HANDBOOK
OF DIFFERENTIAL EQUATIONS

STATIONARY PARTIAL DIFFERENTIAL EQUATIONS

VOLUME III

HANDBOOK OF DIFFERENTIAL EQUATIONS

STATIONARY PARTIAL DIFFERENTIAL EQUATIONS

Volume III

Edited by

M. CHIPOT

Institute of Mathematics, University of Zürich, Zürich, Switzerland

P. QUITTNER

*Department of Applied Mathematics and Statistics, Comenius University,
Bratislava, Slovak Republic*

ELSEVIER

Amsterdam • Boston • Heidelberg • London • New York • Oxford
Paris • San Diego • San Francisco • Singapore • Sydney • Tokyo

North-Holland is an imprint of Elsevier
Radarweg 29, PO Box 211, 1000 AE Amsterdam, The Netherlands
The Boulevard, Langford Lane, Kidlington, Oxford OX5 1GB, UK

First edition 2006

Library of Congress Cataloging-in-Publication Data
A catalog record for this book is available from the Library of Congress

British Library Cataloguing-in-Publication Data
A catalogue record for this book is available from the British Library

ISBN-13: 978-0-444-52846-9
ISBN-10: 0-444-52846-6
Set ISBN: 0 444 51743 x

For information on all North-Holland publications
visit our web site at books.elsevier.com

Printed and bound in The Netherlands

06 07 08 09 10 10 9 8 7 6 5 4 3 2 1

Preface

This handbook is volume III in a series devoted to stationary partial differential equations. Similarly as volumes I and II, it is a collection of self contained, state-of-the-art surveys written by well-known experts in the field.

The topics covered by this handbook include singular and higher order equations, problems near criticality, problems with anisotropic nonlinearities, dam problem, Γ-convergence and Schauder-type estimates. We hope that these surveys will be useful for both beginners and experts and speed up the progress of corresponding (rapidly developing and fascinating) areas of mathematics.

We thank all the contributors for their clearly written and elegant articles. We also thank Arjen Sevenster and Andy Deelen at Elsevier for efficient collaboration.

M. Chipot and P. Quittner

List of Contributors

Antontsev, S., *Departamento de Matematica, Universidade da Beira Interior, 6201-001 Covilha, Portugal* (Ch. 1)

Braides, A., *Dipartimento di Matematica, Università di Roma 'Tor Vergata', Via della Ricerca Scientifica 1, 00133 Roma, Italy* (Ch. 2)

del Pino, M., *Departamento de Ingeniería Matemática and CMM, Universidad de Chile, Casilla 170, Correo 3, Santiago, Chile* (Ch. 3)

Hernández, J., *Departamento de Matemáticas, Universidad Autónoma de Madrid, 28049 Madrid, Spain* (Ch. 4)

Kichenassamy, S., *Laboratoire de Mathématiques, UMR 6056, CNRS and Université de Reims Champagne-Ardenne, Moulin de la Housse, B.P. 1039, F-51687 Reims Cedex 2, France* (Ch. 5)

Lyaghfouri, A., *Mathematical Sciences Department, King Fahd University of Petroleum and Minerals, Dhahran 31261, Saudi Arabia* (Ch. 6)

Mancebo, F.J., *E.T.S.I. Aeronáuticos, Universidad Politécnica de Madrid, Plaza del Cardenal Cisneros 3, 28040 Madrid, Spain* (Ch. 4)

Musso, M., *Departamento de Matemática, Pontificia Universidad Católica de Chile, Avda. Vicuña Mackenna 4860, Macul, Santiago, Chile and Dipartimento di Matematica, Politecnico di Torino, Corso Duca degli Abruzzi, 24, 10129 Torino, Italy* (Ch. 3)

Peletier, L.A., *Mathematical Institute, Leiden University, PB 9512, 2300 RA Leiden, The Netherlands* (Ch. 7)

Shmarev, S., *Departamento de Matematicas, Universidad de Oviedo, Spain* (Ch. 1)

Contents

Preface   v
List of Contributors   vii
Contents of Volume I   xi
Contents of Volume II   xiii

1. Elliptic Equations with Anisotropic Nonlinearity and Nonstandard Growth
    Conditions   1
    S. Antontsev and S. Shmarev
2. A Handbook of Γ-Convergence   101
    A. Braides
3. Bubbling in Nonlinear Elliptic Problems Near Criticality   215
    M. del Pino and M. Musso
4. Singular Elliptic and Parabolic Equations   317
    J. Hernández and F.J. Mancebo
5. Schauder-Type Estimates and Applications   401
    S. Kichenassamy
6. The Dam Problem   465
    A. Lyaghfouri
7. Nonlinear Eigenvalue Problems for Higher-Order Model Equations   553
    L.A. Peletier

Author Index   605

Subject Index   613

Contents of Volume I

Preface   v
List of Contributors   vii

1. Solutions of Quasilinear Second-Order Elliptic Boundary Value Problems via
   Degree Theory   1
   C. Bandle and W. Reichel
2. Stationary Navier–Stokes Problem in a Two-Dimensional Exterior Domain   71
   G.P. Galdi
3. Qualitative Properties of Solutions to Elliptic Problems   157
   W.-M. Ni
4. On Some Basic Aspects of the Relationship between the Calculus of Variations
   and Differential Equations   235
   P. Pedregal
5. On a Class of Singular Perturbation Problems   297
   I. Shafrir
6. Nonlinear Spectral Problems for Degenerate Elliptic Operators   385
   P. Takáč
7. Analytical Aspects of Liouville-Type Equations with Singular Sources   491
   G. Tarantello
8. Elliptic Equations Involving Measures   593
   L. Véron

Author Index   713

Subject Index   721

Contents of Volume II

Preface   v
List of Contributors   vii
Contents of Volume I   xi

1. The Dirichlet Problem for Superlinear Elliptic Equations   1
    T. Bartsch, Z.-Q. Wang and M. Willem
2. Nonconvex Problems of the Calculus of Variations and Differential Inclusions   57
    B. Dacorogna
3. Bifurcation and Related Topics in Elliptic Problems   127
    Y. Du
4. Metasolutions: Malthus versus Verhulst in Population Dynamics. A Dream
    of Volterra   211
    J. López-Gómez
5. Elliptic Problems with Nonlinear Boundary Conditions and the Sobolev
    Trace Theorem   311
    J.D. Rossi
6. Schrödinger Operators with Singular Potentials   407
    G. Rozenblum and M. Melgaard
7. Multiplicity Techniques for Problems without Compactness   519
    S. Solimini

Author Index   601

Subject Index   609

Elliptic Equations with Anisotropic Nonlinearity and Nonstandard Growth Conditions

Stanislav Antontsev

Departamento de Matematica, Universidade da Beira Interior, 6201-001 Covilha, Portugal
E-mail: anton@ubi.pt

Sergey Shmarev

Departamento de Matematicas, Universidad de Oviedo, 33007, Oviedo, Spain
E-mail: shmarev@orion.ciencias.uniovi.es

Contents

1. Introduction 3
   1.1. Assumptions and results 3
   1.2. Physical motivation 6
   1.3. Previous work 10
2. The Lebesgue and Sobolev spaces with variable exponents 12
   2.1. Spaces $L^{p(x)}(\Omega)$ and $W_0^{1,p(x)}(\Omega)$ 12
   2.2. Anisotropic spaces 14
   2.3. Embedding theorems 15
3. Existence theorems 16
   3.1. Generalized $p(x)$-Laplace equation 16
   3.2. Generalized diffusion equation 24
   3.3. Equations with convection terms 32
4. Uniqueness theorems 34
   4.1. Uniqueness of solution of the generalized $p(x)$-Laplace equation 36
   4.2. Uniqueness of solution of the generalized diffusion equation 40
5. Localization caused by the diffusion–absorption balance 43
   5.1. Generalized $p(x)$-Laplace equation 43
   5.2. The energy relation 47
   5.3. The ordinary differential inequality 49
   5.4. Equations with convection terms 55

HANDBOOK OF DIFFERENTIAL EQUATIONS
Stationary Partial Differential Equations, volume 3
Edited by M. Chipot and P. Quittner

6. Directional localization caused by anisotropic diffusion 56
   6.1. Generalized diffusion equation 56
   6.2. Generalized $p(x)$-Laplace equation 68
7. Problems on unbounded domains 72
   7.1. Generalized diffusion equation 72
   7.2. Generalized $p(x)$-Laplace equation 76
8. Systems of elliptic equations 76
   8.1. Existence of solutions 78
   8.2. Localization properties 78
   8.3. Systems of other types 81
9. Examples: localization in borderline cases 82
   9.1. Illustrative examples 82
   9.2. The ordinary differential inequality in the limit case 95
Acknowledgements 97
References 97

1. Introduction

1.1. *Assumptions and results*

This chapter is a contribution to the theory of elliptic equations with nonstandard growth conditions and systems of such equations. We study the Dirichlet problem for the class of elliptic equations

$$
\begin{cases}
-\sum_i D_i\big(a_i(x,u)|D_iu|^{p_i(x)-2}D_iu\big) \\
\quad + c(x,u)|u|^{\sigma(x)-2}u = f(x) & \text{in } \Omega, \\
u = 0 & \text{on } \partial\Omega,
\end{cases}
\tag{1.1}
$$

systems of equations of the same structure

$$
\begin{cases}
-\sum_j D_j\big(a_{ij}(x,\nabla\mathbf{u})\big) = f^{(i)}(x,\mathbf{u}) & \text{in } \Omega, i=1,\ldots,n, \\
\mathbf{u} = 0 & \text{on } \partial\Omega,
\end{cases}
\tag{1.2}
$$

and the Dirichlet problem for the elliptic equations of the type

$$
\begin{cases}
\sum_i D_i\big(a_i(x,u)|u|^{\alpha_i(x)}D_iu\big) \\
\quad + c(x,u)|u|^{\sigma(x)-2}u = f(x) & \text{in } \Omega \subset \mathbb{R}^n, \\
u = 0 & \text{on } \partial\Omega.
\end{cases}
\tag{1.3}
$$

Here and throughout the chapter we use bold characters to denote the vector-valued functions $\mathbf{u} = (u^{(1)},\ldots,u^{(n)})$ and bold capitals for the matrix-valued functions. The notation $\nabla\mathbf{s}$ is used to denote the matrices with the entries $D_ju^{(i)}$. The coefficients a_i, c, a_{ij} and the exponents of nonlinearity p_i, α_i, σ are given functions of their arguments.

A prototype of the differential operators of the form (1.1) is the $p(x)$-Laplacian

$$
\Delta_{p(x)}u \equiv \operatorname{div}\big(|\nabla u|^{p(x)-2}\nabla u\big)
$$

which generalizes the p-Laplacian. By this reason, we term the equations of the structure (1.1) the generalized $p(x)$-Laplace equations. Equations of the type (1.3) are called the generalized diffusion equations.

We discuss the questions of existence, uniqueness and localization of weak solutions to the formulated problems. Anticipating the precise conditions on the structure of the equations under study, let us notice here that the coefficients $a_i(x,u)$ are always assumed separated away from zero so that the possible degeneracy or singularity of equations (1.1) and (1.3) is solely defined by the properties of the nonlinearity exponents $p_i(x)$ and $\alpha_i(x)$. The main feature of equations and systems of the type (1.1) and (1.2) is the gap between the coercivity and monotonicity conditions. Let us write (1.1) in the form

$$
-\operatorname{div} A(x,u,\nabla u) = \Phi(x,u),
$$

and system (1.2) as

$$-\operatorname{div} \mathbf{A}(x, \nabla \mathbf{u}) = \Psi(x, \mathbf{u}),$$

where $A : \Omega \times \mathbb{R} \times \mathbb{R}^n \mapsto \mathbb{R}^n$ is a vector-valued function, and $\mathbf{A}$ is a matrix with the entries $a_{ij} : \Omega \times \mathbb{R}^{n^2} \mapsto \mathbb{R}^{n^2}$. In the present chapter we study the equations and systems whose main parts satisfy the anisotropic nonstandard growth conditions

$$\lambda_1 \sum_i |s_i|^{p_i(x)} \geq A(x, t, \mathbf{s}) \cdot \mathbf{s} \geq \lambda_2 \sum_i |s_i|^{p_i(x)},$$

$$\lambda_1 \sum_{ij} |V_{ij}|^{p_{ij}(x)} \geq \mathbf{A}(x, \mathbf{V}) : \mathbf{V} \geq \lambda_2 \sum_{ij} |V_{ij}|^{p_{ij}(x)}, \qquad \lambda_1, \lambda_2 = \text{const} > 0.$$

For nonconstant $p_i(x)$ and $p_{ij}(x)$ these conditions are usually termed *nonstandard* because of the existing gap between the coercivity and monotonicity assumptions. Unless explicitly stated, we always assume that

$$\Phi(x, u) = -c(x, u)|u|^{\sigma(x)-2}u + F(x)$$

with a nonnegative function $c(x, u)$ and a continuous exponent $\sigma(x) > 1$. In case of a system, the function $\Psi(x, \mathbf{s}) = (\Psi^{(1)}, \ldots, \Psi^{(n)})$ is assumed to satisfy similar growth conditions

$$\begin{cases} \Psi^{(i)}(x, \mathbf{s}) = f^{(i)}(x) - \sum_k c_{ik}(x)|s^{(k)}|^{\sigma_{ik}(x)-2}s^{(k)}, & i = 1, \ldots, n, \\ \sum_{ik} c_{ik}(x)|s^{(k)}|^{\sigma_{ik}(x)-2}s^{(k)}s^{(i)} \geq c_0 \sum_i |s^{(i)}|^{\sigma_i(x)}, & c_0 = \text{const} > 0, \\ \sigma_{ik}(x), \sigma_i(x) > 1, & i, k = 1, \ldots, n, \end{cases} \qquad (1.4)$$

and, additionally, the monotonicity conditions

$$\begin{cases} \forall x \in \Omega, \quad \mathbf{s}, \mathbf{r} \in \mathbb{R}^n, \\ \sum_{ik} c_{ik}(x)\left(|s^{(k)}|^{\sigma_{ik}(x)-2}s^{(k)} - |r^{(k)}|^{\sigma_{ik}(x)-2}r^{(k)}\right)\left(s^{(i)} - r^{(i)}\right) \geq 0. \end{cases}$$

Examples of the systems satisfying all these conditions are furnished by

$$\begin{cases} -\Delta_{p(x)}u_1 + |u_1|^{\delta(x)-2}u_1 - u_2 = f(x), \\ -\Delta_{q(x)}u_2 + u_1 + |u_2|^{\gamma(x)-2}u_2 = g(x), \end{cases}$$

$$\begin{cases} -\Delta_{p(x)}u_1 + |\mathbf{u}|^{\sigma(x)-2}u_1 = f(x), \\ -\Delta_{q(x)}u_2 + |\mathbf{u}|^{\sigma(x)-2}u_2 = g(x) \end{cases}$$

with given exponents $p(x), q(x), \delta(x), \gamma(x), \sigma(x) > 1$.

We prove that under suitable assumptions on the data the afore-formulated problems have weak solutions. The solutions are elements of the function spaces which generalize the Lebesgue and Sobolev spaces. The basic information about these spaces is collected in Section 2. In Section 3 we prove the existence theorems for problems (1.1) and (1.3).

The methods of proof are different for the generalized $p(x)$-Laplace equation and generalized diffusion equations. The solvability of the Dirichlet problem for the generalized $p(x)$-Laplace equation is proved via an adaptation of the Galerkin method. The application of Galerkin's method becomes possible if the coefficients $p_i(x)$ and $\sigma(x)$ are subject to a specific regularity assumption. Namely, it is requested that p_i and σ are continuous with the logarithmic module of continuity. It is known that under this assumption the set of smooth functions is dense in the generalized Sobolev spaces. A weak solution of the generalized diffusion equation is constructed as the limit of a sequence of regularized problems. It is requested that either $\alpha_i(x) \in (-1, \infty)$ and $c(x, u)$ is bounded away from zero, or that simply $\alpha_i(x) \geqslant 0$. In both cases we claim that $\|D_i \alpha_i(x)\|_{2,\Omega}$ are bounded. Boundedness of the weak solution is proved a posteriori in the case of generalized $p(x)$-Laplacian. For the generalized diffusion equation boundedness of the prospected solution is requested for the proof of existence and is established a priori. Systems of the type (1.2) are studied in Section 8.

In Section 4 we establish uniqueness of bounded weak solutions of problems (1.1) and (1.3). In the case of equation (1.1) the proof relies on the monotonicity of the principle part of the differential operator and requires the presence of the lower order term. We claim therefore that $c(x, u) \geqslant c_0 > 0$ and distinguish two cases in dependence of behavior of the lower-order term. We either claim that $\sigma(x) \in (1, 2]$ and $p_i(x) \in (1, \infty)$, or that $\sigma(x) \in (2, \infty)$ but $p_i(x) \in (2n/(n + 2), 2]$. Sticking to the mechanical terminology we may characterize these two cases as the mathematical models of the diffusion–absorption processes of the types (a) slow diffusion–strong absorption, (b) fast diffusion–weak absorption. In the case of the generalized diffusion equation (1.3) it is sufficient to claim that $\alpha_i(x)$ are bounded away from minus one and infinity, $\alpha_i \in L_2(\Omega)$, and that either $c(x, u) \equiv c(x)$ or that $c(x, s)$ is monotone: for all $s, r \in \mathbb{R}$, $x \in \Omega$, $(c(x, s)|s|^{p-2}s - c(x, r)|r|^{p-2}r)(s - r) \geqslant 0$.

The localization properties of weak solutions is the next issue of the study. We show first that the solutions of problems (1.1) and (1.3) possess the same localization properties that are intrinsic for the solutions of nonlinear elliptic equations with constant exponents of nonlinearity and isotropic diffusion. The typical localization property consists in the following: if the right-hand side f of equation (1.1) (or (1.3)) is identically zero in a ball B_r of radius $r > 0$, then one may indicate a concentric ball B_ρ of a smaller radius ρ such that the solution is zero in B_ρ. The radius ρ is defined in terms of the problem data. Under an additional condition on the rate of vanishing of f near the boundary of the ball B_r a stronger localization result is established: the weak solution must be zero in the same ball B_r. Localization of this type is always caused by a suitable balance between the principal part of the differential operator (the diffusion) and the low-order terms (the absorption) and holds for equations with isotropic and anisotropic diffusion.

Such properties were studied first for "model" nonlinear equations of relatively simple structure and the proofs relied on the possibility of comparison of the solution under study with supersolutions of the same equation (see [26] for the details). For more complicated equations, including (1.1) and (1.3) with constant exponents of nonlinearity, such properties were established via the local energy method. This method allows one to reduce the study of the localization properties of solutions to nonlinear PDEs with several independent variables to the analysis of the local energy functions which satisfy certain nonlinear ordinary differential inequalities.

We prove next that the solutions of equations with anisotropic diffusion possess a new property of localization caused by strong anisotropy. Let $u(x)$ be a nonnegative solution of the equation

$$-\Delta_p u + cu^{\sigma-1} = 0 \tag{1.5}$$

in an exterior domain $\Omega \subset \mathbb{R}^n$ with constant exponents of nonlinearity $p > 1$ and $\sigma > 1$. For this equation one may formulate the following alternative:

$$\begin{cases} 1 < p \leqslant \sigma & \Longleftrightarrow & \text{the strong maximum principle holds [69],} \\ 1 < \sigma < p & \Longleftrightarrow & \text{the compact support principle holds [62].} \end{cases} \tag{1.6}$$

It happens, however, that this alternative is wrong for the solutions of equations with anisotropic diffusion operator. We show that for equations with "strong anisotropy" the solutions can be localized in a separate direction even in the absence of the absorption term. The analysis of this effect is carried out in Section 6. Relying on this property one can solve the Dirichlet problem for equations (1.1) and (1.3) posed on unbounded domains without conditions at infinity. The conditions of solvability of these problems are formulated in terms of geometrical restrictions on the problem domain. Roughly speaking, these are conditions on the "asymptotic size" of the problem domain at infinity.

The above-described results extend to the systems of elliptic equations of similar structure. The methods used to study the systems are not specific and the results are obtained via suitable modifications of the arguments explained in the previous sections.

In the concluding section we discuss several borderline cases in which the variable exponent of nonlinearity is allowed to achieve its limit value. In the case of equation (1.5) with $p = 2$ and variable σ this would mean that the function $\sigma(x) - 2$ is not necessarily bounded away from zero. Although in the case of constant exponents $\sigma = p = 2$ the compact support principle is not valid, we show that for solutions of equations with variable exponents the localization property may persist even if $p = 2$ and $\sigma(x) - 2 \to 0$ as $x \to x_0$.

The presentation is partially based on the authors' papers [14–16].

1.2. *Physical motivation*

In the recent years, there has been an increasing interest in the study of equations and systems of equations with nonstandard growth conditions. Let us describe here two kinds of physical processes whose mathematical modeling leads to an equation with variable nonlinearity.

Let us consider first the problem of image recovery. Suppose that the image u_0, defined on a domain $\Omega \subset \mathbb{R}^n$, is the result of a linear transformation A of the true image u to which a random noise n has been added,

$$u_0 = Au + n.$$

It is requested to recover u, knowing u_0. The functions u and u_0 usually are scales of gray. The method of total variation smoothing (see [21] and the further references therein) consists in solving the minimization problem

$$\int_\Omega \left\{ |\nabla u| + \frac{\lambda}{2} |Au - u_0|^2 \right\} dx$$

with a given Lagrangian multiplier $\lambda = \mathrm{const}$. It is assumed that the noise n has zero mean, i.e., $\int_\Omega (Au - u_0)\, dx = 0$, and that the standard deviation σ is given: $\int_\Omega |Au - u_0|^2 dx = \sigma^2$. Another approach to the image recovery consists in minimizing the energy integral $\int_\Omega |\nabla u|^2 dx$ under similar restrictions on the random noise n, which leads to the problem of minimizing the integral

$$\int_\Omega \left\{ |\nabla u|^2 + \frac{\lambda}{2} |Au - u_0|^2 \right\} dx.$$

The former method preserves the edges of the image where $|\nabla u|$ is high. However, the flaw of the method is that it may also create edges due to the presence of the random noise. Using the latter method, one eliminates the noise effect by smoothing the input, but the drawback is that it also destroys small details of the true image. A combination of the two methods consists in minimizing the energy

$$J[u] = \int_\Omega \left\{ |\nabla u|^{p(x)} + \frac{\lambda}{2} |Au - u_0|^2 \right\} dx \quad \text{with } p(x) \in [1, 2],$$

with the exponent $p(x)$ close to 2 where there are likely no edges, and close to 1 where the edges are expected. The approximate location of the edges can be determined by looking for the zones where $|\nabla u|$ is high. The minimizer of the functional $J[u]$ is a solution of the $p(x)$-Laplace equation. A detailed discussion of these and more complicated models in the image restoration problems can be found in [22,57].

A special interest in the study of equations and systems of equations with nonstandard growth conditions is motivated by their applications to the mathematical modeling of non-Newtonian fluids in particular, the electrorheological fluids. This kind of fluids is characterized by their ability to drastically change the mechanical properties under the influence of an external electromagnetic field. A mathematical model of electrorheological fluids was proposed by Rajagopal and Růžička in [63,64]. Let $\mathbf{v}$ stand for the velocity of the fluid, $\mathbf{E}$ denotes the external electromagnetic field and P be the pressure. The system of the modified stationary Navier–Stokes equations reads as [2,3,63,64]

$$\begin{cases} \mathrm{rot}\,\mathbf{E} = 0, \quad \mathrm{div}\,\mathbf{E} = 0, \quad \mathrm{div}\,\mathbf{v} = 0, \\ -\,\mathrm{div}\,a\big(x, \mathbf{E}, D(\mathbf{v})\big) + \nabla P = \mathrm{div}(\mathbf{v} \otimes \mathbf{v}) + \mathbf{f}, \end{cases}$$

where $D(\mathbf{v})$ denotes the symmetric part of the matrix $\nabla \mathbf{v}$, and the function a behaves like

$$a\big(x, \mathbf{E}, D(\mathbf{v})\big) \approx \big(1 + |D(\mathbf{v})|^2\big)^{(p(\mathbf{E})-2)/2} D(\mathbf{v}). \tag{1.7}$$

Once the field $\mathbf{E}$ is defined from Maxwell's equations, the exponent $p(\mathbf{E})$ becomes a function of x. The system for the components of the velocity vector $\mathbf{v}$ transforms then into the system with the following nonstandard growth condition:

$$\begin{cases} a(x, \mathbf{E}, \mathbf{S}) \equiv \tilde{a}(x, \mathbf{S}) : \Omega \times \mathbb{R}^{n^2} \mapsto \mathbb{R}^{n^2}, \\ C_1 |\mathbf{S}|^{p_-} - C_2 \leqslant \tilde{a}(x, \mathbf{S}) : \mathbf{S} \leqslant C_3 |\mathbf{S}|^{p_+} + C_4, \quad C_i = \text{const} > 0. \end{cases}$$

For $p_+ > p_-$, this condition is known in the literature as the *nonstandard growth condition of (p_+, p_-) type*. We refer to the papers by Acerbi and Mingione [2,3] for a discussion of the regularity properties of weak solutions of the systems of equations with this type of nonlinearity (see also the references therein to the previous work on this issue). The regularity of weak solutions to the parabolic counterpart of such systems is studied in [4].

Equations and systems of equations with this type of nonlinearity appear also in the mathematical modeling of stationary thermo-convective flows of non-Newtonian fluids [11,9,71]. In particular, in [11] the stress tensor $\mathbf{S}$ has the form

$$\mathbf{S} = -p\mathbf{I} + \left(\mu(\theta) + \tau(\theta)|\mathbf{D}(\mathbf{u})|^{p(\theta)-2}\right)\mathbf{D}(\mathbf{u}),$$

where $\mathbf{u}$ is the velocity vector, p pressure, θ denotes the temperature and μ, p, τ are known coefficients depending on the temperature. This hypothesis leads to the system of equations with nonstandard growth conditions

$$\begin{cases} (\mathbf{v} \cdot \nabla)\mathbf{v} = \text{div}\left(\mu(\theta) + \tau(\theta)|\mathbf{D}(\mathbf{v})|^{p(\theta)-2}\right)\mathbf{D}(\mathbf{v}) - \nabla p + \mathbf{f}, \\ \text{div } \mathbf{v} = 0, \quad \mathbf{v} \cdot \nabla b(\theta) = \Delta \theta + g(x). \end{cases}$$

The thermistor problem studied in [72] describes the electric current in a conductor under the influence of a nonconstant temperature field and consists in finding the electric potential $u(x)$ and the temperature $\theta(x)$ from the system of equations

$$\text{div}\left(|\nabla u|^{\theta(x)-2}\nabla u\right) = 0, \qquad -\Delta \theta = \lambda |\nabla u|^{\theta(x)} \quad \text{in } \Omega, \lambda > 0.$$

Paper [9] deals with the mathematical model of thermal effects in viscous fluid. In this model the dissipative (absorption) term nonlinearly depends on the temperature and the system of governing equations is of the form

$$\begin{cases} (\mathbf{u} \cdot \nabla)\mathbf{u} = \nu\Delta\mathbf{u} - \nabla p - \delta|\mathbf{u}|^{\sigma(\theta)-2}\mathbf{u}, \\ \mathbf{u} \cdot \nabla \mathcal{C}(\theta) = \Delta\varphi(\theta), \quad \text{div } \mathbf{u} = 0, \end{cases}$$

where $\mathbf{u}$, p and θ denote the velocity, the pressure and the temperature, $\mathcal{C}(\cdot)$ is a prescribed function.

The principal parts of the equations and systems studied in the above-quoted papers can be regarded as generalizations of the *$p(x)$-Laplace equation* which is formally elliptic wherever $0 < |D_i u| < \infty$, degenerates at the points $x \in \Omega$ where either $|D_i u| = 0$ and $p_i(x) > 2$ or $|D_i u| = \infty$ and $p_i(x) < 2$, and becomes singular if either $|D_i u| = 0$ and $p_i(x) < 2$ or $|D_i u| = \infty$ and $p_i(x) > 2$.

As an example, let us consider the motion of a fluid in a porous medium and trace the influence of the hypotheses about the properties of the fluid and the medium on the complexity of the mathematical model of this process. Let us denote by ρ, $\mathbf{v}$ and p the density, velocity and pressure of the fluid. We assume first that the medium is homogeneous and isotropic and that there are no external forces and the mass sinks and sources. The classical Darcy law suggests that the fluid velocity is proportional to the gradient of pressure,

$$\mathbf{v} = -k\nabla p, \quad k = \text{const}. \tag{1.8}$$

For the incompressible fluids the continuity equation holds, $\operatorname{div}\mathbf{v} = 0$. Combining these two equations we conclude that the pressure p is a harmonic function. For the compressible fluids (gases) the continuity equation has the form $\operatorname{div}(\rho\mathbf{v}) = 0$. In barotropic gases the state equation is given by relation

$$p \equiv p(\rho) = \rho^{\gamma}, \quad \gamma = \text{const}, \tag{1.9}$$

which leads to the nonlinear elliptic equation for the pressure

$$\operatorname{div}\left(kp^{1/\gamma}\nabla p\right) = 0. \tag{1.10}$$

If the nonlinear Darcy law is accepted, i.e., either

(a) $\quad \mathbf{v} = -k|\nabla p|^{\lambda-2}\nabla p \quad$ or

(b) $\quad v_i = -k|p_{x_i}|^{\lambda-2}p_{x_i}, \quad i = 1, \ldots, n,$

then the pressure in the incompressible fluid satisfies the equation of the p-Laplace type

$$\operatorname{div}\left(|\nabla p|^{\lambda-2}\nabla p\right) = 0 \quad \text{or} \quad \sum_{i=1}^{n}\left(|p_{x_i}|^{\lambda-2}p_{x_i}\right) = 0.$$

For the barotropic gas the equation for the pressure takes the form

$$\operatorname{div}\left(p^{1/\gamma}|\nabla p|^{\lambda-2}\nabla p\right) = 0 \quad \text{or} \quad \sum_{i=1}^{n}\left(p^{1/\gamma}|p_{x_i}|^{\lambda-2}p_{x_i}\right)_{x_i} = 0.$$

Let now the medium be nonhomogeneous and anisotropic, i.e., the characteristics of the medium may vary in dependence on the direction x_i and the point of the medium: now $\lambda_i \equiv \lambda_i(x)$, $\gamma = \gamma(x)$, and instead of the constant k in (1.8) a diagonal matrix $K(x)$ is used. Let us also admit the presence of the exterior mass forces and sources and sinks of mass which may depend on the point x, the pressure p and its gradient ∇p. Under these assumptions the pressure in the incompressible fluid obeys the anisotropic generalized $p(x)$-Laplace equation of the form

$$\operatorname{div}(\rho\mathbf{v}) \equiv -\sum_{i}^{n} D_i\left(K_i(x)|\nabla p|^{\lambda_i(x)-2}D_i p\right) = h(x, p, \nabla p)$$

or

$$\mathrm{div}(\rho\mathbf{v}) \equiv -\sum_i^n D_i\big(K_i(x)|D_i p|^{\lambda_i(x)-2} D_i p\big) = h(x, p, \nabla p).$$

For the compressible fluids the continuity equation becomes doubly nonlinear,

$$\mathrm{div}(\rho\mathbf{v}) \equiv -\sum_i^n D_i\big(K_i(x)p^{1/\gamma(x)}|D_i p|^{\lambda_i(x)-2} D_i p\big) = h(x, p, \nabla p).$$

The functions $\lambda_i(x)$, $\gamma(x)$, $K_i(x)$ can be either known function given a priori or may implicitly depend on x. For example, in the nonisothermic processes the exponent γ in the state equation is a function of the thermodynamic parameters, say, the temperature $\theta(x)$. Then $p = \rho^{\gamma(\theta(x))}$, where the function $\theta(x)$ has to be defined from some complementary conditions.

The study of the qualitative properties of such equations is very important for applications and indispensable for understanding the mechanism of formation of the stagnation zones and the behavior of the flow around them. We refer to the monographs [18,25,60] for the derivation and thorough analysis of the mathematical models of continuum mechanics.

1.3. *Previous work*

The question of existence of a weak solution of the Dirichlet problem for the isotropic $p(x)$-Laplace equation,

$$-\Delta_{p(x)}u = f(x, u) \quad \text{in } \Omega, \qquad u = 0 \quad \text{on } \partial\Omega, \tag{1.11}$$

is studied by Fan and Zhang in [37]. In most of the cases studied in [37] the solutions are obtained as the limits of minimizing sequences for the functional

$$J(u) = \int_\Omega \frac{1}{p(x)}|\nabla u|^{p(x)}\,\mathrm{d}x - \int_\Omega \int_0^u f(x, s)\,\mathrm{d}s\,\mathrm{d}x.$$

It is shown in [47] that the claim of log-continuity of the exponent $p(x)$ is not necessary for the existence of a minimizer of the functional $J(u)$ and can be substituted by a suitable "jump condition".

Existence of weak solutions to the Dirichlet problem for a system of elliptic equations of the type (1.2) with isotropic nonstandard growth conditions was studied in [44]. Radially symmetrical solutions of a special system of the type (1.2) are constructed in [36].

The variational formulation of the Dirichlet problem for equation (1.11) prompts the natural choice of the function spaces the solutions may belong to. The weak solution of problem (1.11) is sought as an element of the generalized Sobolev space (also called

Orlicz–Sobolev space or Musielak–Orlicz space) $W_0^{1,p(x)}(\Omega)$. The theory of the generalized Sobolev spaces is developed in [30,33,35,40,41,46,55,61,65,70,72,73] (see also the review papers [45,66] for the relevant bibliography).

The solutions of equation (1.1) and system (1.2) belong to the anisotropic analogs of generalized Sobolev spaces. In Section 2 we introduce the anisotropic generalized Lebesgue–Sobolev spaces and collect the already known results on the properties of function spaces $W_0^{1,p(x)}(\Omega)$ and $L^{\sigma(x)}(\Omega)$. The restrictions imposed on the regularity of $\partial\Omega$ and the claim of logarithmic continuity of the exponents of nonlinearity in equations (1.1), (1.2) guarantee the fulfillment of the properties of density and embedding in the generalized Sobolev spaces.

The localization properties of solutions of the p-Laplacian equation with absorption terms and alternative (1.6) are given a detailed discussion in [62]. The study relies on the strong maximum principle for elliptic equations (with constant exponents of nonlinearity). The strong maximum principle asserts that if a nonnegative classical solution of equation (1.5) vanishes at a point $x_0 \in \Omega$, it must be identically zero in Ω, while the compact support principle says that if Ω is an exterior domain in $\mathbb{R}^n$ and $u(x)$ is a classical solution of equation (1.5) such that $u \to 0$ when $|x| \to \infty$, then the support of u is compact. The compact support principle also holds for the weak solutions of general elliptic equations which contain (1.5) as a partial case. This was proved by means of the method of local energy estimates – see [10], Chapter 1, and references therein for the history of the question.

An analog of the strong maximum principle for $p(x)$-Laplace equation is proved in the recent work [42]. Let $u(x) \in W^{1,p(x)}(\Omega)$ be a weak supersolution of the equation

$$-\Delta_{p(x)}u + c(x)|u|^{\sigma(x)-2}u = 0$$

with $p(x) \in C^1(\overline{\Omega})$, $\sigma(x) \in C(\overline{\Omega})$, $c(x) \in L^\infty(\Omega)$ and $c(x) \geqslant 0$ in $\overline{\Omega}$. Assume that the exponents $p(x)$ and $\sigma(x)$ satisfy the condition

$$p(x) \leqslant \sigma(x) < p^*(x) = \begin{cases} \frac{p(x)n}{n-p(x)} & \text{if } p(x) < n, \\ \infty & \text{if } p(x) > n. \end{cases}$$

If $u \geqslant 0$ in $\overline{\Omega}$ and $u(x) \not\equiv 0$ on Ω, then for every nonempty compact subset $K \subset \Omega$ there is a positive constant C such that $u(x) \geqslant C$ a.e. in K.

The possibility of directional localization in solutions of equations with anisotropic nonlinearity in an infinite layer was described in [10], Chapter 2, for the solutions of a special structure.

The validity of the compact support principle (alias the localization property) of weak solutions suggests the possibility of existence of weak solutions to equations and systems of the type (1.1)–(1.2) defined on the whole of $\mathbb{R}^n$ or on noncompact domains in $\mathbb{R}^n$. The existence and multiplicity of solutions for the equation

$$\Delta_{p(x)} + c(x)|u|^{p(x)-2}u = f(x,u) \quad \text{in } \mathbb{R}^n$$

is studied in [34].

In this chapter we do not discuss the regularity of the weak solutions of equations and systems under study. Relevant results can be found in [1–3,6,17,28,29,32,38–40,59] which deal with equations and systems of the type (1.1), (1.2) and their parabolic counterparts.

The questions of existence, uniqueness and qualitative behavior of solutions of elliptic equations of the type (1.1) and (1.3) with constant exponents of nonlinearity, as well as parabolic equations with elliptic parts of similar form, were studied by many authors, see [10,19,23,24,26,27,31,43,49–51,58,68] and the literature cited therein.

Parabolic equations with variable exponents of nonlinearity in the elliptic part were studied in papers [7,12,13], nonlinear parabolic equations with singularly disturbed exponents of nonlinearity near the critical values in the elliptic part were considered in [52–54].

2. The Lebesgue and Sobolev spaces with variable exponents

In this section we introduce the function spaces used throughout the chapter and describe their basic properties. The definitions of the function spaces and the sketch of their properties presented in this subsection follow [48,55,61,67] (see also the works cited in the Introduction).

2.1. Spaces $L^{p(x)}(\Omega)$ and $W_0^{1,p(x)}(\Omega)$

It is always assumed that Ω is a bounded domain in $\mathbb{R}^n$ with Lipschitz-continuous boundary and the function $p(x)$ satisfies the conditions

$$1 < p^- < \inf_{\Omega} p(x) \leqslant p(x) \leqslant \sup_{\Omega} p(x) < p^+ < \infty, \quad p^- < n, \tag{2.1}$$

$$\forall x, y \in \Omega \text{ such that } |x - y| < 1, \quad |p(x) - p(y)| \leqslant \frac{M}{\ln(1/|x - y|)}. \tag{2.2}$$

1. By $L^{p(x)}(\Omega)$ we denote the space of measurable functions $f(x)$ on Ω such that

$$A_{p(\cdot)}(f) = \int_{\Omega} |f(x)|^{p(x)} \, dx < \infty.$$

The space $L^{p(x)}(\Omega)$ equipped with the norm

$$\|f\|_{p(\cdot)} \equiv \|f\|_{L^{p(x)}(\Omega)} = \inf\left\{ \lambda > 0 : A_{p(\cdot)}\left(\frac{f}{\lambda}\right) \leqslant 1 \right\}$$

becomes a Banach space.

2. The following inequalities hold:

$$\begin{cases} \min\left(\|f\|_{p(\cdot)}^{p^-}, \|f\|_{p(\cdot)}^{p^+}\right) \leqslant A_{p(\cdot)}(f) \leqslant \max\left(\|f\|_{p(\cdot)}^{p^-}, \|f\|_{p(\cdot)}^{p^+}\right), \\ \min\left(A_{p(\cdot)}^{1/p^-}, A_{p(\cdot)}^{1/p^+}\right) \leqslant \|f\|_{p(\cdot)} \leqslant \max\left(A_{p(\cdot)}^{1/p^-}, A_{p(\cdot)}^{1/p^+}\right). \end{cases} \tag{2.3}$$

3. Let $f \in L^{p(x)}(\Omega)$, $g \in L^{q(x)}(\Omega)$ with

$$\frac{1}{p(x)} + \frac{1}{q(x)} = 1, \quad 1 < p^- \leqslant p(x) \leqslant p^+ < \infty, 1 < q^- \leqslant q(x) \leqslant q^+ < \infty.$$

Then Hölder's inequality holds

$$\int_\Omega |fg| \, dx \leqslant 2 \|f\|_{p(\cdot)} \|g\|_{q(\cdot)}. \tag{2.4}$$

4. According to (2.4), for every $1 \leqslant q = \text{const} < p^- \leqslant p(x) < \infty$,

$$\|f\|_q \leqslant C \|f\|_{p(\cdot)} \quad \text{with the constant } C = 2\|1\|_{p(\cdot)/(p(\cdot)-q)}. \tag{2.5}$$

It is straightforward to check that, for $|\Omega| < \infty$,

$$\|1\|_{p(\cdot)} \leqslant 2 \max\{|\Omega|^{2/p^-}, |\Omega|^{1/(2p^+)}\}.$$

5. The space $W^{1,p(x)}(\Omega)$, $p(x) \in [p^-, p^+] \subset (1, \infty)$, is defined by

$$W^{1,p(x)}(\Omega) = \{f(x) \in L^{p(x)}(\Omega): |\nabla f(x)| \in L^{p(x)}(\Omega)\}.$$

If condition (2.2) is fulfilled, $W_0^{1,\,p(x)}(\Omega)$ is the closure of the set $C_0^\infty(\Omega)$ with respect to the norm of $W^{1,p(x)}(\Omega)$. If the boundary of Ω is Lipschitz-continuous and $p(x)$ satisfies (2.2), then $C_0^\infty(\Omega)$ is dense in $W_0^{1,\,p(x)}(\Omega)$. The norm in the space $W_0^{1,p(x)}$ is defined by

$$\|u\|_{W_0^{1,p(x)}} = \sum_i \|D_i u\|_{p(\cdot)} + \|u\|_{p(\cdot)}.$$

If the boundary of Ω is Lipschitz and $p(x) \in C^0(\Omega)$, then the norm $\| \cdot \|_{W_0^{1,p(x)}(\Omega)}$ is equivalent to the norm

$$\widetilde{\|u\|}_{W_0^{1,p(x)}(\Omega)} = \sum_i \|D_i u\|_{p(\cdot)}. \tag{2.6}$$

6. If $p(x) \in C^0(\overline{\Omega})$, then $W^{1,p(x)}(\Omega)$ is separable and reflexive.
7. If $p(x), q(x) \in C^0(\overline{\Omega})$,

$$p_*(x) = \begin{cases} \frac{p(x)n}{n-p(x)} & \text{if } p(x) < n, \\ \infty & \text{if } p(x) > n, \end{cases} \quad \text{and} \quad 1 < q(x) \leqslant \sup_\Omega q(x) < \inf_\Omega p_*(x),$$

then the embedding $W_0^{1,p(x)}(\Omega) \hookrightarrow L^{q(x)}(\Omega)$ is continuous and compact.

8. Sobolev's inequality is valid in the following form: if $p(x)$ satisfies conditions (2.1)–(2.2), then there exists a constant $C > 0$ such that, for every $f \in W_0^{1,p(x)}(\Omega)$,

$$\|f\|_{p(\cdot)} \leqslant C\|\nabla f\|_{p(\cdot)}. \tag{2.7}$$

2.2. Anisotropic spaces

Let $p_i(x)$, $\sigma(x)$ and $p(x)$ satisfy (2.1)–(2.2). Define the set

$$V(\Omega) = W_0^{1,p(x)}(\Omega) \cap L^{\sigma(x)}(\Omega)$$

and introduce the norm

$$\|u\|_V = \|\nabla u\|_{p(\cdot)} + \|u\|_{\sigma(\cdot)}.$$

By $\mathbf{V}(\Omega)$ we denote the set of functions

$$u \in L^{\sigma(x)}(\Omega) \cap W_0^{1,1}(\Omega), \qquad D_i u \in L^{p_i(x)}(\Omega), \quad i = 1, \ldots, n.$$

$\mathbf{V}(\Omega)$ equipped with the norm

$$\|u\|_{\mathbf{V}} = \|u\|_{\sigma(\cdot)} + \sum_{i=1}^{n} \|D_i u\|_{p_i(\cdot)} \tag{2.8}$$

becomes a Banach space. By $\mathbf{V}'(\Omega)$ we denote the dual space to $\mathbf{V}(\Omega)$.

1. The spaces $W_0^{1,p(x)}(\Omega)$ and $L^{\sigma(x)}(\Omega)$ are reflexive and separable. According to (2.5) $\mathbf{V}(\Omega) \subset \mathbf{X} = W_0^{1,p^-}(\Omega) \cap L^{\sigma^-}(\Omega)$. $\mathbf{V}(\Omega)$ is reflexive and separable as a closed subspace of $\mathbf{X}$.

2. Set

$$A_{\mathbf{p}(\cdot)}(\nabla u) = \sum_{i=1}^{n} \int_{\Omega} |D_i u|^{p_i(x)}\, \mathrm{d}x, \quad \mathbf{p}(x) = (p_1, p_2, \ldots, p_n).$$

The following counterpart of (2.3) holds:

$$\min\left\{ \sum_i \|D_i u\|_{p_i(\cdot)}^{p^+}, \sum_i \|D_i u\|_{p_i(\cdot)}^{p^-} \right\}$$

$$\leqslant A_{\mathbf{p}(\cdot)}(\nabla u) \leqslant \max\left\{ \sum_i \|D_i u\|_{p_i(\cdot)}^{p^-}, \sum_i \|D_i u\|_{p_i(\cdot)}^{p^+} \right\}. \tag{2.9}$$

3. $\forall \xi, \eta \in \mathbb{R}^n$,

$$\left(\frac{1}{2}\right)^p |\xi - \eta|^p \leqslant \left(|\xi|^{p-2}\xi - |\eta|^{p-2}\eta\right)(\xi - \eta) \quad \text{if } 2 \leqslant p < \infty, \tag{2.10}$$

$$(p-1)|\xi - \eta|^2\left(|\xi|^p + |\eta|^p\right)^{(p-2)/p}$$
$$\leqslant \left(|\xi|^{p-2}\xi - |\eta|^{p-2}\eta\right)(\xi - \eta) \qquad \text{if } 1 < p < 2. \tag{2.11}$$

4. In the case of a system of equations we use the space

$$\mathbf{W}(\Omega) = \left\{\mathbf{u}\colon u^{(i)} \in L^{\sigma_i(x)}(\Omega), D_j u^{(i)} \in L^{p_{ij}(x)}(\Omega), i, j = 1, \ldots, n\right\},$$

$$\|\mathbf{u}\|_{\mathbf{W}} = \sum_{i=1}^{n}\|u^{(i)}\|_{\sigma_i(\cdot)} + \sum_{ij}\|D_j u^{(i)}\|_{p_{ij}(\cdot)},$$

which is a reflexive and separable Banach space.

2.3. *Embedding theorems*

For further convenience, we quote here the embedding and the trace-interpolation theorems in Sobolev spaces which are repeatedly used further. The thorough study of this issue can be found, e.g., in [5]. Let $\Omega \subset \mathbb{R}^n$ be a bounded domain with piecewise smooth boundary Γ.

1. Let $u(x) \in W_0^{1,p}(\Omega)$. Then

$$\|u\|_{q,\Omega} \leqslant C\|\nabla u\|_{p,\Omega} \tag{2.12}$$

for $q < np/(n-p)$ if $n > p$, $q = \infty$ if $n < p$.
2. Let $u(x) \in W^{1,p}(\Omega)$ with $p \in (1, \infty)$. Then

$$\|u\|_{q,\Omega} \leqslant C\left(\|\nabla u\|_{p,\Omega} + C'|\Omega|^{1/p-1/n-1/\gamma}\|u\|_{\gamma,\Omega}\right)^{\theta}\|u\|_{r,\Omega}^{1-\theta} \tag{2.13}$$

with $C, C' = \text{const} > 0$, the exponents $\gamma \in [1, p]$ and
(a) if $p < n$, then

$$q \in \left(\min\left\{r, \frac{np}{n-p}\right\}, \max\left\{r, \frac{np}{n-p}\right\}\right), \qquad \theta = \frac{\frac{1}{r} - \frac{1}{q}}{\frac{1}{r} - \frac{n-p}{np}} \in [0, 1],$$

(b) if $p \leqslant n$, then $q \in [r, \infty)$, $\theta = (1/r - 1/q)/(1/r - (n-p)/(np)) \in [0, 1])$,
(c) if $p > n$, then $q = \infty$, $\theta = np/(np + r(p - n))$.

3. For every $u(x) \in W^{1,p}(\Omega)$,

$$\|u\|_{q,\Gamma} \leqslant C\big(\|\nabla u\|_{p,\Omega} + C'|\Omega|^{1/p-1/n-1/\gamma}\|u\|_{\gamma,\Omega}\big)^{\theta}\|u\|_{r,\Omega}^{1-\theta} \tag{2.14}$$

with the exponents $\theta = (qn - r(n-1))/(p(n+r) - nr)(p/q) \in (0,1)$ and

$$\begin{cases} 1 \leqslant r < \frac{np}{n-p}, & 1 \leqslant q < \frac{p(n-1)}{n-p} & \text{if } n > p, \\ 1 \leqslant r < \infty, & 1 \leqslant q < \infty & \text{if } n = p, \\ 1 \leqslant r \leqslant \infty, & 1 \leqslant q \leqslant \infty & \text{if } n < p. \end{cases}$$

4. In the special case, when $\Omega = B_\rho(x_0) = \{x: |x - x_0| < \rho\}$, $\partial\Omega = \partial B_\rho(x_0)$ inequalities (2.13), (2.14) take the form: $\forall u(x) \in W^{1,p}(B_\rho(x_0))$,

$$\|u\|_{q,B_\rho(x_0)} \leqslant C\big(\|\nabla u\|_{p,B_\rho(x_0)} + \rho^{\delta}\|u\|_{\gamma,B_\rho(x_0)}\big)^{\theta}\|u\|_{r,B_\rho(x_0)}^{1-\theta}, \tag{2.15}$$

$$\|u\|_{q,\partial B_\rho(x_0)} \leqslant C\big(\|\nabla u\|_{p,B_\rho(x_0)} + \rho^{\delta}\|u\|_{\gamma,B_\rho(x_0)}\big)^{\theta}\|u\|_{r,B_\rho(x_0)}^{1-\theta} \tag{2.16}$$

with $\delta = -(1 + (p - \gamma)/(p\gamma)n)$.

5. Young's inequality: for every $a, b \geqslant 0$ and $1 < p < \infty$,

$$ab \leqslant \frac{1}{p}(\varepsilon a)^p + \frac{1}{p'}\left(\frac{b}{\varepsilon}\right)^{p'}, \qquad p' = \frac{p}{p-1}. \tag{2.17}$$

3. Existence theorems

3.1. *Generalized $p(x)$-Laplace equation*

In this section we study solvability of the Dirichlet problem for generalized $p(x)$-Laplace equation

$$\begin{cases} -\sum_i D_i\big(a_i(x,u)|D_iu|^{p_i(x)-2}D_iu\big) \\ \quad + c(x,u)|u|^{\sigma(x)-2}u = \Phi(x) & \text{in } \Omega, \\ u = 0 & \text{on } \partial\Omega, \\ \Phi(x) = f(x) + \operatorname{div}\mathbf{G}(x), \\ f \in L^{\sigma'(x)}(\Omega), \quad \sigma'(x) = \frac{\sigma(x)}{\sigma(x)-1}, & \mathbf{G}(x) \in \mathbf{V}'(\Omega). \end{cases} \tag{3.1}$$

It is assumed that the data of problems (3.1) satisfy the following conditions.

1. $\Omega \subset \mathbb{R}^n$ is a bounded domain with Lipschitz-continuous boundary $\Gamma = \partial\Omega$.

2. The coefficients $a_i(x,r)$ and $c(x,r)$ are Carathéodory functions (measurable in x for all $r \in \mathbb{R}$ and continuous in r for almost all $x \in \Omega$). Unless explicitly stated, we always assume that a_i and c satisfy the conditions

$$\begin{cases} 0 < a_0 \leqslant a_i(x,r) < \infty & \forall x \in \overline{\Omega}, r \in \mathbb{R}, \\ 0 < c_0 \leqslant c(x,r) < \infty \end{cases} \tag{3.2}$$

with some positive constants a_0, c_0.

3. The functions $p_i(x)$ and $\sigma(x)$ are bounded in Ω: it is assumed that there exist constants $p^- \in (1, n)$, $p^+ < \infty$, $\sigma^- > 1$, $\sigma^+ < \infty$ such that, for all $x \in \overline{\Omega}$,

$$\begin{cases} p_i(x) \in (p^-, p^+], & \inf_\Omega p_i(x) > p^-, \quad i = 1, \ldots, n, \\ \sigma(x) \in (\sigma^-, \sigma^+]. \end{cases} \tag{3.3}$$

4. The functions $p_i(x)$ and $\sigma(x)$ are log-continuous in Ω: for all $x, y \in \Omega$ satisfying $|x - y| \leqslant 1$, the inequality

$$\left| \sigma(x) - \sigma(y) \right| + \sum_i \left| p_i(x) - p_i(y) \right| \leqslant \omega\bigl(|x - y|\bigr) \tag{3.4}$$

holds with a function $\omega(\tau)$ satisfying the condition

$$\omega(\tau) \ln \frac{1}{\tau} \leqslant M \quad \text{for } \tau \in [0, 1], \, M = \text{const} > 0.$$

The weak solution to this problem is understood as follows.

DEFINITION 3.1. A locally integrable function $u(x)$ is called *weak solution* of problem (3.1) if

(1) $u \in \mathbf{V}(\Omega)$,
(2) for any test-function $\zeta \in \mathbf{V}(\Omega)$ the integral identity holds:

$$\int_\Omega \left(\sum_{i=1}^n a_i |D_i u|^{p_i(x)-2} D_i u D_i \zeta + c |u|^{\sigma(x)-2} u \zeta - f \zeta - \mathbf{G} \nabla \zeta \right) dx = 0. \tag{3.5}$$

3.1.1. *A model equation.* We begin with the study of the special situation when the coefficients a_i and c do not depend on u and the proof of existence of weak solutions is fairly easy.

THEOREM 3.1. *Let conditions (3.2)–(3.4) be fulfilled, $f \in L^{\sigma'(x)}(\Omega)$, $\mathbf{G} \in \mathbf{V}'(\Omega)$. Assume that $a_i(x, s) \equiv A_i(x)$ and $c(x, s) \equiv C(x)$. Then the problem*

$$\begin{cases} -\sum_{i=1}^n D_i \bigl(A_i(x) |D_i u|^{p_i(x)-2} D_i u \bigr) \\ \quad + C(x) |u|^{\sigma(x)-2} u = \Phi(x) & \text{in } \Omega, \\ u = 0 & \text{on } \Gamma \end{cases} \tag{3.6}$$

has at least one weak solution.

PROOF. Let us introduce the operator $\mathcal{L} : \mathbf{V}(\Omega) \mapsto \mathbf{V}'(\Omega)$,

$$(\mathcal{L} u, \zeta) = \int_\Omega \left(\sum_i A_i(x) |D_i u|^{p_i(x)-2} D_i u D_i \zeta + C(x) |u|^{\sigma(x)-2} u \zeta \right) dx.$$

It is obvious that the mapping $\mathcal{L} : \mathbf{V}(\Omega) \mapsto \mathbf{V}'(\Omega)$ is continuous. According to (2.10)–(2.11) it is monotone.

LEMMA 3.1. *The operator $L : \mathbf{V}(\Omega) \mapsto \mathbf{V}'(\Omega)$ is coercive:*

$$\forall u \in \mathbf{V}(\Omega), \quad (\mathcal{L}u, u) \geqslant C \min \left\{ \left(\frac{\|u\|_{\mathbf{V}}}{n+1} \right)^{p^-}, \left(\frac{\|u\|_{\mathbf{V}}}{n+1} \right)^{p^+} \right\}.$$

PROOF. Let

$$\lambda = \|u\|_{\mathbf{V}}, \qquad \lambda_i = \|D_i u\|_{p_i(\cdot)}, \quad i = 1, \ldots, n, \qquad \lambda_0 = \|u\|_{\sigma(\cdot)}.$$

Then there is $i \in \{0, 1, \ldots, n\}$ such that $\lambda_i \geqslant \lambda/(n+1)$ (we do not loose the generality by assuming that $i = 1$). It follows from (2.3) that

$$(\mathcal{L}u, u) \geqslant C \left(A_{\mathbf{p}(\cdot)}(\nabla u) + A_{\sigma(\cdot)}(u) \right)$$

$$\geqslant C \int_{\Omega} |D_1 u|^{p_1(x)} \, dx$$

$$\geqslant C \begin{cases} \lambda_1^{p^-} & \text{if } \lambda_1 > 1, \\ \lambda_1^{p^+} & \text{if } \lambda_1 \leqslant 1 \end{cases}$$

$$\geqslant C \min \left\{ \left(\frac{\lambda}{n+1} \right)^{p^+}, \left(\frac{\lambda}{n+1} \right)^{p^-} \right\}. \qquad \square$$

The space $\mathbf{V}(\Omega)$ is separable, and the mapping $\mathcal{L} : \mathbf{V}(\Omega) \mapsto \mathbf{V}'(\Omega)$ is monotone, continuous and coercive. By the Browder–Minty theorem [20], Theorem 7.3.2, for every $\Phi \in \mathbf{V}'(\Omega)$ the equation $\mathcal{L}u = \Phi$ has at least one weak solution $u \in \mathbf{V}(\Omega)$. $\qquad \square$

3.1.2. *The general case.*

THEOREM 3.2. *Let conditions (3.2)–(3.4) be fulfilled, $f \in L^{\sigma'(x)}(\Omega)$, $\mathbf{G} \in \mathbf{V}'(\Omega)$. Then problem (3.1) has at least one weak solution satisfying the estimate*

$$\|u\|_{\mathbf{V}(\Omega)} \leqslant K, \tag{3.7}$$

where the constant K depends on $p^{\pm}, \sigma^{\pm}, n, \|f\|_{\sigma'(\cdot)}, \|\mathbf{G}\|_{\mathbf{V}'}$.

PROOF. The solution is obtained as a limit of the sequence of Galerkin's approximations. When constructing the solution we follow the proof of Theorem 9.2 in [56], Chapter 4, Section 9. The needed changes in the arguments are due to the specific form of the functional spaces the solution belongs to.

Define the operator $\mathcal{L}_\Omega : \mathbf{V}(\Omega) \mapsto \mathbf{V}'(\Omega)$,

$$(\mathcal{L}_\Omega u, \zeta) = \int_\Omega \left(\sum_i a_i(x, u) |D_i u|^{p_i(x)-2} D_i u D_i \zeta + c(x, u) |u|^{\sigma(x)-2} u \zeta \right) dx.$$

Since the space $\mathbf{V}(\Omega)$ is separable, there exists a fundamental system $\{\phi_k(x)\} \subset \mathbf{V}(\Omega)$. Let us search Galerkin's approximations $u^{(N)} = \sum_{k=1}^N c_{kN} \phi_k(x)$ as the solutions of the system of equations

$$\mathcal{L}_\Omega\left(u^{(N)}, \phi_k\right) = \int_\Omega \Phi \phi_k \, dx, \quad k = 1, 2, \ldots, N.$$

Let $\mathcal{P}_N$ be the linear span of the system $\{\phi_1, \ldots, \phi_N\}$. The norm in $\mathcal{P}_N$ is defined by

$$\|\eta\|_{\mathcal{P}_N} = \|\eta\|_{\mathbf{V}}.$$

Considering $\mathcal{P}_N$ as the N-dimensional space of vectors $\mathbf{v} = (v_1, \ldots, v_N)$ with the elements $\mathbf{v} = \sum_{i=1}^N v_i \phi_i(x)$ we introduce the scalar product $(\mathbf{v}, \mathbf{w})_{\mathcal{P}_N} = \sum_{i=1}^N v_i w_i$ and the norm $|\mathbf{v}|_N = \sqrt{(\mathbf{v}, \mathbf{v})}$. Then

$$\forall \eta \in \mathcal{P}_N, \quad \mathcal{L}_\Omega\left(u^{(N)}, \eta\right) = \int_\Omega \Phi \eta \, dx. \tag{3.8}$$

For every fixed $u^{(N)}$ the left hand-side of this equality is a linear functional on $\mathcal{P}_N$. By Riesz's theorem there exists $g(u^{(N)}) \in \mathcal{P}_N$ such that

$$\forall \eta \in \mathcal{P}_N, \quad \mathcal{L}_\Omega\left(u^{(N)}, \eta\right) = \left(g\left(u^{(N)}\right) - \Phi, \eta\right)_{\mathcal{P}_N}.$$

LEMMA 3.2. *The operator $\mathcal{L}_\Omega$ is coercive*:

$$\forall v \in \mathbf{V}(\Omega), \quad (\mathcal{L}_\Omega v, v) \geqslant C \min\left\{ \left(\frac{\|v\|_{\mathbf{V}}}{n+1}\right)^{p^-}, \left(\frac{\|v\|_{\mathbf{V}}}{n+1}\right)^{p^+} \right\}.$$

(See the proof of Lemma 3.1.)

LEMMA 3.3. *The operator $g(\cdot) : \mathcal{P}_N \mapsto \mathcal{P}_N$ is continuous.*

PROOF. Let $\{v_k\} \subset \mathcal{P}_N$ and $|v_k - v|_N \to 0$. It follows that $|v_k - v|^{\sigma(x)} + \sum_i |D_i(v_k - v)|^{p_i(x)} \to 0$ almost everywhere in Ω. Given an arbitrary $\varepsilon > 0$ one may choose a set $\Omega_\varepsilon \subset \Omega$ such that

$$|\Omega \setminus \Omega_\varepsilon| < \varepsilon,$$

$$\sup_{\Omega_\varepsilon} |\phi_k| + \sum_i \sup_{\Omega_\varepsilon} |\nabla \phi_k| \leqslant K_\varepsilon = \text{const},$$

$$|v_k - v|^{\sigma(x)} + \sum_i \left|D_i(v_k - v)\right|^{p_i(x)} \to 0 \quad \text{uniformly on } \Omega_\varepsilon.$$

Then $\forall \eta \in \mathcal{P}_N$,

$$\bigl(g(v_k) - g(v), \eta\bigr)_N$$
$$= \bigl(\mathcal{L}_{\Omega_\varepsilon}(v_k, \eta) - \mathcal{L}_{\Omega_\varepsilon}(v, \eta)\bigr) + \bigl(\mathcal{L}_{\Omega \setminus \Omega_\varepsilon}(v_k, \eta) - \mathcal{L}_{\Omega \setminus \Omega_\varepsilon}(v, \eta)\bigr).$$

Because of the absolute continuity of the integral

$$\bigl|\mathcal{L}_{\Omega \setminus \Omega_\varepsilon}(v_k, \eta) - \mathcal{L}_{\Omega \setminus \Omega_\varepsilon}(v, \eta)\bigr|$$
$$\leqslant C\bigl(\|v_k\|_{\mathbf{V}} + \|v\|_{\mathbf{V}}\bigr)\|\eta\|_{\mathbf{V}(\Omega \setminus \Omega_\varepsilon)} \to 0 \quad \text{when } \varepsilon \to 0.$$

On the other hand,

$$\bigl|\mathcal{L}_{\Omega_\varepsilon}(v_k, \eta) - \mathcal{L}_{\Omega_\varepsilon}(v, \eta)\bigr| \to 0 \quad \text{when } k \to \infty \text{ for every fixed } \varepsilon.$$

Thus, for every $\eta \in \mathcal{P}_N$ $(g(v_k) - g(v), \eta)_N \to 0$, which means that $g(v_k) \to g(v)$ weakly in $\mathcal{P}_N$. Since the dimension of $\mathcal{P}_N$ is finite, the weak convergence $g(v_k) \to g(v)$ implies the strong convergence: $|g(v_k) - g(v)|_N \to 0$ when $k \to \infty$. $\qquad\square$

LEMMA 3.4. *For every $N \in \mathbb{N}$, the solution $u^{(N)}$ of problem* (3.8) *satisfies the estimate* $\|u^{(N)}\|_{\mathbf{V}} \leqslant K$ *with a finite constant K independent of N.*

PROOF. Let us take $\eta = u^{(N)}$ for the test-function in (3.8) and then apply Young's inequality: $\forall \varepsilon > 0$,

$$a_0 A_{\mathbf{p}(\cdot)}\bigl(\nabla u^{(N)}\bigr) + c_0 A_{\sigma(\cdot)}\bigl(u^{(N)}\bigr)$$
$$\leqslant \varepsilon A_{\mathbf{p}(\cdot)}\bigl(\nabla u^{(N)}\bigr) + C_1(\varepsilon) \sum_i A_{p_i'(\cdot)}(G_i) + \varepsilon A_{\sigma(\cdot)}\bigl(u^{(N)}\bigr) + C_2(\varepsilon) A_{\sigma'(\cdot)}(f).$$

According to (2.3), this inequality provides the estimate $\|u^{(N)}\|_{\mathbf{V}(\Omega)} \leqslant K$ with a constant K depending only on $p^\pm$, $\sigma^\pm$, $\|\mathbf{G}\|_{\mathbf{V}'}$ and $\|f\|_{\sigma'(\cdot)}$. $\qquad\square$

LEMMA 3.5. *For every $N \in \mathbb{N}$, the equation $g(u^{(N)}) = 0$ has at least one solution in $\mathcal{P}_N$.*

PROOF. Consider the family of operators

$$g_\tau(v) = (1 - \tau)v + \tau g(v), \quad \tau \in [0, 1].$$

By Lemma 3.2, for every $|v|_N > n + 1$,

$$\bigl(g_\tau(v), v\bigr)_N = (1 - \tau)(v, v)_N + \tau \mathcal{L}_\Omega(v, v)$$
$$\geqslant C(1 - \tau)|v|_N^2 + \tau C \left(\frac{|v|_N}{n + 1}\right)^{p^-} > 0.$$

It follows that $(g_\tau(v), v)_N > 0$ for all $\tau \in [0, 1]$ and every $\|v\|_{\mathbf{V}} = |v|_N$ sufficiently large, say, for $|v|_N = R_N \geqslant n + 1$. For $\tau = 0$ the equation $g_0(v) = 0$ has in the ball $B_N = \{v \in \mathcal{P}_N : |v|_N < R_N\}$ only the trivial solution, and for every $\tau \in [0, 1]$, the boundary of the ball B_N does not contain any solution of the equation $g_\tau(v) = 0$. According to Brouwer's fixed point theorem, the equation $g_1(v) \equiv g(v) = 0$ has at least one solution in the ball B_N. $\quad\square$

By Lemma 3.4, the sequence of Galerkin's approximations contains a subsequence $\{u^{(N)}\}$ possessing the following properties: there exist functions $u \in \mathbf{V}(\Omega)$, $A_i(x) \in L^{p_i'(x)}(\Omega)$, $v \in L^{\sigma'(x)}(\Omega)$ such that

$$\begin{cases} u^{(N)} \to u & \text{weakly in } \mathbf{V}(\Omega), \\ a_i\big(x, u^{(N)}\big)\big|D_i u^{(N)}\big|^{p_i(x)-2} D_i u^{(N)} \to A_i(x) & \text{weakly in } L^{p_i'(x)}(\Omega), \\ c\big(x, u^{(N)}\big)\big|u^{(N)}\big|^{\sigma(x)-2} u \to v(x) & \text{weakly in } L^{\sigma'(x)}(\Omega), \\ D_i u^{(N)} \to D_i u & \text{weakly in } L^{p_i(x)}(\Omega), \\ u^{(N)} \to u & \text{weakly in } L^{\sigma(x)}(\Omega), \\ u^{(N)} \to u & \text{a.e. in } \Omega. \end{cases} \quad (3.9)$$

These properties allow one to pass to the limit in (3.8) when $N \to \infty$, which gives

$$\int_\Omega \left[\sum_i A_i(x) D_i \eta + v(x)\eta - \Phi\eta \right] dx = 0, \quad \eta \in \mathbf{V}(\Omega). \tag{3.10}$$

To complete the proof, we have to identify the limits $A_i(x)$ and $v(x)$.

LEMMA 3.6. *For almost all $x \in \Omega$,*

$$\begin{cases} A_i(x) = a_i(x, u)|D_i u|^{p_i(x)-2} D_i u, & i = 1, \ldots, n, \\ v(x) = c(x, u)|u|^{\sigma(x)-2} u. \end{cases} \tag{3.11}$$

PROOF. Let us denote $\tilde{a}_i(x, u, D_i v) \equiv a_i(x, u)|D_i v|^{p_i(x)-2} D_i v$. According to (2.10)–(2.11), for every $\xi \in \mathbf{V}(\Omega)$,

$$\sum_i \int_\Omega \big(\tilde{a}_i\big(x, u^{(N)}, D_i u^{(N)}\big) - \tilde{a}_i\big(x, u^{(N)}, D_i \xi\big)\big) D_i\big(u^{(N)} - \xi\big) dx \geqslant 0. \tag{3.12}$$

Let in (3.8) $\eta = u^{(N)} - \xi$ with $\xi \in \mathcal{P}_N$. Combining (3.8) with (3.12), we obtain

$$\int_\Omega \Big[\big(\Phi - c\big(x, u^{(N)}\big)\big|u^{(N)}\big|^{\sigma(x)-2} u^{(N)}\big)\big(u^{(N)} - \xi\big)$$

$$- \sum_i \tilde{a}_i\big(x, u^{(N)}, D_i \xi\big) D_i\big(u^{(N)} - \xi\big) \Big] dx \geqslant 0.$$

Letting $N \to \infty$ and using (3.9), we have that

$$\int_\Omega \left[(\Phi - v)(u - \xi) - \sum_i \tilde{a}_i(x, u, D_i \xi) D_i(u - \xi) \right] dx \geq 0.$$

Adding this inequality to (3.10) with $\eta = u - \xi$, we have: $\forall \xi \in \mathbf{V}(\Omega)$,

$$\sum_i \int_\Omega \left[A_i(x) - \tilde{a}_i(x, u, D_i \xi) \right](u - \xi)\, dx \geq 0.$$

Since ξ is arbitrary, we may take $\xi = u \pm \varepsilon \zeta$ with $\varepsilon > 0$, $\zeta \in \mathbf{V}(\Omega)$. Simplifying and then letting $\varepsilon \to 0$, we conclude that

$$\forall \zeta \in \mathbf{V}(\Omega), \quad \sum_i \int_\Omega \left[A_i(x) - \tilde{a}_i(x, u, D_i u) \right] D_i \zeta\, dx = 0.$$

This gives the first equality of (3.11). By (3.8) and (3.10), we have then that $\forall \eta \in \mathbf{V}(\Omega)$,

$$\sum_i \int_\Omega \left[\tilde{a}_i\big(x, u^{(N)}, D_i u^{(N)}\big) - \tilde{a}_i(x, u, D_i u) \right] D_i \eta\, dx$$

$$+ \int_\Omega c\big(x, u^{(N)}\big)\big(|u^{(N)}|^{\sigma(x)-2} u^{(N)} - |u|^{\sigma(x)-2} u \big) \eta\, dx$$

$$= \int_\Omega \left[v(x) - c\big(x, u^{(N)}\big) |u|^{\sigma(x)-2} u \right] \eta\, dx.$$

By virtue of (3.9), we may pass to the limit when $N \to \infty$ in all three terms of this equality. The two terms on the left-hand side tend to zero, whence

$$\forall \eta \in \mathbf{V}(\Omega), \quad \int_\Omega \left[v(x) - c(x, u) |u|^{\sigma(x)-2} u \right] \eta\, dx = 0. \qquad \Box$$

Estimate (3.7) is a byproduct of Lemma 3.4. $\qquad \Box$

3.1.3. *Boundedness of weak solutions.*

THEOREM 3.3. *Let in the conditions of Theorem 3.2* $\|\Phi(x)\|_{\infty,\Omega} = K < \infty$ *and* $c_0 > 0$. *Then the weak solution of problem* (3.1) *satisfies the estimate*

$$\|u\|_{\infty,\Omega} \leq \max\left\{ 1; \left(\frac{K}{c_0} \right)^{1/(\sigma^- - 1)} \right\}$$

with the constant c_0 *from condition* (3.2).

PROOF. Let us define the constant μ from the relation $K = c_0 \mu^{\sigma^- - 1}$ and then set $M_\varepsilon = \max\{1; \mu + \varepsilon\}$ with an arbitrary $\varepsilon > 0$. Set $\zeta_\varepsilon = \max\{u - M_\varepsilon, 0\}$ and choose this function for the test-function in the integral identity (3.5). Notice that

$$\zeta_\varepsilon = \begin{cases} 0 & \text{if } u \leqslant M_\varepsilon, \\ u - M_\varepsilon & \text{if } u > M_\varepsilon, \end{cases}$$

and

$$\nabla \zeta = \begin{cases} 0 & \text{if } u \leqslant M_\varepsilon, \\ \nabla u & \text{if } u > M_\varepsilon. \end{cases}$$

Identity (3.5) becomes

$$\int_{\Omega \cap (u \geqslant M_\varepsilon)} \left(\sum_{i=1}^{n} a_i |\nabla u|^{p_i(x)} \right.$$
$$\left. + c|u|^{\sigma(x)-2} u(u - M_\varepsilon) - (f + \operatorname{div} \mathbf{G})(u - M_\varepsilon) \right) dx = 0.$$

Applying (3.2) and using the definition of the constant M_ε, we obtain the inequality

$$\left(c_0 M_\varepsilon^{\sigma^- - 1} - K \right) \int_{\Omega \cap (u \geqslant M_\varepsilon)} (u - M_\varepsilon) \, dx \leqslant 0.$$

It follows that $u \leqslant M_\varepsilon$ for every $\varepsilon > 0$. Likewise we establish that $-u \leqslant M_\varepsilon$, whence

$$\|u\|_{\infty,\Omega} \leqslant M_\varepsilon = \max\left\{ 1; \varepsilon + \left(\frac{K}{c_0} \right)^{1/(\sigma^- - 1)} \right\} \quad \text{for every } \varepsilon > 0. \qquad \square$$

THEOREM 3.4. *Let in the conditions of Theorem 3.2 $c_0 \geqslant 0$*

$$\Phi(x) \equiv f(x) \in L^{(q-1)/q}(\Omega) \quad \text{with } q \in \left(1, \frac{np^-}{n - p^-} \right).$$

Then the solutions of problem (3.1) satisfy the estimate $\|u\|_{\infty,\Omega} \leqslant K$ with a constant M depending on $p^\pm$, $\|f\|$, $1/a_0$ and n.

PROOF. Fix an arbitrary $k \in \mathbb{N}$ and take the function $\zeta(x) = \max\{0, u - k\}$ for the test-function in Definition 3.1. Denote $\Omega_k = \Omega \cap \{x \in \Omega : u(x) > k\}$ and notice that

$$\nabla \zeta(x) = \begin{cases} \nabla u & \text{for } u > k, \\ 0 & \text{for } u \leqslant k. \end{cases}$$

According to (3.1),

$$I_1 \equiv a_0 \sum_i \int_{\Omega_k} |D_i u|^{p_i(x)} \, dx$$

$$\leqslant \int_{\Omega_k} \left(\sum_i a_i(x, u) |D_i u|^{p_i(x)} + c(x, u) |u|^{\sigma(x)-2} u (u - k) \right) dx$$

$$= \int_{\Omega_k} (u - k) f \leqslant \|u - k\|_{q, \Omega_k} \|f\|_{q/(q-1), \Omega} \equiv I_2.$$

Not loosing generality we may assume that $|\Omega| \equiv \operatorname{meas} \Omega < 1$ and $A_{\mathbf{p}(x)}(\nabla u) < 1$. Using (2.3), (2.4) we estimate I_1 as

$$I_1 \geqslant a_0 \sum_i \|D_i u\|_{p_i(\cdot)\Omega_k}^{p^+} \geqslant \left(2|\Omega_k| \right)^{-p^+/p^-} \|\nabla u\|_{p^-, \Omega_k}^{p^+}.$$

On the other hand, applying the embedding theorem (2.12) we have

$$I_2 \leqslant c |\Omega_k|^{1/q - 1/p^- + 1/n} \|\nabla u\|_{p^-, \Omega_k} \|f\|_{q/(q-1), \Omega}.$$

Gathering these estimates we obtain the inequality

$$\|\nabla u\|_{p^-, \Omega_k}^{p^-} \leqslant \left(\frac{c}{a_0} \right)^{p^-/(p^+-1)} |\Omega_k|^{1+(1/q+1/n)p^-/(p^+-1)} \|f\|_{q/(q-1), \Omega}^{p^-/(p^+-1)}.$$

The conclusion follows now from [56], Chapter 2, Lemma 5.3. $\square$

3.2. *Generalized diffusion equation*

Let us consider the problem

$$\begin{cases} - \sum_i D_i \left(a_i(x, u) |u|^{\alpha_i(x)} D_i u \right) \\ \quad + c(x, u) |u|^{\sigma(x)-2} u = f(x) & \text{in } \Omega, \\ u = 0 & \text{on } \partial\Omega. \end{cases} \tag{3.13}$$

About the domain Ω and the coefficients in equation (2.1) we assume the following:

(1) $a_i(x, r)$ and $c(x, r)$ are Carathéodory functions (measurable in x for every $r \in \mathbb{R}$ and continuous in r for a.e. $x \in \Omega$),

$$\forall x \in \Omega, r \in \mathbb{R},$$

$$0 < a_0 \leqslant a_i(x, r) \leqslant A_0 < \infty, \qquad 0 \leqslant c_0 \leqslant c(x, r) \leqslant C_0 < \infty \tag{3.14}$$

with some constants a_0, c_0, A_0, C_0;

(2) $\alpha_i(x)$ and $\sigma(x)$ are continuous functions satisfying the conditions

$$\alpha_i(x) \in \left(\alpha_i^-, \alpha_i^+\right) \subseteq \left(\alpha^-, \alpha^+\right) \subset (-1, \infty), \qquad \sigma(x) \in \left(\sigma^-, \sigma^+\right) \subset (1, \infty),$$

(3.15)

$\alpha_i^\pm, \alpha^\pm, \sigma^\pm$ are known constants;

(3) unless specially indicated, $\Omega \subset \mathbb{R}^n$ is a bounded domain with Lipschitz-continuous boundary Γ.

The solution of problem (3.13) is understood in the following way.

DEFINITION 3.2. A locally integrable in Ω function $u(x)$ is called weak solution of problem (3.13), if

(1) $u \in L^\infty(\Omega)$, $|u|^{\alpha_i(x)/2}|D_i u| \in L^2(\Omega)$, $i = 1, \ldots, n$,

(2) $u = 0$ on Γ in the sense of traces,

(3) for every test-function $\eta \in W_0^{1,2}(\Omega) \cap L^{\sigma(x)}(\Omega)$, the integral identity holds

$$\sum_i \int_\Omega a_i(x, u)|u|^{\alpha_i(x)} D_i u D_i \eta \, dx + \int_\Omega c(x, u)|u|^{\sigma(x)-2} u \eta \, dx = \int_\Omega f \eta \, dx.$$

(3.16)

About the function f we assume that

$$f \in L^p(\Omega) \quad \text{with } p > n/2 \text{ if } \alpha_i(x) \geqslant 0 \text{ in } \Omega,$$

$$f \in L^\infty(\Omega) \quad \text{if } \min_i \inf_\Omega \alpha_i(x) < 0.$$

(3.17)

THEOREM 3.5. *Let conditions* (3.14), (3.15) *be fulfilled and, additionally to these conditions,*

$$\left\| D_i \alpha_i(x) \right\|_{2,\Omega} \leqslant C, \quad i = 1, 2, \ldots, n.$$

(3.18)

Let us assume that either $\alpha_i(x) \geqslant 0$ in Ω, or $\alpha_i(x) > -1$ in Ω and $c_0 > 0$. Then for every right-hand side f, satisfying condition (3.17), *problem* (3.13) *has a.e. bounded weak solution which satisfies the inequalities*

$$\sum_i \left\| |u|^{\alpha_i(x)/2} D_i u \right\|_{2,\Omega} + \sum_i \left\| |u|^{\alpha_i(x)} D_i u \right\|_{2,\Omega} + c_0 \|u\|_{\sigma(\cdot),\Omega} + \|u\|_{\infty,\Omega} \leqslant \Lambda$$

(3.19)

with a constant Λ depending on $\|f\|$, $|\Omega|$, n and the constants in conditions (3.14), (3.15).

REMARK 3.1. As is shown further, one can essentially relax condition (3.18) and claim its fulfillment in the following form:

$$\left\| D_i \alpha_i(x) \right\|_{2,\Omega_i^-} \leqslant C, \quad \Omega_i^- = \left\{ x \in \Omega : \alpha_i(x) \leqslant 0 \right\}.$$

3.2.1. *Regularized problem.*   Let us consider the auxiliary nonlinear elliptic problem

$$\begin{cases} -\sum_i D_i\big(A_i(\varepsilon, M, x, u)D_i u\big) \\ \quad + C(\varepsilon, M, x, u)u = f(x) & \text{in } \Omega, \\ u = 0 & \text{on } \Gamma \end{cases} \tag{3.20}$$

with positive parameters ε, M and the coefficients

$$0 < C_i'\big(\varepsilon, M, \alpha^\pm\big) \leqslant A_i \equiv a_i(x, u)\big(\varepsilon^2 + \min\{u^2, M^2\}\big)^{\alpha_i(x)/2}$$
$$\leqslant C_i''\big(\varepsilon, M, \alpha^\pm\big) < \infty,$$
$$0 \leqslant C'\big(\varepsilon, M, \sigma^\pm\big) \leqslant C \equiv c(x, u)\big(\varepsilon^2 + \min\{u^2, M^2\}\big)^{(\sigma(x)-2)/2}$$
$$\leqslant C''\big(\varepsilon, M, \sigma^\pm\big) < \infty.$$

LEMMA 3.7. *For every $\varepsilon > 0$, $\tau \in [0, 1]$, $f(x) \in L^p(\Omega)$ with $p > n/2$, problem (3.20) has a solution $u \in W_0^{1,2}(\Omega) \cap L^\infty(\Omega) \cap L^{\sigma(x)}(\Omega)$ satisfying the estimate*

$$c_0\|u\|_{\sigma(\cdot)} + \|u\|_{\infty,\Omega} \leqslant C \tag{3.21}$$

with a constant C independent of M and ε.

The existence of a weak solution of problem (3.20) is proved by means of the Schauder fixed point principle [24], Chapter 4, Section 10. Let us consider the linear problem

$$\begin{cases} -\sum_i D_i\big(A_i(\varepsilon, M, x, v)D_i u\big) \\ \quad + C(\varepsilon, M, x, v)u = \tau f(x) & \text{in } \Omega, \tau \in [0, 1], v \in L^2(\Omega), \\ u = 0 & \text{on } \Gamma. \end{cases} \tag{3.22}$$

For every given $v \in L^2(\Omega)$ and every $\varepsilon > 0$, $M \geqslant 1$ and $\tau \in [0, 1]$, problem (3.22) has a unique solution $u \in W_0^{1,2}(\Omega)$ [24], Chapter 3, satisfying the integral identity

$$\forall \eta \in W_0^{1,2}(\Omega), \quad \sum_i \int_\Omega A_i D_i u D_i \eta \, dx + \int_\Omega C u \eta \, dx = \tau \int_\Omega f \eta \, dx. \tag{3.23}$$

Let us denote $\mathcal{B}_R = \{v \colon \|v\|_{L^2(\Omega)} < R\}$. Problem (3.22) defines the mapping $(v, \tau) \mapsto u$ which can be represented in the form

$$u = \tau \Phi(v) \colon \mathcal{B}_R \times [0, 1] \mapsto L^2(\Omega)$$

because of the linearity with respect to τ. The solution of problem (3.20) is then a fixed point of the mapping $\Phi(\cdot)$. According to the Schauder principle the mapping $\Phi(v)$ has at least one fixed point in the ball $\mathcal{B}_R$, if

    (1) the mapping $\Phi(v) \colon \mathcal{B}_R \mapsto \mathcal{B}_R$ is continuous and compact,

(2) for every $\tau \in [0, 1)$ all possible solutions of the equation $u = \tau \Phi(v)$ satisfy the estimate $\|u\|_{2,\Omega} < R$.

LEMMA 3.8. *The mapping* $\Phi(v) : \mathcal{B}_R \mapsto \mathcal{B}_R$ *is continuous and compact.*

PROOF. Since the embedding $W_0^{1,2}(\Omega) \subset L^2(\Omega)$ is compact, the mapping $\Phi(v)$ is compact too. The continuity of Φ follows from the continuity of the coefficients A_i and C with respect to the variable v. $\qquad\square$

LEMMA 3.9. *Let* $\alpha_i(x) \geqslant 0$ *and* $f \in L^p(\Omega)$ *with* $p > n/2$. *Then the solution of the regularized problem* (3.20) *with the parameter* $M \geqslant 1$ *satisfies the inequality*

$$\|u\|_{\infty,\Omega} \leqslant K \equiv \left(1 + \frac{C}{a_0}\|f\|_{p,\Omega}^{2/n-1/p}\right)^{(1+2/n-1/p)/(2/n-1/p)} \tag{3.24}$$

with a constant C *not depending on* M.

PROOF. Take an arbitrary number $k \geqslant 1$ and consider the function $\zeta = \max\{0, u - k\}$. The function ζ is an admissible test-function in identity (3.23), besides

$$\nabla \zeta = \begin{cases} 0 & \text{if } u \leqslant k, \\ \nabla u & \text{if } u > k. \end{cases}$$

Let use denote $\Omega_k = \Omega \cap \{u > k\}$. Substituting ζ into the integral identity (3.23) and taking into account the inequality $u \cdot \zeta \geqslant 0$, we get

$$I := \sum_i \int_{\Omega_k} a_i(x, u)\left(\varepsilon^2 + \min\{u^2, M^2\}\right)^{\alpha_i(x)/2}|D_i u|^2 \, dx \leqslant \int_{\Omega_k} f(u - k) \, dx.$$

By virtue of (3.14) and (3.15), on the set Ω_k the inequality holds

$$\varepsilon^2 + \min\{u^2, M^2\} \geqslant \varepsilon^2 + \min\{k^2, M^2\} \geqslant 1,$$

which is why

$$\left(\varepsilon^2 + \min\{u^2, M^2\}\right)^{\alpha_i(x)/2} \geqslant \frac{1}{k}.$$

Hence

$$\int_{\Omega_k} |\nabla u|^2 \, dx \leqslant \frac{k}{a_0} \int_{\Omega_k} f(u - k) \, dx \equiv \frac{k}{a_0} I_3.$$

Estimating I_3 by Hölder's inequality and then applying the embedding theorem, we arrive at the estimate

$$|I_3| \leqslant |\Omega_k|^{1/2-1/p}\|u - k\|_{2,\Omega_k}\|f\|_{p,\Omega} \leqslant C|\Omega_k|^{1/2+2/n-1/p}\|f\|_{p,\Omega}\|\nabla u\|_{2,\Omega_k}.$$

Thus

$$\int_{\Omega_k} |\nabla u|^2 \, dx \leqslant k^2 \left(\frac{C}{a_0}\right)^2 |\Omega_k|^{1+2/n-2/p} \|f\|^2_{p,\Omega}$$

with a constant C not depending on M. Using [24], Chapter 2, Lemma 5.3, we conclude that for $p > n/2$ the inequality $\|u\|_{\infty,\Omega} \leqslant K$ holds with a constant K which depends on n, p, $1/a_0$ and $\|f\|_{p,\Omega}$ in the following way [24], Chapter 2, Lemma 5.1:

$$1 \leqslant K \leqslant \left(1 + \frac{C}{a_0} \|f\|^{2/n-1/p}_{p,\Omega}\right)^{(1+2/n-1/p)/(2/n-1/p)}. \qquad \square$$

LEMMA 3.10. *Let $c_0 > 0$ and $f \in L^\infty(\Omega)$. Then the solution of the regularized problem* (3.20) *with the parameter $M \geqslant 1$ satisfies the inequality*

$$\|u\|_{\infty,\Omega} \leqslant K \equiv \max\left\{1, \left(\frac{1}{c_0}\|f\|_{\infty,\Omega}\right)^{1/(\sigma^- - 1)}\right\}. \tag{3.25}$$

PROOF. Arguing like in the proof of Lemma 3.9, we take for the test-function $\zeta = \max\{u - k, 0\}$ with the parameter $k \geqslant 1$. Substituting ζ into the integral identity (3.23), we obtain the inequality

$$\sum_i \int_{\Omega_k} A_i |D_i u|^2 \, dx + c_0 \int_{\Omega_k} \min\{k, M\}^{\sigma(x)-1} (u-k) \, dx \leqslant \tau \int_{\Omega_k} f(u-k) \, dx,$$

whence the estimate

$$0 \geqslant \left(c_0 k^{\sigma^- - 1} - \tau \|f\|_{\infty,\Omega}\right) \int_{\Omega_k} (u-k) \, dx.$$

Increasing k, we conclude that necessarily $|\Omega_k| = 0$ for

$$k \geqslant k_0 \equiv \left(\frac{1}{c_0}\|f\|_{\infty,\Omega}\right)^{1/(\sigma^- - 1)}.$$

The inequality $-u \leqslant k_0$ a.e. in Ω is established likewise. $\qquad \square$

The obtained estimates on the maximum of the solution of problem (3.20) do not depend on M, which allows us to choose $M = K$. Then the coefficients in equations (3.20), (3.22) take on the form

$$A_i(\varepsilon, M, x, w) = \left(\varepsilon^2 + w^2\right)^{\alpha_i(x)/2}, \qquad C(\varepsilon, M, x, w) = \left(\varepsilon^2 + w^2\right)^{(\sigma(x)-2)/2},$$

and (3.20) transforms into a problem with the single regularization parameter ε. Not indicating it specially in what follows, we always assume that $M = K$ with the constant K from (3.24) or (3.25).

LEMMA 3.11. *For $c_0 > 0$, the fixed points of the mapping Φ satisfy the inequality*

$$c_0 \int_\Omega |u|^{\sigma(x)}\, dx \leqslant K \|f\|_{\infty,\Omega} |\Omega|$$

with the constant K from (3.25).

PROOF. Let us substitute u into (3.23) as the test-function and then drop the nonnegative term on the left-hand side of the appearing equality

$$c_0 \int_\Omega \left(\varepsilon^2 + u^2\right)^{(\sigma(x)-2)/2} u^2\, dx \leqslant \int_\Omega |f||u|\, dx \leqslant K|\Omega|\,\|f\|_{\infty,\Omega}. \qquad \square$$

The proof of Lemma 3.7 is complete.

3.2.2. *Passage to the limit.*

LEMMA 3.12. *If condition* (3.17) *is fulfilled, then the solution of problem* (3.20) *satisfies the inequalities*

$$\sum_i \left\|\sqrt{A_i}\, D_i u\right\|_{2,\Omega}^2 \leqslant C, \qquad \sum_i \|A_i\, D_i u\|_{2,\Omega}^2 \leqslant C \tag{3.26}$$

with a constant C independent of ε.

PROOF. The former of inequalities (3.26) follows from (3.23) with $\eta = u$. To prove the latter we make use of the following inequality which is a byproduct of (3.24), (3.25):

$$\left(\frac{\varepsilon^2 + u^2}{\varepsilon^2 + K^2}\right)^{\alpha_i(x)/2} \leqslant \left(\frac{\varepsilon^2 + u^2}{\varepsilon^2 + K^2}\right)^{\alpha^-/2} \qquad \text{a.e. in } \Omega.$$

Let us introduce the function

$$\phi(u) \equiv A_0 \int_0^u \left(\varepsilon^2 + s^2\right)^{\alpha^-/2}\, ds \in L^\infty(\Omega) \cap W_0^{1,2}(\Omega).$$

For $\alpha^- > -1$ the function $|\phi(u)|$ is uniformly with respect to ε bounded by the constant $\phi(K)$. Substituting $\phi(u)$ into (3.23) as the test-function, we obtain the inequality

$$\sum_i \int_\Omega A_i \left(\varepsilon^2 + u^2\right)^{\alpha^-/2} |D_i u|^2\, dx + c_0 \int_\Omega \left(\varepsilon^2 + u^2\right)^{(\sigma(x)-2)/2} u\phi(u)\, dx$$

$$\leqslant \int_\Omega |f||\phi(u)|\, dx.$$

Noting that $u\phi(u) \geqslant 0$, we have

$$\int_{\Omega} A_i^2 |D_i u|^2 \, dx \leqslant \sum_i \int_{\Omega} a_i^2(x, u)\big(\varepsilon^2 + u^2\big)^{\alpha_i(x)} |D_i u|^2 \, dx$$

$$\leqslant \phi(K) \begin{cases} |\Omega| \|f\|_{\infty, \Omega} & \text{for } f \in L^{\infty}(\Omega), \\ |\Omega|^{1/p'} \|f\|_{p, \Omega} & \text{for } f \in L^p(\Omega). \end{cases} \qquad \square$$

Denote by $\{u_{\varepsilon}\}$ the sequence of solutions of the regularized problems (3.20). Estimates (3.21) and (3.26) allow us to extract from $\{u_{\varepsilon}\}$ a subsequence (which will be assumed to coincide with the whole of the sequence), possessing the properties:

$$\begin{aligned} A_i D_i u_{\varepsilon} &\to B_i(x) & &\text{weakly in } L^2(\Omega), \\ u_{\varepsilon} &\to u & &\text{a.e. in } \Omega, \\ \big(\varepsilon^2 + u_{\varepsilon}^2\big)^{(\sigma(x)-2)/2} u_{\varepsilon} &\to |u|^{\sigma(x)-2} u & &\text{weakly in } L^2(\Omega). \end{aligned} \qquad (3.27)$$

Taking into account (3.27) and passing to the limit when $\varepsilon \to 0$ in identity (3.23), we have that, for every $\eta \in W_0^{1,2}(\Omega)$,

$$\sum_i \int_{\Omega} B_i(x) D_i \eta \, dx + \int_{\Omega} c(x, u) |u|^{\sigma(x)-2} u \eta \, dx = \int_{\Omega} f \eta \, dx.$$

It remains to show that $B_i(x) = |u|^{\alpha_i(x)} D_i u$. Consider the function

$$G_i(u) = \big(\varepsilon^2 + u^2\big)^{\alpha_i(x)} u.$$

It is easy to calculate that

$$D_i G_i(u) = \left[1 + \alpha_i \frac{u^2}{\varepsilon^2 + u^2}\right] \big(\varepsilon^2 + u^2\big)^{\alpha_i/2} D_i u$$

$$+ \frac{1}{2} u \big(\varepsilon^2 + u^2\big)^{\alpha_i/2} D_i \alpha_i \ln\big(\varepsilon^2 + u^2\big),$$

$$\big|D_i G_i(u)\big|^2 \leqslant C\big[|A_i|^2 |D_i u|^2 + \big|D_i \alpha_i(x)\big|^2\big].$$

By virtue of (3.26), $\|D_i G_i(u_{\varepsilon})\|_{2, \Omega} \leqslant C$ uniformly with respect to ε. This allows us to extract a subsequence $\{u_{\varepsilon}\}$ such that $D_i G_i(u_{\varepsilon}) \to D_i H_i$ weakly in $L^2(\Omega)$. Hence, for every test-function ζ,

$$\begin{aligned} \lim_{\varepsilon \to 0} \int_{\Omega} D_i G_i(u_{\varepsilon}) \zeta \, dx &= \int_{\Omega} D_i H_i \zeta \, dx \\ &= -\lim_{\varepsilon \to 0} G_i(u_{\varepsilon}) D_i \zeta \, dx \\ &= -\int_{\Omega} |u|^{\alpha_i(x)} u D_i \zeta \, dx, \end{aligned}$$

whence the equality

$$D_i H_i = D_i\big(|u|^{\alpha_i(x)}u\big) = \big(1+\alpha_i(x)\big)|u|^{\alpha_i(x)}D_i u + \frac{1}{2}|u|^{\alpha_i(x)}u\big(\ln u^2\big)D_i\alpha_i(x).$$

Further,

$$A_i(\varepsilon, M, x, u_\varepsilon)D_i u_\varepsilon$$
$$= a_i(x, u_\varepsilon)\frac{\varepsilon^2 + u_\varepsilon^2}{(1+\alpha_i(x))u_\varepsilon^2 + \varepsilon^2}$$
$$\times \left[D_i G_i(u_\varepsilon) - \frac{1}{2}u\big(\varepsilon^2 + u_\varepsilon^2\big)^{\alpha_i(x)/2} D_i\alpha_i(x)\ln\big(\varepsilon^2 + u_\varepsilon^2\big)\right],$$

which yields that, for every $\eta \in W_0^{1,2}(\Omega)$,

$$\sum_i \int_\Omega A_i(\varepsilon, M, x, u_\varepsilon)D_i u_\varepsilon D_i\eta\,dx$$
$$\to \sum_i \int_\Omega a_i(x, u)|u|^{\alpha_i(x)}D_i u D_i\eta\,dx \quad \text{as } \varepsilon \to 0.$$

PROOF OF REMARK 3.1. Condition (3.18) was solely used to justify the limit passage in problem (3.20). Let us show that on the set $\{x \in \Omega: \alpha_i(x) > 0\}$ the limit passage is possible without this condition. Not loosing generality, we assume that $\alpha^- > 0$. Let us represent the first term on the left-hand side of (3.23) as

$$I := \sum_i \int_\Omega a_i(x, u_\varepsilon)\big(\varepsilon^2 + u_\varepsilon^2\big)^{\alpha_i(x)/2} D_i u_\varepsilon D_i\eta\,dx = I_{\delta,\varepsilon}^{(1)} + I_{\delta,\varepsilon}^{(2)}$$
$$\equiv \sum_i \int_{\Omega_\delta}\cdots + \sum_i \int_{\Omega\setminus\Omega_\delta}\cdots,$$

where $\Omega_\delta = \{x \in \Omega: |u_\varepsilon| > \delta\}$ and $\delta > 0$ is an arbitrary small number. It is easy to see that if (3.14), (3.15) and (3.26) are fulfilled, then

$$\mu(\delta)\sum_i \int_{\Omega_\delta} |D_i u_\varepsilon|^2\,dx \leqslant \sum_i \int_{\Omega_\delta} a_i(x, u_\varepsilon)\big(\varepsilon^2 + u_\varepsilon^2\big)^{\alpha_i(x)/2}|D_i u_\varepsilon|^2\,dx \leqslant C$$

with a constant $\mu(\delta) > 0$. Hence, the sequences $\{D_i u_\varepsilon\}$ converge weakly in $L^2(\Omega_\delta)$, while the sequences

$$\big\{a_i(x, u_\varepsilon)\big(\varepsilon^2 + u_\varepsilon^2\big)^{\alpha_i(x)/2}\big\}, \quad i = 1, \ldots, n,$$

converge almost everywhere. Therefore, for every fixed $\delta > 0$, there exists

$$\lim_{\varepsilon \to 0} I_{\delta,\varepsilon}^{(1)} = \sum_i \int_{\Omega_\delta} a(x, u)|u|^{\alpha_i(x)} D_i u D_i \eta \, dx.$$

Taking into account (3.14) and (3.26), we estimate $I_{\delta,\varepsilon}^{(2)}$ as

$$\left| I_{\delta,\varepsilon}^{(2)} \right| \leqslant \sqrt{A_0}(\delta^2 + \varepsilon^2)^{\alpha^-/2} \left\| \sqrt{A_i} D_i u_\varepsilon \right\|_{2,\Omega} \| D_i \eta \|_{2,\Omega} \leqslant C(\delta^2 + \varepsilon^2)^{\alpha^-/2}.$$

Thus, $\lim_{\varepsilon \to 0} |I_{\varepsilon,\delta}^{(2)}| \leqslant C\delta^{\alpha^-}$ and, for every $\delta > 0$,

$$\lim_{\varepsilon \to 0} I = \sum_i \int_{\Omega_\delta} a_i(x, u)|u|^{\alpha_i(x)} D_i u D_i \eta \, dx + o(\delta^{\alpha^-}).$$

Plugging estimates (3.26) and applying the Lebesgue dominated convergence theorem, we finally obtain

$$\lim_{\delta \to 0} \lim_{\varepsilon \to 0} |I| = \sum_i \int_{\Omega} a_i(x, u)|u|^{\alpha_i(x)} D_i u D_i \eta \, dx. \qquad \square$$

3.3. *Equations with convection terms*

The above-described scheme of proof of existence of weak solutions is applicable to more general equations of the form

$$-\sum_i \left[D_i \left(a_i(x, u)|D_i u|^{p_i(x)-2} D_i u \right) + b_i(x, u) D_i u \right]$$

$$+ c(x, u)|u|^{\sigma(x)-2} u = \Phi(x) \qquad (3.28)$$

with the right-hand side

$$\Phi(x) = f(x) + \operatorname{div} \mathbf{G}(x) + g(x, u).$$

In equation (3.28) $b_i(x, r)$ and $g(x, r)$ are Carathéodory functions satisfying the following conditions: either

$$\begin{cases} \left| b_i(x, r) r \right|^{p_i'(x)} \leqslant C|u|^\gamma + h(x), & h \in L^1(\Omega), h \geqslant 0, \\ \frac{p(x)}{p(x)-1} \leqslant \gamma < \sigma^- \leqslant \sigma(x), \\ \left| g(x, r) \cdot r \right| \leqslant C|u|^\beta + g_0(x), & g_0 \in L^1(\Omega), g \geqslant 0, \\ 0 \leqslant \beta(x) < \sigma^- \leqslant \sigma(x), \end{cases} \qquad (3.29)$$

with some constant C if $c(x,u) \geqslant c_0 > 0$, or

$$|b_i(x,r)r|^{p_i'(x)} \leqslant C|u|^{\gamma} + h(x), \quad h \in L^1(\Omega), \tag{3.30}$$

$$|g(x,r)r| \leqslant C|u|^{\gamma} + g_0(x), \qquad 0 \leqslant \gamma < p^-, g_0 \in L^1(\Omega), \tag{3.31}$$

if $c(x,u) \geqslant 0$. The principal point of the proof is to insure that conditions (3.29) (or (3.30), (3.31)) guarantee the validity of an estimate of the type (3.7) for Galerkin's approximations and, respectively, for the prospective weak solution. Let us first derive such an estimate under conditions (3.29). For the simplicity of notations we will write u for both the Galerkin approximations and for exact solution. The following inequalities hold:

$$\int_{\Omega} \left[a_0 \sum_i |D_i u|^{p_i(x)} + c_0 |u|^{\sigma(x)} \right] dx$$

$$\leqslant \int_{\Omega} \left[\sum_i a_i(x,u)|D_i u|^{p_i(x)} + c(x,u)|u|^{\sigma(x)} \right] dx$$

$$= \int_{\Omega} \left[fu - \mathbf{G}(x)\nabla u \right] dx + I_1 + I_2, \tag{3.32}$$

where

$$I_1 = \int_{\Omega} gu\, dx, \qquad I_2 = \int_{\Omega} \sum_i b_i(x,u)D_i uu\, dx.$$

All the terms on the right-hand side of this inequality except I_1 and I_2 are already estimated previously (see the proof of Theorem 3.2). Let us estimate I_1 and I_2. Using (3.29) and applying Young's inequality (2.17), we may write

$$|I_1| \leqslant \int_{\Omega} \left[C|u|^{\beta(x)} + g_0(x) \right] dx$$

$$\leqslant \frac{c_0}{4} \int_{\Omega} |u|^{\sigma(x)}\, dx + C\left(1 + \int_{\Omega} g_0(x)\, dx \right), \tag{3.33}$$

$$|I_2| \leqslant \frac{a_0}{2} \int_{\Omega} \sum_i |D_i u|^{p_i(x)}\, dx$$

$$+ \frac{c_0}{4} \int_{\Omega} |u|^{\sigma(x)}\, dx + C\left(1 + \int_{\Omega} h(x)\, dx \right). \tag{3.34}$$

Here C is a constant which may depend on $p^{\pm}$, $\sigma^{\pm}$, c_0, a_0, $\min[\sigma(x) - \gamma]$, $\min[\sigma(x) - \beta(x)]$ and $|\Omega|$. Gathering (3.32)–(3.34) we arrive at the estimate

$$\int_{\Omega} \left[a_0 \sum_i |D_i u|^{p_i(x)} + c_0 |u|^{\sigma(x)} \right] dx \leqslant K. \tag{3.35}$$

Let us turn to the case (3.30). By virtue of (3.31) and (3.32),

$$|I_1| \leqslant \int_\Omega \left[C|u|^\gamma + g_0(x) \right] dx.$$

(3.36)

Applying the embedding theorems and (2.3), (2.5), we obtain the inequality

$$\int_\Omega |u|^\gamma \, dx \leqslant C \left(\sum_i \|D_i u\|_{p^-,\Omega} \right)^\gamma$$
$$\leqslant C \left(\sum_i \|D_i u\|_{p_i,\Omega} \right)^\gamma$$
$$\leqslant C \left[\sum_i \int_\Omega \left(|D_i u|^{p_i(x)} + 1 \right) dx \right]^{\gamma/p^-}$$
$$\leqslant \frac{a_0}{4} \int_\Omega \sum_i |D_i u|^{p_i(x)} \, dx + C$$

(3.37)

with the exponent $\gamma < p^- < np^-/(n - p^-)$. The term I_2 is estimated as

$$|I_2| \leqslant \frac{a_0}{4} \int_\Omega \sum_i |D_i u|^{p_i(x)} \, dx + C \int_\Omega \left(|u|^\gamma + h(x) \right) dx.$$

(3.38)

Combining (3.32), (3.37) and (3.38), we finally have that

$$\int_\Omega \sum_i |D_i u|^{p_i(x)} \, dx \leqslant K.$$

(3.39)

Estimates (3.35), (3.39) hold for all Galerkin's approximations. Because of the linearity of the new terms with respect to $D_i u$ the limit passage in these terms delivers no difficulty.

4. Uniqueness theorems

Before starting the study of uniqueness of weak solutions of equations of the type (1.1), let us mention that the solutions of equations with variables exponents of nonlinearity possess a peculiar property which illustrates the difference between the function spaces $W_0^{1,p}(\Omega)$ and $W_0^{1,p(x)}(\Omega)$. Let us consider the eigenvalue problem for $p(x)$-Laplace equation

$$- \operatorname{div}\left(|\nabla u|^{p(x)-2} \nabla u \right) = \lambda |u|^{p(x)-2} u \quad \text{in } \Omega, \qquad u = 0 \quad \text{on } \Gamma.$$

(4.1)

It is well known that if $p(x)$ is a constant, then the first (minimal) eigenvalue λ_1 must be positive [56]. This need not be true, however, if $p(x)$ is variable: in this case λ_1 may be zero. This means that, for every $0 < \lambda < \lambda_1$, problem (4.1) with $p = \text{const}$ has but trivial solution. In contrast to this situation, for variable $p(x)$ problem (4.1) has the trivial solution if and only if $\lambda = 0$. According to the minimax property of eigenvalues of problem (4.1) the first eigenvalue of the $p(x)$-Laplacian can be defined by the relation

$$\lambda_1 = \inf\left\{ \frac{\int_\Omega |Du|^{p(x)}\, dx}{\int_\Omega |u|^{p(x)}\, dx} : u \in W_0^{1,p(x)}(\Omega), u \neq 0 \right\}.$$

The following example taken from [41] shows that λ_1 may be zero. Let $\Omega = (-2, 2) \subset \mathbb{R}$,

$$p(x) = \begin{cases} 3 & \text{if } 0 \leqslant x \leqslant 1, \\ 4 - |x| & \text{if } 1 \leqslant |x| \leqslant 2, \end{cases}$$

and

$$u(x) = \begin{cases} 1 & \text{for } 0 \leqslant |x| \leqslant 1, \\ 2 - |x| & \text{for } 1 \leqslant |x| \leqslant 2. \end{cases}$$

Obviously, $au(x) \in W_0^{1,p(x)}(\Omega)$ for every $a = \text{const} > 0$. The straightforward computation shows, however, that

$$\int_\Omega |au'(x)|^{p(x)}\, dx = \frac{2a^2}{\ln a}(a - 1), \qquad \int_\Omega |au|^{p(x)}\, dx \geqslant 2a^3,$$

whence

$$\lim_{a \to \infty} \frac{\int_\Omega |Du|^{p(x)}\, dx}{\int_\Omega |u|^{p(x)}\, dx} = 0.$$

This difference between $W_0^{1,p}(\Omega)$ and $W_0^{1,p(x)}(\Omega)$ shows that in $W_0^{1,p(x)}(\Omega)$ the variational problem corresponding to problem (4.1) becomes very complicated. In particular, the standard methods of proof of the existence of trivial solutions need not work in the case of equations of the type (4.1). Let $u(x)$ be a weak solution of equation (4.1). Every weak solution satisfies the energy relation

$$\int_\Omega |\nabla u|^{p(x)}\, dx = \lambda \int_\Omega |u|^{p(x)}\, dx \tag{4.2}$$

which is true for $p = \text{const}$ as well as for $p(x)$ being a function of x. Let $p = \text{const}$. Due to the Poincaré inequality

$$\left(\int_\Omega |u|^p\, dx \right)^{1/p} = \|u\|_{p,\Omega} \leqslant C\|\nabla u\|_{p,\Omega} = C\left(\int_\Omega |\nabla u|^p\, dx \right)^{1/p}, \tag{4.3}$$

whence

$$\left(1 - |\lambda| C^p\right) \int_\Omega |u|^p \, dx \leqslant 0,$$

which means that $\int_\Omega |u|^p \, dx = 0$ for $|\lambda| C^p < 1$.

Let now $p(x)$ be a function of x. Applying the Poincaré inequality and inequality (2.3), we have that

$$\left(\int_\Omega |u|^p \, dx\right)^{1/p^-} \leqslant \|u\|_{p(\cdot),\Omega}$$

$$\leqslant C \|\nabla u\|_{p(\cdot),\Omega}$$

$$\leqslant C \left(\int_\Omega |\nabla u|^p \, dx\right)^{1/p^+}. \tag{4.4}$$

Without loss of generality we may assume that all the norms of less than one. Then

$$\int_\Omega |u|^{p(x)} \, dx \leqslant |\lambda|^{p^-/p^+} C^{p^-} \left(\int_\Omega |u|^{p(x)} \, dx\right)^{p^-/p^+} \tag{4.5}$$

and since $p^- < p^+$ for every variable $p(x)$, the last inequality has a trivial solution iff $\lambda = 0$.

Returning to the Dirichlet problem for equations of the type (1.1), (3.1), let us notice first that in certain simple situations the uniqueness is obvious. For example, the solution of the problem

$$\begin{cases} -\sum_i D_i \left(a_i(x, u) |D_i u|^{p_i(x)-2} D_i u\right) \\ \quad + f(x, u) = 0 \quad \text{in } \Omega, \\ u = g \quad\quad\quad\quad\text{on } \partial\Omega \end{cases} \tag{4.6}$$

is unique if

$$a_i(x, u) = a_i(x) \geqslant a_0 > 0, \qquad \left(f(x, u_1) - f(x, u_2)\right)(u_1 - u_2) \geqslant 0. \tag{4.7}$$

The uniqueness of a weak solution to problem (1.1) with

$$a_i = a_i(x) \geqslant a_0 \geqslant 0, \qquad c = c(x) \geqslant c_0 \geqslant 0, \qquad a_0^2 + c_0^2 > 0,$$

is a direct byproduct of inequalities (2.10)–(2.11).

4.1. *Uniqueness of solution of the generalized $p(x)$-Laplace equation*

We proceed to establish conditions of uniqueness of weak solutions to the generalized $p(x)$-Laplace equation.

Let u_1 and u_2 be two different solutions to problem (3.1). Introduce the function $u = u_1 - u_2$. Taking it for the test-function in the integral identities for u_1 and u_2, we obtain the relation

$$\int_\Omega \left(\sum_i a_{1i} \left(|D_i u_1|^{p_i-2} \nabla u_1 - |D_i u_2|^{p_i-2} \nabla u_2 \right) \nabla u \right) dx$$

$$+ \int_\Omega c_1 \left(|u_1|^{\sigma-2} u_1 - |u_2|^{\sigma-2} u_2 \right) u \, dx$$

$$= \int_\Omega \left(\sum_i (a_{2i} - a_{1i}) |D_i u_2|^{p_i-2} \nabla u_2 \nabla u + (c_2 - c_1) |u_2|^{\sigma-2} u_2 u \right) dx, \quad (4.8)$$

where $a_{ki} = a_i(x, u_k)$, $c_k = c(x, u_k)$ with $k = 1, 2$, $i = 1, \ldots, n$.

THEOREM 4.1. *Let $a_i(x, u) \equiv a_i(x) \geqslant 0$ and $c(x, u) \geqslant c_0 > 0$,*

$$1 < p^- \leqslant p_i^- \leqslant p_i(x) \leqslant p_i^+ \leqslant p^+ < \infty, \quad 1 < \sigma^- \leqslant \sigma(x) \leqslant 2,$$

$$\left| c(x, u_1) - c(x, u_2) \right| \leqslant L |u_1 - u_2|, \quad L = \text{const}, \quad (4.9)$$

and the weak solution u of problem (3.1) satisfies the inequality $\|u\|_{\infty,\Omega} \leqslant M$. Then the solution u is unique if

$$\max_\Omega M^{\sigma(x)-1} < \frac{c_0}{L} \min_\Omega \left[(\sigma(x) - 1) \left(2M^{\sigma(x)} \right)^{(\sigma(x)-2)/\sigma(x)} \right]. \quad (4.10)$$

REMARK 4.1. Notice that under the condition $\sigma(x) \in (1, 2]$ the left-hand side of (4.10) tends to zero when $M \downarrow 0$, while the right-hand side is at least separated away from zero. It follows that the class of uniqueness is not empty because the value of M is controlled through the data of problem (3.1) (see Theorem 3.3).

PROOF OF THEOREM 4.1. Let us write relation (4.8) in the form

$$I := \int_\Omega \left(\sum_i a_i \left(|D_i u_1|^{p_i-2} \nabla u_1 - |D_i u_2|^{p_i-2} \nabla u_2 \right) \nabla u \right) dx$$

$$+ \int_\Omega c_1 \left(|u_1|^{\sigma-2} u_1 - |u_2|^{\sigma-2} u_2 \right) u \, dx$$

$$= \int_\Omega (c_2 - c_1) |u_2|^{\sigma-2} u_2 u \, dx. \quad (4.11)$$

Applying (2.10)–(2.11) we see that the first term on the left-hand side of this inequality is nonnegative. Applying (2.11) and boundedness of u, we obtain the following chain of relations:

$$\lambda_0 \int_\Omega |u|^2 \, dx \leqslant I \leqslant \lambda_1 L \int_\Omega |u|^2 \, dx \quad \Longrightarrow \quad (\lambda_0 - \lambda_1 L) \int_\Omega |u|^2 \, dx \leqslant 0$$

with the constants

$$\lambda_0 = c_0 \min_{\Omega}\left[(\sigma(x) - 1)\left(2M^{\sigma(x)}\right)^{(\sigma(x)-2)/\sigma(x)}\right],$$

and

$$\lambda_1 = \max_{\Omega}\left[M^{\sigma(x)-1}\right].$$

According to (4.10) we come to desired assertion. $\square$

THEOREM 4.2. *Let in equation* (3.1) $a_i(x, u) \equiv a_i(x) \geqslant a_0 > 0$, $c(x, u) \geqslant 0$ *and the exponents* $p_i(x)$, $\sigma(x)$ *satisfy the conditions*

$$\frac{2n}{2+n} < p_i^- \leqslant p_i(x) \leqslant p_i^+ \leqslant 2, \qquad 2 \leqslant \sigma(x) \leqslant \sigma^+ < \infty. \tag{4.12}$$

Let $c(x, t)$ be Lipschitz-continuous with respect to t with the Lipschitz constant L. Then there exists a constant $\varepsilon_0 > 0$, depending on the structural constants in conditions (3.2)–(3.4), *L, n, and the constant in Sobolev's embedding theorem such that the solution of problem* (3.1) *is unique if*

$$\|u\|_{\infty,\Omega} + \|u\|_{\mathbf{V}} < \varepsilon_0.$$

PROOF. Let u_1, u_2 be two different solutions of problem (3.1) satisfying

$$\|u_i\|_{\infty,\Omega} \leqslant M, \qquad \|u_i\|_{\mathbf{V}} \leqslant K, \quad i = 1, 2.$$

Using (4.11), (3.2) and (2.11), we may write

$$a_0(p^- - 1) \sum_i \int_{\Omega} |D_i u|^2 \left(|D_i u_1|^{p_i(x)} + |D_i u_2|^{p_i(x)}\right)^{(p_i(x)-2)/p_i(x)} dx$$

$$\leqslant \lambda_1 L \int_{\Omega} |u|^2 \, dx \tag{4.13}$$

with the constant λ_1 defined in the proof of the previous theorem. Let us make use of the inverse Hölder's inequality: for every constant $q \in (0, 1)$,

$$\left(\int_{\Omega} |g|^{q/(q-1)} \, dx\right)^{(q-1)/q} \left(\int_{\Omega} |f|^q \, dx\right)^{1/q} \leqslant \int_{\Omega} |f||g| \, dx. \tag{4.14}$$

Let us take a constant

$$q \in \left(\frac{p^-}{2}, 1\right) \subseteq \left(\frac{n}{n+2}, 1\right).$$

Gathering (4.14) with (4.13), we obtain the estimate

$$a_0(p^- - 1) \sum_i \left(\int_\Omega |D_i u|^{2q} \, dx \right)^{1/q}$$

$$\times \left(\int_\Omega \left(|D_i u_1|^{p_i(x)} + |D_i u_2|^{p_i(x)} \right)^{(2-p_i(x))q/(p_i(x)(1-q))} dx \right)^{(q-1)/q}$$

$$\leqslant \lambda_1 L \int_\Omega |u|^2 \, dx. \qquad (4.15)$$

In our choice of the constant q,

$$\frac{(2 - p_i(x))q}{p_i(x)(1 - q)} \leqslant 1 \quad \text{in } \Omega \text{ for every } i = 1, \ldots, n.$$

By Hölder's inequality (2.4) and due to (2.3), (3.7),

$$\int_\Omega \left(|D_i u_1|^{p_i(x)} + |D_i u_2|^{p_i(x)} \right)^{(2-p_i(x))q/(p_i(x)(1-q))} dx$$

$$\leqslant C_0(K, p^\pm, q, \Omega) \to 0 \quad \text{as } K \to 0.$$

Since $q < 1$, then for every $i = 1, \ldots, n$,

$$\left(\int_\Omega \left(|D_i u_1|^{p_i(x)} + |D_i u_2|^{p_i(x)} \right)^{(2-p_i(x))q/(p_i(x)(1-q))} dx \right)^{(q-1)/q}$$

$$\geqslant \left\{ (C_0)^{(q-1)/q} \right\} \equiv \delta \to 0 \quad \text{as } K \to 0,$$

which allows us to rewrite (4.15) as

$$\delta a_0(p^- - 1) \sum_i \left(\int_\Omega |D_i u|^{2q} \, dx \right)^{1/q} \leqslant \lambda_1 L \int_\Omega |u|^2 \, dx. \qquad (4.16)$$

Since $2nq/(n - 2q) \geqslant 2$, we may apply Sobolev's embedding theorem

$$\int_\Omega |u|^2 \, dx \leqslant C \left(\int_\Omega |\nabla u|^{2q} \, dx \right)^{1/q} \leqslant C_1 \sum_i \left(\int_\Omega |D_i u|^{2q} \, dx \right)^{1/q}.$$

Substituting this inequality into (4.16) we have that

$$\left[\delta a_0(p^- - 1)C_1^{-1} - \lambda_1 L \right] \int_\Omega |u|^2 \, dx \leqslant 0.$$

Since $\lambda_1 \downarrow 0$ when $M \downarrow 0$, and $\delta \uparrow \infty$ when $K \downarrow 0$, there exists a constant $\varepsilon_0 > 0$ depending on a_0, $p^\pm$, $\sigma^\pm$, n, L and the constant in the embedding theorem, such that

$$\|u\|_{2,\Omega} = 0 \quad \text{if } 0 < M + K < \varepsilon_0. \qquad \square$$

The assertion of Theorem 4.2 can be transformed into a sufficient condition of uniqueness of bounded solution of problem (3.1) formulated as a restriction on the coefficient $c(x, u)$.

COROLLARY 4.1. *Let the exponents $p_i(x)$ and $\sigma(x)$ satisfy the conditions of Theorem 4.2. Then, given a constant $M > 0$, one may indicate a constant ε_0 such that the weak solution of problem (3.1) is unique in the class of functions $\{\|u\|_{\infty,\Omega} \leqslant M\}$ provided that $L < \varepsilon_0$.*

COROLLARY 4.2. *The assertions of Theorems 4.1 and 4.2 remain valid for the weak solutions of equation (3.1) with the right-hand side $\Phi \equiv \phi(x, u) + f(x) + \operatorname{div} \mathbf{G}(x)$, provided that $|\phi(x, u_1) - \phi(x, u_2)| \leqslant L|u_1 - u_2|$, $L = \text{const}$.*

REMARK 4.2. New results about the uniqueness of solutions of problem (4.6) were recently proved in [8] under weaker restrictions on the data, including the case $a_i = a_i(x, u)$ and $f \equiv f(x, u)$.

4.2. *Uniqueness of solution of the generalized diffusion equation*

Let us consider the problem

$$\begin{cases} -\operatorname{div}\big(a(x, u)|u|^{\alpha(x)} \nabla u\big) = f(x) & \text{in } \Omega, \\ u = g & \text{on } \Gamma. \end{cases} \tag{4.17}$$

THEOREM 4.3. *Let the functions $a(x, s)$ and $\alpha(x)$ be such that*

$$\begin{cases} \forall s \in \mathbb{R}, & \nabla_x a(x, s) \in L^2(\Omega), \\ \nabla \alpha(x) \in L^2(\Omega), & -1 < \alpha^- \leqslant \alpha(x) \leqslant \alpha^+ < \infty, \end{cases} \tag{4.18}$$

with $\alpha^\pm = \text{const}$, and

$$0 < a_0 \leqslant a(x, s) \leqslant a_1 < \infty, \qquad \sup_{x \in \Omega, s \in \mathbb{R}} |\nabla_x a(x, s)| = A_1. \tag{4.19}$$

Then the bounded solution of problem (4.17) is unique.

PROOF. Let us introduce the function

$$v(x) = v\big[x, u(x)\big] = \int_0^{u(x)} a(x, s)|s|^{\alpha(x)} \, \mathrm{d}s \tag{4.20}$$

and denote by ψ its inverse, $u = \psi(x, v)$. It is easy to see that

$$\nabla v(x) = a(x, u)|u|^{\alpha(x)} \nabla u$$

$$+ \int_0^{u(x)} \left[\nabla_x a(x, s)|s|^{\alpha(x)} + a(x, s)\nabla\alpha(x)|s|^{\alpha(x)} \ln|s| \right] ds. \tag{4.21}$$

Let u_1 and u_2 be two different solutions of problem (4.17) and v_1, v_2 be the corresponding functions defined by (4.20). Set $w = v_1 - v_2$. The function w is a solution of the problem

$$\begin{cases} \operatorname{div}\left(\nabla w - \int_{u(v_2)}^{u(v_1)} \left[\nabla_x a(x, s)|s|^{\alpha(x)} \right. \right. \\ \qquad\qquad \left. \left. + a(x, s)\nabla\alpha(x)|s|^{\alpha(x)} \ln|s| \right] ds \right) = 0 & \text{in } \Omega, \\ w = 0 & \text{on } \Gamma. \end{cases} \tag{4.22}$$

Let us take an arbitrary function $\phi(x) : \mathbb{R}_+ \mapsto \mathbb{R}_+$ such that

$$\phi(s) > 0 \quad \text{for } s \in (0, s_0) \text{ with some } s_0 > 0,$$

$$\phi(0) = 0, \qquad \int_{0+} \frac{ds}{\sqrt{\phi(s)}} = \infty, \tag{4.23}$$

fix an arbitrary $\varepsilon > 0$ and introduce the functions

$$F_\varepsilon(x) = \begin{cases} \int_\varepsilon^x \frac{ds}{\phi(s)} & \text{for } x > \varepsilon, \\ 0 & \text{otherwise,} \end{cases} \quad \text{and}$$

$$G_\varepsilon(x) = \begin{cases} \int_\varepsilon^x \frac{ds}{\sqrt{\phi(s)}} & \text{for } x > \varepsilon, \\ 0 & \text{otherwise.} \end{cases} \tag{4.24}$$

For every $\varepsilon > 0$,

$$\nabla F_\varepsilon(w) = \begin{cases} \frac{\nabla w}{\phi(w)} & \text{for } w > \varepsilon, \\ 0 & \text{otherwise,} \end{cases} \quad \text{and}$$

$$\nabla G_\varepsilon(w) = \begin{cases} \frac{\nabla w}{\sqrt{\phi(w)}} & \text{for } w > \varepsilon, \\ 0 & \text{otherwise.} \end{cases} \tag{4.25}$$

Let us denote $\Omega_\varepsilon = \Omega \cap \{w > \varepsilon\}$. Multiplying (4.22) by $F_\varepsilon(w)$ and integrating over Ω we obtain the equality

$$\int_{\Omega_\varepsilon} |\nabla G_\varepsilon(w)|^2 \, dx$$

$$= \int_{\Omega_\varepsilon} \frac{|\nabla w|^2}{\phi(w)} \, dx$$

$$= \int_{\Omega_\varepsilon} \left[\int_{u(v_2)}^{u(v_1)} \left[\nabla_x a(x, s)|s|^{\alpha(x)} + a(x, s)\nabla\alpha(x)|s|^{\alpha(x)} \ln|s| \right] ds \right] \frac{\nabla w}{\phi(w)} \, dx,$$

which yields

$$\int_{\Omega_\varepsilon} \left| \nabla G_\varepsilon(w) \right|^2 dx$$

$$\leqslant \int_{\Omega_\varepsilon} \left[\int_{u(v_2)}^{u(v_1)} \left[\nabla_x a(x,s) |s|^{\alpha(x)} + a(x,s) \nabla \alpha(x) |s|^{\alpha(x)} \ln|s| \right] ds \right]^2 \frac{dx}{\phi(x)}.$$

Applying the Poincaré inequality we may write

$$C \int_{\Omega_\varepsilon} \left| G_\varepsilon(w) \right|^2 dx$$

$$\leqslant \int_{\Omega_\varepsilon} \left| \nabla G_\varepsilon(w) \right|^2 dx$$

$$\leqslant \int_\Omega \sup_{s \in \mathbb{R}} a^2(x,s) \left| \nabla \alpha(x) \right|^2 \left(\int_{u(v_2)}^{u(v_1)} |s|^{\alpha(x)} \ln|s| \, ds \right)^2 \frac{dx}{\phi(w)}$$

$$+ \int_\Omega \sup_{s \in \mathbb{R}} \left| \nabla_x a(x,s) \right|^2 \frac{\left(\int_{u_1}^{u_2} |s|^{\alpha(x)} ds \right)^2}{\phi(w)} dx$$

$$\leqslant \left(a_1^2 \left\| \nabla \alpha(x) \right\|_{2,\Omega}^2 + A_1 |\Omega| \right) I, \tag{4.26}$$

with the constants a_1, A_1 from (4.19) and

$$I = \sup_\Omega \frac{\left(\int_{u(v_2)}^{u(v_1)} |s|^{\alpha(x)} \ln|s| \, ds \right)^2 + \left(\int_{u(v_2)}^{u(v_1)} |s|^{\alpha(x)} ds \right)^2}{\phi\left(\left| \int_{u(v_2)}^{u(v_1)} |s|^{\alpha(x)} ds \right| \right)}.$$

By the choice of $\phi(s)$, we have that $G_\varepsilon(w) \to \infty$ as $\varepsilon \to 0$, but for bounded I this is possible only if $|\Omega_\varepsilon| \equiv |\Omega \cap \{w > \varepsilon\}| \to 0$ as $\varepsilon \to 0$. It follows that $w \equiv$ in Ω, whence $u_1 \equiv u_2$ in Ω due to the first condition in (4.19).

Let us check that there exists a function $\phi(s)$ satisfying conditions (4.23) and such that for a finite constant K

$$\left(\int_{u(v_2)}^{u(v_1)} |s|^{\alpha(x)} \ln|s| \, ds \right)^2 + \left(\int_{u(v_2)}^{u(v_1)} |s|^{\alpha(x)} ds \right)^2 \leqslant K^2 \phi\left(\left| \int_{u(v_2)}^{u(v_1)} |s|^{\alpha(x)} ds \right| \right). \tag{4.27}$$

The fulfillment of this inequality is equivalent to the inequality $I \leqslant K^2$. Notice that since u_i are bounded, we may simply assume that $\|u_i\|_{\infty,\Omega} \leqslant 1/e$. Then $\ln|s| \geqslant 1$ and inequality (4.27) follows if

$$\left(\int_{u(v_2)}^{u(v_1)} |s|^{\alpha(x)} \ln|s| \, ds \right)^2 \leqslant \frac{1}{2} K^2 \phi\left(\left| \int_{u(v_2)}^{u(v_1)} |s|^{\alpha(x)} ds \right| \right). \tag{4.28}$$

Set

$$\phi(s) = s^2 \ln^2 s.$$

This function satisfies conditions (4.23). Next,

$$\left| \int_{u_2}^{u_1} |s|^{\alpha(x)} \ln|s| \, ds \right| \geqslant \frac{1}{(1+\alpha)^2} \left| u_1^{1+\alpha} \ln u_1^{1+\alpha} - u_2^{1+\alpha} \ln u_2^{1+\alpha} \right|$$

$$- \frac{1}{(1+\alpha)^2} \left| u_1^{1+\alpha} - u_2^{1+\alpha} \right|,$$

$$\left| \int_{u_2}^{u_1} |s|^{\alpha(x)} \ln|s| \, ds \right| = \frac{1}{1+\alpha} \left| u_1^{1+\alpha} - u_2^{1+\alpha} \right|.$$

Let us denote $x = u_1^{1+\alpha}$, $y = u_2^{1+\alpha}$, and assume without loss of generality that $x \geqslant y$. Inequality (4.28) immediately follows if there exists a constant $K > 0$ such that

$$|x \ln x - y \ln y| \leqslant \frac{K}{2} \left| (x-y) \ln(x-y) \right| \quad \text{for } 0 < y \leqslant x \leqslant \frac{1}{e}.$$

It is easy to see that

$$\left(\frac{1}{x}\right)^x y^y = \left(\frac{1}{x}\right)^{x-y} \left(\frac{y}{x}\right)^y \leqslant \left(\frac{1}{x}\right)^{x-y} \leqslant \left(\frac{1}{x-y}\right)^{x-y} \leqslant (x-y)^{-K(x-y)/2}$$

provided that $K \geqslant 2$. $\qquad \square$

5. Localization caused by the diffusion–absorption balance

5.1. *Generalized $p(x)$-Laplace equation*

In this section we consider the vanishing properties of *local weak solutions* of equation

$$-\sum_i D_i \left(a_i(x,u)|D_i u|^{p_i(x)-2} D_i u \right) + c(x,u)|u|^{\sigma(x)-2} u = f(x). \tag{5.1}$$

By local weak solution we mean a function $u(x)$ locally integrable in Ω, such that $u^{\sigma(x)} \in L^1_{\text{loc}}(\Omega)$, $|D_i u|^{p_i(x)} \in L^1_{\text{loc}}(\Omega)$ for $i = 1, \ldots, n$, and for every test function $\zeta \in \mathbf{V}(\Omega)$ compactly supported in Ω, the integral identity (3.5) holds. Every weak solution of problem (3.1) is a local weak solution in the sense of this definition.

We will consider the local weak solutions with finite energy: for every set $\omega \subseteq \Omega$,

$$\int_\omega \left(\sum_i a_i(x,u)|D_i u|^{p_i(x)} + c(x,u)|u|^{\sigma(x)} \right) dx \leqslant T_0 < \infty. \tag{5.2}$$

Let us introduce the following notations:

$$B_\rho(x_0) = \left\{ x \in \mathbb{R}^N : |x - x_0| < \rho \right\} \subset \Omega,$$

$$S_\rho(x_0) = \partial B_\rho(x_0),$$

$$E(\rho, u) \equiv E(\rho) = \sum_i \int_{B_\rho(x_0)} |D_i u|^{p_i(x)}\, dx,$$

$$b(\rho, u) \equiv b(\rho) = \int_{B_\rho(x_0)} |u|^{\sigma(x)}\, dx,$$

$$\widetilde{E}(\rho) \equiv \sum_i \int_{B_\rho(x_0)} a_i(x, u) |D_i u|^{p_i(x)}\, dx,$$

$$\tilde{b}(\rho) \equiv \int_{B_\rho(x_0)} c(x, u) |u|^{\sigma(x)}\, dx.$$

The functions E and b are called the energy functions associated with the local weak solution $u(x)$ of equation (5.1). We will assume that in addition to conditions (3.2) the coefficients $a_i(x, u)$ and $c(x, u)$ satisfy the conditions

$$\forall x \in \Omega, \quad a_i(x, u) \leqslant a_1 < \infty, \qquad c(x, u) \leqslant c_1 < \infty \tag{5.3}$$

with positive constants a_1, c_1. It follows that

$$\frac{1}{a_1} \widetilde{E}(\rho) \leqslant E(\rho) \leqslant \frac{1}{a_0} \widetilde{E}(\rho), \qquad \frac{1}{c_1} \tilde{b}(\rho) \leqslant b(\rho) \leqslant \frac{1}{c_0} \tilde{b}(\rho).$$

Introducing the system of spherical coordinates with the origin x_0 by the equalities $r = |x - x_0|$, $\omega = (x - x_0)/|x - x_0|$, we may write

$$\widetilde{E}(\rho) = \sum_i \int_{B_\rho(x_0)} a_i(x, u) |D_i u|^{p_i(x)}\, dx$$

$$= \sum_i \int_0^\rho \left(\int_{S_\rho} a_i(x, u) |D_i u|^{p_i(x)}\, ds \right) d\rho.$$

(For the sake of brevity we keep on using the same notation for the functions which are now considered as functions of variables (r, ω).) It follows from the last equality that the function $\widetilde{E}(\rho)$ is monotone nondecreasing and bounded for $u \in V(\Omega)$. Hence, it has the weak derivative $\widetilde{E}'(\rho)$ which possesses the following properties:

$$\widetilde{E}'(\rho) = \sum_i \int_{S_\rho} a_i(x, u) |D_i u|^{p_i(x)}\, ds$$

exists for a.a. $\rho \in (0, R)$, $R = \text{dist}(x_0, \partial \Omega)$,

$$a_0 \int_{S_\rho} |\nabla u|^{p(x)}\, ds \leqslant \widetilde{E}'(\rho) \leqslant a_1 \int_{S_\rho} |\nabla u|^{p(x)}\, ds,$$

$$\widetilde{E}'(\rho) \in L^1(0, R).$$

In the same way we conclude that there exist

$$E'(\rho) = \sum_i \int_{S_\rho} |D_i u|^{p_i(x)}\, ds \in L^1(0, R),$$

$$b'(\rho) = \int_{S_\rho} |u|^{\sigma(x)}\, ds \in L^1(0, R).$$

Without loss of generality we assume that, for every ball $B_\rho \subset \Omega$,

$$E(\rho) + b(\rho) = \int_{B_\rho} \left(\sum_i |D_i u|^{p_i(x)} + |u|^{\sigma(x)} \right) dx \equiv T_0 \leqslant 1. \tag{5.4}$$

According to (2.3),

$$\begin{cases} \sum_i \|D_i u\|_{p_i(\cdot), B_\rho}^{p^+} \leqslant E(\rho) \leqslant \sum_i \|D_i u\|_{p_i(\cdot), B_\rho}^{p^-}, \\ \|u\|_{\sigma(\cdot), B_\rho}^{\sigma^+} \leqslant b(\rho) \leqslant \|u\|_{\sigma(\cdot), B_\rho}^{\sigma^-}. \end{cases}$$

We are now in position to formulate the main results of this section.

THEOREM 5.1. *Let the coefficients $a_i(x, u)$ and $c(x, u)$ in equation (5.1) be Carathéodory functions satisfying (3.2) and (5.3), and $u(x)$ be a local weak solution of equation (5.1) satisfying (5.4). Let us assume that*
  (i) *the oscillation of the exponents $p_i(x)$ is such that*

$$p^+ - p^- < \frac{(p^- - 1)}{n} p^-, \tag{5.5}$$

  (ii) *conditions (3.3), (3.4) hold and*

$$\sigma^+ < p^-, \tag{5.6}$$

  (iii) *$f(x) \equiv 0$ in $B_{\rho_0}(x_0) \subseteq \Omega$ with $\rho_0 \leqslant 1$.*
Then

$$u(x) \equiv 0 \quad a.e.\ in\ B_{\rho_1}(x_0)\ with\ \rho_1^{1+\lambda} = \left[\rho_0^{1+\lambda} - CE^\nu(\rho_0) \right]_+, \tag{5.7}$$

with the positive constants C, λ and ν depending only on $p^\pm$, $\sigma^\pm$, a_0, a_1, c_0, c_1.

REMARK 5.1. If $\rho_1 = 0$, then the above statement offers no information on the vanishing set of $u(x)$. Nonetheless, if the total energy $E(\rho_0)$ is small enough, then always $\rho_1 > 0$ and the vanishing set of $u(x)$ is not empty.

REMARK 5.2. Condition (5.5) means that the oscillation of the function $p(x)$ in the ball $B_{\rho_0}(x_0)$ is small. By virtue of the continuity of $p(x)$ it is always true provided that the ball $B_{\rho_0}(x_0)$ is sufficiently small.

THEOREM 5.2. *Let, under the conditions of Theorem 5.1, $f(x) \not\equiv 0$ in $B_{\rho_0}(x_0)$ but*

$$f(x) \equiv 0 \quad in \ B_{\rho_1}(x_0) \ for \ some \ 0 < \rho_1 < \rho_0. \tag{5.8}$$

There exist positive constants $\gamma > 0$, ε_* and E_*, depending only on $p^\pm$, $\sigma^\pm$, a_0, a_1, c_0, c_1, such that if

$$\int_{B_\rho} |f|^{\sigma'(x)} \, dx \leqslant \varepsilon_*(\rho - \rho_1)_+^\gamma \quad for \ \rho \in (\rho_1, \rho_0), \tag{5.9}$$

and $E(\rho_0) < E_*$, then

$$u(x) \equiv 0 \quad in \ B_{\rho_1}(x_0).$$

The assertions of Theorems 5.1 and 5.2 are illustrated in Figures 1 and 2. Under the conditions of Theorem 5.1, if $f(x) = 0$ in a ball $B_{\rho_0}(x_0)$, the solution may vanish in a ball $B_{\rho_1}(x_0)$ separated away from $\partial\{\mathrm{supp}\, f(x)\}$ by a positive distance. Under the conditions of Theorem 5.2, if $f(x) = 0$ in a ball $B_{\rho_1}(x_0)$ and rapidly decreases in a concentric ball denoted by the dotted line, then the solution is identically zero in the same ball $B_{\rho_1}(x_0)$ which may touch (or coincide with) the set $\partial\{\mathrm{supp}\, f(x)\}$.

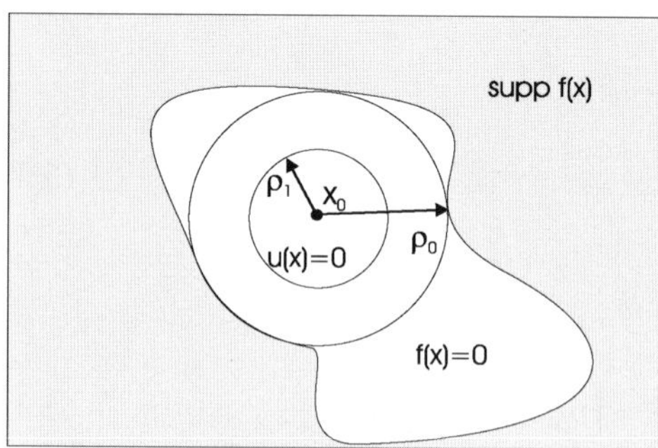

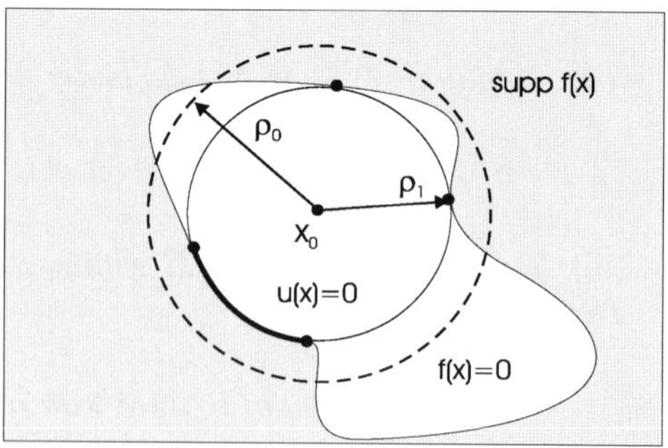

Fig. 1. Theorem 5.1.   Fig. 2. Theorem 5.2.

5.2. *The energy relation*

LEMMA 5.1. *Let $u \in \mathbf{V}(\Omega)$, $\Gamma = \partial\Omega$ be Lipschitz-continuous and $p_i(x)$ satisfy the conditions of Theorem 5.1. Then for every ball $B_\rho \subset \Omega$,*

$$\sum_i \int_{B_\rho} |u|^{p_i(x)} \, dx \leqslant \phi\big(A_{\mathbf{p}(\cdot)}(\nabla u)\big)$$

with a continuous and monotone increasing function $\phi(s)$ such that $\phi(s) \to 0$ when $s \to 0$.

PROOF. In the special case $p_i(x) = p(x)$, the assertion immediately follows from [41], Theorems 2.6 and 2.7. Let us consider the general case. Condition (5.5) means that if $n > p^-$, then

$$p_i(x) < \frac{np^-}{n - p^-}, \quad i = 1, \ldots, n. \tag{5.10}$$

Indeed: since $p_i(x) < p^+$, it suffices to check that

$$p^+ < \frac{np^-}{n - p^-} \equiv p^- + \frac{(p^-)^2}{n - p^-} \quad \Longleftrightarrow \quad p^+ - p^- < \frac{(p^-)^2}{n - p^-};$$

according to (5.5), this is true if

$$p^- \frac{p^- - 1}{n} < \frac{(p^-)^2}{n - p^-} \quad \Longleftrightarrow \quad p^- < (p^-)^2 + n \quad \Longleftarrow \quad p^- > 1.$$

Using (2.3), (2.6) and (5.10), we have

$$\int_{B_\rho} |u|^{p_i(x)} \, dx \leqslant \int_\Omega |u|^{p_i(x)} \, dx$$

$$\leqslant \max\big\{ \|u\|_{p_i(\cdot)}^{p_i^-}, \|u\|_{p_i(\cdot)}^{p_i^+} \big\}$$

$$\leqslant C \max\big\{ \|\nabla u\|_{p^-}^{p_i^-}, \|\nabla u\|_{p^-}^{p_i^+} \big\}$$

$$\leqslant Cn \max\Big\{ \max_i \|D_i u\|_{p^-}^{p_i^-}, \max_i \|D_i u\|_{p^-}^{p_i^+} \Big\}$$

$$\leqslant Cn \max\Big\{ \max_i \|D_i u\|_{p_i(\cdot)}^{p_i^-}, \max_i \|D_i u\|_{p_i(\cdot)}^{p_i^+} \Big\}$$

$$\leqslant Cn \sum_i \big[A_{\mathbf{p}(\cdot)}^{p_i^+/p_i^-}(\nabla u) + A_{\mathbf{p}(\cdot)}^{p_i^-/p_i^+}(\nabla u) \big]$$

$$\leqslant Cn^2 \big[A_{\mathbf{p}(\cdot)}^{p^+/p^-}(\nabla u) + A_{\mathbf{p}(\cdot)}^{p^-/p^+}(\nabla u) \big]. \qquad \square$$

LEMMA 5.2. *Under the conditions of Theorem 5.1, for almost all $\rho \in (0, \rho_0)$ the following energy relation holds:*

$$\int_{B_\rho(x_0)} \left(\sum_i a_i(x, u)|D_i u|^{p_i(x)} + c(x, u)|u|^{\sigma(x)} - fu \right) dx = I(\rho),$$

$$I(\rho) = -\int_{S_\rho(x_0)} u \sum_i a_i(x, u)|D_i u|^{p_i(x)-2} D_i u \cdot \nu_i \, dS, \tag{5.11}$$

where $\nu = (\nu_1, \ldots, \nu_n)$ is the unit outer normal vector to $S_\rho(x_0)$. Moreover, $I(\rho) \in L^1(0, \rho_0)$.

PROOF. We begin with checking the inclusion $I(\rho) \in L^1(0, \rho_0)$. There holds the relation

$$\int_0^{\rho_0} |I(\rho)| \, d\rho \leqslant a_1 \int_0^{\rho_0} \int_{S_\rho(x_0)} |u| \sum_i |D_i u|^{p_i(x)-1} \, dS \equiv a_1 I_1.$$

By Young's inequality,

$$I_1 \leqslant C \sum_i \int_{B_\rho(x_0)} \left(|D_i u|^{p_i(x)} + |u|^{p_i(x)} \right) dx$$

$$= C E(\rho) + C \sum_i \int_{B_\rho} |u|^{p_i(x)} \, dx. \tag{5.12}$$

The first term on the right-hand side of (5.12) is bounded by assumption, while the second one is bounded by virtue of Lemma 5.1, whence $I(\rho) \in L^1(0, \rho_0)$.

Let us introduce the cut-off function

$$\psi_k(r) = \begin{cases} 1 & \text{if } r \in \left[0, \rho - \frac{1}{k}\right], \\ k(\rho - r) & \text{if } r \in \left[\rho - \frac{1}{k}, \rho\right], \\ 0 & \text{if } r \in [\rho, \rho_0], \end{cases} \quad k \in \mathbb{N}. \tag{5.13}$$

It is easy to see that the function

$$\varphi_k(x) \equiv \psi_k(|x - x_0|) u(x)$$

belongs to $\mathbf{V}(B_\rho(x_0))$ and can be taken for the test function in (3.5). Letting in (3.5) $\zeta = \varphi_k$, we have

$$\int_{B_{\rho_0}} \varphi_k(x) \left(\sum_i a_i(x, u)|D_i u|^{p_i(x)} + c(x, u)|u|^{\sigma(x)} - fu \right) dx$$

$$= -\int_{B_{\rho_0}} \sum_i a_i(x, u)|D_i u|^{p_i(x)-2} D_i u \cdot D_i \varphi_k \, dx. \tag{5.14}$$

Passing to the spherical coordinates (r, ω), we have

$$\int_{B_{\rho_0}} u \sum_i a_i(x, u)|D_i u|^{p_i(x)-2} D_i u \cdot D_i \varphi_k \, dx$$

$$= k \int_{\rho-1/k<|x-x_0|<\rho} \left(\int_{S_\rho(x_0)} u \sum_i a_i(r\omega, u)|D_i u|^{p_i(x)-2} \right.$$

$$\left. \times D_i u \cdot \frac{x_i - x_{0i}}{|x - x_0|} \, ds \right) dr$$

$$= k \int_{\rho-1/k<|x-x_0|<\rho} \left(\int_{S_\rho(x_0)} u \sum_i a_i(r\omega, u)|D_i u|^{p_i(x)-2} D_i u \cdot v_i \, ds \right) dr$$

$$= k \int_{\rho-1/k}^{\rho} I(r) \, dr.$$

Since $I(\rho) \in L^1(0, \rho_0)$, it follows from the Lebesgue theorem that, for a.e. $\rho \in (0, \rho_0)$,

$$\lim_{k \to \infty} \int_{B_{\rho_0}} u \sum_i a_i(x, u)|D_i u|^{p_i(x)-2} D_i u \cdot D_i \varphi_k \, dx$$

$$= \int_{S_\rho(x_0)} u \sum_i a_i(x, u)|D_i u|^{p_i(x)-2} D_i u \cdot v_i \, dS$$

$$\equiv -I(\rho),$$

and (5.11) follows. $\qquad\qquad\qquad\qquad\qquad\qquad\qquad\qquad\qquad\qquad\qquad\qquad\quad\square$

5.3. *The ordinary differential inequality*

To derive the ordinary differential inequality for the energy function we adapt the technique developed in [10]. By Hölder's inequality (2.4) and due to (2.3),

$$\left| I(\rho) \right| \leqslant C \sum_i \|u\|_{p_i(\cdot), S_\rho} \left\| |D_i u|^{p_i(x)-1} \right\|_{p_i'(\cdot), S_\rho}$$

$$\leqslant C\|u\|_{p^+, S_\rho} \max\left\{ \sum_i \left(\int_{S_\rho} |D_i u|^{p_i(x)} \right)^{(p_i^- - 1)/p_i^-}, \right.$$

$$\left. \sum_i \left(\int_{S_\rho} |D_i u|^{p_i(x)} \right)^{(p_i^+ - 1)/p_i^+} \right\}$$

$$\equiv C\|u\|_{p^+, S_\rho} \max\left\{ E'^{(p^- - 1)/p^-}, E'^{(p^+ - 1)/p^+} \right\}. \qquad (5.15)$$

We will make use of the trace-interpolation inequality (2.16) written in the form

$$\|u\|_{p^+,S_\rho} \leqslant C\big(\|\nabla u\|_{p^-,B_\rho} + \rho^{-\delta}\|u\|_{\sigma^-,B_\rho}\big)^\theta \|u\|_{\sigma^-,B_\rho}^{1-\theta}$$

$$\leqslant C_1\Big(\sum_i \rho^{n(p_i^- - p^-)/p_i^-}\|D_i u\|_{p_i^-,B_\rho} + \rho^{-\delta}\|u\|_{\sigma^-,B_\rho}\Big)^\theta \|u\|_{\sigma^-,B_\rho}^{1-\theta}$$

$$\leqslant C_2\Big(\sum_i \|D_i u\|_{p_i^-,B_\rho} + \rho^{-\delta}\|u\|_{\sigma^-,B_\rho}\Big)^\theta \|u\|_{\sigma^-,B_\rho}^{1-\theta}, \tag{5.16}$$

where

$$p_i^- = \inf_\Omega p_i(x) \geqslant p^-, \qquad \delta = \frac{n(p^- - \sigma^-) + p^-\sigma^-}{p^-\sigma^-} > 1,$$

$$\theta = \frac{p^-}{p^+}\frac{n(p^+ - \sigma^-) + \sigma^-}{n(p^- - \sigma^-) + p^-\sigma^-}.$$

Notice that $\theta \in (0,1)$ if

$$p^+ > \sigma^- \quad \text{and} \quad p^+ - p^- < \frac{(p^- - 1)}{n}p^-,$$

which is true whenever (5.5) holds. By virtue of (5.4) and (2.3), the inequalities

$$\|u\|_{\sigma^-,B_\rho} \leqslant C\|u\|_{\sigma(\cdot),B_\rho} \leqslant C\Big(\int_{B_\rho} |u|^{\sigma(x)}\,dx\Big)^{1/\sigma^+}$$

$$= Cb^{1/\sigma^+}(\rho) \leqslant Cb^{1/p^-}(\rho) \leqslant Cb^{1/p^+}(\rho),$$

$$\sum_i \|D_i u\|_{p_i^-,B_\rho}$$

$$\leqslant C\sum_i \|D_i u\|_{p_i(\cdot),B_\rho} \leqslant C\sum_i\Big(\int_{B_\rho} |D_i u|^{p_i(x)}\Big)^{1/p^+} = CE^{1/p^+}(\rho)$$

hold, which allows one to rewrite (5.16) in the form

$$\|u\|_{p^+,S_\rho} \leqslant C\rho^{-\delta\theta}\big(E(\rho) + b(\rho)\big)^{\theta/p^+ + (1-\theta)/\sigma^+} \tag{5.17}$$

(recall that $\rho_0 \leqslant 1$ by assumption). Substituting (5.17) into (5.11) and using (5.15), we have

$$E + b \leqslant C\rho^{-\delta\theta}(E + b)^{\theta/p^+ + (1-\theta)/\sigma^+} \max\big\{E'^{(p^- - 1)/p^-}, E'^{(p^+ - 1)/p^+}\big\},$$

whence the differential inequality

$$\rho^{-\delta\theta}(E+b)^{\eta} \leqslant C \max\left\{E'^{(p^--1)/p^-}, E'^{(p^+-1)/p^+}\right\} \tag{5.18}$$

with the exponent

$$0 < \eta = 1 - \left(\frac{\theta}{p^+} + \frac{1-\theta}{\sigma^+}\right) < 1. \tag{5.19}$$

Since $\dfrac{p^-}{p^--1} \geqslant \dfrac{p^+}{p^+-1}$ and $\rho^{\delta\theta}(E+b) \leqslant 1$, then

$$\left(\rho^{\delta\theta}(E+b)^{\eta}\right)^{p^-/(p^--1)}$$
$$= \min\left[\left(\rho^{\delta\theta}(E+b)^{\eta}\right)^{p^-/(p^--1)}, \left(\rho^{\delta\theta}(E+b)^{\eta}\right)^{p^+/(p^+-1)}\right], \tag{5.20}$$

and inequality (5.18) yields the following one:

$$\rho^{\mu}E^{1-\nu} \leqslant \rho^{\mu}(E+b)^{1-\nu} \leqslant CE' \tag{5.21}$$

with the exponents

$$\mu = \delta\theta\,\frac{p^-}{p^--1} > 0, \qquad 1-\nu = \frac{\eta p^-}{p^--1}. \tag{5.22}$$

5.3.1. *Proofs of the main assertions.*

1. If the conditions of Theorem 5.1 are fulfilled, then the local energy function corresponding to the local weak solution of equation (5.1) satisfies inequality (5.21). Let us claim that $1 - \nu \in (0, 1)$. This condition is equivalent to the following inequalities:

$$\begin{cases} 1-\nu > 0 & \Longleftrightarrow \quad \frac{1}{p^-} < \frac{\theta}{p^+} + \frac{1-\theta}{\sigma^+}, \\ 1-\nu < 1 & \Longleftrightarrow \quad 1 > \frac{\theta}{p^+} + \frac{1-\theta}{\sigma^+}. \end{cases}$$

Of these two inequalities, the former is fulfilled provided that $\theta < 1$ and $p^+ > \sigma^+$,

$$\frac{1}{p^-} < \frac{\theta}{p^+} + \frac{1-\theta}{\sigma^+} \quad \Longleftrightarrow \quad \frac{1}{p^-} - \frac{1}{p^+} + \frac{1-\theta}{p^+} < \frac{1-\theta}{\sigma^+} \quad \Longleftarrow \quad p^+ > \sigma^+.$$

The latter holds whenever $\theta > 0$ and $1 < \sigma^+ < p^+$,

$$1 > \frac{\theta}{p^+} + \frac{1-\theta}{\sigma^+} \quad \Longleftrightarrow \quad 1 > \theta\,\frac{\sigma^+ - p^+}{\sigma^+ p^+} + \frac{1}{\sigma^+}.$$

Hence, the condition $1 - \nu \in (0, 1)$ is fulfilled if $\theta \in (0, 1)$ and $p^+ > \sigma^+$, which is guaranteed by (5.5) and (5.6).

Integrating (5.21) over the interval $\rho \in (\rho, \rho_0)$, we arrive at the estimate

$$E^{\nu}(\rho) \leqslant E^{\nu}(\rho_0) - \frac{\nu}{C(\mu + 1)} \left(\rho_0^{\mu+1} - \rho^{\mu+1} \right). \tag{5.23}$$

Since $E(\rho)$ is nonnegative, it is necessary that $E(\rho) \equiv 0$ for all

$$\rho^{\mu+1} \leqslant \frac{C(\mu + 1)}{\nu} E^{\nu}(\rho_0) - \rho_0^{\mu+1}.$$

The proof of Theorem 5.1 is completed.

     2. We proceed to prove Theorem 5.2. Let

$$f(x) \equiv 0 \quad \text{in } B_{\rho_1}(x_0) \text{ with some } 0 < \rho_1 < \rho_0$$

and

$$\int_{B_\rho} |f|^{\sigma'(x)} \, dx \leqslant \varepsilon (\rho - \rho_1)_+^{\gamma} \quad \text{for } \rho \in (\rho_1, \rho_0) \text{ with } \gamma = \frac{1}{1 - \mu}, \tag{5.24}$$

where the constant μ is defined in (5.21). By Young's inequality,

$$\int_{B_\rho} |fu| \, dx \leqslant \int_{B_\rho} \left(\gamma |u|^{\sigma(x)} + \frac{1}{\sigma'(x)} \left(\varepsilon \sigma(x) \right)^{1/(1-\sigma(x))} |f|^{\sigma'(x)} \right) dx. \tag{5.25}$$

Choosing a sufficiently small $\varepsilon > 0$, we have

$$E(\rho) + b(\rho)$$
$$\leqslant C \int_{B_\rho} |f|^{\sigma'(x)} \, dx$$
$$+ C\rho^{-\delta\theta} (E + b)^{\theta/p^+ + (1-\theta)/p^-} \max \left\{ E'^{(p^- - 1)/p^-}, E'^{(p^+ - 1)/p^+} \right\}.$$

Applying Young's inequality to the second term on the right-hand side of this inequality and using (5.24), we arrive at the inequality

$$(E + b) \leqslant C\varepsilon (\rho - \rho_1)_+^{\gamma} + C \left[\rho^{-\delta\theta} \max \left\{ E'^{(p^- - 1)/p^-}, E'^{(p^+ - 1)/p^+} \right\} \right]^{1/\eta}$$
$$\leqslant C(\rho_1) \left[\varepsilon (\rho - \rho_1)_+^{\gamma} + \max \left\{ E'^{(p^- - 1)/p^-}, E'^{(p^+ - 1)/p^+} \right\} \right]^{1/\eta}$$

with the constant $C(\rho_1) = C \max\{1, \rho_1^{-\delta\theta}\}$ and the exponent $\eta \in (0, 1)$ defined in (5.19). Let us write this inequality in the following form:

$$E' \geqslant C_1(\rho_1) \begin{cases} (E + b)^{\eta p^- / (p^- - 1)} - \left[\varepsilon (\rho - \rho_1)_+^{\gamma} \right]^{\eta p^- / (p^- - 1)} & \text{if } E' < 1, \\ (E + b)^{\eta p^+ / (p^+ - 1)} - \left[\varepsilon (\rho - \rho_1)_+^{\gamma} \right]^{\eta p^+ / (p^+ - 1)} & \text{if } E' \geqslant 1. \end{cases}$$

Using (5.20) and noting that

$$\max\left\{\left[\varepsilon(\rho - \rho_1)_+^{\gamma}\right]^{\eta p^-/(p^--1)}, \left[\varepsilon(\rho - \rho_1)_+^{\gamma}\right]^{\eta p^+/(p^+-1)}\right\}$$
$$= \left[\varepsilon(\rho - \rho_1)_+^{\gamma}\right]^{\eta p^+/(p^+-1)}$$

whenever $\varepsilon(\rho - \rho_1)_+^{\gamma} \leqslant 1$, we have that, for all $\rho \in (\rho_1, \rho_0)$,

$$C_3 E^{1-\nu} \leqslant \left[\varepsilon(\rho - \rho_1)_+^{\gamma}\right]^{\eta p^+/(p^+-1)} + E' \tag{5.26}$$

with the exponent $1 - \nu \in (0, 1)$ defined in (5.22). We will need the following lemma.

LEMMA 5.3 ([10], Chapter 1, Lemma 2.4). *Let $E \in W^{1,1}(0, \rho_0)$, $E \geqslant 0$ and $E' \geqslant 0$, satisfy the inequality*

$$E'(\rho) + F(\rho) \geqslant \lambda E^{1-\nu}(\rho) \quad \text{for a.a. } \rho \in (\rho_1, \rho_0)$$

with $\rho_1 \in (0, \rho_0)$, $\nu \in (0, 1)$, $\lambda = \text{const} > 0$ and $F \geqslant 0$. Assume that the integral

$$J(\rho) = \int_{\rho_1}^{\rho} \frac{F(\tau)}{(\tau - \rho_1)^{1/\nu}} \, d\tau$$

is convergent. Then

$$E(\rho_1) \leqslant G(\rho) \equiv E(\rho_0) - (\rho - \rho_1)^{1/\nu}\left((\lambda\nu)^{1/\nu} - J(\rho)\right). \tag{5.27}$$

If the equation $G(\rho) = 0$ has a root $\rho_* \in (\rho_1, \rho_0)$, then $E(\rho) = 0$ for all $\rho \in [0, \rho_1]$.

PROOF. The function $\overline{E}(\rho) = [(\lambda\nu)(\rho - \rho_1)]^{1/\nu}$ is a solution of the problem

$$\lambda\overline{E}^{1-\nu} = \overline{E}' \quad \text{in } (\rho_1, \rho_0), \qquad \overline{E}(\rho_1) = 0.$$

Introduce the notation

$$H(\rho) = \left(E - \overline{E}\right)(\rho) \quad \text{and} \quad \phi(t) = \lambda(1 - \nu)\int_0^1 \frac{d\theta}{(\theta E(\rho) + (1 - \theta)\overline{E}(\rho))^{\nu}}.$$

The function H satisfies the inequality

$$\phi(t)H \leqslant H'(\rho) + F(\rho) \quad \text{in } (\rho_1, \rho_0).$$

Multiplying this inequality by the function $\psi(\rho) = \exp(-\int_{\rho_1}^{\rho} \phi(t)\,dt) \in (0, 1)$, we obtain the inequality

$$(\psi H)' + \psi F \geqslant 0. \tag{5.28}$$

Integrating (5.28) over the interval (ρ_1, ρ) and reversing to the functions E and $\overline{E}$, we obtain

$$E(\rho) \geqslant \overline{E}(\rho) + \frac{E(\rho_1)}{\psi(\rho)} - \int_{\rho_1}^{\rho} F(s) \frac{\psi(s)}{\psi(\rho)}\, ds. \tag{5.29}$$

Consider the function

$$\frac{\psi(s)}{\psi(\rho)} = \exp\left(\int_s^{\rho} \phi(t)\, dt \right).$$

The function E is nonnegative by assumption, whence

$$\begin{aligned}
\int_s^{\rho} \phi(t)\, dt &= \lambda(1 - v) \int_{\rho}^{s} ds \int_0^1 \frac{d\theta}{(\theta E + (1 - \theta)\overline{E})^v} \\
&\leqslant \lambda(1 - v) \int_{\rho}^{s} ds \int_0^1 \frac{d\theta}{(1 - \theta)^v \overline{E}^v} \\
&= \lambda \int_{\rho}^{s} \frac{ds}{\overline{E}^v(s)} \int_0^1 \frac{(1 - v)\, d\theta}{(1 - \theta)^v} \\
&= \lambda \int_{\rho}^{s} \frac{ds}{\overline{E}^v(s)}.
\end{aligned}$$

Using the definition of $\overline{E}$, we continue this inequality as

$$\int_s^{\rho} \phi(t)\, dt \leqslant \lambda \int_{\rho}^{s} \frac{ds}{\overline{E}^v(s)} = \int_s^{\rho} \frac{\overline{E}'(s)}{\overline{E}(s)} = \ln \frac{\overline{E}(\rho)}{\overline{E}(s)}.$$

It follows that

$$\frac{\psi(s)}{\psi(\rho)} \leqslant \frac{\overline{E}(\rho)}{\overline{E}(s)}$$

whence, according to (5.29),

$$E(\rho) \geqslant E(\rho_1) + \overline{E}(\rho)\left(1 - \int_{\rho_1}^{\rho} \frac{F(s)}{\overline{E}(s)}\, ds \right).$$

This inequality coincides with (5.27). Since $E(\rho)$ is monotone, the last inequality can be given the form $E(\rho_1) \leqslant G(\rho)$, whence the second assertion of the lemma. $\qquad\square$

The assertion of Theorem 5.2 immediately follows now from Lemma 5.3 provided that

$$\frac{\gamma \eta p^+}{p^+ - 1} > 1 + \frac{1}{v},$$

which guarantees the convergence of the integral $J(\rho)$, and E^* and ε are so small that the equation $G(\rho) = 0$ has a solution in the interval (ρ_0, ρ_0).

5.4. *Equations with convection terms*

The above-presented arguments admit an immediate extension to certain equations of more complicated structure. Let us consider the equation with convection terms (alias equation (3.28))

$$-\sum_i \left[D_i \left(a_i(x, u)|D_i u|^{p_i(x)-2} D_i u \right) + b_i(x, u) D_i u \right]$$

$$+ c(x, u)|u|^{\sigma(x)-2} u = \Phi(x) \quad \text{in } \Omega, \tag{5.30}$$

with the right-hand side

$$\Phi(x) = f(x) + g(x, u),$$

and the functions $g(x, r)$ and $b_i(x, r)$ subject to the conditions

$$\left| g(x, r)r \right| \leqslant \delta_0 |u|^{\sigma(x)}, \tag{5.31}$$

$$|b_i u|^{p_i'(x)} \leqslant \varepsilon_0 |u|^{\sigma(x)} \tag{5.32}$$

with a positive constants δ_0, ε_0. It is assumed that there exists $\gamma \in (0, 1)$ such that

$$\min\left[c_0 - \delta_0 - \max_{x \in B_\rho, i=1,\dots,n} \frac{p_i(x)\gamma^{-p_i/(p_i-1)}}{p_i(x) - 1} \varepsilon_0^{p_i'(x)}, a_0 - \max_{x \in B_\rho, i=1,\dots,n} \frac{\gamma^{p_i}}{p_i} \right]$$

$$\geqslant C > 0. \tag{5.33}$$

For the sake of simplicity we assume that $f(x) = 0$ in the ball $B_\rho(x_0)$ and that the solution is bounded, $|u| \leqslant 1$.

The local weak solution of equation (5.30) satisfies the energy relation

$$\int_{B_\rho(x_0)} \left(\sum_i a_i(x, u)|D_i u|^{p_i(x)} + c(x, u)|u|^{\sigma(x)} \right) dx = I(\rho) + I_1(\rho) + I_2(\rho),$$

$$I(\rho) = -\int_{S_\rho(x_0)} u \sum_i a_i(x, u)|D_i u|^{p_i(x)-2} D_i u \cdot v_i \, dS,$$

$$I_1(\rho) = \int_{B_\rho(x_0)} g(x, u) u \, dx,$$

$$I_2(\rho) = \int_{B_\rho(x_0)} \sum_i u b_i D_i u \, dx. \tag{5.34}$$

Using assumptions (5.31), (5.32) and Young's inequality (2.17) we estimate the additional terms I_1, I_2 as follows:

$$|I_1(\rho)| \leqslant \delta_0 \int_{B_\rho(x_0)} |u|^{\sigma(x)}\, dx,$$

$$|I_2(\rho)| \leqslant \int_{B_\rho(x_0)} \sum_i \left[\frac{\gamma^{p_i}}{p_i} |D_i u|^{p_i(x)} + \frac{p_i \gamma^{-p_i'}}{p_i - 1} |b_i u|^{p_i'(x)} \right] dx$$

$$\leqslant \int_{B_\rho(x_0)} \sum_i \left[\frac{\gamma^{p_i}}{p_i} |D_i u|^{p_i(x)} + \frac{p_i \gamma^{-p_i'}}{p_i - 1} \varepsilon_0^{p_i'(x)} |u|^{\sigma(x)} \right] dx.$$

Substituting these inequalities into (5.34) and applying (5.33), we arrive at the inequality

$$E + b \leqslant C |I(\rho)|. \tag{5.35}$$

The further study of the localization properties is performed like in the case $g \equiv 0$, $b_i \equiv 0$ (see (5.11)).

Notice that, for $b_i(x, u) \equiv b_i(x)$, condition (5.32) leads to the inequality

$$\frac{p_i(x)}{p_i(x) - 1} \geqslant \sigma(x).$$

Gathering this inequality with (5.6) of Theorem 5.1, we obtain a sufficient condition of localization,

$$\sigma^+ < \min\left[p^-, \frac{p^+}{p^+ - 1} \right].$$

6. Directional localization caused by anisotropic diffusion

6.1. *Generalized diffusion equation*

Let us consider Dirichlet problem for the generalized diffusion equation

$$\begin{cases} -\sum_i D_i\big(a_i(x, u)|u|^{\alpha_i(x)} D_i u\big) \\ \quad + c(x, u)|u|^{\sigma(x)-2} u = f(x) & \text{in } \Omega, \\ u = 0 & \text{on } \partial\Omega, \end{cases} \tag{6.1}$$

assuming that the data of this problem satisfy the conditions of the existence theorem Theorem 3.5.

6.1.1. *The simplest case: $n = 2$.* Let us consider problem (6.1) in the case $n = 2$. We will assume that the bounded simple-connected domain $\Omega \subset \mathbb{R}^2$ is defined by the relations

$$
\begin{cases}
\Omega = \left\{ (x_1, x_2)\colon x_1 \in (0, L),\, x_2 \in \big(g(x_1), h(x_1)\big] \right\}, & L = \mathrm{const}, \\
h, g \in C^{0,1}[0, L], \quad h(t) > g(t) \text{ for } t \in (0, L), \\
h(0) \geqslant g(0), \qquad h(L) \geqslant g(L).
\end{cases}
\tag{6.2}
$$

6.1.2. *The energy identity.* Let in the conditions of Theorem 3.5 the function $f(x)$ be such that

$$
f(x) = 0 \quad \text{for a.a. } x \in \Omega \cap \{x_1 \geqslant l\}
\tag{6.3}
$$

with some constant $0 < l < L$. Let $u(x)$ be a weak solution of problem (3.13). The function u can be taken for the test-function in the integral identity (3.16), which leads to the energy relation

$$
\sum_i \int_\Omega a_i(x, u) |u|^{\alpha_i(x)} |D_i u|^2 \, dx + \int_\Omega c(x, u) |u|^{\sigma(x)} \, dx = \int_\Omega f u \, dx.
\tag{6.4}
$$

THEOREM 6.1. *Let the conditions of Theorem 3.5 and conditions (6.2), (6.3) be fulfilled. Then for a.e. $t > l$,*

$$
- \int_{g(t)}^{h(t)} u |u|^{\alpha_1(x)} D_1 u \big|_{x_1 = t} \, dx_2
$$

$$
= \int_t^L dx_1 \int_{g(x_1)}^{h(x_1)} \left(\sum_i a_i(x, u) |u|^{\alpha_i(x)} |D_i u|^2 \, dx_2 + c(x, u) |u|^{\sigma(x)} \right) dx_2.
$$

$$
\tag{6.5}
$$

PROOF. Let us introduce the function

$$
\psi_k(x_1, t) =
\begin{cases}
1 & \text{for } x_1 > t + \frac{1}{k}, \\
k(x_1 - t) & \text{for } x_1 \in \left[t, t + \frac{1}{k} \right], \quad k = 1, 2, \dots . \\
0 & \text{for } x < t,
\end{cases}
$$

Substituting $\eta = \psi(x_1, t) u(x)$ into (3.16) as the test-function, we have that, for $t > l$,

$$
- I_k(t) \equiv -k \int_t^{t+1/k} dx_1 \int_{g(x_1)}^{h(x_1)} u |u|^{\alpha_1(x)} D_1 u \, dx_2
$$

$$
= \sum_i \int_t^L dx_1 \int_{g(x_1)}^{h(x_1)} a_i(x, u) |u|^{\alpha_i(x)} |D_i u|^2 \, dx_2
$$

$$
+ \int_t^L dx_1 \int_{g(x_1)}^{h(x_1)} c(x, u) \psi(x_1) |u|^{\sigma(x)} \, dx_2.
\tag{6.6}
$$

Let us denote

$$J(r) = \int_{g(r)}^{h(r)} u|u|^{\alpha_1(x)} D_1 u|_{x_1=r} \, dx_2,$$

and

$$I_k(t) = k \int_t^{t+1/k} J(r) \, dr.$$

It follows from the inequality

$$\int_t^L |J(r)| \, dr \leqslant A_0 \int_\Omega |u|^{\alpha_1(x)+1} |D_1 u| \, dx$$

$$\leqslant A_0 \|u\|_{2,\Omega} \big\| |u|^{\alpha_1(x)} D_1 u \big\|_{2,\Omega}$$

$$\leqslant A_0 K \sqrt{|\Omega|} \big\| |u|^{\alpha_1(x)} D_1 u \big\|_{2,\Omega}$$

and Theorem 3.5, that $\|J(r)\|_{1,(l,L)} \leqslant C$. By the Lebesgue dominated convergence theorem, there exists

$$\lim_{k \to \infty} I_k(t) = J(t)$$

and (6.5) follows from (6.6) after the limit passage when $k \to \infty$. $\qquad\square$

6.1.3. *The ordinary differential inequality.* Let us estimate the left-hand side of (6.5)

$$|J(t)| \leqslant \int_{g(t)}^{h(t)} \big(|u|^{(\alpha_1+2)/2}\big)\big(|u|^{\alpha_1/2}|D_1 u|\big) \, ds, \tag{6.7}$$

and make use of the representation

$$|u|^{(\alpha_2^+ +2)/2} = \big(u^2\big)^{(\alpha_2^+ +2)/4}$$

$$= \int_{g(t)}^s D_2\big((u^2)^{(\alpha_2^+ +2)/4}\big) \, d\xi$$

$$= \frac{\alpha_2^+ + 2}{4} \int_{g(t)}^s (u^2)^{(\alpha_2^+ -2)/4} D_2 u^2 \, d\xi$$

$$= \frac{\alpha_2^+ + 2}{2} \int_{g(t)}^s |u|^{(\alpha_2^+ -2)/2} u D_2 u \, d\xi. \tag{6.8}$$

Then for a.e. $t \in (l, L)$ on the truncation $x_1 = t$

$$|u| \leqslant \left(\frac{\alpha_2^+ + 2}{2}\right)^{2/(\alpha_2^+ + 2)} \phi^{1/(2(\alpha_2^+ + 2))}(t) \left(\int_{g(t)}^{h(t)} |u|^{\alpha_2^+} |D_2 u|^2 \, dx_2\right)^{1/(\alpha_2^+ + 2)}.$$

Substituting this inequality into (6.7), and then estimating the right-hand side by Hölder's inequality, we arrive at the inequality

$$|J(t)| \leqslant \int_{g(t)}^{h(t)} \left(\frac{\alpha_2^+ + 2}{2}\right)^{(\alpha_1(x)+2)/(\alpha_2^+ + 2)} \phi^{(\alpha_1(x)+2)/(4(\alpha_2^+ + 2))}(t)$$

$$\times \left(\int_{g(t)}^{h(t)} |u|^{\alpha_2^+} |D_2 u|^2 \, dx_2\right)^{(\alpha_1(x)+2)/(2(\alpha_2^+ + 2))}$$

$$\times \left(|u|^{\alpha_1(x)/2} |D_1 u|\right) dx_2$$

$$\leqslant \left[\int_{g(t)}^{h(t)} \left(\frac{\alpha_2^+ + 2}{2}\right)^{2(\alpha_1(x)+2)/(\alpha_2^+ + 2)} \phi^{(\alpha_1(x)+2)/(2(\alpha_2^+ + 2))}(t)\right.$$

$$\left.\times \left(\int_{g(t)}^{h(t)} |u|^{\alpha_2^+} |D_2 u|^2 \, dx_2\right)^{(\alpha_1(x)+2)/(\alpha_2^+ + 2)} dx_2\right]^{1/2}$$

$$\times \left(\int_{g(t)}^{h(t)} |u|^{\alpha_1(x)} |D_1 u|^2 \, dx_2\right)^{1/2}.$$

Let us introduce the notation

$$\Phi(t) = \int_{g(t)}^{h(t)} |u|^{\alpha_1(x)} |D_1 u|^2 \, dx_2 + \int_{g(t)}^{h(t)} |u|^{\alpha_2^+} |D_2 u|^2 \, dx_2,$$

$$\begin{cases} q(x) = \frac{\alpha_1(x)+2}{2(\alpha_2^+ + 2)}, \quad x = (t, x_2), \\ 0 < q^- < \inf_{\Omega \cap \{t \geqslant l\}} q(x) \leqslant q(x) \leqslant \sup_{\Omega \cap \{t \geqslant l\}} q(x) < q^+ < \infty, \end{cases} \tag{6.9}$$

$$\psi(t) = \left[\int_{g(t)}^{h(t)} \left(\frac{\alpha_2^+ + 2}{2}\right)^{4q(t,x_2)} \phi^{q(t,x_2)}(t) \, dx_2\right]^{1/2}. \tag{6.10}$$

In these notations the estimate on $|J(t)|$ becomes

$$|J(t)| \leqslant \psi(t) \max\left\{\Phi^{q^+}(t), \Phi^{q^-}(t)\right\} \Phi^{1/2}(t).$$

Let us take into consideration the function

$$\Psi(t) = \int_t^L \Phi(s) \, ds, \qquad \Psi'(t) = -\Phi(t) \leqslant 0 \quad \text{for a.e. } t \in (l, L).$$

Taking into account the fact that $|u| \leqslant K$ a.e. in Ω with a constant $K \geqslant 1$, we obtain

$$
\begin{aligned}
E(t) &\equiv \sum_i \int_t^L \mathrm{d}s \int_{g(s)}^{h(s)} |u|^{\alpha_i(x)} |D_i u|^2 \, \mathrm{d}x_2 \\
&\geqslant \int_t^L \mathrm{d}s \int_{g(s)}^{h(s)} |u|^{\alpha_1(x)} |D_1 u|^2 \, \mathrm{d}x_2 \\
&\quad + K^{\alpha_2^- - \alpha_2^+} \int_t^L \mathrm{d}s \int_{g(s)}^{h(s)} |u|^{\alpha_2^+} |D_2 u|^2 \, \mathrm{d}x_2 \\
&\geqslant C(K, \alpha_2^{\pm}) \Psi(t).
\end{aligned}
$$

Gathering this inequality with the estimate on $J(t)$, we arrive at the required assertion.

LEMMA 6.1. *Let the conditions of Theorem 3.5 be fulfilled. Then the solution of problem* (3.13) *satisfies the differential inequality*

$$
C\Psi(t) \leqslant \psi(t) \max\left\{ \left(-\Psi'(t)\right)^{q^+}, \left(-\Psi'(t)\right)^{q^-} \right\} \left(-\Psi'(t)\right)^{1/2} \quad \text{for a.a. } t \in (l, L)
\tag{6.11}
$$

with the constant $C = K^{\alpha_2^- - \alpha_2^+} \in (0, 1]$.

THEOREM 6.2. *Let the conditions of Theorem 3.5 be fulfilled, $f = 0$ for a.e. $x \in \Omega \cap \{x_1 > l\}$, and the exponents $\alpha_i(x)$ be such that*

$$
\alpha_1^- > \alpha_2^+.
\tag{6.12}
$$

Then there exists ε, depending on $\alpha_i^{\pm}$ and properties of the functions $h(t)$, $g(t)$, such that every solution of problem (3.13), *satisfying the condition $E(t) \leqslant \varepsilon$, is localized in the variable x_1: $u(x,t) = 0$ for a.e. $x \in \Omega \cap \{x_1 > t^*\}$ with $t^* \in (l, L)$, defined through ε and $\alpha_i^{\pm}$.*

The assertion of the theorem is illustrated by Figure 3.

PROOF OF THEOREM 6.2. The solutions of problem (3.13) satisfy the differential inequality (6.11). Let us notice that

$$
\max\left\{ \left(-\Psi'(t)\right)^{q^+}, \left(-\Psi'(t)\right)^{q^-} \right\} =
\begin{cases}
\left(-\Psi'(t)\right)^{q^-} & \text{if } \Phi(t) < 1, \\
\left(-\Psi'(t)\right)^{q^+} & \text{if } \Phi(t) \geqslant 1.
\end{cases}
$$

This allows us to rewrite inequality (6.11) in the form

$$
\begin{aligned}
&\min\left\{ \left(C\psi^{-1}(t)\Psi(t)\right)^{2/(1+2q^-)}, \left(C\psi^{-1}(t)\Psi(t)\right)^{2/(1+2q^+)} \right\} \\
&\quad + \Psi'(t) \leqslant 0 \quad \text{for a.e. } t \in (l, L).
\end{aligned}
\tag{6.13}
$$

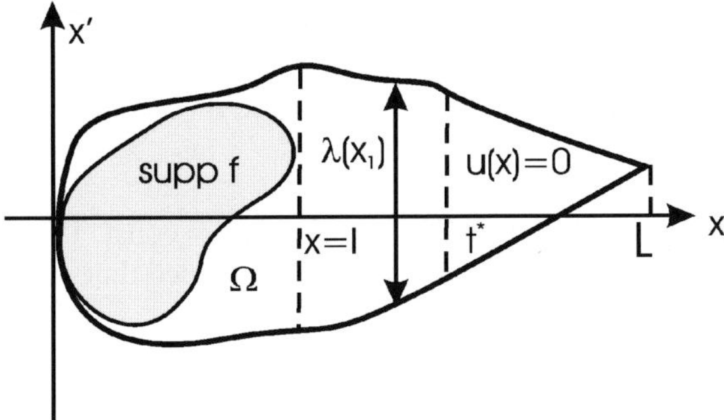

Fig. 3. Localization of the solution in the direction x_1.

Using the inequalities

$$\frac{2}{1+2q^+} \leqslant \frac{2}{1+2q^-} < 1, \qquad \frac{\Psi(t)}{\varepsilon K^{\alpha_2^+ - \alpha_2^-}} \leqslant \frac{E(t)}{\varepsilon} \leqslant 1$$

and taking into account the form of the coefficient C in (6.11), we conclude that

$$\min\left\{\left(C\psi^{-1}(t)\Psi(t)\right)^{2/(1+2q^-)}, \left(C\psi^{-1}(t)\Psi(t)\right)^{2/(1+2q^+)}\right\}$$

$$\geqslant \min\left\{\left(C\varepsilon\psi^{-1}(t)\right)^{2/(1+2q^-)}, \left(C\varepsilon\psi^{-1}(t)\right)^{2/(1+2q^+)}\right\}\left(\frac{\Psi(t)}{\varepsilon K^{\alpha_2^+ - \alpha_2^-}}\right)^{2/(1+2q^-)}$$

$$\geqslant \min\left\{1, \varepsilon^{2/(1+2q^+)-2/(1+2q^-)}\right\}$$

$$\times \min\left\{\psi^{-2/(1+2q^-)}(t), \psi^{-2/(1+2q^+)}(t)\right\}\Psi^{2/(1+2q^-)}(t).$$

Inequality (6.13) can now be continued as follows,

$$D\Psi^{2/(1+2q^-)} + z(t)\Psi'(t) \leqslant 0 \quad \text{for a.e. } t \in (l, L) \tag{6.14}$$

with the coefficients

$$D = \begin{cases} 1 & \text{for } \varepsilon \leqslant 1, \\ \varepsilon^{2/(1+2q^+)-2/(1+2q^-)} & \text{for } \varepsilon > 1, \end{cases}$$

$$z(t) \equiv \max\left\{\psi^{2/(1+2q^-)}(t), \psi^{2/(1+2q^+)}(t)\right\}. \tag{6.15}$$

Integrating (6.14) in the limits (l, t) and making use of the fact that $\Psi(l) \leqslant \varepsilon K^{\alpha_2^+ - \alpha_2^-}$, we obtain the inequality

$$0 \leqslant \Psi^{(2q^- -1)/(2q^- +1)}(t) \leqslant \left(\varepsilon K^{\alpha_2^+ - \alpha_2^-}\right)^{(2q^- -1)/(2q^- +1)} - D\frac{2q^- - 1}{2q^- + 1}\int_l^t \frac{ds}{z(x)}$$

$$\equiv G(t, \varepsilon).$$

By the definition, $\Psi(t)$ is a monotone decreasing function, which is why $\Psi(t) \equiv 0$ for all $t \geqslant t_*$ where t^* is the root of the equation $G(t, \varepsilon) = 0$. This equation always has a solution in the interval (l, L), provided that $\varepsilon < \varepsilon^*$ where ε^* denotes the root of the equation $G(L, \varepsilon^*) = 0$. $\qquad\square$

REMARK 6.1. Let in the conditions of Theorem 6.2, $\alpha_i = $ const. Then the equation $G(t, \varepsilon)$ has a solution in the interval (l, L) for every $\varepsilon > 0$, provided that

$$\int_l^L \frac{\mathrm{d}s}{h(s) - g(s)} = \infty.$$

An example of such a domain is a domain with the corner point at $x_1 = L$: $h(t) - g(t) \sim L - t$ when $t \to L-$.

6.1.4. *The general case $n \geqslant 3$.* Let $n \geqslant 3$. Let us assume that the domain Ω possesses the following properties:

$$\begin{cases} \text{for every } t \in (0, L) \text{ the truncation } \omega(t) = \Omega \cap \{x_1 = t\} \text{ is a simple-connected} \\ \quad \text{domain in } \mathbb{R}^{n-1} \text{ with Lipschitz-continuous boundary } \partial\omega(t), \\ \text{there exists } \kappa \in (0, 1) \text{ such that } \kappa\lambda(t) \leqslant \operatorname{diam}\omega(t) \leqslant \lambda(t), \\ \quad \lambda(0) \geqslant 0, \lambda(L) \geqslant 0, \text{ where } \lambda(t) \text{ is a given continuous function.} \end{cases} \qquad (6.16)$$

By convention, we will denote $x = (x_1, x')$, $x' = (x_2, \ldots, x_n)$. Repeating the proof of Theorem 6.1, it is easy to check the validity of the lemma:

LEMMA 6.2. *Let the conditions of Theorem 3.5 and conditions (6.16) be fulfilled. Assume that $f \equiv 0$ for a.e. $x \in \Omega \cap \{x_1 \geqslant l\}$. Then the solution of problem (3.13) satisfies the energy relation: for a.e. $t \in (l, L)$,*

$$J(t) \equiv - \int_{\omega(t)} u|u|^{\alpha_1(x)} D_1 u \, \mathrm{d}x'$$

$$= \int_t^L \mathrm{d}r \int_{\omega(r)} \left(\sum_i a_i(x, u)|u|^{\alpha_i(x)}|D_i u|^2 \, \mathrm{d}x' + c(x, u)|u|^{\sigma(x)} \right) \mathrm{d}x'.$$

$$(6.17)$$

Let us denote

$$\Phi(t) = \int_{\omega(t)} |u|^{\alpha_1(x)}|D_i u|^2 \, \mathrm{d}x' + \sum_{i \neq 1} \int_{\omega(t)} |u|^\beta |D_i u|^2 \, \mathrm{d}x',$$

$$\Psi(t) = \int_t^L \mathrm{d}r \int_{\omega(r)} |u|^{\alpha_1(x)}|D_i u|^2 \, \mathrm{d}x' + \sum_{i \neq 1} \int_t^L \mathrm{d}r \int_{\omega(r)} |u|^\beta |D_i u|^2 \, \mathrm{d}x,$$

$$\gamma(x) = 2\frac{\alpha_1(x) + 2}{\beta + 2}, \qquad \beta = \max_{i \neq 1} \alpha_i^+,$$

$$\gamma(x) \in \left[\inf_{\Omega}\gamma(x), \sup_{\Omega}\gamma(x)\right] \subset (\gamma^-, \gamma^+).$$

LEMMA 6.3. *Let $n \geqslant 3$. Assume that the conditions of Theorem 3.5 are fulfilled, the domain Ω satisfies conditions (6.16), and the exponents of nonlinearity $\alpha_i(x)$ satisfy the conditions*

$$1 < \frac{\alpha_1^+ + 2}{\beta + 2} \quad for \ n = 3, \qquad 1 < \frac{\alpha_1^+ + 2}{\beta + 2} \leqslant \frac{n-1}{n-3} \quad for \ n > 3. \tag{6.18}$$

Then for the weak solutions of problem (3.13), the inequality

$$C_0 \Psi(t) \leqslant C_1 \psi(t)\sqrt{\Phi(t)}\max\{\Phi^{\gamma^-/4}(t), \Phi^{\gamma^+/4}(t)\},$$

holds, where C_0 and C_1 are absolute constants depending only on the parameters $\alpha_i^{\pm}$, K, a_0 and c_0, and the function $\psi(t)$ has the form

$$\psi(t) = C(K, \alpha_i^{\pm})\|1\|_{\gamma^+/(\gamma^+-\gamma(\cdot)),\omega(t)}^{1/2}$$
$$\times \max\{\lambda^{\gamma^-/2+(n-1)/2(\gamma^-/\gamma^++\gamma^-/2)}(t), \lambda^{\gamma^+/2+(n-1)/2(1+\gamma^+/2)}(t)\}.$$
$$\tag{6.19}$$

PROOF. By Hölder's inequality,

$$|J(t)| \leqslant \left(\int_{\omega(t)}|u|^{\alpha_1(x)+2}\,dx'\right)^{1/2}\left(\int_{\omega(t)}|u|^{\alpha_1(x)}|D_1u|^2\,dx'\right)^{1/2}$$
$$\equiv \sqrt{j_1(t)}\sqrt{j_2(t)}.$$

Let us denote $v = |u|^{(\beta+2)/2}$, and estimate j_1 applying (2.3),

$$j_1(t) \leqslant \int_{\omega(t)}|v|^{\gamma(x)}\,dx' \leqslant 2\|v\|_{\gamma^+/\gamma(\cdot),\omega(t)}\|1\|_{\gamma^+/(\gamma^+-\gamma(\cdot)),\omega(t)}$$
$$\leqslant 2C(|\omega(t)|)\max\{A_{\gamma^+/\gamma(\cdot),\omega(t)}^{\gamma^-/\gamma^+}(|v|^{\gamma(x)}), A_{\gamma^+/\gamma(\cdot),\omega(t)}(|v|^{\gamma(x)})\}$$
$$\leqslant 2C(|\omega(t)|)\max\{A_{\gamma^+,\omega(t)}^{\gamma^-/\gamma^+}(v), A_{\gamma^+,\omega(t)}(v)\}$$
$$\leqslant 2C(|\omega(t)|)\max\{\|v\|_{\gamma^+,\omega(t)}^{\gamma^-}, \|v\|_{\gamma^+,\omega(t)}^{\gamma^+}\}$$

with the constant

$$C(|\omega(t)|) = \|1\|_{\gamma^+/(\gamma^+-\gamma(\cdot)),\omega(t)}$$
$$\leqslant \max\{|\omega(t)|^{\inf\gamma^+/(\gamma^+-\gamma(x))}, |\omega(t)|^{\sup(\gamma^+/(\gamma^+-\gamma(x)))}\}. \tag{6.20}$$

Accept the notation $\tilde{\nabla} v = (D_2 v, \ldots, D_n v)$. Making use of the known interpolation inequalities [24], Chapter 2, Section 2, we estimate

$$\|v\|_{\gamma^+,\omega(t)} \leqslant C\big(\lambda(t)\big)^{1+(n-1)(1/\gamma^+-1/2)} \big\|\tilde{\nabla} v\big\|_{2,\omega(t)}$$

$$\leqslant C\big(\lambda(t)\big)^{1+(n-1)(1/\gamma^+-1/2)} \frac{\beta+2}{2} \Big(\sum_{i\neq 1} \int_{\omega(t)} |u|^\beta |D_i u|^2 \, dx'\Big)^{1/2}$$

$$\leqslant C\big(\lambda(t)\big)^{1+(n-1)(1/\gamma^+-1/2)} \sqrt{\Phi(t)}, \quad C \equiv C(K,\beta),$$

with the function $\lambda(t)$ from condition (6.16). Thus

$$\big|J(t)\big| \leqslant C'\big(K,\alpha_i^\pm\big) \sqrt{\Phi(t)} \max\big\{\Phi^{\gamma^-/4}(t), \Phi^{\gamma^+/4}(t)\big\} \|1\|_{\gamma^+/(\gamma^+-\gamma(\cdot)),\omega(t)}^{1/2}$$

$$\times \max\big\{\lambda^{\gamma^-/2+(n-1)/2(\gamma^-/\gamma^++\gamma^-/2)}(t), \lambda^{\gamma^+/2+(n-1)/2(1+\gamma^+/2)}(t)\big\}. \quad \square$$

By analogy with (6.13), (6.14), we arrive at the inequality

$$C_3 \Psi^{4/(2+\gamma^-)}(t) + \phi(t)\Psi'(t) \leqslant 0, \quad \phi(t) = \psi^{4/(2+\gamma^-)} \text{ for a.e. } t \in (l, L).$$

$$(6.21)$$

Integrating this inequality in t, we prove the following theorem.

THEOREM 6.3. *Let $n \geqslant 3$. Let us assume that the conditions of Theorem 3.5 are fulfilled, the domain Ω satisfies conditions (6.16), $f = 0$ for a.e. $x \in \Omega \cap \{x_1 > l\}$, and the exponents $\alpha_i(x)$ satisfy conditions (6.18). Then there exists ε, depending on $\alpha_i^\pm$ and $\lambda(t)$, such that every solution of problem (3.13), satisfying the conditions $E(t) \leqslant \varepsilon$, $\|u\|_{\infty,\Omega} \leqslant \max\{1, \varepsilon\}$, is localized in the variable x_1: $u(x,t) = 0$ for a.e. $x \in \Omega \cap \{x_1 > t^*\}$ with $t^* \in (l, L)$ defined through ε and $\alpha_i^\pm$.*

The proof repeats the proof of Theorem 6.2 and can be omitted. Let us only notice that the final inequality for the energy function $\Psi(t)$ has the form

$$0 \leqslant \Psi^\sigma(t) \leqslant \big(\varepsilon^{1+\min_{i\neq 1}\alpha_i^- -\beta}\big)^\sigma - C_3 \sigma \int_l^t \frac{ds}{\phi(s)} \equiv \mathcal{F}(\varepsilon, t),$$

$$\sigma = 1 - \frac{4}{2+\gamma^-} \equiv \frac{\gamma^- - 2}{\gamma^- + 2} \in (0,1), \tag{6.22}$$

which guarantees the presence of the localization effect, provided that $\varepsilon < \varepsilon^*$ where for ε^* we choose the solution of the equation $\mathcal{F}(\varepsilon^*, L) = 0$. It is evident also that the one-dimensional localization surely takes place in the case when

$$\int_l^L \frac{ds}{\phi(s)} = \infty.$$

Combining (6.20) with (6.19), it is easy to check that the last condition if fulfilled for the domain satisfying the conditions

$$\lambda(t) \sim (L - t)^{\mu} \quad \text{when } t \to L, \qquad 1 \geqslant \mu \geqslant \frac{\gamma^+}{\gamma^-} \frac{2 + \gamma^-}{2(n-1) - \gamma^+(n-3)}.$$

If $\alpha_i = \mathrm{const}$ and $n = 2$, this condition coincides with the condition imposed in Remark 6.1.

6.1.5. *Influence of the low-order terms.* Condition (6.18) allows us to prove the one-dimensional localization of solutions of equation (3.13) independently of the properties of the low-order terms in the equation. Let us assume that condition (6.18) is false. If this is the case, inequality (6.21) becomes linear and its integration gives no information about the possible localization of the solution. Nonetheless, the effect of one-dimensional localization takes place if the low-order term in equation (3.13) is subject to certain conditions.

Let us introduce the energy functions

$$\Psi(t) = \int_t^L dr \int_{\omega(r)} |u|^{\alpha_1(x)} |D_1 u|^2 \, dx' + \int_t^L dr \int_{\omega(r)} |u|^{\alpha^+} |\widetilde{\nabla} u|^2 \, dx',$$

$$\Lambda(t) = \int_t^L dr \int_{\omega(r)} |u|^{\sigma^+} \, dx', \qquad \Theta(t) = \Phi(t) + \Lambda(t).$$

The functions $\Psi(t)$ and $\Lambda(t)$ are nonnegative and monotone nonincreasing, which is why $\Psi'(t)$ and $\Lambda'(t)$ exist for a.e. $t \in (l, L)$.

LEMMA 6.4. *Let the conditions of Theorem 3.5 and condition (6.16) be fulfilled, and let in condition (3.14) $c_0 > 0$. Let us assume that $f \equiv 0$ for a.e. $x \in \Omega \cap \{x_1 \geqslant l\}$. Then the solutions of problem (3.13) satisfy the inequality: for a.e. $t \in (l, L)$,*

$$\int_{\omega(t)} |u|^{\alpha_1(x)+1} |D_1 u| \, dx' \geqslant \min\{a_0, c_0\} \Theta(t).$$

PROOF. By Lemma 6.2, inequality (6.17) is true. Estimating the right-hand side of (6.17) with the help of (3.14) and the inequality

$$|u|^{\alpha_i(x)} \geqslant K^{\alpha^- - \alpha^+} |u|^{\alpha^+}, \qquad |u|^{\sigma(x)} \geqslant K^{\sigma^- - \sigma^+} |u|^{\sigma^+} \quad \text{for a.e. } x \in \Omega,$$

we obtain the required inequality. $\qquad\square$

LEMMA 6.5. *Let the conditions of Lemma 6.4 be fulfilled, and the exponents of nonlinearity $\alpha_i(x)$ and $\sigma(x)$ be such that*

$$\gamma \equiv \frac{2\sigma^+}{\alpha^+ + 2} \in (1, 2).$$

Then for the solutions of problem (3.13), *the differential inequality holds*

$$\widetilde{C}\Theta^{\beta}(t) + \rho(t)\Theta'(t) \leqslant 0 \quad \text{for a.e. } t \in (l, L), \, \beta^{-1} = \frac{\theta}{2} + \frac{1}{\gamma} + \frac{1}{2}, \tag{6.23}$$

in which $\widetilde{C}$ is an absolute constant,

$$\theta = \frac{1/\gamma - 1/2}{1/\gamma - (n-3)/(2(n-1))} \in (0, 1), \qquad r(x) = \left(\frac{\alpha^+ + 2}{\alpha_1(x) + 2}\right)'$$

and the coefficient $\rho(t)$ has the form $\rho(t) = R^{\beta}(t)$,

$$R(t) = \|1\|_{r(\cdot),\omega(t)}\lambda^{\mu}(t), \quad \mu = \theta + (1-\theta)(n-1)\left(\frac{1}{2} - \frac{1}{\gamma}\right).$$

PROOF. Let us estimate the left-hand side of the inequality obtained in Lemma 6.5, using Hölder's inequality and inequality (2.3):

$$\int_{\omega(t)} |u|^{\alpha_1(x)+1} |D_1 u| \, dx'$$

$$\leqslant \int_{\omega(t)} |u|^{\alpha_1(x)/2+1} \left(|u|^{\alpha_1(x)/2}|D_1 u|\right) dx'$$

$$\leqslant \left(\int_{\omega(t)} |u|^{\alpha_1(x)+2} \, dx'\right)^{1/2} \left(\int_{\omega(t)} |u|^{\alpha_1(x)}|D_1 u|^2 \, dx'\right)^{1/2}$$

$$\leqslant \sqrt{2}\||u|^{\alpha_1(x)+2}\|_{(\alpha^++2)/(\alpha_1(\cdot)+2),\omega(t)}^{1/2}$$

$$\times \|1\|_{((\alpha^++2)/(\alpha_1(\cdot)+2))',\omega(t)}^{1/2} \left(\int_{\omega(t)} |u|^{\alpha_1(x)}|D_1 u|^2 \, dx'\right)^{1/2}$$

$$\leqslant \sqrt{2}\max\left\{\left(\int_{\omega(t)} |u|^{\alpha^++2} \, dx'\right)^{1/2}, \left(\int_{\omega(t)} |u|^{\alpha^++2} \, dx'\right)^{(\alpha^-+2)/(2(\alpha^++2))}\right\}$$

$$\times \|1\|_{r(\cdot),\omega(t)}^{1/2} \left(\int_{\omega(t)} |u|^{\alpha_1(x)}|D_1 u|^2 \, dx'\right)^{1/2}.$$

Denote

$$v = |u|^{(\alpha^++2)/2}, \qquad \int_{\omega(t)} |u|^{\alpha^++2} \, dx' = \int_{\omega(t)} v^2 \, dx'.$$

If $\gamma \in (1, 2)$, we may apply interpolation inequalities (2.12)–(2.13),

$$\|v\|_{2,\omega(t)} \leqslant C\lambda^{\mu}(t)\|\nabla v\|_{2,\omega(t)}^{\theta}\|v\|_{\gamma,\omega(t)}^{1-\theta} \leqslant C\lambda^{\mu}(t)\left(-\Theta'(t)\right)^{\theta/2+1/\gamma}.$$

Gathering the last two inequalities, we arrive at (6.23). $\qquad\square$

REMARK 6.2. In the special case when $\alpha_1(x) = \alpha^+$, in inequality (6.23), $\rho(t) \sim \lambda^\mu(t)$.

REMARK 6.3. If $\alpha_i(x) = 0$, $\sigma(x) \equiv \sigma^+$, then the diffusion part of equation (3.13) becomes linear and the above-imposed conditions on the exponents of nonlinearity transform into the known localization condition [17]: $1 < \sigma^+ < 2$.

Let us denote

$$\varepsilon^* = \sup\left\{\varepsilon > 0 \colon \varepsilon^{1-\beta} < \widetilde{C}(1-\beta) \int_l^L \frac{ds}{R(s)}\right\}.$$

THEOREM 6.4. *Let the conditions of Lemma 6.5 be fulfilled. Then one may indicate* $\delta \equiv \delta(\varepsilon^*, \alpha^\pm, \sigma^+, K)$ *such that*
(1) *$\delta \to \infty$ when $\varepsilon^* \to \infty$,*
(2) *every solution of problem (3.13), satisfying the conditions*

$$\int_l^L dr \int_{\omega(r)} \left(\sum_i |u|^{\alpha_i(x)}|D_iu|^2 + |u|^{\sigma(x)}\right) dx' \leqslant \delta, \quad \|u\|_{\infty,\Omega} \leqslant K,$$

is localized in the variable x_1: $u(x) \equiv 0$ for a.e. $x \in \{x \in \Omega \colon x_1 > T\}$ with some $T \in (l, L)$.

PROOF. It is checked by straightforward calculation that $\beta < 1$ for every $1 < \gamma < 2$. Let us integrate inequality (6.23) in the interval (l, t),

$$\Theta^{1-\beta}(t) \leqslant \Theta^{1-\beta}(l) - \widetilde{C}(1-\beta) \int_l^t \frac{ds}{R(s)}. \tag{6.24}$$

Boundedness of the solution yields the inequality $\Theta(t) \leqslant (K^{\alpha^+-\alpha^-} + K^{\sigma^+-1})\delta$, and for all $t \geqslant l$,

$$\left[\delta\left(K^{\alpha^+-\alpha^-} + K^{\sigma^+-1}\right)\right]^{1-\beta} \geqslant \widetilde{C}(1-\beta) \int_l^t \frac{ds}{R(s)}.$$

Let us choose $\delta = \varepsilon^*(K^{\alpha^+-\alpha^-} + K^{\sigma^+-1})^{-1}$. Then there exists $T \in (l, L)$ such that $\Theta(T) = 0$ and $u(x) = 0$ for a.e. $x \in \Omega \cap \{x_1 > T\}$ due to monotonicity of $\Theta(t)$. $\qquad\square$

REMARK 6.4. The solution of problem (3.13) is always localized in the variable x_1, provided that

$$\int_l^L \frac{ds}{R(s)} = \infty.$$

6.2. *Generalized $p(x)$-Laplace equation*

Let the conditions of Theorems 3.2 and 3.3 be fulfilled and $u(x)$ be a bounded weak solution of the problem

$$
\begin{cases}
-\sum_i D_i\left(a_i(x,u)|D_i u|^{p_i(x)-2} D_i u\right) \\
\quad + c(x,u)|u|^{\sigma(x)-2}u = f & \text{in } \Omega, \\
u = 0 & \text{on } \Gamma.
\end{cases}
\tag{6.25}
$$

The domain Ω is assumed to satisfy the following conditions:

$$
\begin{cases}
\forall s \in (0,L) \text{ the set } \omega(x_1) = \Omega \cap \{x_1 = s\} \text{ is a simple-connected} \\
\quad \text{domain in } \mathbb{R}^{n-1}, \ \partial\omega(t) \text{ is Lipschitz-continuous} \\
\exists \kappa \in (0,1): \ \kappa\lambda(s) \leqslant \operatorname{diam}\omega(s) \leqslant \lambda(s), \ \lambda(0) \geqslant 0, \ \lambda(L) \geqslant 0, \\
\quad \text{where } \lambda(t) \text{ is a given continuous function.}
\end{cases}
\tag{6.26}
$$

Unless specially indicated, we will assume that $L < \infty$.

6.2.1. *The energy relation.*   Set

$$
\phi_k(x,s) =
\begin{cases}
1 & \text{for } x_1 > s + \frac{1}{k}, \\
k(x_1 - s) & \text{for } x \in \left[s, s + \frac{1}{k}\right], \quad k \in \mathbb{N}, \\
0 & \text{for } x_1 < s,
\end{cases}
$$

and choose the function $u(x)\phi_k(x,s)$ for the test-function in the integral identity (3.5). The resulting identity has the form

$$
\sum_{j=1}^{4} I_j(k,s) \equiv \sum_i \int_{\Omega\cap\{x_1>s+1/k\}} a_i |D_i u|^{p_i} \phi_k \, dx
$$

$$
+ k \int_{\Omega\cap\{s<x_1<s+1/k\}} a_1 u |D_1 u|^{p_1-2} D_1 u \, dx \, dt
$$

$$
+ \int_{\Omega\cap\{x_1>s\}} c(x)|u|^{\sigma(x)} \phi_k \, dx - \int_{\Omega\cap\{x_1>s\}} f u \phi_k \, dx = 0. \tag{6.27}
$$

The inclusion $u \in \mathbf{V}(\Omega)$ yields the inclusions $a_i|D_i u|^{p_i}\phi_k, c(x)|u|^{\sigma(x)}\phi_k, fu\phi_k \in L^1(\Omega)$, which allows one to pass to the limit when $k \to \infty$ in I_1, I_3 and I_4,

$$
\lim_{k\to\infty} I_1 = \sum_i \int_{\Omega\cap\{x_1>s\}} a_i |D_i u|^{p_i} \, dx \, dt,
$$

$$
\lim_{k\to\infty} I_3 = \int_{\Omega\cap\{x_1>s\}} c(x)|u|^{\sigma(x)} \, dx,
$$

$$
\lim_{k\to\infty} I_4 = -\int_{\Omega\cap\{x_1>s\}} f u \, dx.
$$

Now notice that by virtue of (6.27) I_2 is bounded uniformly with respect to k provided that so are the integrals I_1, I_2, I_3. Writing I_2 in the form

$$I_2 = k \int_s^{s+1/k} dx_1 \int_{\omega(x_1)} a_i u |D_i u|^{p_i - 2} D_i u \, dx'$$

and applying the Lebesgue theorem, we conclude that there exists

$$\lim_{k \to \infty} I_2(k, s) = \int_{\omega(s)} a_i u |D_i u|^{p_i - 2} D_i u \, dx'.$$

Let us assume that

$$f(x) = 0 \quad \text{a.e. in } \Omega \cap \{x: x_1 > l\} \text{ with some } l \in (0, L). \tag{6.28}$$

Then the energy relation takes on the form

$$\forall s > l, \quad \sum_i \int_s^L dx_1 \int_{\omega(x_1)} a_i |D_i u|^{p_i} \, dx' + \int_s^L \int_{\omega(s)} c|u|^{\sigma(x)} \, dx'$$

$$= - \int_{\omega(s)} a_1 u |D_1 u|^{p_1 - 2} D_1 u \, dx'. \tag{6.29}$$

6.2.2. *The ordinary differential inequality.* Let us introduce the function

$$J \equiv A_1 \int_{\omega(s)} |D_1 u|^{p_1(x) - 1} |u| \, dx' \geqslant \left| \int_{\omega(s)} a_1 u |D_1 u|^{p_1 - 2} D_1 u \, dx' \right|.$$

By Hölder's inequality,

$$J \leqslant A_1 \big\| |D_1 u|^{p_1(x) - 1} \big\|_{p_1(\cdot)/(p_1(\cdot)-1), \omega(s)} \|u\|_{p_1(\cdot), \omega(s)},$$

According to (2.9),

$$\big\| |D_1 u|^{p_1(x) - 1} \big\|_{p_1(\cdot)/(p_1(\cdot)-1), \omega(s)}$$

$$\leqslant \max\big\{ A_{p_1(\cdot), \omega(s)}^{(p_1^- - 1)/p_1^-}(D_1 u), \, A_{p_1(\cdot), \omega(s)}^{(p_1^+ - 1)/p_1^+}(D_1 u) \big\}$$

$$= A_{p_1(\cdot), \omega(s)}^{\gamma_1(s)}(D_1 u) \tag{6.30}$$

with the exponent

$$\gamma_1(s) = \begin{cases} \dfrac{p_1^- - 1}{p_1^-} & \text{if } A_{p_1(\cdot), \omega(s)}(D_1 u) < 1, \\[2ex] \dfrac{p_1^+ - 1}{p_1^+} & \text{otherwise.} \end{cases}$$

The second factor can be estimated by virtue of the embedding theorem,

$$
\|u\|_{p_1(\cdot),\omega(s)}
$$
$$
\leqslant C \max\left\{\lambda^{1+(n-1)/\beta-(n-1)/p_1^-}(s),\, \lambda^{1+(n-1)/\beta-(n-1)/p_1^+}(s)\right\}\|\widetilde{\nabla}u\|_{\beta,\omega(s)}
$$

$$(6.31)$$

with $\beta = \mathrm{const} > 0$ such that $\frac{1}{p_1(x)} > \frac{1}{\beta} - \frac{1}{n-1}$ in Ω if $\beta < n - 1$. Let us claim that $\beta = \min_{j\neq 1} p_j^-$. Then

$$
\|\widetilde{\nabla}u\|_{\beta,\omega(s)} \leqslant 2\sum_{j\neq 1}\|1\|_{p_j(\cdot)/(p_j(\cdot)-\beta),\omega(s)}^{1/\beta}\||D_j u|^\beta\|_{p_j(\cdot)/\beta,\omega(s)}^{1/\beta}
$$

$$
\leqslant 2\sum_{j\neq 1}\max\left\{\lambda(s)^{(n-1)/\beta-(n-1)/p_j^-},\, \lambda(s)^{(n-1)/\beta-(n-1)/p_j^+}\right\}
$$

$$
\times \max\left\{A_{p_j(\cdot),\omega(s)}^{1/p_j^-}(D_j u),\, A_{p_j(\cdot),\omega(s)}^{1/p_j^+}(D_j u)\right\}.
$$

Notice that $\forall j = 2, \ldots, n$,

$$
\max\left\{\lambda(s)^{(n-1)/\beta-(n-1)/p_j^-},\, \lambda(s)^{(n-1)/\beta-(n-1)/p_j^+}\right\}
$$

$$
\leqslant \max\left\{\lambda(s)^{(n-1)/\beta-(n-1)/p^-},\, \lambda(s)^{(n-1)/\beta-(n-1)/p^-}\right\}.
$$

Fix an arbitrary $s \in (l, L)$ and choose $k \in \{2, \ldots, n\}$ such that

$$
A_{p_k(\cdot),\omega(s)}(D_k u) \geqslant A_{p_j(\cdot),\omega(s)}(D_j u) \quad \text{for all } j \geqslant 2.
$$

There are two possibilities: either $A_{p_k(\cdot),\omega(s)}(D_k u) < 1$ or $A_{p_k(\cdot),\omega(s)}(D_k u) \geqslant 1$. In the former case the inequality

$$
\sum_{j\neq 2}A_{p_j(\cdot),\omega(s)}^{1/p_j^-}(D_j u) \leqslant n A_{p_k(\cdot),\omega(s)}^{1/p^+}(D_k u) \leqslant n\left(\sum_{j\neq 2}A_{p_j(\cdot),\omega(s)}(D_j u)\right)^{1/q^+}
$$

holds with $q^+ = \max_{j\neq 1} p_j^+$, otherwise

$$
\sum_{j\neq 2}A_{p_k(\cdot),\omega(s)}^{1/p^-}(D_k u) \leqslant n A_{p_k(\cdot),\omega(s)}^{1/p^-}(D_k u) \leqslant n\left(\sum_{j\neq 2}A_{p_j(\cdot),\omega(s)}(D_j u)\right)^{1/q^-}
$$

with $q^- = \max_{j\neq 1} p_j^-$. Thus, $J \leqslant K\rho(s)n A_{\mathbf{p}(\cdot),\omega(s)}^{\theta(s)}(\nabla u)$ with the exponent and the coefficient

$$
\theta(s) = \gamma_1(s) + \tau_j(s), \quad \tau_j(s) = \begin{cases} \dfrac{1}{q^+} & \text{if } \max_{j\geqslant 2} A_{p_j(\cdot),\omega(s)}(D_j u) < 1, \\[2mm] \dfrac{1}{q^-} & \text{otherwise,} \end{cases}
$$

$$\rho(s) = \max\left\{\lambda(s)^{1+2(n-1)/\beta-2(n-1)/p^-}, \lambda(s)^{1+2(n-1)/\beta-2(n-1)/p^+}\right\},$$

and with an absolute constant K independent of $\lambda(s)$ and $u(x)$. Let us introduce the energy function

$$\Phi(s) \equiv \int_s^L dx_1 \int_{\omega(s)} A_{\mathbf{p}(\cdot),\omega(s)}(\nabla u)\, dx'.$$

Equality (6.29) transforms then into the following inequality:

$$\begin{cases} \forall s > l, & \Phi^{1/\theta(s)}(s) + \psi(s)\Phi'(s) \leq 0, \quad s \in (l, L), \\ \Phi(L) = 0, & \Phi'(s) \leq 0, \quad \phi(s) = \left(\frac{Kn\rho(s)}{a_0}\right)^{1/\theta(s)}. \end{cases} \tag{6.32}$$

LEMMA 6.6. *Let the conditions of Theorems 3.2 and 3.3 be fulfilled, the exponents $p_i(x)$ satisfy the oscillation conditions*

$$\frac{1}{p_1(x)} > \frac{1}{\min_{j \neq 1} p_j^-} - \frac{1}{n-1} \quad \text{if } \min_{j \neq 1} p_j^- < n-1, \tag{6.33}$$

and the function $f(x)$ satisfies condition (6.28). Then for every weak bounded solution $u(x)$, the corresponding energy function $\Phi(s)$ is a solution of the ordinary differential inequality (6.32).

6.2.3. *Directional localization of solutions.* Let the energy function $\Phi(s)$ be uniformly bounded in the interval (l, L) by a finite constant M. Then inequality (6.32) can be written in the form

$$\nu\Phi^\mu + \phi(s)\Phi'(s) \leq 0, \quad \mu = \inf_{(l,L)} \frac{1}{\theta(s)}, \, \nu = \inf_{(l,L)} M^{1/\theta(s)-\mu}. \tag{6.34}$$

Let us claim that $\mu < 1$. This is true if $\theta(s) > 1$, which is guaranteed by the condition $p_1^- > \max_{j \neq 1} p_j^+$.

LEMMA 6.7. *Let the conditions of Lemma 6.6 be fulfilled, and the exponents $p_i(x)$ satisfy the conditions*

$$\begin{cases} 0 < \frac{1}{p_1^+} \leq \frac{1}{p_1^-} < \frac{1}{p_2^+} & \text{if } n = 2, \\ \frac{1}{\min_{j \neq 1} p_j^-} - \frac{1}{n-1} < \frac{1}{p_1^+} \leq \frac{1}{p_1^-} < \sum_{j \neq 1} \frac{1}{\max_{j \neq 1} p_j^+} & \text{if } n \geq 3. \end{cases}$$

Then one may indicate ε_* such that every solution $\Psi(s)$ of (6.25) satisfying the inequality $\Psi(s) \leq \varepsilon_*$ vanishes on an interval $(s_0, L) \subset (l, L)$.

PROOF. Let us integrate inequality (6.34) in the limits (l, s),

$$\Phi^{1-\mu}(s) \leq \varepsilon_*^{1-\mu} - \nu(1-\mu) \int_l^s \frac{dt}{M^{1/\theta(t)}\psi(t)} \tag{6.35}$$

and then take

$$\varepsilon_* < \left(\nu(1-\mu) \int_l^L \frac{dt}{M^{1/\theta(t)}\psi(t)} \right)^{1/(1-\mu)}.$$

The nonnegative and monotone decreasing function $\Phi(s)$ vanishes at a point $s_0 \in (l, L)$, and $\Phi(s) \equiv 0$ for all $s \in [s_0, L]$. $\qquad\square$

THEOREM 6.5. *Let the conditions of Theorems 3.2 and 3.3 and Lemma 6.7 be fulfilled. Then there exists ε_* such that every bounded weak solution of problem (6.25) satisfying $\Phi(l) \leqslant \varepsilon_*$ is localized in the direction x_1: $u(x) = 0$ a.e. in $\Omega \cap \{x_1 > s_0\}$ with some $s_0 \in (l, L)$.*

PROOF. Since the energy function $\Phi(s)$ satisfies the ordinary differential inequality (6.34), it follows from Lemma 6.7 that $\Phi(s) = 0$ for all $s > s_0$, provided that $\Phi(s)$ is sufficiently small. $\qquad\square$

COROLLARY 6.1. *The value of ε_* is arbitrary if*

$$\int_l^L \frac{dt}{\psi(t)} = \infty.$$

If this is the case, every solution with finite total energy is localized in the direction x_1. The conditions of divergence read as conditions on the rate of vanishing of the function $\psi(t)$ as $t \to L$, i.e., on the shape of the problem domain near the point $t = L$.

7. Problems on unbounded domains

7.1. *Generalized diffusion equation*

Let us consider problem (3.13) in the case when the domain Ω is unbounded. Let us assume that $\Omega \subset \mathbb{R}^n$ is simple-connected and condition (6.16) is fulfilled with $L = \infty$:

$$\begin{cases} \text{for every } t \in (0, L) \text{ the truncation } \omega(t) = \Omega \cap \{x_1 = t\} \text{ is a simple-connected} \\ \quad \text{domain in } \mathbb{R}^{n-1} \text{ with Lipschitz-continuous boundary } \partial\omega(t), \\ \forall t \in (0, L), \kappa\lambda(t) \leqslant \operatorname{diam}\omega(t) \leqslant \lambda(t), \lambda(0) \geqslant 0, \lim_{t\to\infty}\lambda(t) \geqslant 0, \\ \kappa \in (0, 1), \lambda(t) \text{ is a given continuous function.} \end{cases} \tag{7.1}$$

Let us assume that

$$f(x) = 0 \quad \text{for a.e. } x \in \Omega \cap \{x_1 > l\}. \tag{7.2}$$

All the above-considered equations admit anisotropic nonlinearity with respect to the variables x_i. The difference between the equations studied in Sections 3–5, consists in the fact that in the derivation of the differential inequality for the energy function we either make use of the essential anisotropy of the principal part of the equation and neglect the low-order term:

$$\alpha_1^- > \max_{i \neq 1} \alpha_i^+ \quad \text{and} \quad \text{in condition (3.14),} \quad c_0 \geqslant 0, \tag{7.3}$$

or rely on the presence of the low-order term and relax the conditions on the character of nonlinearity of the equation

$$\alpha_1^- \geqslant \max_{i \neq 1} \alpha_i^+ \quad \text{and} \quad \text{in condition (2.2),} \quad c_0 > 0.$$

Set

$$\varepsilon^* = \begin{cases} \sup\{\varepsilon > 0: \ \varepsilon^\sigma < \sigma \int_l^\infty \frac{ds}{\phi(s)}\} & \text{for } n \geqslant 3, \\ \sup\{\varepsilon > 0: \ \varepsilon^{(2q^+-1)/(2q^++1)} < \frac{2q^+-1}{2q^++1} \int_l^\infty \frac{ds}{z(s)}\} & \text{for } n = 2, \end{cases}$$

with the functions $\phi(s)$, $z(s)$ and the constants $\sigma \in (0, 1)$, $q^+ > 1/2$ defined in Theorems 6.3 and 6.2.

THEOREM 7.1. *Let $n \geqslant 3$ ($n = 2$), conditions (2.2) and (3.15) be fulfilled, the function $f(x)$ satisfy conditions (7.2) and (3.17), and $\alpha_i(x)$ satisfy conditions (6.18) for $n \geqslant 3$ and (6.12) for $n = 2$. Then one may choose $\delta \equiv \delta(\varepsilon^*)$ such that*

(1) $\delta \to \infty$ when $\varepsilon^* \to \infty$,

(2) problem (3.13) is solvable for every $\| f \|_{s,\Omega} \leqslant \delta$, where $s = p > n/2$ for $\alpha_i(x) \geqslant 0$, $s = \infty$ for $\alpha_i(x) > -1$ and $c_0 > 0$,

(3) problem (3.13) has a solution localized in the variable x_1: $u(x) = 0$ for a.e. $x \in \Omega \cap \{x_1 > T\}$ with some $T \in (l, \infty)$.

PROOF. Let us denote

$$\Omega^{(k)} = \Omega \cap \{t < k\}, \quad k \in \mathbb{N},$$

and consider the auxiliary problem

$$\begin{cases} \sum_i D_i\big(a_i(x, u)|u|^{\alpha_i(x)} D_i u\big) \\ \quad + c(x, u)|u|^{\sigma(x)-2} u = f(x) & \text{in } \Omega^{(k)}, \\ u = 0 & \text{on } \partial\Omega^{(k)}, k = 1, 2, \ldots. \end{cases} \tag{7.4}$$

By Theorem 3.5, for every finite k problem (7.4) has a weak solution $u_k(x)$ in the sense of Definition 1. For this solution, uniform in k estimates (3.19) are fulfilled. This allows us to extract from the sequence $\{u_k\}$ a subsequence (without loss of generality it is assumed to coincide with the whole of the sequence) such that, for every fixed $m \in \mathbb{N}$,

$$u_k \to u \quad \text{a.e. in } \Omega^{(m)}, k \geqslant m,$$

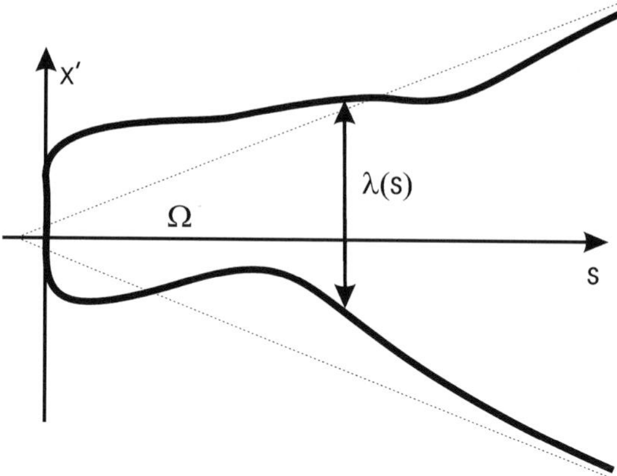

Fig. 4. Unbounded domain $n = 2$.

$$u_k \to u \quad \text{weakly in } L^{\sigma(x)}\big(\Omega^{(m)}\big) \text{ with arbitrary } m,$$

$$|u_k|^{\alpha_i(x)/2} D_i u_k \to |u|^{\alpha_i(x)/2} D_i u \quad \text{weakly in } L^2\big(\Omega^{(m)}\big), i = 1, \ldots, n.$$

Moreover, condition (7.2) means that, for all k beginning with some k_0, the solutions of the auxiliary problem possess the property of localization in the variable x_1: $u_k(x) = 0$ for a.e. $x \in \Omega \cap \{x_1 > l\}$ with some finite l, not depending on k. This allows us to pass to the limit when $k \to \infty$ in the integral identity (3.16) for $u_k(x)$ with every test-function $\eta \in W_0^{1,2}(\Omega) \cap L^{\sigma(x)}(\Omega)$. $\qquad\qquad\qquad\qquad\qquad\qquad\qquad\qquad\qquad\qquad\qquad\qquad\quad\square$

REMARK 7.1. If either $\alpha_i(x) \geqslant 0$, or $\alpha_i(x) > -1$ and $c_0 > 0$, and

$$\int_l^\infty \frac{ds}{\phi(s)} = \infty, \quad n \geqslant 3, \quad \text{or} \quad i(l) = \int_l^\infty \frac{ds}{z(s)} = \infty, \quad n = 2,$$

then problem (3.13) is solvable in the unbounded domain Ω for every right-hand side f satisfying conditions (3.17), (7.2). In the case $n = 2$, the condition of divergence of the integral $i(l)$ has a simple geometrical interpretation: the problem is solvable without additional conditions at infinity, if the domain can be inserted in an angle $\alpha < \pi$ – see Figure 4.

EXAMPLE 1. Let us consider the Dirichlet problem for the equation

$$-\big(2|u|u_{x_1}\big)_{x_1} - u_{x_2 x_2} = f(x_1, x_2)$$

in the domain

$$\Omega = \big\{(x_1, x_2) \colon x_1 > 0, \ x_2 \in \big(-1 - x_1^\mu, 1 + x_1^\mu\big)\big\}, \quad \mu = \text{const} > 0.$$

For $\mu \leqslant 1$ such a problem is solvable for every function $f(x) \in L^p(\Omega)$, $p > 1$, satisfying condition (7.2). For $\mu > 1$ the problem has a localized in x_1 solution if $\|f\|_{p,\Omega}$ is sufficiently small. Let

$$\Omega = \{x_1 \in \mathbb{R}, x_2 \in (0,1)\}, \qquad f(x) = \begin{cases} \varepsilon > 0 & \text{for } x_1 \in (0,1), \\ 0 & \text{for } x_1 \geqslant 1. \end{cases}$$

In this case the solution is localized in the domain $\Omega \cap \{x_1 > 1 + K\varepsilon^{1/5}\}$ with a known constant K.

A similar assertion is true if condition (7.3) of essential anisotropy is false. Let us denote

$$\varepsilon^* = \sup\left\{\varepsilon > 0 \colon \varepsilon^{1-\beta} < \tilde{C}(1-\beta)\int_l^\infty \frac{ds}{\rho(s)}\right\}$$

with the function $\rho(s)$ and the constants β, $\tilde{C}$ defined in the conditions of Theorem 6.4.

THEOREM 7.2. *Let condition* (3.15) *and condition* (3.14) *with* $c_0 > 0$ *be fulfilled, the function* $f(x)$ *satisfy conditions* (3.17) *and* (7.2), *and the exponents* $\alpha_i(x)$ *and* $\sigma(x)$ *be such that*

$$\frac{2\sigma^+}{\alpha^+ + 2} \in (1,2).$$

Then there exists $\delta \equiv \delta(\varepsilon^*)$ *such that*
(1) $\delta \to \infty$ *when* $\varepsilon^* \to \infty$,
(2) problem (3.13) *is solvable for every* $\|f\|_{s,\Omega} \leqslant \delta$, *where* $s = p > n/2$ *for* $\alpha_i(x) \geqslant 0$, *and* $s = \infty$ *for* $\alpha_i(x) > -1$,
(3) the solution of problem (3.13) *is localized in the variable* x_1: $u(x) = 0$ *for a.e.* $x \in \Omega \cap \{x_1 > T\}$ *with some* $T \in (l, \infty)$.

EXAMPLE 2. Let us consider the Dirichlet problem for the equation

$$-\Delta u + |u|^{-1/2 + 1/4 \sin x_1} u = f(x_1, x_2)$$

in the domain

$$\Omega = \left\{(x_1, x_2) \colon x_1 > 0, x_2 \in \left(-1 - x_1^\nu, 1 + x_1^\nu\right)\right\}, \qquad \nu = \text{const} > 0.$$

For this equation $\rho(t) \sim (1 + x_1^\nu)^{14/557}$, which is why the problem is solvable for every right-hand side f if $\nu \leqslant 39\frac{11}{14}$. For $\nu > 39\frac{11}{14}$, a localized in x_1 solution exists if $\|f\|_{\infty,\Omega}$ is sufficiently small.

7.2. *Generalized $p(x)$-Laplace equation*

THEOREM 7.3. *If, in the conditions of Theorem 6.5, $L = \infty$ then there exists $\varepsilon^* > 0$ such that problem (6.25) has a solution localized in the direction x_1 for every right-hand side f satisfying the conditions* (1) $f(x) \equiv 0$ *for* $x_1 > l$, (2) $\|f\| < \varepsilon_*$.

The proof follows the proof of Theorem 7.1 (see also [14]). We consider the sequence $\{u_k\}$ of solution of problems (6.25) posed on bounded domains $\Omega_k = \Omega \cap \{x_1 < k\}$, $k \in \mathbb{N}$. For every finite k this problem has a solution and, for all k from some k_0 on, the solution u_k is localized in the direction x_1: $\operatorname{supp} u_k \subset \Omega_{k_0}$. Passing to the limit as $k \to \infty$ in the integral identity (3.5) for u_k, we obtain a localized weak solution of the problem in the unbounded domain. Notice that the limit value of the total energy ε_* can be arbitrary if $\int_l^\infty \frac{dt}{\psi(t)} = \infty$, which is the condition of the "asymptotic size" of the domain Ω at infinity.

8. Systems of elliptic equations

Let us consider the system of equations

$$
\begin{cases}
-\sum_j D_j\big(a_{ij}(x, \nabla \mathbf{u})\big) \\
\quad + \sum_j c_{ij}(x)\big|u^{(j)}\big|^{\sigma_{ij}(x)-2} u^{(j)} = f^{(i)}(x) & \text{in } \Omega, i = 1, \ldots, n, \\
\mathbf{u} = 0 & \text{on } \partial\Omega.
\end{cases}
\tag{8.1}
$$

We assume that the functions $a_{ij}(x, \mathbf{V})$ are Carathéodory functions, and that a_{ij}, c_{ij} satisfy the following conditions:

$$
\begin{cases}
\forall (x, \mathbf{V}), (x, \mathbf{U}) \in \Omega \times \mathbb{R}^{n^2}, \quad (\mathbf{s}, \mathbf{r}) \in \mathbb{R}^n, \\
\sum_{ij} a_{ij}(x, \mathbf{V}) \cdot V_{ij} \geqslant a_0 \sum_{ij} |V_{ij}|^{p_{ij}(x)}, \quad a_0 = \text{const} > 0, \\
\sum_{ij} \big(a_{ij}(x, \mathbf{V}) - a_{ij}(x, \mathbf{U})\big) \cdot (V_{ij} - U_{ij}) \geqslant 0, \\
\sum_{ij} c_{ij}(x)\big|s^{(j)}\big|^{\sigma_{ij}(x)-2} s^{(j)} s^{(i)} \geqslant c_0 \sum_i \big|s^{(i)}\big|^{\sigma_i(x)}, \quad c_0 = \text{const} > 0, \\
\sum_{ik} c_{ik}(x)\big(\big|s^{(k)}\big|^{\sigma_{ik}(x)-2} s^{(k)} - \big|r^{(k)}\big|^{\sigma_{ik}(x)-2} r^{(k)}\big)\big(s^{(i)} - r^{(i)}\big) \geqslant 0,
\end{cases}
\tag{8.2}
$$

with functions $\sigma_{ij}(x)$, $\sigma_i(x)$ and $p_{ij}(x)$ satisfying the conditions:
1. The functions $a_{ij}(x, \mathbf{V})$ have the form

$$
a_{ij}(x, \mathbf{V}) = A_{ij}(x)|V_{ij}|^{p_{ij}(x)-2} V_{ij}, \quad i, j = 1, \ldots, n,
$$

the coefficients $A_{ij}(x)$ and $c_{ij}(x)$ are bounded in Ω and

$$
\begin{cases}
0 < a_0 \leqslant A_{ij}(x) < \infty, \\
0 < c_0 \leqslant c_{ij}(x) < \infty,
\end{cases}
\quad \forall x \in \overline{\Omega}.
\tag{8.3}
$$

2. The functions $p_i(x)$, $p_{ij}(x)$, $\sigma_i(x)$, $\sigma_{ij}(x)$ and $\sigma(x)$ are bounded in Ω: it is assumed that there exist constants $p^- > 1$, $p^+ < \infty$, $\sigma^+ < \infty$, $\sigma^- > 1$ such that, for all $x \in \overline{\Omega}$,

$$\begin{cases} p_{ij}(x) \in (p^-, p^+], \\ \sigma_i(x), \sigma_{ij}(x) \in (\sigma^-, \sigma^+], \quad i, j = 1, \ldots, n, \\ \inf_\Omega p_{ij}(x) > p^- \in (1, n). \end{cases}$$

3. The functions $p_{ij}(x)$, $\sigma_i(x)$ and $\sigma_{ij}(x)$ are continuous in Ω: for all $x, y \in \Omega$ satisfying $|x - y| \leqslant 1$, the inequality

$$\sum_i \left| \sigma_i(x) - \sigma_i(y) \right| + \sum_{ij} \left(\left| \sigma_{ij}(x) - \sigma_{ij}(y) \right| + \left| p_{ij}(x) - p_{ij}(y) \right| \right)$$
$$\leqslant \omega\big(|x - y|\big). \tag{8.4}$$

holds with a function $\omega(\tau)$ satisfying the condition

$$\omega(\tau) \ln \frac{1}{\tau} \leqslant M \quad \text{for } \tau \in [0, 1], \, M = \text{const} > 0,$$

for every $x, y \in \Omega$ such that $|x - y| \leqslant 1$.
The boundary $\Gamma = \partial\Omega$ is assumed to be Lipschitz-continuous. Let us introduce the function space

$$\mathbf{W}(\Omega) = \left\{ \mathbf{u}: u^{(i)} \in L^{\sigma_i(x)}(\Omega), D_j u^{(i)} \in L^{p_{ij}(x)}(\Omega), i, j = 1, \ldots, n \right\},$$

$$\|\mathbf{u}\|_{\mathbf{W}} = \sum_{i=1}^n \left\| u^{(i)} \right\|_{\sigma_i(\cdot)} + \sum_{ij} \left\| D_j u^{(i)} \right\|_{p_{ij}(\cdot)}.$$

The spaces $\mathbf{W}(\Omega)$ defined in this way are reflexive and separable Banach spaces. Let us introduce the operator $L(\mathbf{s}): \mathbf{W}(\Omega) \mapsto \mathbf{W}'(\Omega)$,

$$\forall \mathbf{h} \in \mathbf{W}(\Omega), \quad \big(L(\mathbf{u}), \mathbf{h}\big) \equiv \int_\Omega \left(\sum_{ij} a_{ij}(x, \nabla \mathbf{u}) \cdot D_j h^{(i)} \right.$$
$$\left. + \sum_{ij} c_{ik}(x) \left| u^{(j)} \right|^{\sigma_{ij}(x)-2} u^{(j)} h^{(i)} \right) dx.$$

DEFINITION 8.1. A vector-valued function $\mathbf{u} = (u_1, \ldots, u_n)$ is said to be a weak solution of problem (8.1) if $\mathbf{u} \in \mathbf{W}(\Omega)$ and, for every test-function $\mathbf{h} \in \mathbf{W}(\Omega)$,

$$\big(L(\mathbf{u}), \mathbf{h}\big) - (\mathbf{f}, \mathbf{h})$$
$$\equiv \int_\Omega \left(\sum_{ij} a_{ij}(x, \nabla \mathbf{u}) \cdot D_j h^{(i)} + \sum_{ik} c_{ik}(x) \left| u^{(k)} \right|^{\sigma_{ik}(x)-2} u^{(k)} h^{(i)} - \mathbf{hf} \right) dx$$
$$= 0. \tag{8.5}$$

8.1. *Existence of solutions*

The proof of existence of a weak solution of problem (8.1) is an adaptation of the proof given in the case of a single scalar equation.

LEMMA 8.1. *The mapping* $L : \mathbf{W}(\Omega) \mapsto \mathbf{W}'(\Omega)$ *is coercive,*

$$
\big(L(\mathbf{u}), \mathbf{u}\big) \geqslant C \min\left\{ \left(\frac{\|\mathbf{u}\|_{\mathbf{W}}}{n(n+1)} \right)^{p^{+}}, \left(\frac{\|\mathbf{u}\|_{\mathbf{W}}}{n(n+1)} \right)^{p^{-}} \right\}.
$$

PROOF. Denote $\lambda = \|\mathbf{u}\|_{\mathbf{W}}$, $\lambda_{0i} = \|u^{(i)}\|_{\sigma_i(\cdot)}$, $\lambda_{jk} = \|D_k u^{(j)}\|_{p_{jk}(\cdot)}$. There exists λ_{ij} such that $n(n+1)\lambda_{ij} \geqslant \lambda$ (we may assume that $i = j = 1$). Then

$$
\big(L(\mathbf{u}), \mathbf{u}\big) \geqslant \min\{a_0, c_0\}\left(\sum_i A_{\sigma_i(\cdot)}\big(u^{(i)}\big) + \sum_{ij} A_{p_{ij}(\cdot)}\big(D_j u^{(i)}\big) \right)
$$

$$
\geqslant C A_{p_{11}(\cdot)}\big(D_1 u^{(1)}\big)
$$

and the conclusion follows like in the proof of Lemma 3.1. $\qquad\square$

LEMMA 8.2. *The mapping* $L : \mathbf{W}(\Omega) \mapsto \mathbf{W}'(\Omega)$ *is continuous and monotone.*

PROOF. The monotonicity immediately follows from (8.2) and (2.10)–(2.11). The continuity is obvious. $\qquad\square$

THEOREM 8.1. *Let the functions* $\sigma_{ij}(x)$, $\sigma_i(x)$ *and* $p_{ij}(x)$ *satisfy conditions* (8.3)–(8.4), *and let conditions* (8.2) *be fulfilled. Then for every* $\mathbf{f} = (f^{(1)}, \ldots, f^{(n)})$ *such that* $f^{(i)} \in L^{\sigma_i'(x)}(\Omega)$, *problem* (8.1) *has a weak solution* $\mathbf{u} \in \mathbf{W}(\Omega)$ *satisfying the estimate*

$$
\int_\Omega \left(a_0 \sum_{ij} |D_j u^{(i)}|^{p_{ij}(x)} + c_0 \sum_i |u^{(i)}|^{\sigma_i(x)} \right) dx
$$

$$
\leqslant C \sum_i \int_\Omega |f^{(i)}|^{\sigma_i'(x)} \, dx. \tag{8.6}
$$

PROOF. Since the Banach space $\mathbf{W}(\Omega)$ is reflexive and separable, and the mapping L is continuous, monotone and coercive, the existence follows by the Browder–Minty theorem [20], Chapter 7. Estimate (8.6) follows from the proof of Lemma 8.1. $\qquad\square$

8.2. *Localization properties*

The study of the localization properties of weak solutions to system (8.1) is an imitation of the analysis given in the case of a single elliptic equation, which is why we concentrate here on the details specific for the case of a system of equations. Given a ball $B_\rho(x_0) \subset \Omega$, we introduce the *energy functions*

$$E(\rho) = \sum_{ij} \int_{B_\rho} |D_j u^{(i)}|^{p_{ij}(x)}\, dx, \qquad b(\rho) = \sum_{i} \int_{B_\rho} |u^{(i)}|^{\sigma_i(x)}\, dx.$$

We will need the following conditions:

$$\begin{cases} \forall (x, \mathbf{V}) \in \Omega \times \mathbb{R}^{n^2}, \quad (\mathbf{s}, \mathbf{r}) \in \mathbb{R}^n, \\ \sum_{ij} |a_{ij}(x, \mathbf{V})| \leqslant a_1 \sum_{ij} |V_{ij}|^{p_{ij}(x)-1}, \\ \sum_{ij} a_{ij}(x, \mathbf{V}) \cdot V_{ij} \leqslant a_1 \sum_{ij} |V_{ij}|^{p_{ij}(x)}, \quad a_1 = \mathrm{const} > 0, \\ \sum_{ij} c_{ij}(x) |s^{(j)}|^{\sigma_{ij}(x)-2} s^{(j)} s^{(i)} \leqslant c_1 \sum_{i} |s^{(i)}|^{\sigma_i(x)}, \quad c_1 = \mathrm{const} > 0. \end{cases} \tag{8.7}$$

THEOREM 8.2. *Let under the conditions of Theorem* 8.1 *the coefficients a_{ij} and c_{ij} satisfy conditions* (8.7). *Let us assume that*
  (i) *the oscillation of the exponents $p_{ij}(x)$ is such that*

$$p^+ - p^- < \frac{(p^- - 1)}{n} p^-,$$

  (ii) *equations* (8.3) *and* (8.4) *hold and*

$$\sigma^+ < p^-,$$

  (iii) $\mathbf{f}(x) \equiv 0$ *in $B_{\rho_0}(x_0) \subseteq \Omega$ with $\rho_0 \leqslant 1$.*
Then every weak solution of problem (8.1) *satisfying*

$$E(\rho_0) + b(\rho_0) \leqslant 1 \tag{8.8}$$

is such that

$$\mathbf{u}(x) \equiv 0 \quad a.e.\ in\ B_{\rho_1}(x_0)\ with\ \rho_1^{1+\lambda} = \left[\rho_0^{1+\lambda} - CE^\nu(\rho_0)\right]_+$$

with the positive constants C, λ and ν depending only on $p^\pm$, $\sigma^\pm$, a_0, a_1, c_0, c_1.

Let the function ψ_k be defined by (5.13). The function

$$\phi_k(x) \equiv \psi_k(|x - x_0|)\mathbf{u}(x),$$

can be taken for the test-function $\mathbf{h}$ in (8.5). Passing to the limit when $k \to \infty$, we obtain the equality

$$\sum_{ij} \int_{B_\rho(x_0)} \left(a_{ij}(x, \nabla \mathbf{u}) \cdot D_j u^{(i)} + c_{ij}(x) |u^{(j)}|^{\sigma_{ij}(x)-2} u^{(j)} u^{(i)} \right) dx = I(\rho),$$

where

$$I(\rho) = -\sum_i \int_{S_\rho(x_0)} u^{(i)} \sum_j a_{ij}(x, \nabla \mathbf{u}) \cdot \nu_j \, dS.$$

Then (see (5.15)) the following inequalities holds:

$$\left| I(\rho) \right| \leqslant C \sum_{ij} \left\| u^{(i)} \right\|_{p_{ij}(\cdot), S_\rho} \left\| \left| D_j u^{(i)} \right|^{p_{ij}(x)-1} \right\|_{p'_{ij}(\cdot), S_\rho}$$

$$\leqslant C \sum_i \left\| u^{(i)} \right\|_{p^+, S_\rho}$$

$$\times \max \left\{ \left(\sum_{ij} \int_{S_\rho} \left| D_j u^{(i)} \right|^{p_{ij}(x)} dx \right)^{(p^+-1)/p^+}, \right.$$

$$\left. \left(\sum_{ij} \int_{S_\rho} \left| D_j u^{(i)} \right|^{p_{ij}(x)} dx \right)^{(p^--1)/p^-} \right\}$$

$$= C \sum_i \left\| u^{(i)} \right\|_{p^+, S_\rho} \max \left\{ \left(E'(\rho) \right)^{(p^+-1)/p^+}, \left(E'(\rho) \right)^{(p^--1)/p^-} \right\},$$

$$\left| I(\rho) \right| \geqslant \min\{a_0, c_0\} \left(E(\rho) + b(\rho) \right)$$

which yield the inequality

$$E(\rho) + b(\rho) \leqslant C \sum_i \left\| u^{(i)} \right\|_{p^+, S_\rho} \max \left\{ \left(E'(\rho) \right)^{(p^+-1)/p^+}, \left(E'(\rho) \right)^{(p^--1)/p^-} \right\}.$$

$$(8.9)$$

According to the trace-interpolation inequality (5.16),

$$\left\| u^{(i)} \right\|_{p^+, S_\rho} \leqslant C \left(\left\| \nabla u^{(i)} \right\|_{p^-, B_\rho} + \rho^{-\delta} \left\| u^{(i)} \right\|_{\sigma_i^-, B_\rho} \right)^\theta \left\| u^{(i)} \right\|_{\sigma_i^-, B_\rho}^{1-\theta}$$

$$\leqslant C' \left(\sum_j \left\| D_j u^{(i)} \right\|_{p_{ij}(\cdot), B_\rho} + \rho^{-\delta} \left\| u^{(i)} \right\|_{\sigma_i(\cdot), B_\rho} \right)^\theta \left\| u^{(i)} \right\|_{\sigma_i(\cdot), B_\rho}^{1-\theta}$$

with the exponents

$$\sigma_i^- = \inf_\Omega \sigma_i(x) > \sigma^-,$$

$$\delta = \frac{n(p^- - \sigma^-) + p^- \sigma^-}{p^- \sigma^-} > 1,$$

$$\theta = \frac{p^-}{p^+} \frac{n(p^+ - \sigma^-) + \sigma^-}{n(p^- - \sigma^-) + p^- \sigma^-}.$$

According to (8.8) and (2.3),

$$\begin{cases} \sum_j \left\| D_j u^{(i)} \right\|_{p_{ij}(\cdot)} \leqslant C \big(\sum_j \int_\Omega \left| D_j u^{(i)} \right|^{p_{ij}(x)} \mathrm{d}x \big)^{1/p^+} = C E^{1/p^+}(\rho), \\ \left\| u^{(i)} \right\|_{\sigma_i(\cdot), B_\rho} \leqslant C b^{1/\sigma^+}(\rho), \end{cases}$$

whence (recall that $\rho \leqslant \rho_0 \leqslant 1$)

$$\sum_i \left\| u^{(i)} \right\|_{p^+, S_\rho} \leqslant C \rho_0^{-\delta\theta} \big(E(\rho) + b(\rho) \big)^{\theta/p^+ + (1-\theta)/\sigma^+}.$$

The last estimate allows us to transform inequality (8.9) for the energy function $E(\rho)$ into the nonlinear ordinary differential inequality (5.18) with the same parameters of nonlinearity, and the assertion of Theorem 8.2 follows exactly like in the case of Theorem 5.1.

Using the derived ordinary differential inequality for the energy function and the arguments used in the proof of Theorem 5.2, we prove the theorem:

THEOREM 8.3. *Let, under the conditions of Theorem 8.2,* $\mathbf{f}(x) \not\equiv 0$ *in* $B_{\rho_0}(x_0)$ *and*

$$\mathbf{f}(x) \equiv 0 \quad \text{in } B_{\rho_1}(x_0) \text{ for some } 0 < \rho_1 < \rho_0.$$

There exist positive constants $\gamma > 0$, ε_* *and* E_*, *depending only on* $p^\pm, \sigma^\pm, a_0, a_1, c_0, c_1,$ *such that if*

$$\sum_i \int_{B_\rho} \left| f^{(i)} \right|^{\sigma_i'(x)} \mathrm{d}x \leqslant \varepsilon (\rho - \rho_1)_+^\gamma \quad \text{for } \rho \in (\rho_1, \rho_0), \sigma_i' = \frac{\sigma_i}{\sigma_i - 1},$$

and $E(\rho_0) < \varepsilon_*$, *then* $\mathbf{u}(x) \equiv 0$ *in* $B_{\rho_1}(x_0)$.

8.3. *Systems of other types*

The assertions of Theorems 8.1–8.3 extend without serious change to the systems of the form

$$\begin{cases} -\sum_j D_j \big(a_{ij}(x, \nabla \mathbf{u}) \big) \\ \quad + c_i(x) |\mathbf{u}|^{\sigma(x)-2} u^{(i)} = f^{(i)}(x) \quad \text{in } \Omega, i = 1, \ldots, n, \\ \mathbf{u} = 0 \quad \text{on } \partial\Omega, \qquad |\mathbf{u}|^2 = \sum_i \left| u^{(i)} \right|^2 \end{cases} \tag{8.10}$$

and

$$\begin{cases} -\sum_j D_j \big(a_{ij}(x) |\nabla \mathbf{u}|^{p(x)-2} D_j u^{(i)} \big) \\ \quad + c_i(x) |\mathbf{u}|^{\sigma(x)-2} u^{(i)} = f^{(i)}(x) \quad \text{in } \Omega, i = 1, \ldots, n, \\ \mathbf{u} = 0 \quad \text{on } \partial\Omega, \qquad |\nabla \mathbf{u}|^2 = \sum_{ij} \left| D_j u^{(i)} \right|^2. \end{cases} \tag{8.11}$$

If the matrices

$$a_{ij}(x, \nabla\mathbf{u}), \qquad a_{ij}(x)|\nabla\mathbf{u}|^{p(x)-2}D_j u^{(i)},$$

$$c_{ij} = \begin{cases} 0 & \text{if } i \neq j, \\ c_i(x)|\mathbf{u}|^{\sigma(x)-2}u^{(i)} & \text{for } i = j, \end{cases}$$

satisfy conditions (8.2) and (8.7), the only difference in the arguments consists in the choice of the functional spaces. For system (8.10), we take

$$\mathbf{W}(\Omega) = \left\{\mathbf{u}: |\mathbf{u}| \in L^{\sigma(x)}, D_j u^{(i)} \in L^{p(x)}\right\},$$

$$\|\mathbf{u}\|_\Omega = \|\,|\mathbf{u}|\,\|_{\sigma(\cdot)} + \sum_{ij} \|D_j u^{(i)}\|_{p_{ij}(\cdot)},$$

and in the case of system (8.11),

$$\mathbf{W}(\Omega) = \left\{\mathbf{u}: |\mathbf{u}| \in L^{\sigma(x)}, |\nabla\mathbf{u}| \in L^{p(x)}\right\},$$

$$\|\mathbf{u}\|_\Omega = \|\,|\mathbf{u}|\,\|_{\sigma(\cdot)} + \|\,|\nabla\mathbf{u}|\,\|_{p(\cdot)}.$$

9. Examples: localization in borderline cases

9.1. *Illustrative examples*

Let us present several examples which illustrate the localization properties of solutions to the equations with nonhomogeneous anisotropic nonlinearity. To avoid technical complications, we select the simplest equations whose solutions display the properties we are interested in. The study of these equations furnishes an explanation of the main steps of application of the energy method and does not require the use of the heavy machinery of embedding and trace-interpolation theorems in the Sobolev spaces.

Let us consider the equation

$$-(a_1 u_x)_x - (a_2 u_y)_y + b_1 u_x + b_2 u_y + c = f(x, y) \tag{9.1}$$

in the half-bounded strip

$$\Omega = \left\{(x, y) \in \mathbb{R}^2: 0 < x < \infty, 0 < y < 1\right\}.$$

It is always assumed that $u(x, 0) = u(x, 1) = 0$ for all $x > 0$. The coefficients

$$a_i \equiv a_i(x, y, u, \nabla u) \geqslant 0, \qquad b_i \equiv b_i(x, y, u, \nabla u), \qquad c \equiv c(x, y, u, \nabla u)$$

are given functions of their arguments and, moreover, $c(x, y, s)s \geqslant 0$ for all $(x, y) \in \Omega$ and $s \in \mathbb{R}$. We use the standard notion of weak solution: a function $u(x, y)$ is called weak solution of equation (9.1) if for every test-function $\eta \in C^{\infty}(\Omega)$, vanishing for $y = 0$, $y = 1$, $x = 0$ and as $x \to \infty$,

$$\int_{\Omega} \left[a_1 u_x \eta_x + a_2 u_y \eta_y + (b_1 u_x + b_2 u_y + c + f)\eta \right] dx \, dy = 0. \tag{9.2}$$

Our conjecture consists in the following: we assume that equation (9.1) admits energy solutions, i.e., the solutions satisfying the following properties.

1. The solution u can be substituted into identity (9.2) as the test-function,

$$\int_{\Omega} \left[a_1 u_x^2 + a_2 u_y^2 + (b_1 u_x + b_2 u_y + c + f)u \right] dx \, dy = 0.$$

2. The energy associated with this solution is bounded,

$$E(t) \equiv \int_t^{\infty} \int_0^1 \left[a_1 u_x^2 + a_2 u_y^2 + cu \right] dx \, dy \leqslant E(0) \leqslant 1. \tag{9.3}$$

3. The solution is uniformly bounded: $|u| \leqslant 1$ in Ω.
4. The integration-by-parts formula holds in the domain $\Omega \cap \{x > t\}, t \geqslant 0$,

$$E(t) + \int_t^{\infty} dx \int_0^1 u\left[(b_1 u_x + b_2 u_y + f) \right] dy = \int_0^1 a_1 u u_x \big|_{x=t} \, dy. \tag{9.4}$$

In the previous sections we established sufficient conditions of existence of such solutions for certain classes of nonlinear equations and then studied their localization properties. Now we want to study the properties of energy solutions assuming their existence a priori.

EXAMPLE 1 (Semilinear equation with nonlinear absorption term). Let a function $u(x, y)$ be the energy solution of the problem

$$\begin{cases} -\Delta u + |u|^{\sigma(x,y)-2}u = f(x, y) & \text{in } \Omega, \\ u(x, 0) = u(x, 1) = 0 & \text{for } 0 < x < \infty \end{cases} \tag{9.5}$$

with the exponent $\sigma(x, y) \in (1, 2]$. Let us recall that the solution and the corresponding energy $E(t)$ defined by (9.3) are assumed to be bounded, $|u| \leqslant 1$ in Ω, $E(t) \leqslant E(0) \leqslant 1$ for $t \geqslant 0$. Convergence of the integral representing the energy $E(t)$ means that necessarily $|\nabla u| + |u| \to 0$ as $x \to 0$. Let us consider first the case when $f(x, y) \equiv 0$ for $x \geqslant 1$. Since $E(t)$ is bounded and monotone, then for a.a. $t > 0$, the function $E(t)$ has the derivative

$$E'(t) = -\int_0^1 \left[u_x^2 + u_y^2 + |u|^{\sigma} \right] \big|_{x=t} \, dy.$$

According to (9.4) the energy function $E(t)$ satisfies the equality

$$E(t) = I(t) \equiv -\int_0^1 u(t, y) u_x(t, y) \, dy \quad \text{for all } t \geqslant 1. \tag{9.6}$$

Because of the assumptions $|u| \leqslant 1$ in Ω, $\sigma \in (1, 2]$, the estimate

$$u^2 = u^\sigma u^{2-\sigma} \leqslant u^\sigma u^{\mu(t)} \tag{9.7}$$

holds with the exponent

$$\mu(t) = \min_{0 \leqslant y \leqslant 1}[2 - \sigma(t, y)] = \left[2 - \max_{0 \leqslant y \leqslant 1} \sigma(t, y)\right] \geqslant 0. \tag{9.8}$$

The inequality

$$|u(t, y)| \equiv \left| \int_0^y u_s(t, s)\, ds \right| \leqslant \left(\int_0^y ds \right)^{1/2} \left(\int_0^y u_y^2\, dy \right)^{1/2} \leqslant \left(\int_0^1 u_y^2\, dy \right)^{1/2}$$

together with (9.7) allows us to estimate $I(t)$ as follows:

$$|I(t)| \leqslant \left[\left(\int_0^1 u_x^2\, dy \right)\Big|_{x=t} \left(\int_0^1 |u|^2\, dy \right)\Big|_{x=t} \right]^{1/2}$$

$$\leqslant \left[\left(\int_0^1 u_x^2\, dy \Big|_{x=t} \right) \left(\int_0^1 |u|^\sigma\, dy \Big|_{x=t} \right) \right]^{1/2} \left(\int_0^1 u_y^2\, dy \Big|_{x=t} \right)^{\mu(t)/4}$$

$$\leqslant \left[\int_0^1 \left(u_x^2 + u_y^2 + |u|^\sigma \right) dy \Big|_{x=t} \right]^{(4+\mu(t))/4}$$

$$= \left(-E'(t) \right)^{(4+\mu(t))/4}. \tag{9.9}$$

Plugging (9.6) into (9.9) we arrive at the ordinary differential inequality

$$\begin{cases} E'(t) + E^{\nu(t)}(t) \leqslant 0 & \text{for } t \geqslant 0, \\ E(0) \leqslant 1, \quad E(\infty) = 0, \quad E'(t) \leqslant 0 \end{cases} \tag{9.10}$$

with the exponent

$$\nu(t) = \frac{4}{4 + \mu(t)} = \frac{4}{6 - \max_{0 \leqslant y \leqslant 1} \sigma(t, y)}. \tag{9.11}$$

REMARK 9.1. The assumption $E(0) \leqslant 1$ does not restrict the generality of our considerations. Let $E(t)$ be a solution of inequality (9.10) with $\infty > E(0) = \sup E(t) > 1$. Let us introduce the function $Z(\tau) = E(\tau/\lambda)/E(0)$ with $\lambda = \text{const}$. The function Z satisfies the conditions

$$Z'_\tau(\tau) = \frac{\lambda}{E(0)} E'(t)\Big|_{t=\lambda\tau} \leqslant -\frac{\lambda}{E(0)} E^{\nu(t)}(t)\Big|_{t=\lambda\tau} = -\lambda E^{\nu(\lambda\tau)-1}(0) Z^{\nu(\lambda\tau)}(\tau).$$

Choosing $\lambda = E^{\min \nu(\lambda\tau)-1}(0)$ and letting $\delta(\tau) = \nu(\lambda\tau)$, we see that the function $Z(\tau)$ satisfies the differential inequality (9.10) with the exponent $\delta(\tau)$ and is bounded: $\sup Z(\tau) = Z(0) \leqslant 1$.

The properties of the functions satisfying inequalities of the type (9.10) with variable exponents $v(t)$ are not completely studied so far which is why in the further considerations we restrict ourselves to the two partial cases.

(1) Let us assume that the exponent $v(t) - 1$ is bounded away from zero,

$$v(t) = \frac{4}{4 + \mu(t)} \leqslant v_0 < 1 \quad \Longleftrightarrow \quad \sigma(t, y) \leqslant \sigma^+ < 2.$$

Since $E(t) \leqslant 1$, this assumption leads to the inequality

$$0 \geqslant E'(t) + E^{v(t)}(t) \geqslant E'(t) + E^{v_0}(t) \quad \text{for } t > 1.$$

Integrating in t the inequality $0 \geqslant E'(t) + E^{v_0}(t)$, we get the estimate

$$E^{1-v_0}(t) \leqslant E^{1-v_0}(0) - (1 - v_0)t.$$

It follows that $E(t^*) = 0$ and, consequently, $u \equiv 0$ in $\Omega \cap \{x > t^*\}$ with

$$t^* = \frac{E^{1-v_0}(0)}{1 - v_0} < \infty. \tag{9.12}$$

(2) The function $v(t) - 1$ is not separated away from zero but is allowed to vanish with a prescribed speed as $t \to \infty$. This situation is studied in detail in Lemma 9.1. Anticipating the further rigorous considerations we announce the required rate of vanishing: for every constant $\gamma \in (0, 1)$ there are finite positive constants C_1, C_2 such that, for all sufficiently large t,

$$C_1 < t^\gamma \left(1 - v(t)\right) \leqslant C_2 \quad \Longleftrightarrow \quad C_1 < t^\gamma \left(2 - \max_{y \in [0,1]} \sigma(t, y)\right) \leqslant C_2. \tag{9.13}$$

If this condition is fulfilled, then $E(t) \equiv 0$ and $u(t, y) \equiv 0$ for all t beginning with some $t_* < \infty$.

REMARK 9.2. The question of admissible behavior of the variable exponent of nonlinearity arises in the study of equations posed on bounded domains in the spaces of arbitrary dimension. For example, let us consider the equation

$$\Delta u + |u|^{\sigma(x)-2} u = 0$$

posed on a bounded domain $\Omega \subset \mathbb{R}^n$. The properties of localization of solutions to this equation are studied in Section 5 for the positive exponents $\sigma(x) - 1$ bounded away from zero. For such $\sigma(x)$ the ordinary differential inequality for the energy function can be reduced to an inequality of similar structure but with constant exponent, and the latter can be studied with the already developed methods. However, one may ask the natural questions:

(1) What conditions on $\sigma(x)$ are necessary and sufficient to guarantee that the solution vanishes on a ball $B_\rho(x_0) = \{|x - x_0| < \rho\} \subset \Omega$?

(2) What happens if $\sigma(x) \to 2-$ when $x \to x_0 \in \Omega$?

For the moment, these questions have been given answers only in some partial situations (see Section 5 and Lemma 9.1). The complete answer is unknown so far even in the case of the simplest equation.

Let us assume now that $f(x, y) \not\equiv 0$ and that

$$\left(\int_t^\infty \int_0^1 |f|^{\sigma'(x,y)} \, dx \, dy \right)^{\nu(t)} \leq \varepsilon C \left[1 - \frac{t}{t_f} \right]_+^\gamma \quad \text{for } t_f > t^* \tag{9.14}$$

with t^* defined by (9.12), $\varepsilon, C = \text{const} > 0$, and an arbitrary $\gamma \geq \nu(t)/(\nu(t) - 1)$ where $\nu(t)$ is defined by (9.11). In this case instead of the energy relation (9.6), we arrive at the equality

$$E(t) = I(t) + \int_t^\infty \int_0^1 uf \, dx \, dy.$$

The term $I(t)$ is estimated with the help of (9.9). The second term on the right-hand side can be estimated by the Young inequality (2.17),

$$\int_t^\infty \int_0^1 uf \, dx \, dy \leq \frac{1}{2} \int_t^\infty \int_0^1 |u|^{\sigma(x,y)} \, dx \, dy + C \int_t^\infty \int_0^1 |f|^{\sigma'(x,y)} \, dx \, dy$$

with a constant C depending on $\min \sigma$. It follows that

$$\frac{1}{2} E(t) + \left(E'(t) \right)^{(4+\mu(t))/4} \leq 2C \int_t^\infty \int_0^1 |f|^{\sigma'(x,y)} \, dx \, dy,$$

whence

$$E'(t) + E^{\nu_0}(t) \leq E'(t) + E^{\nu(t)}(t) \leq \varepsilon C \left[1 - \frac{t}{t_f} \right]_+^\gamma. \tag{9.15}$$

Arguing like in the proof of Lemma 5.3, we obtain sufficient conditions on the data which guarantee the presence of the localization effect: there exists a constant $\varepsilon_* > 0$ such that $E(t) \equiv 0$ for all $t \geq t_f$ provided that $\varepsilon < \varepsilon_*$.

EXAMPLE 2 (Equation of anisotropic diffusion). Let us consider the energy solution of the problem

$$\begin{cases} \left(|u|^{\alpha(x,y)} u_x \right)_x + u_{yy} = 0 & \text{in } \Omega, \\ u(x, 0) = u(x, 1) = 0 & \text{for } x \in \mathbb{R}_+. \end{cases} \tag{9.16}$$

According to (9.4), the energy function $E(t)$ given by (9.3) is expressed by

$$E(t) = I(t) \equiv -\int_0^1 |u(t, y)|^{\alpha(t,y)} u(t, y) u_x(t, y) \, dy. \tag{9.17}$$

Since $|u| \leqslant 1$, the following inequalities hold:

$$|I| \leqslant \left(\int_0^1 |u|^\alpha u_x^2 \, dy \right)^{1/2} \left(\int_0^1 |u|^{\alpha+2} \, dy \right)^{1/2} \leqslant \left(-E' \right)^{1/2} \left(\int_0^1 |u|^{\alpha+2} \, dy \right)^{1/2},$$

$$|u(t, y)|^{2+\alpha(t,y)} \leqslant |u|^{2+\mu(t)} \leqslant \left(\int_0^1 u_y^2(t, y) \, dy \right)^{(2+\mu(t))/2},$$

$$\left(\int_0^1 |u|^{\alpha+2} u_x^2 \, dy \right)^{1/2} \leqslant \left(\int_0^1 |u|^{\mu(t)+2} u_x^2 \, dy \right)^{1/2} \leqslant \left(\int_0^1 |u_y|^2 \, dy \right)^{(\mu(t)+2)/4}$$

with the exponent $\mu(t) = \min_{0 \leqslant y \leqslant 1} \alpha(t, y)$. Gathering these inequalities with (9.17), we conclude that the energy function $E(t)$ must satisfy the already known differential inequality (9.10)

$$E'(t) + E^{\nu(t)}(t) \leqslant 0 \quad \text{with } \nu(t) = \frac{4}{4 + \mu(t)}, \, \mu(t) = \min_{0 \leqslant y \leqslant 1} \alpha(t, y).$$

Like in the previous example we conclude that $E(t) = 0$ and, therefore, $u(t, y) \equiv$ for all $t \geqslant t^*$ with some finite t^* if either $0 < \nu(t) \leqslant \nu_0 < 1$, or if condition (9.13) holds.

The above arguments can be adapted to study the nonlinear diffusion equation with another kind of degeneracy. Let u be the energy solution of the problem

$$\begin{cases} u_{xx} + \left(|u|^{-\beta(x,y)} u_y \right)_y = 0 & \text{in } \Omega, \\ u(x, 0) = u(x, 1) = 0 & \text{for } x > 0, \end{cases} \tag{9.18}$$

with the exponent $\beta(x, y) \in (0, 2)$, and let $E(t) \leqslant E(0) \leqslant 1$ defined by (9.3) be the corresponding energy function. Applying the integration-by-parts formula (9.4) we have that, for every $t > 0$,

$$E(t) = I(t) \equiv -\int_0^1 u u_x|_{x=t} \, dy. \tag{9.19}$$

By the Hölder inequality and by virtue of the definition of $E(t)$,

$$|I(t)| \leqslant \left(\int_0^1 u_x^2|_{x=t} \, dy \right)^{1/2} \left(\int_0^1 |u|^2|_{x=t} \, dy \right)^{1/2}$$

$$\leqslant \sqrt{-E'(t)} \left(\int_0^1 |u|^2|_{x=t} \, dy \right)^{1/2}.$$

The assumption $|u| \leqslant 1$ yields the inequality

$$\left|u(t,y)\right|^{-\beta_0} \leqslant \left|u(t,y)\right|^{-\beta(t)} \quad \text{with the constant } \beta_0 = \min_{0 \leqslant y \leqslant 1} \beta(t,y).$$

Further (see Section 6.1.2),

$$\left|u(x,t)\right| \leqslant C \left(\int_0^1 |u|^{-\beta(t,s)}(t,s) u_y^2(t,s)\,ds \right)^{1/(2-\beta_0)}$$

$$\text{with } C = \max_{0 \leqslant t \leqslant \infty} \left(\frac{2-\beta_0(t)}{2} \right)^{2/(2-\beta_0(t))}.$$

Substituting these relations into (9.19), we obtain the inequality

$$E'(t) + E^{\nu(t)}(t) \leqslant 0 \quad \text{with the exponent } \nu(t) = \frac{2(2-\beta_0)}{4-\beta_0}.$$

REMARK 9.3. To outline the nonstandard character of localization property of solutions to nonlinear equations (9.16) and (9.18), let us recall that unlike this situation for the harmonic functions in the strip the Liouville-type theorems hold: if $\alpha \equiv 0$ ($\beta \equiv 0$ in the case of (9.18)) the function u decays exponentially as $x \to \infty$,

$$C_1 e^{-\nu_1 x} \leqslant \max_{y \in [0,1]} u(x,y) \leqslant C_2 e^{-\nu_2 x},$$

with some positive constants C_i and ν_i.

EXAMPLE 3 (Equations with convective terms). Let us consider the energy solution of the nonlinear degenerate elliptic equation with convective terms

$$\begin{cases} \left(|u|^{\alpha(x,y)} u_x\right)_x + u_{yy} \\ \quad + A(x,y,u)u_x + B(x,y,u)u_y = 0 & \text{in } \Omega, \\ u(x,0) = u(x,1) = 0 & \text{for } x > 0. \end{cases} \tag{9.20}$$

About the exponent $\alpha(x,y)$ we assume that $\alpha(x,y) \in [0,\infty)$. The energy $E(t)$ of this solution satisfies (9.4), which in the present case reads as

$$E(t) = I(t) + L(t) \quad \text{for } t > 0 \tag{9.21}$$

with

$$I(t) \equiv - \int_0^1 |u|^\alpha u u_x|_{x=t}\,dy,$$

$$L(t) = \int_t^\infty \int_0^1 [A u_x u + B u_y u]\,dx\,dy = L_1(t) + L_2(t).$$

First we evaluate $L_1(t)$,

$$|L_1(t)| \leqslant \left(\int_t^\infty \int_0^1 |A|^2 |u|^{2-\alpha} \, dx \, dy \right)^{1/2} \left(\int_t^\infty \int_0^1 |u|^\alpha u_x^2 \, dx \, dy \right)^{1/2}$$

$$\leqslant \left(\int_t^\infty \int_0^1 |A|^2 |u|^{-\alpha} \left| \int_0^1 |u_y(x,s)|^2 \, ds \right| dx \, dy \right)^{1/2} \sqrt{E(t)}.$$

Let us assume that the following estimate holds,

$$\int_0^1 \left| A(x,y,u) \right|^2 |u|^{-\alpha(x,y)} \, dy \leqslant \delta_1^2(x),$$

where $\delta_1(x) \geqslant 0$ is a decreasing function, $\delta_1(\infty) = 0$. Then

$$|L_1(t)| \leqslant \delta_1(t) E(t). \tag{9.22}$$

Further,

$$|L_2(t)| \leqslant \left(\int_t^\infty \int_0^1 |B|^2 |u|^2 \, dx \, dy \right)^{1/2} \left(\int_t^\infty \int_0^1 u_y^2 \, dx \, dy \right)^{1/2}$$

$$\leqslant \left(\int_t^\infty \left(\left| \int_0^1 |u_y(x,s)|^2 \, ds \right| \int_0^1 |B(x,y,u)|^2 \, dy \right) dx \right)^{1/2} \sqrt{E(t)}.$$

If we assume that

$$\int_0^1 \left| B(x,y,u) \right|^2 dy \leqslant \delta_2^2(x)$$

for a decreasing nonnegative function $\delta_2(x)$ such that $\delta_2(x) \to 0$ as $x \to \infty$, then

$$|L_2(t)| \leqslant \delta_2(t) E(t). \tag{9.23}$$

Choosing T so large that, for all $t \geqslant T$ $\delta_2(t) + \delta_2(t) < 1 - 1/C < 1$ with some constant $C > 1$, and gathering (9.21)–(9.23), we arrive at the inequality

$$E(t) \leqslant C I(t) \quad \text{for } t > T.$$

Estimating $I(t)$ with the help of (9.9) we obtain for the energy function $E(t)$ the differential inequality (9.10), which can be studied in the usual way.

EXAMPLE 4 (Anisotropic $p(x)$-Laplacian). Let us consider a nontrivial solution of the problem

$$\begin{cases} \left(|u_x|^{p(x,y)-2} u_x \right)_x + u_{yy} = 0 & \text{in } \Omega, \\ u(x,0) = u(x,1) = 0 & \text{for } x \geqslant 0. \end{cases} \tag{9.24}$$

About the exponent of nonlinearity we assume that $p(x, y) \in (2, \infty)$. According to (9.4)

$$E(t) = -\int_0^1 u|u_x|^{p-2}u_x \, dy \bigg|_{x=t} \equiv I(t) \quad \text{for } t > 0.$$

Let additionally assume that

$$\forall x \in \mathbb{R}_+, \quad \int_0^1 |u_x|^{p(x,y)} \, dy \leqslant 1$$

and accept the notation

$$p^+ \equiv p^+(t) = \max_{y \in [0,1]} p(t, y), \qquad p^+ \equiv p^-(t) = \min_{y \in [0,1]} p(t, y).$$

Using inequalities (2.3)–(2.4) we estimate

$$|I| \leqslant \left(\int_0^1 |u_x|^{q(p(x,y)-1)} \, dy\right)^{1/q} \left(\int_0^1 |u|^{q'} \, dy\right)^{1/q'}$$

$$\leqslant C\left(\|u_x\|_{q(p-1)}^{q(p^--1)}\right)^{1/q} \left(\int_0^1 |u_y|^2 \, dy\right)^{1/2}$$

$$\leqslant C\left(\|u_x\|_{p,(0,1)}^{(p^--1)}\right) \left(\int_0^1 |u_y|^2 \, dy\right)^{1/2}$$

$$\leqslant C\left(\int_0^1 |u_x|^p \, dy\right)^{(p^--1)/p^+} \left(\int_0^1 |u_y|^2 \, dy\right)^{1/2}$$

$$\leqslant C(-E')^{1/v(t)} \quad \text{with } v(t) = \frac{2p^+(t)}{p^+(t) + 2(p^-(t) - 1)}.$$

It follows that $E(t)$ satisfies the ordinary differential inequality

$$E'(t) + E^{v(t)}(t) \leqslant 0 \qquad \text{for } t > 0.$$

The properties of the functions satisfying this inequality are described in Example 1. The energy $E(t)$ and the solution $u(x, y)$ vanish for all $t \geqslant t_*$ with some finite t_* if either $0 < v(t) \leqslant v_0 < 1$, or condition (9.11) is fulfilled. In the present case the former condition reads as

$$0 < p^+(t) - p^-(t) \leqslant \frac{v_0}{2 - v_0}\left(p^-(t) - 2\right).$$

EXAMPLE 5 (Equations of mixed type). Let us consider the equation

$$\left(|u_y|^\alpha u_x\right)_x + u_{yy} = 0. \tag{9.25}$$

This equation can be formally written in the form

$$|u_y|^\alpha u_{xx} + \alpha u_x u_y |u_y|^{\alpha-2} u_{xy} + u_{yy} + \alpha_x u_x |u_y|^\alpha \ln|u_y| = 0.$$

The determinant of the matrix associated with the principal part of this equation is

$$D = |u_y|^\alpha - \frac{1}{4}\alpha^2 |u_x|^2 |u_y|^{2\alpha-2}.$$

The function D may change the sign, which is why the equation cannot belong to any of the definite types. Let u be a nontrivial energy solution of equation (9.25). The energy relation has the form

$$E(t) = \int_t^\infty \int_0^1 \left[|u_y|^\alpha u_x^2 + u_y^2 \right] dx\, dy = - \left. \int_0^1 |u_y|^\alpha u_x u \, dy \right|_{x=t} \equiv I(t).$$

The following inequalities hold:

$$|I(t)| \leqslant \max_{y\in[0,1]} |u(t,y)| \left(\left.\int_0^1 |u_y|^\alpha u_x^2 \, dy \right|_{x=t} \right)^{1/2} \left(\left.\int_0^1 |u_y|^\alpha \, dy \right|_{x=t} \right)^{1/2},$$

$$|u(t,y)| \leqslant \left(\int_0^1 |u_y(t,y)|^2 \, dy \right)^{1/2},$$

$$\left(\int_0^1 |u_y|^\alpha \big|_{x=t} \, dy \right)^{1/2} \leqslant \max\left\{ \|u_y\|_2^{\alpha^-(t)}, \|u_y\|_2^{\alpha^+(t)} \right\}^{1/2}$$

with the exponents

$$0 < \alpha^-(t) = \min_{y\in[0,1]} \alpha(t,y) \leqslant \alpha^+(t) = \max_{y\in[0,1]} \alpha(t,y) \leqslant 2.$$

Gathering these relations we arrive at the inequality

$$\forall t > 0, \quad E(t) \leqslant -E'(t) \max\left\{ \left(-E'(t)\right)^{\alpha^-(t)/4}, \left(-E'(t)\right)^{\alpha^+(t)/4} \right\}.$$

Arguing like in the proof of Theorem 6.2, we derive the standard differential inequality

$$E' + E^{\nu(t)} \leqslant 0 \quad \text{with } \nu(t) = \frac{4}{4+\alpha^-(t)} \in (0,1) \text{ for } \alpha^-(t) \geqslant \alpha_0^- > 0.$$

In the similar way one may consider the more complicated equations of the form

$$\left(|u|^{\alpha_1} |u_y|^{\beta_1} |u_x|^{p_1-2} u_x \right)_x + \left(|u|^{\alpha_2} |u_y|^{p_2-2} |u_x|^{\beta_2-2} u_x \right)_y = 0, \tag{9.26}$$

with given nonnegative exponents α_i, β_i, p_i, $i = 1, 2$.

REMARK 9.4. The above arguments extend without any change to the cases when the equation under study depend on more than two independent variables and is posed on a half-bounded cylinder

$$\Omega = \{(y, t): y \in G \subset \mathbb{R}^n, t \in \mathbb{R}_+\}, \quad |G| < \infty,$$

or on a layer-type domain

$$\Omega = \{(y, t): y \in G \subset \mathbb{R}^n, 0 < t < 1\}.$$

EXAMPLE 6 (Semilinear equation with vanishing absorption). Denote $\Omega = \{(x, y) \in \mathbb{R}^2: x > 0, y \in (0, 1)\}$. Let $u(x, y)$ be a nontrivial energy solution of the semilinear elliptic equation

$$\begin{cases} -\Delta u + a^2(x, y)|u|^{\sigma(x,y)-2}u = 0 & \text{in } \Omega, \\ u(x, 0) = u(x, 1) = 0 & \text{for } 0 < x < \infty. \end{cases} \tag{9.27}$$

We assume that the given functions σ and a satisfy the conditions

$$1 \leqslant \sigma(x, y) \leqslant 2, \qquad 0 \leqslant a(x, y) \leqslant \infty.$$

Introduce the energy function

$$E(t) = \int_t^\infty \int_0^1 \left[u_x^2 + u_y^2 + a^2(x, y)|u|^{\sigma(x,y)}\right] dx \, dy \leqslant E(0) < \infty.$$

In our assumptions, the energy function satisfies the relation

$$E(t) = I(t) := -\int_0^1 u(t, y)u_x(t, y) \, dy. \tag{9.28}$$

Not loosing generality we may assume that $E(0) \leqslant 1$ and $|u| \leqslant 1$ in Ω. Then

$$u^{2-\sigma(t,y)} \leqslant u^{\mu(t)} \quad \text{with } \mu(t) = \min_{0 \leqslant y \leqslant 1} \left[2 - \sigma(t, y)\right] = 2 - \max_{0 \leqslant y \leqslant 1} \sigma(t, y) \geqslant 0.$$

Evidently,

$$|u(t, y)| = \left|\int_0^y u_y \, dy\right| \leqslant \left(\int_0^1 u_y^2 \, dy\right)^{1/2}.$$

Using these relations we find

$$|I| \leqslant \left[\left(\int_0^1 u_x^2 \, dy\right)\left(\int_0^1 a^2|u|^\sigma \frac{1}{a^2}|u|^{2-\sigma} \, dy\right)\right]^{1/2}$$

$$\leqslant \max_{0\leqslant y\leqslant 1} \frac{1}{|a(t,y)|} \left[\left(\int_0^1 u_x^2\, dy\right)\left(\int_0^1 a^2|u|^\sigma\, dy\right)\right]^{1/2} \left(\int_0^1 u_y^2\, dy\right)^{\mu(t)/4}$$

$$\leqslant \frac{1}{\min_{0\leqslant y\leqslant 1}|a(t,y)|} \left[\left(\int_0^1 (u_x^2 + u_y^2 + a^2|u|^\sigma)\, dy\right)\right]^{(4+\mu(t))/4}$$

$$= \frac{1}{\min_{0\leqslant y\leqslant 1}|a(t,y)|} (-E'(t))^{(4+\mu(t))/4}. \tag{9.29}$$

Gathering (9.28) with (9.29), we arrive at the ordinary differential inequality

$$E'(t) + \rho(t)E^{\nu(t)}(t) \leqslant 0, \tag{9.30}$$

with the exponent and the coefficient

$$\nu(t) = \frac{4}{4+\mu(t)} = \frac{4}{6 - \max_{0\leqslant y\leqslant 1}\sigma(t,y)}, \qquad \rho(t) = \min_{0\leqslant y\leqslant 1}|a(t,y)|^{\nu(t)}.$$

The possibility of localization of the weak energy solution depends on the interaction between the weight $\rho(t)$ and the exponent $\nu(t)$. To simplify the matter, let us consider the case $\nu(t) = \nu_0 = \text{const}$. Integrating (9.30) with $\nu(t) \equiv \nu_0$, we obtain the estimate

$$E^{1-\nu_0}(t) \leqslant E^{1-\nu_0}(0) - (1-\nu_0)\int_0^t \min_{0\leqslant y\leqslant 1}|a(t,y)|^{\nu_0}\, ds. \tag{9.31}$$

If

$$\int_0^\infty \min_{0\leqslant y\leqslant 1}|a(t,y)|^{\nu_0}\, ds = \infty,$$

then for any total energy $E(0) < \infty$, there exists $t^* < \infty$ such that $E(t) = 0$ for all $t \geqslant t^*$, whence

$$u(t,y) = 0 \quad \text{for all } t \geqslant t^*, \, y \in [0,1].$$

If

$$\int_0^\infty \min_{0\leqslant y\leqslant 1}|a(s,y)|^{\nu_0}\, ds = C < \infty, \tag{9.32}$$

then there exists $t^* < \infty$ such that

$$E(t) = 0 \quad \text{and} \quad u(t,y) = 0 \quad \text{for all } t \geqslant t^*, \, y \in [0,1],$$

provided that the total energy satisfies $E^{1-\nu_0}(0) < (1-\nu_0)C$.

We have established sufficient conditions of localization of the energy solution to problem (9.27) in a finite part of the unbounded strip Ω. The local vanishing properties of solutions to this problem are already studied in Section 5: if u is a bounded weak solution of the equation

$$\Delta u + |u|^{\sigma(x,y)-2}u = 0 \quad \text{in } B_r(x_0), \tag{9.33}$$

and the total energy of u in $B_r(x_0)$ is sufficiently small, then there exists a concentric ball $B_\rho(x_0)$, $\rho \in (0, r)$, where $u \equiv 0$. Let us show now that the null-set of the energy solution of equation (9.33) in a bounded domain may adjoin the part of the boundary where $u = 0$. Let us denote

$$\Omega_{x,y} = \left\{ (x, y) \in \mathbb{R}^2 : x > 0, -\frac{\pi}{2} < y < \frac{\pi}{2} \right\}$$

and consider the complex-valued function

$$z = x + iy = \Phi(w) = \ln \frac{1+w}{1-w}, \qquad w = \Phi^{-1}(z) = \frac{e^z - 1}{e^z + 1}$$

of the complex variable $w = \xi + i\eta$. It is easy to check that

$$w(z) = re^{i\theta} = \frac{e^x(\cos y + i\sin y) - 1}{e^x(\cos y + i\sin y) + 1}, \quad 0 < r < 1, \theta \in (0, 2\pi),$$

$$w\left(x, \pm i\frac{\pi}{2}\right) = \frac{\pm ie^x - 1}{\pm ie^x + 1} = e^{i\theta}, \qquad \left|w\left(x, \pm i\frac{\pi}{2}\right)\right| = 1,$$

$$\lim_{x \to +\infty} w\left(x, \pm i\frac{\pi}{2}\right) = 1, \qquad w\left(0, \pm i\frac{\pi}{2}\right) = \frac{\pm i - 1}{\pm i + 1} = e^{i\theta_\pm},$$

$$w(0, y) = \frac{(\cos y + i\sin y) - 1}{(\cos y + i\sin y) + 1} = r(y)e^{i\theta(y)}.$$

If $w(0, y)$ is prescribed, the last relation defines a curve $G(r, \theta) = 0$. The concrete form of the function G is unimportant for the further consideration. The function $\Phi^{-1}(w)$ transforms the domain $\Omega_{x,y}$ into a bounded domain ω on the plane of the variables $w = \xi + i\eta = re^{i\theta}$, delimited by the lines

$$\{r = 1, \theta \in (\theta_-, \theta_+)\} \quad \text{and} \quad \{G(r, \theta) = 0, \theta \notin (\theta_-, \theta_+)\}.$$

Assuming $\sigma = \text{const}$, let us choose in (9.27)

$$a^2(x, y) = \left|\frac{dw}{dz}\right|^2 = \left|\frac{2e^z}{(e^z + 1)^2}\right|^2 = \frac{4e^{2x}}{(e^{2x} + 2e^x \cos y + 1)^2}.$$

Then

$$\min_{-\pi/2 \leqslant y \leqslant \pi/2} a^2(x, y) = \frac{4e^{2x}}{(e^{2x} + 2e^x + 1)^2}$$

and

$$\int_0^\infty \min_{-\pi/2 \leqslant y \leqslant \pi/2} |a(t,y)|^{\nu_0} \, ds = \int_0^\infty \left(\frac{4e^{2x}}{(e^{2x} + 2e^x + 1)^2} \right)^{\nu_0} dx = C < \infty.$$

This change of variable establishes the equivalence between problem (9.27) and the Dirichlet problem for the semilinear elliptic equation in the bounded domain ω,

$$\begin{cases} -\Delta_{\xi,\eta} u + |u|^{\sigma-2}u = 0 & \text{in } \omega \, (\Delta_{\xi,\eta} u = u_{\xi\xi} + u_{\eta\eta}), \\ u\big(e^{i\theta}\big) = 0, & \theta \in (\theta_-, \theta_+). \end{cases} \tag{9.34}$$

Arguing in the routine way we conclude that if the total energy,

$$E(0) = \int_0^\infty \int_0^1 \left[u_x^2 + u_y^2 + a^2(x,y)|u|^\sigma \right] dx \, dy = \int_{\Omega_{\xi,\eta}} \left[u_\xi^2 + u_\eta^2 + |u|^\sigma \right] d\xi \, d\eta,$$

is appropriately small, then there exists $\varepsilon^* > 0$ such that $u(\xi, \eta) = 0$ in every domain

$$(\xi - 1)^2 + \eta^2 < \varepsilon^2, \qquad \varepsilon \leqslant \varepsilon^*.$$

The case of variable exponent $\sigma(x, y) \in (1, 2)$ is studied likewise – see Example 1 and the borderline case studied in next subsection.

9.2. *The ordinary differential inequality in the limit case*

As we have seen, a solution of a nonlinear equation vanishes on a set of nonzero measure if the corresponding energy function achieves the limit value inside its domain of definition. This limit value can be either zero (in the case of unbounded domain), or a known constant defined in terms of the problem data (if the domain is bounded). The character of nonlinearity of the equation reveals in the form of the differential relation for the energy function. In all the cases studied the energy function satisfied a nonlinear differential inequality of the type

$$E'(t) + CE^{\nu(t)} \leqslant f(t) \quad \text{for } t \in (t_0, t_1) \subseteq \mathbb{R}_+.$$

If the energy $E(t)$ is uniformly bounded on (t_0, t_1), and if the exponent $\nu(t)$ is bounded away from its limit values, i.e., if

$$0 < \nu_- \leqslant \nu(t) \leqslant \nu_+ < 1 \quad \text{on } (t_0, t_1), \, \nu_\pm = \text{const,}$$

then the differential inequality for $E(t)$ can be reduced to an inequality with the exponent $\nu(t)$ substituted by ν_+, which allows one to study the properties of $E(t)$ with the well-known methods (see [10]). This way of analysis was chosen in most of the studied cases.

A drawback of such a reduction to an inequality typical for equations with constant exponents of nonlinearity consists in the fact that we exclude from consideration the cases of limit behavior of these exponents.

The following lemma gives a partial answer to the question about the behavior of the functions satisfying a nonlinear ordinary differential inequality when the nonlinearity exponent is allowed to reach its limit value.

LEMMA 9.1. *Let $E(t)$ be continuous in $\mathbb{R}_+$ and satisfy the conditions*

$$\begin{cases} E'(t) + E^{\nu(t)}(t) \leqslant 0 & \text{for } t \geqslant 1, \\ 0 \leqslant E(t) \leqslant E_0 \leqslant 1, & E'(t) \leqslant 0. \end{cases} \tag{9.35}$$

If for all t from some $t_0 > 1$ on the exponent $\nu(t)$ can be represented in the form

$$0 < \nu(t) = 1 - \varepsilon(t) \leqslant 1 - Ct^{-\gamma} \quad \text{with a constant } \gamma \in [0, 1), \tag{9.36}$$

then $E(t) \equiv 0$ for

$$t \geqslant I = \int_{-\ln E_0}^{\infty} e^{-C\tau(\tau + \ln E(0))^{-\gamma}} \, d\tau.$$

PROOF. There are two possibilities: $E(0) \equiv 0$ for all $t \geqslant 0$ and $E(t) \not\equiv 0$ in $[1, \infty)$. In the first case the assertion follows with $t^* = 1$. Let us consider the second case. By continuity and monotonicity of $E(t)$, there is an interval $[1, T)$ where $E(t) > 0$. Since $E(t) \leqslant 1$ and $\nu(t) > 0$, then $E(t) \leqslant E^{\nu(t)}(t)$, and by virtue of (9.35),

$$E'(t) + E(t) \leqslant E'(t) + E^{\nu(t)}(t) \leqslant 0. \tag{9.37}$$

This inequality yields $E(t) \leqslant E(0)e^{-t}$, whence

$$t \leqslant \ln \frac{E(0)}{E(t)}. \tag{9.38}$$

Due to (9.38) we may write

$$-\varepsilon(t) \leqslant -C\left(\ln \frac{E(0)}{E(t)}\right)^{-\gamma}$$

and, correspondingly (recall that $E(t) \leqslant 1$),

$$E^{-\varepsilon(t)} \geqslant E^{-C(\ln(E(0)/E(t)))^{-\gamma}}.$$

Now we transform inequality (9.35) into

$$E'(t) + E^{1 - C(\ln(E(0)/E(t)))^{-\gamma}}(t) \leqslant 0. \tag{9.39}$$

The straightforward integration of this inequality over the interval $(1, t)$ with $t \leqslant T$ gives

$$J[E(t)] \equiv \int_{E(0)}^{E(t)} \frac{\mathrm{d}s}{s^{1-C(\ln(E_0/s))^{-\gamma}}} \leqslant -t.$$

Introducing the new independent variable $\tau = -\ln s$ we represent $J[E(t)]$ as

$$J[E(t)] = -\int_{-\ln E_0}^{-\ln E(t)} \frac{\mathrm{d}\tau}{\mathrm{e}^{C\tau(\tau+\ln E_0)^{-\gamma}}}.$$

Let us define t^* by the equality

$$t^* = \int_{-\ln E_0}^{\infty} \frac{\mathrm{d}\tau}{\mathrm{e}^{C\tau(\tau+\ln E_0)^{-\gamma}}} = -J[0].$$

Since $J[s]$ is a monotone decreasing function of s, then $E(t) \equiv 0$ for all $t \geqslant t^* = -J[0]$.
$\square$

Lemma 9.1 allows us to specify the conditions of the exponents of nonlinearity which guarantee the presence of the localization effect in the above examples. It is no longer necessary to assume that the exponents of nonlinearity are bounded away from their limit values.

1. The energy solutions of equation (9.5) possess the localization property if condition (9.13) is fulfilled.
2. For equation (9.16) a sufficient condition of localization of the energy solutions reads as follows: for $t \geqslant 1$,

$$C_1 \leqslant t^\gamma \min_{y \in [0,1]} \alpha(t, y) \leqslant C_2 \quad \text{for some constants } \gamma \in (0, 1), C_i > 0.$$

3. In the case of equation (9.18) the energy solutions are localized if, for $t \geqslant 1$,

$$C_1 \leqslant t^\gamma \min_{y \in [0,1]} \beta(t, y) \leqslant C_2 \quad \text{for some constants } \gamma \in (0, 1), C_i > 0.$$

Acknowledgements

The work of the first author was supported by the Project POCI/MAT/61576/2004, FCT (Portugal). The work of the second author was supported by the Research Project MTM-2004-05417 of the Ministry of Science and Technology, Spain. The authors are deeply grateful to Prof. M. Chipot for stimulating discussions of this work.

References

[1] E. Acerbi and G. Mingione, *Regularity results for a class of functionals with non-standard growth*, Arch. Ration. Mech. Anal. **156** (2001), 121–140.

[2] E. Acerbi and G. Mingione, *Regularity results for electrorheological fluids: The stationary case*, C. R. Math. Acad. Sci. Paris **334** (2002), 817–822.

[3] E. Acerbi and G. Mingione, *Regularity results for stationary electro-rheological fluids*, Arch. Ration. Mech. Anal. **164** (2002), 213–259.

[4] E. Acerbi, G. Mingione and G.A. Seregin, *Regularity results for parabolic systems related to a class of non-Newtonian fluids*, Ann. Inst. H. Poincaré Anal. Non Linéaire **21** (2004), 25–60.

[5] R.A. Adams, *Sobolev Spaces*, Pure Appl. Math., vol. 65, Academic Press, New York–London (1975).

[6] Y. Alkhutov, *The Harnack inequality and the Hölder property of solutions of nonlinear elliptic equations with a nonstandard growth condition*, Differ. Uravn. **33** (1997), 1651–1660.

[7] S. Antontsev, *Nonlinear parabolic equations with a variable exponent of degeneration* (Book of Abstracts), International Conference Nonlinear Partial Differential Equations, Ukraine, Alushta, September 15–21, 2003, Institute of Applied Mathematics and Mechanics of NASU, Donetsk (2003), 12 pp.

[8] S. Antontsev, M. Chipot and Y. Xie, *Uniqueness results for equations of the $p(x)$ Laplacian type*, Preprint 002, Centro de Matemática, Departamento de Matemática, Univ. da Beira Interior (2006).

[9] S.N. Antontsev, J.I. Díaz and H. Oliveira, *Stopping a viscous fluid by a feedback dissipative field: Thermal effects without phase changing*, Progr. Nonlinear Differential Equations Appl. (Proceedings TPDE, Óbidos, Portugal, June 7–10, 2003), vol. 61 (2005), 1–14.

[10] S.N. Antontsev, J.I. Díaz and S. Shmarev, *Energy Methods for Free Boundary Problems: Applications to Non-Linear PDEs and Fluid Mechanics*, Progr. Nonlinear Differential Equations Appl., vol. 48, Birkhäuser, Boston (2002).

[11] S.N. Antontsev and J.F. Rodrigues, *On stationary thermo-rheological viscous flows*, Ann. Univ. Ferrara Sez. VII (N.S.) **51** (2005).

[12] S. Antontsev and S. Shmarev, *Existence and uniqueness of solutions of degenerate parabolic equations with variable exponents of nonlinearity*, Proceedings of the International Conference Differential Equations and Related Topics dedicated to Ivan Petrovskii (2004); Preprint 003, Centro de Matemática, Departamento de Matemática, Univ. da Beira Interior, 2004; Fundamental and Applied Mathematics (2006), to appear.

[13] S. Antontsev and S. Shmarev, *A model porous medium equation with variables exponent of nonlinearity: Existence, uniqueness and localization properties of solutions*, Nonlinear Anal. **60** (2005), 515–545.

[14] S. Antontsev and S. Shmarev, *On localization of solutions of elliptic equations with nonhomogeneous anisotropic degeneracy*, Siberian Math. J. **46** (2005), 765–782.

[15] S. Antontsev and S. Shmarev, *Directional localization of solutions to elliptic equations with nonstandard anisotropic growth conditions*, Preprint 003, Centro de Matemática, Departamento de Matemática, Univ. da Beira Interior, 2006; Proceedings of ISAAC-2005, to appear.

[16] S. Antontsev and S. Shmarev, *Elliptic equations and systems with nonstandard growth conditions: Existence, uniqueness and localization properties of solutions*, Nonlinear Anal. **65** (2006), 722–755.

[17] S.N. Antontsev and V.V. Zhikov, *Higher integrability for parabolic equations of $p(x, t)$-Laplacian type*, Adv. Differential Equations **10** (2005), 1053–1080.

[18] G. Barenblatt, V. Entov and V. Ryzhik, *The Motion of Fluids and Gases in Natural Strata*, Nedra Publishing House, Moscow (1984).

[19] G. Barles, G. Díaz and J.I. Díaz, *Uniqueness and continuum of foliated solutions for a quasilinear elliptic equation with a non-Lipschitz nonlinearity*, Comm. Partial Differential Equations **17** (1992), 1037–1050.

[20] P. Blanchard and E. Brüning, *Variational Methods in Mathematical Physics. A Unified Approach*, Texts Monogr. Phys., Springer-Verlag, Berlin (1992).

[21] A. Chambolle and P.-L. Lions, *Image recovery via total variation minimization and related problems*, Numer. Math. **76** (1997), 167–188.

[22] Y. Chen, S. Levine and M. Rao, *Variable exponent, linear growth functionals in image restoration*, SIAM J. Appl. Math. (2005), to appear. Available at *http://www.mathcs.duq.edu/sel/*.

[23] M. Chipot and M. Michaille, *Uniqueness results and monotonicity properties for strongly nonlinear elliptic variational inequalities*, Ann. Sc. Norm. Super. Pisa Cl. Sci. (4) **16** (1989), 137–166.

[24] M. Chipot and M. Michaille, *Uniqueness results and monotonicity properties for the solutions of some variational inequalities. Existence of a free boundary free boundary problems*, Free Boundary Problems: Theory and Applications, Vol. I (Irsee, 1987), Pitman Res. Notes Math. Ser., vol. 185, Longman, Harlow, (1990), 271–276.

[25] G. de Marsily, *Quantitative Hydrogeology. Groundwater Hydrology for Engineers*, Academic Press (1986).

[26] J.I. Díaz, *Nonlinear Partial Differential Equations and Free Boundaries. Elliptic Equations*. Vol. I, Pitman Res. Notes Math., Vol. 106, Pitman (Advanced Publishing Program), Boston, MA (1985).

[27] J.I. Díaz and J.E. Saá, *Existence et unicité de solutions positives pour certaines équations elliptiques quasilinéaires*, C. R. Acad. Sci. París Sér. I **305** (1987), 521–524.

[28] E. DiBenedetto, *Degenerate Parabolic Equations*, Springer-Verlag, New York (1993).

[29] E. DiBenedetto, J. Urbano and V. Vespri, *Current issues on singular and degenerate evolution equations*, Handbook of Differential Equations. Evolutionary Equations, Vol. 1, North-Holland, Amsterdam (2004), 169–286.

[30] L. Diening, *Maximal function on generalized Lebesgue spaces $L^{p(\cdot)}$*, Math. Inequal. Appl. **7** (2004), 245–253.

[31] L. D'Onofrio and T. Iwaniec, *Notes on p-harmonic analysis*, The p-Harmonic Equation and Recent Advances in Analysis, Contemp. Math., Vol. 370, Amer. Math. Soc., Providence, RI (2005), 25–49.

[32] F. Duzaar and G. Mingione, *Regularity for degenerate elliptic problems via p-harmonic approximation*, Ann. Inst. H. Poincaré Anal. Non Linéaire **21** (2004), 735–766.

[33] D. Edmunds and J. Rákosník, *Sobolev embeddings with variable exponent*, Studia Math. **143** (2000), 267–293.

[34] X. Fan and X. Han, *Existence and multiplicity of solutions for p(x)-Laplacian equations in $\mathbb{R}^N$*, Nonlinear Anal. **59** (2004), 173–188.

[35] X. Fan, J. Shen and D. Zhao, *Sobolev embeddings theorems for spaces $W^{k,p(x)}$*, J. Math. Anal. Appl. **262** (2001), 749–760.

[36] X.-L. Fan, H.-Q. Wu and F.-Z. Wang, *Hartman-type results for p(t)-Laplacian systems*, Nonlinear Anal. **52** (2003), 585–594.

[37] X. Fan and Q. Zhang, *Existence of solutions for p(x)-Laplacian Dirichlet problem*, Nonlinear Anal. **52** (2003), 1843–1852.

[38] X. Fan and D. Zhao, *Regularity of minimizers of variational integrals with continuous p(x)-growth conditions*, Chinese Ann. Math. Ser. A **17** (1996), 557–564.

[39] X. Fan and D. Zhao, *A class of De Giorgi type and Hölder continuity*, Nonlinear Anal. **36** (1999), 295–318.

[40] X. Fan and D. Zhao, *Local $C^{1,0}$ regularity of weak solutions for p(x)-Laplacian equations*, J. Gansu Education College **15** (2001), 1–5.

[41] X. Fan and D. Zhao, *On the spaces $L^{p(x)}(\Omega)$ and $W^{m,p(x)}(\Omega)$*, J. Math. Anal. Appl. **263** (2001), 424–446.

[42] X. L. Fan, Y.Z. Zhao and Q.H. Zhang, *A strong maximum principle for p(x)-Laplace equations*, Chinese Ann. Math. Ser. A **24** (2003), 495–500.

[43] I. Fragalà, F. Gazzola and B. Kawohl, *Existence and nonexistence results for anisotropic quasilinear elliptic equations*, Ann. Inst. H. Poincaré Anal. Non Linéaire **21** (2004), 715–734.

[44] Y. Fu, *The existence of solutions for elliptic systems with nonuniform growth*, Studia Math. **151** (2002), 227–246.

[45] P. Harjulehto and P. Hästö, *An overview of variable exponent Lebesgue and Sobolev spaces*, Future Trends in Geometric Function Theory, Rep. Univ. Jyväskylä Dept. Math. Statist., Vol. 92 (2003), 85–93.

[46] P. Harjulehto and P. Hästö, *A capacity approach to the Poincar'e inequality and Sobolev imbeddings in variable exponent Sobolev spaces*, Rev. Mat. Complut. **17** (2004), 129–146.

[47] P. Harjulehto, P. Hästö, M. Koskenoja and S. Varonen, *The Dirichlet energy integral and variable exponents Sobolev spaces with zero boundary values*, Preprint (2003). Available at *http://www.helsinki.fi/ hasto/pp/*.

[48] H. Hudzik, *On generalized Orlicz–Sobolev space*, Funct. Approx. Comment. Math. **4** (1976), 37–51.

[49] A.V. Ivanov, *Quasilinear Degenerate and Nonuniformly Elliptic and Parabolic Equations of Second Order*, Trudy Mat. Inst. Steklov, vol. 160 (1982) (in Russian); English transl. by J.R. Schulenberger.: Proc. Steklov Inst. Math. (1984).

[50] T. Iwaniec and C. Sbordone, *New and old function spaces in the theory of PDEs and nonlinear analysis*, Orlicz Centenary Volume, Banach Center Publ., vol. 64, Polish Acad. Sci., Warsaw (2004), 85–104.

[51] A.S. Kalashnikov, *Some problems of the qualitative theory of second-order nonlinear degenerate parabolic equations*, Uspekhi Mat. Nauk **42** (1987), 135–176, 287.

[52] A.S. Kalashnikov, *Nonlinear effects in heat propagation in media with sources and sinks that are close to linear*, Differ. Uravn. **31** (1995), 277–288, 366.

[53] A.S. Kalashnikov, *On some problems in the nonlinear theory of heat conduction with data containing a small parameter in the exponents*, Zh. Vychisl. Mat. Mat. Fiz. **35** (1995), 1077–1094.

[54] A.S. Kalashnikov, *On some nonlinear problems in mathematical physics with exponents that are close to critical*, Tr. Semin. im. I.G. Petrovskogo **346** (1996), 73–98.

[55] O. Kováčik and J. Rákosník, *On spaces $L^{p(x)}$ and $W^{k,p(x)}$*, Czechoslovak Math. J. **41** (1991), 592–618.

[56] O.A. Ladyzhenskaya and N.N. Ural'tseva, *Linear and Quasilinear Elliptic Equations*, Academic Press, New York (1968); transl. from the Russian by Scripta Technica, Inc., translation ed. L. Ehrenpreis.

[57] S. Levine, Y. Chen and J. Stanich, *Image restoration via nonstandard diffusion*, Technical Report 04-01, Dept. of Mathematics and Computer Science, Duquesne University (2004).

[58] J.-L. Lions, *Quelques méthodes de résolution des problèmes aux limites non linéaires*, Dunod–Gauthier-Villars, Paris (1969).

[59] P. Marcellini, *Regularity and existence of solutions of elliptic equations with (p,q)-growth conditions*, J. Differential Equations **90** (1991), 1–30.

[60] V. Monakhov, *Boundary-Value Problems with Free Boundaries for Elliptic Systems of Equations*, Transl. Math. Monographs, vol. 57, Amer. Math. Soc., Providence, RI (1983); transl. from Russian.

[61] J. Musielak, *Orlicz Spaces and Modular Spaces*, Lecture Notes Math., Vol. 1034, Springer-Verlag, Berlin (1983).

[62] P. Pucci and J. Serrin, *The strong maximum principle revisited*, J. Differential Equations **196** (2004), 1–66.

[63] K. Rajagopal and M. Růžička, *Mathematical modelling of electro-rheological fluids*, Contin. Mech. Thermodyn. **13** (2001), 59–78.

[64] M. Růžička, *Electrorheological Fluids: Modeling and Mathematical Theory*, Lecture Notes in Math., Vol. 1748, Springer-Verlag, Berlin (2000).

[65] S.G. Samko, *Density $C_0^\infty(\mathbb{R}^n)$ in the generalized Sobolev spaces $W^{m,p(x)}(\mathbb{R}^n)$*, Dokl. Akad. Nauk **369** (1999), 451–454.

[66] S.G. Samko, *On a progress in the theory of Lebesgue spaces with variable exponent: Maximal and singular operators*, Integral Transforms Spec. Funct. **16** (2005), 461–482.

[67] I.I. Šarapudinov, *The topology of the space $\mathcal{L}^{p(t)}([0,1])$*, Mat. Zametki **26** (1979), 613–632, 655.

[68] M. Tsutsumi, *On solutions of some doubly nonlinear degenerate parabolic equations with absorption*, J. Math. Anal. Appl. **132** (1988), 187–212.

[69] J.L. Vázquez, *A strong maximum principle for some quasilinear elliptic equations*, Appl. Math. Optim. **12** (1984), 191–202.

[70] V.V. Zhikov, *On Lavrentiev's phenomenon*, Russ. J. Math. Phys. **3** (1995), 249–269.

[71] V.V. Zhikov, *Meyers type estimates for the solution of a nonlinear Stokes system*, Differ. Equ. **33** (1997), 108–115.

[72] V.V. Zhikov, *On some variational problems*, Russ. J. Math. Phys. **5** (1997), 105–116 (1998).

[73] V.V. Zhikov, *On the density of smooth functions in Sobolev–Orlicz spaces*, Zap. Nauchn. Sem. S.-Peterburg. Otdel. Mat. Inst. Steklov (POMI) **310** (2004), 1–14.

A Handbook of Γ-Convergence

Andrea Braides

Dipartimento di Matematica, Università di Roma 'Tor Vergata',
Via della Ricerca Scientifica 1, 00133 Roma, Italy
E-mail: braides@mat.uniroma2.it

Contents

Preface 103
1. Introduction 103
   Notation 109
2. General theory of Γ-convergence 110
   2.1. The definitions of Γ-convergence 110
   2.2. Γ-convergence and lower semicontinuity 113
   2.3. Computation of Γ-limits 114
   2.4. Properties of Γ-convergence 115
   2.5. Development by Γ-convergence 117
3. Localization methods 117
   3.1. Supremum of measures 118
   3.2. The blow-up technique 119
   3.3. A general compactness procedure 120
   3.4. The "slicing" method 122
4. Local integral functionals on Sobolev spaces 124
   4.1. A prototypical compactness theorem 124
   4.2. Useful technical results 128
   4.3. Convergence of quadratic forms 131
   4.4. Degenerate limits 132
5. Homogenization of integral functionals 134
   5.1. The asymptotic homogenization formula 135
   5.2. The convex case: the cell-problem formula 137
   5.3. Homogenization of quadratic forms 138
   5.4. Bounds on composites 140
   5.5. Homogenization of metrics 142
6. Perforated domains and relaxed Dirichlet problems 144
   6.1. Dirichlet boundary conditions: a direct approach 144
   6.2. Relaxed Dirichlet problems 149

HANDBOOK OF DIFFERENTIAL EQUATIONS
Stationary Partial Differential Equations, volume 3
Edited by M. Chipot and P. Quittner

6.3. Neumann boundary conditions: an extension lemma 151
6.4. Double-porosity homogenization 153
7. Phase-transition problems 155
7.1. Interfacial energies 155
7.2. Gradient theory of phase transitions 158
7.3. A compactness result 164
7.4. Other functionals generating phase-transitions 165
7.5. Some extensions 171
8. Concentration problems 173
8.1. Ginzburg–Landau 173
8.2. Critical-growth problems 176
9. Dimension-reduction problems 180
9.1. The Le Dret–Raoult result 180
9.2. A compactness theorem 183
9.3. Higher-order Γ-limits 185
10. Approximation of free-discontinuity problems 186
10.1. Special functions with bounded variation 187
10.2. The Ambrosio–Tortorelli approximation 190
10.3. Other approximations 193
10.4. Approximation of curvature functionals 196
11. Continuum limits of lattice systems 199
11.1. Continuum energies on Sobolev spaces 200
11.2. Continuum energies on discontinuous functions 205
Note to the bibliography 208
References 208

Preface

The notion of Γ-convergence has become, over the more than thirty years after its introduction by Ennio De Giorgi [93], the commonly-recognized notion of convergence for variational problems, and it would be difficult nowadays to think of any other "limit" than a Γ-limit when talking about asymptotic analysis in a general variational setting (even though special convergences may fit better specific problems, as Mosco-convergence, two-scale convergence, G- and H-convergence, etc.). This short presentation is meant as an introduction to the many applications of this theory to problems in Partial Differential Equations, both as an effective method for solving asymptotic and approximation issues and as a means of expressing results that are derived by other techniques. A complete introduction to the general theory of Γ-convergence is the by-now-classical book by Dal Maso [84], while a user-friendly introduction can be found in my book "for beginners" [46], where also simplified one-dimensional versions of many of the problems in this chapter are treated.

These notes are addressed to an audience of experienced mathematicians, with some background and interest in Partial Differential Equations, and are meant to direct the reader to what I regard as the most interesting features of this theory. The style of the exposition is how I would present the subject to a colleague in a neighboring field or to an interested Ph.D. student: the issues that I think will likely emerge again and link a particular question to others are presented with more detail, while I refer to the main monographs or recent articles in the literature for in-depth knowledge of the single issues. Necessarily, many of the proofs are sketchy, and some expert in the field of the Calculus of Variations might shudder at the liberties I will take in order to highlight the main points without entering in details that can be dealt with only in a more ample and dedicated context.

The choice of the issues presented in these notes has been motivated by their closeness to general questions of Partial Differential Equations. Many interesting applications of Γ-convergence that are a little further from that field, and would need a wider presentation of their motivations are only briefly mentioned (for example, the derivation of low-dimensional theories in Continuum Mechanics [107], functionals on BV and SBV [20,55], the application of Γ-convergence to modeling problems in Mechanics [32,75], etc.), or not even touched at all (for example, nonconvex energies defined on measures [16,42], stochastic Γ-convergence in a continuous or discrete setting [64,86], applications of Γ-convergence to Statistical Mechanics [8,40] and to finite-element methods [72], etc.). The reference to those applications listed here are just meant to be a first suggestion to the interested reader.

1. Introduction

Γ-convergence is designed to express the convergence of minimum problems: it may be convenient in many situations to study the asymptotic behavior of a family of problems

$$m_\varepsilon = \min\{F_\varepsilon(x)\colon x \in X_\varepsilon\} \tag{1.1}$$

not through the study of the properties of the solutions x_ε, but by defining a limit energy F_0 such that, as $\varepsilon \to 0$, the problem

$$m_0 = \min\{F_0(x): x \in X_0\} \tag{1.2}$$

is a "good approximation" of the previous one; i.e., $m_\varepsilon \to m_0$ and $x_\varepsilon \to x_0$, where x_0 is itself a solution of m_0. This latter requirement must be thought upon extraction of a subsequence if the "target" minimum problem admits more than a solution. Note that the convergence problem above can also be stated in the reverse direction: given F_0 for which solutions are difficult to characterize, find approximate F_ε whose solution are more at hand. Of course, in order for this procedure to make sense we must require an *equicoerciveness* property for the energies F_ε; i.e., that we may find a precompact *minimizing sequence* (that is, $F_\varepsilon(x_\varepsilon) \leqslant \inf F_\varepsilon + o(1)$) such that the convergence $x_\varepsilon \to x_0$ can take place.

The existence of such an F_0, the Γ-*limit* of F_ε, is a consequence of the two following conditions:

(i) *Liminf inequality*: for every $x \in X_0$ and for every $x_\varepsilon \to x$, we have

$$F_0(x) \leqslant \liminf_{\varepsilon \to 0} F_\varepsilon(x_\varepsilon). \tag{1.3}$$

In other words, F_0 is a *lower bound* for the sequence F_ε, in the sense that $F_0(x) \leqslant F_\varepsilon(x_\varepsilon) + o(1)$ whenever $x_\varepsilon \to x$. If the family F_ε is equicoercive, then this condition immediately implies one inequality for the minimum problems: if (x_ε) is a minimizing sequence and (upon subsequences) $x_\varepsilon \to x_0$ then

$$\inf F_0 \leqslant F_0(x_0) \leqslant \liminf_{\varepsilon \to 0} F_\varepsilon(x_\varepsilon) = \liminf_{\varepsilon \to 0} \inf F_\varepsilon \tag{1.4}$$

(to be precise, in this argument we take care to start from x_{ε_j} such that $\liminf_{\varepsilon \to 0} F_\varepsilon(x_\varepsilon) = \lim_j F_{\varepsilon_j}(x_{\varepsilon_j})$).

(ii) *Limsup inequality* or *existence of a recovery sequence*: for every $x \in X_0$ we can find a sequence $\bar{x}_\varepsilon \to x$ such that

$$F_0(x) \geqslant \limsup_{\varepsilon \to 0} F_\varepsilon(\bar{x}_\varepsilon). \tag{1.5}$$

Note that if (i) holds then in fact $F_0(x) = \lim_{\varepsilon \to 0} F_\varepsilon(\bar{x}_\varepsilon)$, so that the lower bound is sharp. From (1.5) we get in particular that $F_0(x) \geqslant \limsup_{\varepsilon \to 0} \inf F_\varepsilon$, and since this holds for all x we conclude that

$$\inf F_0 \geqslant \limsup_{\varepsilon \to 0} \inf F_\varepsilon. \tag{1.6}$$

An F_0 satisfying (1.5) is an *upper bound* for the sequence (F_ε) and its computation is usually related to an ansatz leading to the construction of the sequence $\bar{x}_\varepsilon$.

From the two inequalities (1.4) and (1.6) we obtain the convergence of the infima m_ε in (1.1) to the minimum m_0 in (1.2). Not only: we also obtain that every cluster point of a minimizing sequence is a minimum point for F_0. This is the *fundamental theorem of Γ-convergence*, that is summarized by the implication

$$\Gamma\text{-convergence} + \text{equicoerciveness} \implies \text{convergence of minimum problems.}$$

A hidden element in the procedure of the computation of a Γ-limit is the *choice of the right notion of convergence* $x_\varepsilon \to x$. This is actually one of the main issues in the problem: a convergence is not given beforehand and should be chosen in such a way that it implies the equicoerciveness of the family F_ε. The choice of a weaker convergence, with many converging sequences, makes this requirement easier to fulfill, but at the same time makes the liminf inequality more difficult to hold. In the following we will not insist on the motivation of the choice of the convergence, that in most cases will be a strong L^p-convergence (the choice of a separable metric convergence makes life easier). The reader is anyhow advised that this is one of the main points of the Γ-convergence approach. Another related issue is that of the *correct energy scaling*. In fact, in many cases the given functionals F_ε will not give rise to an equicoercive family with respect to a meaningful convergence, but the right scaled functionals, e.g., $\varepsilon^{-\alpha} F_\varepsilon$, will turn out to better describe the behavior of minimum problems. The correct scaling is again usually part of the problem.

Applications of Γ-convergence to Partial Differential Equations can be generally related to the behavior of the Euler–Lagrange equations of some integral energy. The prototype of such problems can be written as

$$m_\varepsilon = \inf\left\{ \int_\Omega f_\varepsilon(x, Du)\, dx - \int_\Omega \langle g, u \rangle\, dx \colon u = \varphi \text{ on } \partial\Omega \right\}. \tag{1.7}$$

In these notes Ω will always stand for an open bounded (sufficiently smooth) subset of $\mathbb{R}^n$, unless otherwise specified. Note that the possibility of defining a Γ-limit related to these problems will not be linked to the properties (or even the existence) of the solutions of the related Euler–Lagrange equations.

It must be noted that the functionals related to (1.7) of which we want to compute the Γ-limit are usually defined on some Sobolev space $W^{1,p}(\Omega; \mathbb{R}^m)$ and can be written as

$$F_\varepsilon^{\varphi,g}(u)$$
$$= \begin{cases} \int_\Omega f_\varepsilon(x, Du)\, dx - \int_\Omega \langle g, u \rangle\, dx & \text{if } u - \varphi \in W_0^{1,p}(\Omega; \mathbb{R}^m), \\ +\infty & \text{otherwise,} \end{cases} \tag{1.8}$$

or, equivalently, we can think of $F_\varepsilon^{\varphi,g}$ as defined on $L^p(\Omega; \mathbb{R}^m)$ extended to $+\infty$ outside $W^{1,p}(\Omega; \mathbb{R}^m)$. We will often use this extension to $+\infty$ in the chapter, leaving it as understood in most cases.

As defined in (1.8), the functional $F_\varepsilon^{\varphi,g}$ depends both on the forcing term g and on the boundary datum φ. Suppose now that the growth conditions on f_ε ensure strong pre-compactness of minimizing sequences in L^p (this is usually obtained by Poincaré's inequality and Rellich's embedding theorem), and that $g \in L^{p'}(\Omega)$. A first important property, following directly from the definition of Γ-convergence is the *stability of*

Γ-convergence with respect to continuous perturbations: if G is a continuous function and $F_0 = \Gamma\text{-}\lim_{\varepsilon \to 0} F_\varepsilon$, then $\Gamma\text{-}\lim_{\varepsilon \to 0}(F_\varepsilon + G) = F_0 + G$. This implies that we can neglect the forcing term in the computation of the Γ-limit of $F_\varepsilon^{\varphi,g}$ and add it a posteriori.

Another, more particular, property is the *compatibility* of boundary conditions, which says that also the boundary condition $u = \varphi$ can be added after computing the Γ-limit of the "free" energy

$$F_\varepsilon(u) = \begin{cases} \int_\Omega f_\varepsilon(x, Du)\, dx & \text{if } u \in W^{1,p}(\Omega; \mathbb{R}^m), \\ +\infty & \text{otherwise.} \end{cases} \tag{1.9}$$

This property of compatibility of boundary conditions is not always true, but holds for a large class of integrals (see Section 4.2.1). Other compatibility properties are available for other types of functionals, as that of volume constraints for Cahn–Hilliard functionals (see Section 7.2.2).

We have hence seen that to characterize the Γ-limit of $F_\varepsilon^{\varphi,g}$ it will be sufficient to compute the Γ-limit of F_ε. We will see that growth conditions on f_ε ensure that the limit always exists (up to subsequences) and can be represented again through an integral functional $F_0(u) = \int_\Omega f_0(x, Du)\, dx$, independently of the regularity and convexity properties of f_ε. This can be done through a general *localization and compactness procedure* due to De Giorgi [90] (see Sections 3.3 and 4.2). As a consequence we obtain a limit problem

$$m_0 = \inf\left\{ \int_\Omega f_0(x, Du)\, dx - \int_\Omega \langle g, u \rangle\, dx : u = \varphi \text{ on } \partial\Omega \right\}, \tag{1.10}$$

with f_0 *independent of the data* g and φ, and also of Ω.

We will see other classes of energies than the integrals as above for which a general approach is possible showing compactness with respect to Γ-convergence and representation results for the Γ-limit. We now briefly outline the description of some specific problems. Many more examples are included in the text.

A "classical" example of Γ-limit is for functionals of the type (1.9) when $f_\varepsilon(x, \xi) = f(x/\varepsilon, \xi)$, and f is a fixed function that is 1-periodic in the first variable (i.e., $f(x + e_i, \xi) = f(x, \xi)$ for the elements e_i of the standard basis of $\mathbb{R}^n$). The Γ-limit is also called the *homogenized functional* of the F_ε (see Section 5).

It is interesting to note how some general issues arise in the study of these functionals. First, one notes that the limit energy f_0, that always exists up to subsequences by what remarked above, is *homogeneous*; i.e., $f_0(x, \xi) = f_0(\xi)$ by the vanishing periodicity of f_ε. At the same time, a general property is the *lower semicontinuity of Γ-limits*, that in this case implies that f_0 is *quasiconvex*, or, what is more important, that the value $f_0(\xi)$ at a fixed $\xi \in \mathbb{M}^{m \times n}$ can be expressed as a minimum problem:

$$f_0(\xi) = \min\left\{ \int_{(0,1)^n} f_0(\xi + D\varphi)\, dx : u \in W_0^{1,p}((0,1)^n; \mathbb{R}^m) \right\}. \tag{1.11}$$

Now the property of convergence of minima provides an ansatz for the function f_0, as the value given by the *asymptotic homogenization formula*

$$f_{\text{hom}}(\xi) = \lim_{\varepsilon \to 0} \min\left\{ \int_{(0,1)^n} f\left(\frac{x}{\varepsilon}, \xi + D\varphi\right) dx : u \in W_0^{1,p}\left((0,1)^n; \mathbb{R}^m\right) \right\}$$

$$= \lim_{T \to +\infty} \frac{1}{T^n} \min\left\{ \int_{(0,T)^n} f(y, \xi + D\varphi) \, dy : u \in W_0^{1,p}\left((0,T)^n; \mathbb{R}^m\right) \right\}.$$

Note that we have made use both of the compatibility of boundary conditions, and of the fact that we have Γ-convergence on all open sets (and in particular on $(0,1)^n$). The problem is thus reduced to showing that this last limit exists, giving a form for the limit independent of a particular subsequence. The derivation of suitable formulas is one of the recurrent issues in the characterization of Γ-limits.

The homogenization theorem is easier in the convex and scalar case. In particular, we can apply it to the study of the behavior of linear equations

$$\begin{cases} -\sum_{i,j=1}^n D_i\left(a_{ij}\left(\frac{x}{\varepsilon}\right)D_j u\right) = g & \text{in } \Omega, \\ u = \varphi & \text{on } \partial\Omega, \end{cases} \tag{1.12}$$

with a_{ij} periodic (given the usual boundedness and uniform ellipticity), that are the Euler equations of the minimum problem

$$m_\varepsilon = \inf\left\{ \int_\Omega \sum_{i,j=1}^n a_{ij}\left(\frac{x}{\varepsilon}\right) D_j u D_i u \, dx - \int_\Omega gu \, dx : u = \varphi \text{ on } \partial\Omega \right\}. \tag{1.13}$$

In this case an additional property of Γ-limits can be used: that Γ-limits of quadratic forms are still quadratic forms, so that we obtain $f_{\text{hom}}(\xi) = \sum_{i,j} q_{ij}\xi_i\xi_j$ with q_{ij} constant coefficients. Since in this case the limit problem has a unique solution we deduce that the solutions u_ε weakly converge in $H^1(\Omega)$ to the solution of the simpler problem

$$\begin{cases} -\sum_{i,j=1}^n q_{ij} D_i D_j u = g & \text{in } \Omega, \\ u = \varphi & \text{on } \partial\Omega. \end{cases} \tag{1.14}$$

Note that the coefficients q_{ij} depend in a nontrivial way on all the coefficients of the matrix a_{ij}, and in particular, differ from their averages $\bar{a}_{ij}$, which give the pointwise limit $\int_\Omega \sum_{ij} \bar{a}_{ij} D_i D_j u \, dx$ of $F_\varepsilon(u)$. Even in the simple case, $a_{ij} \in \{\alpha\delta_{ij}, \beta\delta_{ij}\}$, the characterization of the homogenized matrices is not a trivial task (see Section 5.4). More interesting effects of the form of the Γ-limit are obtained by introducing Dirichlet or Neumann boundary conditions on varying domains (see Section 6).

A feature of Γ-convergence is that it is not linked to a particular assumption of the form of solutions, but relies instead on energetic approaches, tracing the behavior of energies. In this way we could end up with problems of a different nature than those we started with. One of the first examples of this fact, included in an early paper by Modica and

Mortola [123], is the study of the asymptotic behavior of minimizers of

$$m_\varepsilon = \min\left\{\int_\Omega (1-|u|)^2\,dx + \varepsilon^2 \int_\Omega |Du|^2\,dx : u \in H^1(\Omega), \int_\Omega u\,dx = C\right\},$$

$$(1.15)$$

where $|C| < |\Omega|$. This is a problem connected to the Cahn–Hilliard theory of liquid–liquid phase transitions. It is also known as the "scalar Ginzburg–Landau" energy. In this case it is easily seen by energy considerations that minimizers are weakly pre-compact in $L^2(\Omega)$ and that in that topology the Γ-limit is simply $\int_\Omega W^{**}(u)\,dx$ (W^{**} denotes the convex hull of $W(u) = (1-|u|)^2$). Since $W^{**}(u) = 0$ for $|u| \leqslant 1$, we conclude that minimizers converge weakly in $L^2(\Omega)$ to functions with $|u| \leqslant 1$, and that all such functions arise as limit of minimizing sequences. This is clearly not a satisfactory description, and a more meaningful scaling must be performed, considering the functionals

$$F_\varepsilon(u) = \frac{1}{\varepsilon}\int_\Omega (1-|u|)^2\,dx + \varepsilon \int_\Omega |Du|^2\,dx.$$

$$(1.16)$$

These functionals are equicoercive with respect to the strong convergence in $L^1(\Omega)$ and from the first term we may deduce that the limit u of a sequence u_ε equibounded in energy satisfies $|u| = 1$ a.e. in Ω. By using compactness arguments for sets of finite perimeter, we actually may deduce that $E = \{u = 1\}$ is a set of finite perimeter. By the invariance properties and representation theorems we may deduce that $F_0(u) = \sigma \mathcal{H}^{n-1}(\Omega \cap \partial E)$. Once the Γ-limit is computed we may show the *compatibility* of the integral constraint thus expressing the limit of u_ε in terms of a set E that minimizes

$$m_0 = \min\left\{\mathcal{H}^{n-1}(\Omega \cap \partial E) : 2|E| - |\Omega| = C\right\}.$$

$$(1.17)$$

This form of the limit problem is common to many phase-transition energies (see Section 7).

The computation of this Γ-limit shows some remarkable features, such as the one-dimensional nature of minimizers (i.e., their value essentially depends only on the distance to ∂E) that allows for a *slicing procedure* (see Section 3.4), the necessity of finding a "correct scaling" for the energies, the use of Γ-convergence as a selection criterion when we have many solutions to a variational problem (in this case problem (1.15) with $\varepsilon = 0$), and not least, the "change of type" in the limit energy that turns from a "bulk" energy into a "surface" energy.

Other types of limits present an even more dramatic change of type, such as the (complex) Ginzburg–Landau energies (note the different scaling with respect to the "scalar" ones)

$$F_\varepsilon(u_\varepsilon) = \frac{1}{\varepsilon^2|\log\varepsilon|}\int_\Omega (1-|u|)^2\,dx + \frac{1}{|\log\varepsilon|}\int_\Omega |Du|^2\,dx,$$

$$(1.18)$$

where $\Omega \subset \mathbb{R}^2$ and $u : \Omega \to \mathbb{R}^2$. In this problem the relevant objects for a sequence with $F_\varepsilon(u_\varepsilon)$ equibounded for which we have a compactness property are the distributional Jacobians $J(u_\varepsilon)$ that converge (upon subsequences) to measures $\mu = \pi \sum_i d_i \delta_{x_i}$, with $x_i \in \Omega$ and d_i integers, describing the formation of "vortices". The limit energy is then defined as $F_0(\{(x_i, d_i)\}_i) = 2\pi \sum_i |d_i|$ (see Section 8.1).

Other types of problems with concentration can be dealt with in the framework of Γ-convergence. One of those is the *Bernoulli free-boundary problem*: look for an open set A and a function u which is the weak solution of

$$\begin{cases} \Delta u = 0 & \text{in } \Omega \setminus A, \\ u = 1 & \text{on } \partial A, \\ u = 0 & \text{on } \partial\Omega, \\ \frac{\partial u}{\partial v} = q & \text{on } \partial A, \end{cases}$$

and describe the asymptotic behavior of such A and u when $q \to +\infty$. These can be considered as the Euler–Lagrange equations of the following family of variational problems depending on a small parameter $\varepsilon > 0$:

$$S_\varepsilon^V(\Omega) = \max\left\{ |\{u \geq 1\}| : u \in H_0^1(\Omega), \int_\Omega |\nabla u|^2 \, dx \leq \varepsilon^2 \right\}$$

$$= \max\left\{ |A| : \text{Cap}(A, \Omega) \leq \varepsilon^2 \right\} \tag{1.19}$$

(where $\text{Cap}(A, \Omega)$ denotes the capacity of A with respect to Ω) with $q = q_\varepsilon$ a Lagrange multiplier. Again, maximizers show concentration phenomena since maximal sets A will shrink to a point, and can be treated by Γ-convergence (actually, since we have a *maximum* problem we must use the symmetric notion fit for maxima). This asymptotic description can be adapted to treat more general concentration problems (see Section 8.2).

Beside the study of asymptotic properties of minimum problems, we will describe other uses of Γ-convergence. One is the construction of suitable Γ-converging functionals F_ε to a given F_0. This is the case for example of functionals in Computer Vision, such as the Mumford–Shah functional, that are difficult to treat numerically. Their approximation by elliptic functionals (such as the Ambrosio–Tortorelli approximation) provides approximate solutions and numerical schemes, but also many other types of approximating functionals are available (see Section 10). Γ-convergence is particularly suited to such issues, not being linked to a particular form or domain for the energies to be constructed. Moreover, Γ-convergence is also used in the "justification" of physical theories through a limit procedure. One example is the derivation of low-dimensional theories from three-dimensional elasticity (see Section 9), another one is the deduction of properties in Continuum Mechanics from atomistic potentials (see Section 11).

Notation

We will use standard notation and results for Lebesgue, Sobolev and BV spaces (see [20,97,143]). For functionals defined on Sobolev spaces we will identify the distributional derivative Du with its density, so that the same symbol will denote the corresponding L^1-function. We will abandon this identification when we will deal with functions of bounded variation, where the approximate gradient will be denoted by ∇u (Sections 10 and 11).

In the proofs we will often use the letters c and C to indicate unspecified strictly positive constants. A family of objects, even if parameterized by a continuous parameter, will also be often called a "sequence" not to overburden notation.

2. General theory of Γ-convergence

This first section is devoted to an introduction to the main properties of Γ-convergence, in particular to those that are useful in the actual computation of Γ-limits. The reader is referred to [46] for a proof of these results, many of which are nevertheless simple applications of the definitions, and to [84] for a more refined introduction (see also [27]). The original papers on the subject by De Giorgi and collaborators are collected in [91].

In what follows, we will usually compute the Γ-limit of a *family* F_ε of functionals indexed by the positive parameter ε. Within the proofs of the lower bounds it is generally useful to invoke some compactness argument, and consider the problem of computing the Γ-limit of a *sequence* $G_j := F_{\varepsilon_j}$. We will give the definitions for the whole family F_ε. The reader can easily rewrite every definition for functionals depending on a discrete parameter j.

2.1. *The definitions of Γ-convergence*

We have seen in the Introduction how a definition of Γ-convergence can be given in terms of properties of the functions along converging sequences. That one will be the definition we will normally use. For the sake of completeness we now consider the most general case of a family $F_\varepsilon : X \to [-\infty, +\infty]$ defined on a *topological space X*. In that case we say that F_ε *Γ-converges* to $F : X \to [-\infty, +\infty]$ at $x \in X$ as $\varepsilon \to 0$ if we have

$$F(x) = \sup_{U \in \mathcal{N}(x)} \liminf_{\varepsilon \to 0} \inf_{y \in U} F_\varepsilon(y) \left(= \sup_{U \in \mathcal{N}(x)} \sup_{0 < \rho} \inf_{\varepsilon < \rho} \inf_{y \in U} F_\varepsilon(y) \right)$$

$$= \sup_{U \in \mathcal{N}(x)} \limsup_{\varepsilon \to 0} \inf_{y \in U} F_\varepsilon(y) \left(= \sup_{U \in \mathcal{N}(x)} \inf_{0 < \rho} \sup_{\varepsilon < \rho} \inf_{y \in U} F_\varepsilon(y) \right), \qquad (2.1)$$

where $\mathcal{N}(x)$ denotes the family of all neighborhoods of x in X. In this case we say that $F(x)$ is the Γ-*limit* of F_ε at x and we write

$$F(x) = \Gamma\text{-}\lim_{\varepsilon \to 0} F_\varepsilon(x). \qquad (2.2)$$

If (2.2) holds for all $x \in X$ then we say that F_ε *Γ-converges* to F (on the whole X).

Note that we sometime will consider families of functionals $F_\varepsilon : X_\varepsilon \to \mathbb{R}$, where the domain may depend on ε. In this case it is understood that we identify such functionals with

$$\widetilde{F}_\varepsilon(x) = \begin{cases} F_\varepsilon(x) & \text{if } x \in X_\varepsilon, \\ +\infty & \text{if } X \setminus X_\varepsilon, \end{cases}$$

where X is a space containing all X_ε where the convergence takes place.

The definition in (2.1) makes sense in any topological space and is rather suggestive as it shows how these limits are constructed from the elementary operations of "sup" and "inf". This definition is sometimes handy to prove general properties as compactness or lower semicontinuity; however it is less suited to most direct computations. In applications we will usually deal with metric spaces (as L^p spaces) or metrizable spaces (as bounded subsets of Sobolev spaces or of spaces of measures, equipped with the weak topology), that in addition are also separable. For such spaces the previous definitions are simplified as follows.

THEOREM 2.1 (Equivalent definitions of Γ-convergence). *Let X be a metric space and let F_ε, $F : X \to [-\infty, +\infty]$. Then the Γ-convergence of F_ε to F at x is equivalent to any of the following conditions*:
   (a) *we have*

$$F(x) = \inf\left\{ \liminf_{\varepsilon \to 0} F_\varepsilon(x_\varepsilon) \colon x_\varepsilon \to x \right\} = \inf\left\{ \limsup_{\varepsilon \to 0} F_\varepsilon(x_\varepsilon) \colon x_\varepsilon \to x \right\}; \qquad (2.3)$$

   (b) *we have*

$$F(x) = \min\left\{ \liminf_{\varepsilon \to 0} F_\varepsilon(x_\varepsilon) \colon x_\varepsilon \to x \right\} = \min\left\{ \limsup_{\varepsilon \to 0} F_\varepsilon(x_\varepsilon) \colon x_\varepsilon \to x \right\}; \qquad (2.4)$$

   (c) (sequential Γ-convergence) *we have*
      (i) (liminf inequality) *for every sequence* (x_ε) *converging to* x,

$$F(x) \leqslant \liminf_{\varepsilon \to 0} F_\varepsilon(x_\varepsilon); \qquad (2.5)$$

      (ii) (limsup inequality) *there exists a sequence* (x_ε) *converging to* x *such that*

$$F(x) \geqslant \limsup_{\varepsilon \to 0} F_\varepsilon(x_\varepsilon); \qquad (2.6)$$

   (d) *the liminf inequality* (c)(i) *holds and*
      (ii)$'$ (existence of a recovery sequence) *there exists a sequence* (x_ε) *converging to* x *such that*

$$F(x) = \lim_{\varepsilon \to 0} F_\varepsilon(x_\varepsilon). \qquad (2.7)$$

(e) *the liminf inequality* (c)(i) *holds and*

(ii)″ (approximate limsup inequality) *for all $\eta > 0$ there exists a sequence (x_ε) converging to x such that*

$$F(x) \geqslant \limsup_{\varepsilon \to 0} F_\varepsilon(x_\varepsilon) - \eta. \tag{2.8}$$

Moreover, the Γ-convergence of F_ε to F on the whole X is equivalent to
(f) (limits of minimum problems) *inequality*

$$\inf_U F \geqslant \limsup_{\varepsilon \to 0} \inf_U F_\varepsilon \tag{2.9}$$

holds for all open sets U and inequality

$$\inf_K F \leqslant \sup\left\{ \liminf_{\varepsilon \to 0} \inf_U F_\varepsilon \colon U \supset K, U \text{ open} \right\} \tag{2.10}$$

holds for all compact sets K.

Finally, if d denotes a distance on X and we have a uniform lower bound $F_\varepsilon(x) \geqslant -c(1 + d(x, x_0)^p)$ for some $p > 0$ and $x_0 \in X$, then the Γ-convergence of F_ε to F on the whole X is equivalent to
(g) (convergence of Moreau–Yosida transforms) *for all x we have*

$$F(x) = \sup_{\lambda \geqslant 0} \liminf_{\varepsilon \to 0} \inf_{y \in X} \left\{ F_\varepsilon(y) + \lambda d(x, y)^p \right\}$$

$$= \sup_{\lambda \geqslant 0} \limsup_{\varepsilon \to 0} \inf_{y \in X} \left\{ F_\varepsilon(y) + \lambda d(x, y)^p \right\}. \tag{2.11}$$

PROOF. The equivalence between (a)–(e) is easily proved, as their equivalence with definition (2.1). In particular note that (2.7) is equivalent to (2.6) since (2.5) holds, and (2.8) is equivalent to (2.7) by the arbitrariness of η and a diagonal argument. The proofs of (f) and (g) are only a little more delicate (see [46], Section 1.4, for details). $\qquad\square$

The two inequalities in (c) are usually taken as the definition of Γ-convergence (for first-countable spaces). Note the asymmetry between inequalities with upper and lower limits, due to Γ-convergence being "unbalanced" towards minimum problems.

REMARK 2.2. From these definitions we can make some immediate but interesting observation:

(1) *Stability under continuous perturbations*: if (F_ε) Γ-converges to F and $G: X \to [-\infty, +\infty]$ is a d-continuous function then $(F_\varepsilon + G)$ Γ-converges to $F + G$. This is an immediate consequence of the definition (e.g. from condition (d));

(2) *Γ-limit of a constant sequence*: Γ-convergence does not enjoy the property that a constant family $F_\varepsilon = F$ converges to F. In fact if this were true, then from the liminf inequality we would have $F(x) \leqslant \liminf_{\varepsilon \to 0} F(x_\varepsilon)$ for all $x_\varepsilon \to x$; i.e., F would be *lower semicontinuous* (which is not always the case);

(3) *Comparison with uniform and pointwise limits.* The previous observation in particular shows that we cannot deduce the existence of the Γ-limit from pointwise convergence. If F_ε converge to G pointwise and $F = \Gamma\text{-}\lim_{\varepsilon\to 0} F_\varepsilon$ then $F \leqslant G$. However, if F_ε converge uniformly to a *continuous* F on an open set U then we easily see that F_ε Γ-converge to F;

(4) *Dependence on the metric.* Note that the value and the existence of the Γ-limit depend on the metric d. If we want to highlight the role of the metric, we can add the dependence on the distance d and write $\Gamma(d)$-limit, $\Gamma(d)$-convergence, and so on. When two distances d and d' are comparable then from (2.3) we get an inequality between the two Γ-limits (if they exist); in general, note that the existence of the Γ-limit in one metric does not imply the existence of the Γ-limit in the second.

2.1.1. *Upper and lower Γ-limits.* As for usual limits, it is convenient to define quantities that always exist (as upper and lower limits) and rephrase the existence of a Γ-limit as an equality between those two quantities. Theorem 2.1(a) suggests the following definition of *lower and upper Γ-limits*:

$$\Gamma\text{-}\liminf_{\varepsilon\to 0} F_\varepsilon(x) = \inf\left\{\liminf_{\varepsilon\to 0} F_\varepsilon(x_\varepsilon)\colon x_\varepsilon \to x\right\}, \tag{2.12}$$

$$\Gamma\text{-}\limsup_{\varepsilon\to 0} F_\varepsilon(x) = \inf\left\{\limsup_{\varepsilon\to 0} F_\varepsilon(x_\varepsilon)\colon x_\varepsilon \to x\right\}, \tag{2.13}$$

respectively. In this way the existence of the $\Gamma\text{-}\lim_{\varepsilon\to 0} F_\varepsilon(x) = F(x)$ is stated as

$$\Gamma\text{-}\liminf_{\varepsilon\to 0} F_\varepsilon(x) = \Gamma\text{-}\limsup_{\varepsilon\to 0} F_\varepsilon(x) = F(x). \tag{2.14}$$

REMARK 2.3. If (F_{ε_k}) is a subsequence of (F_ε) then

$$\Gamma\text{-}\liminf_{\varepsilon\to 0} F_\varepsilon \leqslant \Gamma\text{-}\liminf_{k} F_{\varepsilon_k}, \qquad \Gamma\text{-}\limsup_{k} F_{\varepsilon_k} \leqslant \Gamma\text{-}\limsup_{\varepsilon\to 0} F_\varepsilon.$$

In particular, if $F = \Gamma\text{-}\lim_{\varepsilon\to 0} F_\varepsilon$ exists then, for every infinitesimal sequence (ε_k), $F = \Gamma\text{-}\lim_{k} F_{\varepsilon_k}$.

2.2. *Γ-convergence and lower semicontinuity*

As noted in Remark 2.2(2), the Γ-limit of a constant family $F_\varepsilon = F$ does not converge to F. This is true, however, if F is d-lower semicontinuous. More, the class of lower-semicontinuous functions provides a "stable class" for Γ-convergence. This is summarized in the following propositions.

PROPOSITION 2.4 (Lower semicontinuity of Γ-limits). *The Γ-upper and lower limits of a family F_ε are d-lower-semicontinuous functions.*

PROPOSITION 2.5 (Γ-limits and lower-semicontinuous envelopes – relaxation).

   (1) *The Γ-limit of a constant sequence $F_\varepsilon = F$ is equal to*

$$\overline{F}(x) = \liminf_{y \to x} F(y), \tag{2.15}$$

that is, the lower-semicontinuous envelope of F, defined as the largest lower-semicontinuous function not greater than F. This operation is also called relaxation.

   (2) *The Γ-limit is stable by substituting F_ε by its lower-semicontinuous envelope $\overline{F}_\varepsilon$; i.e., we have*

$$\Gamma\text{-}\liminf_{\varepsilon \to 0} F_\varepsilon = \Gamma\text{-}\liminf_{\varepsilon \to 0} \overline{F}_\varepsilon, \qquad \Gamma\text{-}\limsup_{\varepsilon \to 0} F_\varepsilon = \Gamma\text{-}\limsup_{\varepsilon \to 0} \overline{F}_\varepsilon. \tag{2.16}$$

REMARK 2.6. If $F_\varepsilon \to F$ pointwise then $\Gamma\text{-}\limsup_{\varepsilon \to 0} F_\varepsilon \leqslant F$, and hence, taking both lower-semicontinuous envelopes, also $\Gamma\text{-}\limsup_{\varepsilon \to 0} F_\varepsilon \leqslant \overline{F}$.

REMARK 2.7. The following observations are often useful in computations:

   (i) the supremum of a family (not necessarily finite or countable) of lower-semicontinuous functions is itself lower semicontinuous;

   (ii) if $f : X \to \overline{\mathbb{R}}$ is bounded from below and $p > 0$; then for all $x \in X$,

$$\bar{f}(x) = \sup_{\lambda \geqslant 0} \inf_{y \in X} \left\{ f(y) + \lambda d(x, y)^p \right\}.$$

In particular, every lower-semicontinuous function bounded from below is the supremum of an increasing family of Lipschitz functions.

2.3. *Computation of Γ-limits*

For some classes of functionals, a common compactness procedure has been formalized (see Section 3), but in general the computation of the Γ-limit of a family (F_ε) is usually divided into the computation of a separate lower and an upper bound. A *lower bound* is a functional G such that $G \leqslant \Gamma\text{-}\liminf_{\varepsilon \to 0} F_\varepsilon$; i.e., such that

$$G(x) \leqslant \liminf_{j} F_{\varepsilon_j}(x_j) \quad \text{for all } \varepsilon_j \to 0 \text{ and } x_j \to x. \tag{2.17}$$

The lower semicontinuity of Γ-limit allows us to limit our search for lower bounds to the class of lower-semicontinuous G. If we can characterize a large enough family $\mathcal{G}$ of G satisfying (2.17) then the optimal lower bound is obtained as $\overline{G}(x) := \sup\{G(x) \colon G \in \mathcal{G}\}$. Note that this function is lower semicontinuous, being the supremum of a family of lower-semicontinuous functions.

   The optimization of the lower bound usually suggests an ansatz (or more) to approximate a target element $x \in X$ by a family $\bar{x}_\varepsilon \to x$, thus defining $H(x) := \lim_{\varepsilon \to 0} F_\varepsilon(\bar{x}_\varepsilon)$. By definition, $H \geqslant \Gamma\text{-}\limsup_{\varepsilon \to 0} F_\varepsilon$, so that H is an *upper bound* for the Γ-limit. If we use more ansätze, we obtain a family $\mathcal{H}$ and then a candidate optimal upper bound as $\overline{H}(x) = \inf\{H(x): H \in \mathcal{H}\}$. The existence (and computation) of the Γ-limit is then expressed in the equality $\overline{G} = \overline{H}$.

REMARK 2.8 (A density argument). The lower semicontinuity of the Γ-limsup can be used to reduce its computation to a dense class. Let d' be a distance on X inducing a topology which is not weaker than that induced by d; i.e., $d'(x_\varepsilon, x) \to 0$ implies $d(x_\varepsilon, x) \to 0$, and suppose that

  (i) $\mathcal{D}$ is a dense subset of X for d';
  (ii) we have $\Gamma\text{-}\limsup_{\varepsilon \to 0} F_\varepsilon(x) \leqslant F(x)$ on $\mathcal{D}$, where F is a function which is continuous with respect to d;

then we have $\Gamma\text{-}\limsup_{\varepsilon \to 0} F_\varepsilon \leqslant F$ on X.

To check this, it suffices to note that if $d'(x_k, x) \to 0$ and $x_k \in \mathcal{D}$ then

$$\Gamma\text{-}\limsup_{\varepsilon \to 0} F_\varepsilon(x) \leqslant \liminf_k \left(\Gamma\text{-}\limsup_{\varepsilon \to 0} F_\varepsilon(x_k) \right)$$

$$\leqslant \liminf_k F(x_k) = F(x).$$

This method will be used very often in the following, without repeating these arguments. It will be applied for example for integral functionals with d the L^p-topology and d' the strong $W^{1,p}$-topology.

2.4. *Properties of Γ-convergence*

DEFINITION 2.9. We will say that a sequence $F_\varepsilon : X \to \overline{\mathbb{R}}$ is *equicoercive* if for all $t \in \mathbb{R}$ there exists a compact set K_t such that $\{F_\varepsilon \leqslant t\} \subset K_t$.

If needed, we will also use the more general definition: for all $\varepsilon_j \to 0$ and x_j such that $F_{\varepsilon_j}(x_j) \leqslant t$, there exist a subsequence of j (not relabeled) and a converging sequence x'_j such that $F_{\varepsilon_j}(x'_j) \leqslant F_{\varepsilon_j}(x_j) + \mathrm{o}(1)$.

We can state the main convergence result of Γ-convergence, whose proof has been shown in the Introduction.

THEOREM 2.10 (Fundamental theorem of Γ-convergence). *Let (X, d) be a metric space, let (F_ε) be a equicoercive sequence of functions on X and let $F = \Gamma\text{-}\lim_{\varepsilon \to 0} F_\varepsilon$. Then*

$$\exists \min_X F = \lim_{\varepsilon \to 0} \inf_X F_\varepsilon. \tag{2.18}$$

Moreover, if (x_ε) is a precompact sequence such that $\lim_{\varepsilon \to 0} F_\varepsilon(x_\varepsilon) = \lim_{\varepsilon \to 0} \inf_X F_\varepsilon$, then every limit of a subsequence of (x_ε) is a minimum point for F.

REMARK 2.11 (Γ-convergence as a choice criterion). If in the theorem above all functions F_ε admit a minimizer x_ε then, up to subsequences, x_ε converge to a minimum point of F. The converse is clearly not true: we may have minimizers of F which are not limits of minimizers of F_ε. A trivial example is $F_\varepsilon(t) = \varepsilon t^2$ on the real line. This situation is not exceptional; on the contrary: we may often view some functional as a Γ-limit of some particular perturbations, and single out from its minima those chosen as limits of minimizers.

Γ-limits of monotone sequences.   We can state some simple but important cases when the Γ-limit does exist, and it is easily computed.

REMARK 2.12. (i) If $F_{j+1} \leqslant F_j$ for all $j \in \mathbb{N}$, then

$$\Gamma\text{-}\lim_j F_j = \overline{\left(\inf_j F_j\right)} = \overline{\left(\lim_j F_j\right)}. \tag{2.19}$$

In fact, as $F_j \to \inf_k F_k$ pointwise, by Remark 2.6 we have $\Gamma\text{-}\limsup_j F_j \leqslant \overline{(\inf_k F_k)}$, while the other inequality comes trivially from the inequality $\overline{(\inf_k F_k)} \leqslant \inf_k F_k \leqslant F_j$;
   (ii) if $F_j \leqslant F_{j+1}$ for all $j \in \mathbb{N}$, then

$$\Gamma\text{-}\lim_j F_j = \sup_j \overline{F}_j = \lim_j \overline{F}_j; \tag{2.20}$$

in particular, if F_j is l.s.c. for every $j \in \mathbb{N}$, then

$$\Gamma\text{-}\lim_j F_j = \lim_j F_j. \tag{2.21}$$

In fact, since $\overline{F}_j \to \sup_k \overline{F}_k$ pointwise,

$$\Gamma\text{-}\limsup_j F_j = \Gamma\text{-}\limsup_j \overline{F}_j \leqslant \sup_k \overline{F}_k$$

by Remark 2.6. On the other hand, $\overline{F}_k \leqslant F_j$ for all $j \geqslant k$ so that the converse inequality easily follows.

Γ-limits and pointwise properties.

PROPOSITION 2.13. *If each element of the family (F_ε) is positively homogeneous of degree d (resp., convex, a quadratic form) then their Γ-limit is F_0 is positively homogeneous of degree d (resp., convex, a quadratic form).*

PROOF. The proof follows directly from the definition. The only care is to note that $F : X \to [0, +\infty]$ is a quadratic form if and only if $F(0) = 0$, $F(x + x') + F(x - x') \leqslant 2(F(x) + F(x'))$ and $F(tx) \leqslant t^2 F(x)$ for all $x, x' \in X$ and $t > 0$. $\square$

2.4.1. *Topological properties of Γ-convergence.*

PROPOSITION 2.14 (Compactness). *Let (X, d) be a separable metric space, and for all $j \in \mathbb{N}$ let $F_j : X \to \overline{\mathbb{R}}$ be a function. Then there is an increasing sequence of integers (j_k) such that the Γ-$\lim_k F_{j_k}(x)$ exists for all $x \in X$.*

PROOF. The proof follows easily from the topological definition (2.1) by extracting a subsequence such that $\inf_{y \in U} F_{j_k}(y)$ converges in a countable basis of open sets U. $\qquad\square$

PROPOSITION 2.15 (Urysohn property). *We have Γ-$\lim_j F_j = F$ if and only if for every subsequence (f_{j_k}) there exists a further subsequence which Γ-converges to F.*

REMARK 2.16 (Metrizability). Γ-convergence on spaces of lower-semicontinuous functions satisfying some uniform equicoerciveness properties is metrizable (see [84], Chapter 10, for more detailed statements). This property is often useful, for example in the definition of "diagonal" Γ-converging sequences.

2.5. *Development by Γ-convergence*

In many cases a first Γ-limit provides a functional with a lot of minimizers. In this case a further "Γ-limit of higher order", with a different scaling, may bring more information, as formalized in this result by Anzellotti and Baldo [25] (see also [11,26]).

THEOREM 2.17 (Development by Γ-convergence). *Let $F_\varepsilon : X \to \overline{\mathbb{R}}$ be a family of d-equicoercive functions and let $F^0 = \Gamma(d)$-$\lim_{\varepsilon \to 0} F_\varepsilon$. Let $m_\varepsilon = \inf F_\varepsilon$ and $m^0 = \min F^0$. Suppose that, for some $\delta_\varepsilon > 0$ with $\delta_\varepsilon \to 0$, there exists the Γ-limit*

$$F^1 = \Gamma(d')\text{-}\lim_{\varepsilon \to 0} \frac{F_\varepsilon - m^0}{\delta_\varepsilon}, \tag{2.22}$$

and that the sequence $F^1_\varepsilon = (F_\varepsilon - m^0)/\delta_\varepsilon$ is d'-equicoercive for a metric d' which is not weaker than d. Define $m^1 = \min F^1$ and suppose that $m^1 \neq +\infty$; then we have that

$$m_\varepsilon = m^0 + \delta_\varepsilon m^1 + o(\delta_\varepsilon) \tag{2.23}$$

and from all sequences (x_ε) such that $F_\varepsilon(x_\varepsilon) - m_\varepsilon = o(\delta_\varepsilon)$ (in particular this holds for minimizers, if any) there exists a subsequence converging in (X, d') to a point x which minimizes both F^0 and F^1.

3. Localization methods

The abstract compactness properties of Γ-convergence (Proposition 2.14) always ensure the existence of a Γ-limit, upon passing to a subsequence, but in general the limit function defined in this way remains an abstract object, that needs more information to be satisfactorily identified. In applications to minimum problems of the Calculus of Variations, we often encounter functionals as volume or surface integrals depending on the "local" behavior of some function u; e.g., *integral functionals* of the form

$$F_\varepsilon(u) = \int_\Omega f_\varepsilon(x, Du(x))\,\mathrm{d}x, \quad u \in W^{1,p}(\Omega; \mathbb{R}^m) \tag{3.1}$$

(see Section 4), or *free-discontinuity energies*

$$F_\varepsilon(u) = \int_\Omega f_\varepsilon(x, Du(x))\,\mathrm{d}x + \int_{S(u)} \varphi(x, u^+ - u^-, v_u)\,\mathrm{d}\mathcal{H}^{n-1} \tag{3.2}$$

for $u \in SBV(\Omega; \mathbb{R}^m)$ (see Section 10). These energies can be *localized* on open subsets A of Ω; i.e., we may define

$$F_\varepsilon(u, A) = \int_A f_\varepsilon(x, Du(x))\,\mathrm{d}x, \quad u \in W^{1,p}(\Omega; \mathbb{R}^m), \tag{3.3}$$

and

$$F_\varepsilon(u, A) = \int_A f_\varepsilon(x, \nabla u(x))\,\mathrm{d}x + \int_{S(u) \cap A} \varphi_\varepsilon(x, u^+ - u^-, v_u)\,\mathrm{d}\mathcal{H}^{n-1} \tag{3.4}$$

($u \in SBV(\Omega; \mathbb{R}^m)$), respectively.

The essential property defining *local functionals* $F(u, A)$ is that

$$F(u, A) = F(v, A) \quad \text{if } u = v \text{ a.e. on } A. \tag{3.5}$$

For the rest of the section, with the examples above in mind we suppose to have a sequence of functionals $F_j(u)$ (i.e., we may fix a subsequence (F_{ε_j}) of some (F_ε)) that may be "localized" by defining $F_j(u, A)$ for all open subsets of some open set Ω.

3.1. *Supremum of measures*

The localization methods can be used to simplify the computation of lower bounds. A simple but useful observation is that if F_j are local, then, for fixed u, the set function $A \mapsto F'(u, A) := \Gamma\text{-}\liminf_j F_j(u, A)$ is a superadditive set function on open sets with disjoint compact closures; i.e.,

$$F'(u, A \cup B) \geqslant F'(u, A) + F'(u, B)$$

if $\overline{A} \cap \overline{B} = \emptyset$, $\overline{A} \cup \overline{B} \subset\subset \Omega$. This inequality directly derives from the definition of Γ-liminf since test functions for $F'(u, A \cup B)$ can be used as test functions for both $F'(u, A)$ and $F'(u, B)$.

If we have a family of lower bounds of the form

$$F'(u, A) \geqslant G_i(u, A) =: \int_A \psi_i \, d\lambda,$$

where λ is a positive measure and ψ_i are positive Borel functions, then we can apply to $\mu(A) = F'(u, A)$ the following general lemma (see, e.g., [46], Lemma 15.2).

LEMMA 3.1 (Supremum of a family of measures). *Let μ be a function defined on the family of open subsets of Ω which is superadditive on open sets with disjoint compact closures, let λ be a positive measure on Ω, let ψ_i be positive Borel functions such that $\mu(A) \geqslant \int_A \psi_i \, d\lambda$ for all open sets A and let $\psi(x) = \sup_i \psi_i(x)$. Then $\mu(A) \geqslant \int_A \psi \, d\lambda$ for all open sets A.*

3.2. *The blow-up technique*

The procedure described in Section 3.1 highlights that for local functionals the liminf inequality can be itself localized on open subsets. Another type of localization argument is by the "blow-up" technique introduced by Fonseca and Müller [103] (see also [43]). It applies to the lower estimate along a sequence $F_j(u_j)$ with $u_j \to u$, for energies that for fixed j can be written as measures; i.e., $F_j(u_j, A) = \mu_j(A)$. For the functionals in (3.3) and (3.4), we have $\mu_j = f_{\varepsilon_j}(x, Du_j)\mathcal{L}^n$ and $\mu_j = f_{\varepsilon_j}(x, \nabla u_j)\mathcal{L}^n + \varphi_{\varepsilon_j}(x, u_j^+ - u_j^-, \nu_{u_j})\mathcal{H}^{n-1} \llcorner S(u_j)$, respectively.

Step 1: Definition of a limit measure. If $\liminf_j F_j(u_j)$ is finite (which is the nontrivial case) then we deduce that the family of measures (μ_j) is finite and hence, up to subsequences, we may suppose it converges weakly* to some measure μ. We fix some measure λ (whose choice is driven by the target function u) and consider the decomposition $\mu = (d\mu/d\lambda)\lambda + \mu^s$ in a part absolutely continuous with respect to λ and a singular part. In the case of Sobolev functionals and $u \in W^{1,p}(\Omega; \mathbb{R}^m)$ we expect the limit to be again an integral of the same type and we choose $\lambda = \mathcal{L}^n$; for free-discontinuity energies instead we expect the limit to have an additional term concentrated on $S(u)$, and we may choose $\lambda = \mathcal{L}^n$ or $\lambda = \mathcal{H}^{n-1} \llcorner S(u)$.

Step 2: Local analysis. We fix a "meaningful" x_0; i.e., such that x_0 is a Lebesgue point for μ with respect to λ, i.e.,

$$\frac{d\mu}{d\lambda}(x_0) = \lim_{\rho \to 0} \frac{\mu(x_0 + \rho D)}{\lambda(x_0 + \rho D)},$$

where D is a suitable open set properly chosen for the problem. In the case when $\lambda = \mathcal{L}^n$, for example, we may choose $D = (-1/2, 1/2)^n$ so that $\lambda(x_0 + \rho D) = \rho^n$. Note that for all ρ except for a countable set, we have $\mu(\partial(x_0 + \rho D)) = 0$, and hence $\mu(x_0 + \rho D) = \lim_j \mu_j(x_0 + \rho D) = \lim_j F_j(u_j, x_0 + \rho D)$; for $\lambda = \mathcal{H}^{n-1} \llcorner S(u)$ we choose D a cube with one face orthogonal to the normal vector $\nu_u(x_0)$ to $S(u)$ at x_0, so that $\lambda(x_0 + \rho D) = \rho^{n-1} + o(\rho^{n-1})$.

Step 3: *Blow up.* We choose $\rho_j \to 0$ such that we still have

$$\frac{d\mu}{d\lambda}(x_0) = \lim_j \frac{F_j(u_j, x_0 + \rho_j D)}{\lambda(x_0 + \rho_j D)},$$

and change our variables obtaining functionals

$$G_j(v_j, D) = \lambda(x_0 + \rho_j D)^{-1} F_j(u_j, x_0 + \rho_j D).$$

Up to a proper choice of scaling we may suppose that v_j converges to a meaningful v_0. In the case $\lambda = \mathcal{L}^n$, we have $v_j(x) := 1/\rho_j (u_j(x_0 + \rho_j x) - u_j(x_0)) \to \langle \nabla u(x_0), x - x_0 \rangle =: v_0(x)$; in the case $\lambda = \mathcal{H}^{n-1} \llcorner S(u)$, we choose $v_j(x) = u_j(x_0 + \rho_j x)$, converging to the function taking the only two values $u^\pm(x_0)$ jumping across the linear hyperplane orthogonal to $\nu_u(x_0)$.

Step 4: *Local estimates.* At this point we only have to estimate the limit of the scaled energies $G_j(v_j, D)$ when v_j converges to a simple target v_0. This is done in different ways depending on the type of energies, obtaining then an inequality $\frac{d\mu}{d\lambda}(x_0) \geqslant \varphi^\lambda(x_0)$, and some formulas linking $\varphi^\lambda(x_0)$ to the local behavior of u at x_0. In the case $\lambda = \mathcal{L}^n$, $\varphi^\lambda(x_0) = f_0(x, \nabla u(x_0))$; if $\lambda = \mathcal{H}^{n-1} \llcorner S(u)$ then $\varphi^\lambda(x_0) = \varphi_0(x, u^+(x_0) - u^-(x_0), \nu_u(x_0))$.

Step 5: *Global estimates.* We integrate the local estimates in Step 4. For integral functionals, for example, we then conclude that

$$F_0(u, \Omega) = \mu(\Omega) \geqslant \int_\Omega \frac{d\mu}{d\lambda} d\lambda = \int_\Omega f_0(x, \nabla u(x_0)) dx. \tag{3.6}$$

3.3. *A general compactness procedure*

If the functionals F_ε Γ-converge to some F, we may look at the behavior of the localized limit functionals $F(u, A)$ both with respect to u and A, and deduce enough information to give a description of F (e.g., that it is itself an integral in the cases deriving from functionals as in (3.3), (3.4)). The great advantage of this piece of information is that it reduces the computation of a particular Γ-limit within that class to the pointwise characterization of its energy densities. We briefly describe a procedure introduced by De Giorgi [90] that leads, e.g., to compactness results for classes of integrals in (3.3), (3.4) (see [54,84] for more details).

Step 1: *Compactness.* The first step is to apply the compactness result (not only on Ω, but also) on a dense countable family $\mathcal{V}$ of open subsets of Ω. For example, we can choose as $\mathcal{V}$ the family of all unions of open polyrectangles with rational vertices. Since $\mathcal{V}$ is countable, by a diagonal argument, upon extracting a subsequence, we can suppose that all $F_j(\cdot, A)$ Γ-converge for $A \in \mathcal{V}$. We denote by $F_0(\cdot, A)$ the Γ-limit, whose form may a priori depend on A.

Step 2: *Inner regularization.* The next idea is to consider $F_0(u, \cdot)$ as a set function and prove some properties that lead to some (integral) representation. The first property is "inner regularity" (see below). In general there may be exceptional sets where this property is not valid; hence, in place of F_0, we define the set function $\overline{F}_0(u, \cdot)$, the *inner-regular envelope* of F_0, on all open subsets of Ω by setting

$$\overline{F}_0(u, A) = \sup\{F_0(u, A'): A' \in \mathcal{V}, A' \subset\subset A\}.$$

In this way, $\overline{F}_0(u, \cdot)$ is automatically *inner regular*; i.e., $\overline{F}_0(u, A) = \sup\{\overline{F}_0(u, A'): A' \subset\subset A\}$. An alternative approach is directly proving that $F_0(u, \cdot)$ can be extended to an inner-regular set function (which is not always the case).

Step 3: *Subadditivity.* A crucial property (see Step 4 below) of $\overline{F}_0$ is subadditivity; i.e., that

$$\overline{F}_0(u, A \cup B) \leqslant \overline{F}_0(u, A) + \overline{F}_0(u, B)$$

(which is enjoyed for example by nonnegative integral functionals). This is usually the most technical part to prove that may involve a complex analysis of the behavior of the functionals F_j.

Step 4: *Measure property.* The next step is to prove that $\overline{F}_0(u, \cdot)$ is the restriction of a finite Borel measure to the open sets of Ω. To this end it is customary to use the *De Giorgi–Letta measure criterion* (see below).

Step 5: *Integral representation.* Since $\overline{F}_0(u, \cdot)$ is (the restriction of) a measure, we may write it as an integral. For example, in the case of Sobolev spaces if $\overline{F}_0(u, \cdot)$ is absolutely continuous with respect to the Lebesgue measure, then it can be written as

$$\overline{F}_0(u, A) = \int_A f_u(x)\, dx;$$

subsequently, by combining the properties of $\overline{F}_0$ as a set function with those with respect to u we deduce that indeed we may write $f_u(x) = f(x, Du(x))$.

This step is usually summarized in separate integral representation theorems that state that a local functional $F_0(u, A)$ satisfying suitable growth conditions, that is lower semi-continuous in u and that is a measure in A, can be written as an integral functional (see Section 4.1.1 for the case of functionals on Sobolev spaces).

Step 6: *Recovery of the Γ-limit.* The final step is to check that, taking $A = \Omega$, indeed $\overline{F}_0(u, \Omega) = F_0(u, \Omega)$ so that the representation we have found holds for the Γ-limit (and not for its "inner regularization"). This last step is an inner regularity result on Ω and for some classes of problems is sometime directly proved in Step 2.

REMARK 3.2 (Fundamental estimate). The subadditivity property in Step 3 is often derived by showing that the sequence F_j satisfies the so-called *fundamental estimate*. In the case of Γ-limits with respect to the L^p-convergence this is stated as follows: for all A, A', B open subsets of Ω with $A' \subset\subset A$, and for all $\sigma > 0$, there exists $M > 0$ such that for all u, v in the domain of F_j one may find a function w such that $w = u$ in A', $w = v$ on $B \setminus A$ such that

$$F_j(w, A' \cup B)$$
$$\leqslant (1+\sigma)\bigl(F_j(u, A) + F_j(v, B)\bigr) + M \int_{(A \cap B) \setminus A'} |u - v|^p \, dx + \sigma, \tag{3.7}$$

and

$$\|u - w\|_{L^p} + \|v - w\|_{L^p} \leqslant C \|u - v\|_{L^p}.$$

In the case of functionals on Sobolev spaces, w is usually of the form $\varphi u + (1 - \varphi)v$ with $\varphi \in C_0^\infty(A; [0, 1])$, $\varphi = 1$ in A', but can also be constructed differently (e.g., solving some auxiliary minimum problem in $A \setminus A'$ with data 0 on ∂A and 1 on $\partial A'$).

The subadditivity of $\overline{F}_0$ in Step 3 above is easily proved from this property by directly using the definition of Γ-convergence.

The proof of the following characterization of measures can be found in [94] (see also [54]).

LEMMA 3.3 (De Giorgi–Letta measure criterion). *If a set function α defined on all open subsets of a set Ω satisfies*
    (i) $\alpha(A) \leqslant \alpha(B)$ *is $A \subset B$ (α is increasing)*;
    (ii) $\alpha(A) = \sup\{\alpha(B) \colon B \subset\subset A\}$ *(α is inner regular)*;
    (iii) $\alpha(A \cup B) \leqslant \alpha(A) + \alpha(B)$ *(α is subadditive)*;
    (iv) $\alpha(A \cup B) \geqslant \alpha(A) + \alpha(B)$ *if $A \cap B = \emptyset$ (α is superadditive)*,
then α is the restriction to all open sets of Ω of a regular Borel measure.

3.4. *The "slicing" method*

In this subsection we describe a fruitful method to recover the liminf a inequality for Γ-limits through the study of one-dimensional problems by a "sectioning" argument. An example of application of this procedure will be given by the proof of Theorem 7.3.

The main idea of this method is the following. Let F_ε be a sequence of functionals defined on a space of functions with domain a fixed open set $\Omega \subset \mathbb{R}^n$. Then we may examine the behavior of F_ε on one-dimensional sections as follows: for each $\xi \in S^{n-1}$ we consider the hyperplane

$$\Pi_\xi := \bigl\{ z \in \mathbb{R}^n \colon \langle z, \xi \rangle = 0 \bigr\} \tag{3.8}$$

passing through 0 and orthogonal to ξ. For each $y \in \Pi_\xi$ we then obtain the one-dimensional set

$$\Omega_{\xi,y} := \{t \in \mathbb{R}: \ y + t\xi \in \Omega\}, \tag{3.9}$$

and for all u defined on Ω, we define the one-dimensional function

$$u_{\xi,y}(t) = u(y + t\xi) \tag{3.10}$$

defined on $\Omega_{\xi,y}$. We may then give a lower bound for the Γ-liminf of F_ε by looking at the limit of some functionals "induced by F_ε" on the one-dimensional sections.

Step 1. *We "localize" the functional F_ε highlighting its dependence on the set of integration. This is done by defining functionals $F_\varepsilon(\cdot, A)$ for all open subsets $A \subset \Omega$ as in* (3.3) *and* (3.4).

Step 2. *For all $\xi \in S^{n-1}$ and for all $y \in \Pi_\xi$, we find functionals $F_\varepsilon^{\xi,y}(v, I)$, defined for $I \subset \mathbb{R}$ and $v \in L^1(I)$, such that setting*

$$F_\varepsilon^\xi(u, A) = \int_{\Pi_\xi} F_\varepsilon^{\xi,y}(u_{\xi,y}, A_{\xi,y}) \, d\mathcal{H}^{n-1}(y) \tag{3.11}$$

we have $F_\varepsilon(u, A) \geqslant F_\varepsilon^\xi(u, A)$. This is usually an application of Fubini's theorem.

Step 3. *We compute the Γ-$\liminf_{\varepsilon \to 0} F_\varepsilon^{\xi,y}(v, I) = F^{\xi,y}(v, I)$ and define*

$$F^\xi(u, A) = \int_{\Pi_\xi} F^{\xi,y}(u_{\xi,y}, A_{\xi,y}) \, d\mathcal{H}^{n-1}(y). \tag{3.12}$$

Step 4. *Apply Fatou's lemma. If $u_\varepsilon \to u$, we have*

$$\liminf_{\varepsilon \to 0^+} F_\varepsilon(u_\varepsilon, A) \geqslant \liminf_{\varepsilon \to 0^+} F_\varepsilon^\xi(u_\varepsilon, A)$$

$$= \liminf_{\varepsilon \to 0^+} \int_{\Pi_\xi} F_\varepsilon^{\xi,y}\big((u_\varepsilon)_{\xi,y}, A_{\xi,y}\big) \, d\mathcal{H}^{n-1}(y)$$

$$\geqslant \int_{\Pi_\xi} \liminf_{\varepsilon \to 0^+} F_\varepsilon^{\xi,y}\big((u_\varepsilon)_{\xi,y}, A_{\xi,y}\big) \, d\mathcal{H}^{n-1}(y)$$

$$\geqslant \int_{\Pi_\xi} F^{\xi,y}(u_{\xi,y}, A_{\xi,y}) \, d\mathcal{H}^{n-1}(y) = F^\xi(u, A).$$

Hence, we deduce that Γ-$\liminf_{\varepsilon \to 0^+} F_\varepsilon(u, A) \geqslant F^\xi(u, A)$ for all $\xi \in S^{n-1}$.

Step 5. *Describe the domain of the limit.* From the estimates above and some directional characterization of function spaces we deduce the domain of the Γ-liminf. For example, if

$F^\xi(u, A) \geq \int_A |\langle Du, \xi\rangle|^p \, dx$ for $p > 1$ for ξ in a basis of $\mathbb{R}^n$ we deduce that the limit is a $W^{1,p}$-function.

Step 6. *Optimize the lower estimate.* We finally deduce that $F'(u, A) \geq \sup\{F^\xi(u, A): \xi \in S^{n-1}\}$. If the latter supremum is obtained also by restricting to a countable family $(\xi_i)_i$ of directions, if possible we use Lemma 3.1 to get an explicit form of a lower bound.

4. Local integral functionals on Sobolev spaces

The most common functionals encountered in the treatment of Partial Differential Equations are integral functionals defined on some subset of a Sobolev space; i.e., of the form (we limit to first derivatives)

$$F_\varepsilon(u) = \int_\Omega g_\varepsilon(x, u, Du) \, dx.$$

It must be previously noted that if we can isolate the explicit dependence on u; i.e., if we can write

$$F_\varepsilon(u) = \int_\Omega f_\varepsilon(x, Du) \, dx + \int_\Omega h(x, u) \, dx,$$

and h is a Carathéodory function satisfying a growth condition of the form $|h(x, u)| \leq c(\alpha(x) + |u|^p)$ with $\alpha \in L^1(\Omega)$, then the second integral is a continuous functional on L^p, and can therefore be dropped when computing the Γ-limit in the L^p topology. Note that this observation in particular applies to the continuous linear perturbations

$$u \mapsto \int_\Omega \langle \psi, u \rangle \, dx,$$

where $\psi \in L^{p'}(\Omega; \mathbb{R}^m)$.

4.1. *A prototypical compactness theorem*

Integral functionals of the form

$$F(u) = \int_\Omega f(x, Du) \, dx, \quad u \in W^{1,p}(\Omega; \mathbb{R}^m), \tag{4.1}$$

$$c_1\left(|\xi|^p - 1\right) \leq f(x, \xi) \leq c_2\left(|\xi|^p + 1\right) \tag{4.2}$$

($f \colon \Omega \times \mathbb{M}^{m \times n} \to \mathbb{R}$, Borel function) with Ω a bounded open subset of $\mathbb{R}^n$, $p > 1$ and $c_i > 0$, represent a "classical" class of energies for which the general compactness procedure in Section 3.3 can be applied. For minimum problems of the form

$$\min_{u = \varphi \text{ on } \partial\Omega} F(u), \qquad \min\left\{ F(u) - \int_{\Omega} \langle \psi, u \rangle \, dx \colon u \in W^{1,p}(\Omega; \mathbb{R}^m) \right\} \tag{4.3}$$

(with the usual conditions of Ω connected and $\int_{\Omega} \psi \, dx = 0$ in the second case) we easily infer that minimizers satisfy some $W^{1,p}$ bound depending only on c_i, φ or ψ. Hence the correct convergence to use in the computation of the Γ-limit is the weak convergence in $W^{1,p}$ or, equivalently by Rellich's theorem, the strong L^p convergence.

REMARK 4.1 (Growth conditions). The growth conditions (4.2) can be relaxed to cover more general cases. The case $p = 1$ can be dealt with in the similar way, but in this case the natural domain for these energies is the space of functions with bounded variation $BV(\Omega; \mathbb{R}^m)$ on which the relaxation of F takes a more complex form (see [20]), whose description goes beyond the scopes of this presentation. Moreover, we may also deal with conditions of the form

$$c_1\left(|\xi|^p - 1\right) \leqslant f(x, \xi) \leqslant c_2\left(|\xi|^q + 1\right) \tag{4.4}$$

if the gap between p and q is not too wide. The methods in this chapter work exactly the same if $q < p^*$ by the Sobolev embedding theorem, but wider gaps can also be treated (see the book by Fonseca and Leoni [102]). Outside the convex context, a long-standing conjecture that I learned from De Giorgi is that it should be sufficient to deal with the class of f that satisfy a condition

$$c_1\left(\psi(\xi) - 1\right) \leqslant f(x, \xi) \leqslant c_2\left(\psi(\xi) + 1\right), \tag{4.5}$$

where ψ is such that $\Psi(u) = \int_{\Omega} \psi(Du) \, dx$ is lower semicontinuous. This is a completely open problem, in particular because in general we do not have a characterization of good dense sets in the domain of Ψ.

THEOREM 4.2 (Compactness of local integral energies). *Let $p > 1$ and let f_j be a sequence of Borel functions uniformly satisfying the growth condition (4.2). Then there exist a subsequence of f_j (not relabeled) and a Carathéodory function f_0 satisfying the same condition (4.2) such that, if we set*

$$F_0(u, A) = \int_A f_0(x, Du) \, dx, \quad u \in W^{1,p}(A; \mathbb{R}^m), \tag{4.6}$$

and the localized functionals defined by

$$F_j(u, A) = \int_A f_j(x, Du) \, dx, \quad u \in W^{1,p}(A; \mathbb{R}^m), \tag{4.7}$$

then $F_j(\cdot, A)$ converges to $F_0(\cdot, A)$ with respect to the $L^p(A; \mathbb{R}^m)$ convergence for all A open subsets of Ω.

PROOF. To prove the theorem above the steps in Section 3.3 can be followed. In this case the Γ-limit can be proved to be inner regular (this can be done by using the argument in Section 4.2.1, which also proves the subadditivity property) and we may use the integral representation result in the next section. $\qquad\square$

REMARK 4.3 (Convergence of minimum problems with Neumann boundary conditions). In the class above we immediately obtain the convergence of minimum problems as the second one in (4.3). The only thing to check is equicoerciveness, which follows from the Poincaré–Wirtinger inequality. Note in fact that we may reduce to the case $\int u \, dx = 0$ upon a translation of a constant vector.

4.1.1. *An integral representation result.* The prototype of the integral representation results is the following classical theorem in Sobolev spaces due to Buttazzo and Dal Maso (see [54,66]).

THEOREM 4.4 (Sobolev integral representation theorem). *If $F = F(u, A)$ is a functional defined for $u \in W^{1,p}(\Omega; \mathbb{R}^m)$ and A open subset of Ω satisfying*

    (i) *(lower semicontinuity) $F(\cdot, A)$ is lower semicontinuous with respect to the L^p convergence;*

    (ii) *(growth estimate) $0 \leqslant F(u, A) \leqslant C \int_A (1 + |Du|^p) \, dx$;*

    (iii) *(measure property) $F(u, \cdot)$ is the restriction of a regular Borel measure;*

    (iv) *(locality) F is local: $F(u, A) = F(v, A)$ if $u = v$ a.e. on A, then there exists a Borel function f such that*

$$F(u, A) = \int_A f(x, Du) \, dx.$$

REMARK 4.5. Note that $f(x, \xi)$ can be obtained by derivation for all $\xi \in \mathbb{M}^{m \times n}$ and a.a. $x \in \Omega$ as

$$f(x, \xi) = \lim_{\rho \to 0^+} \frac{F(u_\xi, B_\rho(x))}{|B_\rho(x)|}, \tag{4.8}$$

where $u_\xi(y) = \xi y$.

A remark by Dal Maso and Modica shows that this formula can be used to give an indirect description of f_0 from f_j in Theorem 4.2, by introducing the functions

$$M_j(x, \xi, \rho) := \min_{w \in W_0^{1,p}(B_\rho(x); \mathbb{R}^m)} \int_{B_\rho(x)} f_j(y, \xi + \nabla w) \, dy.$$

Then F_j Γ-converges to F_0 if and only if

$$f_0(x,\xi) = \liminf_{\rho \to 0} \liminf_j \frac{M_j(x,\xi,\rho)}{|B_\rho(x)|} = \limsup_{\rho \to 0} \limsup_j \frac{M_j(x,\xi,\rho)}{|B_\rho(x)|}$$

for almost every $x \in \Omega$ and every $\xi \in \mathbb{M}^{m \times n}$.

This fact can be proved by a blow-up argument, upon using the argument in Section 4.2.1 to match the boundary conditions.

REMARK 4.6 (Other classes of integral functionals). The compactness procedure above can also be applied to energies of the form

$$F_\varepsilon(u) = \int_\Omega f_\varepsilon(x, u, Du)\,\mathrm{d}x, \tag{4.9}$$

with f_ε satisfying analogous growth conditions. It must be remarked that more complex integral-representation results must be used, for which we refer to the book by Fonseca and Leoni [102].

4.1.2. *Convexity conditions.* It is useful to note that the Borel function f_0 in the compactness theorem enjoys some convexity properties in the gradient variable due to the fact that the Γ-limit is a lower-semicontinuous functional (more precisely, sequentially lower semicontinuous with respect to the weak convergence in $W^{1,p}$). The following theorem by Acerbi and Fusco [4] shows that the function $f_0(x, \cdot)$ satisfies Morrey's *quasiconvexity* condition [126].

THEOREM 4.7 (Quasiconvexity and lower semicontinuity). *Let f be a Borel function satisfying* (4.2) *and let $F(u) = \int_\Omega f(x, Du)\,\mathrm{d}x$ be defined on $W^{1,p}(\Omega; \mathbb{R}^m)$. Then F is sequentially weakly lower semicontinuous on $W^{1,p}(\Omega; \mathbb{R}^m)$ if and only if $f(x, \cdot)$ is quasiconvex; i.e., we have*

$$|D| f(x,\xi) = \min\left\{ \int_D f\big(x, \xi + D\varphi(y)\big)\,\mathrm{d}y : \varphi \in W_0^{1,p}(D; \mathbb{R}^m) \right\}, \tag{4.10}$$

where D is any open subset of $\mathbb{R}^n$.

REMARK 4.8. (1) In the scalar case, $m = 1$ or for curves $n = 1$ quasiconvexity reduces to the usual convexity and (4.10) to Jensen's inequality; in all other cases convexity is a restrictive condition. A family of nonconvex quasiconvex functions is that of *polyconvex* functions (i.e., convex functions of the minors of ξ; for example, $\det \xi$ if $m = n$) (see [29,81] and [54], Chapter 5).

(2) Quasiconvexity implies convexity on rank-1 lines, in particular, in the coordinate directions. This, together with the growth condition (4.2) implies a locally uniform Lipschitz condition on f_0 with a Lipschitz constant growing like $|\xi|^{p-1}$ (see [54], Remark 5.15).

(3) Condition (4.10) can be equivalently stated with the more general condition of φ periodic (and D a periodicity set for φ).

(4) The integrand of the lower-semicontinuous envelope of a functional as in (4.1) is given by the *quasiconvex envelope Qf of f*

$$Qf(x,\xi) = \min\left\{ |D|^{-1} \int_D f\big(x,\xi + D\varphi(y)\big)\,dy \colon \varphi \in W_0^{1,p}(D;\mathbb{R}^m) \right\} \qquad (4.11)$$

(see, e.g., [4,54,81]). Note that here x acts as a parameter.

4.2. *Useful technical results*

4.2.1. *The De Giorgi method for matching boundary values.*   From the Γ-convergence of functionals F_j to F_0 as in Theorem 4.2, we do not immediately deduce the convergence of minimum problems with Dirichlet boundary conditions. In fact, to do so we must prove the *compatibility* of the condition $u = \varphi$ on $\partial\Omega$; i.e., that the functionals,

$$F_j^\varphi(u) = \begin{cases} F_j(u) = \int_\Omega f_j(x, Du)\,dx & \text{if } u = \varphi \text{ on } \partial\Omega, \\ +\infty & \text{otherwise,} \end{cases}$$

Γ-converge to F_0^φ analogously defined. The liminf inequality is an immediate consequence of the stability of the boundary condition under weak $W^{1,p}$-convergence. It remains to prove an approximate limsup inequality; i.e., that for all $u \in \varphi + W_0^{1,p}$ and fixed $\eta > 0$ there exists a sequence $u_j \in \varphi + W_0^{1,p}$ such that $\limsup_j F_j(u_j) \leqslant F_0(u) + \eta$.

The proof of this fact will use a method introduced by De Giorgi [90]. From the Γ-convergence of the F_j we know that there exists a sequence $v_j \to u$ such that $F_0(u) = \lim_j F_j(v_j)$; in particular, the $W^{1,p}$-norms of v_j are equibounded. We wish to modify v_j close to the boundary. For simplicity suppose that the Lebesgue measure of $\partial\Omega$ is zero and all f_j are positive. We consider functions $w_j = \phi v_j + (1 - \phi)u$, where $\phi = 0$ on $\partial\Omega$ and $\phi(x) = 1$ if $\mathrm{dist}(x, \partial\Omega) > \eta$. Such functions tend to u and satisfy the desired boundary condition. However, we may obtain only the estimate

$$F_j(w_j) \leqslant F_j(v_j) + C \int_{\Omega^\eta} \left(|Du|^p + |Dv_j|^p + \frac{1}{\eta^p} |u - v_j|^p \right) dx, \qquad (4.12)$$

where $\Omega^\eta = \{x \in \Omega \colon \mathrm{dist}(x, \partial\Omega) < \eta\}$, and passing to the limit we get

$$\limsup_j F_j(w_j) \leqslant F_0(u) + C \int_{\Omega^\eta} \big(|Du|^p + |Dv_j|^p \big)\,dx, \qquad (4.13)$$

which is not sufficient to conclude since we do not know if we can choose $\int_{\Omega^\eta} |Dv_j|^p\,dx$ arbitrarily small. Note that this would be the case if $(|Dv_j|^p)$ were an equiintegrable sequence (see Section 4.2.2).

This choice of w_j can be improved by fixing an integer N and considering ϕ_k (for $k = 1, \ldots, N$) such that $\phi_k = 0$ on $\Omega^{(N+k-1)\eta/N}$, $\phi_k = 1$ on $\Omega \setminus \Omega^{(N+k)\eta/N}$ and $|D\phi_k| \leqslant N/\eta$. With this ϕ_k in place of ϕ we define w_j^k and obtain for each j the estimate

$$F_j\big(w_j^k\big) \leqslant F_j\big(v_j, \Omega \setminus \Omega^\eta\big) + C F_j\big(u, \Omega^{2\eta}\big)$$

$$+ C \int_{D_\eta^k} \left(|Dv_j|^p + \frac{N^p}{\eta^p} |u - v_j|^p \right) \mathrm{d}x, \tag{4.14}$$

where $D_\eta^k = \Omega^{(N+k)\eta/N} \setminus \Omega^{(N+k-1)\eta/N}$. Now, summing up for $k = 1, \ldots, N$, we obtain

$$\sum_{k=1}^{N} F_j\big(w_j^k\big) \leqslant N F_j\big(v_j, \Omega \setminus \Omega^\eta\big) + N C F_j\big(u, \Omega^{2\eta}\big)$$

$$+ C \int_{\Omega^{2\eta} \setminus \Omega^\eta} \left(|Dv_j|^p + \frac{N^p}{\eta^p} |u - v_j|^p \right) \mathrm{d}x, \tag{4.15}$$

and we may choose $k = k_j$ such that

$$F_j\big(w_j^k\big) \leqslant F_j\big(v_j, \Omega \setminus \Omega^\eta\big) + C F_j\big(u, \Omega^\eta\big)$$

$$+ \frac{C}{N} \int_{\Omega^{2\eta} \setminus \Omega^\eta} \left(|Dv_j|^p + \frac{N^p}{\eta^p} |u - v_j|^p \right) \mathrm{d}x. \tag{4.16}$$

If we define $u_j = w_j^{k_j}$ we then obtain

$$\limsup_j F_j(u_j) \leqslant F_0(u) + C \int_{\Omega^{2\eta}} \big(1 + |Du|^p\big) \, \mathrm{d}x + \frac{C}{N} \sup_j \int_{\Omega} |Dv_j|^p \, \mathrm{d}x, \tag{4.17}$$

which proves the approximate limsup inequality.

REMARK 4.9. (1) The idea of the method above consists in finding suitable "annuli", where the energy corresponding to $|Du_j|^p$ does not concentrate, and then taking *cut-off* functions with gradient supported in those annuli to "join" u_j and u through a convex combination. An alternative way to construct such annuli would be to consider sets that are not charged by the weak* limit of the measures $\mu_j = |Dv_j|^p \mathcal{L}^n$ (this method is for instance used in the book by Evans [96]).

(2) (Proof of the inner regularity and fundamental estimate.) We can use the method above to prove the inner regularity in Step 2 of Section 3.3. Note that we have $F_0(u, B) \leqslant c_2 \int_B (1 + |Du|^p) \, \mathrm{d}x$ for all B. Fix an open set A, $\eta > 0$ and set $A' = A \setminus A^\eta$ in the notation above. Then from inequality (4.16) with A in the place of Ω and (u_j) an optimal sequence

for $F_0(u, A')$, we have

$$F_0(u, A) \leqslant \limsup_j F_j(u_j, A)$$

$$\leqslant F_0(u, A') + C \int_{A^{2\eta}} (1 + |Du|^p) \, dx + \frac{C}{N} \tag{4.18}$$

so that we obtain the inner regularity of F_0 by the arbitrariness of η and N.

In a similar fashion we may use the same argument to "join" recovery sequences on sets A' and B and prove the fundamental estimate (Remark 3.2), and hence the subadditivity property of Step 3 of Section 3.3.

(3) The method described above is very general and can be extended also to varying domains (see, e.g., Lemma 6.1). In the scalar case $m = 1$ and with a fixed Ω a simpler truncation argument can be used (see [46], Section 2.7).

REMARK 4.10 (Convergence of minimum problems with Dirichlet boundary conditions). The compatibility of boundary conditions immediately implies the convergence of problems with Dirichlet boundary conditions from the Γ-convergence of the energies, as remarked in the Introduction.

4.2.2. *An equiintegrability lemma.*　As remarked in the proof exhibited in the previous section, sequences of functions with $|Du_j|^p$ equiintegrable are often easier to handle. A method introduced by Acerbi and Fusco [4] shows that this is essentially always the case, as stated by the following theorem due to Fonseca, Müller and Pedregal [104].

THEOREM 4.11 (Equivalent sequences with equiintegrability properties). *Let (u_j) be a sequence weakly converging to u in $W^{1,p}(\Omega; \mathbb{R}^m)$; then there exist a subsequence of (u_j), not relabeled, and a sequence (w_j) with $(|Dw_j|^p)$ equiintegrable such that*

$$\lim_j \left| \{ x \in \Omega : u_j(x) \neq w_j(x) \} \right| = 0$$

and still converging to u.

Thanks to this result we can limit our analysis to such (w_j) with $(|Dw_j|^p)$ equiintegrable. In fact, set $\Omega_j = \{u_j = w_j\}$ and note that $Du_j = Dw_j$ a.e. on Ω_j and $\lim_j \int_{\Omega \setminus \Omega_j} |Dw_j|^p \, dx = 0$ by the equiintegrability property. Then we have (for simplicity assume $f_j \geqslant 0$)

$$\liminf_j \int_\Omega f_j(x, Dw_j) \, dx$$

$$\leqslant \liminf_j \int_{\Omega_j} f_j(x, Dw_j) \, dx + \lim_j \int_{\Omega \setminus \Omega_j} c_2(1 + |Dw_j|^p) \, dx$$

$$= \liminf_j \int_{\Omega_j} f_j(x, Dw_j) \, dx = \liminf_j \int_{\Omega_j} f_j(x, Du_j) \, dx$$

$$\leqslant \liminf_j \int_\Omega f_j(x, Du_j) \, dx,$$

with c_2 given by (4.2), so that if we give a lower bound for such (w_j) we obtain a lower bound also for a general (u_j).

4.2.3. *Higher-integrability results.* When using the characterization of Γ-limits of integral functionals via Moreau–Yosida transforms with respect to the $L^p(\Omega; \mathbb{R}^m)$ convergence it is often useful to resort to some regularity properties of solutions of variational problems stated as follows.

THEOREM 4.12 (Meyers regularity theorem). *Let f be as in (4.2), let A be a bounded open set with smooth boundary and $\bar{u} \in C^\infty(\overline{A}; \mathbb{R}^m)$. Then there exists $\eta = \eta(c_1, c_2, A, \bar{u}) > 0$ such that for all $\lambda > 0$ any solution u_λ of*

$$\min\left\{ \int_A f(x, Du + D\bar{u})\,dx + \lambda \int_A |u|^p\,dx: u \in W^{1,p}(A; \mathbb{R}^m) \right\} \tag{4.19}$$

belongs to $W^{1,p+\eta}(A; \mathbb{R}^m)$, and there exists $C = C(\lambda, c_1, c_2, \Omega, \bar{u})$ such that

$$\|u_\lambda\|_{W^{1,p+\eta}(A;\mathbb{R}^m)} \leqslant C. \tag{4.20}$$

This theorem shows that for fixed λ minimizers of the Moreau–Yosida transforms related to a family F_ε as in (4.1) (see (2.11)) satisfy a uniform bound (4.20) independent of ε.

4.3. *Convergence of quadratic forms*

From the stability property of quadratic forms (Proposition 2.13), we have the following particular case of the compactness Theorem 4.2 (for simplicity we treat the scalar case $n = 1$ only).

THEOREM 4.13 (Compactness of quadratic forms). *Let $A_j : \Omega \to \mathbb{M}^{m \times n}$ be a sequence of symmetric matrix-valued measurable functions, and suppose that $\alpha, \beta > 0$ exist such that $\alpha \mathrm{Id} \leqslant A_j \leqslant \beta \mathrm{Id}$ for all j. Then there exist a subsequence of A_j, not relabeled, and a matrix-valued function A satisfying the same conditions, such that*

$$\int_\Omega \langle A(x)Du, Du\rangle\,dx = \Gamma\text{-}\lim_j \int_\Omega \langle A_j(x)Du, Du\rangle\,dx \tag{4.21}$$

with respect to the $L^2(\Omega)$-convergence, for all $u \in H^1(\Omega)$.

As a consequence of this theorem we have a result of convergence for the related Euler equations. Note that all functionals are strictly convex, so that the solutions to minimum problems with Dirichlet boundary conditions are unique.

COROLLARY 4.14 (*G-convergence*). *If A_j, A are as before then, for all $\varphi \in H^1(\Omega)$ and $f \in L^2(\Omega)$, the solutions u_j of*

$$\begin{cases} -\operatorname{div}(A_j Du_j) = f & \text{in } \Omega, \\ u_j - \varphi \in H_0^1(\Omega) \end{cases}$$

weakly converge in $H^1(\Omega)$ to the solution u of

$$\begin{cases} -\operatorname{div}(A Du) = f & \text{in } \Omega, \\ u - \varphi \in H_0^1(\Omega). \end{cases}$$

This is usually referred to as the G-convergence of the differential operators $G_j(v) = -\operatorname{div}(A_j Dv)$.

4.4. Degenerate limits

If the growth conditions of order p are not uniformly satisfied, then the limit of a family of integral functionals may take a different form and, in particular, lose the locality property. In this subsection we give two examples of such a case.

4.4.1. *Functionals of the sup norm.* A simple example of a family of functionals not satisfying uniformly a p-growth condition is the following:

$$F_\varepsilon(u) = \varepsilon \int_\Omega |a(x)Du|^{1/\varepsilon}\, dx, \quad u \in W^{1,1/\varepsilon}(\Omega), \tag{4.22}$$

where $a \in L^\infty(\Omega)$ and $\inf a > 0$. These functionals can be thought to be defined on $W^{1,1}(\Omega)$, and each F_ε satisfies a $1/\varepsilon$-growth condition. Limits of problems involving these functionals are described by the following result by Garroni, Nesi and Ponsiglione [113].

THEOREM 4.15. (i) *The functionals F_ε Γ-converge with respect to the L^1-convergence to the functional F_0 given by*

$$F_0(u) = \begin{cases} 0 & \text{if } \|a\,Du\|_\infty \leqslant 1, \\ +\infty & \text{otherwise.} \end{cases} \tag{4.23}$$

(ii) *The functionals G_ε given by $G_\varepsilon(u) = (F_\varepsilon(u))^\varepsilon$ Γ-converge with respect to the L^1-convergence to the functional G_0 given by*

$$G_0(u) = \|a\,Du\|_\infty. \tag{4.24}$$

PROOF. (i) The liminf inequality follows by noticing that if $\sup_\varepsilon F_\varepsilon(u_\varepsilon) < +\infty$ and

$u_\varepsilon \to u$ in L^1, then actually $u_\varepsilon \rightharpoonup u$ in $W^{1,q}(\Omega)$ for each $q > 1$, so that

$$\left|\{|a\,Du| > t\}\right| t^q \leqslant \int_\Omega |a\,Du|^q \, dx \leqslant \liminf_{\varepsilon \to 0} \int_\Omega |a\,Du_\varepsilon|^q \, dx$$

$$\leqslant \liminf_{\varepsilon \to 0} \left(\frac{1}{\varepsilon} F_\varepsilon(u)\right)^{\varepsilon q} = 1,$$

and we get $\left|\{|a\,Du| > t\}\right| = 0$ for all $t > 1$. A recovery sequence is trivially $u_\varepsilon = u$.

(ii) follows from the increasing convergence of $\varepsilon^{-\varepsilon} G_\varepsilon$ to G_0. $\qquad\square$

We can apply the result above to minimum problems of the form

$$m_\varepsilon = \min\left\{\int_\Omega |a(x)Du|^{1/\varepsilon} \, dx : u = \varphi \text{ on } \partial\Omega\right\}; \tag{4.25}$$

note however that the limit minimum problem (of the scaled functionals)

$$\min\left\{\|a(x)Du\|_\infty : u = \varphi \text{ on } \partial\Omega\right\} \tag{4.26}$$

possesses many solutions. Hence, the limit of the (unique) solutions of m_ε should be characterized otherwise (see, e.g., [30,80]).

For generalizations of this result we refer, e.g., to the paper by Champion, De Pascale and Prinari [73].

4.4.2. *The closure of quadratic forms.* We consider now the problem of characterizing the closure of all quadratic forms when the coefficients do not satisfy uniform bounds from above and below as in Theorem 4.13. To this end we have to introduce some definitions (for details see the book by Fukushima [109]).

DEFINITION 4.16 (Dirichlet form). A quadratic form F on $L^2(\Omega)$ is called a *Dirichlet form* if

(i) it is *closed*; i.e., its domain $\mathrm{Dom}(F)$ (where $F(u) = B(u,u) < +\infty$, B a bilinear form) endowed with the scalar product $(u,v)_F = B(u,v) + \int_\Omega uv \, dx$ is a Hilbert space;

(ii) it is *Markovian* (or *decreasing by truncature*); i.e., $F((u \vee 0) \wedge 1) \leqslant F(u)$ for all $u \in L^2(\Omega)$.

The following remark helps to get an intuition of a general Dirichlet form.

REMARK 4.17 (Deny–Beurling integral representation). A *regular Dirichlet form F* is such that $\mathrm{Dom}(F) \cap C^0(\Omega)$ is dense in both $C^0(\Omega)$ with respect to the uniform norm and $\mathrm{Dom}(F)$. Such F admits the representation

$$F(u) = \sum_{i,j} \int_\Omega D_i u\, D_j u\, d\mu_{ij} + \int_\Omega |u|^2 \, d\nu + \int_{\Omega \times \Omega} (u(x) - u(y))^2 \, d\mu \tag{4.27}$$

for $u \in \mathrm{Dom}(F) \cap C_0^1(\Omega)$, where μ_{ij}, ν and μ are Radon measures such that $\mu(\{(x,x): x \in \Omega\}) = 0$ and $\sum_{i,j} z_i z_j \mu_{ij}(K) \geqslant 0$ for all compact subsets K of Ω and $z \in \mathbb{R}^n$.

For the use of Dirichlet form for the study of asymptotic problems we refer to Mosco [127]. The following theorem is due to Camar-Eddine and Seppecher [68].

THEOREM 4.18 (Closure of quadratic forms). *Let $n \geqslant 3$. The closure with respect to the $L^2(\Omega)$-convergence of isotropic quadratic forms of diffusion type, i.e., of the form*

$$F_\alpha(u) = \int_\Omega \alpha(x)|Du|^2 \, dx, \quad u \in H^1(\Omega), \tag{4.28}$$

where $0 < \inf \alpha \leqslant \sup \alpha < +\infty$ (but not equibounded) is the set of all Dirichlet form that are objective; i.e., $F(u+c) = F(u)$ for all constants c.

We do not give a proof of this result, referring to the paper [68]. We only remark that the density of isotropic quadratic forms in all (coercive) quadratic forms can be obtained by local homogenization (see Remark 5.8). The prototype of a nonlocal term is $\mu = \delta_{x_0, y_0}$ in the Deny–Beurling formula; this can be reached by taking $\alpha \to +\infty$ on a set composed of two balls centered on x_0 and y_0 and a tubular neighborhood of the segment joining the two points with suitable (vanishing) radius. Note that the use of this construction is not possible in dimension two.

5. Homogenization of integral functionals

An important case of limits of integral functionals is that of energies within the theory of *homogenization*; i.e., when we want to take into account fast-oscillating inhomogeneities. The simpler way to model such a behavior is to consider a function $f: \mathbb{R}^n \times \mathbb{M}^{m \times n} \to \mathbb{R}$ *periodic* in the first variable (up to a change of basis), we may suppose it is T-periodic (if not otherwise specified $T = 1$); i.e.,

$$f(x + Te_i, \xi) = f(x, \xi) \quad \text{for all } x, \xi,$$

for all vectors e_i of the standard basis of $\mathbb{R}^n$, and examine the asymptotic behavior of energies

$$F_\varepsilon(u) = \int_\Omega f\left(\frac{x}{\varepsilon}, Du\right) dx, \quad u \in W^{1,p}(\Omega; \mathbb{R}^m). \tag{5.1}$$

Note that we may apply the general compactness Theorem 4.2 to any sequence (f_j), where $f_j(x, \xi) = f(x/\varepsilon_j, \xi)$, thus obtaining the existence of Γ-converging subsequences. The main issues here are:

(i) prove that the whole family (F_ε) Γ-converges;

(ii) give a description of the energy density of the Γ-limit in terms of the properties of f.

In this section we give a simple account of the main features of this problem referring to the book [54] for more details.

The natural ansatz for the Γ-limit of (F_ε) is that it is "homogeneous"; i.e., its energy density does not depend on x, so that it takes the form

$$F_{\mathrm{hom}}(u) = \int_\Omega f_{\mathrm{hom}}(Du)\,dx, \quad u \in W^{1,p}(\Omega;\mathbb{R}^m). \tag{5.2}$$

If such a Γ-limit exists it will be called the *homogenized* functional of F_ε. As a consequence of the theorem on the convergence of minimum problems, we obtain that families of problems with solutions with highly oscillating gradients are approximated by solutions of simpler problems with F_{hom} in place of F_ε, where oscillations are "averaged out".

5.1. *The asymptotic homogenization formula*

From the localization methods we can easily derive an ansatz for a formula describing f_{hom}. As a first remark, recall that f_{hom} is quasiconvex, so that it can be expressed as a minimum problem; e.g., choosing $D = (0,1)^n$ in (4.10), for all $\xi \in \mathbb{M}^{m \times n}$ we may write

$$f_{\mathrm{hom}}(\xi) = \min\left\{ \int_{(0,1)^n} f_{\mathrm{hom}}(\xi + D\varphi)\,dy \colon \varphi \in W_0^{1,p}((0,1)^n;\mathbb{R}^m) \right\}. \tag{5.3}$$

Now, from the convergence of minima and the compatibility of addition of boundary conditions, we obtain

$$f_{\mathrm{hom}}(\xi) = \lim_{\varepsilon \to 0} \inf\left\{ \int_{(0,1)^n} f\left(\frac{y}{\varepsilon}, \xi + D\varphi\right) dy \colon \varphi \in W_0^{1,p}((0,1)^n;\mathbb{R}^m) \right\}. \tag{5.4}$$

The final *asymptotic homogenization formula* is obtained from this by the change of variables $y = \varepsilon x$ that isolates the dependence on ε in a scaling argument (here we set $T = 1/\varepsilon$)

$$f_{\mathrm{hom}}(\xi) = \lim_{T \to +\infty} \frac{1}{T^n} \inf\left\{ \int_{(0,T)^n} f(x, \xi + D\varphi)\,dx \colon \varphi \in W_0^{1,p}((0,T)^n;\mathbb{R}^m) \right\}. \tag{5.5}$$

To make this ansatz into a theorem we need just to prove that the candidate f_{hom} is indeed homogeneous, and that the limit on the right-hand side exists. In this way we can use the compactness theorem, and prove that the limit is independent of the sequence (ε_j), being characterized by formula (5.4).

THEOREM 5.1 (Homogenization theorem). *Let f be as above. If F_ε are defined as in (5.1), then $\Gamma\text{-}\lim_{\varepsilon \to 0} F_\varepsilon = F_{\mathrm{hom}}$, given by (5.2) and (5.5) with respect to the $L^2(\Omega;\mathbb{R}^m)$ convergence.*

PROOF. The "homogeneity" of an $f_{\text{hom}} = f_{\text{hom}}(x, \xi)$ given by the compactness theorem can be easily obtained thanks to the ε-periodicity of F_ε, that ensures that $F_\varepsilon(u, A) = F_\varepsilon(u_y, y + A)$ for all open sets A and $y \in \varepsilon\mathbb{Z}^n$, where $u_y(x) = u(x - y)$. This implies, by a translation and approximation argument, that $F_{\text{hom}}(u, A) = F_{\text{hom}}(u_y, y + A)$ for all open sets A and $y \in \mathbb{R}^n$, so that $f_{\text{hom}}(x, \xi) = f_{\text{hom}}(x + y, \xi)$ by derivation (see (4.8)). The existence of the limit in (5.5) can be derived from the scaling argument in Remark 5.3. $\square$

PROPOSITION 5.2 (Asymptotic behavior of subadditive functions). *Let g be a function defined on finite unions of cubes of $\mathbb{R}^n$ which is subadditive (i.e., $g(A \cup B) \leqslant g(A) + g(B)$ if $|A \cap B| = 0$) such that $g(z + A) = g(A)$ for all $z \in \mathbb{Z}^n$ and $g(A) \leqslant c|A|$. Then there exists the limit*

$$\lim_{T \to +\infty} \frac{g((0, T)^n)}{T^n}. \tag{5.6}$$

PROOF. It suffices to check that if $S > T$ then we have $g((0, S)^n) \leqslant (S/T)^n g((0, T)^n) + C(T, S)$, with $\lim_{T \to +\infty} \lim_{S \to +\infty} C(T, S) = 0$, and then take the limsup in S first and eventually the liminf in T. $\square$

REMARK 5.3. To prove the existence of the limit in (5.5) it suffices to apply the previous proposition to

$$g(A) = \inf\left\{ \int_A f(x, \xi + D\varphi)\, dx \colon \varphi \in W_0^{1,p}(A; \mathbb{R}^m) \right\}. \tag{5.7}$$

5.1.1. *A periodic formula.* We can easily derive alternative formulas for f_{hom}, for instance taking periodic minimum problems. The following *asymptotic periodic formula* is due to Müller [128] (we suppose that f is 1-periodic)

$$f_{\text{hom}}(\xi) = \inf_{k \in \mathbb{N}} \frac{1}{k^n} \inf\left\{ \int_{(0,k)^n} f(x, \xi + D\varphi)\, dx \colon \varphi \in W_\#^{1,p}((0, k)^n; \mathbb{R}^m) \right\}, \tag{5.8}$$

where $W_\#^{1,p}((0, k)^n; \mathbb{R}^m)$ denotes the space of k-periodic functions in $W_{\text{loc}}^{1,p}(\mathbb{R}^n; \mathbb{R}^m)$. Note that since $W_\#^{1,p}((0, k)^n; \mathbb{R}^m) \subset W_\#^{1,p}((0, lk)^n; \mathbb{R}^m)$ for all $l \in \mathbb{N}$ with $l \geqslant 1$, the $\inf_k$ is actually a $\lim_k$. Moreover, since $W_0^{1,p}((0, k)^n; \mathbb{R}^m) \subset W_\#^{1,p}((0, k)^n; \mathbb{R}^m)$, the right-hand side of formula (5.8) is not greater than the value for $f_{\text{hom}}(\xi)$ given by (5.5). It remains to prove the opposite inequality. To this end we will make use of the following lemma, which is a fundamental tool for dealing with oscillating energies.

LEMMA 5.4 (Riemann–Lebesgue lemma). *Let g be an L_{loc}^1 periodic function of period Y, let $g_\varepsilon(x) = g(x/\varepsilon)$ and let $\bar{g} = |Y|^{-1} \int_Y g\, dy$. Then $g_\varepsilon \rightharpoonup \bar{g}$. In particular, $\int_D g_\varepsilon\, dx \to \bar{g}|D|$ for all bounded open subsets D.*

To conclude the proof of formula (5.8) it suffices to fix k and φ a test function for the corresponding minimum problem. Define $u_\varepsilon(x) = \xi x + \varepsilon \varphi(x/\varepsilon)$, so that $u_\varepsilon \to \xi x$. We can use these functions in the liminf inequality to get

$$|\Omega| f_{\text{hom}}(\xi) = F_{\text{hom}}(\xi x, \Omega) \leqslant \liminf_{\varepsilon \to 0} F_\varepsilon(u_\varepsilon, \Omega)$$

$$= \lim_{\varepsilon \to 0} \int_\Omega f\left(\frac{x}{\varepsilon}, \xi + D\varphi\left(\frac{x}{\varepsilon}\right)\right) dx$$

$$= |\Omega| \frac{1}{k^n} \int_{(0,k)^n} f(y, \xi + D\varphi) \, dy,$$

where we have used Lemma 5.4 with $Y = (0, k)^n$ and $g(\tilde{y}) = f(y, \xi + D\varphi(y))$. By the arbitrariness of φ and k the desired inequality is proved.

5.2. *The convex case: the cell-problem formula*

In the convex case (i.e., $f(x, \cdot)$ convex for a.a. x) the formula for f_{hom} is further simplified. In fact, in this case we have a single *cell-problem formula* (we suppose that f is 1-periodic):

$$f_{\text{hom}}(\xi) = \inf\left\{ \int_{(0,1)^n} f(x, \xi + D\varphi) \, dx : \varphi \in W_\#^{1,p}\left((0,1)^n; \mathbb{R}^m\right) \right\}. \tag{5.9}$$

To check this, by (5.8), it is sufficient to prove that

$$\frac{1}{k^n} \inf\left\{ \int_{(0,k)^n} f(x, \xi + D\varphi) \, dx : \varphi \in W_\#^{1,p}\left((0,k)^n; \mathbb{R}^m\right) \right\}$$

$$\geqslant \inf\left\{ \int_{(0,1)^n} f(x, \xi + D\varphi) \, dx : \varphi \in W_\#^{1,p}\left((0,1)^n; \mathbb{R}^m\right) \right\}$$

for all k, the converse inequality being trivial since $W_\#^{1,p}((0,1)^n; \mathbb{R}^m) \subset W_\#^{1,p}((0,k)^n; \mathbb{R}^m)$. Now, take φ a k-periodic test function and define

$$\tilde{\varphi}(x) = \frac{1}{k^n} \sum_{j \in \{1,\dots,k\}^n} \varphi(x + j).$$

Then $\tilde{\varphi}$ is 1-periodic and it is a convex combination of periodic translations of φ. By the convexity of f, then

$$\int_{(0,1)^n} f(x, \xi + D\tilde{\varphi}) \, dx = \frac{1}{k^n} \int_{(0,k)^n} f(x, \xi + D\tilde{\varphi}) \, dx$$

$$\leqslant \frac{1}{k^n} \sum_j \frac{1}{k^n} \int_{j+(0,k)^n} f(x, \xi + D\varphi) \, dx$$

$$= \frac{1}{k^n} \int_{(0,k)^n} f(x, \xi + D\varphi) \, dx,$$

that proves the inequality.

REMARK 5.5 (Homogenization and convexity conditions). Note that convexity is preserved by Γ-convergence also in the vectorial case; i.e., f_{hom} is convex if $f(x, \cdot)$ is convex for a.a. x. On the contrary, it can be seen that the same fails for the condition that $f(x, \cdot)$ be polyconvex (see [46]).

5.2.1. *Müller's counterexample.* Formula (5.9) proves the ansatz that recovery sequences for convex homogenization problems can be though locally periodic of minimal period (ε in the case above). Note that in the case $n = 1$ or $m = 1$ convexity is not a restrictive hypothesis, since we may consider the lower-semicontinuous envelope of F_ε in its place, whose integrand is convex.

The local-periodicity ansatz is false if the problem is vectorial, as shown by a counterexample by Müller [54,128]. We do not enter in the detail of the example, but try to give an interpretation of the physical idea behind the construction: the function f is defined in the periodicity cell $(0, 1)^3$ as

$$f(x, \xi) = \begin{cases} f_1(\xi) & \text{if } x \in B_{1/4}\left(\tfrac{1}{2}, \tfrac{1}{2}\right) \times (0, 1), \\ f_0(\xi) & \text{otherwise,} \end{cases}$$

where f_1 is a (suitable) polyconvex function and f_0 is a "weak" convex energy; e.g., $f_0(\xi) = \delta |\xi|^p$ with δ small enough (we may think f_0 being 0, even though that case is not covered by our results). We may interpret the energy F_ε as describing a periodic array or thin vertical bars. For suitable polyconvex f_1, we will have buckling instabilities and the array of thin bars will sustain much less vertical compression than the single bar in the periodicity cell. This corresponds to the inequality

$$f_{\mathrm{hom}}(\xi) < \inf\left\{ \int_{(0,1)^n} f(x, \xi + D\varphi)\,\mathrm{d}x \colon \varphi \in W_{\#}^{1,p}\big((0, 1)^n; \mathbb{R}^m\big) \right\} \tag{5.10}$$

for $\xi = -e_3 \otimes e_3$.

5.3. *Homogenization of quadratic forms*

As remarked in Theorem 4.13, quadratic forms are closed under Γ-convergence. In the case of homogenization we can give the following characterization (for simplicity we deal with the scalar case $m = 1$ only).

THEOREM 5.6 (Homogenization of quadratic forms). *Let $A : \mathbb{R}^n \to \mathbb{M}^{m \times n}$ be a 1-periodic symmetric matrix-valued measurable function, and suppose that $\alpha, \beta > 0$ exist such that $\alpha\mathrm{Id} \leqslant A \leqslant \beta\mathrm{Id}$ for all j. Then we have*

$$\int_\Omega \langle A_{\mathrm{hom}} Du, Du \rangle \,\mathrm{d}x = \Gamma\text{-}\lim_{\varepsilon \to 0} \int_\Omega \left\langle A\left(\frac{x}{\varepsilon}\right) Du, Du \right\rangle \mathrm{d}x \tag{5.11}$$

with respect to the $L^2(\Omega)$-convergence, for all $u \in H^1(\Omega)$, where the constant matrix A_{hom} is given by

$$\langle A_{\mathrm{hom}}\xi, \xi\rangle = \inf\left\{\int_{(0,1)^n} \langle A(x)(\xi + D\varphi), \xi + D\varphi\rangle\,\mathrm{d}x : \varphi \in H^1_{\#}\big((0,1)^n\big)\right\}.$$
(5.12)

REMARK 5.7 (One-dimensional homogenization). In the one-dimensional case, when we simply have $\langle A(x)\xi, \xi\rangle = a(x)\xi^2$ with $a : \mathbb{R} \to [\alpha, \beta]$ 1-periodic, the limit energy density is of the simple form $a_{\mathrm{hom}}\xi^2$. The coefficient a_{hom} is easily computed and is the *harmonic mean* of a

$$a_{\mathrm{hom}} = \underline{a} := \left(\int_0^1 \frac{1}{a(y)}\,\mathrm{d}y\right)^{-1}.$$
(5.13)

REMARK 5.8 (Laminates). We can easily compute the homogenized matrix for an A of the form $a(\langle x, v\rangle)\mathrm{Id}$

$$a(s) = \begin{cases} \alpha & \text{if } 0 < s - [s] < t, \\ \beta & \text{if } t < s - [s] < 1, \end{cases} \qquad [s] \text{ is the } integer\ part \text{ of } s.$$
(5.14)

The corresponding energy density is called a *lamination* of the two energies densities $\alpha|\xi|^2$ and $\beta|\xi|^2$ in direction v with *volume fractions* t and $1 - t$, respectively.

It is not restrictive, upon a rotation, to consider $v = e_1$. By a symmetry argument, we see that A_{hom} is diagonal. By (5.12) the coefficient a_{kk} of A_{hom} is computed by considering the minimum problem

$$a_{kk} = \inf\left\{\int_{(0,1)^n} a(x_1)|e_k + D\varphi|^2\,\mathrm{d}x : \varphi \in H^1_{\#}\big((0,1)^n\big)\right\}.$$
(5.15)

For $k = 1$ the solution is $\varphi(x) = \varphi_1(x_1)$, where φ_1 is the solution of the one-dimensional problem with coefficient a; hence $a_{11} = \underline{a}$ as defined in (5.13). For $k > 1$ we may easily see that the solution is $\varphi(x) = 0$, so that, in conclusion,

$$a_{11} = \frac{\alpha\beta}{t\beta + (1 - t)\alpha}, \qquad a_{kk} = t\alpha + (1 - t)\beta \quad \text{for } k > 1.$$

Note that we have obtained a nonisotropic matrix by homogenization of isotropic ones. Of course, by varying v we obtain all symmetric matrices with the same eigenvalues.

The same computation can be performed for $A(x) = \prod_{k=1}^n a_i(x_i)|\xi|^2$, with $\alpha \leq a_i(y) \leq \beta$. Note that if we vary α and β and a_i then we may obtain all symmetric matrices as homogenization of isotropic ones.

5.4. *Bounds on composites*

The example of lamination in Remark 5.8 shows that mixtures of two simple energies can give rise to more complex ones. A general question is to describe all possible mixtures of a certain number of "elementary" energies. This is a complex task giving rise to numerous types of questions, most of which still open (see [121]). Here we want to highlight a few connections with the theory of homogenization as presented in previous sections, by considering only the case of mixtures of two isotropic energies $\alpha|\xi|^2$ and $\beta|\xi|^2$.

A "mixture" will be given by a choice of measurable sets $E_j \subset \Omega$. We will consider energies of the form

$$F_j(u) = \alpha \int_{E_j} |Du|^2 \, dx + \beta \int_{\Omega \setminus E_j} |Du|^2 \, dx. \tag{5.16}$$

Note that we can rewrite $F_j(u) = \int_\Omega a_j(x)|Du|^2 \, dx$ and apply Theorem 4.13, thus obtaining, upon subsequences, a Γ-limit of the form

$$F_0 = \int_\Omega \langle A_0(x)Du, Du \rangle \, dx. \tag{5.17}$$

The problem is to give the best possible description of the possible reachable A_0.

The limit local (statistical) description of the behavior of (E_j) is given by the weak* limit of χ_{E_j}, which will be denoted by θ and called the *local volume fraction* of the energy α. Note that, by Remark 5.8, the knowledge of θ is not sufficient to describe the limit of F_j (since we can take laminates in two different directions with the same $\theta = t$ but different limit energies).

5.4.1. *The localization principle.*

With fixed $\bar\theta$ we can consider the set of all *matrices obtained by homogenization* of energies α and β with volume fraction $\bar\theta$ of α, corresponding by (5.12) to matrices A satisfying

$$\langle A\xi, \xi \rangle = \inf\left\{ \alpha \int_E |\xi + D\varphi|^2 \, dx + \beta \int_{(0,1)^n \setminus E} |\xi + D\varphi|^2 \, dx \colon \varphi \in H^1_\#\big((0,1)^n\big) \right\} \tag{5.18}$$

for some measurable $E \subset (0,1)^n$ with $|E| = \bar\theta$. We will denote by $\mathcal{H}(\bar\theta)$ the set of all such matrices; E is called an *underlying microgeometry* of such A.

The matrices A_0 in (5.17) are characterized by a *localization principle* [134,141].

PROPOSITION 5.9 (Localization principle). $A_0(x) \in \mathcal{H}(\theta(x))$ *for almost all $x \in \Omega$.*

PROOF. We only sketch the main points of the proof. Let $\bar x$ be a Lebesgue point for $\theta(x)$. Upon a translation argument we can suppose that $\bar x = 0$. For all open sets U, the functional defined by $\int_U \langle A_0(0)Du, Du \rangle \, dx$ is the Γ-limit of $\int_U \langle A_0(\rho x)Du, Du \rangle \, dx$ as $\rho \to 0$ since

$A_0(\rho x)$ converges to $A_0(0)$ in L^1 on U. Let $Q_\rho(x)$ denote the coordinate cube centered at x and with side length ρ. We can then infer that, for any fixed ξ,

$$\langle A_0(\bar{x})\xi, \xi \rangle = \min\left\{ \int_{Q_1(0)} \langle A_0(\rho x)(\xi + D\varphi), (\xi + D\varphi) \rangle \colon \varphi \text{ 1-periodic} \right\} + o(1)$$

$$= \rho^{-n} \min\left\{ F_0(\xi + D\varphi, Q_\rho(0)) \colon \varphi \ \rho\text{-periodic} \right\} + o(1)$$

$$= \rho^{-n} \min\left\{ F_j(\xi + D\varphi, Q_\rho(0)) \colon \varphi \ \rho\text{-periodic} \right\} + o(1)$$

as $\rho \to 0$ and $j \to +\infty$. Upon scaling, the formula in the last limit is of type (5.18) for some θ_ρ^j tending to $\theta(\bar{x})$ as $\rho \to 0$ and $j \to +\infty$, and the proposition is proved, upon remarking that the limit of matrices in $\mathcal{H}(\theta_\rho^j)$ belongs to $\mathcal{H}(\theta(\bar{x}))$. $\qquad\square$

The previous proposition reduces the problem of characterizing all $A_0(x)$ to that of studying the sets $\mathcal{H}(\theta)$ for fixed $\theta \in [0, 1]$.

REMARK 5.10 (Set of all reachable matrices). From the trivial one-dimensional estimates we have

$$\frac{\alpha\beta}{\theta\beta + (1-\theta)\alpha} \leqslant \lambda_i \leqslant \theta\alpha + (1-\theta)\beta, \tag{5.19}$$

where λ_i denote the eigenvalues of the matrices in $\mathcal{H}(\theta)$.

In the two-dimensional case we deduce that all such matrices have eigenvalues satisfying

$$\frac{\alpha\beta}{\alpha + \beta - \lambda_1} \leqslant \lambda_2 \leqslant \alpha + \beta - \frac{\alpha\beta}{\lambda_1}, \tag{5.20}$$

and actually all matrices satisfying (5.20) belong to some $\mathcal{H}(\theta)$.

5.4.2. *Optimal bounds.* The computation of $\mathcal{H}(\theta)$ is obtained by exhibiting "optimal bounds"; it is due to Murat and Tartar (see [141]; see also the derivation of Cherkaev and Lurie in the two-dimensional case [119]). It is not based on Γ-convergence arguments, so for completeness we only include the (two-dimensional) optimal bounds, which only constrain the eigenvalues λ_1, λ_2 of the macroscopic conductivity tensor A_0. The formula is

$$\begin{cases} \frac{1}{\lambda_1 - \alpha} + \frac{1}{\lambda_2 - \alpha} \leqslant \frac{1}{\bar{a}(\theta) - \alpha} + \frac{1}{\underline{a}(\theta) - \alpha}, \\ \frac{1}{\beta - \lambda_1} + \frac{1}{\beta - \lambda_2} \leqslant \frac{1}{\beta - \bar{a}(\theta)} + \frac{1}{\beta - \underline{a}(\theta)}, \end{cases} \tag{5.21}$$

where $\underline{a}(\theta)$ and $\bar{a}(\theta)$ are the harmonic and arithmetic means of α and β with proportion θ:

$$\underline{a}(\theta) = \frac{\alpha\beta}{\theta\beta + (1-\theta)\alpha}, \qquad \bar{a}(\theta) = \theta\alpha + (1-\theta)\beta.$$

Note that the two "extremal" geometries are given by laminates.

5.5. *Homogenization of metrics*

We conclude this chapter with some observation regarding another type of homogenization, that of functionals of the type

$$F_\varepsilon(u) = \int_\Omega f\left(\frac{u}{\varepsilon}, Du\right) dx, \tag{5.22}$$

with f periodic in the first variable and satisfying the usual growth conditions. In this case, by Remark 4.6, we can carry over the compactness procedure and also represent the limit as an integral of the usual form

$$f_{\text{hom}}(\xi) = \lim_{T\to+\infty} \frac{1}{T^n}$$
$$\times \inf\left\{\int_{(0,T)^n} f(u + \xi y, Du + \xi)\,dy \colon u \in W_0^{1,p}\big((0,T)^n; \mathbb{R}^m\big)\right\}. \tag{5.23}$$

We do not treat the general case (for details see [54], Chapter 15), but briefly outline two applications.

5.5.1. *The closure of Riemannian metrics.* We consider the one-dimensional integrals, related to distances on a periodic isotropic Riemannian manifold

$$F_\varepsilon(u) = \int_0^1 a\left(\frac{u}{\varepsilon}\right)|u'|^2\,dt, \quad u \in H^1\big((0,1); \mathbb{R}^m\big). \tag{5.24}$$

The following result states that the limit of such energies corresponds to a *homogeneous Finsler metric* (see [1]) and the converse is also true; i.e., every homogeneous Finsler metric can be approximated by homogenization of (isotropic) Riemannian metrics (see [47]).

THEOREM 5.11 (Closure of Riemannian metrics by homogenization). (i) *Let a be a 1-periodic function satisfying $0 < \alpha \leqslant a \leqslant \beta < +\infty$; then the Γ-limit of F_ε is*

$$F_{\text{hom}}(u) = \int_0^1 f_{\text{hom}}(u')\,dt, \tag{5.25}$$

where

$$f_{\text{hom}}(z) = \lim_{T\to+\infty} \frac{1}{T} \inf\left\{\int_0^1 a(v)|v'|^2\,dt \colon v(0) = 0,\, v(T) = Tz\right\}; \tag{5.26}$$

(ii) *for all $\psi \colon \mathbb{R}^m \to [0, +\infty)$ even, convex, positively homogeneous of degree two and such that $\alpha|z|^2 \leqslant \psi(z) \leqslant \beta|z|^2$, and for all $\eta > 0$ there exists f_{hom} as above such that $|f_{\text{hom}}(z) - \psi(z)| \leqslant \eta|z|^2$ for all $z \in \mathbb{R}^m$.*

PROOF. (i) can be achieved by repeating the compactness procedure (see Remark 4.6). The formula follows from the representation of $f_{\text{hom}}(z)$ as a minimum problem.

(ii) Let (ν_i) be a sequence of rational directions (i.e., such that for all i there exists $T_i \in \mathbb{R}$ such that $T_j \nu_j \in \mathbb{Z}^m$) dense in S^{m-1}, fix M and define a^M as

$$a^M(s) = \begin{cases} \psi(\nu_i) & \text{if } s \in (\mathbb{Z}^m + \nu_i \mathbb{R}) \setminus \bigcup_{j \neq i, 1 \leqslant j \leqslant M} (\mathbb{Z}^m + \nu_j \mathbb{R}), \\ & 1 \leqslant i \leqslant M, \\ \beta & \text{otherwise.} \end{cases} \tag{5.27}$$

The coefficient a^M is β except on a $\mathbb{Z}^m$ periodic set of lines in the directions ν_i. Then from formula (5.26) for the corresponding f_{hom}^M we easily get that $f_{\text{hom}}^M(\nu_i) = \psi(\nu_i)$ on all ν_i for $i \leqslant M$. Note in fact that for all v we have $\int_0^1 a^M(v)|v'|^2 \, dt \geqslant \int_0^1 \psi(v') \, dt$ and equality holds on the functions $v(t) = t\nu_i$. Since ψ is positively homogeneous of degree two and convex, this implies that $f_{\text{hom}}^M \to \psi$ uniformly on S^{m-1} as $M \to +\infty$, as desired. $\quad\square$

This result has been generalized to the approximation of arbitrary (nonhomogeneous) Finsler metrics by Davini [89].

5.5.2. *Homogenization of Hamilton–Jacobi equations.* The solution u_ε of a Hamilton–Jacobi equation of the form

$$\begin{cases} \frac{\partial u_\varepsilon}{\partial t} + H\left(\frac{x}{\varepsilon}, Du_\varepsilon(x,t)\right) = 0 & \text{in } \mathbb{R}^m \times [0,+\infty), \\ u_\varepsilon(x,0) = \varphi(x) & \text{in } \mathbb{R}^m, \end{cases} \tag{5.28}$$

where H is a quadratic *Hamiltonian* and φ is a smooth bounded initial datum, is given by the *Lax formula*

$$u_\varepsilon(x) = \inf\{\varphi(y) + S_\varepsilon(x,t;y,s) : y \in \mathbb{R}^m, 0 \leqslant s < t\},$$

$$S_\varepsilon(x,t;y,s) = \inf\left\{\int_s^t L\left(\frac{u}{\varepsilon}, u'\right) d\tau : u(s) = y, u(t) = x\right\},$$

$$L(x,z) = \sup\{\langle z, z'\rangle - H(x,z') : z' \in \mathbb{R}^m\}$$

(the *Legendre transform* of H). We can apply the results outlined in (5.22) and (5.23) to the integrals in the minimum problem defining ε, obtaining that the pointwise limit of S_ε is given by

$$S_{\text{hom}}(x,t;y,s) = \inf\left\{\int_s^t L_{\text{hom}}(u') \, d\tau : u(s) = y, u(t) = x\right\}$$

$$= (t-s) L_{\text{hom}}\left(\frac{x-y}{t-s}\right),$$

where L_{hom} is obtained through formula (5.23), and u_ε converge uniformly on compact sets to the corresponding u. As a conclusion we may prove that u satisfies the *homogenized Hamilton–Jacobi equation*

$$\begin{cases} \frac{\partial u}{\partial t} + H_{\mathrm{hom}}\big(Du(x,t)\big) = 0 & \text{in } \mathbb{R}^m \times [0, +\infty), \\ u(x,0) = \varphi(x) & \text{in } \mathbb{R}^m, \end{cases} \tag{5.29}$$

where H_{hom} is given by

$$H_{\mathrm{hom}}(x,z) = \sup\{\langle z, z'\rangle - L_{\mathrm{hom}}(x,z') \colon z' \in \mathbb{R}^m\}.$$

Details can be found in [46], Section 3.4 (see also [98]).

6. Perforated domains and relaxed Dirichlet problems

A class of problems that cannot be directly framed within the class of integral functionals considered in the previous sections are those defined on varying domains. The prototype of these domains are *perforated domains*; i.e., obtained from a fixed Ω by removing some periodic set, the simplest of which is a periodic array of closed sets,

$$\Omega_\varepsilon = \Omega \setminus \bigcup_{i \in \mathbb{Z}^n} (\varepsilon i + \delta_\varepsilon K). \tag{6.1}$$

We suppose that the set K is a bounded closed set. On the boundary of Ω_ε (or on the boundary of Ω_ε interior to Ω) we can consider various types of conditions. We will examine Dirichlet and Neumann boundary conditions, leading to different relevant scales for δ_ε and technical issues.

6.1. *Dirichlet boundary conditions: a direct approach*

We first treat the model case of Ω_ε as in (6.1) and $u = 0$ on $\partial\Omega_\varepsilon$, with in mind minimum problems of the form

$$\min\left\{ \int_\Omega |Du|^2 \, dx - 2\int_\Omega gu \, dx \colon u = 0 \text{ on } \partial\Omega_\varepsilon \right\}. \tag{6.2}$$

The results of this section can be extended to vector u, to different boundary conditions on $\partial\Omega$ (provided we introduce a "safe zone" close to $\partial\Omega$ vanishing with ε where the perforation is absent in order not to make the boundary conditions interact) and to general integrands satisfying the growth conditions of Section 4.

The observation that for suitable δ_ε the solutions u_ε of the equations

$$\begin{cases} -\Delta u_\varepsilon = g & \text{in } \Omega_\varepsilon, \\ u_\varepsilon = 0 & \text{on } \partial\Omega_\varepsilon, \end{cases} \tag{6.3}$$

extended to 0 inside the perforation, may converge to a function u satisfying

$$\begin{cases} -\Delta u + Cu = g & \text{in } \Omega, \\ u = 0 & \text{on } \partial\Omega, \end{cases} \tag{6.4}$$

for some $C > 0$, goes back to Marchenko and Khruslov [120], and was subsequently recast in a variational framework by Cioranescu and Murat [74]. We want to reread this phenomenon on problems in (6.2), of which (6.3) is the Euler equation.

We will extend our functions to the whole Ω by setting $u = 0$ on the perforation. As usual, we will neglect the continuous part $2 \int_\Omega gu \, dx$ since it commutes with the Γ-limit, and consider the functionals

$$F_\varepsilon(u) = \begin{cases} \int_\Omega |Du|^2 \, dx & \text{if } u \in H_0^1(\Omega) \text{ and } u = 0 \text{ on } \Omega \setminus \Omega_\varepsilon, \\ +\infty & \text{otherwise.} \end{cases} \tag{6.5}$$

It remains to understand the role of δ_ε. We have to understand the meaningful scalings of the energy and possibly an optimal minimum formula describing the limit. To this end, consider a sequence $u_\varepsilon \to u$. We make the assumption that the energy "far from the perforation" gives a term of Dirichlet form that can be dealt with separately, and focus on the contribution "close to the perforation". We also assume that the energy due to each set $\varepsilon i + \delta_\varepsilon K$ can be dealt with separately. Suppose for the time being that u is continuous; since $u_\varepsilon \to u$ close to $\varepsilon i + \delta_\varepsilon K$ the function u_ε will be close to the limit value $u(\varepsilon i)$. Assume that this is true (and then that we may directly suppose $u_\varepsilon = u(\varepsilon i)$) on the boundary of some ball $B_{\varepsilon R}(\varepsilon i)$ containing $\varepsilon i + \delta_\varepsilon K$. We have

$$\int_{B_{\varepsilon R}(\varepsilon i)} |Du_\varepsilon|^2 \, dx$$

$$\geq \min\left\{ \int_{B_{\varepsilon R}(0)} |Dv|^2 \, dx : v = 0 \text{ on } \delta_\varepsilon K, v = u(\varepsilon i) \text{ on } \partial B_{\varepsilon R}(0) \right\}$$

$$= \delta_\varepsilon^{n-2} \min\left\{ \int_{B_{\varepsilon R/\delta_\varepsilon}(0)} |Dv|^2 \, dy : v = 0 \text{ on } K, v = u(\varepsilon i) \text{ on } \partial B_{\varepsilon R/\delta_\varepsilon}(0) \right\}$$

$$\geq \delta_\varepsilon^{n-2} \inf_T \min\left\{ \int_{B_T(0)} |Dv|^2 \, dy : v = 0 \text{ on } K, v = u(\varepsilon i) \text{ on } \partial B_T(0) \right\}$$

$$= \delta_\varepsilon^{n-2} |u(\varepsilon i)|^2 \inf_T \min\left\{ \int_{B_T(0)} |Dv|^2 \, dy : v = 0 \text{ on } K, v = 1 \text{ on } \partial B_T(0) \right\}. \tag{6.6}$$

For the sake of simplicity we suppose that $n > 2$; in that case the last minimum problem is the *capacity* of the set K (with respect to $\mathbb{R}^n$), that we will denote by $\mathrm{Cap}(K)$. We then have a lower estimate on the contribution "close to the perforation" of the form

$$\sum_i \int_{B_{\varepsilon R}(\varepsilon i)} |Du_\varepsilon|^2 \, dx \geq \mathrm{Cap}(K) \sum_i \delta_\varepsilon^{n-2} |u(\varepsilon i)|^2. \tag{6.7}$$

The last is a Riemann sum provided that $\delta_\varepsilon^{n-2} = M\varepsilon^n + o(\varepsilon)$. This given a guess for the correct meaningful scaling for which the limit is influenced by the perforation (we may suppose $M = 1$ upon scaling K)

$$\delta_\varepsilon = \varepsilon^{n/(n-2)}. \tag{6.8}$$

We will see that all other scaling can be reduced to this one by a comparison argument.

The argument above needs some refinement if $n = 2$, due to the scaling-invariance properties of the Dirichlet integral. In that case, the minimum problem on B_T in (6.6) scales as $(\log T)^{-1}$, so that taking the infimum in T would give a trivial lower bound. Instead, to obtain an inequality as in (6.7) from the first inequality in (6.6), we choose δ_ε so that $(\log T)^{-1} = \varepsilon^2$ $(T = \varepsilon R/\delta_\varepsilon)$. This choice gives the correct scaling $\delta_\varepsilon = e^{-c/\varepsilon^2}$. Note that the dependence on K in the limit disappears. In this section we will always assume that $n \geqslant 3$.

6.1.1. *A joining lemma on perforated domains.*   In the argument in (6.6) we have supposed that it is not restrictive to vary the value of a sequence u_ε on some sets surrounding the perforation. This can be obtained easily if the family $(|Du_\varepsilon|^2)$ is equiintegrable. Unfortunately, Theorem 4.11 cannot be directly used since the modified sequence might violate the constraint $u_\varepsilon = 0$ on the perforation. Nevertheless, we can modify De Giorgi's method to match boundary conditions and obtain the following technical lemma proved by Ansini and Braides. We suppose that $K \subset B_1(0)$ for simplicity.

LEMMA 6.1. *Let (u_ε) converge weakly to u in $H^1(\Omega)$. Let $k \in \mathbb{N}$ be fixed and $R < 1/2$. Let Z_ε be the set of all $i \in \mathbb{Z}^n$ with $\mathrm{dist}(\varepsilon i, \partial\Omega) > n\varepsilon$. For each such i there exists $k_i \in \{0, \dots, k-1\}$ such that, having set*

$$C_i^\varepsilon = \left\{ x \in \Omega : 2^{-k_i-1} R\varepsilon < |x - \varepsilon i| < 2^{-k_i} R\varepsilon \right\}, \tag{6.9}$$

$$u_\varepsilon^i = \frac{1}{|C_i^\varepsilon|} \int_{C_i^\varepsilon} u_\varepsilon \, dx \quad and \quad \rho_\varepsilon^i = \frac{3}{4} 2^{-k_i} R\varepsilon \tag{6.10}$$

(the mean value of u_ε on C_i^ε and the middle radius of C_i^ε, respectively), there exists a sequence (w_ε), with $w_\varepsilon \rightharpoonup u$ in $H^1(\Omega)$ such that

$$w_\varepsilon = u_\varepsilon \quad on \ \Omega \setminus \bigcup_{i \in Z_\varepsilon} C_i^\varepsilon, \qquad w_\varepsilon(x) = u_\varepsilon^i \quad if \ |x - \varepsilon i| = \rho_\varepsilon^i \tag{6.11}$$

and

$$\int_\Omega \left| |Dw_\varepsilon|^2 - |Du_\varepsilon|^2 \right| dx \leqslant c\frac{1}{k}. \tag{6.12}$$

PROOF. The proof of the lemma follows the idea of the De Giorgi method for matching boundary values. In this case the value to match is u_ε^i, and the choice where to operate the cut-off procedure is between the annuli C_i^ε, $i \in \{1, \ldots, N\}$. The proof is a little more complex since we have to use Poincaré's inequality on C_i^ε to estimate the excess of energy due to this process (note that the annuli are all homothetic in order to control the Poincaré constant by the scaling ratio). We refer to [22] for the details of the proof. $\qquad\square$

With this lemma, it is relatively easy to describe the Γ-limit of F_ε.

THEOREM 6.2. *Let $n > 2$ and let F_ε be given by (6.5) and $\delta_\varepsilon = \varepsilon^{n/(n-2)}$. Then the Γ-limit of F_ε with respect to the $L^2(\Omega)$ convergence is given by*

$$F_0(u) = \int_\Omega |Du|^2 \, dx + \mathrm{Cap}(K) \int_\Omega |u|^2 \, dx \tag{6.13}$$

on $H_0^1(\Omega)$.

PROOF. By Lemma 6.1, we can use the argument in (6.6) with u_ε^i in place of $u(\varepsilon i)$ to give a lower bound on the contribution close to the perforation with $\mathrm{Cap}(K) \sum_{i \in Z_\varepsilon} |u_\varepsilon^i|^2$, which converges to $\mathrm{Cap}(K) \int_\Omega |u|^2 \, dx$. As for the contribution away from the perforation, we can write it as $\int_\Omega |Dz_\varepsilon|^2 \, dx$, where z_ε is the $H^1(\Omega)$-extension of w_ε which is constant on each ball $B_{\rho_\varepsilon^i}(\varepsilon i)$. The limit of z_ε is still u so that we have the inequality $\liminf_{\varepsilon \to 0} \int_\Omega |Dz_\varepsilon|^2 \, dx \geq \int_\Omega |Du|^2 \, dx$, completing the lower bound.

The upper bound can be achieved by a direct construction. We first show it for $u = 1$ constant (even though this does not satisfy the boundary condition $u \in H_0^1(\Omega)$). In this case we simply choose $T > 0$ and v_T minimizing the last minimum problem in (6.6) and define

$$u_\varepsilon(x) = \begin{cases} v_T\left(\varepsilon^{n/(2-n)}(x - \varepsilon i)\right) & \text{on } \varepsilon^{n/(n-2)} B_T(\varepsilon i), \\ 1 & \text{otherwise.} \end{cases} \tag{6.14}$$

Then $u_\varepsilon \to u$ and $\lim_{\varepsilon \to 0} F_\varepsilon(u_\varepsilon) = |\Omega| \int_{B_T(0)} |Dv_T|^2 \, dx$, which proves the approximate limsup inequality. For $u \in C_0^\infty(\Omega)$ we can use the recovery sequence $\tilde{u}_\varepsilon = u_\varepsilon u$ with u_ε as in (6.14), and for $u \in H_0^1(\Omega)$ use a density argument. $\qquad\square$

REMARK 6.3 (Other limits). (i) As a first remark, note that we may consider perforations with locally varying size. For example, we can fix a smooth bounded function $g : \mathbb{R}^n \to [0, +\infty)$ and take

$$\Omega_\varepsilon = \Omega \setminus \bigcup_{i \in \mathbb{Z}^n} \left(\varepsilon i + \varepsilon^{n/(n-2)} g(\varepsilon i) K\right). \tag{6.15}$$

We can follow word for word the previous proof, noting that a term $(g(x))^{n-2}$ appears in (6.6), and obtain the limit functional

$$F_0(u) = \int_\Omega |Du|^2 \, dx + \int_\Omega a(x)|u|^2 \, dx \tag{6.16}$$

on $H_0^1(\Omega)$, where $a(x) = \mathrm{Cap}(K)(g(x))^{n-2}$. By approximation, in this way we may obtain any $a \in L_{\mathrm{loc}}^1(\mathbb{R}^n)$ in the limit functional.

(ii) The "noncritical cases" can be easily dealt with by comparison. If we take a perforation as in (6.1) with

$$\lim_{\varepsilon \to 0} \frac{\delta_\varepsilon}{\varepsilon^{n/(n-2)}} = 0 \tag{6.17}$$

then we have an upper bound for the Γ-limit by any functional of the form (6.16) with a any fixed constant, so that we may take $a = 0$ and obtain that the contribution of the perforation disappears leaving only the Dirichlet integral (the lower bound is trivial in this case by the lower semicontinuity of the Dirichlet integral). Conversely, if

$$\lim_{\varepsilon \to 0} \frac{\delta_\varepsilon}{\varepsilon^{n/(n-2)}} = +\infty, \tag{6.18}$$

then the functionals (6.16) with a any fixed constant are a lower bound and in the limit we obtain that the functional is finite (and its value is 0) only on the constant 0 (for which the upper bound is trivial).

(iii) We can extend the method outlined above to cover other cases. For example, we can require the unilateral condition $u \geqslant 0$ on the perforation. In this case the same proof gives a limit of the form

$$F_0(u) = \int_\Omega |Du|^2 \, dx + \mathrm{Cap}(K) \int_\Omega |u^-|^2 \, dx, \tag{6.19}$$

where the sole *negative part* of u contributes to the extra term. It is sufficient to note that

$$\min\left\{\int_{B_T(0)} |Dv|^2 \, dy \colon v \geqslant 0 \text{ on } K, \, v = u(\varepsilon i) \text{ on } B_T(0)\right\}$$

$$= |u^-(\varepsilon i)|^2 \min\left\{\int_{B_T(0)} |Dv|^2 \, dy \colon v = 0 \text{ on } K, \, v = 1 \text{ on } B_T(0)\right\}$$

in the last equality of (6.6).

REMARK 6.4 (Perforated domains as degenerate quadratic forms). We note that, for $n \geqslant 3$ at fixed ε, the functional in (6.5) can be seen as the Γ-limit of a family of usual quadratic integral functionals on $H_0^1(\Omega)$. We can easily check this by a double-limit procedure. We first fix $\rho > 0$ and consider the set $\mathcal{G}_\rho = \{x \colon \mathrm{dist}(x, \mathcal{G}) \leqslant \rho\}$, where $\mathcal{G}$ is the "integer grid" $\mathcal{G} = \{x \in \mathbb{R}^n \colon \#\{i \colon x_i \notin \mathbb{Z}\} \leqslant 1\}$. It is not restrictive to suppose that $0 \in K$ and K connected, so that the set $C_\rho = \mathcal{G}_\rho \cup \bigcup_i (\varepsilon i + \delta_\varepsilon K)$ is periodic and connected (and, in particular, connected with $\partial\Omega$). We can then define

$$a_n^\rho(x) = \begin{cases} n & \text{if } x \in C_\rho, \\ 1 & \text{otherwise,} \end{cases} \qquad F_n^\rho(u) = \int_\Omega a_n^\rho |Du|^2 \, dx, \quad u \in H_0^1(\Omega). \tag{6.20}$$

As $n \to +\infty$, F_n^ρ converge increasingly to the functionals F^ρ defined by the Dirichlet integral with zero boundary conditions on $\partial\Omega \cup C_\rho$ (note that we may use Remark 2.12(ii) to deduce that F^ρ is also the Γ-limit of F_n^ρ). We now let $\rho \to 0$ so that F^ρ converge decreasingly to F_ε as defined in (6.5) and use Remark 2.12(i) to deduce their Γ-convergence. Note that here we use that $\mathcal{G}$ has zero capacity (this fact is not true for $n = 2$). A diagonal sequence (that we may construct thanks to Remark 2.16) does the job.

6.2. *Relaxed Dirichlet problems*

The problem of the computation of the Γ-limit for an arbitrary family of perforations needs a general setting including both the original constraint $u = 0$ on some E, and the limit case obtained in the previous section with the "extra term" $\int |u|^2 \, dx$. To this end, note that both energies can be written as

$$F(u) = \int_\Omega |Du|^2 \, dx + \int_\Omega |u|^2 \, d\mu,$$

where the Borel measure μ is defined either as $\mu = \mathrm{Cap}(K)\mathcal{L}^n$ or as $\mu = \infty_E$, where

$$\infty_E(B) = \begin{cases} 0 & \text{if } \mathrm{Cap}(B \setminus E) = 0, \\ +\infty & \text{otherwise,} \end{cases} \tag{6.21}$$

corresponding to the zero condition on the set E. The notion of capacity is naturally linked to H^1-functions, that are defined up to sets of zero capacity, so that in the second case the condition $F(u) < +\infty$ can be equivalently read as $u \in H^1(\Omega)$ and $u = 0$ (up to a set of zero capacity) on E.

DEFINITION 6.5 (Relaxed Dirichlet problems). We denote by $\mathcal{M}_0$ the set of all (possibly nonfinite) nonnegative Borel measures μ on $\mathbb{R}^n$ such that $\mu(B) = 0$ for every Borel set $B \subset \mathbb{R}^n$ of zero capacity.

A *relaxed Dirichlet problem* is a minimum problem of the form

$$\min\left\{ \int_\Omega |Du|^2 \, dx + \int_\Omega |u|^2 \, d\mu - 2 \int_\Omega gu \, dx : u \in H_0^1(\Omega) \right\},$$

where $\mu \in \mathcal{M}_0$ and $g \in L^2(\Omega)$; the solution to this minimum problem solves the problem

$$\begin{cases} -\Delta u + u\mu = g, \\ u \in H_0^1(\Omega). \end{cases}$$

We define the γ-*convergence* of μ_j to μ as the Γ-convergence of the functionals defined on $H_0^1(A)$ by

$$F_{\mu_j}(u, A) = \int_A |Du|^2 \, dx + \int_A |u|^2 \, d\mu_j$$

to the corresponding

$$F_\mu(u, A) = \int_A |Du|^2 \, dx + \int_A |u|^2 \, d\mu$$

for all A bounded open subset of $\mathbb{R}^n$.

For the class $\mathcal{M}_0$ we have a compactness and density result as follows.

THEOREM 6.6 (Closure of relaxed Dirichlet problems). (i) *For every sequence (μ_j) in $\mathcal{M}_0$ there exist a subsequence, not relabeled, and μ in $\mathcal{M}_0$ such that μ_j γ-converge to μ.*

(ii) *For every $\mu \in \mathcal{M}_0$ there exists a sequence (K_j) of compact subsets of $\mathbb{R}^n$ such that the measures $\mu_j = \infty_{K_j}$ (defined as in (6.21) with $E = K_j$) γ-converge to μ.*

PROOF. For a complete proof we refer to the paper by Dal Maso and Mosco [87], Section 4. Here we want to highlight that for the proof of (i) (a variation of) the compactness method in Section 3.3 can be applied. In this case the limit F_0 of the functionals F_{μ_j}, obtained by a compactness and localization argument, can be written as $\int_A |Du|^2 \, dx + G(u, A)$, and a suitable representation theorem for G by Dal Maso (see [82]) shows that $G(u, A) = \int_A g(x, u) \, d\mu$. Eventually, as F_0 is a quadratic form, we deduce that we may take $g(x, u) = |u|^2$, so that $F_0 = F_\mu$.

As for (ii) note that the case $\mu = a(x)\mathcal{L}^n$ with $a \in L^\infty$ is taken care in Remark 6.3(i). In the general case, one proceeds by approximation. $\qquad\square$

REMARK 6.7 (Computation of the limit of perforated domains). The construction in the previous section shows that in the case of functionals $F_{\infty_{K_\varepsilon}}$, where $K_\varepsilon = \bigcup_i \varepsilon i + \varepsilon^{n/(n-2)} K$ the measure μ in the limit energy F_μ can be computed as the weak* limit of the measures $\sum_i \varepsilon^n \, \text{Cap}(K)\delta_{\varepsilon i}$ (here $\delta_{\varepsilon i}$ stands for the Dirac mass at εi), and the effect of the capacity of the set K_ε can be decomposed as the sum of the capacities of each $\varepsilon i + \varepsilon^{n/(n-2)} K$. This is not the case in general, since the capacity is not an additive set function. Nevertheless, a formula for the limit of a family F_{μ_j} can be proved (in particular, we may have $\mu_j = \infty_{K_j}$, with K_j an arbitrary perforation): the limit is F_μ if a Radon measure ν exists and $f : \mathbb{R}^n \to [0, +\infty]$ is such that $f(x) < +\infty$ up to a set of zero capacity and

$$f(x) = \liminf_{\rho \to 0} \liminf_j \frac{\text{Cap}_{\mu_j}(B_\rho(x), B_{2\rho}(x))}{\nu(B_\rho(x))}$$

$$= \liminf_{\rho \to 0} \limsup_j \frac{\text{Cap}_{\mu_j}(B_\rho(x), B_{2\rho}(x))}{\nu(B_\rho(x))},$$

where the μ_j-*capacity* is defined as

$$\text{Cap}_{\mu_j}(E, A) = \min\left\{\int_A |Du|^2 \, dx + \int_E u^2 \, d\mu_j : u - 1 \in H_0^1(A)\right\} \qquad (6.22)$$

then we have $\mu = fv$ (see [67], Theorem 5.2).

In particular, if $\mu_j = \infty_{K_j}$, then we have to compute the behavior of

$$\mathrm{Cap}_{\mu_j}(E, A) = \mathrm{Cap}(E \cap K_j, A)$$

$$= \min\left\{\int_A |Du|^2 \, dx \colon u - 1 \in H_0^1(A), \ u = 1 \text{ on } K_j \cap E\right\}.$$

This formula describes the local behavior of the energies due to a perforation in terms of the μ_j-capacities.

Another way to express the measure μ is as the least superadditive set function satisfying

$$\mu(A) \geqslant \inf_{\substack{U \text{ open} \\ A \subseteq U}} \ \sup_{\substack{B \text{ compact} \\ B \subseteq U}} \ \limsup_j \mathrm{Cap}(K_j \cap B, \Omega)$$

for every Borel subset $A \subseteq \Omega$ (see [83]).

REMARK 6.8 (Limits of obstacle problems). As noted in Remark 6.3(iii) problems on perforated domains can be extended to problems with (unilateral or bilateral) obstacles. In particular, the condition $u = 0$ on the perforation can be seen as a particular case of bilateral obstacle. We refer to the paper by Dal Maso [82,85] for the treatment of limits of such problems, and in particular for their integral representation, which is used to represent limits of relaxed Dirichlet problems.

REMARK 6.9 (Closure of quadratic forms with Dirichlet boundary conditions). Theorem 4.18 shows that the closure of quadratic forms of *diffusion type* are all objective Dirichlet forms. On the other hand, Remark 6.4 has shown that functionals on perforated domains, and hence also all functionals of relaxed Dirichlet problems by Remark 6.8, can be obtained as limits of quadratic forms of diffusion type on $H_0^1(\Omega)$. Note that relaxed Dirichlet problems possess the missing nonobjective part in the Deny–Beurling formula. In fact, a result by Camar-Eddine and Seppecher shows that *the closure of quadratic forms of diffusion type are all Dirichlet forms*. We refer to [68] for details.

6.3. *Neumann boundary conditions: an extension lemma*

The issues in the treatment of Neumann boundary conditions are different; the first one being which convergence to use in the definition of Γ-limit; the second one being the most general hypothesis under which a limit exists and defines a nondegenerate functional. The fundamental tool to answer these questions is an *extension lemma* by Acerbi, Chiadò Piat, Dal Maso and Percivale [3] (see also [54], the Appendix).

LEMMA 6.10 (Extension lemma). *Let E be a periodic, connected, open subset of $\mathbb{R}^n$, with Lipschitz boundary. Given a bounded open set $\Omega \subset \mathbb{R}^n$, and a real number $\varepsilon > 0$, there* *exist a linear and continuous extension operator* $T_\varepsilon : W^{1,p}(\Omega \cap \varepsilon E) \to W^{1,p}_{\text{loc}}(\Omega)$ *and three constants* $k_0, k_1, k_2 > 0$, *such that*

$$T_\varepsilon u = u \quad a.e. \ in \ \Omega \cap \varepsilon E, \tag{6.23}$$

$$\int_{\Omega(\varepsilon k_0)} |T_\varepsilon u|^p \, \mathrm{d}x \leqslant k_1 \int_{\Omega \cap \varepsilon E} |u|^p \, \mathrm{d}x, \tag{6.24}$$

$$\int_{\Omega(\varepsilon k_0)} \left| D(T_\varepsilon u) \right|^p \, \mathrm{d}x \leqslant k_2 \int_{\Omega \cap \varepsilon E} |Du|^p \, \mathrm{d}x, \tag{6.25}$$

where we use the notation $A(\lambda)$ *for the* retracted *set* $\{x \in A: \ \mathrm{dist}\,(x, \partial A) > \lambda\}$, *for every* $u \in W^{1,p}(\Omega \cap \varepsilon E)$. *The constants* k_0, k_1, k_2 *depend on* E, n, p, *but are independent of* ε *and* Ω.

With this result in mind, we can look at the behavior of solutions to problems of the form

$$m_\varepsilon = \min\left\{ \int_{\Omega \cap \varepsilon E} f\left(\frac{x}{\varepsilon}, Du\right) \mathrm{d}x - \int_{\Omega \cap \varepsilon E} gu \, \mathrm{d}x : \ u = \varphi \ on \ \partial\Omega \right\}. \tag{6.26}$$

In fact, if u_ε is a solution to m_ε, then we can consider $T_\varepsilon u_\varepsilon$ as defined above (componentwise if $u : \Omega \cap \varepsilon E \to \mathbb{R}^m$ with $m > 1$). If f satisfies the growth condition of Theorem 4.2 then we infer that $(T_\varepsilon u_\varepsilon)$ is locally bounded in $W^{1,p}(\Omega; \mathbb{R}^m)$ so that a limit $u \in W^{1,p}_{\text{loc}}(\Omega; \mathbb{R}^m)$ exists up to subsequences. Actually, it is easily seen that we obtain a uniform bound on $|Du|^p$ on each $\Omega(\lambda)$ independent of λ, so that $u \in W^{1,p}(\Omega; \mathbb{R}^m)$. The notion of Γ-convergence in this case must follow this compactness result.

THÉOREM 6.11 (Homogenization of perforated domains). *Let* f *satisfy the hypotheses of Theorem 5.1, let* E *be a periodic, connected, open subset of* $\mathbb{R}^n$, *with Lipschitz boundary and let*

$$F_\varepsilon(u) = \int_{\Omega \cap \varepsilon E} f\left(\frac{x}{\varepsilon}, Du\right) \mathrm{d}x, \quad u \in W^{1,p}(\Omega \cap \varepsilon E; \mathbb{R}^m). \tag{6.27}$$

Then F_ε Γ-*converge with respect to the weak convergence in* $W^{1,p}_{\text{loc}}(\Omega; \mathbb{R}^m)$ *to the functional defined on* $W^{1,p}(\Omega; \mathbb{R}^m)$ *by* $F_{\text{hom}}(u) = \int_\Omega f_{\text{hom}}(Du)\,\mathrm{d}x$, *with* f_{hom} *still satisfying a growth condition as* f, *given by*

$$f_{\text{hom}}(\xi) = \lim_{T \to +\infty} \frac{1}{T^n}$$

$$\times \inf\left\{ \int_{(0,T)^n \cap E} f(x, \xi + D\varphi)\,\mathrm{d}x : \ \varphi \in W^{1,p}_0\big((0,T)^n \cap E; \mathbb{R}^m\big) \right\}. \tag{6.28}$$

This formula can be simplified to a cell-problem formula if $f(y, \cdot)$ is convex. Furthermore, if Ω has a Lipschitz boundary then problems m_ε converge to

$$m_{\text{hom}} = \min\left\{ \int_\Omega f_{\text{hom}}(Du)\, dx - C \int_\Omega gu\, dx : u = \varphi \text{ on } \partial\Omega \right\}, \qquad (6.29)$$

where $C = |E \cap (0,1)^n|$.

It must be noted that the result of Γ-convergence still holds if we only suppose that E is connected and contains a periodic connected set with Lipschitz boundary (for example, we can take E as the complement of a periodic array of "cracks"; i.e., of $(n-1)$-dimensional closed sets). Of course, in this case in general, the solutions to m_ε cannot be extended to $W^{1,p}_{\text{loc}}$-functions in Ω. For details we refer to [54], Chapter 20.

6.4. *Double-porosity homogenization*

The homogenization of perforated media presents an interesting variant when the "holes" are not "empty", but the energy density therein has a different scaling. The prototype of such problems is of the form

$$m_\varepsilon = \min\left\{ \int_{\Omega \cap \varepsilon E} |Du|^2\, dx \right.$$

$$\left. + \varepsilon^2 \int_{\Omega \setminus \varepsilon E} |Du|^2\, dx - \int_\Omega gu\, dx : u = \varphi \text{ on } \partial\Omega \right\}, \qquad (6.30)$$

where E is a periodic open subset of $\mathbb{R}^n$, with Lipschitz boundary, not necessarily connected. Contrary to (6.26) on the part $\Omega \setminus \varepsilon E$ we consider a "weak" energy scaling as ε^2 (other scalings as usual can be considered giving less interesting results). Note moreover that now the forcing term $\int_\Omega gu\, dx$ is considered on the whole Ω.

A first observation is that if E is also connected then the Γ-limit with respect to the weak convergence in $W^{1,p}_{\text{loc}}(\Omega; \mathbb{R}^m)$ of $\int_{\Omega \cap \varepsilon E} |Du|^2\, dx + \varepsilon^2 \int_{\Omega \setminus \varepsilon E} |Du|^2\, dx$ is the same as that of $\int_{\Omega \cap \varepsilon E} |Du|^2\, dx$ by Theorem 6.11 and a simple comparison argument. However, it must be noted that we cannot derive from this result the convergence of problems m_ε, as we cannot obtain a bound on the L^2 norms of the gradients of solutions (u_ε).

A way to overcome this lack of compactness is by considering only the part of Ω where we can apply the compactness argument in the previous section: we define a limit u considering only the limit of $T_\varepsilon u_\varepsilon$. In this way the contribution of $\varepsilon^2 \int_{\Omega \setminus \varepsilon E} |Du_\varepsilon|^2\, dx - \int_{\Omega \setminus E} gu_\varepsilon\, dx$ can be considered as a perturbation. In order to understand its effect, suppose that $K = (0,1)^n \setminus E$ is compactly contained in $(0,1)^n$ and g is continuous.

We focus on the energy contained on a set $\varepsilon i + \varepsilon K$. Since $u_\varepsilon \to u$, we may suppose that $u_\varepsilon = u(\varepsilon i)$ on $\partial(\varepsilon i + \varepsilon K)$, so that we may estimate the contribution

$$\varepsilon^2 \int_{\varepsilon i + \varepsilon K} |Du_\varepsilon|^2 \, dx - \int_{\varepsilon i + \varepsilon K} g u_\varepsilon \, dx$$

$$\geq \inf\left\{ \varepsilon^2 \int_{\varepsilon K} |Dv|^2 \, dx - \int_{\varepsilon K} g(\varepsilon i + x) v \, dx : v = u(\varepsilon i) \text{ on } \partial \varepsilon K \right\}$$

$$= \varepsilon^n \inf\left\{ \int_K |Dv|^2 \, dx - \int_K g(\varepsilon i + \varepsilon x) v \, dx : v = u(\varepsilon i) \text{ on } \partial K \right\}.$$

If we set

$$\phi(x, u) = \inf\left\{ \int_K |Dv|^2 \, dx - g(x) \int_K v \, dx : v = u \text{ on } \partial K \right\}$$

then we deduce a lower estimate of the limit of the contributions on $\Omega \setminus \varepsilon E$ by $\int_\Omega \phi(x, u(x)) \, dx$.

This argument can be carried over rigorously and also removing the assumption that E consists of a single connected component, as stated in the next section.

6.4.1. *Multiphase limits.* Let $E = \bigcup_{i=1}^N E_i$, where E_i are periodic connected open subsets of $\mathbb{R}^n$ with Lipschitz boundary and such that $\overline{E}_i \cap \overline{E}_j = \emptyset$ for $i \neq j$. We also set $E_0 = \mathbb{R}^n \setminus E$. Let the extension operators T_ε^j corresponding to $\Omega \cap E_j$ be defined as in Theorem 6.10. We define the convergence on $H^1(\Omega)$ as the L^2_{loc} convergence of these extensions. Namely, we will write that $u_\varepsilon \to (u_1, \dots, u_N)$ if $T_\varepsilon^j u_\varepsilon \to u_j$ for all $j = 1, \dots, N$, or equivalently, if

$$\lim_{\varepsilon \to 0} \sum_{i=1}^N \int_{\Omega \cap \varepsilon E_i} |u_\varepsilon - u_i|^2 \, dx = 0. \tag{6.31}$$

We define the energies

$$F_\varepsilon(u) = \int_{\Omega \cap \varepsilon E} |Du|^2 \, dx + \varepsilon^2 \int_{\Omega \cap \varepsilon E_0} |Du|^2 \, dx + \int_\Omega |u|^2 \, dx \tag{6.32}$$

for $u \in H^1(\Omega)$. For simplicity we only consider the quadratic perturbation $\int_\Omega |u|^2 \, dx$. By Theorem 6.10, F_ε are equicoercive with respect to the convergence above.

Note that by the closure of quadratic forms there exist A_{hom}^j constant matrices such that

$$\int_\Omega \langle A_{\text{hom}}^j Du, Du \rangle \, dx = \Gamma\text{-}\lim_{\varepsilon \to 0} \int_{\Omega \cap \varepsilon E_j} |Du|^2 \, dx, \tag{6.33}$$

in the sense of Theorem 6.11. Moreover, we define

$$\phi(z_1,\ldots,z_N) = \min\left\{\int_{E_0\cap(0,1)^n} \left(|Dv|^2 + |v|^2\right)\,dy:\right.$$

$$\left. v \in H^1_\#\left((0,1)^n\right), v = z_j \text{ on } E_j, j = 1,\ldots,N\right\}. \qquad (6.34)$$

The following theorem is a particular case of a result by Braides, Chiadò Piat and Piatnitski [50], where general integrands in the vector case are considered.

THEOREM 6.12. *If* $|\partial\Omega| = 0$ *then the functionals* F_ε *defined by* (6.32) Γ*-converge with respect to the convergence* (6.31) *to the functional* F_{hom} *with domain* $H^1(\Omega; \mathbb{R}^N)$ *defined by*

$$F_{\mathrm{hom}}(u_1,\ldots,u_N)$$

$$= \sum_{j=1}^{N} \int_\Omega \left(\langle A^j_{\mathrm{hom}} Du_j, Du_j\rangle + C_j |u_j|^2\right)\,dx + \int_\Omega \phi(u_1,\ldots,u_N)\,dx, \qquad (6.35)$$

where $C_j = |E_j \cap (0,1)^n|$ *and* A^j_{hom} *and* ϕ *are given by* (6.33) *and* (6.34), *respectively.*

7. Phase-transition problems

In the previous sections we have examined sequences of functionals defined on Sobolev spaces, whose minimizers satisfy some weak compactness properties, so that the limit is automatically defined on a Sobolev space, even though the actual form of the limit takes into account oscillations and compactness effects. In this section we will consider families of functionals whose minimizers tend to generate sharp interfaces between zones where they are approximately constant. In the limit we expect the relevant properties of such minimizers to be described by energies, whose domain are partitions of the domain Ω into sets (the *phase domains*).

7.1. *Interfacial energies*

The types of energies we have in mind are functionals defined on partitions of a reference set Ω into sets, which take into account some measure of the interface between those sets. The simplest of such functionals is the "perimeter functional", suitably defined to suit problems in the Calculus of Variations.

7.1.1. *Sets of finite perimeter.* The simplest way to have a definition of *perimeter* which is lower semicontinuous by the L^1-convergence of the sets is by *lower semicontinuity*: if

$E \subset \mathbb{R}^n$ is of class C^1 define the perimeter $\mathcal{P}(E, \Omega)$ of the set E inside the open set Ω in a classical way, and then for an arbitrary set, define

$$\mathcal{P}(E, \Omega) = \inf\left\{\liminf_j \mathcal{P}(E_j, \Omega) \colon \chi_{E_j} \to \chi_E \text{ in } L^1(\Omega), \, E_j \text{ of class } C^1\right\}.$$

Another choice leading to the same definition is to start with E_j of polyhedral type.

If $\mathcal{P}(E, \Omega) < +\infty$, then we say that E is a *set of finite perimeter* or *Caccioppoli set* in Ω. For such sets it is possible to define a notion of measure-theoretical boundary, where a normal is defined, so that we may heuristically picture those sets as having a smooth boundary. In order to make these concepts more precise we recall the definition of the *k-dimensional Hausdorff measure* (in this context we will limit ourselves to $k \in \mathbb{N}$). If E is a Borel set in $\mathbb{R}^n$, then we define

$$\mathcal{H}^k(E) = \sup_{\delta > 0} \frac{\omega_k}{2^k} \inf\left\{\sum_{i \in \mathbb{N}} (\text{diam } E_i)^k \colon \text{diam } E_i \leqslant \delta, \, E \subseteq \bigcup_{i \in \mathbb{N}} E_i\right\},$$

where ω_k is the Lebesgue measure of the unit ball in $\mathbb{R}^k$.

We say that $x \in E$ is a *point of density* $t \in [0, 1]$ if the limit $\lim_{\rho \to 0+} (\omega_n)^{-1} \rho^{-n} |E \cap B_\rho(x)| = t$ exists. The set of all points of density t will be denoted by E_t. If E is a set of finite perimeter in Ω then the De Giorgi's *essential boundary* of E, denoted by $\partial^* E$, is defined as the set of points $x \in \Omega$ with density $1/2$.

THEOREM 7.1 (De Giorgi's rectifiability theorem). *Let $E \subset \mathbb{R}^n$ be a set of finite perimeter in Ω. Then $\partial^* E$ is rectifiable; i.e., there exists a countable family (Γ_i) of graphs of C^1 functions of $(n-1)$ variables such that $\mathcal{H}^{n-1}(\partial^* E \setminus \bigcup_{i=1}^\infty \Gamma_i) = 0$. Moreover, the perimeter of E in $\Omega' \subseteq \Omega$ is given by*

$$\mathcal{P}(E, \Omega') = \mathcal{H}^{n-1}(\partial^* E \cap \Omega').$$

By the previous theorem and the implicit function theorem a *internal normal* $\nu = \nu_E(x)$ to E is defined at $\mathcal{H}^{n-1}$-almost all points x of $\partial^* E$ as the normal of the corresponding Γ_i. A *generalized Gauss–Green formula* holds, which states that the distributional derivative of χ_E is a vector measure given by

$$D\chi_E(B) = \int_B \nu_E \, d\mathcal{H}^{n-1}.$$

In particular, we have $\mathcal{P}(E, \Omega) = |D\chi_E|(\Omega)$, the total variation of the measure $D\chi_E$ on Ω, so that χ_E is a *function with bounded variation*.

A finite *Caccioppoli partition*; i.e., a partition of Ω into sets of finite perimeter $E_1, \ldots, E_M$ can be identified with an element $u \in BV(\Omega; T)$, where $\#T = M$. In this case we will also use the notation $S(u)$ for $\bigcup_i \partial^* E_i$, which is the *jump set* of u. This notation also holds if $u = \chi_E$.

7.1.2. *Convexity and subadditivity conditions.* From the characterization above we easily see that the characteristic functions of a sequence of sets with equibounded perimeter are bounded in BV, so that we may extract a converging subsequence in L^1, and $\mathcal{P}(E, \Omega) = |D\chi_E|(\Omega) \leqslant \liminf_j |D\chi_{E_j}|(\Omega) = \liminf_j \mathcal{P}(E_j, \Omega)$ by the lower semicontinuity of the total variation, so that $\mathcal{P}(\cdot, \Omega)$ is a lower-semicontinuous functional.

Actually, it may be easily seen that functionals of the form

$$F(E) = \int_{\Omega \cap \partial^* E} \varphi(v) \, d\mathcal{H}^{n-1} \tag{7.1}$$

are L^1-lower semicontinuous if and only if the positively homogeneous extension of degree 1 of φ to $\mathbb{R}^n$ is convex.

7.1.3. *Integral representation.* The application of the localization methods often necessitates the representation of functionals defined on sets of finite perimeter or on (finite) Caccioppoli partitions. An analogue of the representation theorem for integral functionals is the following, of which a simple proof can be obtained from that in the paper by Braides and Chiadò Piat [49], Section 3, where it is directly proved for infinite Caccioppoli partitions.

THEOREM 7.2 (Integral representation on Caccioppoli partitions). *Let T be a finite set and $F : BV(\Omega; T) \times \mathcal{B}(\Omega) \to [0, +\infty)$ be a function defined on pairs Caccioppoli partition/Borel subset of Ω, satisfying*
  (i) *$F(u, \cdot)$ is a measure for every $u \in BV(\Omega; T)$;*
  (ii) *F is local on open sets; i.e., $F(u, A) = F(v, A)$ whenever $u = v$ a.e. in A;*
  (iii) *$F(\cdot, A)$ is L^1-lower semicontinuous for all open sets A;*
  (iv) *there exist constants $c_1, c_2 > 0$ such that $c_1 \mathcal{H}^{n-1}(B \cap S(u)) \leqslant F(u, B) \leqslant c_2 \mathcal{H}^{n-1}(B \cap S(u))$.*
Then there exist Borel functions $\varphi_{ij} : \Omega \times S^{n-1} \to [0, +\infty)$ such that

$$F(u, B) = \sum_{i \neq j} \int_{B \cap \partial^* E_i \cap \partial^* E_j} \varphi_{ij}(x, v_j) \, d\mathcal{H}^{n-1} \tag{7.2}$$

for all $u \in BV(\Omega; T)$ identified with the partition $E_1, \ldots, E_M$ with inner normal v_j to E_j and every Borel subset B of Ω.

For Caccioppoli partitions lower-semicontinuity conditions are more complex than the simple convexity: for homogeneous functionals F as in (7.2) of the form

$$F(E_1, \ldots, E_M) = \sum_{i < j} \int_{\Omega \cap \partial^* E_i \cap \partial^* E_j} \varphi_{ij}(v_i) \, d\mathcal{H}^{n-1}, \tag{7.3}$$

where v_i is the interior normal to E_i, necessary conditions are the convexity of each φ_{ij}, and their *subadditivity*: $\varphi_{ij}(v) \leqslant \varphi_{ik}(v) + \varphi_{kj}(v)$ for all v. These two combined conditions are not sufficient, and a more complex condition called *BV-ellipticity*, that mirrors the notion of quasiconvexity, turns out to be necessary and sufficient [18].

7.1.4. *Energies depending on curvature terms.* In the literature other types of energies defined on boundaries of sets have been introduced, especially for Computer Vision models. One type of energy (in a two-dimensional setting) is the *elastica functional* (see [130])

$$F(E) = \int_{\partial E} \left(1 + \kappa^2\right) d\mathcal{H}^1,\tag{7.4}$$

defined on sets with $W^{2,2}$ boundary, where κ denotes the curvature of ∂E. Note that on one hand the $W^{2,2}$ bounds ensure easier compactness properties for sets with equibounded energy, while on the other hand parts of the boundary of such sets may "cancel" in the limit, giving rise to sets with cusp singularities. As a consequence the functional is not lower semicontinuous and its relaxation exhibits complex nonlocal effects that have been studied by Bellettini, Dal Maso and Paolini [33] and more recently by Bellettini and Mugnai [34].

7.2. *Gradient theory of phase transitions*

It is well known that the minimization of a nonconvex energy often leads to minimizing sequences with oscillations, highlighted by a relaxation of the energy. This is not the case if we add a singular perturbation with a gradient term. We will be looking at the behavior of minimum problems

$$\min\left\{\int_\Omega W(u)\,dx + \varepsilon^2 \int_\Omega |Du|^2\,dx : \int_\Omega u\,dx = C\right\},\tag{7.5}$$

where $u : \Omega \to \mathbb{R}$, and W is a nonconvex energy. Upon an affine translation of u, that does not change the minimizers of problem (7.5), is not restrictive to suppose that

$$W \geqslant 0 \quad \text{and} \quad W(u) = 0 \quad \text{only if } u = 0, 1\tag{7.6}$$

(or two other points). W is called a *double-well energy* and the energies above are related to the Cahn–Hilliard theory of liquid–liquid phase transitions.

Note that under the hypotheses above, if $0 < C < |\Omega|$ then the minimum of $\int_\Omega W(u)\,dx$ is 0, and is achieved on any $u = \chi_E$ with $|E| = C$. The gradient term, however, forbids such configurations, and we expect the creation of interfaces to be penalized by the second integral. A heuristic scaling argument can be performed in dimension one to understand the scale of this penalization: if the transition of u is on an interval I of size δ, where the gradient is of the order $1/\delta$, we have

$$\int_I W(u)\,dx + \varepsilon^2 \int_I |u'|^2\,dx \approx \delta + \frac{\varepsilon^2}{\delta}.\tag{7.7}$$

The minimization in δ gives $\delta = \varepsilon$ and a contribution of order ε. This argument suggests a scaling of the problem and to consider

$$m_\varepsilon = \min\left\{\frac{1}{\varepsilon}\int_\Omega W(u)\,dx + \varepsilon \int_\Omega |Du|^2\,dx : \int_\Omega u\,dx = C\right\},\tag{7.8}$$

whose minimizers are clearly the same as the problem in (7.5).

7.2.1. *The Modica–Mortola result.* The Γ-limit of the energy in (7.8) is one of the first examples in the literature and is due to Modica and Mortola [123] (see also [5,41,45,122, 140]). In this subsection we will examine the behavior of the energies

$$F_\varepsilon(u) = \frac{1}{\varepsilon} \int_\Omega W(u)\, dx + \varepsilon \int_\Omega |Du|^2\, dx, \quad u \in H^1(\Omega), \tag{7.9}$$

with W as in (7.6) and such that $W(u) \geqslant c(|u|^2 - 1)$.

Note that the *volume constraint* $\int_\Omega u\, dx = C$ is not continuous, so that it cannot be simply added to the Γ-limit, so that a separate argument must, and will, be used.

THEOREM 7.3 (Modica–Mortola's theorem). *The functionals in (7.9) Γ-converge with respect to the $L^1(\Omega)$ convergence to the functional*

$$F_0(u) = \begin{cases} c_W \mathcal{P}(\{u = 1\}, \Omega) \\ \quad = c_W \mathcal{H}^{n-1}(\partial^*\{u = 1\} \cap \Omega) & \text{if } u \in \{0, 1\} \text{ a.e.,} \\ +\infty & \text{otherwise,} \end{cases} \tag{7.10}$$

where $c_W = 2\int_0^1 \sqrt{W(s)}\, ds$.

PROOF. *The one-dimensional case.* Suppose that $\lim_j F_{\varepsilon_j}(u_j) < +\infty$. Fix some $\eta > 0$ and consider an interval I such that u_j takes the values η and $1 - \eta$ at the endpoints of the interval. We can use the following *Modica–Mortola trick* to estimate

$$\int_I \left(\frac{1}{\varepsilon_j} W(u_j) + \varepsilon_j |u'|^2\right) dt \geqslant 2 \int_I \sqrt{W(u_j)}\, |u'_j|\, dt$$

$$\geqslant 2 \int_\eta^{1-\eta} \sqrt{W(s)}\, ds =: C_\eta \tag{7.11}$$

(we have simply used the algebraic inequality $a^2 + b^2 \geqslant 2ab$ and the change of variables $s = u_j(t)$). From this inequality we easily deduce that the number of transitions between η and $1 - \eta$ is equibounded. Since $\int_\Omega W(u_j)\, dt \leqslant \varepsilon C$, we also deduce that $u_j \to \{0, 1\}$ in measure, so that we have (up to subsequences) $u_j \to u$, where u is a piecewise-constant function taking values in $\{0, 1\}$. If we denote by $S(u)$ the set of discontinuity points of u the inequality above yields

$$\liminf_j F_{\varepsilon_j}(u_j) \geqslant C_\eta \#\big(S(u)\big), \tag{7.12}$$

and then the lower bound is achieved by the arbitrariness of η.

To prove the limsup inequality, take v the solution of

$$v'(s) = \sqrt{W(v)}, \quad v(0) = \frac{1}{2} \tag{7.13}$$

(suppose for simplicity that we have a global solution to this problem), and define $v_\varepsilon(t) = v(t/\varepsilon)$. Note that v_ε tends to $H = \chi_{[0,+\infty)}$ (the Heaviside function with jump in 0) and it optimizes the inequality in (7.11): $\frac{1}{\varepsilon}W(v_\varepsilon) + \varepsilon|v_\varepsilon'|^2 = 2\sqrt{W(v_\varepsilon)}|v_\varepsilon'|$ so that it gives a recovery sequence for $u(t) = H(t)$. For a general $u \in \{0, 1\}$ we easily construct a recovery sequence by suitably gluing the functions $v_\varepsilon((\bar{t} \pm t)/\varepsilon)$, where $\bar{t} \in S(u)$.

The n-dimensional case. In order to apply the "slicing procedure" we will need a result characterizing sets of finite perimeter through their sections. For a piecewise-constant function u on an open set of $\mathbb{R}$ we use the notation $S(u)$ for its set of discontinuity points (if u is thought as an L^1 function we mean its essential discontinuity points). We use the notation for one-dimensional sections introduced in Section 3.4.

THEOREM 7.4 (Sections of sets of finite perimeter). (a) *Let E be a set of finite perimeter in a smooth open set $\Omega \subset \mathbb{R}^n$ and let $u = \chi_E$. Then for all $\xi \in S^{n-1}$ and for $\mathcal{H}^{n-1}$-a.a. $y \in \Pi_\xi$, the function $u_{\xi,y}$ is piecewise constant on $\Omega_{\xi,y}$. Moreover, for such y we have $S(u_{\xi,y}) = \{t \in \mathbb{R}: y + t\xi \in \Omega \cap \partial^*E\}$, and for all Borel functions g,*

$$\int_{\Pi_\xi} \sum_{t \in S(u_{\xi,y})} g(t)\, d\mathcal{H}^{n-1}(y) = \int_{\Omega \cap \partial^*E} g(x)\big|\langle \nu_E, \xi\rangle\big|\, d\mathcal{H}^{n-1}. \tag{7.14}$$

(b) *Conversely, if $E \subset \Omega$ and for all $\xi \in \{e_1, \dots, e_n\}$ and for $\mathcal{H}^{n-1}$-a.a. $y \in \Pi_\xi$ the function $u_{\xi,y}$ is piecewise constant in each interval of $\Omega_{\xi,y}$ and*

$$\int_{\Pi_\xi} \#\big(S(u_{\xi,y})\big)\, d\mathcal{H}^{n-1}(y) < +\infty, \tag{7.15}$$

then E is a set of finite perimeter in Ω.

We follow the steps outlined in Section 3.4.

Step 1. The localized functionals are

$$F_\varepsilon(u, A) = \frac{1}{\varepsilon}\int_A W(u)\, dx + \varepsilon \int_A |Du|^2\, dx. \tag{7.16}$$

Step 2. We choose

$$F_\varepsilon^{\xi,y}(v, I) = \frac{1}{\varepsilon}\int_I W(v)\, dt + \varepsilon \int_I |v'|^2\, dt \tag{7.17}$$

(in this case $F_\varepsilon^{\xi,y}$ is independent of y). We then have, by Fubini's theorem,

$$F_\varepsilon^\xi(u, A) = \frac{1}{\varepsilon}\int_A W(u)\, dx + \varepsilon \int_A |\langle \xi, Du\rangle|^2\, dx. \tag{7.18}$$

Note that $F_\varepsilon^\xi \leqslant F_\varepsilon$.

Step 3. By the one-dimensional proof the Γ-limit

$$F^{\xi,y}(v, I) := \Gamma\text{-}\lim_{\varepsilon \to 0} F_\varepsilon^{\xi,y}(v, I) = c_W \#\big(S(v)\big) \quad \text{if } v \in \{0, 1\} \text{ a.e. on } I \qquad (7.19)$$

$(+\infty$ otherwise). We define F^ξ as in (3.12). Note that $F^\xi(u, A)$ is finite if and only if $u \in \{0, 1\}$ a.e. in Ω, $u_{\xi,y}$ is piecewise constant on $A_{\xi,y}$ for $\mathcal{H}^{n-1}$-a.a. $y \in \Pi_\xi$, and (7.15) holds.

Step 4. From Fatou's lemma we deduce that

$$\Gamma\text{-}\liminf_{\varepsilon \to 0^+} F_\varepsilon(u, A) \geqslant \Gamma\text{-}\liminf_{\varepsilon \to 0^+} F^\xi_\varepsilon(u, A) \geqslant F^\xi(u, A)$$

for all $\xi \in S^{n-1}$.

Step 5. By Step 3 and Theorem 7.4(b), we deduce that the Γ-lower limit $F'(u, A) = \Gamma\text{-}\liminf_{\varepsilon \to 0^+} F_\varepsilon(u, A)$ is finite only if $u = \chi_E$ for some set E of finite perimeter in A. Moreover, $C > 0$ exists such that $F'(u, A) \geqslant C\mathcal{H}^{n-1}(A \cap \partial^*E)$.

Step 6. If $u = \chi_E$, for some set E of finite perimeter from Theorem 7.4(a), we have

$$F^\xi(u, A) = c_W \int_{A \cap \partial^*E} \big|\langle \xi, v_u \rangle\big| \, d\mathcal{H}^{n-1}(y). \qquad (7.20)$$

Hence

$$F'(u, A) \geqslant c_W \int_{A \cap \partial^*E} \big|\langle \xi, v_u \rangle\big| \, d\mathcal{H}^{n-1}(y). \qquad (7.21)$$

Step 7. Since all F_ε are local, then if $u = \chi_E$ for some set E of finite perimeter the set function $\mu(A) = F'(u, A)$ is superadditive on disjoint open sets. From Theorem 3.1 applied with $\lambda = \mathcal{H}^{n-1} \llcorner \partial^*E$, and $\psi_i(x) = \chi_{\partial^*E}|\langle \xi_i, v_u \rangle|$, where (ξ_i) is a dense sequence in S^{n-1}, we conclude that

$$F'(u, A) \geqslant c_W \int_{S(u) \cap \partial^*E} \sup_i \{|\langle \xi_i, v \rangle|\} \, d\mathcal{H}^{n-1}. \qquad (7.22)$$

The liminf inequality follows noticing that $\sup_i\{|\langle \xi_i, v \rangle|\} = 1$.

The liminf inequality is sharp for functions with a "unidimensional" profile; i.e., that on lines orthogonal to ∂^*E follow the one-dimensional recovery sequences. This argument can be easily carried over if ∂^*E is smooth; the general case will then be achieved by approximation via the following result (whose proof easily follows from Sard's theorem by using the coarea formula).

PROPOSITION 7.5 (Density of smooth sets). *If Ω is a Lipschitz set and E is a set of finite perimeter in Ω, then there exists a sequence (E_j) of sets of finite perimeter in Ω, such that $\lim_j |E \triangle E_j| = 0$, $\lim_j \mathcal{P}(E_j, \Omega) = \mathcal{P}(E, \Omega)$, and for every open set Ω' with $\Omega \subset\subset \Omega'$ there exist sets E'_j of class C^∞ in Ω' and such that $E'_j \cap \Omega = E_j$.*

It remains to exhibit a recovery sequence when ∂E is smooth. In that case it suffices to take

$$u_\varepsilon(x) = v\left(\frac{d(x)}{\varepsilon}\right), \tag{7.23}$$

where $d(x) = \mathrm{dist}(x, \Omega \setminus E) - \mathrm{dist}(x, E)$ is the *signed distance function* to ∂E. A simple computation using the coarea formula in the form

$$\int_\Omega f(x)|Dd|\,\mathrm{d}x = \int_{-\infty}^{+\infty} \int_{\{d=t\}\cap A} f(y)\,\mathrm{d}\mathcal{H}^{n-1}(y)\,\mathrm{d}t \tag{7.24}$$

valid if f is a Borel function (recalling that $|Dd| = 1$ a.e.) gives the desired estimate. $\qquad\square$

REMARK 7.6 (Optimal profile problem). We may rewrite the constant c_W as the minimum problem

$$c_W = \min\left\{ \int_{-\infty}^{+\infty} \left(W(v) + |v'|^2\right) \mathrm{d}t \colon u(-\infty) = 0,\ u(+\infty) = 1 \right\}. \tag{7.25}$$

By the proof before, we get that the function v defined in (7.13) is a solution to this minimum problem (*optimal profile* problem). The proof of the limsup inequality shows that a recovery sequence is obtained by scaling an optimal profile.

REMARK 7.7 (Generalizations). By some easy convexity arguments we can adapt the proof of Theorem 7.3 to the case when we substitute $|Du|^2$ by a more general $\varphi^2(Du)$ with φ convex and positively homogeneous of degree one (see [45], Section 4.1.2), in which case the limit is given by

$$F_0(u) = c_W \int_{\Omega \cap \partial^* E} \varphi(v)\,\mathrm{d}\mathcal{H}^{n-1} \tag{7.26}$$

if $u = \chi_E$. As a particular case, we may take $\varphi(v) = \sqrt{\langle Av, v \rangle}$.

7.2.2. *Addition of volume constraints.* As for boundary conditions for integral functionals, to apply Theorem 7.3 to the convergence of the minimum problems m_ε we have to prove that the volume constraint $\int_\Omega u\,\mathrm{d}x = C$ is *compatible* with the Γ-limit. Clearly, the constraint is closed under $L^1(\Omega)$-convergence. By taking the density argument into account, the compatibility then amounts to proving the following.

PROPOSITION 7.8 (Compatibility of volume constraints). *Let E be a set with smooth boundary; then there exist $\bar{u}_\varepsilon \to \chi_E$ such that $\int_\Omega \bar{u}_\varepsilon \, dx = |E \cap \Omega|$ and $\lim_{\varepsilon \to +\infty} F_\varepsilon(\bar{u}_\varepsilon) = c_W \mathcal{H}^{n-1}(\partial E \cap \Omega)$.*

PROOF. The proof of this proposition can be easily achieved by adding to the recovery sequence constructed above u_ε a suitable perturbation. For example, if W is smooth in 0 and 1, we can choose a ball B contained in E (or $\Omega \setminus E$; it is not restrictive to suppose that such B exists by approximation) and $\phi_\varepsilon \in C_0^\infty(B)$ with $0 \leqslant \phi_\varepsilon \leqslant 1$, $\phi_\varepsilon \to 1$ and $\int_B (|\phi_\varepsilon|^2 + \varepsilon^2 |D\phi_\varepsilon|^2) \, dx = O(1)$, and consider $\bar{u}_\varepsilon = u_\varepsilon + c_\varepsilon \phi_\varepsilon$ where $c_\varepsilon \to 0$ are such that $\int_\Omega \bar{u}_\varepsilon \, dx = |E \cap \Omega|$. $\qquad\square$

With this proposition the proof of the Γ-limit of the functionals in (7.8) is complete. Note moreover that the sequence is equicoercive by the proof of Theorem 7.3. We then obtain the convergence result as follows.

COROLLARY 7.9 (Convergence to minimal sharp interfaces). *Let u_ε be a minimizer for problem m_ε as defined in (7.8). Then, we have $m_\varepsilon \to m$ and, up to subsequences, $u_\varepsilon \to u$, where $u = \chi_E$ and E is a minimizer of the problem*

$$m = \min\{c_W \mathcal{P}(E; \Omega) : |E| = C\}. \tag{7.27}$$

7.2.3. A selection criterion: minimal interfaces. The result in Corollary 7.9 can also be read as a result on the convergence of the original minimum problems in (7.5). Note that we have

$$\int_\Omega W^{**}(u) \, dx = \Gamma\text{-}\lim_{\varepsilon \to 0} \int_\Omega \left(W(u) + \varepsilon^2 |Du|^2 \right) dx \tag{7.28}$$

with respect to the weak-$L^2(\Omega)$ convergence, so that the limit of (7.5) can be expressed as

$$\min\left\{ \int_\Omega W^{**}(u) \, dx : \int_\Omega u \, dx = C \right\}, \tag{7.29}$$

where W^{**} is the convex envelope of W. In our case $W^{**} = 0$ on $[0, 1]$ so that the first minimum is 0 and is achieved on all test functions with $0 \leqslant u \leqslant 1$.

In other words, sequences (u_ε) with $\int_\Omega (W(u_\varepsilon) + \varepsilon^2 |Du_\varepsilon|^2) \, dx = o(1)$ may converge weakly in $L^2(\Omega)$ to any $0 \leqslant u \leqslant 1$. On the contrary, if $\int_\Omega (W(u_\varepsilon) + \varepsilon^2 |Du_\varepsilon|^2) \, dx = o(\varepsilon)$, then $u_\varepsilon \to \chi_E$, where E is a set with minimal perimeter. Since the minimum in (7.29) coincides with

$$\min\left\{ \int_\Omega W(u) \, dx : \int_\Omega u \, dx = C \right\}, \tag{7.30}$$

the addition of the singular perturbation represents a choice criterion between all minimizers of this nonconvex variational problem.

7.2.4. *Addition of boundary values.* It is interesting to note that boundary values are trivially *not compatible* with this Γ-limit, as the limit energy is defined only on characteristic functions. Nevertheless the limit of an energy of the form

$$F_\varepsilon^\varphi(u) = \begin{cases} \frac{1}{\varepsilon}\int_\Omega W(u)\,\mathrm{d}x + \varepsilon \int_\Omega |Du|^2\,\mathrm{d}x & \text{if } u = \varphi \text{ on } \partial\Omega, \\ +\infty & \text{otherwise,} \end{cases} \tag{7.31}$$

can be easily computed.

To check this, we may first consider the one-dimensional case, with $\Omega = (0, 1)$ and the boundary condition $u(0) = u_0$. The same line of proof as before shows that the Γ-limit is again finite only if $u \in \{0, 1\}$ and is piecewise constant, but we have the additional boundary term $\phi(u_0, u(0+))$, where

$$\phi(s, t) = 2 \left| \int_s^t \sqrt{W(\tau)}\,\mathrm{d}\tau \right|, \tag{7.32}$$

and $u(0+)$ is the (approximate) right-hand side limit of u at 0. This term accounts for the boundary mismatch of the two wells from the boundary condition, and is derived again using the Modica–Mortola trick. The same applies with a boundary condition $u(1) = u_1$.

In the general n-dimensional case, we may obtain results of the following form.

THEOREM 7.10 (Relaxed boundary data). *Let Ω be a set with smooth boundary and φ be a continuous function. Then the Γ-limit of the functionals F_ε^φ in (7.31) is given by*

$$F_0^\varphi(u) = c_W \mathcal{P}(\{u = 1\}; \Omega) + \int_{\partial\Omega} \phi\big(\varphi(y), u(y)\big)\,\mathrm{d}\mathcal{H}^{n-1}(y), \tag{7.33}$$

where $u(y)$ for $y \in \partial\Omega$ is understood as the inner trace of u at y.

7.3. *A compactness result*

As done for integral functionals in Section 4.1, the gradient theory of phase transitions described in Section 7.2 can be framed in a more abstract framework by proving a compactness theorem via the localization methods. This has been done for example by Ansini, Braides and Chiadò Piat [23], to get a result as follows.

Let $W : \mathbb{R} \to [0, +\infty)$ be a continuous function satisfying the hypothesis of Theorem 7.3, let $V_\varepsilon : \mathbb{R}^n \times \mathbb{R} \to [0, +\infty)$ be functions satisfying $c_1 W(u) \leqslant V_\varepsilon(x, u) \leqslant c_2 W(u)$ and let f_ε be a sequence of integrands as in Theorem 4.2. We will consider the functionals $G_\varepsilon : L^1_{\mathrm{loc}}(\mathbb{R}^n) \times \mathcal{A} \to [0, +\infty]$ defined by

$$G_\varepsilon(u, A) = \int_A \left(\frac{V_\varepsilon(x, u)}{\varepsilon} + \varepsilon f_\varepsilon(x, Du) \right) \mathrm{d}x, \quad u \in H^1(A) \tag{7.34}$$

(extended to $+\infty$ elsewhere), where $\mathcal{A}$ denotes the family of bounded open subsets of $\mathbb{R}^n$.

THEOREM 7.11 (Compactness by Γ-convergence). *For every sequence (ε_j) converging to 0, there exist a subsequence (not relabeled) and a functional $G : L^1_{\mathrm{loc}}(\mathbb{R}^n) \times \mathcal{A} \to [0, +\infty]$, such that (G_{ε_j}) Γ-converges to G for every A bounded Lipschitz open set, and for every $u \in L^1_{\mathrm{loc}}(\mathbb{R}^n)$ such that $u = \chi_E$ with E a set of finite perimeter, with respect to the strong topology of $L^1(A)$. Moreover, there exists a Borel function $\varphi : \mathbb{R}^n \times S^{n-1} \to [0, +\infty)$ such that*

$$G(u, A) = \int_{\partial^* E \cap A} \varphi(x, v) \, d\mathcal{H}^{n-1} \tag{7.35}$$

for every open set A.

PROOF. We may follow the localization method in Section 3.3. Note that we may follow the same line as in Section 4.2.1 to prove the fundamental estimate, with some finer technical changes. Moreover, by a simple comparison argument we obtain $c_1 c_W \mathcal{P}(E, A) \leqslant G(u, A) \leqslant c_2 c_W \mathcal{P}(E, A)$, so that we may apply the representation theorem in Section 7.1.3. Details are found in [23]. $\qquad\square$

REMARK 7.12 (A formula for the interfacial energy density). Note that a simple derivation formula for φ as in Theorem 4.4 does not hold; in particular, φ is not determined by the behavior of $G(u, A)$ when $S(u)$ is an hyperplane (this would be the analogue of an affine function in a Sobolev setting). To see this it is sufficient to take $\varphi(x, v) = 2 - \chi_{\partial B_1(0)}(x)$, which gives a lower-semicontinuous G; in this case, $G(u, A) = 2\mathcal{H}^{n-1}(A \cap S(u))$ if $S(u)$ is an hyperplane, but for example, $G(\chi_{B_1(0)}, A) = \mathcal{H}^{n-1}(A \cap \partial B_1(0))$. Nevertheless, if G is translation invariant then $\varphi = \varphi(v)$ and hence by convexity,

$$\varphi(v) = \min\{G(u, \overline{Q}_v) : u = \chi_E \ v^\perp\text{-periodic}, u = 1 \text{ on } Q_v^+, u = 0 \text{ on } Q_v^-\}, \tag{7.36}$$

where Q_v is a cube with center 0, side length 1 and $Q_v^\pm = \partial Q_v \cap \{\langle x, v \rangle = \pm 1/2\}$ are the two faces of Q_v orthogonal to v. The $v^\perp$-periodicity of u must be understood as periodicity in the $n - 1$ directions given by the edges of the cube Q_v other than v, the values on $Q_v^\pm$ are taken in the sense of traces, and the functional G is extended to a measure on all Borel sets.

The formula above is useful to derive a characterization of φ in terms of the approximating G_ε simply by applying the theorem on convergence of minimum problems after Γ-convergence, in the same spirit of the derivation of the homogenization formula in Section 5.1.

7.4. *Other functionals generating phase-transitions*

In this section we present some other types of functionals whose Γ-limit is a phase-transition energy, briefly highlighting differences and analogies with the gradient theory outlined in Section 7.2.

7.4.1. *A nonlocal model.* Another class of energies giving rise to phase transitions, and linked to some models deriving from Ising systems, have been studied by Alberti and Bellettini [7] (see also [8]). They have the form

$$F_\varepsilon(u) := \frac{1}{\varepsilon} \int_\Omega W\big(u(x)\big)\,\mathrm{d}x + \frac{\varepsilon}{4} \int_{\Omega \times \Omega} J_\varepsilon\big(x' - x\big) \left(\frac{u(x') - u(x)}{\varepsilon} \right)^2 \mathrm{d}x\,\mathrm{d}x',$$

where $J_\varepsilon(y) := 1/\varepsilon^N J(y/\varepsilon)$, and J is an even positive L^1 kernel with $\int_{\mathbb{R}^n} J(h)|h|\,\mathrm{d}h < +\infty$. Note that F_ε can be obtained from the functional studied by Modica and Mortola (Section 7.2.1) by replacing the term $|Du(x)|$ in the second integral in (7.9) with the average of the finite differences $1/\varepsilon |u(x + \varepsilon h) - u(x)|$ with respect to the measure $J_\varepsilon \mathcal{L}^n$. An equicoerciveness property for the functionals F_ε in $L^2(\Omega)$ can be proved. The Γ-limit is finite only on characteristic functions of sets of finite perimeter; its characterization is the following.

THEOREM 7.13. *The Γ-limit of F_ε in the $L^2(\Omega)$ topology is given by*

$$F(u) = \int_{\Omega \cap \partial^* E} g(\nu)\,\mathrm{d}\mathcal{H}^{n-1} \tag{7.37}$$

on $u = \chi_E$ characteristic functions of sets of finite perimeter, where the anisotropic phase-transition energy density g is defined as

$$g(\nu) = \lim_{T \to +\infty} \frac{1}{T^{n-1}}$$

$$\times \inf\left\{ \int_{TQ_\nu} W(u)\,\mathrm{d}x + \int_{TQ_\nu \times \mathbb{R}^n} J(h)\big(u(x+h) - u(x)\big)^2 \mathrm{d}x\,\mathrm{d}h : \right.$$

$$\left. u\, T\nu^\perp\text{-}periodic,\, u(x) = 1 \text{ if } \langle x, \nu \rangle \geqslant \frac{T}{2},\, u = 0 \text{ if } \langle x, \nu \rangle \leqslant -\frac{T}{2} \right\}$$

(we use the notation of Remark 7.12).

PROOF. Even though the functions are nonlocal, the limit contribution to the surface energy can be computed by using the arguments of Remark 7.12 (which explains the form of g). A particular care must be used to deal with the boundary data. Details can be found in [7]. $\qquad\square$

7.4.2. *A two-parameter model.* An intermediate model between the local Cahn–Hilliard model and the nonlocal above deriving from Ising systems, can be obtained by considering energies depending on one more parameter v, of the form

$$F_\varepsilon(u, v) = \frac{1}{\varepsilon} \int_\Omega W(u)\,\mathrm{d}x + \frac{\alpha}{\varepsilon} \int_\Omega (u - v)^2\,\mathrm{d}x + \varepsilon \int_\Omega |Dv|^2\,\mathrm{d}x \tag{7.38}$$

for $u \in L^2(\Omega)$ and $v \in H^1(\Omega)$. These functionals arise independently in the study of thin bars, with the additional variable taking into account the deviation from one-dimensional deformations (see [136]), and their Γ-limit has been studied by Solci and Vitali [139]. In the case $\alpha = +\infty$, we recover the Modica–Mortola functionals.

Note that the second term in (7.38) forces $u = v$ as $\varepsilon \to 0$ and the first one gives $u \in \{0, 1\}$. Note however that the variable u may be discontinuous at fixed $\varepsilon > 0$. An equi-coerciveness theorem can be proved for the family F_ε with respect to the L^2-convergence, as well as that the Γ-limit is finite only on characteristic functions of sets of finite perimeter. The characterization of the Γ-limit (which is finite for $u = v = \chi_E$) is the following theorem.

THEOREM 7.14. *Let F_ε be as above. Then $\Gamma\text{-}\lim_{\varepsilon \to 0} F_\varepsilon(u, v) = F^\alpha(u)$, where*

$$F^\alpha(u) = c_W^\alpha \mathcal{H}^{n-1}\big(\Omega \cap \partial^* E\big) \tag{7.39}$$

if $u = \chi_E$, and c_W^α is defined as

$$c_W^\alpha = \sqrt{\alpha}\,\inf\bigg\{ \int_{\mathbb{R}} W(\varphi)\,\mathrm{d}x + \frac{\alpha^2}{4} \int_{\mathbb{R}^2} e^{-\alpha|x-y|}\big(\varphi(x) - \varphi(y)\big)^2 \,\mathrm{d}x\,\mathrm{d}y :$$

$$\varphi(-\infty) = 0,\, \varphi(+\infty) = 1 \bigg\}.$$

Furthermore, we have $\lim_{\alpha \to +\infty} c_W^\alpha = 2 \int_0^1 \sqrt{W(s)}\,\mathrm{d}s$.

PROOF. The idea of the proof is to reduce to the one-dimensional case by slicing, and then minimize the effect of v for fixed u. In this way we recover a nonlocal one-dimensional functional as in Theorem 7.13. Details can be found in [139]. $\qquad\square$

Note that we may recover anisotropic functionals by considering terms of the form $g^2(Dv)$ in the place of $|Dv|^2$, with g a norm.

7.4.3. *A perturbation with the $H^{1/2}$ norm.* Energies similar to those in Section 7.4.1 are the following ones studied by Alberti, Bouchitté and Seppecher [9]:

$$G_\varepsilon(u) = \int_\Omega W(u)\,\mathrm{d}t + \varepsilon^2 \int_{\Omega \times \Omega} \left| \frac{u(t) - u(s)}{t - s} \right|^2 \mathrm{d}t\,\mathrm{d}s \tag{7.40}$$

on a one-dimensional set $\Omega = (a, b)$. In this case, a different scaling is needed. Adapting the argument in Section 7.2 we can argue that the first term forces $u \in \{0, 1\}$, and we look at a transition from 0 to 1 taking place on an interval $[t, t + \delta]$. We then have

$$G_\varepsilon(u) \geqslant C\delta + 2\varepsilon^2 \int_{(a,t) \times (t+\delta, b)} \left| \frac{1}{t - s} \right|^2 \mathrm{d}t\,\mathrm{d}s \geqslant C\delta - 2\varepsilon^2(\log \delta + C).$$

By optimizing the last expression, we get $\delta = 2\varepsilon^2/C$ and hence $G_\varepsilon(u) \geqslant 4\varepsilon^2|\log\varepsilon| + O(\varepsilon^2)$. We are then led to the scaled energies

$$F_\varepsilon(u) = \frac{1}{\varepsilon^2|\log\varepsilon|} \int_\Omega W(u)\, dt + \frac{1}{|\log\varepsilon|} \int_{\Omega\times\Omega} \left| \frac{u(t) - u(s)}{t - s} \right|^2 dt\, ds, \qquad (7.41)$$

for which we have the following Γ-convergence result (see [9], to which we also refer for the proof of their equicoerciveness).

THEOREM 7.15 (Phase transitions generated by an $H^{1/2}$-singular perturbation). *The Γ-limit F_0 of F_ε with respect to the L^1-convergence is finite only on piecewise-constant functions, for which $F_0(u) = 4\#(S(u))$.*

PROOF. The crucial point is a compactness and rearrangement argument that allows to reduce to the case where u is close to 0 or 1 except for a finite number of intervals, to which the computation above can be applied. A recovery sequence is obtained by taking $u_\varepsilon = u$ except on intervals of length ε^2 around $S(u)$. $\qquad\square$

It must be noted that contrary to the energies considered until now, here we do not have an "equipartition" of the energy in the two terms of F_ε, but the whole lower bound is due to the double integral. As a consequence we do not obtain an optimal profile problem by scaling the energy. The loss of such a property makes the problem more difficult and will be found again for Ginzburg–Landau energies (see Section 8.1).

REMARK 7.16. By renaming ε the scaling factor $1/|\log\varepsilon|$, we obtain the Γ-convergence of the energies

$$H_\varepsilon(u) = \lambda_\varepsilon \int_\Omega W(u)\, dt + \varepsilon \int_{\Omega\times\Omega} \left| \frac{u(t) - u(s)}{t - s} \right|^2 dt\, ds, \qquad (7.42)$$

to $2K\#(S(u))$ whenever $\varepsilon\log\lambda_\varepsilon \to K$.

REMARK 7.17 (An application: the line-tension effect). The result above has been applied by Alberti, Bouchitté and Seppecher [10] to the study of energies defined on $\Omega \subset \mathbb{R}^3$ with smooth boundary by

$$\mathcal{F}_\varepsilon(u) = \frac{1}{\varepsilon} \int_\Omega W_1(u)\, dx + \lambda_\varepsilon \int_{\partial\Omega} W_2(u)\, d\mathcal{H}^2 + \varepsilon \int_\Omega |Du|^2\, dx, \qquad (7.43)$$

where W_i are two double-well potentials.

We can give a heuristic derivation of the limit of such energies, and for the sake of simplicity we suppose that $\Omega \subset \mathbb{R}^2$ (the case treated in [10] uses further blow-up and slicing arguments). If $\mathcal{F}_\varepsilon(u_\varepsilon) < +\infty$, then $u_\varepsilon \to u$ in Ω and, if v_ε is the trace of u_ε on $\partial\Omega$, then (up to subsequences) $v_\varepsilon \to v$ as $\varepsilon \to 0$. Note that we can rewrite $\mathcal{F}_\varepsilon$ as $\mathcal{F}_\varepsilon^1 + \mathcal{F}_\varepsilon^2$, where

$$\mathcal{F}_\varepsilon^1(u) = \frac{1}{\varepsilon} \int_\Omega W_1(u)\, dx + \delta\varepsilon \int_\Omega |Du|^2\, dx, \tag{7.44}$$

$$\mathcal{F}_\varepsilon^2(u) = \lambda_\varepsilon \int_{\partial\Omega} W_2(u)\, d\mathcal{H}^1 + (1-\delta)\varepsilon \int_\Omega |Du|^2\, dx. \tag{7.45}$$

This latter term will give the contribution due to v. Upon a blow-up close to each essential discontinuity points of v in $\partial\Omega$ and change of variables argument we can reduce to treat the case $\{x_1^2 + x_2^2 < r, x_2 > 0\}$ a half-disk and $(0,0)$ is a discontinuity point of v on $(-r, r)$, for which a lower bound for this latter functional is

$$\mathcal{G}_\varepsilon^2(v_\varepsilon) = \lambda_\varepsilon \int_{-r}^{r} W_2(v_\varepsilon)\, dx_1$$

$$+ \frac{1}{2\pi}(1-\delta)\varepsilon \int_{(-r,r)^2} \left| \frac{v_\varepsilon(t) - v_\varepsilon(s)}{t - s} \right|^2 dt\, ds \tag{7.46}$$

(the last term obtained by minimization at fixed v_ε).

If $\varepsilon \log \lambda_\varepsilon \to K$, the use of the previous result (see Remark 7.16) adapted to $\mathcal{G}_\varepsilon^2$ gives $v \in \{0, 1\}$ and provides a term in the limit energy of the form $(1-\delta)K\#(S(v))/\pi$, where $S(v)$ denotes the essential discontinuity points of v in $\partial\Omega$. Since $u_\varepsilon \to u$ in Ω with limit relaxed boundary condition v (see Section 7.2.4), from the limit of $\mathcal{F}_\varepsilon^1$ we get that $u \in \{0, 1\}$ and a term in the limit energy of the form

$$\sqrt{\delta}\left(c_{W_1} \mathcal{H}^1\big(S(u) \cap \Omega\big) + 2 \int_{\partial\Omega} \left| \int_u^v \sqrt{W_1(s)}\, ds \right| d\mathcal{H}^1 \right). \tag{7.47}$$

Note that this last term can also be written as $c_{W_1} \mathcal{H}^1(\partial\Omega \cap \{u \neq v\})$ taking into account that both u and v may only take the value 0 and 1, but gives a general form if the wells of W_i differ. We can use Lemma 3.1 to optimize the role of $\sqrt{\delta}$ and $(1-\delta)$ separately. We refer to [10] for the construction of a recovery sequence which optimizes these lower bounds.

In the three-dimensional case and for W_2 with wells α and β possibly different from 0 and 1 we have the Γ-limit of the form

$$\mathcal{F}_0(u, v) = K \frac{(\beta - \alpha)^2}{\pi} \mathcal{H}^1\big(S(v)\big) + c_{W_1} \mathcal{H}^2\big(S(u) \cap \Omega\big)$$

$$+ c_{W_1} 2 \int_{\partial\Omega} \left| \int_u^v \sqrt{W_1(s)}\, ds \right| d\mathcal{H}^2. \tag{7.48}$$

Note that to get a more formally correct statement we should identify $\mathcal{F}_\varepsilon$ with the functional defined on $H^1(\Omega) \times H^{1/2}(\partial\Omega)$ by

$$\mathcal{F}_\varepsilon(u, v) = \begin{cases} \mathcal{F}_\varepsilon(u) & \text{if } v \text{ is the trace of } u, \\ +\infty & \text{otherwise,} \end{cases} \tag{7.49}$$

where now $S(v)$ denotes the essential boundary of $\{v = \alpha\}$ on $\partial\Omega$.

These functionals have applications in the study of capillarity phenomena. Similar functionals arise in the study of dislocations, where additional difficulties are related to the presence of an infinite-wells potential (see [112]).

7.4.4. *A phase transition with Gibbs' phenomenon.* A one-dimensional perturbation problem deriving from a nonlinear model of a shell-membrane transition has been studied by Ansini, Braides and Valente [24]. The energies F_ε take the form

$$F_\varepsilon(u) = \frac{1}{\varepsilon^3} \int_0^1 \left(\int_0^t u(u-1)\, ds \right)^2 dt + \varepsilon \int_0^1 |u'|^2\, dt, \quad u \in H^1(0,1). \tag{7.50}$$

Note that this energy can be compared with the corresponding Modica–Mortola functional

$$\frac{1}{\varepsilon} \int_0^1 \left(u(u-1)\right)^2 dt + \varepsilon \int_0^1 |u'|^2\, dt, \quad u \in H^1(0,1), \tag{7.51}$$

with $W(s) = s^2(s-1)^2$, which shows a different scaling in ε. Another feature of these energies is that it is not true that $F_\varepsilon((0 \vee u) \wedge 1) \leqslant F_\varepsilon(u)$ so that truncation arguments are not applicable (see Remark 7.19).

THEOREM 7.18. *The functionals F_ε Γ-converge to F with respect to the L^1-convergence, whose domain are piecewise-constant functions taking only the values 0 and 1, and for such functions $F(u) = C\#(S(u))$, where*

$$C = \inf_{T>0} \min \left\{ \int_{-T}^T \left(\int_{-T}^t \varphi(\varphi-1)\, ds \right)^2 dt + \int_{-T}^T |\varphi|^2\, dt : \right.$$

$$\left. \varphi \in H^1(-T,T), \varphi(-T) = 0, \varphi(T) = 1, \int_{-T}^T \varphi(\varphi-1)\, ds = 0 \right\},$$

$$\tag{7.52}$$

and $S(u)$ denotes the set of essential discontinuity points of u.

PROOF. The proof of the equicoerciveness, and hence of the liminf inequality of the functionals is particularly tricky, since the function $u(u-1)$ in the double integral may change sign. As a consequence, the truncation arguments that make computations easier in the Γ-limits considered previously do not hold, and in particular recovery sequences do not satisfy $0 \leqslant u_\varepsilon \leqslant 1$ (and hence are not monotone). This is a kind of Gibbs' phenomenon (see the remark below). Note that the conditions for φ in (7.52) make it easy to construct a recovery sequence. It is a much more technical issue to prove that we may always reduce to sequences (u_ε) such that $\int_{t-\varepsilon T_\varepsilon}^{t+\varepsilon T_\varepsilon} u_\varepsilon(u_\varepsilon - 1)\, ds = 0$ and $u_\varepsilon(t \pm \varepsilon T_\varepsilon) \in \{0,1\}$ for some T_ε, from which derives the possibility of localization of the computation of the Γ-limit on $S(u)$. For details we refer to [24]. $\square$

REMARK 7.19 (Gibbs' phenomenon as a scaling effect). Note that the integral condition $\int_{-T}^{T} \varphi(\varphi - 1)\,ds = 0$ in (7.52) cannot be satisfied if $0 \leqslant \varphi \leqslant 1$; this implies that the same observation is valid for recovery sequences u_ε. More precisely, if $u_\varepsilon \to u$, $0 \leqslant u_\varepsilon \leqslant 1$ and u is not constant, then we have

$$\liminf_{\varepsilon \to 0} \varepsilon^{1/6} F_\varepsilon(u_\varepsilon) > 0.$$

This shows that the addition of the constraint $0 \leqslant u_\varepsilon \leqslant 1$ not only is not compatible with the construction of recovery sequences, but even gives a different scaling of the energy.

7.5. *Some extensions*

7.5.1. *The vector case: multiple wells.* The vector case of the Modica–Mortola functional when $u : \Omega \to \mathbb{R}^m$ and $W : \mathbb{R}^m \to [0, +\infty)$ possesses a finite number of wells (or a discrete set of zeros) can be dealt with similarly. Note that such a setting is necessary when dealing with phases parameterizing mixtures of more than two fluids. In this case, we may suppose that W is continuous, with superlinear growth at infinity and $\{W = 0\} = \{\alpha_1, \ldots, \alpha_M\}$. The L^1-limit u of a sequence with equibounded energy can therefore be identified with a partition (E_i) with $E_i = \{u = \alpha_i\}$, and the Γ-limit is described by the following theorem (see [28]).

THEOREM 7.20 (Multiple phase transitions). *The Γ-limit of the energies*

$$F_\varepsilon(u) = \frac{1}{\varepsilon} \int_\Omega W(u)\,dx + \varepsilon \int_\Omega |Du|^2\,dx, \quad u \in H^1(\Omega; \mathbb{R}^m), \tag{7.53}$$

is described by the functional

$$F(E_1, \ldots, E_M) = \sum_{i > j} c_{ij} \mathcal{H}^{n-1}\left(\Omega \cap \partial^* E_i \cap \partial^* E_j\right)$$

$$c_{ij} = \inf\left\{\int_{\mathbb{R}} \left(W(u) + |u'|^2\right)\,dt : u(-\infty) = \alpha_i, u(+\infty) = \alpha_j\right\}. \tag{7.54}$$

PROOF. We may follow the line of the proof of the scalar case through the localization procedure. The Γ-limit can be represented as in (7.3). By (7.36) we obtain that

$$\varphi_{ij}(\nu) = \lim_{\varepsilon \to 0} \min\left\{\frac{1}{\varepsilon} \int_\Omega W(u)\,dx + \varepsilon \int_\Omega |Du|^2\,dx :\right.$$

$$\left. u\ \nu^\perp\text{-periodic}, u = \alpha_i \text{ on } Q_\nu^+, u = \alpha_j \text{ on } Q_\nu^-\right\}$$

$$\geqslant \lim_{\varepsilon \to 0} \min\left\{\int_{-1/(2\varepsilon)}^{1/(2\varepsilon)} \left(W(\phi) + |\phi'|^2\right)\,dt : \phi\left(-\frac{1}{2\varepsilon}\right) = \alpha_i, \phi\left(\frac{1}{2\varepsilon}\right) = \alpha_j\right\},$$

the last inequality obtained by testing on $u(x) = \phi(\langle x, v \rangle / \varepsilon)$. Taking the limit as $\varepsilon \to 0$ we obtain the inequality $\varphi_{ij}(v) \geqslant c_{ij}$. The converse inequality is obtained by a direct construction with one-dimensional scalings of optimal profiles. $\qquad\square$

Note that the constants c_{ij} automatically satisfy the *wetting condition* $c_{ij} \leqslant c_{ik} + c_{kj}$ corresponding to the necessary subadditivity constraint.

7.5.2. *Solid–solid phase transitions.* The Cahn–Hilliard theory accounts for liquid–liquid phase transitions. The inclusion of functionals of elastic problems into this framework would translate into the Γ-limit of energies of the form

$$F_\varepsilon(u) = \frac{1}{\varepsilon} \int_\Omega W(Du)\, dx + \varepsilon \int_\Omega \left|D^2 u\right|^2 dx, \quad u \in H^2(\Omega; \mathbb{R}^n), \tag{7.55}$$

where $u : \Omega \to \mathbb{R}^n$ represents a deformation, and W is an energy density possessing at least two minimizers A and B in $\mathbb{M}^{m \times n}$.

If W represents a hyperelastic free energy, it must be remarked that the physical assumption of *frame-indifference* would actually force W to vanish on the set $SO(n)A \cup SO(n)B$, where $SO(n)$ is the set of *rotations* in $\mathbb{R}^n$. Nonaffine weak solutions for the limiting problem may exist if the two wells are *rank-one connected (Hadamard's compatibility condition)*; i.e., there exist $R, R' \in SO(n)$ and vectors a, v such that $RA - R'B = a \otimes v$.

We state the Γ-convergence result only in a simplified version obtained by neglecting the frame-indifference constraint, as in the following theorem by Conti, Fonseca and Leoni [76].

THEOREM 7.21 (Solid–solid phase transitions). *Let Ω be a convex set of $\mathbb{R}^n$, let A and B be $n \times n$ matrices and suppose that vectors a, v exist such that $A - B = a \otimes v$. We suppose that W is continuous, positive, growing more than linearly at infinity and vanishing exactly on $\{A, B\}$. Then the Γ-limit F of F_ε is finite only on continuous piecewise-affine functions u such that $Du \in \{A, B\}$ almost everywhere. If $S(Du)$ denotes the set of discontinuity points for Du then $S(Du)$ is the union of parallel hyperplanes orthogonal to v, and*

$$F(u) = c_{A,B}\mathcal{H}^{n-1}\big(S(Du)\big), \tag{7.56}$$

where

$$c_{A,B} = \inf\left\{\liminf_{\varepsilon \to 0} F_\varepsilon(u_\varepsilon, Q_v) : u_\varepsilon \to u_{A,B}\right\} \tag{7.57}$$

and

$$u_{A,B}(x) = \begin{cases} Ax & \text{if } \langle x, v \rangle \geqslant 0, \\ Bx & \text{if } \langle x, v \rangle < 0 \end{cases} \tag{7.58}$$

(i.e., $c_{A,B} = \Gamma\text{-}\liminf_{\varepsilon \to 0} F_\varepsilon(u_{A,B}, Q_v)$).

PROOF. Functions u with finite energy for the limiting problem are necessarily piecewise-affine deformations, whose interfaces are hyperplanes with normal v (see the notes by Müller [129]). Despite the simple form of the limit deformations, a number of difficulties arises in the construction of a recovery sequence. In particular, De Giorgi's trick to glue together low-energy sequences does not work as such, and more properties of those sequences must be exploited such as that no rotations of the gradient are allowed. We refer to [76] for details. $\qquad\square$

To tackle the physical case more refined results must be used such as rigidity properties of low-energy sequences (see Theorem 9.8). We refer to the work of Conti and Schweizer [77] for a detailed proof.

REMARK 7.22. The computation of the Γ-limit in the higher-order scalar case, when we consider energies of the form

$$F_\varepsilon(u) = \frac{1}{\varepsilon} \int_\Omega \left(1 - |Du|^2\right) dx + \varepsilon \int_\Omega |D^2u|^2 dx, \quad u \in W^{1,1}(\Omega), \tag{7.59}$$

where the "rigidity" of the gradient is missing, is an interesting open problem. For this problem we have a equicoerciveness property and lower estimates (see [19,95]). A key observation is that the "Modica–Mortola trick" as such is not applicable, but other lower bounds can be obtained in the same spirit.

8. Concentration problems

In the previous sections we have examined the behavior of functionals defined on Sobolev spaces, first (Sections 4–6) in some cases where the Γ-limit is automatically defined on some Sobolev space and its form must be described (with some notable exception when the "dimension" of the domain of the limit increases as in Section 6.4), and then for phase-transition limit energies, when the weak coerciveness on Sobolev spaces fails and the limit can be defined on (functions equivalent to) sets of finite perimeter. It must be noted that the actual object on which the final phase-transition energies depend is the measure $\mathcal{H}^{n-1} \llcorner \partial^*E$ which can be seen as the limit of the gradients of the recovery sequences. In this section we examine other cases where concentration occurs in a more evident way. In such cases, the limit relevant objects can and will again be thought as measures, and their connection to the corresponding limit functions will be less relevant than in the previous cases.

8.1. *Ginzburg–Landau*

We consider the simplified Ginzburg–Landau energy

$$\frac{1}{\varepsilon^2} \int_\Omega \left(|u| - 1\right)^2 dx + \int_\Omega |Du|^2 dx, \tag{8.1}$$

where $\Omega \subset \mathbb{R}^n$ and $u : \Omega \to \mathbb{R}^2$. The formal analogy with the Cahn–Hilliard model (up to scaling) is apparent, but it must be immediately noted that in this case the zeroes of the potential $W(u) = (|u| - 1)^2$ are the whole set S^1 and that the functional coincides with the Dirichlet integral on the space $H^1(\Omega; S^1)$. This latter space is not trivial contrary to the space $H^1(\Omega; \{-1, 1\})$. However, in some problems (e.g., when a boundary datum is added with nonzero degree) a further scaling of this energy is necessary. We will give a heuristic derivation of this scaling and a description of the Γ-limit of the scaled energies in dimension two, and give an idea of the extension to higher dimensions.

8.1.1. *The two-dimensional case.* An in-depth study of the behavior of minimizers for the energy above subject to nontrivial boundary data is contained in the book by Bethuel, Brezis and Hélein [35] (we also refer to the monograph by Sandier and Serfaty [138] for the case with magnetic field). Some of the results therein can be rephrased in the language of Γ-convergence. The first step towards an energetic interpretation is the derivation of the correct scaling of the energies. The heuristic idea is that a boundary datum with nonzero degree will force minimizing sequences (u_ε) to create a finite number of singularities $\{x_i\}_i$ in the interior of Ω so that their limit will belong to $H^1_{\mathrm{loc}}(\Omega \setminus \{x_i\}_i; S^1)$, and the nonzero degree condition on $\partial\Omega$ will be balanced by some "vortices" centered at x_i.

We now fix a sequence (u_ε) with fixed nonzero degree on $\partial\Omega$. We note that for fixed $\delta < 1$ the set $T_\varepsilon = \{|u_\varepsilon| < \delta\}$ is not empty (otherwise we could "project" u_ε on S^1 obtaining a homotopy with a map with zero degree). The limit of such sets T_ε is a candidate for the set $\{x_i\}_i$ above. In order to estimate the scale of the energy contribution of such a sequence we assume that T_ε is composed of disks $B^\varepsilon_i = B_{\rho^\varepsilon_i}(x_i)$ on the boundary of which the degree of u_ε is some nonzero integer d_i. We also assume that the gradient $|Du_\varepsilon|$ is of order $1/\rho^\varepsilon_i$ in B^ε_i. Note that the restriction of u_ε to $\partial B_r(x_i)$ for each $0 < r < R$, with R such that $B_R(x_i)$ are pairwise disjoint and contained in Ω, has degree d_i so that the integral of the square of its tangential derivative on $\partial B_r(x_i)$ is at least $2\pi d_i^2/r$. We then obtain

$$\frac{1}{\varepsilon^2} \int_\Omega (|u_\varepsilon| - 1)^2 \, dx + \int_\Omega |Du_\varepsilon|^2 \, dx$$

$$\geqslant \sum_i \left(\pi (\rho^\varepsilon_i)^2 \left(\frac{(1-\delta)^2}{\varepsilon^2} + \frac{C}{(\rho^\varepsilon_i)^2} \right) + 2\pi d_i^2 \int_{\rho^\varepsilon_i}^R \frac{dr}{r} \right)$$

$$\approx \pi \left(C \#(\{x_i\}_i) + \sum_i \left((1-\delta)^2 \frac{(\rho^\varepsilon_i)^2}{\varepsilon^2} + 2d_i^2 \left(|\log \rho^\varepsilon_i| + \log R \right) \right) \right).$$

Optimizing in ρ^ε_i gives $\rho^\varepsilon_i = d_i \varepsilon/(1 - \delta)$, so that we get a lower bound with $2\pi |\log \varepsilon| \times \sum_i d_i^2 + O(1)$. This estimate suggests the desired scaling, and gives the functionals

$$F_\varepsilon(u) = \frac{1}{\varepsilon^2 |\log \varepsilon|} \int_\Omega (|u_\varepsilon| - 1)^2 \, dx + \frac{1}{|\log \varepsilon|} \int_\Omega |Du_\varepsilon|^2 \, dx. \tag{8.2}$$

Note that, as in the case of Section 7.4.3, the leading part of the energy is logarithmic and is due to the "far-field" away from the singularities.

At this point, we have to define the correct notion of convergence, that will give in the limit the "vortices" x_i. To this end, it must be noted (see [115]) that the relevant quantity "concentrating" at x_i is the *distributional Jacobian* defined as $Ju = D_1(u^1 D_2 u^2) - D_2(u^1 D_1 u^2)$, which coincides with the usual Jacobian determinant $\det(Du)$ if $u \in H^1(\Omega; \mathbb{R}^2)$, and is a measure of the form $\sum_i \pi d_i \delta_{x_i}$ if u is regular outside a finite set $\{x_i\}$ with degree d_i around x_i. The notion of convergence that makes our functionals equicoercive is the *flat convergence* (i.e., testing against C^1 functions). If (u_ε) is such that $\sup_\varepsilon F_\varepsilon(u_\varepsilon) < +\infty$ then, up to subsequences, Ju_ε converge flat to some measure μ of the form $\sum_i \pi d_i \delta_{x_i}$. This defines the convergence $u_\varepsilon \to \{(x_i, d_i)\}_i$, for which we have the following result.

THEOREM 8.1 (Energy of vortices). *The Γ-limit of the functionals in* (8.2) *is given by*

$$F_0\big(\{(x_i, d_i)\}_i\big) = 2\pi \sum_i |d_i|. \tag{8.3}$$

PROOF. We do not include the details of the proof for which we refer to [35]. We only remark that the lower bound above is almost sharp, giving $\sum_i d_i^2$ in the place of $\sum_i |d_i|$. This is easily made optimal by approximating a vortex of degree d_i with d_i vortices of degree $\mathrm{sign}(d_i)$ in the limsup inequality. Moreover, for each vortex of degree ± 1, a recovery sequence is obtained by mollifying $(x - x_i)/|x - x_i|$. $\qquad\square$

REMARK 8.2 (Interaction of vortices). Note that this result implies that minimizers with boundary datum of degree $d \neq 0$ will generate $|d|$ vortices of degree $\mathrm{sign}(d)$, but does not give any information about the location of such vortices. To this end we have to look for the behavior of the *renormalized energies* (in the terminology of [35]).

In terms of higher-order Γ-limits we fix a boundary datum $g: \partial\Omega \to S^1$ with $\deg(g) > 0$, and consider the functionals

$$G_\varepsilon(u) = \begin{cases} \frac{1}{\varepsilon^2} \int_\Omega \big(|u_\varepsilon| - 1\big)^2 \, dx \\ \qquad + \int_\Omega |Du_\varepsilon|^2 \, dx - 2\pi d \,|\log\varepsilon| & \text{if } u = g \text{ on } \partial\Omega, \\ +\infty & \text{otherwise.} \end{cases} \tag{8.4}$$

Then the functionals G_ε Γ-converge with respect to the convergence defined above to a limit G_0 that describes the interactions between the vortices. We refer to [35], Theorem 1.7, for the description of this renormalized limit energy via Green's function of some auxiliary boundary value problem.

Finally, we note that Γ-convergence has also been applied to the asymptotic study of the gradient flows of Ginzburg–Landau energies by Sandier and Serfaty [137].

8.1.2. *The higher-dimensional case.* The analogue of Theorem 8.1 in three dimensions (or higher) is more meaningful, as we expect the distributional Jacobian to give rise to a limit with a more complex geometry than a set of points. Indeed, it can be seen that these problems have a two-dimensional character that forces concentration on sets of co-dimension two (hence, lines in three dimensions). Actually, we expect limit objects with some multiplicity defined, taking the place of the degree in dimension two. These objects are indeed *currents*, whose treatment is beyond the scope of these notes and for which we refer to the introductory book by Morgan [125]. We just mention the following result due to Jerrard and Soner [116] and Alberti, Baldo and Orlandi [6].

THEOREM 8.3. *The Γ-limit of the functionals F_ε with respect to the flat convergence of currents is defined on integer (up to a factor π) rectifiable currents T and is equal to $F_0(T) = 2\pi\|T\|$, where $\|T\|$ is the mass of the current T.*

8.2. *Critical-growth problems*

Another class of variational problems where concentration occurs are problems related to the critical growth for the Sobolev embedding. It is well known that the best constant in the Sobolev inequality is not achieved on domains different from the whole space, due to a scaling-invariance property that implies that optimal sequences concentrate at a point. The techniques of *concentration–compactness* type of Lions are the classical tool to study such phenomena for a wide class of variational problems of the same nature. Some of these concentration phenomena can be also treated within the theory of Γ-convergence. We give some applications to a large class of variational problems that exhibit concentration and include critical-growth problems.

We will study the behavior of the family of maximum problems depending on a small parameter $\varepsilon > 0$,

$$S_\varepsilon^\Psi(\Omega) = \varepsilon^{-2^*} \sup\left\{ \int_\Omega \Psi(\varepsilon u)\,dx \colon u \in H_0^1(\Omega),\ \int_\Omega |\nabla u|^2\,dx \leqslant 1 \right\}, \tag{8.5}$$

where Ω is a bounded open set in $\mathbb{R}^n$ with $n \geqslant 3$ and $2^* = 2n/(n-2)$ is the usual critical *Sobolev exponent*, through some limit of the corresponding functionals

$$F_\varepsilon(u) = \begin{cases} \varepsilon^{-2^*} \int_\Omega \Psi(\varepsilon u)\,dx & \text{if } \int_\Omega |Du|^2\,dx \leqslant 1, \\ 0 & \text{otherwise in } L^{2^*}(\Omega). \end{cases} \tag{8.6}$$

We assume $0 \leqslant \Psi(t) \leqslant c|t|^{2^*}$ for every $t \in \mathbb{R}$; $\Psi \not\equiv 0$ and upper semicontinuous and, in order to simplify the exposition, we also assume that the following two limits

$$\Psi_0(t) = \lim_{t \to 0} \frac{\Psi(t)}{|t|^{2^*}} \quad \text{and} \quad \Psi_\infty(t) = \lim_{t \to \infty} \frac{\Psi(t)}{|t|^{2^*}}$$

exist.

REMARK 8.4. (1) Within this class we recover the *capacity* problem (see the Introduction), with

$$\Psi(t) = \begin{cases} 0 & \text{if } t < 1, \\ 1 & \text{if } t \geqslant 1. \end{cases} \tag{8.7}$$

(2) If Ψ is smooth and $\Psi' = \psi$ then the functional is linked to the study of the asymptotic properties of solutions of

$$\begin{cases} -\Delta u = \lambda \psi(u) & \text{in } \Omega, \\ u = 0 & \text{on } \partial\Omega, \end{cases} \tag{8.8}$$

where $\lambda \to +\infty$ and $0 \leqslant \psi(t) \leqslant c|t|^{2^*-1}$.

Note that the functionals F_ε are weakly equicoercive in $H_0^1(\Omega)$. If $Du_\varepsilon \rightharpoonup Du$, we may also assume that there exists a measure $\mu \in \mathcal{M}(\overline{\Omega}) := (C(\overline{\Omega}))'$ such that $|Du_\varepsilon|^2 \mathcal{L}^n \rightharpoonup^* \mu$ in $\mathcal{M}(\overline{\Omega})$. In general, by the lower semicontinuity of the norm, we get $\mu \geqslant |Du|^2 \mathcal{L}^n$. Thus we can isolate the atoms of μ, $\{x_i\}_{i \in J}$, and rewrite μ as follows

$$\mu = |Du|^2 \mathcal{L}^n + \sum_{i \in J} \mu_i \delta_{x_i} + \tilde{\mu}, \tag{8.9}$$

where μ_i denotes the positive weight of the atom x_i and $\tilde{\mu}$ is the nonatomic part of $\mu - |Du|^2 \mathcal{L}^n$.

In general, we say that a sequence u_ε converges to (u, μ) if

$$u_\varepsilon \rightharpoonup u \quad \text{in } H_0^1(\Omega) \quad \text{and} \quad |Du_\varepsilon|^2 \mathcal{L}^n \rightharpoonup^* \mu \quad \text{in } \mathcal{M}(\overline{\Omega}). \tag{8.10}$$

In view of the study of the asymptotic behavior of the maxima of F_ε and the corresponding maximizing sequences we introduce the notion of Γ^+-convergence, symmetric to the notion of Γ-convergence used until now for minimum problems.

DEFINITION 8.5. Let $F_\varepsilon : X \to \overline{\mathbb{R}}$, be a family of functionals. We say that the sequence F_ε Γ^+-*converges* to the functional $F : X \to \overline{\mathbb{R}}$ if the following two properties are satisfied for all $x \in X$
   (i) for every sequence $x_\varepsilon \to x$ we have that $\limsup_{\varepsilon \to 0} F_\varepsilon(x_\varepsilon) \leqslant F(x)$;
   (ii) for every $x \in X$, there exists a sequence x_ε such that $x_\varepsilon \to x$ and

$$\liminf_{\varepsilon \to 0} F_\varepsilon(x_\varepsilon) \geqslant F(x).$$

To define the Γ^+-limit we denote by $S^\Psi := S_1^\Psi(\mathbb{R}^n)$ the *ground-state energy*; i.e.,

$$S^\Psi = \sup \left\{ \int_{\mathbb{R}^n} \Psi(u) \, dx : u \in D^{1,2}(\mathbb{R}^n), \int_{\mathbb{R}^n} |Du|^2 \, dx \leqslant 1 \right\}, \tag{8.11}$$

where $D^{1,2}$ is the closure of C^1 with respect to the L^2-norm of the gradient. We then have the following result proved in [17].

THEOREM 8.6 (Concentration by Γ-convergence). *The sequence F_ε Γ^+-converges with respect to the convergence given by (8.10) to the functional*

$$F(u, \mu) := \Psi_0 \int_\Omega |u|^{2^*} \, dx + S^\Psi \sum_{i \in J} (\mu_i)^{2^*/2}. \tag{8.12}$$

As a consequence we get the result of concentration known for this type of problem (obtained by Lions in the smooth case and Flucher and Müller in the general case). In fact, by a scaling argument one can see that $S^* \Psi_0 \leqslant S^\Psi$ (S^* denotes the best Sobolev constant) and that $S_\varepsilon^\Psi \to S^\Psi$, hence by the Γ^+-convergence we get $S^\Psi = \max F(u, \mu)$. On the other hand, by the convexity of the function $|t|^{2^*/2}$, we get

$$F(u, \mu) = \Psi_0 \int_\Omega |u|^{2^*} \, dx + S^\Psi \sum_{i \in J} (\mu_i)^{2^*/2}$$

$$\leqslant \Psi_0 S^* \left(\int_\Omega |Du|^2 \, dx \right)^{2^*/2} + S^\Psi \sum_{i \in J} (\mu_i)^{2^*/2}$$

$$\leqslant S^\Psi \left[\left(\int_\Omega |Du|^2 \, dx \right)^{2^*/2} + \sum_{i \in J} (\mu_i)^{2^*/2} \right] \leqslant S^\Psi \mu(\overline{\Omega}) \leqslant S^\Psi. \tag{8.13}$$

Since the Sobolev constant is not attained in Ω, the first inequality is strict unless $u = 0$ and the third inequality is strict unless $\mu = \delta_{x_0}$ for some $x_0 \in \overline{\Omega}$. In other words,

$$F(u, \mu) = S^\Psi = \max F \quad \Longleftrightarrow \quad (u, \mu) = (0, \delta_{x_0}),$$

which corresponds to the concentration of a maximizing sequence at x_0.

PROOF OF THEOREM 8.6. The Γ^+-"limsup inequality" (i) in Definition 8.5 is essentially the so-called concentration–compactness lemma in its generalized version proved by Flucher and Müller (see [99] for details), where the asymptotic behavior of the sequence $\varepsilon^{-2^*} \Psi(\varepsilon u_\varepsilon)$ is given in terms of the limit (u, μ). The optimization of the upper bound is easily achieved on pairs $(0, \delta_x)$ by arguing as in (8.13) above, while if μ does not contain an atomic part it is derived from the strong convergence in L^{2^*} due to the concentration–compactness lemma. For details we refer to [17]. $\qquad \square$

We can apply the result above to the case of the capacity, with the choice of Ψ as in Remark 8.4(1). The maximum problem S_ε^Ψ can be rewritten as (V stands for "volume")

$$S_\varepsilon^V(\Omega) = \max\{|A|: \mathrm{Cap}(A, \Omega) \leqslant \varepsilon^2\}, \tag{8.14}$$

after identifying any open set A with the level set $\{v \geqslant 1\}$ of its *capacitary potential* (see (1.19)). We have that the maximizing sets A_ε concentrate at a single point x_0 in $\overline{\Omega}$; i.e., the corresponding capacitary potentials (divided by ε) converge to $(0, \delta_{x_0})$ in the sense of (8.10).

As for the two-dimensional Ginzburg–Landau energies, a classical question in these problems of concentration is the identification of the concentration points. The answer can also be given in terms of the Γ^+-convergence of F_ε suitably scaled. A key role is played by the diagonal of the regular part of Green's function of Ω for the Laplacian and the Dirichlet problem. This plays the same role as the renormalized energy for the Ginzburg–Landau functionals. It is called the *Robin function* τ_Ω and it is given by

$$\tau_\Omega(x) = H_\Omega(x, x),$$

where $H_\Omega(x, y)$ is the regular (harmonic) part of Green's function $G_\Omega(x, y)$. In the case of the capacity we have the following result (see [100] and [17]).

THEOREM 8.7 (Identification of concentration points). *Let Ψ be as in Remark 8.4(1); then the sequence*

$$\frac{F_\varepsilon(u) - S^V}{\varepsilon^2},$$

where $S^V := S_1^V(\mathbb{R}^n)$, Γ^+-converges with respect to the convergence given by (8.10) to the functional

$$F^1(u, \mu) = \begin{cases} -\dfrac{n}{n-2} S^V \tau_\Omega(x_0) & \text{if } (u, \mu) = (0, \delta_{x_0}), \\ -\infty & \text{otherwise.} \end{cases}$$

PROOF. The proof of the Γ^+-limsup inequality is based on an asymptotic formula for the capacity of small sets involving the Robin function (see [110,111]). To prove the converse inequality, if v_ε is the capacitary potential of the level set of Green's function $A_\varepsilon = \{G_\Omega(x_0, \cdot) > \varepsilon^{-2}\}$, for which it is not difficult to prove that

$$|A_\varepsilon| \geqslant \varepsilon^{2^*} S^V \left(1 - \frac{n}{n-2} \tau_\Omega(x_0) \varepsilon^2 + o(\varepsilon^2)\right),$$

then a recovery sequence is given by $u_\varepsilon = v_\varepsilon / \varepsilon$. $\qquad\square$

As a consequence of this Γ^+-convergence result we deduce that all sequences of *almost maximizers* of problem (8.14); i.e., satisfying

$$\varepsilon^{2^*} |A_\varepsilon| = S_\varepsilon^V(\Omega) + o(\varepsilon^2), \tag{8.15}$$

concentrate at a minimum point of the Robin function (a *harmonic center* of Ω).

More in general, one can compute the asymptotic behavior of

$$\frac{F_\varepsilon(u) - S^\psi}{\varepsilon^2}$$

under a condition which rules out dilation-invariant problems; e.g., $S^\psi > \max\{\Psi_0, \Psi_\infty\} S^*$.

THEOREM 8.8. *If $S^\psi > \max\{\Psi_0, \Psi_\infty\} S^*$, then all sequences of almost maximizers for F_ε; i.e., satisfying*

$$\varepsilon^{2^*} \int_\Omega \Psi(\varepsilon u_\varepsilon)\, dx = S_\varepsilon^\psi(\Omega) + o(\varepsilon^2), \tag{8.16}$$

up to a subsequence, concentrate at a harmonic center of Ω.

9. Dimension-reduction problems

The small parameter ε entering the definition of the functionals F_ε may be some times related to some small dimension of the domain of integration. This happens for example in the theory of thin films, rods and shells. As $\varepsilon \to 0$ some energy defined on a lower-dimensional set is expected to arise as the Γ-limit. We will illustrate some aspects of this passage to the limit for "thin films" (i.e., for n-dimensional domains whose limit is $(n-1)$-dimensional, with the case $n = 3$ in mind).

9.1. *The Le Dret–Raoult result*

We begin with an illuminating result for homogeneous functionals. We consider a domain of the form

$$\Omega_\varepsilon = \omega \times (0, \varepsilon), \quad \omega \text{ bounded open subset of } \mathbb{R}^{n-1}, \tag{9.1}$$

and an energy

$$F_\varepsilon(u) = \frac{1}{\varepsilon} \int_{\Omega_\varepsilon} f(Du)\, dx, \quad u \in W^{1,p}(\Omega_\varepsilon; \mathbb{R}^m), \tag{9.2}$$

where f satisfies a standard growth condition of order p. Note that up to a relaxation argument, we can suppose that f be quasiconvex. The normalization factor $1/\varepsilon$ is a simple scaling proportional to the measure of Ω_ε.

In order to understand in what sense a Γ-limit of F_ε can be defined, we first identify F_ε with a functional defined on a fixed domain

$$G_\varepsilon(v) = \int_\Omega f\left(D_\alpha v, \frac{1}{\varepsilon} D_n v\right) dx, \quad v \in W^{1,p}(\Omega; \mathbb{R}^m), \tag{9.3}$$

where $x_\alpha = (x_1, \ldots, x_{n-1})$ and $D_\alpha v = (D_1 v, \ldots, D_{n-1} v)$, $\Omega = \omega \times (0, 1)$ and v is obtained from u by the scaling

$$v(x_\alpha, x_n) = u(x_\alpha, \varepsilon x_n). \tag{9.4}$$

Note that G_ε satisfies a degenerate growth condition

$$G_\varepsilon(v) \geqslant C \int_\Omega \left(|D_\alpha v|^p + \frac{1}{\varepsilon^p} |D_n v|^p \right) dx - C. \tag{9.5}$$

From this condition we deduce first that a family with equibounded energies $G_\varepsilon(v_\varepsilon)$ and bounded in L^p is weakly precompact in $W^{1,p}(\Omega; \mathbb{R}^m)$; second, that if $v_\varepsilon \rightharpoonup v$ (up to subsequences) then

$$\int_\Omega |D_n v|^p \, dx \leqslant \liminf_{\varepsilon \to 0} \int_\Omega |D_n v_\varepsilon|^p \leqslant \liminf_{\varepsilon \to 0} \varepsilon^p C = 0, \tag{9.6}$$

so that v is actually independent of the nth variable and can therefore be identified with a function $u \in W^{1,p}(\omega; \mathbb{R}^m)$.

With this compactness result in mind, we can define a convergence $u_\varepsilon \to u$ of functions $u_\varepsilon \in W^{1,p}(\Omega_\varepsilon; \mathbb{R}^m)$ to $u \in W^{1,p}(\omega; \mathbb{R}^m)$ if the corresponding v_ε defined above converges to the function $v(x_\alpha, x_n) = u(x_\alpha)$. The functionals F_ε are equicoercive with respect to this convergence. We then have the following convergence result [118].

THEOREM 9.1 (Le Dret–Raoult thin-film limit). *Let F_ε be defined above, and let*

$$F_0(u) = \int_\omega Q\bar{f}(D_\alpha u) \, dx_\alpha, \quad u \in W^{1,p}(\omega; \mathbb{R}^m), \tag{9.7}$$

where Q denotes the quasiconvex envelope *(for energy densities on the space of $m \times (n-1)$ matrices) and*

$$\bar{f}(\xi) = \inf\{ f(\xi|b) \colon b \in \mathbb{R}^m \}. \tag{9.8}$$

(Here $(\xi|b) \in \mathbb{M}^{m \times n}$ is a matrix whose first $n - 1$ columns coincide with the $m \times (n-1)$ matrix ξ, and the last column with the vector b.)

This result gives an easy way to characterize the energy density of the limit that highlights the superposition of two optimality processes: first, the minimization in the dependence on the nth variable disappearing in the limit. This gives the function $\bar{f}$, that may be not $(m \times (n-1))$-quasiconvex even though f is $(m \times n)$-quasiconvex; hence a second optimization in oscillations in the $n - 1$ "planar" coordinates must be taken into account (expressed by the quasiconvexification process).

PROOF OF THEOREM 9.1. The lower bound is easily achieved since

$$\liminf_{\varepsilon \to 0} G_\varepsilon(v_\varepsilon) = \liminf_{\varepsilon \to 0} \int_\Omega f\left(D_\alpha v_\varepsilon, \frac{1}{\varepsilon} D_n v_\varepsilon\right) dx$$

$$\geq \liminf_{\varepsilon \to 0} \int_\Omega \bar{f}(D_\alpha v_\varepsilon) \, dx$$

$$\geq \liminf_{\varepsilon \to 0} \int_\Omega Q\bar{f}(D_\alpha v_\varepsilon) \, dx$$

$$\geq \int_\Omega Q\bar{f}(D_\alpha v) \, dx$$

$$= \int_\omega Q\bar{f}(D_\alpha u) \, dx_\alpha,$$

the last inequality due to the lower semicontinuity of $\int_\Omega Q\bar{f}(D_\alpha v) \, dx$.

The limsup inequality needs to be shown only for piecewise-affine function by a density argument. It is sufficient to exhibit a recovery sequence for the target function $u(x_\alpha) = \xi x_\alpha$. We may suppose that f be quasiconvex and hence continuous (locally Lipschitz), so that we easily get the existence of $b \in \mathbb{R}^m$ such that $\bar{f}(\xi) = f(\xi|b)$. Note that if $\bar{f}$ were quasiconvex then

$$u_\varepsilon(x) = \xi x_\alpha + \varepsilon b x_n \tag{9.9}$$

would be a recovery sequence for $F_0(u)$. In general we have to improve this argument: we can fix a 1-periodic smooth $\varphi(x_\alpha)$ such that

$$\int_{(0,1)^{n-1}} \bar{f}\big(\xi + D\varphi(x_\alpha)\big) \, dx_\alpha \leq Q\bar{f}(\xi) + \eta.$$

Let $b = b(x_\alpha)$ be such that $\bar{f}(\xi + D\varphi(x_\alpha)) = f(\xi + D\varphi(x_\alpha), b(x_\alpha))$. Then a recovery sequence is given by

$$u_\varepsilon(x) = \xi x_\alpha + \varphi(x_\alpha) + \varepsilon x_n b(x_\alpha).$$

Note a few technical issues: the existence of b may be proven by suitable measurable-selection criteria; if b is not differentiable then a mollification argument must be used; we get an extra term $\varepsilon x_n D_\alpha b(x_\alpha)$ in $D_\alpha u_\varepsilon$ that can be neglected due to the local Lipschitz continuity of f. $\quad\square$

REMARK 9.2 (The convex case). Note that if f is convex then $\bar{f}$ is convex, and the quasiconvexification process is not necessary. In this case recovery sequences for affine functions are simply given by (9.9).

9.2. *A compactness theorem*

The localization arguments in Section 3.3 can be adapted to sequences of functionals defined on thin films. It must be noted that in order to obtain a limit functional defined on $W^{1,p}(\omega; \mathbb{R}^m)$ the localization argument must be performed on open sets of $\mathbb{R}^{n-1}$, or equivalently, on cylindrical sets $A \times (0, 1)$ with A open set of $\mathbb{R}^{n-1}$. With this observation in mind the following theorem holds by Braides, Fonseca and Francfort [56], which is an analogue of Theorem 4.2.

THEOREM 9.3 (Compactness theorem for thin films). *Given a family of Borel functions* $f_\varepsilon : \Omega_\varepsilon \times \mathbb{M}^{m \times n} \to [0, +\infty)$, *satisfying growth condition* (4.2), *there exists a Carathéodory function* $f_0 : \omega \times \mathbb{M}^{m \times (n-1)} \to [0, +\infty)$, *satisfying the same growth conditions, such that, up to subsequences,*

$$\int_A f_0(x_\alpha, D_\alpha u)\, dx_\alpha = \Gamma\text{-} \lim_{\varepsilon \to 0} \frac{1}{\varepsilon} \int_{A \times (0, \varepsilon)} f_\varepsilon(x, Du)\, dx \tag{9.10}$$

with respect to the convergence $u_\varepsilon \to u$ *as in Theorem* 9.1, *for all open subsets A of* ω.

PROOF. Follow the arguments in Section 3.3 applied to the energies

$$G_\varepsilon(v, A) = \int_{A \times (0, 1)} f_\varepsilon\left(x_\alpha, \varepsilon x_n, D_\alpha v, \frac{1}{\varepsilon} D_n v\right) dx.$$

Note that we can use De Giorgi's argument to match boundary conditions on $(\partial A) \times (0, 1)$ to prove the fundamental estimate, provided we choose the cut-off functions independent of the nth variable. $\qquad\square$

REMARK 9.4 (Convergence of minimum problems). Minimum problems of the type

$$m_\varepsilon = \min\left\{\frac{1}{\varepsilon} \int_{\Omega_\varepsilon} f_\varepsilon(x, Du)\, dx : u = \phi \text{ on } \partial\omega \times (0, 1)\right\} \tag{9.11}$$

can be treated by the usual arguments provided that $\phi = \phi(x_\alpha)$, so that

$$m_\varepsilon = \min\left\{\int_\Omega f_\varepsilon\left(x_\alpha, \varepsilon x_n, D_\alpha v, \frac{1}{\varepsilon} D_n v\right) dx : v = \phi \text{ on } \partial\omega \times (0, 1)\right\} \tag{9.12}$$

converge to

$$m = \min\left\{\int_\omega f_0(x_\alpha, D_\alpha u)\, dx_\alpha : u = \phi \text{ on } \partial\omega\right\}. \tag{9.13}$$

Again, to prove this, it suffices to use the arguments in Section 4.2.1.

In the same way we may treat the case $\omega = (0, 1)^{n-1}$ and u satisfying 1-periodic conditions in the x_α variables.

REMARK 9.5 (Alternate formula for the Le Dret–Raoult result). In the case of integrands that are "homogeneous" and independent of ε (i.e., $f_\varepsilon(x,\xi) = f(\xi)$) as in Theorem 9.1, we may easily infer that the function f_0 given by the compactness Theorem 9.3 is itself homogeneous (and quasiconvex) so that

$$
\begin{aligned}
f_0(\xi) &= \min\left\{ \int_{(0,1)^{n-1}} f_0(\xi + D_\alpha u)\,\mathrm{d}x_\alpha: \ u = 0 \text{ on } \partial(0,1)^{n-1} \right\} \\[2mm]
&= \min\left\{ \int_{(0,1)^n} f_0(\xi + D_\alpha u)\,\mathrm{d}x: \ u = 0 \text{ on } \left(\partial(0,1)^{n-1}\right) \times (0,1) \right\} \\[2mm]
&= \lim_{\varepsilon \to 0} \min\left\{ \int_{(0,1)^{n-1}\times(0,1)} f\left(\xi + D_\alpha v, \frac{1}{\varepsilon} D_n v\right)\,\mathrm{d}x: \right. \\[2mm]
&\qquad\qquad\qquad \left. v = 0 \text{ on } \left(\partial(0,1)^{n-1}\right) \times (0,1) \right\} \\[2mm]
&= \lim_{T \to +\infty} \frac{1}{T^{n-1}} \min\left\{ \int_{(0,T)^{n-1}\times(0,1)} f(\xi + D_\alpha v, D_n v)\,\mathrm{d}y: \right. \\[2mm]
&\qquad\qquad\qquad \left. v = 0 \text{ on } \left(\partial(0,T)^{n-1}\right) \times (0,1) \right\},
\end{aligned}
$$

where we have performed the change of variables $y_\alpha = \varepsilon x_\alpha$ and set $T = 1/\varepsilon$. This is a formula of *homogenization type*, and can be easily extended to cover the case $f_\varepsilon(x,\xi) = f(x_\alpha/\varepsilon, \xi)$ with f 1-periodic in the x_α directions (see [56]). Note moreover that the zero boundary condition can be replaced by periodicity.

REMARK 9.6 (Thin films with oscillating profiles). It is interesting to note that the fact that the dependence on the nth variable disappears allows to consider more complex geometries for the sets Ω_ε; for example, we can consider

$$
\Omega_\varepsilon = \left\{ (x_\alpha, x_n): \ 0 < x_n < \varepsilon \psi_\varepsilon(x_\alpha) \right\}. \tag{9.14}
$$

The compactness argument must be adequately extended since now Ω_ε cannot be rescaled to a single set Ω. In order not to have a degenerate behavior we assume that $0 < c \leqslant \psi_\varepsilon \leqslant 1$ uniformly. The scaling in the nth variable brings Ω_ε into $\Omega^\varepsilon = \{(x_\alpha, x_n): \ 0 < x_n < \psi_\varepsilon(x_\alpha)\}$. If $F_\varepsilon(u_\varepsilon)$ is equibounded we can apply the compactness argument to the scaled sequence v_ε on $\omega \times (0,c)$ and define a limit $u \in W^{1,p}(\omega; \mathbb{R}^m)$. We can easily see that we indeed have

$$
\lim_{\varepsilon \to 0} \int_{\Omega^\varepsilon} |v_\varepsilon - u|^p\,\mathrm{d}x = \lim_{\varepsilon \to 0} \int_\omega \int_0^{\psi_\varepsilon(x_\alpha)} |v_\varepsilon - u|^p\,\mathrm{d}x_n\,\mathrm{d}x_\alpha = 0
$$

by a simple use of Poincaré–Wirtinger's inequality since $v - u_\varepsilon \to 0$ on $\omega \times (0,c)$, and we have a bound for $1/\varepsilon\, D_n v_\varepsilon = 1/\varepsilon\, D_n(v_\varepsilon - u)$. The sequence F_ε is then equicoercive with respect to this convergence, and the compactness Theorem 9.3 can be proven using the same arguments as before.

Note that we have used the fact that the sections of Ω_ε in the nth direction are connected. This cannot be dropped, as shown by an example by Bhattacharya and Braides [31] (with Ω_ε with an increasing number of small cracks as $\varepsilon \to 0$), showing that otherwise we may have a limit in which the dependence on the nth variable does not disappear.

REMARK 9.7 (Equiintegrability for thin films). In the computation of the Γ-limits for thin films, as for the case of n-dimensional objects, the possibility of reducing to sequences with some equiintegrability property is very useful. A result of Bocea and Fonseca [39] shows that, for any converging sequence (u_ε) such that $\sup_\varepsilon \int_\Omega (|\nabla_\alpha u_\varepsilon|^p + \frac{1}{\varepsilon^p}|\nabla_n u_\varepsilon|^p)\,dx < +\infty$, there exists an "equivalent sequence" v_ε such that the sequence $(|\nabla_\alpha v_\varepsilon|^p + \frac{1}{\varepsilon^p}|\nabla_n v_\varepsilon|^p)$ is equiintegrable on Ω. An alternative proof and the generalization to any co-dimension of this result can be found in the paper by Braides and Zeppieri [65].

Note that the general approach outlined above may be generalized to cover the cases when the "thin directions" are more than one, and in the limit we get objects of co-dimension more than one. In the case of low-dimensional theories of rods, it must be noted that the one-dimensional nature of the final objects easily allows for more general growth conditions for the energies (see [2]).

9.3. *Higher-order Γ-limits*

The analysis carried over in the first part of this chapter can be applied to derive low-dimensional theories from three-dimensional (finite) elasticity, where $m = n = 3$ and the function $f : \mathbb{M}^{3\times 3} \to [0, +\infty]$, the *elastic stored energy* of the material, is continuous and *frame indifferent*; i.e., $f(R\xi) = f(\xi)$ for every rotation R and every $\xi \in \mathbb{M}^{3\times 3}$, where $R\xi$ denotes the usual product of 3×3 matrices. We assume that f vanishes on the set $SO(3)$ of rotations in $\mathbb{R}^3$, is of class C^2 in a neighborhood of $SO(3)$, and satisfies the inequality

$$f(\xi) \geqslant C\big(\mathrm{dist}\big(\xi, SO(3)\big)\big)^2 \quad \text{for every } \xi \in \mathbb{M}^{3\times 3}, \tag{9.15}$$

with a constant $C > 0$.

For these energies other scalings than that considered previously provide a variational justification of a number of low-dimensional theories commonly used in Mechanics (see [107]). In this subsection we briefly focus on a derivation of plate theory. In this case the energies F_ε must be further scaled by $1/\varepsilon^2$ obtaining the functionals $\mathcal{F}_\varepsilon$ defined by

$$\mathcal{F}_\varepsilon(v) := \frac{1}{\varepsilon^2} \int_\Omega f\left(D_1 v | D_2 v | \frac{1}{\varepsilon} D_3 v\right) dx, \quad v \in W^{1,2}(\Omega; \mathbb{R}^3),$$

where again $v(x_1, x_2, x_3) = u(x_1, x_2, \varepsilon x_3)$, $u : \Omega_\varepsilon \to \mathbb{R}^3$ is the deformation of Ω_ε and $\xi = (\xi_1|\xi_2|\xi_3)$ represents a matrix ξ via its columns.

The Γ-limit of $\mathcal{F}_\varepsilon$ with respect to the L^2-convergence turns out to be finite on the set $\Sigma(\omega; \mathbb{R}^3)$ of all *isometric embeddings* of ω into $\mathbb{R}^3$ of class $W^{2,2}$; i.e., $v \in \Sigma(\omega; \mathbb{R}^3)$ if and only if $v \in W^{2,2}(\omega; \mathbb{R}^3)$ and $(Dv)^\top Dv = I$ a.e. on ω. As before, elements of $\Sigma(\omega; \mathbb{R}^3)$ are also regarded as maps from Ω into $\mathbb{R}^3$ independent of x_3. To describe the Γ-limit we introduce the quadratic form $\mathcal{Q}_3$ defined on $\mathbb{M}^{3\times3}$ by

$$\mathcal{Q}_3(\xi) := \frac{1}{2}D^2 f(I)[\xi, \xi],$$

which is the density of the linearized energy for the three-dimensional problem, and the quadratic form $\mathcal{Q}_2$ defined on the space of symmetric 2×2 matrices by

$$\mathcal{Q}_2 \begin{pmatrix} a_{11} & a_{12} \\ a_{12} & a_{22} \end{pmatrix} := \min_{(b_1,b_2,b_3)\in\mathbb{R}^3} \mathcal{Q}_3 \begin{pmatrix} a_{11} & a_{12} & b_1 \\ a_{12} & a_{22} & b_2 \\ b_1 & b_2 & b_3 \end{pmatrix}.$$

The Γ-limit of $\mathcal{F}_\varepsilon$ is the functional $\mathcal{F}: L^2(\Omega; \mathbb{R}^3) \to [0, +\infty]$ defined by

$$\mathcal{F}(v) := \frac{1}{12} \int_\Omega \mathcal{Q}_2\big(A(v)\big)\,dx \quad \text{if } v \in \Sigma\big(\omega; \mathbb{R}^3\big),$$

where $A(v)$ denotes the second fundamental form of v; i.e.,

$$A_{ij}(v) := -D_i D_j v \cdot v, \tag{9.16}$$

with normal vector $v := D_1 v \wedge D_2 v$. The proof of this fact can be found in the paper by Friesecke, James and Müller [105].

Equicoerciveness for problems involving the functionals $\mathcal{F}_\varepsilon$ in $L^2(\Omega; \mathbb{R}^3)$ is not trivial; it follows from (9.15) through the following lemma which is due to Friesecke, James and Müller (see [106]).

LEMMA 9.8 (Geometric rigidity estimate). *Let* $\Omega \subset \mathbb{R}^n$ *be a Lipschitz domain; then there exists a constant* $C(\Omega)$ *such that*

$$\min_{R\in SO(n)} \int_\Omega |Du - R|^2\,dx \leqslant C(\Omega) \int_\Omega \big(\text{dist}(Du, SO(3))\big)^2\,dx$$

for all $u \in H^1(\Omega; \mathbb{R}^n)$.

10. Approximation of free-discontinuity problems

"Free-discontinuity problems", following a terminology introduced by De Giorgi, are those problems in the Calculus of Variations where the unknown is a pair (u, K), with K varying in a class of (sufficiently smooth) closed hypersurfaces contained in a fixed open set $\Omega \subset \mathbb{R}^n$ and $u : \Omega \setminus K \to \mathbb{R}^m$ belonging to a class of (sufficiently smooth) functions. Such problems are usually of the form

$$\min\{E_v(u, K) + E_s(u, K) + \text{"lower-order terms"}\}, \tag{10.1}$$

with E_v, E_s being interpreted as *volume* and *surface* energies, respectively. Several examples can be described in this setting, among which: image and signal reconstruction problems (linked to the Mumford and Shah functional, see Section 10.1.2), fracture in brittle hyperelastic media (where E_v denotes the elastic energy, and E_s the surface energy due to the crack), equilibrium problems for drops of liquid crystals, or even simply prescribed curvature problems (for which $u = \chi_E$ and $K = \partial E$, $E_v(u, K) = \int_E g(x) \, dx$ and $E_s(u, K) = \mathcal{H}^{n-1}(\partial E)$).

Despite the existence theory developed in *SBV*-spaces, functionals arising in free-discontinuity problems present some serious drawbacks. First, the lack of differentiability in any reasonable norm implies the impossibility of flowing these functionals, and dynamic problems can be tackled only in an indirect way. Moreover, numerical problems arise in the detection of the unknown discontinuity surface. To bypass these difficulties, a considerable effort has been spent recently to provide variational approximations of free discontinuity problems, and in particular of the Mumford–Shah functional *MS* defined in (10.8), with differentiable energies defined on smooth functions.

10.1. *Special functions with bounded variation*

The treatment of free-discontinuity problems following the direct methods of the Calculus of Variations presents many difficulties, due to the dependence of the energies on the surface K. Unless topological constraints are added, it is usually not possible to deduce compactness properties from the only information that such kind of energies are bounded. An idea of De Giorgi has been to interpret K as the set of discontinuity points of the function u, and to set the problems in a space of discontinuous functions. The requirements on such a space are of two kinds:

(a) *structure properties*: if we define K as the set of discontinuity points of the function u then K can be interpreted as an hypersurface, and u is "differentiable" on $\Omega \setminus K$ so that bulk energy depending on ∇u can be defined;

(b) *compactness properties*: it is possible to apply the direct method of the Calculus of Variations, obtaining compactness of sequences of functions with bounded energy.

The answer to the two requirements above has been De Giorgi and Ambrosio's space of *special functions of bounded variation*: a function u belongs to $SBV(\Omega)$ if and only if its distributional derivative Du is a bounded measure that can be split into a bulk and a surface term. This definition can be further specified: if $u \in SBV(\Omega)$ and $S(u)$ (the *jump set* or *discontinuity set* of u) stands for the complement of the set of the Lebesgue points for u then a *measure-theoretical normal* ν_u to $S(u)$ can be defined $\mathcal{H}^{n-1}$-a.e. on $S(u)$, together with the *traces* $u^{\pm}$ on both sides of $S(u)$; moreover, the *approximate gradient* ∇u exists a.e. on Ω, and we have

$$Du = \nabla u \mathcal{L}_n + \left(u^+ - u^-\right)\nu_u \mathcal{H}^{n-1} \llcorner S(u). \tag{10.2}$$

Replacing the set K by $S(u)$ we obtain a weak formulation for a class of free-discontinuity problems, whose energies take the general form

$$\int_{\Omega} f(x, u, \nabla u)\,dx + \int_{S(u)} \vartheta\left(x, u^+, u^-, \nu_u\right)d\mathcal{H}^{n-1}. \tag{10.3}$$

An existence theory for problems involving these kinds of energies has been developed by Ambrosio. Various regularity results show that for a wide class of problems the weak solution u in $SBV(\Omega)$ provides a solution to the corresponding free-discontinuity problem, taking $K = \overline{S(u)}$ (see [20,88,124]).

10.1.1. *A density result in SBV.* It is useful to ensure the existence of "dense" sets in *SBV* spaces to use a density argument in the computation of the limsup inequality. To this end, we can show that "piecewise-smooth functions" are dense as in the following lemma (see [49]) derived from the regularity results in [92].

We recall that $\mathcal{M}^{n-1}(E)$ is the $(n-1)$-dimensional Minkowski content of E, defined by

$$\mathcal{M}^{n-1}(E) = \lim_{\varepsilon \to 0} \frac{1}{2\varepsilon}\left|\left\{x \in \mathbb{R}^n \colon \operatorname{dist}(x, E) < \varepsilon\right\}\right|, \tag{10.4}$$

whenever the limit in (10.4) exists.

LEMMA 10.1. *Let Ω be a bounded open set with Lipschitz boundary. Let $u \in SBV(\Omega) \cap L^{\infty}(\Omega)$ with $\int_{\Omega} |\nabla u|^2\,dx + \mathcal{H}^{n-1}(S(u)) < +\infty$ and let Ω' be a bounded open subset of $\mathbb{R}^n$ such that $\Omega \subset\subset \Omega'$. Then u has an extension $z \in SBV(\Omega') \cap L^{\infty}(\Omega')$ such that $\int_{\Omega'} |\nabla z|^2\,dx + \mathcal{H}^{n-1}(S(z)) < +\infty$, $\mathcal{H}^{n-1}(S(z) \cap \partial\Omega) = 0$, and $\|z\|_{L^{\infty}(\Omega')} = \|u\|_{L^{\infty}(\Omega)}$. Moreover, there exists a sequence (z_k) in $SBV(\Omega') \cap L^{\infty}(\Omega')$ such that (z_k) converges to z in $L^1(\Omega')$, (∇z_k) converges to ∇z in $L^2(\Omega'; \mathbb{R}^n)$, $\|z_k\|_{L^{\infty}(\Omega')} \leqslant \|u\|_{L^{\infty}(\Omega)}$,*

$$\lim_k \mathcal{H}^{n-1}\left(S(z_k)\right) = \mathcal{H}^{n-1}\left(S(z)\right), \tag{10.5}$$

$$\lim_k \mathcal{H}^{n-1}\left(S(z_k) \cap \overline{\Omega}\right) = \mathcal{H}^{n-1}\left(S(z) \cap \Omega\right) = \mathcal{H}^{n-1}\left(S(u)\right), \tag{10.6}$$

$\mathcal{H}^{n-1}(\overline{S(z_k)} \setminus S(z_k)) = 0$ *and* $\mathcal{H}^{n-1}(S(z_k) \cap K) = \mathcal{M}^{n-1}(S(z_k) \cap K)$ *for every compact set $K \subseteq \Omega'$.*

REMARK 10.2 (Approximations with polyhedral jump sets). The result above has been further refined in a very handy way by Cortesani and Toader [78], who have proved that a dense class of *SBV*-functions are those which jump set $S(u)$ is composed of a finite number of polyhedral sets.

10.1.2. *The Mumford–Shah functional.* The prototype of the energies in (10.3) derives from a model in Image Reconstruction due to Mumford and Shah [131], where we minimize

$$\int_{\Omega\setminus K} |\nabla u|^2 \, dx + c_1 \int_{\Omega\setminus K} |u - g|^2 \, dx + c_2 \mathcal{H}^1(K) \tag{10.7}$$

on an open set $\Omega \subset \mathbb{R}^2$. In this case g is interpreted as the input picture taken from a camera, u is the "cleaned" image and K is the relevant contour of the objects in the picture. The constants c_1 and c_2 are contrast parameters. Note that the problem is meaningful also adding the constraint $\nabla u = 0$ outside K, in which case we have a minimal partitioning problem. Note moreover that very similar energies are linked to Griffith's theory of fracture, where $|\nabla u|^2$ is substituted by a linear elastic energy.

In the framework of $SBV(\Omega)$ functions, with Ω a bounded open subset of $\mathbb{R}^n$, the *Mumford–Shah functional* is written as

$$MS(u) = \alpha \int_{\Omega} |\nabla u|^2 \, dx + \beta \, \mathcal{H}^{n-1}\big(S(u)\big). \tag{10.8}$$

Comparing with (10.7) note that we drop the term $\int |u - g|^2 \, dx$ since it is continuous with respect to the $L^2(\Omega)$-convergence.

Weak solutions for problems involving energies (10.7) are obtained by applying the following compactness and lower-semicontinuity theorem by Ambrosio [20].

THEOREM 10.3 (*SBV compactness*). *Let (u_j) be a sequence of functions in $SBV(\Omega)$ such that $\sup_j MS(u_j) < +\infty$ and $\sup \|u_j\|_\infty < +\infty$. Then, up to subsequences, there exists a function $u \in SBV(\Omega)$ such that $u_j \to u$ in $L^2(\Omega)$ and $\nabla u_j \rightharpoonup \nabla u$ weakly in $L^2(\Omega; \mathbb{R}^n)$. Moreover $\mathcal{H}^{n-1}(S(u)) \leqslant \liminf_j \mathcal{H}^{n-1}(S(u_j))$.*

REMARK 10.4 (Existence of optimal segmentation). Let $g \in L^\infty(\Omega)$ and consider the minimum problem

$$\min\left\{ MS(u) + \int_{\Omega} |u - g|^2 \, dx \colon u \in SBV(\Omega) \right\}. \tag{10.9}$$

If (v_j) is a minimizing sequence for this problem, note that such is also the truncated sequence $u_j = ((-\|g\|_\infty) \vee v_j) \wedge \|g\|_\infty$. To this sequence we can apply the compactness result above to obtain a solution.

If $(u_\varepsilon) \to u$ in $L^1(\Omega)$, $u_\varepsilon \in SBV(\Omega)$ and $\sup_\varepsilon MS(u_\varepsilon) < +\infty$ then u need not be a SBV-function. To completely describe the domain of MS we should introduce the following extension of $SBV(\Omega)$.

DEFINITION 10.5 (Generalized special functions of bounded variation). A function u belongs to $GSBV(\Omega)$ if all its truncations $u_T = (u \vee (-T)) \wedge T$ belong to $SBV(\Omega)$.

In the rest of the section it will not be restrictive to describe our functionals on *SBV* since we will always be able to reduce to that space by truncations.

10.1.3. *Two asymptotic results for the Mumford–Shah functional.*

(1) The "complete" Mumford–Shah functional (10.7) with given g may be scaled to understand the role of the contrast parameter c_1, c_2. This has been done by Rieger and Tilli [135], who have shown that an interesting scaling gives the energy

$$F_\varepsilon(u) = \frac{1}{\varepsilon^2} \int_\Omega \left(|\nabla u|^2 + |u - g|^2 \right) dx + \varepsilon \mathcal{H}^1 \left(S(u) \right). \tag{10.10}$$

In this case the convergence taken into account is the weak* convergence of measures applied to the measures $(\mathcal{H}^1(S(u)))^{-1} \mathcal{H}^1 \llcorner S(u)$, so that the limit is defined on probability measures μ on $\overline{\Omega}$. If $g \in H^1(\Omega)$ in this space it takes the form

$$F_0(\mu) = \left(\frac{9}{16} \int_\Omega \frac{|\nabla g|^2}{(d\mu/dx)^2} \, dx \right)^{1/3}, \tag{10.11}$$

where $d\mu/dx$ is the density of the absolutely continuous part of μ with respect to the Lebesgue measure.

(2) The presence of two competing terms with different growth and scaling properties makes asymptotic problems for energies of the Mumford–Shah type technically more challenging. The homogenization of functionals of this form has been computed in [55], while a very interesting variant of the Γ-limit of functionals on perforated domains as in Section 6, with the Mumford–Shah functional in place of the Dirichlet integral in (6.5), has been performed by Focardi and Gelli [101]. In that case, the scaling is given by $\delta_\varepsilon = \varepsilon^{n/(n-1)}$ and the limit is represented as

$$F_0(u) = \alpha \int_\Omega |\nabla u|^2 \, dx + \beta \mathcal{H}^{n-1} \left(S(u) \right) + \beta' |\{u \neq 0\}|, \tag{10.12}$$

with β' depending only on K and β and defined through the minimization of the perimeter among sets containing K (see [101] for details and generalizations).

10.2. *The Ambrosio–Tortorelli approximation*

The first approximation by Γ-convergence of the Mumford–Shah functional was given by Ambrosio and Tortorelli in [21]. Following the idea developed by Modica and Mortola for the approximation of the perimeter functional by elliptic functionals, they introduced an approximation procedure of $MS(u)$ with an auxiliary variable v, which in the limit approaches $1 - \chi_{S(u)}$. As the approximating functionals are elliptic, even though nonconvex, numerical methods can be applied to them. It is clear, though, that the introduction of an extra variable v can be very demanding from a numerical viewpoint.

10.2.1. *An approximation by energies on set-function pairs.* A common pattern in the approximation of free-discontinuity energies (often, in the liminf inequality) is the substitution of the sharp interface $S(u)$ by a "blurred" interface, of small but finite Lebesgue measure, shrinking to $S(u)$. In this subsection we present a simple result in that direction, which will be used in the proof of the Ambrosio–Tortorelli result in the next one, and formalizes an intermediate step present in many approximation procedures (see, e.g., [44]). The result mixes Sobolev functions and sets of finite perimeter and is due to Braides, Chambolle and Solci [48].

THEOREM 10.6 (Set-function approximation). *Let δ_ε be a sequence of positive numbers converging to 0, and let F_ε be defined on pairs (u, E), where E is a set of finite perimeter and u is such that $u(1 - \chi_E) \in SBV(\Omega)$ and $S(u(1 - \chi_E)) \subseteq \partial^* E$ (i.e., $u \in H^1(\Omega \setminus E)$ if E is smooth) by*

$$F_\varepsilon(u, E) = \begin{cases} \alpha \int_{\Omega \setminus E} |\nabla u|^2 \, dx + \frac{1}{2}\beta \mathcal{H}^{n-1}(\partial^* E) & \text{if } |E| \leq \delta_\varepsilon, \\ +\infty & \text{otherwise.} \end{cases} \tag{10.13}$$

Then F_ε Γ-converges to the functional (equivalent to MS)

$$F_0(u, E) = \begin{cases} MS(u) & \text{if } |E| = 0, \\ +\infty & \text{otherwise,} \end{cases} \tag{10.14}$$

if $u \in SBV(\Omega)$, with respect to the $L^2(\Omega)$-convergence of u and χ_E. Clearly, the functional F_0 is equivalent to MS as far as minimum problems are concerned.

PROOF. The proof is achieved by slicing. The one-dimensional case is easily achieved as follows: if $u_\varepsilon \to u$ and $\sup_\varepsilon F_\varepsilon(u_\varepsilon, E_\varepsilon) < +\infty$, then $\chi_{E_\varepsilon} \to 0$ and E_ε are composed by an equibounded number of segments. We can therefore assume, up to subsequences, that E_ε shrink to a finite set S. Since $u_\varepsilon \rightharpoonup u$ in $H^1_{\text{loc}}(\Omega \setminus S)$ and $\int_{\Omega \setminus S} |\nabla u|^2 \, dx \leq \liminf_{\varepsilon \to 0} \int_{\Omega \setminus E_\varepsilon} |\nabla u_\varepsilon|^2 \, dx$ we deduce that $u \in SBV(\Omega)$ and $S(u) \subset S \cap \Omega$. Since for each point $x \in S \cap \Omega$ we have at least two families of points in ∂E_ε converging to x, we deduce that $2\#(S(u)) \leq \liminf_{\varepsilon \to 0} \#(\partial E_\varepsilon)$, and then the liminf inequality. The limsup inequality is then achieved by taking $u_\varepsilon = u$ and E_ε any sequence of open sets containing $S(u)$ with $|E_\varepsilon| \leq \delta_\varepsilon$. The n-dimensional case follows exactly the proof of Theorem 7.3 with the due changes. $\square$

Note that in the previous result we may restrict the domain of F_ε to pairs (u, E) with E with Lipschitz boundary and $u \in H^1(\Omega)$, up to a smoothing argument for the recovery sequence.

10.2.2. *Approximation by elliptic functionals.* We now can introduce the Ambrosio–Tortorelli approximating energies

$$F_\varepsilon(u, v) = \alpha \int_\Omega v^2 |\nabla u|^2 \, dx + \frac{\beta}{2} \int_\Omega \left(\varepsilon |\nabla v|^2 + \frac{1}{4\varepsilon}(1 - v)^2 \right) dx, \tag{10.15}$$

defined on functions u, v such that $v \in H^1(\Omega)$, $uv \in H^1(\Omega)$ and $0 \leqslant v \leqslant 1$. The idea behind the definition of F_ε is that the variable v tends to take the value 1 almost everywhere due to the last term in the second integral, and the value 0 on $S(u)$ in order to make the first integral finite. As a result, the interaction of the two terms in the second integral will give a surface energy concentrating on $S(u)$. We present a proof which uses the result in the previous section; in a sense, v^2 is replaced by $1 - \chi_E$ and the second integral by $\mathcal{H}^{n-1}(\Omega \cap \partial^* E)$.

THEOREM 10.7 (Ambrosio–Tortorelli approximation). *The functionals F_ε defined in (10.15) Γ-converge as $\varepsilon \to 0$ with respect to the $(L^1(\Omega))^2$-topology to the functional*

$$F(u, v) = \begin{cases} MS(u) & \text{if } v = 1 \text{ a.e. on } \Omega, \\ +\infty & \text{otherwise,} \end{cases} \tag{10.16}$$

defined on $(L^1(\Omega))^2$.

PROOF. Let $u_\varepsilon \to u$ and $v_\varepsilon \to v$ be such that $F_\varepsilon(u_\varepsilon, v_\varepsilon) \leqslant c < +\infty$. We first observe that $v = 1$ a.e. since $\int_\Omega (v_\varepsilon - 1)^2 \, dx \leqslant C\varepsilon$. For every A open subset of Ω we denote by $F_\varepsilon(\cdot, \cdot; A)$ the energy obtained by computing both integrals on A. The "Modica–Mortola trick" and an application of the coarea formula gives

$$F_\varepsilon(u_\varepsilon, v_\varepsilon; A) \geqslant \alpha \int_A v_\varepsilon^2 |\nabla u_\varepsilon|^2 \, dx + \beta \int_A |1 - v_\varepsilon||\nabla v_\varepsilon| \, dx$$

$$\geqslant \alpha \int_A v_\varepsilon^2 |\nabla u_\varepsilon|^2 \, dx + \beta \int_0^1 (1 - s)\mathcal{H}^{n-1}\big(\partial\{v_\varepsilon < s\} \cap A\big) \, ds.$$

Now, we fix $\delta \in (0, 1)$. The mean value theorem ensures the existence of $t_\varepsilon^\delta \in (\delta, 1)$ such that

$$\int_\delta^1 (1 - s)\mathcal{H}^{n-1}\big(\partial\{v_\varepsilon < s\} \cap A\big) \, ds \geqslant \int_\delta^1 (1 - s) \, ds \, \mathcal{H}^{n-1}\big(\partial E_\varepsilon^\delta \cap A\big)$$

$$= \frac{1}{2}(1 - \delta)^2 \mathcal{H}^{n-1}\big(\partial E_\varepsilon^\delta \cap A\big),$$

where $E_\varepsilon^\delta = \{v_\varepsilon < t_\varepsilon^\delta\}$; hence

$$F_\varepsilon(u_\varepsilon, v_\varepsilon; A) \geqslant \alpha\delta^2 \int_{A\setminus E_\varepsilon^\delta} |\nabla u_\varepsilon|^2 \, dx + \frac{\beta}{2}(1 - \delta)^2 \mathcal{H}^{n-1}\big(\partial E_\varepsilon^\delta \cap A\big). \tag{10.17}$$

An application of Theorem 10.6 gives

$$\liminf_{j \to +\infty} \left(\alpha\delta^2 \int_{A\setminus E_\varepsilon^\delta} |\nabla u_\varepsilon|^2 \, dx + \frac{\beta}{2}(1 - \delta)^2 \mathcal{H}^{n-1}\big(\partial E_\varepsilon^\delta \cap A\big) \right)$$

$$\geqslant \alpha\delta^2 \int_A |\nabla u|^2 \, dx + \beta(1 - \delta)^2 \mathcal{H}^{n-1}\big(S(u) \cap A\big).$$

An easy application of Lemma 3.1 with $\lambda = \mathcal{L}^n + \mathcal{H}^{n-1} \llcorner S(u)$ gives the lower bound.

By Lemma 10.1 it suffices to prove the limsup inequality when the target function u is smooth and $S(u)$ is essentially closed with Minkowski content equal to its $\mathcal{H}^{n-1}$ measure. The proof of the optimality of the lower bound is obtained through the construction of recovery sequences $(u_\varepsilon, v_\varepsilon)$ with $u_\varepsilon = u$ and

$$
v_\varepsilon(x) = \begin{cases} v\left(\frac{\text{dist}(x, S(u)) - \varepsilon^2}{\varepsilon}\right) & \text{if } \text{dist}(x, S(u)) \geq \varepsilon^2, \\ 0 & \text{if } \text{dist}(x, S(u)) < \varepsilon^2, \end{cases}
$$

with a "one-dimensional" structure, as in the proof of the Modica–Mortola result. The function v_ε equals to 0 in a neighborhood of $S(u)$ so that the first integral converges to $\int_\Omega |\nabla u|^2 \, dx$. The function v is chosen to be an optimal profile giving the equality in the Modica–Mortola trick, with datum $v(0) = 0$. In the case above the computation is explicit, giving $v(t) = 1 - e^{-t/2}$. $\qquad\square$

REMARK 10.8 (Approximate solutions of the Mumford–Shah problem). A little variation must be made to obtain coercive functionals approximating MS, by adding a perturbation of the form $k_\varepsilon \int_\Omega |\nabla u|^2 \, dx$ to $G_\varepsilon(u, v)$ with $0 < k_\varepsilon = o(\varepsilon)$. In this way, with fixed $g \in L^\infty(\Omega)$, for each ε we obtain a solution to

$$
m_\varepsilon = \min\left\{ F_\varepsilon(u, v) + k_\varepsilon \int_\Omega |\nabla u|^2 \, dx + \int_\Omega |u - g|^2 \, dx \colon u, v \in H^1(\Omega) \right\},
$$

(10.18)

and up to subsequences, these solutions $(u_\varepsilon, v_\varepsilon)$ converge to $(u, 1)$, where u is a minimizer of (10.9).

10.3. *Other approximations*

In this subsection we briefly illustrate some alternative ideas, for which details can be found in [45]. A simpler approach to approximate MS is to try an approximation by means of local integral functionals of the form

$$
\int_\Omega f_\varepsilon(\nabla u(x)) \, dx,
$$

(10.19)

defined in the Sobolev space $H^1(\Omega)$. It is clear that such functionals cannot provide any variational approximation for MS. In fact, if an approximation existed by functionals of this form, the functional $MS(u)$ would also be the Γ-limit of their lower-semicontinuous envelopes; i.e., the convex functionals

$$
\int_\Omega f_\varepsilon^{**}(\nabla u(x)) \, dx,
$$

(10.20)

where f_ε^{**} is the convex envelope of f_ε (see Proposition 2.5(2) and Remark 4.8(4)), in contrast with the lack of convexity of *MS*. However, functionals of the form (10.19) can be a useful starting point for a heuristic argument. We can begin by requiring that for every $u \in SBV(\Omega)$ with ∇u and $S(u)$ sufficiently smooth we have

$$\lim_{\varepsilon \to 0^+} \int_\Omega f_\varepsilon\big(\nabla u_\varepsilon(x)\big)\, dx = \alpha \int_\Omega |\nabla u|^2\, dx + \beta \mathcal{H}^{n-1}\big(S(u)\big)$$

if we choose u_ε very close to u, except in an ε-neighborhood of $S(u)$ (where the gradient of u_ε tends to be very large). It can be easily seen that this requirement is fulfilled if we choose f_ε of the form

$$f_\varepsilon(\xi) = \frac{1}{\varepsilon} f\big(\varepsilon |\xi|^2\big), \quad \text{with } f'(0) = \alpha \text{ and } \lim_{t \to +\infty} f(t) = \frac{\beta}{2}; \tag{10.21}$$

the simplest such f is $f(t) = \alpha t \wedge \frac{\beta}{2}$.

10.3.1. *Approximation by convolution functionals.* Nonconvex integrands of the form (10.21) can be used for an approximation, provided we slightly modify the functionals in (10.19). This can be done in many ways. For example, the convexity constraint in ∇u can be removed by considering approximations of the form

$$F_\varepsilon(u) = \frac{1}{\varepsilon} \int_\Omega f\left(\varepsilon \fint_{B_\varepsilon(x) \cap \Omega} |\nabla u(y)|^2\, dy \right) dx, \tag{10.22}$$

defined for $u \in H^1(\Omega)$, where f is a suitable nondecreasing continuous (nonconvex) function and $\fint_B$ denotes the *average* on B. These functionals, proposed by Braides and Dal Maso, are nonlocal in the sense that their energy density at a point $x \in \Omega$ depends on the behavior of u in the whole set $B_\varepsilon(x) \cap \Omega$ or, in other words, on the value of the convolution of u with $\frac{1}{|B_\varepsilon|}\chi_{B_\varepsilon}$. More general convolution kernels with compact support may be considered. Note that, even if the term containing the gradient is not convex, the functional F_ε is weakly lower semicontinuous in $H^1(\Omega)$ by Fatou's lemma. These functionals Γ-converge, as $\varepsilon \to 0$, to the Mumford–Shah functional *MS* in (10.8) if f satisfies the limit conditions in (10.21) (see [52]).

The proof of this result is rather technical in the liminf part, reducing to a "nonlocal slicing procedure". It must be remarked that, on the other side, the limsup inequality is easily obtained as the Γ-limit coincides with the limit (at least for regular-enough $S(u)$) of mollified u_ε with a mollifier with support on a scale finer than ε. In fact for such u_ε,

$$\lim_{\varepsilon \to 0} F_\varepsilon(u_\varepsilon) = \lim_{\varepsilon \to 0} \left(\frac{1}{\varepsilon} \int_{\Omega \cap \{\mathrm{dist}(x, S(u)) > \varepsilon\}} f\left(\varepsilon \fint_{B_\varepsilon(x) \cap \Omega} |\nabla u_\varepsilon(y)|^2\, dy \right) dx \right.$$

$$\left. + \frac{1}{\varepsilon} \int_{\Omega \cap \{\mathrm{dist}(x, S(u)) \leq \varepsilon\}} f\left(\varepsilon \fint_{B_\varepsilon(x) \cap \Omega} |\nabla u_\varepsilon(y)|^2\, dy \right) dx \right)$$

$$= \lim_{\varepsilon \to 0} \left(\frac{1}{\varepsilon} \int_{\Omega \cap \{\mathrm{dist}(x, S(u)) > \varepsilon\}} \alpha \varepsilon \fint_{B_\varepsilon(x) \cap \Omega} \left| \nabla u(y) \right|^2 dy \, dx \right.$$

$$\left. + \beta \frac{\left| \Omega \cap \{\mathrm{dist}(x, S(u)) \leqslant \varepsilon\} \right|}{2\varepsilon} \right)$$

$$= \lim_{\varepsilon \to 0} \left(\int_{\Omega} \alpha \fint_{B_\varepsilon(x) \cap \Omega} \left| \nabla u(y) \right|^2 dy \, dx \right.$$

$$\left. + \beta \frac{\left| \Omega \cap \{\mathrm{dist}(x, S(u)) \leqslant \varepsilon\} \right|}{2\varepsilon} \right)$$

$$= \alpha \int_{\Omega} \left| \nabla u(y) \right|^2 dy + \beta \mathcal{H}^{n-1}\left(\Omega \cap S(u) \right).$$

10.3.2. *A singular-perturbation approach.* A different path can be followed by introducing a second-order singular perturbation. Dealing for simplicity with the one-dimensional case, we have functionals of the form (f again as in (10.21))

$$F_\varepsilon(u) = \frac{1}{\varepsilon} \int_{\Omega} f\left(\varepsilon |u'|^2 \right) dx + \varepsilon^3 \int_{\Omega} |u''|^2 \, dx \tag{10.23}$$

on $H^2(\Omega)$. Note that the Γ-limit of these functionals would be trivial without the last term, and that the convexity in u'' assures the weak lower semicontinuity of F_ε in $H^2(\Omega)$. Alicandro, Braides and Gelli [14] have proven that the family (F_ε) Γ-converges to the functional defined on $SBV(\Omega)$ by

$$F(u) = \alpha \int_{\Omega} |u'|^2 \, dx + C \sum_{t \in S(u)} \sqrt{|u^+(t) - u^-(t)|}, \tag{10.24}$$

with C explicitly computed from β through the optimal profile problem

$$C = \min_{T > 0} \min \left\{ \beta T + \int_{-T}^{T} |v''|^2 \, dt : \ v(\pm T) = \pm \frac{1}{2}, \ v'(\pm T) = 0 \right\} \tag{10.25}$$

corresponding to minimizing the contribution of the part of the energy $F_\varepsilon(u_\varepsilon)$ concentrating on an interval where $f(\varepsilon |u'_\varepsilon|^2) = \beta/2$, centered in some x_ε, after the usual scaling $v(t) = u_\varepsilon(x_\varepsilon + \varepsilon t)$.

In contrast with those in (10.15), functionals (10.23) possess a particularly simple form, with no extra variables. The form of the approximating functionals gets more complex if we want to use this approach to recover in the limit other surface energies (as, for example, in the Mumford–Shah functional) in which case we must substitute f by more complex f_ε.

10.3.3. *Approximation by finite-difference energies.* A sequence of functionals proposed by De Giorgi provides another type of nonlocal approximation of the Mumford–Shah functional, proved by Gobbino [114], namely,

$$F_\varepsilon(u) = \frac{1}{\varepsilon^{n+1}} \int_{\Omega \times \Omega} f\left(\frac{(u(x) - u(y))^2}{\varepsilon}\right) e^{-|x-y|^2/\varepsilon} \, dx \, dy, \tag{10.26}$$

defined on $L^1(\Omega)$, with f as in (10.21). In this case the constants α and β in MS must be replaced by two other constants A and B.

The idea behind the energies in (10.26) derives from an analogous result by Chambolle [70] in a discrete setting, giving an anisotropic version of the Mumford–Shah functional (see Section 11.2.2); F_ε are designed to eliminate such anisotropy. This procedure is particularly flexible, allowing for easy generalizations, to approximate a wide class of functionals. The main drawback of this approach is the difficulty in obtaining coerciveness properties. In this case the Γ-limit coincides with the pointwise limit. In particular, if $u(x) = x_1$ we get

$$\lim_{\varepsilon \to 0} F_\varepsilon(u) = \lim_{\varepsilon \to 0} \frac{1}{\varepsilon^{n+1}} \int_{\Omega \times \Omega} f\left(\frac{(x_1 - y_1)^2}{\varepsilon}\right) e^{-|x-y|^2/\varepsilon} \, dx \, dy$$

$$= \alpha |\Omega| \int_{\mathbb{R}^n} |\xi_1|^2 e^{-|\xi|^2} \, d\xi,$$

and similarly, if we choose $u(x) = \mathrm{sign}(x_1)$, we get

$$\lim_{\varepsilon \to 0} F_\varepsilon(u) = \lim_{\varepsilon \to 0} \frac{1}{\varepsilon^{n+1}} \int_{\Omega \times \Omega} f\left(\frac{(\mathrm{sign}(x_1) - \mathrm{sign}(y_1))^2}{\varepsilon}\right) e^{-|x-y|^2/\varepsilon} \, dx \, dy$$

$$= \beta \mathcal{H}^{n-1}\big(\Omega \cap S(u)\big) \int_{\mathbb{R}^n} |\xi| e^{-|\xi|^2} \, d\xi,$$

from which we obtain the value of the constants A and B.

10.4. *Approximation of curvature functionals*

We close this section with a brief description of the approximation for another type of functional used in Visual Reconstruction; namely,

$$\mathcal{G}(u, C, P) = \#(P) + \int_C (1 + \kappa^2) \, d\mathcal{H}^1 + \int_{\Omega \setminus (C \cup P)} |\nabla u|^2 \, dx$$

(to which a "fidelity term" $\int_\Omega |u - g|^2 \, dx$ can be added), where C is a family of curves, P is the set of the endpoints of the curves in C, and $\#(P)$ is the number of points in P. The functional $\mathcal{G}$ has been proposed in this form by Anzellotti, and has a number of connections with similar energies proposed in Computer Vision. Existence results for minimizers of $\mathcal{G}$ can be found in a paper by Coscia [79]. They are relatively simple, since energy bounds imply a bound on the number of components of C and on their norm as $W^{2,2}$-functions.

We now illustrate an approximation due to Braides and March [63]. This result will use at the same time many of the arguments introduced before. One will be the introduction of intermediate energies defined on set-function triplets (as in Section 10.2.1), another will be the iterated use of a gradient approach as in the Modica–Mortola case to generate energies on sets of different dimensions, and finally the modification of the Ambrosio–Tortorelli construction to obtain recovery sequences in H^2.

We first show the idea for an approximation by energies on triplets function-sets. The first step is to construct a variational approximation of the functional $\#(P)$ that simply counts the number of the points of a set P by another functional on sets, whose minimizers are discs of small radius ε (additional conditions will force these discs to contain the target set of points). Such a functional is given by

$$\mathcal{E}_{\varepsilon}^{(1)}(D) = \frac{1}{4\pi} \int_{\partial D} \left(\frac{1}{\varepsilon} + \varepsilon \kappa^2(x) \right) d\mathcal{H}^1(x),$$

where κ denotes the curvature of ∂D. The number $1/(4\pi)$ is a normalization factor that derives from the fact that minimizers of $\mathcal{E}_{\varepsilon}^{(1)}(D)$ are given by balls of radius ε. This functional may be interpreted, upon scaling, as a singular perturbation of the perimeter functional by a curvature term.

The next step is then to construct another energy defined on sets, that approximates the functional $\int_C (1 + \kappa^2) d\mathcal{H}^1$, where C is a (finite) union of $W^{2,2}$-curves with endpoints contained in P. To this end we approximate C away from D by sets A, whose energy is defined as

$$\mathcal{E}_{\varepsilon}^{(2)}(A, D) = \frac{1}{2} \int_{(\partial A) \setminus D} \left(1 + \kappa^2 \right) d\mathcal{H}^1, \quad |A| \leqslant \delta_{\varepsilon},$$

where $0 < \delta_{\varepsilon} = o(1)$ play the same role as in Theorem 10.6, and force A to shrink to C. As in Section 10.2.1, the factor $1/2$ depends on the fact that, as A tends to C, each curve of C is the limit of two arcs of ∂A.

The intermediate function-set approximation is thus constructed by assembling the pieces above and the simpler terms that account for u

$$\mathcal{E}_{\varepsilon}(u, A, D) = \mathcal{E}_{\varepsilon}^{(1)}(D) + \mathcal{E}_{\varepsilon}^{(2)}(A, D) + \int_{\Omega \setminus (A \cup D)} |\nabla u|^2 dx, \quad |A| \leqslant \delta_{\varepsilon},$$

defined for A and D compactly contained in Ω. Note that $A \cup D$ contains the singularities of u. By following a recovery sequence for the Γ-limit of $\mathcal{E}_{\varepsilon}$, a triplet (u, C, P) is approximated by means of triplets $(u_{\varepsilon}, A_{\varepsilon}, D_{\varepsilon})$.

To obtain an energy defined on functions, we again use a gradient-theory approach as by Modica and Mortola, where it is shown that the perimeter measures $\mathcal{H}^1 \llcorner \partial A$ and $\mathcal{H}^1 \llcorner \partial D$ are approximated by the measures $\mathcal{H}^1_\varepsilon(s, \nabla s)\,dx$ and $\mathcal{H}^1_\varepsilon(w, \nabla w)\,dx$, where

$$\mathcal{H}^1_\varepsilon(s, \nabla s) = \zeta_\varepsilon |\nabla s|^2 + \frac{s^2(1-s)^2}{\zeta_\varepsilon}$$

and

$$\mathcal{H}^1_\varepsilon(w, \nabla w) = \zeta_\varepsilon |\nabla w|^2 + \frac{w^2(1-w)^2}{\zeta_\varepsilon},$$

$\zeta_\varepsilon \to 0$ as $\varepsilon \to 0$, and s and w are optimal-profile functions approximating $1 - \chi_A$ and $1 - \chi_D$, respectively. We define the curvature of s and w as

$$\kappa(\nabla s) = \begin{cases} \operatorname{div}\left(\frac{\nabla s}{|\nabla s|}\right) & \text{if } \nabla s \neq 0, \\ 0 & \text{otherwise,} \end{cases} \qquad \kappa(\nabla w) = \begin{cases} \operatorname{div}\left(\frac{\nabla w}{|\nabla w|}\right) & \text{if } \nabla w \neq 0, \\ 0 & \text{otherwise,} \end{cases}$$

respectively. The next step is formally to replace the characteristic functions $1 - \chi_A$ and $1 - \chi_D$ by functions s and w. The terms $\mathcal{E}^{(1)}_\varepsilon(D)$ and $\mathcal{E}^{(2)}_\varepsilon(A, D)$ are then substituted by

$$\mathcal{G}^{(1)}_\varepsilon(w) = \int_\Omega \left(\frac{1}{\varepsilon} + \varepsilon\kappa^2(\nabla w)\right)\mathcal{H}^1_\varepsilon(w, \nabla w)\,dx,$$

$$\mathcal{G}^{(2)}_\varepsilon(s, w) = \int_\Omega w^2\left(1 + \kappa^2(\nabla s)\right)\mathcal{H}^1_\varepsilon(s, \nabla s)\,dx,$$

respectively, and the constraint that $|A| \leqslant a_\varepsilon$ by an integral penalization

$$\mathcal{I}_\varepsilon(s, w) = \frac{1}{\mu_\varepsilon} \int_\Omega \left((1-s)^2 + (1-w)^2\right)dx$$

(where $\mu_\varepsilon \to 0$) that forces s and w to be equal to 1 almost everywhere in the limit as $\varepsilon \to 0$, so that we construct a candidate functional

$$\mathcal{G}_\varepsilon(u, s, w) = \frac{1}{4\pi b_0}\mathcal{G}^{(1)}_\varepsilon(w) + \frac{1}{2b_0}\mathcal{G}^{(2)}_\varepsilon(s, w) + \int_\Omega s^2|\nabla u|^2\,dx + \mathcal{I}_\varepsilon(s, w),$$

where b_0 is a normalization constant.

The following result shows that these elliptic energies are indeed variational approximations of the energy $\mathcal{G}$, for a suitable choice of ζ_ε and μ_ε.

THEOREM 10.9 (Approximation of curvature functionals). *The functionals* $\mathcal{G}_\varepsilon$ *Γ-converge to* $\mathcal{G}$ *as* $\varepsilon \to 0^+$ *with respect to the convergence* $(u_\varepsilon, s_\varepsilon, w_\varepsilon) \to (u, C, P)$ *defined as the convergence of a.a. level sets (see* [63] *for a precise definition).*

A technical but important difference with the Ambrosio–Tortorelli approach is that there the double-well potential $s^2(1-s)^2$ in the approximation of the perimeter is replaced by the single-well potential $(1-s)^2$. This modification breaks the symmetry between 0 and 1 and forces automatically s to tend to 1 as $\varepsilon \to 0^+$. Unfortunately, it also forbids recovery sequences to be bounded in H^2: with this substitution the curvatures terms in $\mathcal{G}_\varepsilon^{(1)}$ and $\mathcal{G}_\varepsilon^{(2)}$ would necessarily be unbounded. In our case, the necessary symmetry breaking is obtained by adding the "lower-order" term $\mathcal{I}_\varepsilon$. We note that the complex form of the functionals $\mathcal{G}_\varepsilon$, in particular of $\mathcal{G}_\varepsilon^{(1)}$, seems necessary despite the simple form of the target energy. Indeed, in order to describe an energy defined on points in the limit, it seems necessary to consider degenerate functionals. Our approach may be compared with that giving vortices in the Ginzburg–Landau theory or concentration of energies for functionals with critical growth in Section 8.

11. Continuum limits of lattice systems

In this last section we touch a subject of active research, with connections with many issues in Statistical Mechanics, Theoretical Physics, Computer Vision, computational problems, approximation schemes, etc. Namely, that of the passage from a variational problem defined on a discrete set to a corresponding problem on the continuum as the number of the points of the discrete set increases. Some of these problems naturally arise in an atomistic setting, or as finite-difference numerical schemes. An overview of scale problems for atomistic theories can be found in the review paper by Le Bris and Lions [117], an introduction to some of their aspects in Computational Materials Science can be found in that by Blanc, Le Bris and Lions [37]. An introduction to the types of problems treated in this section in a one-dimensional setting can be found in the notes by Braides and Gelli [60] (see also [46], Chapters 4 and 11).

The setting for discrete problems in which we have a fairly complete set of results is that of *central interactions* for *lattice systems*; i.e., systems where the reference positions of the interacting points lie on a prescribed lattice, whose parameters change as the number of points increases, and each point of the lattice interacts separately with each other point. In more precise terms, we consider an open set $\Omega \subset \mathbb{R}^n$ and take as reference lattice $Z_\varepsilon = \Omega \cap \varepsilon \mathbb{Z}^n$. The general form of a *pair-potential* energy is then

$$F_\varepsilon(u) = \sum_{i,j \in Z_\varepsilon} f_{ij}^\varepsilon\big(u(i), u(j)\big), \tag{11.1}$$

where $u : Z_\varepsilon \to \mathbb{R}^m$. The analysis of energies of the form (11.1) has been performed under various hypotheses on f_{ij}. The first simplifying assumption is that F_ε is invariant under translations (in the target space); that is,

$$f_{ij}^\varepsilon(u, v) = g_{ij}^\varepsilon(u - v). \tag{11.2}$$

Furthermore, an important class is that of homogeneous interactions (i.e., invariant under translations in the reference space); this condition translates into

$$f_{ij}^{\varepsilon}(u, v) = g_{(i-j)/\varepsilon}^{\varepsilon}(u, v). \tag{11.3}$$

If both conditions are satisfied, we may rewrite the energies F_{ε} above as

$$F_{\varepsilon}(u) = \sum_{k \in \mathbb{Z}^n} \sum_{i, j \in Z_{\varepsilon}, i-j=\varepsilon k} \varepsilon^n \psi_k^{\varepsilon}\left(\frac{u(i) - u(j)}{\varepsilon}\right), \tag{11.4}$$

where $\psi_k^{\varepsilon}(\xi) = \varepsilon^{-n} g_k^{\varepsilon}(\varepsilon \xi)$. In this new form the interactions appear through the (discrete) difference quotients of the function u. Upon identifying each function u with its piecewise-constant interpolation (extending the definition of u arbitrarily outside Ω), we can consider F_{ε} as defined on (a subset of) $L^1(\Omega; \mathbb{R}^m)$, and hence compute the Γ-limit with respect to the L_{loc}^1-topology. Under some coerciveness conditions the computation of the Γ-limit will give a continuous approximate description of the behavior of minimum problems involving the energies F_{ε} for ε small.

11.1. *Continuum energies on Sobolev spaces*

Growth conditions on energy densities ψ_k^{ε} imply correspondingly boundedness conditions on gradient norms of piecewise-affine interpolations of functions with equibounded energy. The simplest type of growth condition that we encounter is on *nearest neighbors*; i.e., for $|k| = 1$. If $p > 1$ exists such that

$$c_1 |z|^p - c_2 \leqslant \psi_k^{\varepsilon}(z) \leqslant c_2\big(1 + |z|^p\big) \tag{11.5}$$

$(c_1, c_2 > 0$ for $|k| = 1)$, and if $\psi_k^{\varepsilon} \geqslant 0$ for all k then the energies are *equicoercive*: if (u_{ε}) is a bounded sequence in $L^1(\Omega; \mathbb{R}^m)$ and $\sup_{\varepsilon} F_{\varepsilon}(u_{\varepsilon}) < +\infty$, then from every sequence (u_{ε_j}) we can extract a subsequence converging to a function $u \in W^{1,p}(\Omega; \mathbb{R}^m)$. In this section we will consider energies satisfying this assumption. Hence, their Γ-limits are defined in the Sobolev space $W^{1,p}(\Omega; \mathbb{R}^m)$.

First, we remark that the energies F_{ε} can also be seen as an integration with respect to measures concentrated on Dirac deltas at the points of $Z_{\varepsilon} \times Z_{\varepsilon}$. If each ψ_k^{ε} satisfies a growth condition $\psi_k^{\varepsilon}(z) \leqslant c_k^{\varepsilon}(1 + |z|^p)$, then we have

$$F_{\varepsilon}(u) \leqslant \int_{\Omega \times \Omega} \big(1 + |u(x) - u(y)|^p\big) \, d\mu_{\varepsilon},$$

where

$$\mu_{\varepsilon} = \sum_{k=1}^{\infty} \sum_{\substack{i-j=\varepsilon k, \\ i, j \in Z_{\varepsilon}}} c_k^{\varepsilon} \frac{1}{\varepsilon^p} \delta_{(i,j)}.$$

A natural condition for the finiteness of the limit of F_ε is the equiboundedness of these measures (as $\varepsilon \to 0$) for every fixed set Ω. However, under such assumption, we can have a nonlocal Γ-limit of the form

$$F(u) = \int_\Omega f\big(Du(x)\big)\,dx + \int_{\Omega \times \Omega} \psi\big(u(x) - u(y)\big)\,d\mu(x, y),$$

where μ is the weak* limit of the measures μ_ε outside the "diagonal" of $\mathbb{R}^n \times \mathbb{R}^n$ (just as we obtain Dirichlet forms from degenerate quadratic functionals). Under some decay conditions, such long-range behavior may be ruled out. The following compactness result proved by Alicandro and Cicalese [15] shows that a wide class of discrete systems possesses a "local" continuous limit (an analogue for linear difference operators can be found in [133]). We state it in a general "space-dependent" case.

THEOREM 11.1 (Compactness for discrete systems). *Let $p > 1$ and let ψ_k^ε satisfy:*

(i) (coerciveness on nearest neighbors) *there exists $c_1, c_2 > 0$ such that, for all $(x, z) \in \Omega \times \mathbb{R}^m$ and $i \in \{1, \ldots, n\}$,*

$$c_1 |z|^p - c_2 \leqslant \psi_{e_i}^\varepsilon (x, z), \tag{11.6}$$

(ii) (decay of long-range interactions) *for all $(x, z) \in \Omega \times \mathbb{R}^m$ and $k \in \mathbb{Z}^n$,*

$$\psi_k^\varepsilon (x, z) \leqslant c_k^\varepsilon \big(1 + |z|^p\big), \tag{11.7}$$

where c_k^ε satisfy
(H1) $\limsup_{\varepsilon \to 0^+} \sum_{k \in \mathbb{Z}^n} c_k^\varepsilon < +\infty$;
(H2) *for all $\delta > 0$, $M_\delta > 0$ exists such that* $\limsup_{\varepsilon \to 0^+} \sum_{|k| > M_\delta} c_k^\varepsilon < \delta$.
Let F_ε be defined by

$$F_\varepsilon(u) = \sum_{k \in \mathbb{Z}^n} \sum_{i \in R_\varepsilon^k} \varepsilon^n \psi_k^\varepsilon \left(i, \frac{u(i + \varepsilon k) - u(i)}{\varepsilon |k|}\right),$$

where $R_\varepsilon^k := \{i \in Z_\varepsilon : i + \varepsilon k \in Z_\varepsilon\}$. Then for every sequence (ε_j) of positive real numbers converging to 0, there exist a subsequence (not relabeled) and a Carathéodory function $f : \Omega \times \mathbb{M}^{m \times n} \to \mathbb{R}$ satisfying

$$c\big(|\xi|^p - 1\big) \leqslant f(x, \xi) \leqslant C\big(|\xi|^p + 1\big),$$

with $0 < c < C$, such that (F_{ε_j}) Γ-converges with respect to the $L^p(\Omega; \mathbb{R}^m)$-topology to the functional F defined as

$$F(u) = \int_\Omega f(x, Du)\,dx \quad \textit{if } u \in W^{1,p}(\Omega; \mathbb{R}^m). \tag{11.8}$$

PROOF. The proof of this theorem follows the general compactness procedure in Section 3.3. It must be remarked, though, that discrete functionals are nonlocal by nature, so that all arguments where locality is involved must be carefully adapted. The nonlocality disappears in the limit thanks to condition (ii). For a detailed proof we refer to [15]. $\quad\square$

REMARK 11.2 (Homogenization). In the case of energies defined by a scaling process; i.e., when

$$\psi_k^\varepsilon(x, z) = \psi_k\left(\frac{z}{\varepsilon}\right), \tag{11.9}$$

then the limit energy density $\varphi(M) = f(x, M)$ is independent of x and of the subsequence, and is characterized by the *asymptotic homogenization formula*

$$\varphi(M) = \lim_{T \to +\infty} \frac{1}{T^N} \min\{\mathcal{F}_T(u), u|_{\partial Q_T} = Mi\}, \tag{11.10}$$

where $Q_T = (0, T)^N$,

$$\mathcal{F}_T(u) = \sum_{k \in \mathbb{Z}^n} \sum_{i \in R_1^k(Q_T)} \psi_k\left(\frac{u(i+k) - u(i)}{|k|}\right)$$

and $u|_{\partial Q_T} = Mi$ means that "close to the boundary" of Q_T the function u is the discrete interpolation of the affine function Mx (see [15] for further details). In the one-dimensional case this formula was first derived in [61], and it is the discrete analogue of the nonlinear asymptotic formula for the homogenization of nonlinear energies of the form $\int_\Omega f(x/\varepsilon, Du)\,dx$.

Note, however, that the two formulas differ in two important aspects: the first is that (11.10) transforms functions depending on difference quotients (hence, vectors or scalars) into functions depending on gradients (hence, matrices or vectors, respectively); the second one is that the boundary conditions in (11.10) must be carefully specified, since we have to choose whether considering or not interactions that may "cross the boundary" of Q_T.

It must be noted that formula (11.10) does not simplify even in the simplest case of three levels of interactions in dimension one, thus showing that this effect, typical of nonlinear homogenization, is really due to the lattice interactions and not restricted to vector-valued functions as in the case of homogenization on the continuum.

It is worth examining formula (11.10) in some special cases. First, if all ψ_k are *convex* then, apart from a possible lower-order boundary contribution, the solution in (11.10) is simply $u_i = Mi$. In this case the Γ-limit coincides with the pointwise limit. Note that convexity in a sense always "trivializes" discrete systems, in the sense that their continuous counterpart, obtained by simply substituting difference quotients with directional derivatives is already lower semicontinuous, and hence provides automatically an optimal lower bound.

Next, if only *nearest-neighbor interactions* are present then it reduces to

$$\varphi(M) = \sum_{i=1}^{n} \psi_i^{**}(Me_i),$$

where $\psi_i = \psi_{e_i}$ and ψ_i^{**} denotes the lower-semicontinuous and convex envelope of ψ_i. Note that convexity is not a necessary condition for lower semicontinuity at the discrete level: this convexification operation should be interpreted as an effect due to oscillations at a "mesoscopic scale" (i.e., much larger than the "microscopic scale" ε but still vanishing as $\varepsilon \to 0$). If not only nearest neighbors are taken into account then the mesoscopic oscillations must be coupled with microscopic ones (see [61,132,46] and the next section).

Finally, note that also in the nonconvex case (the relaxation of) the pointwise limit always gives an upper bound for the Γ-limit and is not always trivial (see, e.g., the paper by Blanc, Le Bris and Lions [36]).

11.1.1. *Microscopic oscillations: the Cauchy–Born rule.* One issue of interest in the study of discrete-to-continuous problems is whether to a "macroscopic" gradient there corresponds at the "microscopic" scale a "regular" arrangement of lattice displacement. For energies deriving from a scaling process as in (11.9) this can be translated into the asymptotic study of minimizers for the problems defining $\varphi(M)$; in particular, whether $u_i = Mi$ is a minimizer (in which case we say that the (strict) *Cauchy–Born rule* holds at M), or if minimizers tend to a periodic perturbation of Mi; i.e., ground states are periodic (in which case we say that the *weak Cauchy–Born rule* holds at M). Note that the strict Cauchy–Born rule can be translated into the equality

$$\varphi(M) = \sum_{k \in \mathbb{Z}^N} \psi_k\left(\frac{Mk}{|k|}\right), \tag{11.11}$$

and that it always holds if all ψ_k are convex, as remarked above.

A simple example in order to understand how the validity and failure of the Cauchy–Born rule can be understood in terms of the form of φ is given by the one-dimensional case with *next-to-nearest neighbors*; i.e., when only ψ_1 and ψ_2 are nonzero. In this case $\varphi = \psi^{**}$, where

$$\psi(z) = \psi_2(2z) + \frac{1}{2} \min\{\psi_1(z_1) + \psi_1(z_2) : z_1 + z_2 = 2z\}. \tag{11.12}$$

The second term, obtained by minimization, is due to oscillations at the microscopic level: nearest neighbors rearrange so as to minimize their interaction coupled with that between second neighbors (see [46] for a simple treatment of these one-dimensional problems). In this case we can read the microscopic behavior as follows (for the sake of simplicity we suppose that the minimum problem in (11.12) has a unique solution, upon changing z_1 into z_2).

(i) First case: ψ *is convex at* z (i.e., $\psi(z) = \varphi(z)$). We have the two cases:

(a) $\psi(z) = \psi_1(z) + \psi_2(z)$; in this case $z = z_1 = z_2$ minimizes the formula giving φ and (11.11) holds; hence, the strict Cauchy–Born rule applies;

(b) $\psi(z) < \psi_1(z) + \psi_2(z)$; in this case we have a 2-periodic ground state with "slopes" z_1 and z_2, and the weak Cauchy–Born rule applies.

(ii) Second case: ψ *is not convex at* z (i.e., $\psi(z) > \varphi(z)$). In this case the Cauchy–Born rule is violated, but a finer analysis (see below) shows that minimizers are fine mixtures of states satisfying the conditions above; hence the condition holds "locally".

For energies in higher dimensions this analysis is more complex. A similar argument as in the one-dimensional case is used by Friesecke and Theil [108] to show the nonvalidity of the Cauchy–Born rule even for some types of very simple lattice interactions in dimension two, with nonlinearities of geometrical origin.

11.1.2. *Higher-order developments: phase transitions.* In the case of failure of the Cauchy–Born rule, nonuniform states may be preferred as minimizers, and surface energies must be taken into account in their description. A first attempt to rigorously describe these phenomena can be found in Braides and Cicalese [51], again in the simplest nontrivial case of next-to-nearest neighbor interactions of the form independent of ε. In that case, we may infer that (under some technical assumptions) the discrete systems are equivalent to the perturbation of a nonconvex energy on the continuum, of the form

$$\int_\Omega \psi(u')\, dt + \varepsilon^2 C \int_\Omega |u''|^2\, dt,$$

thus recovering the well-known formulation of the gradient theory of phase transitions. This result shows that a surface term (generated by the second gradient) penalizes high oscillations between states locally satisfying some Cauchy–Born rule.

11.1.3. *Homogenization of networks.* We have stated above that convex discrete problems are "trivial" since they can immediately be translated into local integral functionals. However, in some cases constraints are worse expressed in the continuous translation rather than in the original lattice notation, so that a direct treatment of the discrete system is easier. A striking and simple example is the computation of bounds for composite linear conducting networks in dimension two, as shown by Braides and Francfort [57]. This is the discrete analogue of the problem presented in Section 5.4, that translates into the computation of homogenized matrices given by

$$\langle A\xi, \xi \rangle = \frac{1}{N^2} \min \left\{ \sum_{i \in \{1,\dots,N\}^2} h_i \big(\xi_1 + \varphi(i_1 + 1, i_2) - \varphi(i_1, i_2)\big)^2 \right.$$

$$+ \sum_{i \in \{1,\dots,N\}^2} v_i \big(\xi_2 + \varphi(i_1, i_2 + 1) - \varphi(i_1, i_2)\big)^2 :$$

$$\left. \varphi : \mathbb{Z}^2 \to \mathbb{R} \ N\text{-periodic} \right\},$$

where $h_i, v_i \in \{\alpha, \beta\}$ (h_i stands for the horizontal connection at i, v_i for vertical), N is arbitrary, and the percentage θ of α-connections is given; i.e., $\#\{i\colon v_i = \alpha\} + \#\{i\colon h_i = \alpha\} = 2N^2\theta$. The set of such matrices $\mathcal{H}_d(\theta)$ is the analogue of the set $\mathcal{H}(\theta)$ in Section 5.4.1.

We do not discuss the computation of optimal bounds for $\mathcal{H}(\theta)$, for which we refer to [57], but just remark that the additional microscopic dimension brings new microgeometries. In fact, it is easily seen that $\mathcal{H}(\theta) \subset \mathcal{H}_d(\theta)$ since coefficients h_i, v_i such that $h_i = v_i$ correspond to discretizing continuous coefficients, but we can also construct discrete laminates at the same time in the horizontal and vertical directions, providing a much larger set of homogenized matrices. As an example, in the discrete case the set of all reachable matrices (see Remark 5.10) contains all diagonal matrices with eigenvalues λ_1, λ_2 with $\alpha \leqslant \lambda_i \leqslant \beta$, while in the continuous case we are restricted by the bounds (5.20).

11.2. *Continuum energies on discontinuous functions*

In many cases discrete potentials related to atomic theories do not satisfy the hypotheses of the compactness result in Theorem 11.1 and the limits are defined on spaces of discontinuous functions.

11.2.1. *Phase transitions in discrete systems.*   The easiest example of a discrete system exhibiting a phase transition is that of nearest-neighbor interactions for an elementary Ising system. We can consider energies of the form

$$- \sum_{|i-j|=\varepsilon} u_i u_j \qquad (\textit{ferromagnetic interactions}), \tag{11.13}$$

defined on functions $u\colon \varepsilon\mathbb{Z}^n \cap \Omega \to \{-1, 1\}$, and $u_i = u(i)$. Upon a scaling of the energies we can equivalently consider the functionals

$$F_\varepsilon(u) = \sum_{|i-j|=\varepsilon} \varepsilon^{n-1}(1 - u_i u_j). \tag{11.14}$$

Note that we have $1 - u_i u_j = 0$ if $u_i = u_j$ and $1 - u_i u_j = 2$ if $u_i \neq u_j$. With this observation in mind we may identify each u with the function which takes the value u_i in the coordinate cube Q_i^ε with center εi and side length ε. In this way we can rewrite

$$F_\varepsilon(u) = 2 \int_{S(u)\cap\Omega} \|v\|_1 \, d\mathcal{H}^{n-1} + \mathrm{O}(\varepsilon) \tag{11.15}$$

(the error term comes from the cubes intersecting the boundary of Ω), where $\|v\|_1 = \sum_i |v_i|$. Here we have taken into account that the interface between $\{u = 1\}$ and $\{u = -1\}$ is composed of the common boundaries of neighboring cubes Q_i^ε and Q_j^ε where $u_i \neq u_j$ and that these boundaries are orthogonal to some coordinate direction (note that $\|v\|_1$ is the greatest convex and positively homogeneous function of degree one with value 1 on the coordinate directions). This heuristic derivation can be turned into a theorem (see the paper by Alicandro, Braides and Cicalese [13]). Note that the symmetries of the limit energy density are derived from those of the square lattice, and that this reasoning also provides an equicoerciveness result.

THEOREM 11.3 (Continuum limit of a binary lattice system). *The Γ-limit in the L^1-convergence of the functional F_ε defined in (11.14) is given by*

$$F_0(u) = 2 \int_{\Omega \cap S(u)} \|v\|_1 \, d\mathcal{H}^{n-1} \tag{11.16}$$

on characteristic functions of sets of finite perimeter.

PROOF. The argument above gives the lower bound by the lower semicontinuity of F_0, an upper bound is obtained by first considering smooth interfaces, for which we may define $u_\varepsilon = u$, and then reason by density. $\square$

Similar results can also be obtained for long-range interactions, and a more subtle way to define the limit phases must be envisaged in the case of long-range *anti-ferromagnetic* interactions (i.e., when we change sign in (11.13), so that microscopic oscillations are preferred to uniform states). Details are found in [13].

11.2.2. *Free-discontinuity problems deriving from discrete systems.* The pioneering example for this case is due to Chambolle [69,70], who treated the limit of some finite-difference schemes in Computer Vision (see [38]), producing as the continuum counterpart the one-dimensional version of the Mumford–Shah functional.

THEOREM 11.4 (Blake–Zisserman approximation of the Mumford–Shah functional). *Let F_ε be defined by the* truncated quadratic energy

$$F_\varepsilon(u) = \sum_{|i-j|=\varepsilon} \varepsilon^n \min\left\{ \left(\frac{u_i - u_j}{\varepsilon}\right)^2, \frac{1}{\varepsilon} \right\} \tag{11.17}$$

defined on functions $u : \varepsilon\mathbb{Z}^n \cap \Omega \to \mathbb{R}$; then the Γ-limit in the L^1-convergence of F_ε is the anisotropic Mumford–Shah functional

$$F_0(u) = \int_\Omega |\nabla u|^2 \, dx + \int_{\Omega \cap S(u)} \|v\|_1 \, d\mathcal{H}^{n-1}. \tag{11.18}$$

on $GSBV(\Omega)$.

PROOF. In the one-dimensional case it suffices to identify each u with a discontinuous piecewise-affine interpolation v for which $F_\varepsilon(u) = F_0(v)$ (see [46], Section 8.3). A more complex interpolation can be used in the two-dimensional case (see [70]), while the general case can be achieved by adapting the slicing method (see [58,71]). Note that for functions $u_i \in \{-1, 1\}$ the functional coincides (up to a factor 2) with the one studied in the previous section. $\square$

11.2.3. *Lennard–Jones potentials.* The case of more general convex–concave potentials such as Lennard–Jones or Morse potentials brings additional problems due to the degenerate behavior at infinity. In the one-dimensional case we consider energies

$$F_\varepsilon(u) = \sum_i \varepsilon J\left(\frac{u_i - u_{i-1}}{\varepsilon}\right), \quad J(z) = \frac{1}{z^{12}} - \frac{1}{z^6}, \tag{11.19}$$

defined on $u : \varepsilon\mathbb{Z} \cap (0,1) \to \mathbb{R}$ with the constraint $u_i > u_{i-1}$ (*noninterpenetration*). It must be noted that the Γ-limit of such an energy is equal to

$$F_0(u) = \int_0^1 J^{**}(u')\,dt, \quad u \text{ increasing} \tag{11.20}$$

(u' here denotes the absolutely continuous part of the distributional derivative Du). Note that the singular part of Du does not appear here so that u may in particular have infinitely-many (increasing) jumps at "no cost".

Note that $J^{**}(z)$ takes the constant value $\min J$ for $z \geqslant z^* := \arg\min J = 2^{1/6}$, and hence $F_0(u) = \min J$ for all increasing u with $u' \geqslant z^*$. Some other method must be used to derive more information. These problems can be treated in different ways, which we mention briefly.

1. *Γ-developments.* We can consider the scaled functionals $F_\varepsilon^1(u) = 1/\varepsilon(F_\varepsilon(u) - \min J)$. In this case we obtain as a Γ-limit the functional whose domain are increasing (discontinuous) piecewise-affine functions with $u' = z^*$, on which the limit is $F^1(u) = C\#(S(u))$, and $C = -\min J$. This limit function can be interpreted as a *fracture energy* for a rigid body (see [142] and [46], Section 11.4 for a complete proof).

2. *Scaling of convex–concave potentials.* The study of these types of energies have been initiated in a paper by Braides, Dal Maso and Garroni [53], who consider potentials of convex–concave type and express the limit in terms of different scalings of the two parts, expressing the limit in the space of functions with bounded variation. In mechanical terms the limit captures softening phenomena and size effects. The general case of nearest-neighbor interactions has been treated by Braides and Gelli in [59]. The fundamental issue here is the *separation of scale effect* (see also [46], Chapter 11, for a general presentation).

3. *Renormalization-group approach.* This suggests a different scaling of the energy, and to consider

$$F_\varepsilon'(u) = \sum_i \left(J\left(z^* + \frac{u_i - u_{i-1}}{\sqrt{\varepsilon}}\right) - \min J \right). \tag{11.21}$$

The Γ-limit of F_ε' has been studied by Braides, Lew and Ortiz [62], and can be reduced to the case studied in Theorem 11.4, obtaining as a Γ-limit

$$F'(u) = \alpha \int_0^1 |u'|^2\,dt + \beta\#(S(u)), \quad \text{with } u^+ > u^- \text{ on } S(u), \tag{11.22}$$

which is interpreted as a Griffith fracture energy with a unilateral condition on the jumps. Here α is the curvature of J at its minimum and $\beta = -\min J$.

All these approaches can be extended to long-range interactions, but are more difficult to repeat in higher dimension.

11.2.4. *Boundary value problems.* For the sake of completeness it must be mentioned that according to the variational nature of the approximations all these convergence results lead to the study of convergence of minimum problems. To this regard we have to remark that in the case of the so called "long-range interactions" for functional allowing for fracture (that is, when the limit energy presents a nonzero surface part) more than one type of boundary-value problem can be formulated and an effect of boundary layer also occurs. The problem was first studied by Braides and Gelli in [58] where two types of problems where treated. The first one is to define discrete functions on the whole $\varepsilon \mathbb{Z}^N$ and to fix the values on the nodes outside the domain Ω equal to a fixed function φ; in this case the interactions "across the boundary of Ω" give rise to an additional boundary term in the limit energy of the type

$$\int_{\partial\Omega} \mathcal{G}\big(\gamma(u) - \varphi, \nu_{\partial\Omega}\big)\, d\mathcal{H}^{n-1}, \tag{11.23}$$

where $\gamma(u)$ is the inner trace of u on $\partial\Omega$.

The second method consists in considering the functions as fixed only on $\partial\Omega$, that is, only a proper subset of pairwise interactions are linked with the constraint; in this case, the boundary term gives a different contribution, corresponding to a boundary-layer effect. Indeed, the additional term is still of type (11.23) but with the surface density $\mathcal{G}'$ strictly less than $\mathcal{G}$, the gap magnifying with the range of interaction considered.

Boundary-layer effects also appear in the definition of the surface energies for fracture in the case of long-range interactions (see [51,62]) and in the study of *discrete thin films* [12].

Note to the bibliography

The following list of references contains only the works directly quoted in the text, as it would be impossible to write down an exhaustive bibliography. We refer to [54,84] and [46] for a guide to the literature.

References

[1] E. Acerbi and G. Buttazzo, *On the limits of periodic Riemannian metrics*, J. Anal. Math. **43** (1983), 183–201.

[2] E. Acerbi, G. Buttazzo and D. Percivale, *A variational definition for the strain energy of an elastic string*, J. Elasticity **25** (1991), 137–148.

[3] E. Acerbi, V. Chiadò Piat, G. Dal Maso and D. Percivale, *An extension theorem from connected sets, and homogenization in general periodic domains*, Nonlinear Anal. **18** (1992), 481–496.

[4] E. Acerbi and N. Fusco, *Semicontinuity problems in the calculus of variations*, Arch. Ration. Mech. Anal. **86** (1984), 125–145.

[5] G. Alberti, *Variational models for phase transitions, an approach via Γ-convergence*, Calculus of Variations and Partial Differential Equations, L. Ambrosio and N. Dancer, eds, Springer-Verlag, Berlin (2000), 95–114.

[6] G. Alberti, S. Baldo and G. Orlandi, *Variational convergence for functionals of Ginzburg–Landau type*, Indiana Univ. Math. J. **54** (2005), 1411–1472.

[7] G. Alberti and G. Bellettini, *A non-local anisotropic model for phase transitions: Asymptotic behaviour of rescaled energies*, European J. Appl. Math. **9** (1998), 261–284.

[8] G. Alberti, G. Bellettini, M. Cassandro and E. Presutti, *Surface tension in Ising systems with Kac potentials*, J. Statist. Phys. **82** (1996), 743–796.

[9] G. Alberti, G. Bouchitté and P. Seppecher, *A singular perturbation result involving the $H^{1/2}$ norm*, C. R. Acad. Sci. Paris Ser. I **319** (1994), 333–338.

[10] G. Alberti, G. Bouchitté and P. Seppecher, *Phase transitions with the line-tension effect*, Arch. Ration. Mech. Anal. **144** (1998), 1–46.

[11] G. Alberti and S. Müller, *A new approach to variational problems with multiple scales*, Comm. Pure Appl. Math. **54** (2001), 761–825.

[12] R. Alicandro, A. Braides and M. Cicalese, *Continuum limits of discrete thin films with superlinear growth densities*, Preprint SNS, Pisa (2005).

[13] R. Alicandro, A. Braides and M. Cicalese, *Phase and anti-phase boundaries in binary discrete systems: A variational viewpoint*, Networks and Heterogeneous Media **1** (2006), 85–107.

[14] R. Alicandro, A. Braides and M.S. Gelli, *Free-discontinuity problems generated by singular perturbation*, Proc. Roy. Soc. Edinburgh Sect. A **128** (1998), 1115–1129.

[15] R. Alicandro and M. Cicalese, *Representation result for continuum limits of discrete energies with superlinear growth*, SIAM J. Math. Anal. **36** (2004), 1–37.

[16] M. Amar and A. Braides, *Γ-convergence of non-convex functionals defined on measures*, Nonlinear Anal. TMA **34** (1998), 953–978.

[17] M. Amar and A. Garroni, *Γ-convergence of concentration problems*, Ann. Scuola Norm. Sup. Pisa Cl. Sci. (5) **2** (2003), 151–179.

[18] L. Ambrosio and A. Braides, *Functionals defined on partitions of sets of finite perimeter, I and II*, J. Math. Pures Appl. **69** (1990), 285–305, 307–333.

[19] L. Ambrosio, C. De Lellis and C. Mantegazza, *Line energies for gradient vector fields in the plane*, Calc. Var. Partial Differential Equations **9** (1999), 327–255.

[20] L. Ambrosio, N. Fusco and D. Pallara, *Functions of Bounded Variation and Free Discontinuity Problems*, Oxford Univ. Press, Oxford (2000).

[21] L. Ambrosio and V.M. Tortorelli, *Approximation of functionals depending on jumps by elliptic functionals via Γ-convergence*, Comm. Pure Appl. Math. **43** (1990), 999–1036.

[22] N. Ansini and A. Braides, *Asymptotic analysis of periodically-perforated nonlinear media*, J. Math. Pures Appl. **81** (2002), 439–451; **84** (2005), 147–148 (Erratum).

[23] N. Ansini, A. Braides and V. Chiadò Piat, *Gradient theory of phase transitions in inhomogeneous media*, Proc. Roy. Soc. Edinburgh Sect. A **133** (2003), 265–296.

[24] N. Ansini, A. Braides and V. Valente, *Multi-scale analysis by Γ-convergence of a shell-membrane transition*, Preprint SNS, Pisa (2005); SIAM J. Math. Anal., to appear.

[25] G. Anzellotti and S. Baldo, *Asymptotic development by Γ-convergence*, Appl. Math. Optim. **27** (1993), 105–123.

[26] G. Anzellotti, S. Baldo and D. Percivale, *Dimensional reduction in variational problems, asymptotic developments in Γ-convergence, and thin structures in elasticity*, Asymptotic Anal. **9** (1994), 61–100.

[27] H. Attouch, *Variational Convergence for Functions and Operators*, Pitman, Boston, MA (1984).

[28] S. Baldo, *Minimal interface criterion for phase transitions in mixtures of Cahn–Hilliard fluids*, Ann. Inst. H. Poincaré Anal. Non Linéaire **7** (1990), 67–90.

[29] J.M. Ball, *Convexity conditions and existence theorems in nonlinear elasticity*, Arch. Ration. Mech. Anal. **63** (1977), 337–403.

[30] E.N. Barron, *Viscosity solutions and analysis in L^∞*, Nonlinear Analysis, Differential Equations and Control, F.H. Clarke and R.J. Stern, eds, Kluwer, Dordrecht (1999).

[31] K. Bhattacharya and A. Braides, *Thin films with many small cracks*, Proc. Roy. Soc. London Ser. A **458** (2002), 823–840.

[32] K. Bhattacharya and R.D. James, *The material is the machine*, Science **307** (2005), 53–54.

[33] G. Bellettini, G. Dal Maso and M. Paolini, *Semicontinuity and relaxation properties of a curvature depending functional in 2D*, Ann. Scuola Norm. Sup. Pisa Cl. Sci. **20** (1993), 247–299.

[34] G. Bellettini and L. Mugnai, *Characterization and representation of the lower semicontinuous envelope of the elastica functional*, Ann. Inst. H. Poincaré Anal. Non Linéaire **21** (2004), 839–880.

[35] F. Bethuel, H. Brezis and F. Hélein, *Ginzburg–Landau Vortices*, Birkhäuser, Boston (1994).

[36] X. Blanc, C. Le Bris and P.-L. Lions, *From molecular models to continuum models*, Arch. Ration. Mech. Anal. **164** (2002), 341–381.

[37] X. Blanc, C. Le Bris and P.-L. Lions, *Atomistic to continuum limis for computational materials science*, Math. Mod. Numer. Anal., to appear.

[38] A. Blake and A. Zisserman, *Visual Reconstruction*, MIT Press, Cambridge (1987).

[39] M. Bocea and I. Fonseca, *Equi-integrability results for 3D–2D dimension reduction problems*, ESAIM: Cont. Optim. Calc. Var. **7** (2002), 443–470.

[40] T. Bodineau, D. Ioffe and Y. Velenik, *Rigorous probabilistic analysis of equilibrium crystal shapes*, J. Math. Phys. **41** (2000), 1033–1098.

[41] G. Bouchitté, *Singular perturbations of variational problems arising from a two-phase transition model*, Appl. Math. Optim. **21** (1990), 289–315.

[42] G. Bouchitté and G. Buttazzo, *Integral representation of nonconvex functionals defined on measures*, Ann. Inst. H. Poincaré Anal. Non Linéaire **9** (1992), 101–117.

[43] G. Bouchitté, I. Fonseca, G. Leoni and L. Mascarenhas, *A global method for relaxation in $W^{1,p}$ and in SBV^p*, Arch. Ration. Mech. Anal. **165** (2002), 187–242.

[44] B. Bourdin and A. Chambolle, *Implementation of an adaptive finite-element approximation of the Mumford–Shah functional*, Numer. Math. **85** (2000), 609–646.

[45] A. Braides, *Approximation of Free-Discontinuity Problems*, Lecture Notes in Math., Vol. 1694, Springer-Verlag, Berlin (1998).

[46] A. Braides, *Γ-Convergence for Beginners*, Oxford Univ. Press, Oxford (2002).

[47] A. Braides, G. Buttazzo and I. Fragalà, *Riemannian approximation of Finsler metrics*, Asymptot. Anal. **31** (2002), 177–187.

[48] A. Braides, A. Chambolle and M. Solci, *A relaxation result for energies defined on pairs set-function and applications*, Preprint SNS, Pisa (2005).

[49] A. Braides and V. Chiadò Piat, *Integral representation results for functionals defined on $SBV(\Omega; \mathbb{R}^m)$*, J. Math. Pures Appl. **75** (1996), 595–626.

[50] A. Braides, V. Chiadò Piat and A. Piatnitski, *A variational approach to double-porosity problems*, Asymptot. Anal. **39** (2004), 281–308.

[51] A. Braides and M. Cicalese, *Surface energies in nonconvex discrete systems*, Preprint SNS, Pisa (2004).

[52] A. Braides and G. Dal Maso, *Nonlocal approximation of the Mumford–Shah functional*, Calc. Var. Partial Differential Equations **5** (1997), 293–322.

[53] A. Braides, G. Dal Maso and A. Garroni, *Variational formulation of softening phenomena in fracture mechanics: The one-dimensional case*, Arch. Ration. Mech. Anal. **146** (1999), 23–58.

[54] A. Braides and A. Defranceschi, *Homogenization of Multiple Integrals*, Oxford Univ. Press, Oxford (1999).

[55] A. Braides, A. Defranceschi and E. Vitali, *Homogenization of free discontinuity problems*, Arch. Ration. Mech. Anal. **135** (1996), 297–356.

[56] A. Braides, I. Fonseca and G.A. Francfort, *3D–2D asymptotic analysis for inhomogeneous thin films*, Indiana Univ. Math. J. **49** (2000), 1367–1404.

[57] A. Braides and G.A. Francfort, *Bounds on the effective behavior of a square conducting lattice*, Proc. Roy. Soc. London Ser. A **460** (2004), 1755–1769.

[58] A. Braides and M.S. Gelli, *Limits of discrete systems with long-range interactions*, J. Convex Anal. **9** (2002), 363–399.

[59] A. Braides and M.S. Gelli, *Continuum limits of discrete systems without convexity hypotheses*, Math. Mech. Solids **7** (2002), 41–66.

[60] A. Braides and M.S. Gelli, *From discrete systems to continuous variational problems: An introduction*, Topics on Concentration Phenomena and Problems with Multiple Scales, A. Braides and V. Chiadò Piat, eds, Lec. Notes Unione Mat. Ital., Vol. 2, Springer-Verlag, Berlin (2006), 3–78.

[61] A. Braides, M.S. Gelli and M. Sigalotti, *The passage from non-convex discrete systems to variational problems in Sobolev spaces: The one-dimensional case*, Proc. Steklov Inst. Math. **236** (2002), 408–427.

[62] A. Braides, A.J. Lew and M. Ortiz, *Effective cohesive behavior of layers of interatomic planes*, Arch. Ration. Mech. Anal. **180** (2006), 151–182.

[63] A. Braides and R. March, *Approximation by Γ-convergence of a curvature-depending functional in Visual Reconstruction*, Comm. Pure Appl. Math. **58** (2006), 71–121.

[64] A. Braides and A. Piatnitski, *Overall properties of a discrete membrane with randomly distributed defects*, Preprint SNS, Pisa (2004).

[65] A. Braides and C. Zeppieri, *A note on equi-integrability in dimension reduction problems*, Calc. Var. Partial Differential Equations, to appear.

[66] G. Buttazzo, *Semicontinuity, Relaxation and Integral Representation in the Calculus of Variations*, Pitman, London (1989).

[67] G. Buttazzo, G. Dal Maso and U. Mosco, *A derivation theorem for capacities with respect to a Radon measure*, J. Funct. Anal. **71** (1987), 263–278.

[68] M. Camar-Eddine and P. Seppecher, *Closure of the set of diffusion functionals with respect to the Mosco-convergence*, Math. Models Methods Appl. Sci. **12** (2002), 1153–1176.

[69] A. Chambolle, *Un theoreme de Γ-convergence pour la segmentation des signaux*, C. R. Acad. Sci. Paris Ser. I **314** (1992), 191–196.

[70] A. Chambolle, *Image segmentation by variational methods: Mumford and Shah functional and the discrete approximations*, SIAM J. Appl. Math. **55** (1995), 827–863.

[71] A. Chambolle, *Finite-differences discretizations of the Mumford–Shah functional*, M2AN Math. Model. Numer. Anal. **33** (1999), 261–288.

[72] A. Chambolle and G. Dal Maso, *Discrete approximation of the Mumford–Shah functional in dimension two*, M2AN Math. Model. Numer. Anal. **33** (1999), 651–672.

[73] T. Champion, L. De Pascale and F. Prinari, *Γ-convergence and absolute minimizers for supremal functionals*, ESAIM Control Optim. Calc. Var. **10** (2004), 14–27.

[74] D. Cioranescu and F. Murat, *Un terme étrange venu d'ailleur. Nonlinear Partial Differential Equations and their Applications*, Pitman Research Notes in Math., Vol. 60, Pitman, London (1982), 98–138.

[75] S. Conti, A. De Simone, G. Dolzmann, S. Müller and F. Otto, *Multiscale modeling of materials – The role of Analysis*, Trends in Nonlinear Analysis, Springer-Verlag, Berlin (2003), 375–408.

[76] S. Conti, I. Fonseca and G. Leoni, *A Γ-convergence result for the two-gradient theory of phase transitions*, Comm. Pure Appl. Math. **55** (2002), 857–936.

[77] S. Conti and B. Schweizer, *A sharp-interface limit for a two-well problem in geometrically linear elasticity*, Arch. Ration. Mech. Anal. **179** (2006), 413–452.

[78] G. Cortesani and R. Toader, *A density result in SBV with respect to non-isotropic energies*, Nonlinear Anal. **38** (1999), 585–604.

[79] A. Coscia, *On curvature sensitive image segmentation*, Nonlinear Anal. **39** (2000), 711–730.

[80] M.G. Crandall, L.C. Evans and R.F. Gariepy, *Optimal Lipschitz extensions and the infinity Laplacian*, Calc. Var. Partial Differential Equations **13** (2001), 123–139.

[81] B. Dacorogna, *Direct Methods in the Calculus of Variations*, Springer-Verlag, Berlin (1989).

[82] G. Dal Maso, *Asymptotic behaviour of minimum problems with bilateral obstacles*, Ann. Mat. Pura Appl. **129** (1981), 327–366.

[83] G. Dal Maso, *Γ-convergence and μ-capacities*, Ann. Scuola Norm. Sup. Pisa Cl. Sci. **14** (1987), 423–464.

[84] G. Dal Maso, *An Introduction to Γ-Convergence*, Birkhäuser, Boston (1993).

[85] G. Dal Maso, *Asymptotic behaviour of solutions of Dirichlet problems*, Boll. Unione Mat. Ital. Sez. A **11** (1997), 253–277.

[86] G. Dal Maso and L. Modica, *Nonlinear stochastic homogenization and ergodic theory*, J. Reine Angew. Math. **368** (1986), 28–42.

[87] G. Dal Maso and U. Mosco, *Wiener's criterion and Γ-convergence*, Appl. Math. Optim. **15** (1987), 15–63.

[88] G. David, *Singular Sets of Minimizers for the Mumford–Shah Functional*, Progr. Math., Vol. 233, Birkhäuser, Basel (2005).

[89] A. Davini, *Smooth approximation of weak Finsler metrics*, Differential Integral Equations **18** (2005), 509–530.

[90] E. De Giorgi, *Sulla convergenza di alcune successioni di integrali del tipo dell'area*, Rend. Mat. Appl. (6) **8** (1975), 277–294.

[91] E. De Giorgi, *Selected Works*, Springer-Verlag, Berlin (2006).

[92] E. De Giorgi, M. Carriero and A. Leaci, *Existence theorem for a minimum problem with free discontinuity set*, Arch. Ration. Mech. Anal. **108** (1989), 195–218.

[93] E. De Giorgi and T. Franzoni, *Su un tipo di convergenza variazionale*, Atti Accad. Naz. Lincei Cl. Sci. Fis. Mat. Natur. Rend. **58** (1975), 842–850.

[94] E. De Giorgi and G. Letta, *Une notion générale de convergence faible pour des fonctions croissantes d'ensemble*, Ann. Scuola Norm. Sup. Pisa Cl. Sci. **4** (1977), 61–99.

[95] A. De Simone, S. Müller, R.V. Kohn and F. Otto, *A compactness result in the gradient theory of phase transitions*, Proc. Roy. Soc. Edinburgh Sect. A **131** (2001), 833–844.

[96] L.C. Evans, *Weak Convergence Methods in Nonlinear Partial Differential Equations*, Amer. Math. Soc., Providence, RI (1990).

[97] L.C. Evans and R.F. Gariepy, *Measure Theory and Fine Properties of Functions*, CRC Press, Boca Raton, FL (1992).

[98] L.C. Evans and D. Gomes, *Effective Hamiltonians and averaging for Hamiltonian dynamics. I*, Arch. Ration. Mech. Anal. **157** (2001), 1–33.

[99] M. Flucher, *Variational Problems with Concentration*, Progr. Nonlinear Differential Equations Appl., Vol. 36, Birkhäuser, Basel (1999).

[100] M. Flucher, A. Garroni and S. Müller, *Concentration of low energy extremals: Identification of concentration points*, Calc. Var. Partial Differential Equations **14** (2002), 483–516.

[101] M. Focardi and M.S. Gelli, *Asymptotic analysis of the Mumford–Shah functional in periodically-perforated domains*, Preprint Dip. Mat., Pisa (2005); Interfaces Free Boundaries, to appear.

[102] I. Fonseca and G. Leoni, *Modern Methods in the Calculus of Variations with Applications to Nonlinear Continuum Physics*, Springer-Verlag, Berlin, to appear.

[103] I. Fonseca and S. Müller, *Quasi-convex integrands and lower semicontinuity in L^1*, SIAM J. Math. Anal. **23** (1992), 1081–1098.

[104] I. Fonseca, S. Müller and P. Pedregal, *Analysis of concentration and oscillation effects generated by gradients*, SIAM J. Math. Anal. **29** (1998), 736–756.

[105] G. Friesecke, R.D. James and S. Müller, *Rigorous derivation of nonlinear plate theory and geometric rigidity*, C. R. Acad. Sci Paris Ser. I **334** (2002), 173–178.

[106] G. Friesecke, R.D. James and S. Müller, *A theorem on geometric rigidity and the derivation of nonlinear plate theory from three-dimensional elasticity*, Comm. Pure Appl. Math. **55** (2002), 1461–1506.

[107] G. Friesecke, R.D. James and S. Müller, *A hierarchy of plate models derived from nonlinear elasticity by Gamma-convergence*, Preprint MPI, Leipzig (2005).

[108] G. Friesecke and F. Theil, *Validity and failure of the Cauchy–Born hypothesis in a two-dimensional mass-spring lattice*, J. Nonlinear Sci. **12** (2002), 445–478.

[109] M. Fukushima, *Dirichlet Forms and Markov Processes*, North-Holland Mathematical Library, Vol. 23, North-Holland, Amsterdam (1980).

[110] A. Garroni, *Γ-Convergence for Concentration Problems*, Topics on Concentration Phenomena and Problems with Multiple Scales, A. Braides and V. Chiadò Piat, eds, Lec. Notes Unione Mat. Ital., Vol. 2, Springer-Verlag, Berlin (2006), 233–266.

[111] A. Garroni and S. Müller, *Concentration phenomena for the volume functional in unbounded domains: Identification of concentration points*, J. Funct. Anal. **199** (2003), 386–410.

[112] A. Garroni and S. Müller, *Line tension model for dislocations*, Arch. Ration. Mech. Anal. **180** (2006), 183–236.

[113] A. Garroni, V. Nesi and M. Ponsiglione, *Dielectric breakdown: Optimal bounds*, Proc. Roy. Soc. London Ser. A **457** (2001), 2317–2335.

[114] M. Gobbino, *Finite difference approximation of the Mumford–Shah functional*, Comm. Pure Appl. Math. **51** (1998), 197–228.

[115] R.L. Jerrard, *Lower bounds for generalized Ginzburg–Landau functionals*, SIAM J. Math. Anal. **30** (1999), 721–746.

[116] R.L. Jerrard and H.M. Soner, *Limiting behavior of the Ginzburg–Landau functional*, J. Funct. Anal. **192** (2002), 524–561.

[117] C. Le Bris and P.-L. Lions, *From atoms to crystals: A mathematical journey*, Bull. Amer. Math. Soc. **42** (2005), 291–363.

[118] H. Le Dret and A. Raoult, *The nonlinear membrane model as variational limit of nonlinear three-dimensional elasticity*, J. Math. Pures Appl. **74** (1995), 549–578.

[119] K. Lurie and A.V. Cherkaev, *Exact estimates of conductivity of mixtures composed of two materials taken in prescribed proportion (plane problem)*, Dokl. Akad. Nauk SSSR **5** (1982), 1129–1130.

[120] A.V. Marchenko and E.Ya. Khruslov, *Boundary Value Problems in Domains with Fine-Granulated Boundaries*, Naukova Dumka, Kiev (1974) (in Russian).

[121] G.W. Milton, *The Theory of Composites*, Cambridge Univ. Press, Cambridge (2002).

[122] L. Modica, *The gradient theory of phase transitions and the minimal interface criterion*, Arch. Ration. Mech. Anal. **98** (1987), 123–142.

[123] L. Modica and S. Mortola, *Un esempio di Γ-convergenza*, Boll. Unione Mat. Ital. Sez. B **14** (1977), 285–299.

[124] J.M. Morel and S. Solimini, *Variational Models in Image Segmentation*, Birkhäuser, Boston (1995).

[125] F. Morgan, *Geometric Measure Theory*, Academic Press, San Diego, CA (1988).

[126] C.B. Morrey, *Quasiconvexity and the semicontinuity of multiple integrals*, Pacific J. Math. **2** (1952), 25–53.

[127] U. Mosco, *Composite media and asymptotic Dirichlet forms*, J. Funct. Anal. **123** (1994), 368–421.

[128] S. Müller, *Homogenization of nonconvex integral functionals and cellular elastic materials*, Arch. Ration. Mech. Anal. **99** (1987), 189–212.

[129] S. Müller, *Variational Models for Microstructure and Phase Transitions*, Calculus of Variations and Geometric Evolution Problems, Cetraro, 1996, Lecture Notes in Math., Vol. 1713, Springer-Verlag, Berlin (1999), 85–210.

[130] D. Mumford, *Elastica and Computer Vision*, Algebraic Geometry and Its Applications, C.L. Bajaj, ed., Springer-Verlag, Berlin (1993).

[131] D. Mumford and J. Shah, *Optimal approximation by piecewise smooth functions and associated variational problems*, Comm. Pure Appl. Math. **17** (1989), 577–685.

[132] S. Pagano and R. Paroni, *A simple model for phase transitions: From the discrete to the continuum problem*, Quart. Appl. Math. **61** (2003), 89–109.

[133] A. Piatnitski and E. Remy, *Homogenization of elliptic difference operators*, SIAM J. Math. Anal. **33** (2001), 53–83.

[134] U. Raitums, *On the local representation of G-closure*, Arch. Ration. Mech. Anal. **158** (2001), 213–234.

[135] M.O. Rieger and P. Tilli, *On the Γ-limit of the Mumford–Shah functional*, Calc. Var. Partial Differential Equations **23** (2005), 373–390.

[136] R.C. Rogers and L. Truskinovsky, *Discretization and hysteresis*, Physica B **233** (1997), 370–375.

[137] E. Sandier and S. Serfaty, *Gamma-convergence of gradient flows and application to Ginzburg–Landau*, Comm. Pure Appl. Math. **57** (2004), 1627–1672.

[138] E. Sandier and S. Serfaty, *Vortices in the Magnetic Ginzburg–Landau Model*, Birkhäuser, to appear.

[139] M. Solci and E. Vitali, *Variational models for phase separation*, Interfaces Free Bound **5** (2003), 27–46.

[140] P. Sternberg, *The effect of a singular perturbation on nonconvex variational problems*, Arch. Ration. Mech. Anal. **101** (1988), 209–260.

[141] L. Tartar, *Estimations fines de coefficients homogénéisés*, Ennio De Giorgi Colloquium, P. Krée, ed., Pitman Research Notes Math., Vol. 125, Pitman, London (1985), 168–187.

[142] L. Truskinovsky, *Fracture as a phase transition*, Contemporary Research in the Mechanics and Mathematics of Materials, R.C. Batra and M.F. Beatty, eds, CIMNE, Barcelona (1996), 322–332.

[143] W. Ziemer, *Weakly Differentiable Functions*, Springer-Verlag, Berlin (1989).

CHAPTER 3

Bubbling in Nonlinear Elliptic Problems Near Criticality

Manuel del Pino

*Departamento de Ingeniería Matemática and CMM, Universidad de Chile,
Casilla 170, Correo 3, Santiago, Chile*

Monica Musso

*Departamento de Matemática, Pontificia Universidad Católica de Chile,
Avda. Vicuña Mackenna 4860, Macul, Santiago, Chile*
and
Dipartimento di Matematica, Politecnico di Torino, Corso Duca degli Abruzzi, 24, 10129 Torino, Italy

Contents

1. Introduction 217
2. Nearly critical bubbling: the proof of Theorem 1.1 225
   2.1. Ansatz and scheme of the proof 225
   2.2. Variational reduction and conclusion of the proof 233
3. Solvability of slightly supercritical problems and the topology of the domain 255
   3.1. The case of a small hole 255
   3.2. Bubbling under symmetries 266
4. The Brezis–Nirenberg problem in dimension $N = 3$: the proof of Theorem 1.2 269
   4.1. Energy expansion of single bubbling 270
   4.2. The method of proof 273
   4.3. The linear problem 276
   4.4. Solving the nonlinear problem 277
   4.5. Variational formulation of the reduced problem for $k = 1$ 278
   4.6. Proof of Theorem 1.2, part (a): single bubbling 279
   4.7. Proof of Theorem 1.2, part (b): multiple bubbling 280
5. Liouville-type equations 285
   5.1. Proof of Theorem 1.3 285
   5.2. A related 2-d problem involving nonlinearity with large exponent 309
Acknowledgement 312
References 312

HANDBOOK OF DIFFERENTIAL EQUATIONS
Stationary Partial Differential Equations, volume 3
Edited by M. Chipot and P. Quittner

1. Introduction

The purpose of this paper is to review some recent results concerning asymptotic analysis and construction of solutions of semilinear elliptic boundary value problems near criticality in $\mathbb{R}^N$. When the nonlinearity has a power-like behavior, it is well known that the exponent $\frac{N+2}{N-2}$, $N \geqslant 3$, sets a threshold where the structure of the solution set may suffer dramatic change. In particular, the effect of lower-order terms in the nonlinearity or topology–geometry of the domain becomes crucial in the solvability issue. Criticality has been a subject broadly treated in the PDE literature for more than two decades. While highly nontrivial understanding has been achieved, this effect still hides many mysterious aspects. In particular, understanding of supercritical problems appears as a vastly open subject.

Most of our discussion will be centered on the boundary value problem

$$\begin{cases} \Delta u + \lambda u + u^q = 0 & \text{in } \Omega, \\ u > 0 & \text{in } \Omega, \\ u = 0 & \text{on } \partial\Omega, \end{cases} \tag{1.1}$$

where $\Omega \subset \mathbb{R}^N$, $N \geqslant 3$, is a bounded domain with smooth boundary $\partial\Omega$, $q > 1$ and $\lambda \in \mathbb{R}$.

When $\lambda = 0$, this equation is sometimes called Lane–Emden–Fowler equation. It was used first in the mid-19th century in the study of internal structure of stars, see [20], on the other hand it constitutes a basic model equation for steady states of reaction–diffusion systems and nonlinear Schrödinger equations. The case $q = \frac{N+2}{N-2}$ is especially meaningful in geometry, versions of this problem on manifolds correspond to the well-known problem of finding conformal metrics with prescribed scalar curvature, in particular the well-known Yamabe problem.

Testing (1.1) against a first eigenfunction for the problem $\Delta\phi_1 + \lambda_1\phi_1 = 0$ with zero Dirichlet boundary condition, readily yields that a necessary condition for solvability is $\lambda < \lambda_1$. On the other hand, if $\lambda < \lambda_1$ and $q < \frac{N+2}{N-2}$ if $N \geqslant 3$, a solution may be found by minimizing the Rayleigh quotient

$$Q_\lambda(u) \equiv \frac{\int_\Omega |\nabla u|^2 - \lambda \int_\Omega |u|^2}{(\int_\Omega |u|^{q+1})^{2/(q+1)}}, \qquad u \in H_0^1(\Omega) \setminus \{0\}. \tag{1.2}$$

In fact, the quantity

$$S_\lambda \equiv \inf_{u \in H_0^1(\Omega)\setminus\{0\}} Q_\lambda(u)$$

is achieved thanks to compactness of Sobolev embeddings for $q < \frac{N+2}{N-2}$. A suitable scalar multiple of a minimizer turns out to be a solution of (1.1). The case $q \geqslant \frac{N+2}{N-2}$ is considerably more delicate: for $q = \frac{N+2}{N-2}$ compactness of the embedding is lost while for $q > \frac{N+2}{N-2}$ there is no such an embedding. This obstruction is not just technical for the solvability question, but essential. Pohozaev [75] showed that if Ω is strictly star shaped then no solution of (1.1) exists if $\lambda \leqslant 0$ and $q \geqslant \frac{N+2}{N-2}$.

Let $\mathcal{S}_N$ be the best constant in the critical Sobolev embedding,

$$\mathcal{S}_N = \inf_{u \in C_0^1(\mathbb{R}^N) \setminus \{0\}} \frac{\int_{\mathbb{R}^N} |\nabla u|^2}{\left(\int_{\mathbb{R}^N} |u|^{2N/(N-2)}\right)^{(N-2)/N}}. \tag{1.3}$$

Let us consider the case $q = \frac{N+2}{N-2}$ in Q_λ in (1.2) and the number

$$\lambda_* \equiv \inf\{\lambda > 0 \colon S_\lambda < \mathcal{S}_N\}. \tag{1.4}$$

In their well-known paper [14], Brezis and Nirenberg established that $\lambda_* = 0$ for $N \geqslant 4$, $0 < \lambda^* < \lambda_1$ for $N = 3$ and that S_λ is achieved whenever $\lambda^* < \lambda < \lambda_1$, hence (1.1) is solvable in this range. When Ω is a ball and $N = 3$ they find that $\lambda^* = \lambda_1/4$ and that no solution exists for $\lambda \leqslant \lambda^*$.

Thus $\lambda > 0$ taken at the appropriate range makes compactness restored and solvability holds. Pohozaev's result, on the other hand, puts in evidence the central role of topology or geometry in the domain for solvability if, say, $\lambda = 0$. For instance, Kazdan and Warner observed in [52] that problem (1.1) is actually solvable for any $p > 1$ if Ω is a radial annulus. In fact, compactness in the Rayleigh quotient Q_λ is gained within the class of radially symmetric functions, hence a radial extremal always exists. On the other hand, Coron in [24] found via a variational method that (1.1) is solvable for $\lambda = 0$ at the critical exponent $p = \frac{N+2}{N-2}$ whenever Ω is a domain exhibiting a small hole. Substantial improvement of this result was found by Bahri and Coron [7], proving that if $\lambda = 0$ and some homology group of Ω with coefficients in $\mathbf{Z}_2$ is not trivial, then (1.1) has at least one solution for p critical, in particular, in any three-dimensional domain which is not contractible to a point. Examples showing that this condition is actually not necessary for solvability at the critical exponent were found by Dancer [26], Ding [37] and Passaseo [71] (see also Passaseo [73] for the same issue in the supercritical range). A question due to Rabinowitz, collected in Brezis' survey [12], was whether a solution to (1.1) with $\lambda = 0$ existed also for *supercritical* p, namely $p > \frac{N+2}{N-2}$. It is important to observe that the use of variational arguments in this range becomes less obvious since the nonlinearity falls off the natural energy space $H_0^1(\Omega)$. The general answer to this question is negative. Passaseo in [72] used a Pohozaev-type identity to exhibit a torus-like domain in $\mathbb{R}^N$, $N \geqslant 4$, for which no solution to (1.1) for $\lambda = 0$ exists if $p > \frac{N+1}{N-3}$. Still the question of existence remained open for a power super-critical, but close to critical. Dancer conjectured that solvability of (1.1) in a domain with nontrivial topology persists for small $\varepsilon > 0$.

The change of structure of solution set taking place at the critical exponent is strongly linked to the presence of unbounded sequences of solutions or *bubbling solutions*. By a *bubbling solution* for (1.1) near the critical exponent we mean an unbounded sequence of solutions u_n of (1.1) for $\lambda = \lambda_n$ bounded, and $q = q_n \to \frac{N+2}{N-2}$. Setting

$$M_n \equiv \alpha^{-1} \max_{\Omega} u_n = \alpha^{-1} u_n(x_n) \to +\infty$$

with $\alpha > 0$ to be chosen, we see then that the scaled function

$$v_n(y) \equiv M_n^{-1} u_n\left(x_n + M_n^{-(q_n-1)/2} y\right),$$

satisfies

$$\Delta v_n + v_n^{q_n} + M_n^{-(q_n-1)}\lambda_n v_n = 0$$

in the expanding domain $\Omega_n = M_n^{(q_n-1)/2}(\Omega - x_n)$. Assuming for instance that x_n stays away from the boundary of Ω, elliptic regularity implies that locally over compacts around the origin, v_n converges up to subsequences to a positive solution of

$$\Delta w + w^{(N+2)/(N-2)} = 0$$

in entire space, with $w(0) = \max w = \alpha$. It is known, see [16], that for the convenient choice $\alpha = \alpha_N \equiv (N(N-2))^{(N-2)/4}$, this solution is explicitly given by

$$w(y) = \alpha_N \left(\frac{1}{1+|y|^2}\right)^{(N-2)/2}$$

which corresponds precisely to an extremal of the Sobolev constant $\mathcal{S}_N$, see [5,84]. Coming back to the original variable, we expect then that "near x_n" the behavior of $u_n(x)$ can be approximated as

$$u_n(x) = \alpha_N \left(\frac{1}{1 + M_n^{4/(N-2)}|x - x_n|^2}\right)^{(N-2)/2} M_n\left(1 + o(1)\right). \tag{1.5}$$

A natural problem is that of constructing solutions exhibiting this property around one or several points of the domain. Let us consider the special case of problem (1.1) given by

$$\begin{cases} \Delta u + u^{(N+2)/(N-2)-\varepsilon} = 0 & \text{in } \Omega, \\ u > 0 & \text{in } \Omega, \\ u = 0 & \text{on } \partial\Omega, \end{cases} \tag{1.6}$$

where $\varepsilon > 0$ is a small number. A solution is given by a minimizer u_ε of the Rayleigh quotient (1.2) for $\lambda = 0$ and $q = \frac{N+2}{N-2} - \varepsilon$. Clearly u_ε cannot remain bounded as $\varepsilon \downarrow 0$, since otherwise Sobolev's constant $\mathcal{S}_N$ would be achieved by a function supported in Ω. The bubbling asymptotic behavior of u_ε was first described in the radial case by Brezis and Peletier [15] and in the general case by Han [48] and Rey [78]. The conclusion is that u_ε has asymptotically just a single maximum point x_ε and that asymptotics (1.5) holds globally in Ω with $M_\varepsilon \sim \varepsilon^{-1/2}$. Moreover, x_ε approaches a critical point of Robin's function $H(x,x)$. Here $H(x,y) = c_N|y-x|^{2-N} - G(x,y)$ is the regular part of Green's function $G(x,y)$ of the problem

$$-\Delta_y G(x,y) = \delta_x(y), \quad y \in \Omega,$$

$$G(x,y) = 0, \qquad\qquad y \in \partial\Omega. \tag{1.7}$$

Wei [87] found that x_ε actually approaches a global minimizer of $H(x, x)$. Rey [78] established furthermore the following: Given a nondegenerate critical point of $H(x, x)$, a family of bubbling solutions u_ε of (1.6) around this point exists as $\varepsilon \downarrow 0$.

In [8] this conclusion was refined to the case of solutions u_ε exhibiting bubbling at multiple points. Their result can be phrased in the following terms: Given a *nondegenerate critical point* of the functional of $(\xi, \lambda) = (\xi_1, \ldots, \xi_k, \lambda_1, \ldots, \lambda_k) \in \Omega^k \times \mathbb{R}_+^k$,

$$
\Psi_k^-(\xi, \lambda) = \sum_{j=1}^{k} H(\xi_j, \xi_j)\lambda_j^{N-2}
$$
$$
- 2 \sum_{i<j} G(\xi_i, \xi_j)\lambda_i^{(N-2)/2}\lambda_j^{(N-2)/2} - 2\log(\lambda_1 \cdots \lambda_k), \tag{1.8}
$$

there exists a k-bubble solution u_ε to problem (1.6) with centers near points ξ_i, with (uneven) heights of order $O(\varepsilon^{-1/2})$ modified by the scalars λ_i. Needless to say, this functional may not have any critical points if $k > 1$ as it is the case of a ball, where only one solution exists.

This result opens up important questions: Is there any solution at all other than the least energy (minimizer of Rayleigh quotient)? Even in topological situations where one can predict existence of critical points for the above functional their nondegeneracy is an assumption hard to check, therefore it is important to lift this requirement in order to obtain (concrete) classes of domains where more than one solution exists. As a model situation let us consider a domain Ω formed by two fixed disjoint domains Ω_1 and Ω_2 connected by a narrow cylindrical channel. It is then easy to see that in such a situation, if the width of the channel is taken small enough, then Robin's function $H(x, x)$ will exhibit a two-well situation: there are sets $\overline{\mathcal{D}}_i \subset \Omega_i$, such that

$$
\inf_{\mathcal{D}_i} H(x, x) > \inf_{\Omega \setminus \mathcal{D}_1 \cup \mathcal{D}_2} H(x, x), \quad i = 1, 2,
$$

and hence two local minimizers and a mountain pass for Robin's function is present: as we will see, as a consequence of our main results, in such a situation indeed three bubbling solutions to (1.6) exist. More than this, if besides the channel is *sufficiently long* (or *sufficiently thin*), a critical point for functional (1.8) for $k = 2$ is also present, and associated to it there is a two bubble solution with centers inside the $\mathcal{D}_i$'s. Details on these examples are provided in Remark 2.3.

The second question deals with the possibility of solving problems of this type *above* the critical exponent. We consider now the equation

$$
\begin{cases}
\Delta u + u^{(N+2)/(N-2)+\varepsilon} = 0 & \text{in } \Omega, \\
u > 0 & \text{in } \Omega, \\
u = 0 & \text{on } \partial\Omega,
\end{cases} \tag{1.9}
$$

where $\varepsilon > 0$. What we discover is that bubbling solutions to this problem can be found with a criterion *dual* to that described above. It turns out that *nontrivial critical points of* the functional

$$\Psi_k^+(\xi,\lambda) = \sum_{j=1}^{k} H(\xi_j,\xi_j)\lambda_j^{N-2}$$

$$- 2\sum_{i<j} G(\xi_i,\xi_j)\lambda_i^{(N-2)/2}\lambda_j^{(N-2)/2} + 2\log(\lambda_1\cdots\lambda_k), \tag{1.10}$$

inherit corresponding bubbling solutions for the slightly supercritical problem (1.9).

We say that Ψ_k^+ (resp., Ψ_k^-) exhibits a *nontrivial critical point situation* in $\mathcal{D}$, open and bounded set with

$$\overline{\mathcal{D}} \subset \left\{ (\xi,\lambda) \in \Omega^k \times \mathbb{R}_+^k : \xi_i \neq \xi_j, \text{ if } i \neq j \right\},$$

if there exists a $\delta > 0$ such that for any $g \in C^1(\overline{\mathcal{D}})$ with $\|g\|_{C^1(\overline{\mathcal{D}})} < \delta$, a critical point for $\Psi_k^+ + g$ (resp., $\Psi_k^- + g$) in $\mathcal{D}$ exists. This notion was introduced first by Li in [53,54], in the analysis of a different singular perturbation problem.

The following general result holds true.

THEOREM 1.1. *Let $k \geq 1$ be given and assume that the function Ψ_k^+ has a nontrivial critical point situation in some set $\mathcal{D}$. Then for all sufficiently small ε there are points*

$$(\xi_\varepsilon,\lambda_\varepsilon) = (\xi_{1\varepsilon},\ldots,\xi_{k\varepsilon},\lambda_{1\varepsilon},\ldots,\lambda_{k\varepsilon}) \in \mathcal{D}$$

and a solution u_ε of problem (1.9) of the form

$$u_\varepsilon(x) = \alpha_N \sum_{j=1}^{k} \left[\frac{\lambda_{j\varepsilon}}{(\lambda_{j\varepsilon})^2 + \varepsilon^{-2/(N-2)}|x - \xi_{j\varepsilon}^\pm|^2} \right]^{(N-2)/2} \varepsilon^{-1/2} + o(1),$$

where $o(1) \to 0$ uniformly in $\overline{\Omega}$. Moreover,

$$\nabla\Psi_k^+(\xi_\varepsilon,\lambda_\varepsilon) \to 0 \quad as\ \varepsilon \to 0.$$

The same conclusions hold for problem (1.6) with Ψ_k^+ replaced by Ψ_k^-.

This result is essentially contained in [32,34]. A main implication of this result in the slightly supercritical problem (1.9) is given in Theorem 3.1, originally proved in [32], which states that in a domain with a small hole, like that in Coron's result [24], problem (1.9) has a two-bubble solution. More precisely, let

$$\Omega = \mathcal{D} \setminus \omega, \tag{1.11}$$

where $\mathcal{D}$ and ω are bounded domains with smooth boundary. Then if $\omega \subset \overline{B}(0,\rho) \subset \mathcal{D}$, and ρ is fixed sufficiently small, then problem (1.9) is solvable, with a solution like in

Theorem 1.1 for $k = 2$. In fact, in such a situation, a nontrivial critical point situation for Ψ_2^+ can be described. More generally, if several spherical holes are drilled, a solution obtained by gluing of several two bubbles can be found, see Theorem 3.5 and [34].

Two-bubble solutions are the simplest to be obtained: single-bubble solutions for problem (1.9) *do not exist*, see [11]. While possible, it is harder to obtain three-bubble solutions in the case of the small spherical hole, see Pistoia and Rey [74]. We conjecture that actually the size of the perforation is irrelevant to the existence issue. We believe that in Bahri–Coron's situation, a noncontractible domain Ω, a k-bubble solution of (1.9) exists for all k sufficiently large. This has been proven if Ω is an annulus, or more generally, if it enjoys certain rotational symmetries, see Theorem 3.6 and also Molle and Passaseo [63,64]. There is a strong analogy between bubbling above and below critical and that in Brezis–Nirenberg problem for $N \geqslant 4$,

$$
\begin{cases}
\Delta u + \lambda u + u^{(N+2)/(N-2)} = 0 & \text{in } \Omega, \\
u > 0 & \text{in } \Omega, \\
u = 0 & \text{on } \partial\Omega.
\end{cases}
\tag{1.12}
$$

In fact, in [78] Rey finds single bubbling at nondegenerate critical points of Robin's function $H(x, x)$ as $\lambda \to 0$ with $\lambda > 0$. In [66], on the other hand it is found that a two-bubble solution exists in this situation if $\lambda \to 0$ and $\lambda < 0$ in the case of the domain with small hole (1.11), see Theorem 3.3.

In the line of thought of existence condition for critical case "seems to imply" presence of bubbling solutions above critical, we may wonder about existence in the slightly super-critical Brezis–Nirenberg problem

$$
\begin{cases}
\Delta u + \lambda u + u^{(N+2)/(N-2)+\varepsilon} = 0 & \text{in } \Omega, \\
u > 0 & \text{in } \Omega, \\
u = 0 & \text{on } \partial\Omega,
\end{cases}
\tag{1.13}
$$

where $\varepsilon > 0$. In [30], the following result has been found for the case $N = 3$.

THEOREM 1.2. (a) *Assume that $\lambda^* < \lambda < \lambda_1$, where λ^* is the number given by (1.4). Then there exists a number $\varepsilon_1 > 0$ such that problem (1.13) is solvable for any $\varepsilon \in (0, \varepsilon_0)$.*

(b) *Assume that Ω is a ball and that $\lambda^* = \lambda_1/4 < \lambda < \lambda_1$. Then, given $k \geqslant 1$ there exists a number $\varepsilon_k > 0$ such that problem (2.1) has at least k radial solutions for any $\varepsilon \in (0, \varepsilon_k)$.*

While the result of part (a) resembles that by Brezis and Nirenberg when $q = 5$, in reality the solution we find has a very different nature: it blows up as $\varepsilon \downarrow 0$ developing a *single bubble* around a certain point inside the domain. We *do not know* if the solution found in [14] actually persists. The other solutions predicted by part (b) blow up only at the origin but exhibit *multiple bubbling*. More precisely, given $k \geqslant 1$, there exists for all sufficiently small $\varepsilon > 0$ a solution u_ε of problem (1.13) of the form

$$
u_\varepsilon(x) = \sum_{j=1}^{k} \frac{3^{1/4} M_{j\varepsilon}}{\sqrt{1 + M_{j\varepsilon}^4 |x|^2}} + \mathrm{o}(1),
$$

where $o(1) \to 0$ uniformly in $\overline{\Omega}$ and for $j = 1, \ldots, k$,

$$M_{j\varepsilon} \sim \varepsilon^{1/2-j}.$$

In other words this solution is built as a *tower of bubbles* of different blow-up orders. In higher dimensions, $N \geqslant 4$, this type of solutions has been found in the radial case in [29] provided that λ lies in the range $\lambda \sim \varepsilon^{(N-2)/(N-4)}$ for $N \geqslant 5$. Recently Ge, Jing and Pacard [45] (see also [50]) have found the presence of these towers in this situation, without symmetries, sitting near a nondegenerate critical point of certain functional of points of the domain. In [43], Felli and Terracini found solutions exhibiting super position of bubbles for a related nonlinear elliptic equation with critical growth and Hardy-type potential.

The results above do have two-dimensional analogues. It seems that a good model for criticality, or for the loss of compactness associated to the critical exponent is given by exponential nonlinearity. Let us consider the problem

$$\begin{cases} \Delta u + \varepsilon^2 e^u = 0 & \text{in } \Omega, \\ u = 0 & \text{on } \partial\Omega, \end{cases} \tag{1.14}$$

where Ω is a smooth bounded domain in $\mathbb{R}^2$ and $\varepsilon > 0$ is a small parameter. Sometimes called *Liouville equation* [59], this problem and qualitatively similar ones have attracted great attention over the last decades. In a two-dimensional domain or a compact manifold this type of equation arises in a broad range of applications, in particular in astrophysics and combustion theory, see [20,46,51,62] and references, the prescribed Gaussian curvature problem [21,22], mean field limit of vortices in Euler flows [18], and vortices in the relativistic Maxwell–Chern–Simons–Higgs theory [10,17,56,82,85].

It is a standard fact that problem (1.14) does not admit any solutions for large ε, as testing against a first eigenfunction of the Laplacian readily shows, while for small ε a solution close to zero exists, which represents a strict local minimizer of the energy functional

$$E(u) = \frac{1}{2} \int_\Omega |\nabla u|^2 - \varepsilon^2 \int_\Omega e^u. \tag{1.15}$$

Moreover, Trudinger–Moser embedding yields necessary compactness to apply in this range of ε the mountain pass lemma thus getting a second solution, which clearly becomes unbounded as $\varepsilon \downarrow 0$. This second, "large" solution of (1.14) was found in simply-connected domains in [88], see also [25] for earlier work on existence. The behavior of blowing-up families of solutions to problem (1.14) has become understood after the works [13,55,60, 67,83]. It is known that if u_ε is an unbounded family of solutions for which $\varepsilon^2 \int_\Omega e^{u_\varepsilon}$ remains uniformly bounded, then necessarily

$$\lim_{\varepsilon \to 0} \varepsilon^2 \int_\Omega e^{u_\varepsilon} = 8m\pi \tag{1.16}$$

for some integer $m \geqslant 1$. Moreover, there are m-tuples of distinct points of Ω, $(x_1^\varepsilon, \ldots, x_m^\varepsilon)$, separated at uniformly positive distance from each other and from $\partial\Omega$ as $\varepsilon \to 0$ for which u_ε remains uniformly bounded on $\Omega \setminus \bigcup_{j=1}^m B_\delta(x_i^\varepsilon)$ and

$$\sup_{B_\delta(x_i^\varepsilon)} u_\varepsilon \to +\infty \tag{1.17}$$

for any $\delta > 0$.

An obvious question is the reciprocal, namely existence of solutions of problem (1.14) with the property (1.16). Here we prove that such a family indeed exists if Ω is not simply connected.

THEOREM 1.3 [35]. *Assume that Ω is not simply connected. Then given any $m \geqslant 1$ there exists a family of solutions u_ε to (1.14) with*

$$\lim_{\varepsilon \to 0} \varepsilon^2 \int_\Omega e^{u_\varepsilon} = 8m\pi.$$

In case of existence, location of blowing-up points is well understood: it is established in [67,83] that the m-tuple $(x_1^\varepsilon, \ldots, x_m^\varepsilon)$ in (1.17) converges, up to subsequences, to a critical point of the functional

$$\varphi_m(y_1, \ldots, y_m) = \sum_{j=1}^m H(y_j, y_j) - \sum_{i \neq j} G(y_i, y_j), \tag{1.18}$$

where $G(x, y)$ is Green's function (1.7) and now

$$H(x, y) = \frac{1}{2\pi} \log \frac{1}{|x - y|} - G(x, y). \tag{1.19}$$

Obvious question is the reciprocal, namely presence of multiple-bubbling solutions with concentration at a critical point of φ_m.

Baraket and Pacard [9] established that for any *nondegenerate critical point* of φ_m, a family of solutions u_ε concentrating at this point as $\varepsilon \to 0$ does exist. As remarked in [9], their construction, based on a very precise approximation of the actual solution and an application of Banach fixed point theorem, uses nondegeneracy in essential way. This assumption, however, is hard to check in practice. We will sketch a construction of blowing-up families of solutions of (1.14) which lifts the nondegeneracy assumption of [9], and it is in particular enough for the proof of Theorem 1.3. Now the solutions u_ε will look, near each ξ_j like

$$u_\varepsilon(x) \sim u_j(x) = \log \frac{8\mu_j^2}{(\mu_j^2 \varepsilon^2 + |x - \xi_j|^2)^2}$$

for certain ε-independent numbers μ_j. Observe that u_j satisfies in entire $\mathbb{R}^2$

$$\Delta u_j + \varepsilon^2 e^{u_j} = 0,$$

in fact up to scaling and translation invariance, these are the only solutions with $\int e^u < +\infty$.

The rest of this paper will be devoted to the proofs of the above mentioned results. In Section 2 we will give a detailed proof of Theorem 1.1, while in Section 3 we will study the above described applications to derive results on bubbling solutions under suitable topological features of the domain. In Section 4 we will analyze, without reference to topology, the Brezis–Nirenberg problem in dimension $N = 3$, proving Theorem 1.2. Finally, in Section 5 we will consider Liouville-type equations, proving Theorem 1.3.

2. Nearly critical bubbling: the proof of Theorem 1.1

2.1. *Ansatz and scheme of the proof*

We will only carry out the proof for the supercritical case, since the other is completely analogous. We will first describe the shape of a bubbling solution to the slightly supercritical problem,

$$\begin{cases} \Delta u + u^{(N+2)/(N-2)+\varepsilon} = 0 & \text{in } \Omega, \\ u > 0 & \text{in } \Omega, \\ u = 0 & \text{on } \partial\Omega, \end{cases} \tag{2.1}$$

where Ω is a general bounded smooth domain in $\mathbb{R}^N$, with $N \geqslant 3$, and ε is a small positive parameter. Furthermore we explain the general strategy we follow to construct such solutions. With minor changes, the same construction goes through for the slightly subcritical problem (1.5). To simplify the exposition, we just treat the supercritical case.

If we consider problem (2.1) in the enlarged domain

$$\Omega_\varepsilon = \varepsilon^{-1/(N-2)}\Omega, \quad \varepsilon > 0,$$

with the following change of variable

$$v(y) = \varepsilon^{\frac{1}{2+\varepsilon(N-2)/2}} u\big(\varepsilon^{1/(N-2)}y\big), \quad y \in \Omega_\varepsilon,$$

it is straightforward to see that u solves (2.1) if and only if v satisfies

$$\begin{cases} \Delta v + v^{(N+2)/(N-2)+\varepsilon} = 0 & \text{in } \Omega_\varepsilon, \\ v > 0 & \text{in } \Omega_\varepsilon, \\ v = 0 & \text{on } \partial\Omega_\varepsilon. \end{cases} \tag{2.2}$$

Since Ω_ε is expanding to the whole $\mathbb{R}^N$, and all positive solutions of the "limit" problem

$$\Delta v + v^{(N+2)/(N-2)} = 0 \quad \text{in } \mathbb{R}^N$$

are given by the functions

$$\overline{U}(x) = \alpha_N \left(\frac{1}{1+|x|^2} \right)^{(N-2)/2} \quad \text{and} \quad \overline{U}_{\lambda,y}(x) = \lambda^{-(N-2)/2} \overline{U}\left(\frac{x-y}{\lambda} \right)$$

with $\alpha_N = (N(N-2))^{(N-2)/4}$, $y \in \mathbb{R}^N$ and $\lambda > 0$ (see [5,84]), it is natural to look for solutions v to (2.2) of the form

$$v(y) \sim \sum_{j=1}^{k} \overline{U}_{\lambda_j,\xi'_j}(y) \tag{2.3}$$

for certain set of points ξ_j in Ω and numbers $\lambda_j > 0$, where from now on we use the letter ξ to denote a point in Ω and

$$\xi' = \varepsilon^{-1/(N-2)}\xi \in \Omega_\varepsilon.$$

In the original domain, a solution of the form (2.3) has the shape of a smooth function which has k maximum points, which are close to the ξ_i's, where the size of the maximum is of order $\varepsilon^{-1/2}$. In the literature, these maximum are called peaks, or bubbles, and solutions to (2.1) which provides peaks are called (multi)peak solutions, or (multi)bubbling solutions, as already explained in the Introduction.

Given an integer k, the location of the k peaks ξ_i's in Ω and the size of the dilation parameters λ_i's of a k peak solution are not arbitrary. As already mentioned in the Introduction, they are related with the existence of a nontrivial critical point situation of the function Ψ_k^+ given by (1.10) for the slightly supercritical case (respectively Ψ_k^- given by (1.8) for the slightly subcritical case). Namely, if the domain Ω is such that Ψ_k^+ (or Ψ_k^-) has a nontrivial critical point situation, a solution to (2.1) (or (1.5)) of the form (2.3), exists. As we will see later, in order to guarantee that $\Psi_k^\pm$ has a critical point the topology and geometry of Ω may play a crucial role, as we will see in Section 4. What we want to do now is to describe how the existence of a nontrivial critical point situation for this function enters in the construction of a solution to (2.1).

Observe first that the approximation given in (2.3) does not take into account of the boundary condition a solution to (2.2) has to satisfy. In fact, a better approximation in (2.3) should be obtained by using the orthogonal projections onto $H_0^1(\Omega_\varepsilon)$ of the functions $\overline{U}_{\lambda,\xi'}$, denoted by $V_{\lambda,\xi'}$, namely the unique solution of the equation

$$-\Delta V_{\lambda,\xi'} = \overline{U}_{\lambda,\xi'}^{(N+2)/(N-2)} \quad \text{in } \Omega_\varepsilon,$$

$$V_{\lambda,\xi'} = 0 \qquad\qquad\qquad \text{on } \partial\Omega_\varepsilon,$$

so that the function $\phi_{\lambda,\xi'}$, defined as $\phi_{\lambda,\xi'} = \overline{U}_{\lambda,\xi'} - V_{\lambda,\xi'}$, will satisfy the equation

$$-\Delta\phi_{\lambda,\xi'} = 0 \quad \text{in } \Omega_\varepsilon,$$
$$\phi_{\lambda,\xi'} = \overline{U}_{\lambda,\xi'} \quad \text{on } \partial\Omega_\varepsilon.$$

Since, for $x \in \partial\Omega_\varepsilon$,

$$\phi_{\lambda,\xi'}(x) = \varepsilon\Gamma\big(\varepsilon^{1/(N-2)}x - \xi\big)\lambda^{(N-2)/2}\int_{\mathbb{R}^N}\overline{U}^{(N+2)/(N-2)} + \mathrm{o}(\varepsilon),$$

by harmonicity we get

$$\phi_{\lambda,\xi'}(x) = \varepsilon H\big(\varepsilon^{1/(N-2)}x, \xi\big)\lambda^{(N-2)/2}\int_{\mathbb{R}^N}\overline{U}^{(N+2)/(N-2)} + \mathrm{o}(\varepsilon), \tag{2.4}$$

uniformly on compact sets of Ω_ε. On the other hand,

$$V_{\lambda,\xi'}(x) = \varepsilon G\big(\varepsilon^{1/(N-2)}x, \xi\big)\lambda^{(N-2)/2}\int_{\mathbb{R}^N}\overline{U}^{(N+2)/(N-2)} + \mathrm{o}(\varepsilon), \tag{2.5}$$

uniformly for x on each compact subset of $\Omega_\varepsilon \setminus \{\xi'\}$. Here G and H are respectively the Green function of the Laplacian with Dirichlet boundary condition on Ω and its regular part.

We write

$$\overline{U}_i = \overline{U}_{\lambda_i,\xi'_i}, \qquad V_i = V_{\lambda_i,\xi'_i}, \tag{2.6}$$

and

$$\overline{V} = \sum_i \overline{U}_i, \qquad V = \sum_i V_i. \tag{2.7}$$

Our goal is to find a solution v of problem (2.2) of the form

$$v = V + \phi \tag{2.8}$$

which for suitable points ξ and scalars λ will have the remainder term ϕ of small order all over Ω_ε, in fact with magnitude not exceeding $\mathrm{O}(\varepsilon)$ in any reasonable norm over Ω_ε.

For notational convenience from now on we denote $p = \frac{N+2}{N-2}$.

In terms of ϕ, problem (2.2) becomes

$$\begin{cases} L(\phi) = -R_\varepsilon - N_\varepsilon(\phi) & \text{in } \Omega_\varepsilon, \\ \phi = 0 & \text{on } \partial\Omega_\varepsilon. \end{cases} \tag{2.9}$$

Here L is the linear operator defined by

$$L(\phi) = \Delta\phi + (p+\varepsilon)V^{p+\varepsilon-1}\phi. \tag{2.10}$$

The term R_ε is defined as follows

$$R_\varepsilon(y) = R_\varepsilon(\xi', \lambda)(y) = \Delta V(y) + V^{p+\varepsilon}(y), \qquad y \in \Omega_\varepsilon. \tag{2.11}$$

It is a function defined on Ω_ε that depends on the points ξ_i' and the parameters λ_i. It represents the error for V to be an actual solution of problem (2.2).

The term $N_\varepsilon(\phi)$ is the function defined in Ω_ε given by

$$N_\varepsilon(\phi)(y) = N_\varepsilon(\xi', \lambda, \phi)(y)$$
$$= (V + \phi)_+^{p+\varepsilon}(y) - V^{p+\varepsilon}(y) - (p+\varepsilon)V^{p+\varepsilon-1}(y)\phi(y), \qquad y \in \Omega_\varepsilon. \tag{2.12}$$

It is quadratic in ϕ and it depends on the points ξ_i' and the parameters λ_i.

In order to solve problem (2.2), or equivalently problem (2.9), we first develop a solvability theory for the linearized operator L under suitable orthogonality conditions.

For $j = 1, \ldots, N$, let us consider the functions

$$\overline{Z}_{ij} = \frac{\partial \overline{U}_i}{\partial \xi'_{ij}}, \qquad \overline{Z}_{iN+1} = \frac{\partial \overline{U}_i}{\partial \lambda_i} = (x - \xi_i') \cdot \nabla \overline{U}_i + (N-2)\overline{U}_i,$$

namely

$$\overline{Z}_{ij}(x) = \alpha_N (N-2)\lambda_i^{(N-2)/2} \frac{(x - \xi_i')_j}{(\lambda_i^2 + |x - \xi_i'|^2)^{N/2}}$$

and

$$\overline{Z}_{iN+1}(x) = \alpha_N \frac{N-2}{2} \lambda_i^{(N-4)/2} \frac{|x - \xi_i'|^2 - \lambda_i^2}{|x - \xi_i'|^2 + \lambda_i^2}.$$

Define the Z_{ij}'s to be their respective $H_0^1(\Omega_\varepsilon)$-projections, namely the unique solutions of

$$\Delta Z_{ij} = \Delta \overline{Z}_{ij} \quad \text{in } \Omega_\varepsilon,$$
$$Z_{ij} = 0 \qquad\quad \text{on } \partial \Omega_\varepsilon.$$

A direct argument shows that

$$Z_{ij}(x) = \overline{Z}_{ij}(x) - \varepsilon^{(N-1)/(N-2)}\lambda_i^{(N-2)/2}\left(\int_{\mathbb{R}^N} \overline{U}^p\right)\frac{\partial}{\partial \xi_{ij}} H\big(\varepsilon^{1/(N-2)}x, \xi_i\big)$$
$$+ \mathrm{o}\big(\varepsilon^{(N-1)/(N-2)}\big)$$

and

$$Z_{iN+1}(x) = \overline{Z}_{iN+1}(x) - \varepsilon \frac{N-2}{2}\lambda_i^{(N-4)/2}\left(\int_{\mathbb{R}^N} \overline{U}^p\right)H\big(\varepsilon^{1/(N-2)}x, \xi_i\big) + \mathrm{o}(\varepsilon)$$

uniformly for x in compact sets of Ω_ε.

The first step to deal with problem (2.9) is to study the following linear problem: Given points $\xi_i \in \Omega$, positive parameters λ_i and $h \in C^\alpha(\overline{\Omega}_\varepsilon)$, find a function ϕ such that

$$
\begin{cases}
L\phi = h + \sum_{i,j} c_{ij} V_i^{p-1} Z_{ij} & \text{in } \Omega_\varepsilon, \\
\phi = 0 & \text{on } \partial\Omega_\varepsilon, \\
\int_{\Omega_\varepsilon} V_i^{p-1} Z_{ij} \phi = 0 & \text{for all } i, j,
\end{cases}
\tag{2.13}
$$

for certain constants c_{ij}, $i = 1, \ldots, k$, $j = 1, \ldots, N+1$. Obviously, both the function ϕ and the constants c_{ij} do depend on ξ_i and λ_i.

In order to perform an invertibility theory for L subject to the above orthogonality conditions, we introduce $L_*^\infty(\Omega_\varepsilon)$ and $L_{**}^\infty(\Omega_\varepsilon)$ to be respectively the spaces of functions defined on Ω_ε with finite $\|\cdot\|_*$-norm (respectively $\|\cdot\|_{**}$-norm), where

$$
\|\psi\|_* = \sup_{x \in \Omega_\varepsilon} \left|\omega^{-\beta}(x)\psi(x)\right| + \left|\omega^{-(\beta+1/(N-2))}(x)D\psi(x)\right|,
$$

with

$$
\omega(x) = \sum_{j=1}^{k} \left(1 + |x - \xi_j'|^2\right)^{-(N-2)/2},
$$

$\beta = 1$ if $N = 3$ and $\beta = \frac{2}{N-2}$ if $N \geqslant 4$. Similarly we define, for any dimension $N \geqslant 3$,

$$
\|\psi\|_{**} = \sup_{x \in \Omega_\varepsilon} \left|\omega^{-4/(N-2)}(x)\psi(x)\right|.
$$

Indeed, if the points ξ_i are far away from the boundary of Ω and far away from each other, and if the parameters λ_i are uniformly bounded below from 0 and above, the operator L is uniformly invertible with respect to the above weighted L^∞-norm, for all ε small enough. This fact will be proved in Proposition 2.1.

The second step in dealing with problem (2.9) consists in solving the following auxiliary problem: Given points ξ_i in Ω and positive parameters λ_i, find a function ϕ and constants c_{ij} solution of (2.13) with $h = -R_\varepsilon - N_\varepsilon(\phi)$ (see (2.11) and (2.12)). Namely, given ξ_i and λ_i, find a function ϕ and constants c_{ij}, all depending on ξ_i and λ_i such that

$$
\begin{cases}
L\phi = -R_\varepsilon - N_\varepsilon(\phi) + \sum_{i,j} c_{ij} V_i^{p-1} Z_{ij} & \text{in } \Omega_\varepsilon, \\
\phi = 0 & \text{on } \partial\Omega_\varepsilon, \\
\int_{\Omega_\varepsilon} V_i^{p-1} Z_{ij} \phi = 0 & \text{for all } i, j.
\end{cases}
\tag{2.14}
$$

In fact, one obtains existence and uniqueness of such solutions in a certain range of functions and constants, as consequence of a fixed point argument. This will be done in Proposition 2.3.

Once the auxiliary problem (2.14) has been solved, one gets that $v = V + \phi$ is a solution for (2.2) if and only if the constants c_{ij} that appears in (2.14) are all zero. Taking into account the dependence of the constants c_{ij} on the points ξ_i and the parameters λ_i, one proves that this is equivalent to finding a critical point of a reduced functional depending on ξ_i and λ_i (see Lemma 2.2). It is exactly at this point where the nontrivial critical point situation of the function defined in (1.10) enters. Indeed the reduced functional is given by

$$(\xi, \lambda) \to J(\xi, \lambda) = I_\varepsilon(V + \phi),$$

where

$$I_\varepsilon(u) = \frac{1}{2} \int_{\Omega_\varepsilon} |Du|^2 - \frac{1}{p+1+\varepsilon} \int_{\Omega_\varepsilon} u^{p+1+\varepsilon}. \tag{2.15}$$

Not too surprisingly, this function $I_\varepsilon(V + \phi)$ coincides with $I_\varepsilon(V)$ at main order (see Proposition 2.4). On the other hand, one can easily compute the explicit asymptotic expansion of $I_\varepsilon(V)$, which, as shown in the next lemma, coincides at first order with the function introduced in (1.10).

LEMMA 2.1. *Let us fix a small number $\delta > 0$. The following expansion holds*

$$I_\varepsilon(V) = kC_N + \varepsilon\big[\gamma_N + \omega_N \Psi_k^+(\xi, \lambda)\big] + o(\varepsilon) \tag{2.16}$$

uniformly with respect to (ξ, λ) satisfying

$$|\xi_i - \xi_j| > \delta \quad \text{if } i \neq j, \qquad \text{dist}(\xi_i, \partial\Omega) > \delta, \quad \delta < \lambda_i < \delta^{-1}. \tag{2.17}$$

Here Ψ_k^+ is the function defined in (1.10), while

$$C_N = \frac{1}{2} \int_{\mathbb{R}^N} |D\overline{U}|^2 - \frac{1}{p+1} \int_{\mathbb{R}^N} |\overline{U}|^{p+1},$$

$$\gamma_N = \left\{ \frac{k}{p+1} \omega_N - \frac{k}{p+1} \int_{\mathbb{R}^N} \overline{U}^{p+1} \log \overline{U} \right\} \tag{2.18}$$

and $\omega_N = \frac{1}{p+1} \int_{\mathbb{R}^N} \overline{U}^{p+1}$.

PROOF. We first write

$$I_\varepsilon(V) = I_0(V) + \frac{1}{p+1} \int_{\Omega_\varepsilon} V^{p+1} - \frac{1}{p+1+\varepsilon} \int_{\Omega_\varepsilon} V^{p+1+\varepsilon}, \tag{2.19}$$

where

$$I_0(V) = \frac{1}{2} \int_{\Omega_\varepsilon} |DV|^2 - \frac{1}{p+1} \int_{\Omega_\varepsilon} V^{p+1}.$$

Let us first estimate $I_0(V)$. We have

$$
I_0(V) = I_0\left(\sum_{j=1}^{k} V_j\right)
$$

$$
= \sum_{j=1}^{k}\left[\frac{1}{2}\int_{\Omega_\varepsilon}|DV_j|^2 - \frac{1}{p+1}\int_{\Omega_\varepsilon}|V_j|^{p+1}\right]
$$

$$
+ \sum_{i\neq j}\int_{\Omega_\varepsilon} DV_i\,DV_j - \frac{1}{p+1}\int_{\Omega_\varepsilon}\left[\left(\sum_{j=1}^{k}V_j\right)^{p+1} - \sum_{j=1}^{k}V_j^{p+1}\right].
$$

$$(2.20)$$

Arguing like in [6–8,33], and taking into account (2.4) and (2.5), one can prove that

$$
\int_{\Omega_\varepsilon}|DV_i|^2 = \int_{\mathbb{R}^N}|D\overline{U}|^2 - \left(\int_{\mathbb{R}^N}\overline{U}^p\right)^2 H(\xi_i,\xi_i)\lambda_i^{N-2}\varepsilon + \mathrm{o}(\varepsilon),
\tag{2.21}
$$

$$
\int_{\Omega_\varepsilon} DV_i\,DV_j = \left(\int_{\mathbb{R}^N}\overline{U}^p\right)^2 G(\xi_i,\xi_j)\lambda_i^{(N-2)/2}\lambda_j^{(N-2)/2}\varepsilon + \mathrm{o}(\varepsilon),
\tag{2.22}
$$

$$
\frac{1}{p+1}\int_{\Omega_\varepsilon}V_i^{p+1} = \frac{1}{p+1}\int_{\mathbb{R}^N}\overline{U}^{p+1} - \left(\int_{\mathbb{R}^N}\overline{U}^p\right)^2 H(\xi_i,\xi_i)\lambda_i^{N-2}\varepsilon + \mathrm{o}(\varepsilon)
\tag{2.23}
$$

and finally, for $i\neq j$,

$$
\frac{1}{p+1}\int_{\Omega_\varepsilon}\left[\left(\sum_{j=1}^{k}V_j\right)^{p+1} - \sum_{j=1}^{k}V_j^{p+1}\right]
$$

$$
= 2\left(\int_{\mathbb{R}^N}\overline{U}^p\right)^2 G(\xi_i,\xi_j)\lambda_i^{(N-2)/2}\lambda_j^{(N-2)/2}\varepsilon + \mathrm{o}(\varepsilon).
\tag{2.24}
$$

From (2.20)–(2.24) we conclude that

$$
I_0(V) = kC_N + \frac{1}{2}\left(\int_{\mathbb{R}^N}\overline{U}^p\right)^2\left\{\sum_{j=1}^{k}H(\xi_j,\xi_j)\lambda_j^{N-2}\right.
$$

$$
\left. - 2\sum_{i<j}G(\xi_i,\xi_j)\lambda_i^{(N-2)/2}\lambda_j^{(N-2)/2}\right\} + \mathrm{o}(\varepsilon).
$$

Let us consider now the quantity

$$I_\varepsilon(V) - I_0(V) = \frac{\varepsilon}{(p+1)^2} \int_{\Omega_\varepsilon} V^{p+1} - \frac{\varepsilon}{p+1} \int_{\Omega_\varepsilon} V^{p+1} \log V + o(\varepsilon). \quad (2.25)$$

First we see that

$$\int_{\Omega_\varepsilon} V^{p+1} = k \int_{\mathbb{R}^N} \overline{U}^{p+1} + o(1).$$

On the other hand, for a number $\rho > 0$ we can write

$$\int_{\Omega_\varepsilon} V^{p+1} \log V = \sum_{j=1}^{k} \int_{|x-\xi_j'|<\rho} V^{p+1} \log V + o(\varepsilon).$$

For any index j we have

$$\int_{|x-\xi_j'|<\rho} V^{p+1} \log V$$

$$= -\frac{N-2}{2} \log \lambda_j \int_{|x-\xi_j'|<\rho} V^{p+1}$$

$$\quad + \int_{|x-\xi_j'|<\rho} V^{p+1} \log\big((\lambda_j)^{(N-2)/2} V_j + (\lambda_j)^{(N-2)/2}(V - V_j)\big)$$

$$= -\frac{N-2}{2} \log \lambda_j \left(\int_{\mathbb{R}^N} \overline{U}^{p+1} + o(\varepsilon) \right) + \int_{\mathbb{R}^N} \overline{U}^{p+1} \log \overline{U} + o(1).$$

Then we conclude

$$\int_{\Omega_\varepsilon} V^{p+1} \log V$$

$$= -\frac{N-2}{2} \log(\lambda_1 \cdots \lambda_k) \left(\int_{\mathbb{R}^N} \overline{U}^{p+1} \right) + k \int_{\mathbb{R}^N} \overline{U}^{p+1} \log \overline{U} + o(1);$$

hence from (2.25) and the previous computation we get

$$I_\varepsilon(V) - I_0(V) = \varepsilon \left[\frac{k}{(p+1)^2} \int_{\mathbb{R}^N} \overline{U}^{p+1} + \frac{N-2}{2} \log(\lambda_1 \cdots \lambda_k) \left(\int_{\mathbb{R}^N} \overline{U}^{p+1} \right) \right.$$

$$\left. - \frac{k}{p+1} \int_{\mathbb{R}^N} \overline{U}^{p+1} \log \overline{U} \right] + o(\varepsilon). \qquad \square$$

REMARK 2.1. The quantity $o(\varepsilon)$ in the expansion of (2.16) is actually also of that size in the C^1-norm as a function of ξ and λ in the considered region.

2.2. *Variational reduction and conclusion of the proof*

In this section we prove Theorem 1.1. As mentioned in the previous section, in order to do so we first have to show the invertibility of the linear operator L (see (2.10)) subject to the orthogonality conditions given in (2.13). We will do it next for a slight modification of the linear operator L, which will be useful for the section devoted to the Brezis–Nirenberg problem.

Indeed, we consider problem (2.13) with the linear operator L given by

$$L(\phi) = \Delta\phi + \mu\varepsilon^{2/(N-2)}\phi + (p+\varepsilon)V^{p+\varepsilon-1}\phi, \tag{2.26}$$

where μ is less than μ_1, the first eigenvalue of the Laplace operator with Dirichlet boundary condition on Ω.

We start with the resolubility of problem (2.13), keeping in mind that L is now given by (2.26).

PROPOSITION 2.1. *Let $\delta > 0$ be fixed. There are numbers $\varepsilon_0 > 0$, $C > 0$, such that, for points ξ in Ω and parameters λ satisfying (2.17) problem (2.13) admits a unique solution $\phi \equiv T(h)$ for all $0 < \varepsilon < \varepsilon_0$ and all $h \in C^\alpha(\overline{\Omega}_\varepsilon)$. Besides,*

$$\left\|T(h)\right\|_* \leqslant C\|h\|_{**} \tag{2.27}$$

and

$$|c_{ij}| \leqslant C\|h\|_{**}. \tag{2.28}$$

PROOF. The proof of this proposition consists of 2 steps.

STEP 1. Assume there exists a sequence $\varepsilon = \varepsilon_n \to 0$ such that there are functions ϕ_ε and h_ε with $\|h_\varepsilon\|_{**} = o(1)$, such that

$$\begin{cases} L(\phi_\varepsilon) = h_\varepsilon + \sum_{i,j} c_{ij} V_i^{p-1} Z_{ij} & \text{in } \Omega_\varepsilon \\ \phi_\varepsilon = 0 & \text{on } \partial\Omega_\varepsilon, \\ \int_{\Omega_\varepsilon} V_i^{p-1} Z_{ij}\phi_\varepsilon \, dx = 0 & \text{for all } i, j, \end{cases}$$

for certain constants c_{ij}, depending on ε. Then

$$\|\phi_\varepsilon\|_* \to 0.$$

We shall establish first the slightly weaker assertion that

$$\|\phi_\varepsilon\|_\rho = \sup_{x\in\Omega_\varepsilon} \left|\omega^{-(\beta-\rho)}\phi_\varepsilon(x)\right| + \left|\omega^{-(\beta-\rho+1/(N-2))} D\phi_\varepsilon(x)\right| \to 0$$

with $\rho > 0$ a small fixed number. To do this, we assume the opposite, so that with no loss of generality we may take $\|\phi_\varepsilon\|_\rho = 1$. Testing the above equation against Z_{lh}, integrating by parts twice we get that

$$\sum c_{ij} \int_{\Omega_\varepsilon} V_i^{p-1} Z_{ij} Z_l = \int_{\Omega_\varepsilon} \left[\Delta Z_{lh} + (p+\varepsilon) V^{p-1+\varepsilon} Z_{lh} \right] \phi_\varepsilon$$

$$+ \mu \varepsilon^{2/(N-2)} \int_{\Omega_\varepsilon} \phi_\varepsilon Z_{lh} - \int_{\Omega_\varepsilon} h_\varepsilon Z_{lh}. \tag{2.29}$$

This defines a linear system in the c_{ij} which is "almost diagonal" as ε approaches zero, since we have, for $h = 1, \ldots, N$,

$$\int_{\Omega_\varepsilon} V_i^{p-1} Z_{ij} Z_{lh} = \delta_{i,l} \delta_{j,h} \int_{\mathbb{R}^N} \overline{U}_{\Lambda_i}^{p-1} \left(\frac{\partial \overline{U}_{\Lambda_i,0}}{\partial x_h} \right)^2 + o(1) \tag{2.30}$$

and for $h = N + 1$,

$$\int_{\Omega_\varepsilon} V_i^{p-1} Z_{ij} Z_{l(N+1)} = \delta_{i,l} \delta_{j,N+1} \int_{\mathbb{R}^N} \overline{U}_{\Lambda_i}^{p-1} \left(x \overline{U}_{\Lambda_i} + (N-2) \overline{U}_{\Lambda_i} \right)^2 + o(1) \tag{2.31}$$

for suitable $\Lambda_i > 0$.

On the other hand, it is easy to see that, for $l = 1, \ldots, k$, we have $[\Delta Z_{lh} + (p + \varepsilon) V^{p+\varepsilon-1} Z_{lh}] \leqslant C \omega^{(N+2)/(N-2)}(x)$ for $h = 1, \ldots, N$ and $[\Delta Z_{lh} + (p+\varepsilon) V^{p+\varepsilon-1} Z_{lh}] \leqslant C \omega^{(N+3)/(N-2)}(x)$ for $h = N + 1$. Hence, we get

$$\int_{\Omega_\varepsilon} \left[\Delta Z_{lh} + (p+\varepsilon) V^{p+\varepsilon-1} Z_{lh} \right] \phi_\varepsilon = o(1) \|\phi_\varepsilon\|_\rho, \tag{2.32}$$

after noticing that $\Delta Z_{lh} + p \overline{V}_l^{p-1} Z_{lh} = 0$, and an application of dominated convergence.

Let us now estimate the second term in the right-hand side of (2.29). If $N \geqslant 4$ and $h = N + 1$, then

$$\mu \varepsilon^{2/(N-2)} \left| \int_{\Omega_\varepsilon} \phi_\varepsilon Z_{lN+1} \right| \leqslant C \varepsilon^{2/(N-2)} \|\phi_\varepsilon\|_\rho \int_{\Omega_\varepsilon} \omega^{-(\beta-\rho)} \frac{1}{(1 + |x - \xi_j'|)^{N-2}}$$

$$\leqslant C \varepsilon^{2/(N-2)} \|\phi_\varepsilon\|_\rho \int_{\Omega_\varepsilon} \frac{1}{(1 + |x - \xi_j'|)^{N-\rho}}$$

$$\leqslant C \varepsilon^{(2-\rho)/(N-2)} \|\phi_\varepsilon\|_\rho.$$

Hence, in this case we conclude that, provided ρ is small,

$$\mu \varepsilon^{2/(N-2)} \int_{\Omega_\varepsilon} \phi_\varepsilon Z_{lN+1} = o(1) \|\phi_\varepsilon\|_\rho.$$

If $N \geqslant 4$ and $h = 1, \ldots, N$, similar computation yields to

$$\mu \varepsilon^{2/(N-2)} \int_{\Omega_\varepsilon} \phi_\varepsilon Z_{lh} = O(1) \varepsilon^{2/(N-2)} \|\phi_\varepsilon\|_\rho = o(1) \|\phi_\varepsilon\|_\rho.$$

Assume now that $N = 3$ and $h = N + 1$. Then

$$\mu \varepsilon^2 \left| \int_{\Omega_\varepsilon} \phi_\varepsilon Z_{lN+1} \right| \leqslant C \varepsilon^2 \|\phi_\varepsilon\|_\rho \int_{\Omega_\varepsilon} \frac{1}{(1 + |x - \xi_j'|)^{2-\rho}} \leqslant C \|\phi_\varepsilon\|_\rho \varepsilon^{1-\rho}.$$

Hence we conclude that, provided ρ is small,

$$\mu \varepsilon^2 \int_{\Omega_\varepsilon} \phi_\varepsilon Z_{lN+1} = o(1) \|\phi_\varepsilon\|_\rho.$$

Analogously, one proves that, for $N = 3$ and $h = 1, \ldots, N$,

$$\mu \varepsilon^2 \int_{\Omega_\varepsilon} \phi_\varepsilon Z_{lh} = o(1) \|\phi_\varepsilon\|_\rho.$$

Finally, for the last term in the right-hand side of (2.29), we have

$$\left| \int_{\Omega_\varepsilon} h_\varepsilon Z_{lh} \right| \leqslant C \|h_\varepsilon\|_{**}.$$

Thus, we conclude that

$$|c_{ij}| \leqslant C \|h_\varepsilon\|_{**} + o(1) \|\phi_\varepsilon\|_\rho \tag{2.33}$$

so that $c_{ij} = o(1)$.

Taking into account that $\mu < \mu_1$, we can rewrite the equation in the following form

$$\phi_\varepsilon(x) - (p + \varepsilon) \int_{\Omega_\varepsilon} G_\varepsilon(x, y) V^{p+\varepsilon-1} \phi_\varepsilon \, dy - \mu \varepsilon^{2/(N-2)} \int_{\Omega_\varepsilon} G_\varepsilon(x, y) \phi_\varepsilon$$

$$= - \int_{\Omega_\varepsilon} G_\varepsilon(x, y) h_\varepsilon \, dy - \sum c_{ij} \int_{\Omega_\varepsilon} V_i^{p-1} Z_{ij} G_\varepsilon(x, y) \, dy, \quad x \in \Omega_\varepsilon, \tag{2.34}$$

where G_ε denotes Green's function of Ω_ε. Furthermore, the function ϕ_ε is of class C^1 and

$$\partial_{x_j} \phi_\varepsilon(x) - (p + \varepsilon) \int_{\Omega_\varepsilon} \partial_{x_j} G_\varepsilon(x, y) V^{p+\varepsilon-1} \phi_\varepsilon \, dy$$

$$- \mu \varepsilon^{2/(N-2)} \int_{\Omega_\varepsilon} \partial_{x_j} G_\varepsilon(x, y) \phi_\varepsilon$$

$$= - \int_{\Omega_\varepsilon} \partial_{x_j} G_\varepsilon(x, y) h_\varepsilon \, dy$$

$$- \sum c_{ij} \int_{\Omega_\varepsilon} V_i^{p-1} Z_{ij} \partial_{x_j} G_\varepsilon(x, y) \, dy, \quad x \in \Omega_\varepsilon. \tag{2.35}$$

We make now the following observation

$$\int_{\Omega_\varepsilon} \big|G_\varepsilon(x,y)h_\varepsilon\big|\,\mathrm{d}y \leqslant \|h_\varepsilon\|_{**} C \int_{\mathbb{R}^N} \Gamma(x-y)\omega^{4/(N-2)}(x)\,\mathrm{d}y$$

$$\leqslant C\|h_\varepsilon\|_{**}\left(\sum_i (1+|x-\xi_i'|^2)^{-(N-2)/2}\right)^\beta. \tag{2.36}$$

Indeed, one has

$$\int_{\mathbb{R}^N} \Gamma(x-y)\omega^{4/(N-2)}(x)\,\mathrm{d}y \leqslant C\sum_{j=1}^{k}\int_{\mathbb{R}^N} \Gamma(x-y)\big(1+|y-\xi_j'|^2\big)^{-2}\,\mathrm{d}y$$

$$\leqslant C\sum_{j=1}^{k}(I_{1j}+I_{2j}+I_{3j}),$$

where

$$I_{1j} = \int_{B(x,|x-\xi_j'|/2)} \Gamma(x-y)\big(1+|y-\xi_j'|^2\big)^{-2}\,\mathrm{d}y,$$

$$I_{2j} = \int_{B(\xi_j',|x-\xi_j'|/2)} \Gamma(x-y)\big(1+|y-\xi_j'|^2\big)^{-2}\,\mathrm{d}y$$

and

$$I_{3j} = \int_{\mathbb{R}^N} \Gamma(x-y)\big(1+|y-\xi_j'|^2\big)^{-2}\,\mathrm{d}y - I_{1j} - I_{2j}.$$

Now, (2.36) follows from

$$I_{1j} \leqslant \frac{C}{(1+|x-\xi_j'|^2)^2}\int_0^{|x-\xi_j'|/2} t\,\mathrm{d}t \leqslant \frac{C}{(1+|x-\xi_j'|^2)},$$

$$I_{2j} \leqslant \frac{C}{(1+|x-\xi_j'|^2)^{(N-2)/2}}\int_0^{|x-\xi_j'|/2} \frac{t^{N-1}}{(1+t^2)^2}\,\mathrm{d}t$$

$$\leqslant \frac{C}{(1+|x-\xi_j'|^2)^{(N-2)/2}}\int_1^{|x-\xi_j'|/2} t\big(t^2+1\big)^{(N-6)/2}\,\mathrm{d}t$$

$$\leqslant \frac{C}{(1+|x-\xi_j'|^2)}$$

and $I_{3j} \leqslant C I_{2j}$.

Analogously we get

$$\int_{\Omega_\varepsilon} \left| \partial_{x_j} G_\varepsilon(x, y) h_\varepsilon \right| dy$$

$$\leqslant \|h_\varepsilon\|_{**} C \sum_j \int_{\mathbb{R}^N} \frac{1}{|x-y|^{N-1}} \left(1 + |y - \xi_j'|^2\right)^{-2} dy$$

$$\leqslant C \|h_\varepsilon\|_{**} \omega^{\beta + 1/(N-2)}(x).$$

On the other hand, we have

$$\left| \sum c_{ij} \int_{\Omega_\varepsilon} V_i^{p-1} Z_{ij} G_\varepsilon(x, y) \, dy \right|$$

$$\leqslant C \left(\|\phi_\varepsilon\|_\rho + \|h_\varepsilon\|_{**} \right) \sum_i \int_{\mathbb{R}^N} \Gamma(x-y) \left(\left(1 + |y - \xi_i'|^2\right)^{-(N+3)/2} \right)$$

$$\leqslant C \left(\|\phi_\varepsilon\|_\rho + \|h_\varepsilon\|_{**} \right) \left(\sum_i \left(1 + |x - \xi_i'|^2\right)^{-(N-2)/2} \right)$$

and

$$\left| \sum c_{ij} \int_{\Omega_\varepsilon} V_i^{p-1} Z_{ij} \partial_{x_j} G_\varepsilon(x, y) \, dy \right|$$

$$\leqslant C \left(\|\phi_\varepsilon\|_\rho + \|h_\varepsilon\|_{**} \right) \sum \int_{\mathbb{R}^N} \frac{1}{|x-y|^{N-1}} \left(\left(1 + |y - \xi_i'|^2\right)^{-(N+3)/2} \right)$$

$$\leqslant C \left(\|\phi_\varepsilon\|_\rho + \|h_\varepsilon\|_{**} \right) \omega^{\beta + 1/(N-2)}(x).$$

Similarly, we obtain

$$\int_{\Omega_\varepsilon} \left| G_\varepsilon(x, y) V^{p+\varepsilon-1} \phi_\varepsilon \right| dy$$

$$\leqslant C \|\phi_\varepsilon\|_\rho \left(\sum_i \left(1 + |x - \xi_i'|^2\right)^{-(N-2)/2} \right)^\beta$$

and

$$\int_{\Omega_\varepsilon} \left| \partial_{x_j} G_\varepsilon(x, y) V^{p+\varepsilon-1} \phi_\varepsilon \right| dy$$

$$\leqslant C \|\phi_\varepsilon\|_\rho \omega^{\beta + 1/(N-2)}(x).$$

Let us now estimate $\mu \varepsilon^{2/(N-2)} \int_{\Omega_\varepsilon} G_\varepsilon(x,y)\phi_\varepsilon$. First assume that $N \geqslant 4$. Then

$$\mu \varepsilon^{2/(N-2)} \int_{\Omega_\varepsilon} \left| G_\varepsilon(x,y)\phi_\varepsilon \right|$$

$$\leqslant C \varepsilon^{2/(N-2)} \|\phi_\varepsilon\|_\rho \int_{\Omega_\varepsilon} \Gamma(x-y) \left(\sum_i (1+|y-\xi_i'|^2)^{-(N-2)/2} \right)^{\beta-\rho}$$

$$\leqslant C \varepsilon^{2/(N-2)} \|\phi_\varepsilon\|_\rho \sum_i \int_{\Omega_\varepsilon} \frac{1}{|x-y|^{N-2}} \frac{1}{(1+|y-\xi_i'|)^{2-\rho}} \, dy$$

$$= C \varepsilon^{2/(N-2)} \|\phi_\varepsilon\|_\rho \sum_i \left(\int_{B(x,|x-\xi_i'|/2)} + \int_{B(\xi_i',|x-\xi_i'|/2)} \right.$$

$$\left. + \int_{\Omega_\varepsilon \backslash (B(x,|x-\xi_i'|/2) \cup B(\xi_i',|x-\xi_i'|/2))} \right)$$

$$= C \varepsilon^{2/(N-2)} \|\phi_\varepsilon\|_\rho (A+B+C).$$

Now, first we have

$$\varepsilon^{2/(N-2)} \|\phi_\varepsilon\|_\rho A \leqslant C \varepsilon^{2/(N-2)} \|\phi_\varepsilon\|_\rho \sum_i \frac{1}{(1+|y-\xi_i'|)^{2-\rho}} \int_0^{|x-\xi_i'|/2} t \, dt$$

$$\leqslant C \|\phi_\varepsilon\|_\rho \sum_i \frac{1}{(1+|y-\xi_i'|)^{2-\rho}}.$$

Second,

$$\varepsilon^{2/(N-2)} \|\phi_\varepsilon\|_\rho B \leqslant C \varepsilon^{2/(N-2)} \|\phi_\varepsilon\|_\rho \sum_i \frac{1}{(1+|y-\xi_i'|)^{N-2}} \int_1^{|x-\xi_i'|/2} t^{N-3+\rho} \, dt$$

$$\leqslant C \|\phi_\varepsilon\|_\rho \sum_i \frac{\varepsilon^{2/(N-2)} |x-\xi_i'|^{N-2+\rho}}{(1+|x-\xi_i'|)^{N-2}}$$

$$\leqslant C \|\phi_\varepsilon\|_\rho \sum_i \frac{1}{(1+|y-\xi_i'|)^{2-\rho}}.$$

Taking into account that the integral C can be estimated by the integral B, we conclude that, for $N \geqslant 4$,

$$\mu \varepsilon^{2/(N-2)} \int_{\Omega_\varepsilon} \left| G_\varepsilon(x,y)\phi_\varepsilon \right| \leqslant C \|\phi_\varepsilon\|_\rho \omega^{\beta-\rho}(x).$$

If $N = 3$, then one has

$$\mu\varepsilon^2 \int_{\Omega_\varepsilon} |G_\varepsilon(x, y)\phi_\varepsilon| \leqslant C\|\phi_\varepsilon\|_\rho \omega^\beta(x).$$

Similarly, one gets

$$\mu\varepsilon^2 \int_{\Omega_\varepsilon} |\partial_{x_j} G_\varepsilon(x, y)\phi_\varepsilon| \leqslant \begin{cases} C\|\phi_\varepsilon\|_\rho \omega^{\beta-\rho+1/(N-2)}(x) & \text{if } N \geqslant 4, \\ C\|\phi_\varepsilon\|_\rho \omega^{\beta+1/(N-2)}(x) & \text{if } N = 3. \end{cases}$$

Equations (2.34) and (2.35) and the above estimates imply that

$$|\phi_\varepsilon(x)| \leqslant \begin{cases} C\big(\|\phi_\varepsilon\|_\rho + \|h_\varepsilon\|_{**}\big)\omega^{\beta-\rho}(x) & \text{if } N \geqslant 4, \\ C\big(\|\phi_\varepsilon\|_\rho + \|h_\varepsilon\|_{**}\big)\omega^\beta(x) & \text{if } N = 3 \end{cases} \tag{2.37}$$

and

$$|\partial_{x_j}\phi_\varepsilon(x)| \leqslant \begin{cases} C\big(\|\phi_\varepsilon\|_\rho + \|h_\varepsilon\|_{**}\big)\omega^{\beta-\rho+1/(N-2)}(x) & \text{if } N \geqslant 4, \\ C\big(\|\phi_\varepsilon\|_\rho + \|h_\varepsilon\|_{**}\big)\omega^{\beta+1}(x) & \text{if } N = 3. \end{cases} \tag{2.38}$$

In particular, we have that

$$\omega^{-(\beta-2\rho)}(x)|\phi_\varepsilon(x)| \leqslant C\omega^\rho(x).$$

Since ρ is arbitrarily small and $\|\phi_\varepsilon\|_\rho = 1$, it follows the existence of a radius $R > 0$ and a number $\gamma > 0$, both independent of ε such that $\|\phi_\varepsilon\|_{L^\infty(B_R(\xi_i'))} > \gamma$ for some i. Assume this happens for $i = 1$. Then local elliptic estimates and the bound (2.37) yield that, up to a subsequence, $\tilde{\phi}_\varepsilon(x) = \phi_\varepsilon(x - \xi_1')$ converges uniformly over compacts of $\mathbb{R}^N$ to a nontrivial solution $\tilde{\phi}$ of

$$\Delta\tilde{\phi} + p\overline{U}_{\Lambda,0}^{p-1}\tilde{\phi} = 0 \tag{2.39}$$

for some $\Lambda > 0$, which besides satisfies

$$|\tilde{\phi}(x)| \leqslant C|x|^{(2-N)(\beta-\rho)}. \tag{2.40}$$

Hence, for $N = 3$ we have

$$|\tilde{\phi}(x)| \leqslant C|x|^{2-N}.$$

Now, since $\tilde{\phi}$ satisfies (2.39) and estimate (2.40) holds, a bootstrap argument leads to

$$|\tilde{\phi}(x)| \leqslant C|x|^{2-N} \quad \text{for any } N > 3.$$

It is well known that this implies that $\tilde{\phi}$ is a linear combination of the functions $\frac{\partial \overline{U}_{\Lambda,0}}{\partial x_j}$, $x \cdot \nabla \overline{U}_{\Lambda,0} + (N-2)\overline{U}_{\Lambda,0}$, see for instance [78]. On the other hand, we recall that

$$\int_{\Omega_\varepsilon} \phi_\varepsilon V_i^{p-1} Z_{ij} = 0 \quad \text{for all } i, j.$$

By dominated convergence, this relation is easily seen to be preserved up to the limit, hence

$$\int_{\mathbb{R}^N} \tilde{\phi} \, \overline{U}_{\Lambda,0}^{p-1} \frac{\partial \overline{U}_{\Lambda,0}}{\partial x_j} = \int_{\mathbb{R}^N} \tilde{\phi} \, \overline{U}_{\Lambda,0}^{p-1} \left(x \cdot \nabla \overline{U}_{\Lambda,0} + (N-2)\overline{U}_{\Lambda,0} \right) = 0$$

for all j. Hence the only possibility is that $\tilde{\phi} \equiv 0$, which is a contradiction which yields the proof of $\|\phi_\varepsilon\|_\rho \to 0$. Finally, from estimate (2.37), we observe that

$$\|\phi_\varepsilon\|_* \leqslant C \left(\|h_\varepsilon\|_{**} + \|\phi_\varepsilon\|_\rho \right),$$

hence $\|\phi_\varepsilon\|_* \to 0$, and the proof is thus complete.

STEP 2. Now we are in a position to prove Proposition 2.1. To do this, let us consider the space

$$H = \left\{ \phi \in H_0^1(\Omega_\varepsilon) \; \middle| \; \int_{\Omega_\varepsilon} V_i^{p-1} Z_{ij} \phi = 0 \; \forall i, j \right\}$$

endowed with the usual inner product $[\phi, \psi] = \int_{\Omega_\varepsilon} \nabla\phi \nabla\psi$. Denote with $\langle f, g \rangle$ the inner product in $L^2(\Omega_\varepsilon)$, namely $\langle f, g \rangle = \int_{\Omega_\varepsilon} fg$ for any $f, g \in L^2(\Omega_\varepsilon)$. Problem (2.13) expressed in weak form is equivalent to that of finding a $\phi \in H$ such that

$$[\phi, \psi] = \left\langle \left(\mu\varepsilon^{2/(N-2)}\phi + (p+\varepsilon)V^{p+\varepsilon-1}\phi - h \right), \psi \right\rangle \quad \forall \psi \in H.$$

With the aid of Riesz's representation theorem, this equation gets rewritten in H in the operational form

$$\phi = K(\phi) + \tilde{h} \tag{2.41}$$

with certain $\tilde{h} \in H$ which depends linearly in h and where K is a compact operator in H since $\mu < \mu_1$. Fredholm's alternative guarantees unique solvability of this problem for any h provided that the homogeneous equation

$$\phi = K(\phi)$$

has only the zero solution in H. Let us observe that this last equation is equivalent to

$$\begin{cases} \Delta\phi + \mu\varepsilon^{2/(N-2)}\phi + (p+\varepsilon)V^{p-1+\varepsilon}\phi = \sum_{i,j} c_{ij} V_i^{p-1} Z_{ij} & \text{in } \Omega_\varepsilon, \\ \phi = 0 & \text{on } \partial\Omega_\varepsilon, \\ \langle \phi, V_i^{p-1} Z_{ij} \rangle = 0 \end{cases} \tag{2.42}$$

for certain constants c_{ij}. Assume it has a nontrivial solution $\phi = \phi_\varepsilon$, which with no loss of generality may be taken so that $\|\phi_\varepsilon\|_* = 1$. But this makes the previous step applicable, so that necessarily $\|\phi_\varepsilon\|_* \to 0$. This is certainly a contradiction that proves that this equation only has the trivial solution in H. We conclude then that for each h, problem (2.13) admits a unique solution. We check that

$$\|\phi\|_* \leqslant C\|h\|_{**}.$$

We assume again the opposite. In doing so, we find a sequence h_ε with $\|h_\varepsilon\|_{**} = o(1)$ and solutions $\phi_\varepsilon \in H$ of problem (2.13) with $\|\phi_\varepsilon\|_* = 1$. Again this makes the previous step applicable, and a contradiction has been found. This proves estimate (2.27). Estimate (2.28) follows from this and relation (2.33). This concludes the proof of the proposition. $\qquad\square$

It is important for later purposes to understand the differentiability of the operator T with respect to the variables $\xi' \in \Omega_\varepsilon^k$, $\lambda \in \mathbb{R}_+^k$ which satisfy constraints (2.17). Consider the L_*^∞ (resp. L_{**}^∞) of functions defined on Ω_ε with finite $\|\cdot\|_*$ norm (resp. $\|\cdot\|_{**}$ norm). We consider the map

$$\left(\xi', \lambda, h\right) \mapsto S\left(\xi', \lambda, h\right) \equiv T(h), \tag{2.43}$$

as a map with values in $L_*^\infty \cap H_0^1(\Omega_\varepsilon)$. We have the following result:

PROPOSITION 2.2. *Under the conditions of Proposition* 2.1, *the map S is of class C^1. Besides, we have*

$$\left\|\nabla_{\xi',\lambda} S\left(\xi', \lambda, h\right)\right\|_* \leqslant C\|h\|_{**}.$$

PROOF. Let us consider differentiation with respect to the variable ξ'_{hl}, $h = 1, \ldots, k$, $l = 1, \ldots, N$. For notational simplicity we write $\frac{\partial}{\partial \xi'_{ij}} = \partial_{\xi'}$. Let us set, $\phi = S(\xi', \lambda, h)$ and, still formally, $Z = \partial_{\xi'}\phi$. We seek for an expression for Z. Then Z satisfies the following equation:

$$\Delta Z + \mu \varepsilon^{2/(N-2)} Z + (p + \varepsilon) V^{p+\varepsilon-1} Z$$
$$= -(p + \varepsilon)\partial_{\xi'}\left(V^{p-1+\varepsilon}\right)\phi + \sum_{i,j} d_{ij} V_i^{p-1} Z_{ij} + c_{ij}\partial_{\xi'}\left(V_i^{p-1} Z_{ij}\right) \quad \text{in } \Omega_\varepsilon.$$

Here $d_{ij} = \partial_{\xi'} c_{ij}$. Besides, from differentiating the orthogonality condition $\langle \phi, V_i^{p-1} Z_{ij}\rangle = 0$ we further obtain the relations

$$\left\langle \phi, \partial_{\xi'}\left(V_i^{p-1} Z_{ij}\right)\right\rangle + \left\langle Z, V_i^{p-1} Z_{ij}\right\rangle = 0.$$

Let us consider constants b_{ij} such that

$$\left\langle Z - \sum_{l,h} b_{lh} Z_{lh}, V_i^{p-1} Z_{ij}\right\rangle = 0.$$

These relations amount to

$$\sum_{l,h} b_{lh}\langle Z_{lh}, V_i^{p-1} Z_{ij}\rangle = \langle \phi, \partial_{\xi'} V_i^{p-1} Z_{ij}\rangle. \tag{2.44}$$

Since this system is diagonal dominant with uniformly bounded coefficients, we see that it is uniquely solvable and that

$$b_{lh} = O(\|\phi\|_*)$$

uniformly on ξ', λ in the considered region. Now, we easily see that

$$\left\|\phi \partial_{\xi'}\left(V^{p-1+\varepsilon}\right)\right\|_{**} \leqslant C\|\phi\|_*.$$

Recall now that, from Proposition 2.1, $c_{ij} = O(\|h\|_{**})$. On the other hand,

$$\left|\partial_{\xi'}\left(V_i^{p-1} Z_{ij}(x)\right)\right| \leqslant C\left|x - \xi_i'\right|^{-N-4},$$

hence

$$\left\|c_{ij}\partial_{\xi'} V_i^{p-1} Z_{ij}\right\|_{**} \leqslant C\|h\|_{**}.$$

Let us now set $\eta = Z - \sum_{i,j} b_{ij} Z_{ij}$. Then, summing up the estimates above and using that $\|\phi\|_* \leqslant C\|h\|_{**}$, we get that η satisfies the relation

$$\Delta\eta + \mu\varepsilon^{2/(N-2)}\eta + (p+\varepsilon)V^{p-1+\varepsilon}\eta = f + \sum_{i,j} d_{ij} V_i^{p-1} Z_{ij} \quad \text{in } \Omega_\varepsilon, \tag{2.45}$$

where

$$f = \sum_{i,j} b_{ij}\left(-\left(\Delta + \mu\varepsilon^{2/(N-2)} + (p+\varepsilon)V^{p-1+\varepsilon}\right)\right)Z_{ij} + c_{ij}\partial_{\xi'}\left(V_i^{p-1} Z_{ij}\right)$$

$$- (p+\varepsilon)\partial_{\xi'}\left(V^{p-1+\varepsilon}\right)\phi, \tag{2.46}$$

so that

$$\|f\|_{**} \leqslant C\|h\|_{**}.$$

Since besides $\eta \in H_0^1(\Omega_\varepsilon)$ and

$$\langle \eta, V_i^{p-1} Z_{ij}\rangle = 0 \quad \text{for all } i, j, \tag{2.47}$$

we have that $\eta = T(f)$. Reciprocally, if we now define

$$Z = T(f) + \sum_{i,j} b_{ij} Z_{ij}$$

with b_{ij} given by relations (2.44) and f by (2.46), then it is a matter of routine to check that indeed $Z = \partial_{\xi'}\phi$. In fact, Z depends continuously on the parameters ξ', Λ and h for the norm $\|\cdot\|_*$, and $\|Z\|_* \leqslant C\|h\|_{**}$ for points in the considered region. The corresponding result for differentiation with respect to the λ_i's follow similarly. This concludes the proof. $\qquad\square$

REMARK 2.2. We can also state the above result by saying that the map $(\xi', \lambda) \mapsto T$ is of class C^1 in $\mathcal{L}(L^\infty_{**}, L^\infty_*)$ and, for instance,

$$(D_{\xi'}T)(h) = T(f) + \sum_{i,j} b_{ij} Z_{ij}, \tag{2.48}$$

where f is given by (2.46) and b_{ij} by (2.44).

Let us now go back to problem (2.2) and consider $\mu = 0$. Next step in the proof of Theorem 1.1 is the finite-dimensional reduction: we consider the nonlinear problem of finding a function ϕ such that for some constants c_{ij} the following equation holds

$$\begin{cases} \Delta(V + \phi) + (V + \phi)_+^{p+\varepsilon} = \sum_{i,j} c_{ij} V_i^{p-1} Z_{ij} & \text{in } \Omega_\varepsilon, \\ \phi = 0 & \text{on } \partial\Omega_\varepsilon, \\ \int_{\Omega_\varepsilon} \phi V_i^{p-1} Z_{ij} = 0 & \text{for all } i, j. \end{cases} \tag{2.49}$$

Let us rewrite the first equation in (2.49) in the following form

$$\Delta\phi + (p + \varepsilon)V^{p+\varepsilon-1}\phi = -R_\varepsilon - N_\varepsilon(\phi) + \sum_{i,j} c_{ij} V_i^{p-1} Z_{ij} \quad \text{in } \Omega_\varepsilon,$$

where, in this case, R_ε and $N_\varepsilon(\phi)$ are defined respectively by (2.11) and (2.12).

To estimate the $\|\cdot\|_{**}$-norm of $N_\varepsilon(\eta)$, it is convenient, and sufficient for our purposes, to assume $\|\eta\|_* < 1$. Note that

$$N_\varepsilon(\eta) = \frac{(p + \varepsilon)(p - 1 + \varepsilon)}{2}(V_1 + V_2 + t\eta)^{p-2+\varepsilon}\eta^2 \tag{2.50}$$

with $t \in (0, 1)$. If $N \leqslant 6$ then $p \geqslant 2$, and we can estimate

$$\left| \overline{V}^{-4/(N-2)} N_\varepsilon(\eta) \right| \leqslant C\overline{V}^{(p-2)\beta - 4/(N-2) + 2\beta} \|\eta\|_*^2,$$

hence

$$\left\| N_\varepsilon(\eta) \right\|_{**} \leqslant C\|\eta\|_*^2.$$

Assume now that $N > 6$. In the region where $\mathrm{dist}(y, \partial\Omega_\varepsilon) \geqslant \delta\varepsilon^{-1/(N-2)}$ for some $\delta > 0$, then $V(y) \geqslant \alpha_\delta \overline{V}(y)$ for some $\alpha_\delta > 0$; hence in this region, we have

$$\left| \overline{V}^{-4/(N-2)} N_\varepsilon(\eta) \right| \leqslant C\overline{V}^{2\beta-1} \|\eta\|_*^2 \leqslant C\varepsilon^{2\beta-1} \|\eta\|_*^2.$$

On the other hand, when $\mathrm{dist}(y, \partial\Omega_\varepsilon) \leqslant \delta\varepsilon^{-1/(N-2)}$, the following facts occur: $\overline{V}(y)$, $V(y) = O(\varepsilon)$ and, as $y \to \partial\Omega_\varepsilon$, $V(y) = C\varepsilon^{(N-1)/(N-2)}\,\mathrm{dist}(y, \partial\Omega_\varepsilon) + o(\varepsilon)$. This second assertion is a consequence of the fact that

$$V(y) = \varepsilon^{1/2} \sum_{j=1}^{2} U_{\Lambda_j \varepsilon^{1/(N-2)}, \xi_j}\left(\varepsilon^{1/(N-2)} y\right)$$

and hence, taking into account that the Green function of the domain Ω vanishes linearly with respect to $\mathrm{dist}(x, \partial\Omega)$ as $x \to \partial\Omega$,

$$V(y) = C\varepsilon\,\mathrm{dist}\left(\varepsilon^{1/(N-2)} y, \partial\Omega\right) + o(\varepsilon) = \varepsilon^{(N-1)/(N-2)}\,\mathrm{dist}(y, \partial\Omega_\varepsilon) + o(\varepsilon).$$

These facts imply that, if $\mathrm{dist}(y, \partial\Omega_\varepsilon) \leqslant \delta\varepsilon^{-1/(N-2)}$, then

$$\left|\overline{V}^{-4/(N-2)} N_\varepsilon(\eta)\right|$$
$$\leqslant \overline{V}^{-4/(N-2)} V^{p-2}\|\eta\|^2$$
$$\leqslant C\overline{V}^{-4/(N-2)}\left(\varepsilon^{(N-1)/(N-2)}\,\mathrm{dist}(y, \partial\Omega_\varepsilon)\right)^{p-2}\mathrm{dist}(y, \partial\Omega_\varepsilon)^2\left\|D\eta(y)\right\|^2$$
$$\leqslant C\varepsilon^{-4/(N-2)+((N-1)/(N-2))(p-2)-p/(N-2)+2\beta+2/(N-2)}\|\eta\|_*^2 \leqslant C\varepsilon^{2\beta-1}\|\eta\|_*^2.$$

If $|\eta| \leqslant \frac{1}{2}\overline{V}$ then relation (2.50) yields that

$$\left|\overline{V}^{-4/(N-2)} N_\varepsilon(\eta)\right| \leqslant C\overline{V}^{2\beta-1}\|\eta\|_*^2 \leqslant C\varepsilon^{2\beta-1}\|\eta\|_*^2.$$

In the other case, we see directly from (2.50) that $|N_\varepsilon(\eta)| \leqslant C|\eta|^p$ and hence

$$\left|\overline{V}^{-4/(N-2)} N_\varepsilon(\eta)\right| \leqslant \overline{V}^{p\beta-4/(N-2)}\|\eta\|_*^p \leqslant C\varepsilon^{-(2-p)\beta}\|\eta\|_*^p.$$

Combining these relations we get

$$\|N_\varepsilon(\eta)\|_{**} \leqslant \begin{cases} C\|\eta\|_*^2 & \text{if } N \leqslant 6, \\ C\left(\varepsilon^{2\beta-1}\|\eta\|_*^2 + \varepsilon^{-(2-p)\beta}\|\eta\|_*^p\right) & \text{if } N > 6. \end{cases} \tag{2.51}$$

Next we estimate the term R_ε. We have

$$|R_\varepsilon| \leqslant \left|\overline{V}_i^{p+\varepsilon} - \overline{V}_i^p\right| + o\left(\varepsilon^{(N+2)/(N-2)}\right) \leqslant \varepsilon C\overline{V}_i^p\left|\log\overline{V}_i\right|(x) + o\left(\varepsilon^{(N+2)/(N-2)}\right)$$

in the regions where $|x - \xi_i'| \leqslant \bar{\delta}\varepsilon^{-1/(N-2)}$, for small $\bar{\delta} > 0$. Taking into account that $|R_\varepsilon| \leqslant C\varepsilon^{(N+2)/(N-2)}$ in the complement of these two regions, we get

$$\|R_\varepsilon\|_{**} \leqslant C\varepsilon. \tag{2.52}$$

PROPOSITION 2.3. *Assume the conditions of Proposition 2.1 are satisfied. Then there is a $C > 0$, such that for all small ε there exists a unique solution*

$$\phi = \phi(\xi', \lambda) = \tilde{\phi} + \psi$$

to problem (2.49) with ψ defined by

$$\psi = -T(R_\varepsilon). \tag{2.53}$$

Besides, the map $(\xi', \lambda) \to \phi(\xi', \lambda)$ is of class C^1 for $\|\cdot\|_*$-norm and

$$\|\phi\|_* \leqslant C\varepsilon, \tag{2.54}$$

$$\|\nabla_{(\xi', \lambda)}\phi\|_* \leqslant C\varepsilon. \tag{2.55}$$

PROOF. Problem (2.49) is equivalent to solving a fixed point problem. Indeed $\phi = \tilde{\phi} + \psi$ is a solution of (2.49) if and only if

$$\tilde{\phi} = -T\left(N_\varepsilon(\tilde{\phi} + \psi)\right) \equiv A_\varepsilon(\tilde{\phi})$$

taking into account that $\psi = -T(R_\varepsilon)$.

Then we need to prove that the operator A_ε defined above is a contraction inside a properly chosen region.

First observe that, from the definition of ψ, from (2.52) and from Proposition 2.1, we infer that

$$\|\psi\|_{**} \leqslant C\varepsilon$$

and, by (2.51), for $\|\eta\|_* \leqslant 1$,

$$\left\|N_\varepsilon(\psi + \eta)\right\|_{**}$$
$$\leqslant \begin{cases} C\left(\|\eta\|_*^2 + \varepsilon^2\right) & \text{if } N \leqslant 6, \\ C\left(\varepsilon^{2\beta-1}\|\eta\|_*^2 + \varepsilon^{-(2-p)\beta}\|\eta\|_*^p + \varepsilon^{p\beta+1}\right) & \text{if } N > 6. \end{cases} \tag{2.56}$$

Let us set

$$\mathcal{F} = \left\{\eta \in H_0^1 : \|\eta\|_* \leqslant \varepsilon\right\}.$$

From Proposition 2.1 and (2.56) we conclude that, for ε sufficiently small and any $\eta \in \mathcal{F}_r$, we have

$$\left\|A_\varepsilon(\eta)\right\|_* = \left\|L_\varepsilon\left(N_\varepsilon(\eta + \psi)\right)\right\|_* \leqslant C\left\|N_\varepsilon(\eta + \psi)\right\|_{**}$$
$$\leqslant \begin{cases} C\varepsilon^2 \leqslant \varepsilon & \text{if } N \leqslant 6, \\ C\left(\varepsilon^{2\beta+1} + \varepsilon^{p\beta+1}\right) \leqslant \varepsilon & \text{if } N > 6, \end{cases}$$

where the last inequality holds provided that ε is sufficiently small. Now we will show that the map A_ε is a contraction, for any ε small enough. That will imply that A_ε has a unique fixed point in $\mathcal{F}$ and hence problem (2.49) has a unique solution.

For any η_1, η_2 in $\mathcal{F}_r$ we have

$$\big\| A_\varepsilon(\eta_1) - A_\varepsilon(\eta_2) \big\|_* \leqslant C \big\| N_\varepsilon(\psi + \eta_1) - N_\varepsilon(\psi + \eta_2) \big\|_{**},$$

hence we just need to check that N_ε is a contraction in its corresponding norms. By definition of N_ε,

$$D_{\bar{\eta}} N_\varepsilon(\bar{\eta}) = (p + \varepsilon)\big[(V + \bar{\eta})_+^{p+\varepsilon-1} - V^{p+\varepsilon-1}\big].$$

Hence we get

$$\big| N_\varepsilon(\psi + \eta_1) - N_\varepsilon(\psi + \eta_2) \big| \leqslant C \overline{V}^{p-2} |\bar{\eta}| |\eta_1 - \eta_2|$$

for some $\bar{\eta}$ in the segment joining $\psi + \eta_1$ and $\psi + \eta_2$. Hence, we get for small enough $\|\bar{\eta}\|_*$,

$$\overline{V}^{-4/(N-2)} \big| N_\varepsilon(\psi + \eta_1) - N_\varepsilon(\psi + \eta_2) \big| \leqslant C \overline{V}^{2\beta-1} \|\bar{\eta}\|_* \|\eta_1 - \eta_2\|_*.$$

We conclude

$$\big\| N_\varepsilon(\psi + \eta_1) - N_\varepsilon(\psi + \eta_2) \big\|_{**}$$
$$\leqslant \begin{cases} \overline{V}^{2\beta-1}\big(\|\eta_1\|_* + \|\eta_2\|_* + \|\psi\|_*\big) \|\eta_1 - \eta_2\|_* & \text{if } N \leqslant 6, \\ \varepsilon^{2\beta-1}\big(\|\eta_1\|_* + \|\eta_2\|_* + \|\psi\|_*\big)\|\eta_1 - \eta_2\|_* \\ \qquad \leqslant \varepsilon^{\min\{2\beta,1\}} \|\eta_1 - \eta_2\|_* & \text{if } N > 6, \end{cases}$$

and hence A_ε is a contraction mapping for the $\|\cdot\|_*$-norm inside $\mathcal{F}_r$.

Let us now analyze the differentiability properties of the function $\phi(\xi', \lambda)$.

We recall that ϕ is defined through the relation

$$B\big(\xi', \lambda, \phi\big) \equiv \phi + T\big(N_\varepsilon(\phi + \psi)\big) = 0.$$

Write $N(\xi', \lambda, \bar{\phi}) = N_\varepsilon(\bar{\phi})$, namely

$$N\big(\xi', \lambda, \bar{\phi}\big) = (V + \bar{\phi})_+^{p+\varepsilon} - V^{p+\varepsilon} - (p + \varepsilon)V^{p+\varepsilon-1}\bar{\phi}.$$

Then

$$D_{\bar{\phi}} N\big(\xi', \lambda, \bar{\phi}\big) = (p + \varepsilon)\big[(V + \bar{\phi})_+^{p+\varepsilon-1} - V^{p+\varepsilon-1}\big]$$

and

$$D_{\xi'} N\big(\xi', \lambda, \bar{\phi}\big)$$
$$= (p + \varepsilon)\big[(V + \bar{\phi})_+^{p+\varepsilon-1} - V^{p+\varepsilon-1} - (p + \varepsilon - 1)V^{p+\varepsilon-2}\bar{\phi}\big]D_{\xi'}V, \quad (2.57)$$

similarly for $D_\lambda N(\xi', \lambda, \bar{\phi})$. We have that

$$D_\phi B\big(\xi', \lambda, \phi\big)[\theta] = \theta + T\big(\theta D_{\bar{\phi}} N_\varepsilon(\phi + \psi)\big) \equiv \theta + M(\theta).$$

Now,

$$\big\| M(\theta) \big\|_* \leqslant C \big\| \theta D_{\bar{\phi}} N_\varepsilon(\phi + \psi) \big\|_{**} \leqslant C \big\| V^{-4/(N-2)+\beta} D_{\bar{\phi}} N_\varepsilon(\phi + \psi) \big\|_\infty \| \theta \|_*.$$

We have

$$\overline{V}^{-4/(N-2)+\beta} \big| D_{\bar{\phi}} N_\varepsilon(\phi + \psi) \big|$$

$$\leqslant C \begin{cases} \overline{V}^{2\beta - 1} \| \phi + \psi \|_* & \text{if } N \leqslant 6, \\ \varepsilon^{2\beta - 1} \| \phi + \psi \|_* & \text{if } N > 6, \end{cases}$$

$$\leqslant C \varepsilon^{\min\{2\beta, 1\}}.$$

It follows that for small ε, the linear operator $D_\phi B(\xi', \lambda, \phi)$ is invertible in L_*^∞, with uniformly bounded inverse. It also depends continuously on its parameters.

Now, let us consider differentiability with respect to the (ξ', λ) variables. We have

$$D_{\xi'} B\big(\xi', \lambda, \phi\big)$$
$$= (D_{\xi'} L)\big(N_\varepsilon(\phi + \psi)\big)$$
$$+ \Big[L\big((D_{\xi'} N)\big(\xi', \lambda, \phi + \psi\big)\big) + L\big((D_{\bar{\phi}} N)\big(\xi', \lambda, \phi + \psi\big) D_{\xi'} \psi\big) \Big].$$

Here $D_{\xi'} L$ is the operator defined by the expression (2.48) and the second quantity by (2.57). Observe also that

$$D_{\xi'} \psi = (D_{\xi'} L)(R_\varepsilon) + L(D_{\xi'} R_\varepsilon). \tag{2.58}$$

Also

$$D_{\xi_1'} R_\varepsilon = (p + \varepsilon) V^{p+\varepsilon-1} D_{\xi_1'} V_1 - p \overline{V}_1^{p-1} D_{\xi_1'} \overline{V}_1. \tag{2.59}$$

These expressions also depend continuously on their parameters. We have a similar expression for the derivative with respect to Λ.

The implicit function theorem then applies to yield that $\phi(\xi', \lambda)$ indeed defines a C^1 function into L_*^∞. Moreover, we have for instance

$$D_{\xi'} \phi = -\big(D_\phi B(\xi, \lambda, \phi)\big)^{-1} \Big[(D_{\xi'} L)\big(N_\varepsilon(\phi + \psi)\big)$$
$$+ \Big[L\big(D_{\xi'}\big[N\big(\xi', \lambda, \phi + \psi\big)\big]\big)$$
$$+ L\big((D_{\bar{\phi}} N)\big(\xi', \lambda, \phi + \psi\big) D_{\xi'} \psi\big) \Big] \Big].$$

Hence,

$$\|D_{\xi'}\phi\|_* \leq C\big(\|N_\varepsilon(\phi+\psi)\|_{**}$$
$$+ \|D_{\xi'}N(\xi',\lambda,\phi+\psi)\|_{**} + \|D_{\bar\phi}N(\xi',\lambda,\psi+\phi)D_{\xi'}\psi\|_{**}\big),$$

where we have used Remark 2.2. From (2.56), we get

$$\|N_\varepsilon(\phi+\psi)\|_{**} \leq \begin{cases} C\varepsilon^2 & \text{if } N \leq 6, \\ C\varepsilon^{p\beta+1} & \text{if } N > 6. \end{cases} \tag{2.60}$$

On the other hand, from (2.57) we have

$$\big|(D_{\xi'}N)(\xi',\lambda,\bar\phi)\big|$$
$$\leq C\overline{V}^{(N-1)/(N-2)}\big|(V+\bar\phi)_+^{p+\varepsilon-1} - V^{p+\varepsilon-1} - (p+\varepsilon-1)V^{p+\varepsilon-2}\bar\phi\big|$$
$$\leq C \begin{cases} \overline{V}^{5/(N-2)+\varepsilon+\beta}\|\bar\phi\|_* & \text{if } N \leq 6, \\ \varepsilon^{5/(N-2)+\varepsilon+\beta}\|\bar\phi\|_* & \text{if } N > 6. \end{cases}$$

Hence,

$$\big\|(D_{\xi'}N)(\xi',\lambda,\psi+\phi)\big\|_{**} \leq C\|\phi+\psi\|_* \leq C\varepsilon.$$

In similar way we get that

$$\big\|D_{\bar\phi}N(\xi',\lambda,\psi+\phi)D_\xi\psi\big\|_{**} \leq C\varepsilon.$$

Hence, we finally get

$$\|D_{\xi'}\phi\|_* \leq C\varepsilon,$$

as desired. A similar estimate holds for differentiation with respect to the λ_i's. This concludes the proof. $\qquad\square$

For what we are going to do next, it is more convenient to recast the parameter λ_i into the parameters Λ_i given by

$$\lambda_i^{N-2} = a_N \Lambda_i^2 \tag{2.61}$$

with

$$a_N = \frac{1}{p+1}\frac{\int_{\mathbb{R}^N}\overline{U}^{p+1}}{(\int_{\mathbb{R}^N}\overline{U}^p)^2}. \tag{2.62}$$

With this in mind, let us consider points (ξ,Λ) which satisfy constraints

$$|\xi_i - \xi_j| > \delta, \qquad \mathrm{dist}(\xi_i,\partial\Omega_\varepsilon) > \delta, \qquad \delta < \Lambda_i < \delta^{-1},$$

for some small fixed $\delta > 0$. Let $\phi(y) = \phi(\xi', \Lambda)(y)$ be the unique solution of problem

$$\begin{cases} \Delta(V + \phi) + (V + \phi)_+^{p+\varepsilon} = \sum_{i,j} c_{ij} V_i^{p-1} Z_{ij} & \text{in } \Omega_\varepsilon, \\ \phi = 0 & \text{on } \partial\Omega_\varepsilon, \\ \int_{\Omega_\varepsilon} \phi V_i^{p-1} Z_{ij} = 0 & \text{for all } i, j \end{cases} \tag{2.63}$$

given by Proposition 2.3. Let us consider the functional

$$J(\xi, \Lambda) = I_\varepsilon(V + \phi), \tag{2.64}$$

where I_ε was defined in (2.15). The definition of ϕ yields that

$$I_\varepsilon'(V + \phi)[\eta] = 0$$

for all η which vanishes on $\partial\Omega_\varepsilon$ and such that

$$\int_{\Omega_\varepsilon} \eta V_i^{p-1} Z_{ij} = 0 \quad \text{for all } i, j.$$

It is easy to check that

$$\frac{\partial V}{\partial \xi_{ij}} = Z_{ij} + o(1), \quad \frac{\partial V}{\partial \Lambda_i} = Z_{i(N+1)} + o(1),$$

with o(1) small as $\varepsilon \to 0$. This fact, together with the last part of Proposition 2.3, give the validity of the following lemma.

LEMMA 2.2. $v = V + \phi$ *is a solution of problem* (2.2), *namely* $c_{ij} = 0$ *in* (2.63) *for all* i, j, *if and only if* (ξ, Λ) *is a critical point of* J.

Next step is then to give an asymptotic estimate for $J(\xi, \Lambda)$. We see next that this functional and $I_\varepsilon(V)$ coincide up to order $o(\varepsilon)$.

PROPOSITION 2.4. *We have the expansion,*

$$J(\xi, \Lambda) = kC_N + \varepsilon\left[\bar{\gamma}_N + w_N \Psi_k(\xi, \Lambda) + o(1)\right], \tag{2.65}$$

where $o(1) \to 0$ *as* $\varepsilon \to 0$ *in the uniform* C^1*-sense with respect to* (ξ, Λ) *satisfying* (2.17). *Here,*

$$\Psi_k(\xi, \Lambda) = \frac{1}{2}\left\{ \sum_{j=1}^k H(\xi_j, \xi_j)\Lambda_j^2 - 2\sum_{i<j} G(\xi_i, \xi_j)\Lambda_i\Lambda_j \right\} + \log(\Lambda_1 \cdots \Lambda_h), \tag{2.66}$$

and the constants ω_N and C_N are those in Lemma 2.1 and

$$\bar{\gamma}_N = \left\{ \frac{k}{p+1}\omega_N + \frac{k}{2}\omega_N \log a_N - \frac{k}{p+1}\int_{\mathbb{R}^N} \overline{U}^{p+1}\log\overline{U}\right\}$$

with a_N given by (2.62).

PROOF. We start showing that

$$J(\xi,\Lambda) - I_\varepsilon(V) = o(\varepsilon) \tag{2.67}$$

and

$$\nabla_{\xi,\Lambda}\big[J(\xi,\Lambda) - I_\varepsilon(V)\big] = o(\varepsilon). \tag{2.68}$$

Taking into account that $0 = DI_\varepsilon(V + \psi + \tilde{\phi})[\tilde{\phi}]$, a Taylor expansion gives

$$I_\varepsilon(V+\psi) - I(\xi,\Lambda) = \int_0^1 t\,dt\, D^2 I_\varepsilon(V+\psi+t\tilde{\phi})[\tilde{\phi},\tilde{\phi}],$$

$$\int_0^1 t\,dt\left[\int_{\Omega_\varepsilon}|\nabla\tilde{\phi}|^2 - (p+\varepsilon)(V+\psi+t\tilde{\phi})^{p+\varepsilon-1}\tilde{\phi}^2\right]$$

$$= \int_0^1 t\,dt\left(\int_{\Omega_\varepsilon} N_\varepsilon(\tilde{\phi}+\psi)\tilde{\phi}\right.$$

$$\left. + \int_{\Omega_\varepsilon}(p+\varepsilon)\big[V^{p+\varepsilon-1} - (V+\psi+t\tilde{\phi})^{p+\varepsilon-1}\big]\tilde{\phi}^2\right). \tag{2.69}$$

Since $\|\tilde{\phi}\|_* = O(\varepsilon)$, we get

$$J(\xi,\Lambda) - I_\varepsilon(V+\psi) = O(\varepsilon^2). \tag{2.70}$$

Differentiating with respect to ξ variables we get form (2.69) that

$$D_\xi\big[I_\varepsilon(V+\psi) - J(\xi,\Lambda)\big]$$

$$= \varepsilon^{-1/(N-2)}\int_0^1 t\,dt\left(\int_{\Omega_\varepsilon} D_{\xi'}\big[(N_\varepsilon(\tilde{\phi}+\psi))\tilde{\phi}\big]\right.$$

$$+ (p+\varepsilon)\int_{\Omega_\varepsilon} D_{\xi'}\big[((V+\psi+t\tilde{\phi})^{p+\varepsilon-1}$$

$$\left. - (V+\psi)^{p+\varepsilon-1})\tilde{\phi}^2\big]\right). \tag{2.71}$$

Using the computations in the proof of Proposition 2.3 we get that the first integral in relation (2.71) can be estimated by $O(\varepsilon^2)$, so does the second; hence

$$D_\xi\big[J(\xi,\Lambda) - \mathcal{I}_\varepsilon(V+\psi)\big] = O\big(\varepsilon^{2-1/(N-2)}\big).$$

Now, since $D\mathcal{I}_\varepsilon(V)[\psi] = \int_{\Omega_\varepsilon} R_\varepsilon \psi$,

$$\mathcal{I}_\varepsilon(V+\psi) - \mathcal{I}_\varepsilon(V)$$

$$= \left\{\int_0^1 (1-t)\,dt\left[(p+\varepsilon)\int_{\Omega_\varepsilon}\big((V+t\psi)^{p+\varepsilon-1} - V^{p+\varepsilon-1}\big)\psi^2\right]\right.$$

$$\left. - 2\int_{\Omega_\varepsilon} R_\varepsilon\psi\right\}. \tag{2.72}$$

Since $\|\psi\|_* + \|R_\varepsilon\|_{**} = O(\varepsilon)$, the above term is $O(\varepsilon^2)$; then (2.67) follows from (2.70) and (2.72). Using again (2.72), we see that

$$D_\xi\big[\mathcal{I}_\varepsilon(V+\psi) - \mathcal{I}_\varepsilon(V)\big]$$

$$= \varepsilon^{-1/(N-2)}D_{\xi'}\left\{\int_0^1 (1-t)\,dt\left[(p+\varepsilon)\int_{\Omega_\varepsilon}\big((V+t\psi)^{p+\varepsilon-1} - V^{p+\varepsilon-1}\big)\psi^2\right]\right.$$

$$\left. - 2\int_{\Omega_\varepsilon} R_\varepsilon\psi\right\}.$$

Since from Proposition 2.3 it follows that $\|D_{\xi'}\psi\|_* = O(\varepsilon)$, we get

$$D_\xi\big[\mathcal{I}_\varepsilon(V+\psi) - \mathcal{I}_\varepsilon(V)\big]$$

$$= O\big(\varepsilon^2\big) - 2\varepsilon^{-1/(N-2)}D_{\xi'}\left(\int_{\Omega_\varepsilon} R_\varepsilon\psi\right).$$

The desired result will follow if we prove that

$$\varepsilon^{-1/(N-2)}D_{\xi'}\left(\int_{\Omega_\varepsilon} R_\varepsilon\psi\right) = o(\varepsilon). \tag{2.73}$$

First, if $N > 3$, Proposition 2.3 yields

$$\varepsilon^{-1/(N-2)}D_{\xi'}\left(\int_{\Omega_\varepsilon} R_\varepsilon\psi\right) = \begin{cases} O\big(\varepsilon^{2-1/(N-2)}\big) & \text{if } N = 4, 5, \\ O\big(\varepsilon^{7/4}|\log\varepsilon|\big) & \text{if } N = 6, \\ O\big(\varepsilon^{(N+1)/(N-2)}\big) & \text{if } N \geqslant 7. \end{cases}$$

Let us consider now the case $N = 3$. We have that

$$D_{\xi'_1}\left(\int_{\Omega_\varepsilon} R_\varepsilon\psi\right) = \int_{\Omega_\varepsilon} (D_{\xi'_1} R_\varepsilon)\psi + \int_{\Omega_\varepsilon} (D_{\xi'_1}\psi) R_\varepsilon = \varepsilon^2(I + II).$$

Let us estimate first *II*. Our first observation is that, locally, around ξ_1',

$$\varepsilon^{-1} R_\varepsilon(\xi_1 + x) \to V_0^5 \log V_0 + c V_0^4$$

uniformly over compacts, for certain constant c. Here $V_0(|x|) = \overline{U}_{\lambda,0}$ for some $\lambda > 0$. We also set $Z_0 = x \cdot \nabla V_0 + V_0$. Hence, $\varepsilon^{-1} \psi(x + \xi_1) \to w(|x|)$ where w is the unique radial solution of

$$\Delta w + p V_0^4 w = V_0^5 \log V_0 + c V_0^4 + b V_0^4 Z_0$$

which goes to zero at ∞, and is such that

$$\int_{\mathbb{R}^N} V_0^4 Z_0 w = 0.$$

The constant b is precisely that making the integral of the right-hand side of the above equation against Z_0 equal to zero. In a similar way,

$$\varepsilon^{-1} D_{\xi_1} \psi(x + \xi_1) \to w'(|x|) \frac{x}{|x|}.$$

After a suitable application of dominated convergence, we get that

$$
\begin{aligned}
II &= \varepsilon^{-2} \int_{\Omega_\varepsilon} (D_{\xi_1} \psi) R_\varepsilon \\
&\to \int_{\mathbb{R}^N} \left(V_0^5 \log V_0 + c V_0^4 + b V_0^4 Z_0 \right) (|x|) w'(|x|) \frac{x}{|x|} = 0,
\end{aligned}
$$

by symmetry. The term *I* can we dealt with in a similar manner. We conclude that $I \to 0$. Hence relation (2.73) has been established, and this proves the result in what concerns to derivatives with respect to ξ. Derivatives with respect to Λ can be handle in a simpler way, since the term $\varepsilon^{-1/(N-2)}$ does not appear in the differentiation. The validity of (2.73) thus follows.

From Lemma 2.1, we can finally conclude that

$$J(\xi, \Lambda) = k C_N + \varepsilon \left[\bar{\gamma}_N + w_N \Psi_k(\xi, \Lambda) \right] + o(\varepsilon). \tag{2.74}$$

On the other hand, as a consequence of (2.68) and Remark 2.2 we also get

$$\nabla J(\xi, \Lambda) = \varepsilon \frac{1}{p+1} \left(\int_{\mathbb{R}^N} \overline{U}^{p+1} \right) \left(\nabla \Psi_k(\xi, \Lambda) + o(1) \right). \tag{2.75}$$

The proof is complete. $\qquad\square$

We are now ready to give the proof of Theorem 1.1.

PROOF OF THEOREM 1.1. For simplicity we give the proof of the theorem only for the supercritical case. As mentioned before, the subcritical case works exactly the same.

Lemma 2.2 guarantees that $v = V + \phi$, where V is given by (2.7) and ϕ is the solution given by Proposition 2.3, is a solution to problem (2.2) if and only if the point (ξ, Λ) is a critical point for $J(\xi, \Lambda) = I_\varepsilon(V + \phi)$ (see (2.64)). Hence we need to find a critical point for J, or equivalently, a critical point for

$$\widetilde{J}(\xi, \Lambda) = \omega_N \left[\varepsilon^{-1} \left(J(\xi, \Lambda) - kC_N \right) - \bar{\gamma}_N \right]$$

(see Proposition 2.4). Proposition 2.4 implies that

$$\widetilde{J}(\xi, \Lambda) - \Psi_k(\xi, \Lambda) = o(1), \tag{2.76}$$

where $o(1)$ is in C^1 sense as $\varepsilon \to 0$ and Ψ_k is given by (2.66).

The assumption of Theorem 1.1 is that Ψ_k has a nontrivial critical point situation in $\mathcal{D}$. Hence, taking into account (2.76), there exists an $\bar{\varepsilon} > 0$ such that, for any $\varepsilon \in (0, \bar{\varepsilon})$, there exists a critical point $(\xi_\varepsilon, \Lambda_\varepsilon)$ in $\mathcal{D}$ of $\widetilde{J}(\xi, \Lambda)$ such that $\nabla \Psi_k(\xi_\varepsilon, \Lambda_\varepsilon) \to 0$ as $\varepsilon \to 0$. The qualitative behavior of the solution predicted by Theorem 1.1 follows directly by construction from the definition of V and of ϕ. $\qquad \square$

REMARK 2.3. In the slightly subcritical problem (1.5), existence of bubbling solution is governed by existence of nontrivial critical point situations for Ψ_k^- (see (1.8)).

Here we want to be more precise with the examples of contractible domains described in the Introduction on which problem (1.5) admits respectively multiple-bubbling solutions at one point and bubbling solutions at multiple points. These examples are contained in [65].

The first example is the following: take $\Omega_0 = \Omega_1 \cup \Omega_2$, where Ω_1 and Ω_2 are two smooth bounded domains such that $\overline{\Omega}_1 \cap \overline{\Omega}_2 = \emptyset$. Assume that

$$\Omega_1 \subset \left\{ (x_1, x') \in \mathbb{R} \times \mathbb{R}^{N-1} \,\big|\, 0 < a \leqslant x_1 \leqslant b \right\}$$

and

$$\Omega_2 \subset \left\{ (x_1, x') \in \mathbb{R} \times \mathbb{R}^{N-1} \,\big|\, -b \leqslant x_1 \leqslant -a < 0 \right\}.$$

For any $\delta > 0$ let

$$C_\delta = \left\{ (x_1, x') \in \mathbb{R} \times \mathbb{R}^{N-1} \,\big|\, x_1 \in (-b, b), \, |x'| \leqslant \delta \right\}.$$

Let Ω_δ be a smooth connected domain such that

$$\Omega_0 \subset \Omega_\delta \subset \Omega_0 \cup C_\delta. \tag{2.77}$$

Let G_D and H_D denote respectively the Green function relative to a set D and its regular part. It is straightforward to show that

$$\lim_{\delta \to 0} H_{\Omega_\delta}(x) = H_{\Omega_0}(x) \tag{2.78}$$

C^1-uniformly on compact sets of Ω_0, and

$$\lim_{\delta \to 0} G_{\Omega_\delta}(x, y) = G_{\Omega_0}(x, y) \tag{2.79}$$

C^1-uniformly on compact sets of $\Omega_0 \times \Omega_0 \setminus \{x = y\}$. Hence the number of nontrivial critical point situations of H_{Ω_δ} can be estimated from above by the sum of the numbers of nontrivial critical point situations of H_{Ω_1} and of H_{Ω_2}. Observe now that both H_{Ω_1} has a strict and H_{Ω_2} have a strict minimum point respectively in Ω_1 and in Ω_2. Furthermore, a strict minimum point is an example of nontrivial critical point situation. Let Ω_δ be defined as in (2.77). By (2.78) we deduce that if δ is small enough H_{Ω_δ} has two different strict minimum points and the claim is proved. A more general example with several disjoint sets linked together by a thin tube can be built with minor changes.

We next construct an example of contractible domain where (1.5) admits a bubbling solution with multiple, say $k > 1$ bubbles.

Let $\Omega_0 = \Omega_1 \cup \cdots \cup \Omega_k$, where, $\Omega_1, \ldots, \Omega_k$ are k smooth bounded domains such that $\overline{\Omega}_i \cap \overline{\Omega}_j = \emptyset$ if $i \neq j$. Denote by $\Psi_k^{\Omega_0}$ the function Ψ_k^- relative to Ω_0. It is easy to check that this function has a strict minimum point in the connected component $\Omega_1 \times \cdots \times \Omega_k \times \mathbb{R}_+^k$ of the set $\Omega_0^k \times \mathbb{R}_+^k$.

Assume that

$$\Omega_i \subset \left\{ (x_1, x') \in \mathbb{R} \times \mathbb{R}^{N-1} \,\big|\, a_i \leqslant x_1 \leqslant b_i \right\} \quad \text{with } b_i < a_{i+1}, i = 1, \ldots, k.$$

For any $\delta > 0$ let

$$C_\delta = \left\{ (x_1, x') \in \mathbb{R} \times \mathbb{R}^{N-1} \,\big|\, x_1 \in (a_1, b_k), |x'| \leqslant \delta \right\}.$$

Let Ω_δ be a smooth connected domain such that $\Omega_0 \subset \Omega_\delta \subset \Omega_0 \cup C_\delta$.

As before we can prove that

$$\lim_{\delta \to 0} H_{\Omega_\delta}(x) = H_{\Omega_0}(x)$$

C^1-uniformly on compact sets of Ω_0, and

$$\lim_{\delta \to 0} G_{\Omega_\delta}(x, y) = G_{\Omega_0}(x, y)$$

C^1-uniformly on compact sets of $\Omega_0 \times \Omega_0 \setminus \{x = y\}$. Therefore we deduce that Ψ_k^- relative to Ω_δ converges C^1-uniformly on compact sets of $\Omega_0^k \times (\mathbb{R}^+)^k$ to $\Psi_k^{\Omega_0}$. Using again that a strict minimum point is an example of nontrivial critical point situation, we can conclude that if δ is small enough the function $\Psi_k^{\Omega_\delta}$ has a strict minimum point, and hence problem (1.5) has a solution with k bubbles in Ω_δ.

3. Solvability of slightly supercritical problems and the topology of the domain

In this section we describe several concrete situations where problem (2.1) has a bubbling solution, or equivalently, where the function Ψ_k (see (2.66)) does have a nontrivial critical point situation. It is here where the topology of the domain plays a fundamental role: it guarantees the existence of nontrivial critical point situation for Ψ_k and hence a bubbling solution for (2.1).

3.1. *The case of a small hole*

The first result we want to show is the case in which the domain Ω has a hole. Coron [24] proved that the problem at the critical exponent, namely when $\varepsilon = 0$ in (2.1), has a solution whenever the domain Ω has a sufficiently small hole. We can prove that the same result holds true for the slightly supercritical case: namely when $\varepsilon > 0$ in (2.1) is sufficiently small and Ω is a domain with a sufficiently small hole, problem (2.1) has a solution. Not only that: we can describe this solution. It is a two-peak solution, whose maximum points are located close to the little hole dropped. Let us mention two facts. First, it is hopeless finding a solution which exhibits just one peak, as rigorously proved by [11]. Roughly speaking, this is due to the fact that the function $\xi \to H(\xi, \xi)$ cannot have a negative critical point, since it always assumes positive values. Second, the solution we find for the slightly supercritical case cannot be obtained from Coron's solution by a perturbation argument, since our solution uniformly converges to 0, as $\varepsilon \to 0$, on compact sets of Ω minus two points.

The first result of this section is the following.

THEOREM 3.1. *Assume that*

$$\Omega = \mathcal{D} \setminus \omega, \tag{3.1}$$

where $\mathcal{D}$ and ω are bounded domains with smooth boundary in $\mathbb{R}^N$, $N \geqslant 3$, with the property that $\omega \subset \overline{B}(0, \rho) \subset \mathcal{D}$.

There exists a $\rho_0 > 0$ such that, if $0 < \rho < \rho_0$ is fixed and Ω is given by (3.1), then there exists $\bar{\varepsilon} > 0$ such that, for any $0 < \varepsilon < \bar{\varepsilon}$, problem (2.1) has a solution u_ε. Furthermore, there are two points ξ_1 and ξ_2 in Ω, with $\xi_1 \neq \xi_2$, such that, for any $\delta > 0$ and as $\varepsilon \to 0$,

$$\sup_{x \notin B(\xi_j, \delta)} u_\varepsilon(x) \to 0 \quad \forall j = 1, 2$$

and

$$\sup_{x \in B(\xi_j, \delta)} u_\varepsilon(x) \sim \varepsilon^{-1/2} \quad \forall j = 1, 2.$$

In view of Theorem 1.1, the proof of Theorem 3.1 consists in setting up a min–max scheme to find a critical point of the function Ψ_2 and to show that this is a nontrivial critical point situation.

The function Ψ_2 takes the explicit form

$$\Psi_2(\xi_1, \xi_2, \Lambda_1, \Lambda_2) = \frac{1}{2}\Lambda^{\mathrm{T}}M(\xi)\Lambda + \log(\Lambda_1\Lambda_2), \tag{3.2}$$

where

$$\Lambda^{\mathrm{T}}M(\xi)\Lambda = H(\xi_1, \xi_1)\Lambda_1^2 + H(\xi_2, \xi_2)\Lambda_2^2 - 2G(\xi_1, \xi_2)\Lambda_1\Lambda_2 \tag{3.3}$$

for $\xi = (\xi_1, \xi_2) \in \Omega \times \Omega \setminus \{\xi_1 = \xi_2\}$, $\Lambda = (\Lambda_1, \Lambda_2) \in \mathbb{R}^+$.

In order to show that Ψ_2 has a critical point, we first need to prove some properties of the Green function and its regular part. Let $\mathcal{D}$ be a smooth bounded domain in $\mathbb{R}^N$ containing the origin. We shall emphasize the dependence of Green's function on the domain by writing it as $G_{\mathcal{D}}(x, y)$, and similarly for its regular part $H_{\mathcal{D}}(x, y)$. Let us consider a number $\rho > 0$ and the domain

$$\mathcal{D}_\rho = \mathcal{D} \setminus \overline{B}(0, \rho).$$

We denote by G_ρ, H_ρ respectively its Green's function and regular part.

LEMMA 3.1. *The following result holds*

$$\lim_{\rho \to 0} H_\rho(x, y) = H_{\mathcal{D}}(x, y),$$

uniformly on x, y in compact subsets of $\overline{\mathcal{D}} \setminus \{0\}$.

PROOF. The maximum principle yields

$$H_\rho(x, y) \leqslant H_{\mathcal{D}}(x, y),$$

hence the family of functions $H_\delta(x, y)$ is uniformly bounded as $\rho \to 0$ on each compact subset of $\overline{\mathcal{D}} \setminus \{0\} \times \overline{\mathcal{D}} \setminus \{0\}$, and strictly increasing in ρ. By harmonicity, its pointwise limit as $\rho \to 0$ is therefore uniform on compacts of $\mathcal{D} \setminus \{0\}$. Since the resulting limit $H(x, y)$ is harmonic in x and bounded, it extends smoothly as a harmonic function in all of $\mathcal{D}$. H therefore satisfies equation

$$\Delta_x H(x, y) = 0, \qquad x \in \mathcal{D},$$
$$H(x, y) = \Gamma(x - y), \quad x \in \partial\mathcal{D},$$

and is thus equal to $H_{\mathcal{D}}$. $\qquad\qquad\square$

Consider now our domain Ω given by (3.1). Denote by G and H its Green's function and regular part, and consider the function

$$\varphi(\xi) = \varphi(\xi_1, \xi_2) = H^{1/2}(\xi_1, \xi_1)H^{1/2}(\xi_2, \xi_2) - G(\xi_1, \xi_2) \tag{3.4}$$

defined on $\Omega \times \Omega \setminus \{\xi_1 = \xi_2\}$. The sign of the function $\varphi(\xi)$ is related to the sign of the determinant of the matrix $M(\xi)$ defined in (3.3).

Next two lemmas provide properties of the function φ when Ω is given by (3.1), which will be useful in the sequel. More precisely, the first lemma says that, if the hole in Ω is sufficiently small, then there is a region around the hole where φ assumes negative values; the second lemma says that, if the points (ξ_1, ξ_2) are sufficiently far away from the boundary of Ω then one can find a direction along which the gradient of $\varphi(\xi_1, \xi_2)$ is not zero.

LEMMA 3.2. *For any (fixed) sufficiently small number $\sigma > 0$ there is a $\rho_0 > 0$ such that if ω is any domain with $\omega \subset \overline{B}(0, \rho)$ and $\rho < \rho_0$, then*

$$\sup_{|\xi_1|=|\xi_2|=\sigma} \varphi(\xi_1, \xi_2) < 0.$$

PROOF. We have that $H_{\mathcal{D}}$ is smooth near $(0, 0)$ while $G_{\mathcal{D}}$ becomes unbounded, hence for any $\sigma > 0$

$$\sup_{|\xi_1|=|\xi_2|=\sigma} \tilde{\varphi}(\xi_1, \xi_2) < 0,$$

where $\tilde{\varphi}$ is defined by

$$\tilde{\varphi}(\xi_1, \xi_2) = H_{\mathcal{D}}^{1/2}(\xi_1, \xi_1) H_{\mathcal{D}}^{1/2}(\xi_2, \xi_2) - G_{\mathcal{D}}(\xi_1, \xi_2).$$

On the other hand, for this σ, it follows from the previous lemma that H and hence G become uniformly close to $H_{\mathcal{D}}$ and $G_{\mathcal{D}}$ on $|\xi_1| = |\xi_2| = \sigma$ as ρ gets smaller. The desired conclusion then readily follows. $\square$

Let now $\Omega_\delta = \{x \in \Omega : \operatorname{dist}(x, \partial\Omega) > \delta\}$. We have the lemma.

LEMMA 3.3. *Given $c < 0$ there exists a sufficiently small number $\delta > 0$ with the following property: If $(\bar{\xi}_1, \bar{\xi}_2) \in \partial(\Omega_\delta \times \Omega_\delta)$ is such that $\varphi(\bar{\xi}_1, \bar{\xi}_2) = c$, then there is a vector τ, tangent to $\partial(\Omega_\delta \times \Omega_\delta)$ at the point $(\bar{\xi}_1, \bar{\xi}_2)$, so that*

$$\nabla\varphi(\bar{\xi}_1, \bar{\xi}_2) \cdot \tau \neq 0. \tag{3.5}$$

The number δ does not depend on c.

PROOF. Consider, for small δ, the modified domain

$$\tilde{\Omega} = \delta^{-1}\Omega,$$

and observe that for this domain, its associated Green's function and regular part are given by

$$\tilde{G}(x_1, x_2) = \delta^{N-2} G(\delta x_1, \delta x_2), \qquad \tilde{H}(x_1, x_2) = \delta^{N-2} H(\delta x_1, \delta x_2).$$

Then $\varphi(\delta x_1, \delta x_2) = c$ translates into $\tilde{\varphi}(x_1, x_2) = c\delta^{N-2}$, where

$$\tilde{\varphi}(x_1, x_2) = \tilde{H}^{1/2}(x_1, x_1)\tilde{H}^{1/2}(x_2, x_2) - \tilde{G}(x_1, x_2).$$

Assume that $\mathrm{dist}(\delta x_1, \partial\Omega) = \delta$, namely that $\mathrm{dist}(x_1, \partial\tilde{\Omega}) = 1$. After a rotation and a translation, we assume that the closest point of the boundary to x_1 is the origin, that $x_1 = (\mathbf{0}, 1)$, where $\mathbf{0} = 0_{\mathbb{R}^{N-1}}$ and that as $\delta \to 0$ the domain $\tilde{\Omega}$ becomes the half-space $x^N > 0$. In order to make the relation $\tilde{\varphi}(x_1, x_2) = c\delta^{N-2}$ remain, as $\delta \to 0$, we claim that necessarily we must have $d = |x_1 - x_2| = O(1)$ as $\delta \to 0$. In fact, otherwise we will have

$$\tilde{H}^{1/2}(x_1, x_1)\tilde{H}^{1/2}(x_2, x_2) \geqslant Cd^{-(N-2)/2}$$

while

$$\tilde{G}(x_1, x_2) \leqslant Cd^{-(N-2)}.$$

Hence, for large d,

$$Cd^{-(N-2)/2} \leqslant \tilde{\varphi}(x_1, x_2) = c\delta^{N-2}$$

which is impossible since $c < 0$. We observe that this conclusion does not depend on the value of c, but on the fact c is negative. By assumption, we also have $|x_1 - x_2| \geqslant 1$. Then we let $\delta \to 0$ and then assume that the point x_2 converges to some $\bar{x}_2 = (\bar{x}_2', \bar{x}_2^N)$, where $\bar{x}_2^N \geqslant 1$. We also set, consistently $\bar{x}_1 = (\mathbf{0}, 1)$. The functions $\tilde{H}(x, y)$ and $\tilde{G}(x, y)$ converge to the corresponding ones $\widehat{H}$ and $\widehat{G}$ in the half-space $x_N > 0$, namely to

$$\widehat{H}(x, y) = \frac{b_N}{|x - \hat{y}|^{N-2}}$$

and

$$\widehat{G}(x, y) = b_N\left(\frac{1}{|x - y|^{N-2}} - \frac{1}{|x - \hat{y}|^{N-2}}\right).$$

Here, for $y = (y', y_N)$ we denote $\hat{y} = (y', -y_N)$. Similarly, $\nabla\tilde{\varphi}$ converges to $\nabla\hat{\varphi}$, where

$$\hat{\varphi}(x_1, x_2) = \widehat{H}^{1/2}(x_1, x_1)\widehat{H}^{1/2}(x_2, x_2) - \widehat{G}(x_1, x_2).$$

We have that

$$\hat{\varphi}(\bar{x}_1, \bar{x}_2) = 0.$$

Assume first that $\bar{x}_2' \neq 0$. Then

$$\nabla_{x_2'}\hat{\varphi}(\bar{x}_1, \bar{x}_2) = -\nabla_{x_2'}\widehat{G}(\bar{x}_1, \bar{x}_2)$$

$$= -(N-2)b_N\left(\frac{1}{|\bar{x}_2 - \bar{x}_1|^{N-2}} - \frac{1}{|\bar{x}_2 - \hat{\bar{x}}_1|^{N-2}}\right)x_2' \neq 0$$

since clearly $|\bar{x}_2 - \bar{x}_1| < |\bar{x}_2 - \hat{\bar{x}}_1|$. The vector of $\mathbb{R}^N \times \mathbb{R}^N$ $(0', 0, x_2', 0)$ is clearly tangent to the boundary of the restriction $x_1^N > 0$, where we are assuming the considered point lies. Assume now that $x_2' = 0$, case in which otherwise $\bar{x}_2 = (0', a_0)$ with $a_0 \geqslant 2$, and then

$$b_N^{-1}\hat{\varphi}\big(\bar{x}_1, \bar{x}_2 + (a - a_0)\bar{x}_1\big)$$

$$= \frac{1}{2^{(N-2)/2}} \frac{1}{(2a)^{(N-2)/2}} - \left(\frac{1}{(a-1)^{N-2}} - \frac{1}{(1+a)^{N-2}} \right).$$

Differentiating with respect to a we get

$$b_N^{-1}\nabla_{x_2^N}\hat{\varphi}(\bar{x}_1, \bar{x}_2)$$

$$= -(N-2)\big[2^{-(N-1)}a_0^{-N/2} - (a_0 - 1)^{-(N-1)} + (a_0 + 1)^{-(N-1)}\big].$$

This combined with the relation $\hat{\varphi}(\bar{x}_1, \bar{x}_2) = 0$ yields

$$b_N^{-1}\nabla_{x_2^N}\hat{\varphi}(\bar{x}_1, \bar{x}_2)$$

$$= (N-2)\big[(a_0 - 1)^{-(N-1)} - 2^{-(N-1)}a_0^{-N/2} - (a_0 + 1)^{-(N-1)}\big] > 0.$$

Indeed, since $\hat{\varphi}(\bar{x}_1, \bar{x}_2) = 0$ we have

$$\frac{1}{2^{N-1}a_0^{N/2}} = \frac{1}{2a_0} \frac{1}{2^{N-2}a_0^{(N-2)/2}} < \left[\frac{1}{(a_0 - 1)^{N-1}} - \frac{1}{(a_0 + 1)^{N-1}} \right].$$

So we can conclude that

$$b_N^{-1}\nabla_{x_2^N}\hat{\varphi}(\bar{x}_1, \bar{x}_2) > 0. \qquad \square$$

In what follows, we prove the existence of a critical point for Ψ_2 when Ω is given by (3.1). In order to avoid the singularity of Ψ_2 and φ over the diagonal $\{\xi_1 = \xi_2\}$ we consider $M > 0$, we define

$$G_M(\xi) = \begin{cases} G(\xi) & \text{if } G(\xi) \leqslant M, \\ M & \text{if } G(\xi) > M, \end{cases} \tag{3.6}$$

and we replace G with G_M in the definition of Ψ_2 and φ. The choice of M will be made later on.

Let σ be fixed as in Lemma 3.2. A direct consequence of Lemma 3.2 is that, if $l_1(M(\xi))$ denotes the first eigenvalue of $M(\xi)$, then

$$(\xi_1, \xi_2) \in \sigma S^{N-1} \times \sigma S^{N-1} \Rightarrow l_1(M(\xi)) < 0.$$

Call

$$\mathcal{S} = \sigma S^{N-1} \times \sigma S^{N-1}.$$

For every $\xi \in \mathcal{S}$ we let $e(\xi) = (e_1(\xi), e_2(\xi)) \in \mathbb{R}^2$ be the negative direction of the quadratic form defined in (3.3). It is not restrictive to assume that $|e(\xi)| = 1$. Furthermore, it is easy to see that there is a constant $c > 0$ such that

$$c < e_1(\xi)e_2(\xi) < c^{-1} \tag{3.7}$$

for all $\xi \in \mathcal{S}$.

Let $I_0 = [s_0, s_0^{-1}]$ for some $s_0 > 0$ and define

$$b_0 = \max_{\xi \in \mathcal{S}, s \in I_0} \Psi_2(\xi, se(\xi)), \qquad a_0 = \max_{\xi \in \mathcal{S}, s \in \partial I_0} \Psi_2(\xi, se(\xi)).$$

Then one easily finds that, if s_0 is small, then

$$b_0 > a_0.$$

Define

$$W_\delta^{\varphi_0} = \left\{ \xi \in \Omega \times \Omega : \varphi(\xi) < -\varphi_0 \right\} \cap (\Omega_\delta \times \Omega_\delta) \tag{3.8}$$

with $\varphi_0 < \min\{\frac{1}{2}e^{-2b_0-1}, -\frac{1}{2}\max_{\xi \in \mathcal{S}} \varphi(\xi)\}$. It is easy to see that, for δ small, $\mathcal{S} \subset W_\delta^{\varphi_0}$. Furthermore, we have that

$$W_\delta^{\varphi_0} = \left\{ \xi \in \Omega \times \Omega : l_1(M(\xi)) < -l_0 \right\} \cap (\Omega_\delta \times \Omega_\delta)$$

for an explicit $l_0 > 0$. In particular, if we define

$$a = -\frac{s_0^2}{2}l_0 + \log c^{-1} + 2\log s_0,$$

then $a_0 < a$.

Let $I = \{\Lambda = (\Lambda_1, \Lambda_2) \in \mathbb{R}_+^2 : \Lambda_1 \Lambda_2 = 1\}$. Then, taking s_0 smaller if necessary, the following inequalities hold true

$$b_0 \geqslant \min_{\xi \in \mathcal{S}, \Lambda \in I} \Psi_2(\xi, \Lambda) \geqslant \min_{\xi \in W_\delta^l, \Lambda \in I} \Psi_2(\xi, \Lambda) > a. \tag{3.9}$$

Indeed, in (3.9) the first inequality follows if we take $s_0 < c$, with c given by (3.7); the second inequality is trivial; the third inequality follows if we choose $a < -M$, which can always be obtained provided s_0 is sufficiently small.

THEOREM 3.2. *There exists a critical level for Ψ_2 between a and b_0.*

PROOF. We first prove that for every sequence $\{(\xi_n, \Lambda_n)\} \subset W_\delta^{\varphi_0} \times \mathbb{R}_+^2$ such that $(\xi_n, \Lambda_n) \to (\bar{\xi}, \bar{\Lambda}) \in \partial(W_\delta^{\varphi_0} \times \mathbb{R}_+^2)$ and $\Psi_2(\xi_n, \Lambda_n) \in [a, b_0]$ there is a vector T, tangent to $\partial(W_\delta^{\varphi_0} \times \mathbb{R}_+^2)$ at $(\bar{\xi}, \bar{\Lambda})$, such that

$$\nabla \Psi_2(\bar{\xi}, \bar{\Lambda}) \cdot T \neq 0. \tag{3.10}$$

In order to prove (3.10) we first observe that if $\Lambda_n \to \overline{\Lambda} \in \partial\mathbb{R}^2_+$ then $\Psi_2(\xi_n, \Lambda_n) \to -\infty$. Thus we can assume that $\overline{\Lambda} \in \mathbb{R}^2_+$, $\bar{\xi} \in \overline{\Omega}_\delta \times \overline{\Omega}_\delta$ and $\varphi(\bar{\xi}) \leqslant -\varphi_0$. Two cases arise, if $\nabla_\Lambda \Psi_2(\bar{\xi}, \overline{\Lambda}) \neq 0$ then T can be chosen parallel to $\nabla_\Lambda \Psi_2(\bar{\xi}, \overline{\Lambda})$. Otherwise, when $\nabla_\Lambda \Psi_2(\bar{\xi}, \overline{\Lambda}) = 0$ we have that $\overline{\Lambda}$ satisfies

$$\overline{\Lambda}_1^2 = -\frac{H(\bar{\xi}_2, \bar{\xi}_2)^{1/2}}{H(\bar{\xi}_1, \bar{\xi}_1)^{1/2}\varphi(\bar{\xi}_1, \bar{\xi}_2)}$$

and

$$\overline{\Lambda}_2^2 = -\frac{H(\bar{\xi}_1, \bar{\xi}_1)^{1/2}}{H(\bar{\xi}_2, \bar{\xi}_2)^{1/2}\varphi(\bar{\xi}_1, \bar{\xi}_2)},$$

and $\bar{\xi}$ satisfies $\varphi(\bar{\xi}) < 0$. Substituting back in Ψ_2, we get

$$\Psi_2(\bar{\xi}_1, \bar{\xi}_2, \overline{\Lambda}_1, \overline{\Lambda}_2) = -\frac{1}{2} + \frac{1}{2}\log\frac{1}{|\varphi(\bar{\xi}_1, \bar{\xi}_2)|}$$

and then $\varphi(\bar{\xi}) \leqslant -\exp(-2b_0 - 1) \leqslant -2\varphi_0 < -\varphi_0$, so that necessarily $\bar{\xi} \in \partial(\Omega_\delta \times \Omega_\delta)$. At this point we choose M: we take $\delta > 0$ as in Lemma 3.3, then we let $H_\delta = \max\{H(\xi_1, \xi_1)/\xi_1 \in \Omega_\delta\}$ and consider $M \geqslant \exp(2a - 1) + H_\delta$. We observe then that $G(\bar{\xi}_1, \bar{\xi}_2) \leqslant M$. Thus, we can apply (3.5) to complete the proof of (3.10).

Second we prove the Palais–Smale condition in $W_\delta^{\varphi_0} \times \mathbb{R}^2_+$ for Ψ_2 in the range $[a, b_0]$, that is, if $\{(\xi_n, \Lambda_n)\} \subset W_\delta^{\varphi_0} \times \mathbb{R}^2_+$ satisfies $\Psi(\xi_n, \Lambda_n) \in [a, b_0]$ and $\nabla\Psi(\xi_n, \Lambda_n) \to 0$ then $\{(\xi_n, \Lambda_n)\}$ has a subsequence converging to some $(\bar{\xi}, \overline{\Lambda}) \in W_\delta^{\varphi_0} \times \mathbb{R}^2_+$. In fact, it can be shown that the sequence Λ_n remains bounded. Then we conclude using (3.10).

Assume now by contradiction that there are no critical value in the interval $[a, b_0]$ for Ψ_2. Because of what established above we can define an appropriate negative gradient flow that will remain in $W_\delta^{\varphi_0} \times \mathbb{R}^2_+$ in $[a, b_0]$.

Hence there exists a continuous deformation

$$\eta : [0, 1] \times \Psi_2^{b_0} \to \Psi_2^{b_0}$$

such that for some $a' < a$,

$$\eta(0, u) = u \quad \forall u \in \Psi_2^{b_0},$$

$$\eta(t, u) = u \quad \forall u \in \Psi_2^{a'},$$

$$\eta(1, u) \in \Psi_2^{a'}.$$

Let us call

$$\mathcal{A} = \left\{(\xi, \Lambda) \in W_\delta^{\varphi_0} \times \mathbb{R}^2_+ \,\middle|\, \xi \in \mathcal{S}, \Lambda = re(\xi), r \in I_0\right\},$$

$$\partial\mathcal{A} = \left\{(\xi, \Lambda) \in W_\delta^{\varphi_0} \times \mathbb{R}^2_+ \,\middle|\, \xi \in \mathcal{S}, \Lambda = re(\xi), r \in \partial I_0\right\},$$

$$\mathcal{C} = W_\delta^{\varphi_0} \times I.$$

From (3.9) we deduce that $\mathcal{A} \subset \Psi_2^{b_0}$, $\partial \mathcal{A} \subset \Psi_2^{a'}$ and $\Psi_2^{a'} \cap \mathcal{C} = \emptyset$. Therefore

$$
\begin{aligned}
\eta(0, u) &= u \quad \forall u \in \mathcal{A}, \\
\eta(t, u) &= u \quad \forall u \in \partial \mathcal{A}, \\
\eta(1, \mathcal{A}) &\cap \mathcal{C} = \emptyset.
\end{aligned}
\tag{3.11}
$$

For any $(\xi, \Lambda) \in \mathcal{A}$ and for any $t \in [0, 1]$, we denote

$$
\eta\big(t, (\xi, \Lambda)\big) = \big(\tilde{\xi}(\xi, \Lambda, t), \tilde{\Lambda}(\xi, \Lambda, t)\big) \in W_\delta^{\varphi_0} \times \mathbb{R}_+^2.
$$

We define the set

$$
\mathcal{B} = \big\{(\xi, \Lambda) \in \mathcal{A} \,\big|\, \tilde{\Lambda}(\xi, \Lambda, 1) \in I\big\}.
$$

Since $\eta(1, \mathcal{A}) \cap \mathcal{C} = \emptyset$ it holds $\mathcal{B} = \emptyset$. Now let $\mathcal{U}$ be a neighborhood of $\mathcal{B}$ in $W_\delta^{\varphi_0} \times \mathbb{R}_+^2$ such that $H^*(\mathcal{U}) = H^*(\mathcal{B})$. If $\pi : \mathcal{U} \to \mathcal{S}$ denotes the projection, arguing like in Lemma 7.1 of [33] we can show that

$$
\pi^* : H^*(\mathcal{S}) \to H^*(\mathcal{U}) \quad \text{is a monomorphism.}
$$

This condition provides a contradiction, since $H^*(\mathcal{U}) = \{0\}$ and $H^*(\mathcal{S}) \neq \{0\}$. $\square$

Now we are in a position to complete the proof of Theorem 3.1, proving that the reduced functional J has a critical point.

PROOF OF THEOREM 3.1 COMPLETED. Given the result of Lemma 2.2, we are left to prove the existence of a critical point $(\xi_\varepsilon, \Lambda_\varepsilon)$ for J, for all ε small enough. Equivalently, we are thus interested in critical points of $\tilde{J}$ defined by

$$
\tilde{J}(\xi, \Lambda) = \omega_N^{-1}\big[J(\xi, \Lambda) - 2C_N - \varepsilon \bar{\gamma}_N\big].
$$

We consider the domain $D_{r,R} = W_\delta^{\varphi_0} \times [r, R]^2$, with $W_\delta^{\varphi_0}$ given by (3.8) and r, R to be chosen later. The functional $\tilde{J}$ is well defined on $D_{r,R}$ except on the set $\Delta_\rho = \{\xi \in \Omega_\rho \times \Omega_\rho \mid |\xi_1 - \xi_2| < \rho\}$. Taking into account Proposition 2.4 and proceeding as in (3.6) with the function G, we can extend J to all $D_{r,R}$, keeping the relations (3.5) and (3.9) over $D_{r,R}$.

Arguing like in the proof of Theorem 3.2 and taking into account (2.74) and (2.75), there are positive numbers r, R, $\bar{\varepsilon}$ and $\bar{\alpha}$ such that, for all $\varepsilon \in (0, \bar{\varepsilon})$, for all $\alpha \in (0, \bar{\alpha})$, the function $\tilde{J}(\xi, \Lambda)$ satisfies the Palais–Smale condition in $D_{r,R}$ in the range $[a - \alpha, b_0 + \alpha]$. Furthermore, if $(\xi_n, \Lambda_n) \to (\bar{\xi}, \overline{\Lambda}) \in \partial D_{r,R}$ with $\tilde{J}(\xi_n, \Lambda_n) \in [a - \alpha, b_0 + \alpha]$ then there exists a vector T, tangent to $D_{r,R}$ at $(\bar{\xi}, \overline{\Lambda})$ such that $\nabla \tilde{J}(\bar{\xi}, \overline{\Lambda}) \cdot T \neq 0$.

Direct consequence of these two facts is that an appropriate negative gradient flow for $\tilde{J}(\xi, \Lambda)$ can be defined and it remains in $D_{r,R}$ in the range $[a - \alpha, b_0 + \alpha]$. Using again the C^1-closeness between Ψ_2 and $\tilde{J}$ given by relations (2.74) and (2.75), one can prove estimates similar to (3.9) for $\tilde{J}$. Arguing as in the last part of the proof of Theorem 3.2, one shows the existence, for all ε small enough, of a critical point $(\xi_\varepsilon, \Lambda_\varepsilon) = (\xi_{1\varepsilon}, \xi_{2\varepsilon}, \Lambda_{1\varepsilon}, \Lambda_{2\varepsilon})$ for $\tilde{J}$.

Hence, the function

$$u_\varepsilon(x) = \varepsilon^{-\frac{1}{2+\varepsilon(N-2)/2}} \left[V_{\lambda_{1\varepsilon}, \xi'_{1\varepsilon}} + V_{\lambda_{2\varepsilon}, \xi'_{2\varepsilon}} + \phi(\xi'_\varepsilon, \Lambda_\varepsilon) \right] \left(\varepsilon^{-1/(N-2)} x \right)$$

(see (2.6)) with $\lambda_{i\varepsilon} = (a_N \Lambda_{i\varepsilon}^2)^{1/(N-2)}$, is the solution to (2.1) we were looking for. In fact, one easily gets that this solution u_ε has the following qualitative behavior

$$u_\varepsilon(x) = \alpha_n \sum_{j=1}^{2} \left[\frac{\lambda_{j\varepsilon}}{\lambda_{j\varepsilon}^2 + \varepsilon^{-2/(N-2)} |x - \xi_{j\varepsilon}|^2} \right]^{(N-2)/2} \varepsilon^{-1/2} + o(1),$$

where $o(1)$ is uniform over compact sets of Ω. $\square$

Let us mention a problem where similar phenomena to those described take place: In [66], the authors studied problem (1.12) with μ negative and small, say $\mu = -\varepsilon^{(N-4)/(N-2)}$ for dimensions $N \geqslant 5$, namely

$$\begin{cases} \Delta u - \varepsilon^{(N-4)/(N-2)} u + u^{(N+2)/(N-2)} = 0 & \text{in } \Omega, \\ u > 0 & \text{in } \Omega, \\ u = 0 & \text{on } \partial\Omega, \end{cases} \tag{3.12}$$

with Ω a smooth bounded domain of $\mathbb{R}^N$, with $N \geqslant 5$.

Despite of being critical and not supercritical, problem (3.12) shares some common patterns with (2.1). Indeed, if Ω is star shaped, Pohozaev's identity shows that (3.12) does not have any nontrivial solution. On the other hand, if Ω is an annulus, then (3.12) has nontrivial positive solutions.

In [66], a result similar to the one obtained in Theorem 3.1 for problem (2.1) holds true for (3.12). Indeed it is proven that in a general domain Ω with a sufficiently small hole then problem (3.12) has a positive solution u_ε which, as $\varepsilon \to 0$, concentrates as the sum of two Dirac deltas around points ξ_1 and ξ_2 of the domain Ω. More precisely, for any $\delta > 0$,

$$\sup_{x \notin B(\xi_j, \delta)} u_\varepsilon(x) \to 0 \quad \forall j = 1, 2$$

and

$$\sup_{x \in B(\xi_j, \delta)} u_\varepsilon(x) \sim \varepsilon^{-(N-2)/(2(N-4))} \quad \forall j = 1, 2.$$

In this case, the location of the two points of concentration is defined through the critical points of the function

$$\widehat{\Psi}_2(\xi_1, \xi_2, \Lambda_1, \Lambda_2) = \frac{1}{2}\left[H(\xi_1, \xi_1)\Lambda_1^2 + H(\xi_2, \xi_2)\Lambda_2^2 - 2G(\xi_1, \xi_2)\Lambda_1\Lambda_2\right] + \frac{1}{2}\left[\Lambda_1^{4/(N-2)} + \Lambda_2^{4/(N-2)}\right]. \tag{3.13}$$

An argument similar to the one used to prove the existence of a nontrivial critical point situation for Ψ_2 given by in the slightly supercritical Bahri–Coron setting of Theorem 3.1 can be reproduced here, thus yielding to the following theorem.

THEOREM 3.3. *Assume that*

$$\Omega = \mathcal{D} \setminus \omega, \tag{3.14}$$

where $\mathcal{D}$ and ω are bounded domains with smooth boundary in $\mathbb{R}^N$, $N \geqslant 5$, with the property that $\omega \subset \overline{B}(0, \rho) \subset \mathcal{D}$.

There exists a $\rho_0 > 0$ such that, if $0 < \rho < \rho_0$ is fixed and Ω is given by (3.1), then there exists $\bar{\varepsilon} > 0$ such that, for any $0 < \varepsilon < \bar{\varepsilon}$, problem (3.12) has a solution u_ε, which has the following form

$$u_\varepsilon(x) = \left[\sum_{j=1}^{2}\left(\frac{\lambda_j^\varepsilon}{(\lambda_j^\varepsilon)^2 + \varepsilon^{-2/(N-4)}|x - \xi_j^\varepsilon|^2}\right)^{(N-2)/2}\right]\varepsilon^{-(N-2)/(2(N-4))} + o(1),$$

where $o(1)$ *is in* C^0*-sense over compact sets of Ω as $\varepsilon \to 0$. Here $(\lambda_i^\varepsilon)^{N-2} = \frac{\int_{\mathbb{R}^N} \overline{U}^2}{(\int_{\mathbb{R}^N} \overline{U}^p)^2}(\Lambda_i^\varepsilon)^2$ and $(\xi_1^\varepsilon, \xi_2^\varepsilon, \Lambda_1^\varepsilon, \Lambda_2^\varepsilon)$ converges (up to subsequences), as $\varepsilon \to 0$, to a critical point of the function $\widehat{\Psi}_2$ defined in (3.13).*

The result obtained in Theorem 3.1 (and also in Theorem 3.3) is consequence of the following more general theorem, which we state for the slightly supercritical problem (2.1) and which has been proved in details in [33].

THEOREM 3.4. *Let Ω be a bounded domain with smooth boundary in $\mathbb{R}^N$, with the following property: There exist a compact manifold $\mathcal{M} \subset \Omega$ and an integer $d \geqslant 1$ such that $\varphi < 0$ on $\mathcal{M} \times \mathcal{M}$, $\iota^* : H^d(\Omega) \to H^d(\mathcal{M})$ is nontrivial and either d is odd or $H^{2d}(\Omega) = 0$. Here φ is the function defined in (3.4).*

Then there exists $\varepsilon_0 > 0$ such that, for any $0 < \varepsilon < \varepsilon_0$, problem (2.1) has at least one solution u_ε. Moreover, let $\mathcal{C}$ be the component of the set where $\varphi < 0$ which contains $\mathcal{M} \times \mathcal{M}$. Then, given any sequence $\varepsilon = \varepsilon_n \to 0$, there is a subsequence, which we denote in the same way, and a critical point $(\xi_1, \xi_2) \in \mathcal{C}$ of the function φ such that $u_\varepsilon(x) \to 0$ on compact subsets of $\Omega \setminus \{\xi_1, \xi_2\}$ and such that, for any $\delta > 0$,

$$\sup_{|x - \xi_i| < \delta} u_\varepsilon(x) \to +\infty, \quad i = 1, 2,$$

as $\varepsilon \to 0$.

In view of Theorem 3.4, observe that the perforation in Coron's situation contained in Theorem 3.1 need not to be symmetric or contained in a small ball. For instance in $\mathbb{R}^3$, the result holds true in a domain with a torus with narrow section excised.

A natural question now concerns the issue of solvability of problem (2.1) in a domain exhibiting multiple holes. Next result asserts that in such a situation, multipeak solutions exist, consisting of gluing of double-spikes associated to each of the holes. More precisely, we have the theorem.

THEOREM 3.5. *Let $\mathcal{D}$ be a bounded, smooth domain in $\mathbb{R}^N$, $N \geqslant 3$, and $P_1, \ldots, P_m$ points of $\mathcal{D}$. Let us consider the domain*

$$\Omega = \mathcal{D} \setminus \bigcup_{j=1}^{m} \overline{B}(P_j, \rho), \tag{3.15}$$

where $\rho > 0$.

There exists a $\rho_0 > 0$, which depends on $\mathcal{D}$ and the points $P_1, \ldots, P_m$ such that if $0 < \rho < \rho_0$ is fixed and Ω is the domain given by (3.15), then the following holds: Given a number $1 \leqslant k \leqslant m$, there exists $\bar{\varepsilon} > 0$ and a family of solutions, u_ε, $0 < \varepsilon < \bar{\varepsilon}$ of (2.1), with the following properties: u_ε has exactly k pairs of local maximum points $(\xi_{j1}^\varepsilon, \xi_{j2}^\varepsilon) \in \Omega^2$, $j = 1, \ldots, k$, with $c\rho < |\xi_{ji}^\varepsilon - P_j| \leqslant C\rho$ for certain constants c, C independent of ρ, and such that, for each small $\delta > 0$,

$$\sup_{\{|x - \xi_{ji}^\varepsilon| > \delta \ \forall i, j\}} u_\varepsilon(x) \to 0$$

and

$$\sup_{\{|x - \xi_{ji}^\varepsilon| < \delta\}} u_\varepsilon(x) \to +\infty \quad \forall i, j$$

as $\varepsilon \to 0$.

The result contained in the previous theorem is proved in detail in [32]. We will skip the proof of it and we refer the reader to [32].

We just want to mention few facts here. Thanks to the fact that the holes in (3.15) are spherical, we can be extremely precise in the location of each couple of peaks: indeed, as explained in the theorem, they are located at distance of order ρ from the hole. Not only that. From the proof it follows that they are antipodal with respect to the hole.

On the other hand, from the proof of this result it is clear that, in order to get the result on existence of multispike solutions, there is no need for the small excised domains to be balls of the same radii. We consider only this case for notational simplicity.

Let us also observe that by re-labeling the points $P_1, \ldots, P_m$, the above result actually yields that, for each $1 \leqslant k \leqslant m$ and any set of indices $i_1, \ldots, i_k$ in $\{1, \ldots, m\}$, a solution exhibiting double-spikes simultaneously near the points $P_{i_1}, \ldots, P_{i_k}$ exists. This in particular yields the existence of at least $2^m - 1$ solutions of the problem whenever ε is sufficiently small.

3.2. *Bubbling under symmetries*

As it was natural to expect, Theorem 3.5 shows that multiplicity of solution to (2.1) appears in presence of multiple holes. What we want to show next is that the concentration phenomena involved in problem (2.1) when ε is sufficiently small, is in fact much richer than it might be a priori suspected, even in domains with just one hole, at least in the case of domains exhibiting symmetries. In particular, we find the presence of a large number of geometrically distinct solutions to problem (2.1) when Ω is an annulus,

$$A_a^b = \{x \mid a < |x| < b\}, \tag{3.16}$$

for given $0 < a < b$, provided that $\varepsilon > 0$ is sufficiently small. More precisely, we find that a k-spike solution of (2.1) exists for any k sufficiently large. The same result holds true in the solid torus T_a^b, $0 < a < b$ in $\mathbb{R}^3$, obtained after rotating around the x_3-axis a disk of radius $(b-a)/2$ centered at the point $((a+b)/2, 0, 0)$. Our main result contained in Theorem 3.6 covers these examples, and also the case of a solid of revolution, symmetric on the variable x_3, which does not contain the origin. See also [23,49,61] for related problems.

For notational simplicity, we write $x \in \mathbb{R}^N$ as $x = (z, x')$ with $z \in \mathbb{R}^2$ which we identify with an element of the complex plane, and $x' \in \mathbb{R}^{N-2}$.

THEOREM 3.6. *Assume that the domain Ω in $\mathbb{R}^N$, $N \geqslant 3$, is such that $0 \notin \Omega$ and satisfies the following properties:*
(a) *For any z, x',*

$$(z, x') \in \Omega \quad \text{implies} \quad (e^{i\theta} z, x') \in \Omega \quad \text{for all } \theta \in [0, 2\pi].$$

(b) *For each $3 \leqslant i \leqslant N$,*

$$(z, x_3, \ldots, x_i, \ldots, x_N) \in \Omega \quad \text{implies} \quad (z, x_3, \ldots, -x_i, \ldots, x_N) \in \Omega.$$

For $k \in \mathbb{N}$, let us denote $P_j = (e^{2\pi i j/k}, 0')$, $j = 1, \ldots, k$. Then there is a k_0, such that for $k \geqslant k_0$ there exists a positive number ε_0 such that for each $0 < \varepsilon < \varepsilon_0$ there are numbers λ_ε and ρ_ε and a solution u_ε to problem (2.1) of the form

$$u_\varepsilon(x) = \alpha_N \sum_{j=1}^k \left(\frac{\lambda_\varepsilon}{\lambda_\varepsilon^2 + \varepsilon^{-2/(N-2)} |x - \rho_\varepsilon P_j|^2} \right)^{(N-2)/2} \varepsilon^{-1/2} + o(1), \tag{3.17}$$

where $o(1) \to 0$ uniformly as $\varepsilon \to 0$ and $\alpha_N = (N(N-2))^{(N-2)/4}$. Moreover, λ_ε is bounded above and below away from zero and $\rho_\varepsilon P_j$ remains uniformly away from the boundary of Ω.

As it will become clear from the construction, the number k may in principle need to be chosen very large for a k-peak solution to exist.

According to the results of Lemma 2.2 in the previous section, our task reduces to finding critical points of the functional of (ξ, Λ) given by $J(\xi, \Lambda) = I_\varepsilon(V + \phi)$.

Let (a, b) be any maximal interval so that $\rho P_1 \in \Omega$ for all $\rho \in (a, b)$. The setting of Theorem 3.6 suggests us to restrict ourselves to seek for critical points of the form

$$\xi_j = \rho P_j, \qquad \Lambda_j = \mu, \quad \text{for all } j = 1, \ldots, k, \tag{3.18}$$

where $\rho \in (a, b)$ and $\mu > 0$. Let us set

$$\widetilde{J}(\rho, \mu) = J\big(\rho(P_1, \ldots, P_k), \mu(1, \ldots, 1)\big). \tag{3.19}$$

LEMMA 3.4. *Under the assumptions of Theorem 3.6, if (ρ, μ) is a critical point of $\widetilde{J}$, then $(\xi, \Lambda) = (\rho(P_1, \ldots, P_k), \mu(1, \ldots, 1))$ is a critical point of J.*

PROOF. We first investigate the symmetries inherited by the function $\phi(\xi', \lambda)$ in the variables $(y_1, \ldots, y_N)$ when the points (ξ, λ) belong to Ω. We recall that ϕ satisfies

$$\begin{cases} \Delta(V + \phi) + (V + \phi)_+^{p+\varepsilon} = \sum_{ij} c_{ij} V_i^{p-1} Z_{ij} & \text{in } \Omega_\varepsilon, \\ \phi = 0 & \text{on } \partial\Omega_\varepsilon, \\ \int_{\Omega_\varepsilon} \phi V_i^{p-1} Z_{ij} = 0 & \text{for all } i, j. \end{cases} \tag{3.20}$$

Now, because of the symmetry of the data and the domain, we see that V and each of the V_i's (see (2.7)) are even with respect to each of the variables $y_2, \ldots, y_N$. It then follows that $c_{ij} = 0$ for all $1 \leqslant i \leqslant k$ and for all $2 \leqslant j \leqslant N$. In fact, for instance for $j = 3$, the function $\phi(y_1, y_2, -y_3, \ldots, y_N)$ satisfies an equation of the same form, with (possibly) different constants c_{ij}. Uniqueness of the solution ϕ to the above problem yields that ϕ is even with respect to y_3. For $i = l$ then only the terms $c_{l1} V_1^{p-1} Z_{l1} + c_{l(N+1)} V_1^{p-1} Z_{l(N+1)}$ are present in the summation on the right-hand side of (3.20). On the other hand, the invariance under rotations assumed implies that actually ϕ is symmetric with respect to the line $z = te^{2\pi i l/k}$ in the space of the first two variables. Call

$$\widetilde{Z}_l = Z_{l1} \cos \frac{2\pi l}{k} + Z_{l2} \sin \frac{2\pi l}{k},$$

namely the H_0^1-projection of the directional derivative of $\overline{V}_l$ in the direction of the above line. Then by the symmetry, we obtain that the right-hand side of (3.20) reduces simply to an expression of the form $\sum_l c_1 V_l^{p-1} \widetilde{Z}_l + c_2 Z_{l,N+1}$, where $c_i = c_i(\rho, \mu)$, $i = 1, 2$. Thus according to the previous section, $(\xi, \Lambda) = (\rho(P_1, \ldots, P_k), \mu(1, \ldots, 1))$ is a critical point of J provided that $c_i(\rho, \mu) = 0$ for $i = 1, 2$. We claim that these relations hold true if (ρ, μ) is a critical point of $\widetilde{J}$. In fact, $\nabla \widetilde{J} = 0$ means

$$DI_\varepsilon(V + \phi)\left[\frac{\partial}{\partial \rho}(V + \phi)\right] = 0 = DI_\varepsilon(V + \phi)\left[\frac{\partial}{\partial \mu}(V + \phi)\right]. \tag{3.21}$$

Now,

$$\frac{\partial}{\partial \rho}(V + \phi) = \sum_{l=1}^{k} Z_l + o(1), \qquad \frac{\partial}{\partial \mu}(V + \phi) = \sum_{l=1}^{k} Z_{l(N+1)} + o(1)$$

with $o(1) \to 0$ uniformly as $\varepsilon \to 0$. So we have that relations (3.21) are respectively equivalent to

$$(1 - o(1))c_1 + o(1)c_2 = 0, \qquad (1 - o(1))c_2 + o(1)c_1 = 0,$$

hence $c_1 = c_2 = 0$, and the proof is concluded. $\square$

PROOF OF THEOREM 3.6. From Lemma 3.4, we need to find a critical point of the functional $\widetilde{J}(\rho, \mu)$. This is equivalent to finding a critical point of

$$\Psi_\varepsilon(\rho, \mu) \equiv \omega_N^{-1}\left(\varepsilon^{-1}\left(\widetilde{J}(\rho, \mu) - kC_N\right) - \bar{\gamma}_N\right) \tag{3.22}$$

for the constants introduced in expansion given by Proposition 2.4. Now, again from Proposition 2.4,

$$\Psi_\varepsilon(\rho, \mu) = \Psi_0(\rho, \mu) + o(1),$$

where $o(1)$ is small in ε, in the C^1-sense, uniformly on compact subsets of $(a, b) \times (0, \infty)$, and

$$\Psi_0(\rho, \mu) = \Psi\left(\rho(P_1, \ldots, P_k), \mu(1, \ldots, 1)\right) = k\left\{\frac{\mu^2}{2} F_k(\rho) + \log \mu\right\} \tag{3.23}$$

with

$$F_k(\rho) = H(\xi_1, \xi_1) - \sum_{i=2}^{k} G(\xi_1, \xi_i)$$

and $\xi_j = \rho P_j$. Since Robin's function $H(\xi, \xi)$ tends to $+\infty$ as ξ approaches $\partial\Omega$, it follows that for any integer k, $\lim_{\rho \to a} F_k(\rho) = \lim_{\rho \to b} F_k(\rho) = +\infty$. On the other hand, if $\xi_j = \frac{a+b}{2} P_j$, then

$$G(\xi_1, \xi_2) = \gamma_N |\xi_1 - \xi_2|^{2-N} + O(1),$$

where the quantity $O(1)$ is bounded independently of k, hence $G(\xi_1, \xi_2) \geqslant Ak^{N-2}$ for all large k, with A independent of k. Now, $H(\xi_1, \xi_1) \leqslant B$ with B independent of k. It follows that

$$F_k\left(\frac{a+b}{2}\right) \leqslant k\left(B - Ak^{N-2}\right) < 0$$

for all sufficiently large k.

Then there are numbers $a < \bar{a} < \bar{b} < b$ for which $F_k'(\bar{a}) < 0$, $F_k'(\bar{b}) > 0$ and $F_k(\rho) < 0$ hold for all $\rho \in (\bar{a}, \bar{b})$. Then if δ is fixed and sufficiently small we see that the following relations hold

$$\frac{\partial}{\partial \mu} \Psi_0(\rho, \delta) > 0, \qquad \frac{\partial}{\partial \mu} \Psi_0(\rho, \delta^{-1}) < 0 \quad \text{for all } \rho \in \left[\bar{a}, \bar{b}\right] \tag{3.24}$$

and

$$\frac{\partial}{\partial \rho} \Psi_0(\bar{a}, \mu) < 0, \qquad \frac{\partial}{\partial \rho} \Psi_0(\bar{b}, \mu) > 0 \quad \text{for all } \mu \in \left[\delta, \delta^{-1}\right]. \tag{3.25}$$

Let us set $\mathcal{R} = (\bar{a}, \bar{b}) \times (\delta, \delta^{-1})$ and let (d_1, d_2) be the center point of this rectangle. Let us consider the homotopy

$$H_t(\rho, \mu) = t \nabla \Psi_0(\rho, \mu) + (1 - t)(\rho - d_1, -(\mu - d_2)), \quad t \in [0, 1].$$

Then, from relations (3.24) and (3.25), we see that the degree $\deg(H_t, \mathcal{R}, 0)$ is well defined and constant for $t \in [0, 1]$. It follows then that $\deg(\nabla \Psi_0, \mathcal{R}, 0) = -1$. Since $\nabla \Psi_\varepsilon$ is a small uniform perturbation of $\nabla \Psi_0$ on $\mathcal{R}$, we conclude that $\deg(\nabla \Psi_\varepsilon, \mathcal{R}, 0) = -1$ for all sufficiently small ε. Hence a critical point $(\rho_\varepsilon, \mu_\varepsilon) \in \mathcal{R}$ of Ψ_ε indeed exists for all sufficiently small ε and the proof of the theorem is thus concluded. $\qquad \square$

REMARK 3.1. In the limit $(\rho_\varepsilon, \mu_\varepsilon) \to (\rho, \mu)$ as $\varepsilon \to 0$ we find a critical point of the function Ψ_0 in (3.23). Computing directly, we see that

$$\mu^2 = -\frac{1}{k F_k(\rho)}; \tag{3.26}$$

relation that makes sense only if $F_k(\rho) < 0$. Moreover, a refinement of the region $\mathcal{R}$ in the above proof yields that ρ can actually be chosen to be a minimum of F_k on (a, b).

4. The Brezis–Nirenberg problem in dimension $N = 3$: the proof of Theorem 1.2

In this section we construct single and multiple-bubbling solutions for problem (1.13) in dimension $N = 3$. Unlike the case of higher dimensions, the results of [15] concerning asymptotic analysis of radial solutions in a ball when the exponent approaches critical from below, suggests that the object ruling the location of blowing up in single-bubble solutions of (2.1) is *Robin's function* g_λ defined as follows. Let $\lambda < \lambda_1$ and consider Green's function $G_\lambda(x, y)$, solution for any given $x \in \Omega$ of

$$-\Delta_y G_\lambda - \lambda G_\lambda = \delta_x, \quad y \in \Omega,$$

$$G_\lambda(x, y) = 0, \qquad\qquad y \in \partial\Omega.$$

Let $H_\lambda(x, y) = \Gamma(y - x) - G_\lambda(x, y)$ with $\Gamma(z) = \frac{1}{4\pi|z|}$, be its regular part. In other words, $H_\lambda(x, y)$ can be defined as the unique solution of the problem

$$\Delta_y H_\lambda + \lambda H_\lambda = \lambda \Gamma(x - y), \quad y \in \Omega,$$
$$H_\lambda = \Gamma(x - y), \qquad\qquad y \in \partial\Omega.$$

Let us consider Robin's function of G_λ, defined as

$$g_\lambda(x) \equiv H_\lambda(x, x).$$

It turns out that $g_\lambda(x)$ is a smooth function which goes to $+\infty$ as x approaches $\partial\Omega$. Its minimum value is not necessarily positive. In fact this number is strictly decreasing in λ. It is strictly positive when λ is close to 0 and approaches $-\infty$ as $\lambda \uparrow \lambda_1$. It is suggested in [15] and proven by Druet in [38] that the number λ^* given by (1.4) can be characterized as

$$\lambda^* = \sup\left\{\lambda > 0 : \min_\Omega g_\lambda > 0\right\}. \tag{4.1}$$

The minimum value of g_λ is thus negative whenever $\lambda_* < \lambda < \lambda_1$. In what follows we shall denote simply by $\mathcal{D}$ the subset of Ω where g_λ is negative and $q = 5 + \varepsilon$. Further we shall present general scheme of the proof. Full details can be found in [30].

4.1. *Energy expansion of single bubbling*

Given a point $\zeta \in \mathbb{R}^3$ and a positive number μ, we denote in what follows

$$w_{\mu,\zeta}(y) \equiv \frac{3^{1/4}}{\sqrt{1 + \mu^{-2}|y - \zeta|^2}} \mu^{-1/2},$$

which correspond to all positive solutions of the problem

$$\Delta w + w^5 = 0 \quad \text{in } \mathbb{R}^3.$$

The solutions we look for in Theorem 1.2, part (a) have the form $u(y) \sim w_{\mu,\zeta}(y)$, where $\zeta \in \Omega$ and μ is a very small number. It is natural to correct this initial approximation by a term that provides Dirichlet boundary conditions. We assume in all what follows that $0 < \lambda < \lambda_1$. We define $\pi_{\mu,\zeta}(y)$ to be the unique solution of the problem

$$\Delta\pi_{\mu,\zeta} + \lambda\pi_{\mu,\zeta} = -\lambda w_{\mu,\zeta} \quad \text{in } \Omega \quad \text{with}$$
$$\pi_{\mu,\zeta} = -w_{\mu,\zeta} \qquad\qquad \text{on } \partial\Omega. \tag{4.2}$$

Fix a small positive number μ and a point $\zeta \in \Omega$. We consider as a first approximation of the solution one of the form

$$U_{\mu,\zeta}(y) = w_{\mu,\zeta} + \pi_{\mu,\zeta}. \tag{4.3}$$

Observe that $U = U_{\mu,\zeta}$ satisfies then the equation

$$\Delta U + \lambda U + w_{\mu,\zeta}^5 = 0 \quad \text{in } \Omega,$$
$$U = 0 \qquad\qquad\qquad \text{on } \partial\Omega.$$

Classical solutions to (1.13) correspond to critical points of the energy functional

$$E_{q,\lambda}(u) \equiv \frac{1}{2}\int_\Omega |Du|^2 - \frac{\lambda}{2}\int_\Omega |u|^2 - \frac{1}{q+1}\int_\Omega |u|^{q+1}. \tag{4.4}$$

Here and in what follows we denote $q = 5 + \varepsilon$. If there was a solution very close to U_{μ^*,ζ^*} for a certain pair (μ^*, ζ^*), then we would formally expect $E_{q,\lambda}$ to be nearly stationary with respect to variations of (μ, ζ) on $U_{\mu,\zeta}$ around this point. Under this intuitive basis, it seems important to understand critical points of the functional $(\mu, \zeta) \mapsto E_{q,\lambda}(U_{\mu,\zeta})$. We have the following explicit asymptotic expression for this functional. For q close to 5.

LEMMA 4.1. *Consider $U_{\zeta,\mu}$ and $E_{q,\lambda}$ defined respectively by (4.3) and (4.4). Then, as $\mu \to 0$,*

$$E_{q,\lambda}(U_{\mu,\zeta}) = a_0 + a_1\mu g_\lambda(\zeta) + a_2\mu^2\lambda - a_3\mu^2 g_\lambda^2(\zeta)$$
$$+ (q-5)[a_4 \log \mu + a_5]$$
$$+ (q-5)^2\theta_1(\zeta,\mu,q) + \mu^{3-\sigma}\theta_2(\zeta,\mu,q), \tag{4.5}$$

where for $j = 0, 1, 2,\ i = 0, 1,\ i + j \leqslant 2,\ l = 1, 2,$

$$\mu^j \frac{\partial^{i+j}}{\partial\zeta^i \partial\mu^j}\theta_l(\zeta,\mu,q)$$

is bounded uniformly on all small μ, $\varepsilon = q - 5$ small and ζ in compact subsets of Ω. Here $a_0, \ldots, a_5$ are explicit constants.

Let us consider then $q = 5 + \varepsilon$ and assume that $\zeta \in \mathcal{D}$, i.e., $g_\lambda(\zeta) > 0$. It is convenient to consider Λ defined by

$$\mu \equiv -\varepsilon \frac{a_4}{a_1} \frac{1}{g_\lambda(\zeta)} \Lambda, \tag{4.6}$$

where a_4 and a_1 are the constants in the expansion (4.5).

LEMMA 4.2. *In the situation of Theorem 1.2, part (a), for μ given by (4.6), consider a functional of the form*

$$\psi_\varepsilon(\Lambda, \zeta) \equiv E_{5+\varepsilon,\mu}(U_{\mu,\zeta}) + \varepsilon\theta_\varepsilon(\Lambda, \zeta)$$

for $\Lambda > 0$ and $\zeta \in \Omega$. Denote $\nabla = (\partial_\Lambda, \partial_\zeta)$ and assume that

$$|\theta_\varepsilon| + |\nabla \theta_\varepsilon| + |\nabla \partial_\Lambda \theta_\varepsilon| \to 0 \qquad (4.7)$$

uniformly on (ζ, Λ) as $\varepsilon \to 0$, with

$$\delta < \Lambda < \delta^{-1}, \qquad g_\lambda(\zeta) < -\delta,$$

for any given δ. Then ψ_ε has a critical point $(\Lambda_\varepsilon, \zeta_\varepsilon)$ with $\zeta_\varepsilon \in \mathcal{D}$,

$$\Lambda_\varepsilon \to 1, \qquad g_\lambda(\zeta_\varepsilon) \to \min_\Omega g_\lambda.$$

PROOF. The expansion given in Lemma 4.1 implies

$$\psi_\varepsilon(\Lambda, \zeta) \equiv a_0 + a_4\varepsilon\left[-\Lambda + \log \Lambda + \log\left(-\frac{1}{g_\lambda(\zeta)}\right)\right]$$
$$+ a_4\varepsilon\left[\log\left(\frac{a_4}{a_1}\right) + \log \varepsilon + \frac{a_5}{a_4}\right] + \varepsilon\theta_\varepsilon(\Lambda, \zeta),$$

where θ_ε still satisfies (4.7). The main term in the above expansion is the functional

$$\psi_0(\Lambda, \zeta) = -\Lambda + \log \Lambda + \log\left(-\frac{1}{g_\lambda(\zeta)}\right),$$

which obviously has a critical point since it has a nondegenerate maximum in Λ at $\Lambda = 1$. Consider the equation

$$\partial_\Lambda \psi_\varepsilon(\Lambda, \zeta) = 0,$$

which has the form

$$\Lambda = 1 + o(1)\theta_\varepsilon(\Lambda, \zeta),$$

where the function θ_ε has a continuous, uniformly bounded derivative in (Λ, ζ) in the considered region. It then follows that for each $\zeta \in \mathcal{D}$ there exists a unique $\Lambda = \Lambda_\varepsilon(\zeta)$, function of class C^1 satisfying the above equation which has the form

$$\Lambda_\varepsilon(\zeta) = 1 + o(1)\beta_\varepsilon(\zeta),$$

where β_ε and its derivative are uniformly bounded in the considered region. Clearly we get a critical point of ψ_ε if we have one of the functional $\zeta \mapsto \psi_\varepsilon(\Lambda_\varepsilon(\zeta), \zeta)$. Observe that on $\mathcal{D}$

$$\psi_\varepsilon(\Lambda_\varepsilon(\zeta), \zeta) = c_\varepsilon + a_4\varepsilon\left[\log\left(-\frac{1}{g_\lambda(\zeta)}\right) + o(1)\right],$$

where o(ε) is small uniformly on $\mathcal{D}$ in the C^1-sense and c_ε is a constant. The linking structure is thus preserved, and a critical point $\zeta_\varepsilon \in \mathcal{D}$ of the above functional with the desired properties thus exists. This concludes the proof. $\qquad\Box$

4.2. *The method of proof*

Our purpose in what follows is to find in each of the situations stated in the theorems, solutions with single or multiple bubbling for some well chosen $\zeta \in \Omega$, which at main order look like

$$U = \sum_{i=1}^{k} (w_{\mu_i,\zeta} + \pi_{\mu_i,\zeta}) \tag{4.8}$$

with μ_1 small and, in case $k \geq 1$, also with $\mu_{i+1} \ll \mu_i$. This requires the understanding of the linearization of the equation around this initial approximation. It is convenient and natural, especially in what concerns multiple bubbling to recast the problem using spherical coordinates around the point ζ and a transformation which takes into account the natural dilation invariance of the equation at the critical exponent. This transformation is a variation of the so-called Emden–Fowler transformation, see [44].

Let ζ be a point in Ω. We consider spherical coordinates $y = y(\rho, \Theta)$ centered at ζ given by

$$\rho = |y - \zeta| \quad \text{and} \quad \Theta = \frac{y - \zeta}{|y - \zeta|},$$

and the transformation $\mathcal{T}$ defined by

$$v(x, \Theta) = \mathcal{T}(u)(x, \Theta) \equiv 2^{1/2} e^{-x} u\left(\zeta + e^{-2x}\Theta\right). \tag{4.9}$$

Denote by D the ζ-dependent subset of $S = \mathbb{R} \times S^2$ where the variables (x, Θ) vary. After these changes of variables, problem (2.1) becomes

$$\begin{cases} 4\Delta_{S^2} v + v'' - v + 4\lambda e^{-4x} v + c_q e^{(q-5)x} v^q = 0 & \text{in } D, \\ v > 0 & \text{in } D, \\ v = 0 & \text{on } \partial D \end{cases} \tag{4.10}$$

with

$$c_q \equiv 2^{-(q-5)/2}.$$

Here and in what follows, "$'$" $= \frac{\partial}{\partial x}$. We observe then that

$$\mathcal{T}(w_{\mu,\zeta})(x, \Theta) = W(x - \xi),$$

where

$$W(x) \equiv (12)^{1/4} e^{-x} \left(1 + e^{-4x}\right)^{-1/2} = 3^{1/4} \left[\cosh(2x)\right]^{-1/2}$$

and $\mu = e^{-2\xi}$. The function W is the unique solution of the problem

$$\begin{cases} W'' - W + W^5 = 0 & \text{on } (-\infty, \infty), \\ W'(0) = 0, \\ W > 0, \quad W(x) \to 0 & \text{as } x \to \pm\infty. \end{cases}$$

We see also that setting

$$\Pi_{\xi,\zeta} \equiv T(\pi_{\mu,\zeta}) \quad \text{with } \mu = e^{-2\xi},$$

then $\Pi = \Pi_{\xi,\zeta}$ solves the boundary value problem

$$-\left(4\Delta_{S^2}\Pi + \Pi'' - \Pi + 4\lambda e^{-4x}\Pi\right) = 4\lambda e^{-4x} W(x - \xi) \quad \text{in } D,$$

$$\Pi = -W(x - \xi) \qquad\qquad\qquad\qquad \text{on } \partial D.$$

An observation useful to fix ideas is that this transformation leaves the energy functional associated invariant. In fact associated to (4.10) is the energy

$$J_{q,\lambda}(v) \equiv 2 \int_D |\nabla_\Theta v|^2 + \frac{1}{2} \int_D \left[|v'|^2 + |v|^2\right]$$

$$- 2\lambda \int_D e^{-4x} v^2 - \frac{c_q}{q+1} \int_D e^{(q-5)x} |v|^{q+1}. \tag{4.11}$$

If $v = T(u)$ we have the identity

$$4E_{q,\lambda}(u) = J_{q,\lambda}(v).$$

Let $\zeta \in \Omega$ and consider the numbers $0 < \xi_1 < \xi_2 < \cdots < \xi_k$. Set

$$W_i(x) = W(x - \xi_i), \qquad \Pi_i = \Pi_{\xi_i,\zeta}, \qquad V_i = W_i + \Pi_i, \qquad V = \sum_{i=1}^{k} V_i.$$

We observe then that $V = T(U)$ where U is given by (4.8) and $\mu_i = e^{-2\xi_i}$. Thus finding a solution of (2.1) which is a small perturbation of U is equivalent to finding a solution of (4.10) of the form $v = V + \phi$ where ϕ is small in some appropriate sense. Then solving (4.10) is equivalent to finding ϕ such that

$$\begin{cases} L(\phi) = -N(\phi) - R, \\ \phi = 0 \quad \text{on } \partial D, \end{cases}$$

where

$$L(\phi) \equiv 4\Delta_{S^2}\phi + \phi'' - \phi + 4\lambda e^{-4x}\phi + q c_q e^{(q-5)x} V^{q-1}\phi, \tag{4.12}$$

$$N(\phi) \equiv c_q e^{(q-5)x}\left[(V+\phi)_+^q - V^q - q V^{q-1}\phi\right] \tag{4.13}$$

and

$$R \equiv c_q e^{(q-5)x} V^q - \sum_{i=1}^{k} W_i^5. \tag{4.14}$$

Rather than solving (4.10) directly, we consider first the following intermediate problem: Given points $\xi = (\xi_1, \ldots, \xi_k) \in \mathbb{R}^k$ and a point $\zeta \in \Omega$, find a function ϕ such that for certain constants c_{ij},

$$\begin{cases} L(\phi) = -N(\phi) - R + \sum_{i,j} c_{ij} Z_{ij} & \text{in } D, \\ \phi = 0 & \text{on } \partial D, \\ \int_D Z_{ij}\phi\, dx\, d\Theta = 0 & \text{for all } i, j, \end{cases} \tag{4.15}$$

where the Z_{ij} span an "approximate kernel" for L. They are defined as follows.

Let $\mathbf{z}_{ij}$ be given by $\mathbf{z}_{ij}(x, \Theta) = \mathcal{T}(z_{ij})$, $i = 1, \ldots, k$, $j = 1, \ldots, 4$, where z_{ij} are respectively given by

$$z_{ij}(y) = \frac{\partial}{\partial \zeta_j} w_{\mu_i,\zeta}(y), \qquad j = 1, \ldots, 3,$$

$$z_{i4}(y) = \mu_i \frac{\partial}{\partial \mu_i} w_{\mu_i,\zeta}(y), \quad i = 1, \ldots, k,$$

with $\mu_i = e^{-2\xi_i}$. We recall that for each i, the functions z_{ij} for $j = 1, \ldots, 4$, span the space of all bounded solutions of the linearized problem

$$\Delta z + 5 w_{\mu_i,\zeta}^4 z = 0 \quad \text{in } \mathbb{R}^3.$$

This implies that the $\mathbf{z}_{ij}$'s satisfy

$$4\Delta_{S^2}\mathbf{z}_{ij} + \mathbf{z}_{ij}'' - \mathbf{z}_{ij} + 5 W_i^4 \mathbf{z}_{ij} = 0.$$

Explicitly, we find that setting

$$Z(x) = 12^{1/4} e^{-3x}\left(1 + e^{-4x}\right)^{-3/2} = 3^{1/4} 2^{-1}\left[\cosh(2x)\right]^{-3/2},$$

we get

$$\mathbf{z}_{ij} = Z(x - \xi_i)\Theta_j, \quad j = 1, 2, 3, \qquad \mathbf{z}_{i4} = W'(x - \xi_i).$$

Observe that

$$\int_{\mathbb{R}\times S^2} \mathbf{z}_{ij}\mathbf{z}_{il} = 0 \quad \text{for } l \neq j.$$

The Z_{ij} are corrections of $\mathbf{z}_{ij}$ which vanish for very large x. Let $\eta_M(s)$ be a smooth cut-off function with

$$\eta_M(s) = 0 \quad \text{for } s < M, \qquad \eta_M(s) = 1 \quad \text{for } s > M+1.$$

We define

$$Z_{ij} = \big(1 - \eta_M(x - \xi_i)\big)\mathbf{z}_{ij},$$

where $M > 0$ is a large fixed number. We will see that with these definitions, problem (4.15) is uniquely solvable if the points ξ_i, ζ satisfy appropriate constrains and q is close enough to 5. After this is done, the remaining task is to adjust the parameters ζ and ξ_i in such a way that all constants $c_{ij} = 0$. We will see that this is indeed possible under the different assumptions of the theorems.

4.3. *The linear problem*

In order to solve problem (4.15) it is necessary to understand first its linear part. Given a function h we consider the problem of finding ϕ such that for certain real numbers c_{ij} the following is satisfied

$$\begin{cases} L(\phi) = h + \sum_{i,j} c_{ij} Z_{ij} & \text{in } D, \\ \phi = 0 & \text{on } \partial D, \\ \int_D Z_{ij}\phi = 0 & \text{for all } i, j. \end{cases} \tag{4.16}$$

Recall that L defined by (4.12) takes the expression

$$L(\phi) = 4\Delta_{S^2}\phi + \phi'' - \phi + 4\lambda e^{-4x}\phi + qc_q e^{(q-5)x} V^{q-1}\phi.$$

We need uniformly bounded solvability in proper functional spaces for problem (4.16), for a proper range of the ξ_i's and ζ. To this end, it is convenient to introduce the following norm. Given an arbitrarily small but fixed number $\sigma > 0$, we define

$$\|f\|_* = \sup_{(x,\Theta)\in D} \omega(x)^{-1}\big|f(x,\Theta)\big| \quad \text{with } \omega(x) = \sum_{i=1}^{k} e^{-(1-\sigma)|x-\xi_i|}.$$

We shall denote by $\mathcal{C}_*$ the set of continuous functions f on $\overline{D}$ such that $\|f\|_*$ is finite.

PROPOSITION 4.1. *Fix a small number $\delta > 0$ and take the cut-off parameter $M > 0$ of Section 4.2 large enough. Then there exist positive numbers ε_0, δ_0, R_0 and a constant $C > 0$ such that if $|q - 5| < \varepsilon_0$,*

$$0 \leqslant \lambda \leqslant \lambda_1 - \delta, \qquad \operatorname{dist}(\zeta, \partial\Omega) > \delta_0, \tag{4.17}$$

and the numbers $0 < \xi_1 < \xi_2 < \cdots < \xi_k$ satisfy

$$R_0 < \xi_1, \qquad R_0 < \min_{1 \leqslant i < k} (\xi_{i+1} - \xi_i) \tag{4.18}$$

with $\xi_k < \frac{\delta_0}{|q-5|}$ if $q \neq 5$, then for any $h \in C^\alpha(D)$ with $\|h\|_* < +\infty$, problem (4.16) admits a unique solution $\phi \equiv T(h)$. Besides,

$$\left\| T(h) \right\|_* \leqslant C \|h\|_* \quad \text{and} \quad |c_{ij}| \leqslant C \|h\|_*.$$

4.4. *Solving the nonlinear problem*

In this section we will solve problem (4.15). We assume that the conditions in Proposition 4.1 hold. We have the following result.

LEMMA 4.3. *Under the assumptions of Proposition 4.1 there exist numbers $c_0 > 0$, $C_1 > 0$ such that if ξ and ζ are additionally such that $\|R\|_* < c_0$, then problem (4.15) has a unique solution ϕ which satisfies*

$$\|\phi\|_* \leqslant C_1 \|R\|_*.$$

PROOF. In terms of the operator T defined in Proposition 4.1, problem (4.15) becomes

$$\phi = T\big(N(\phi) + R\big) \equiv A(\phi), \tag{4.19}$$

where $N(\phi)$ and R where defined in (4.13) and (4.14). For a given R, let us consider the region

$$\mathcal{F}_\gamma \equiv \big\{ \phi \in C(\overline{D}) : \|\phi\|_* \leqslant \gamma \|R\|_* \big\}$$

for some $\gamma > 0$, to be fixed later. From Proposition 4.1, we get

$$\left\| A(\phi) \right\|_* \leqslant C_0 \big[\left\| N(\phi) \right\|_* + \|R\|_* \big].$$

On the other hand, we can represent

$$N(\phi) = c_q e^{(q-5)x} q(q-1) \int_0^1 (1 - t)\, dt\, [V + t\phi]^{q-2} \phi^2,$$

so that (making $q - 5$ smaller if necessary) $|N(\phi)| \leqslant C_1 |\phi|^2$, and hence $\|N(\phi)\|_* \leqslant C_1 \|\phi\|_*^2$. It is also easily checked that N satisfies, for $\phi_1, \phi_2 \in \mathcal{F}_\gamma$,

$$\left\| N(\phi_1) - N(\phi_2) \right\|_* \leqslant C_2 \gamma \|R\|_* \|\phi_1 - \phi_2\|_*.$$

Hence for a constant C_3 depending on C_0, C_1, C_2, we get

$$\left\| A(\phi) \right\|_* \leqslant C_3 \left[\gamma^2 \|R\|_* + 1 \right] \|R\|_*,$$
$$\left\| A(\phi_1) - A(\phi_2) \right\|_* \leqslant C_3 \gamma \|R\|_* \|\phi_1 - \phi_2\|_*.$$

With the choices

$$\gamma = 2C_3, \qquad \|R\|_* \leqslant c_0 = \frac{1}{4C_3^2},$$

we get that A is a contraction mapping of $\mathcal{F}_\gamma$, and therefore a unique fixed point of A exists in this region. This concludes the proof. $\qquad\square$

After problem (4.15) has been solved, we will find solutions to the full problem (4.10) if we manage to adjust the pair (ξ, ζ) in such a way that $c_{ij}(\xi, \zeta) = 0$ for all i, j. This is the *reduced problem*. A nice feature of this system of equations is that it turns out to be equivalent to finding critical points of a functional of the pair (ξ, ζ) which is close, in appropriate sense, to the energy of the single or multiple bubble V. We make this precise in the next section for the case of single bubbling, $k = 1$.

4.5. *Variational formulation of the reduced problem for $k = 1$*

Next we assume $k = 1$ in problem (4.15). We omit the subscript $i = 1$ in c_{ij}, Z_{ij} and ξ_i. Then in order to obtain a solution of (4.10) we need to solve the system of equations

$$c_j(\xi, \zeta) = 0 \quad \text{for all } j = 1, \dots, 4. \tag{4.20}$$

If (4.20) holds, then $v = V + \phi$ will be a solution to (4.10). This system turns out to be equivalent to a variational problem, as we discuss next.

Let us consider the functional $J_{q,\lambda}$ in (4.11), the energy associated to problem (4.10). Let us define

$$F(\mu, \zeta) \equiv J_{q,\lambda}(V + \phi), \quad \mu = e^{-2\xi}, \tag{4.21}$$

where $\phi = \phi(\xi, \zeta)$ is the solution of problem (4.15) given by Proposition 4.1. Critical points of F correspond to solutions of (4.20) under a mild assumption that will be satisfied in the proofs of the theorems, as we shall see further.

LEMMA 4.4. *Under the assumptions of Proposition 4.1, the functional $F(\zeta,\xi)$ is of class C^1. Assume additionally that R in (4.14) satisfies that $\|R\|_* \leqslant \mu^{8\sigma}$ where $\sigma > 0$ is the number in the definition of the $*$-norm. Then for all $\mu > 0$ sufficiently small, if $\nabla F(\xi,\zeta) = 0$ then (ξ,ζ) satisfies system (4.20).*

We have now all the elements for the proof of our main results regarding single bubbling.

4.6. *Proof of Theorem 1.2, part* (a): *single bubbling*

We choose μ as in (4.6),

$$\mu = -\varepsilon \frac{a_4}{a_1} \frac{1}{g_\lambda(\zeta)} \Lambda,$$

where $\varepsilon = q - 5$. We have to find a critical point of the functional $F(\mu,\zeta)$ in (4.21) for $q = 5 + \varepsilon$. Consider

$$R = c_q e^{(q-5)x} \left(W(x - \xi) + \Pi_\xi(x, \Theta) \right)^{5+\varepsilon} - W(x - \xi)^5,$$

where $e^{-2\xi} = \mu$. We write as usual $W_1 = W(x - \xi)$, $V = W_1 + \Pi_\xi$. Then we can decompose $R = R_1 + R_2 + R_3 + R_4$, where

$$R_1 \equiv e^{\varepsilon x}\left(V^{5+\varepsilon} - V^5\right), \qquad R_2 \equiv V^5\left(e^{\varepsilon x} - 1\right),$$

$$R_3 \equiv V^5 - W_1^5, \qquad R_4 \equiv (c_q - 1)e^{\varepsilon x} V^{5+\varepsilon}.$$

We have

$$R_1 = \varepsilon e^{\varepsilon x} \int_0^1 (1 - t)\, dt \left(V^{5+t\varepsilon} \log V\right),$$

from where it follows that

$$|R_1| \leqslant C\varepsilon e^{\varepsilon \xi} e^{\varepsilon |x - \xi|} V^{4+1/2} \leqslant C\varepsilon V^4.$$

Since $|\Pi_\xi| \leqslant C e^{-(x+\xi)} \leqslant C e^{-|x-\xi|}$, we get $|R_1| \leqslant C\varepsilon e^{-4|x-\xi|}$ and hence $\|R_1\|_* \leqslant C\varepsilon$. Direct differentiation of the above expression, using the bounds for derivatives of Π_ξ yields as well

$$\left\|\partial_{\xi^2}^2 R_1\right\|_* + \left\|\partial_{\zeta\xi}^2 R_1\right\|_* + \left\|\partial_\zeta R_1\right\|_* \leqslant \varepsilon.$$

Let us denote $\nabla = [\partial_\xi, \partial_\zeta]$. Thus we have

$$\|R_1\|_* + \|\nabla R_1\|_* + \|\nabla \partial_\xi R_1\|_* \leqslant C\varepsilon.$$

Observe that the same estimate is also valid for R_4. On the other hand, we have

$$R_2 = V^5 \left(e^{\varepsilon x} - 1 \right) = \varepsilon x V^5 \int_0^1 e^{t\varepsilon x} \, dt.$$

Since $\xi \sim c \log(1/\varepsilon)$, we obtain for R_2 and derivatives the bounds

$$\|R_2\|_* + \|\nabla R_2\|_* + \|\nabla \partial_\xi R_2\|_* \leqslant C\varepsilon |\log \varepsilon|.$$

Finally, for

$$R_3 = 5 \int_0^1 (1-t) \, dt \, (W_1 + t\Pi_\xi)^4 \Pi_\xi,$$

we find the bound

$$|R_3| \leqslant C e^{-\xi - x - 4|x-\xi|} \leqslant C e^{-2\xi - |x-\xi|},$$

and similarly for derivatives. We get, recalling that $e^{-2\xi} \equiv \mu \leqslant C\varepsilon$,

$$\|R_3\|_* + \|\nabla R_3\|_* + \|\nabla \partial_\xi R_3\|_* \leqslant C e^{-2\xi}.$$

Concerning R_4, a direct computation gives $|R_4| \leqslant C\varepsilon e^{-5|x-\xi|}$. Thus for full R we have

$$\|R\|_* + \|\nabla R\|_* + \|\nabla \partial_\xi R\|_* \leqslant C\varepsilon |\log \varepsilon|.$$

It follows from Lemma 5.20 that for this choice of μ,

$$F(\xi, \zeta) = J_{q,\lambda}(V) + \mu^2 |\log \mu|^2 \theta(\xi, \zeta)$$

with $|\theta| + |\nabla \partial_\xi \theta| + |\nabla \theta| \leqslant C$. Define $\psi_\varepsilon(\Lambda, \zeta) = F(\frac{1}{2} \log \frac{1}{\mu}, \zeta)$ with μ as above. A critical point for ψ_ε is in correspondence with one of F. We conclude that

$$\psi_\varepsilon(\Lambda, \zeta) = 4 E_{5+\varepsilon,\mu}(U_{\mu,\zeta}) + \varepsilon \theta_\varepsilon(\Lambda, \zeta)$$

with θ_ε as in Lemma 4.2. The lemma thus applies to predict a critical point of ψ_ε and the proof of part (a) is complete.

4.7. *Proof of Theorem* 1.2, *part* (b): *multiple bubbling*

Let us consider the solution $\phi(\xi, \zeta)$ of (4.15) given by Proposition 4.1 where $\xi = (\xi_1, \xi_2, \ldots, \xi_k)$. Choosing $\zeta = 0$ makes ϕ symmetric in the Θ_i variables, which automatically yields $c_{ij} = 0$ for all $i = 1, \ldots, k$ and $j = 1, 2, 3$. Thus we just need to adjust ξ in such a way that $c_{i4} = 0$ for $i = 1, \ldots, k$. Arguing exactly as in the proof of Lemma 4.4 we get that this is equivalent to finding a critical point of the functional $F(\xi) = J_{q,\lambda}(V + \phi)$, where ζ has been fixed to be zero. Similarly as before, we find now that

$$F(\xi) = J_{q,\lambda}(V) + \left(\|R\|_*^2 + \|\partial_\xi R\|_*^2 \right)\theta(\xi),$$

where θ and its first derivative are continuous and uniformly bounded in large ξ.

In what remains of this section we fix a number $\delta > 0$, set $\varepsilon = q - 5 > 0$ and choose $\mu_i = e^{-2\xi_i}$ in order that

$$\mu_1 = \varepsilon \Lambda_1, \qquad \mu_{j+1} = \mu_j (\Lambda_{j+1}\varepsilon)^2, \quad j = 1, \ldots, k-1, \tag{4.22}$$

with

$$\delta < \Lambda_j < \delta^{-1}, \quad j = 1, \ldots, k. \tag{4.23}$$

Let us measure the size of $\|R\|_*$ and $\|\partial_\xi R\|_*$ for this ansatz. We can now decompose $R = R_1 + R_2 + R_3 + R_4 + R_5$ where

$$R_1 \equiv e^{\varepsilon x}\left(V^{5+\varepsilon} - V^5\right), \qquad R_2 \equiv V^5\left(e^{\varepsilon x} - 1\right),$$

$$R_3 \equiv V^5 - \left(\sum_i W_i\right)^5, \qquad R_4 \equiv \left(\sum_i W_i\right)^5 - \sum_i W_i^5,$$

$$R_5 \equiv (c_q - 1)e^{\varepsilon x} V^{5+\varepsilon}.$$

We can estimate

$$|R_4| \leqslant C \sum_{i=1}^{k-1} e^{-(\xi_{i+1}-\xi_i)} e^{-3|x-\xi_i|},$$

hence $\|R_4\|_* \leqslant C\varepsilon$, a similar bound being valid for its derivatives in ξ_i's. The quantities R_j for $j = 1, 2, 3, 5$ can be estimated in exactly the same way as in the proof of Theorem 1.2, part (a). Thus $\|R\|_* + \|\partial_\xi R\|_* \leqslant C\varepsilon|\log \varepsilon|$. Let us set $\Lambda = (\Lambda_1, \ldots, \Lambda_k)$ and define $\psi_\varepsilon(\Lambda) = F(\xi)$ with ξ given by (4.22). We need to find a critical point of ψ_ε. We have proven that

$$\psi_\varepsilon(\Lambda) = J_{q,\lambda}(V) + O\left(\varepsilon^2|\log \varepsilon|^2\right)\theta_\varepsilon(\Lambda), \tag{4.24}$$

where θ_ε and its first derivative are uniformly bounded. We have the validity of the following fact, whose proof we postpone for the moment

$$\frac{1}{4} J_{q,\lambda}(V) = ka_0 + \left[\psi_*(\Lambda) + o(1)\right]\varepsilon + \frac{1}{2}k(k+1)a_4\varepsilon|\log \varepsilon|, \tag{4.25}$$

where

$$\psi_*(\Lambda) = a_1 g_\lambda(0)\Lambda_1 + ka_4 \log \Lambda_1 + \sum_{j=2}^{k}\left[(k - j + 1)a_4 \log \Lambda_j - a_6\Lambda_j\right]$$

and the term o(1) as $\varepsilon \to 0$ is uniformly small in C^1-sense on parameters Λ_j satisfying (4.23). Here the constants a_0, a_1, a_4 are the same as those in Lemma 4.1, while $a_6 = 16\pi\sqrt{3}$. The assumption $g_\lambda(0) < 0$ implies the existence of a unique critical point Λ_* which can easily be solved explicitly. It follows that o(1) C^1 perturbation of ψ_* will have a critical point located at o(1) distance of Λ_*. After this observation, the combination of relations (4.25) and (4.24) give the existence of a critical point of ψ_ε close to Λ_* which translates exactly as the result of Theorem 1.2, part (a).

It only remains to establish the validity of expansion (4.25). We recall that

$$V = \sum_{i=1}^{k} V_i = \sum_{i=1}^{k} W_i + \Pi_i = \mathcal{T}(U),$$

where $U = \sum_{i=1}^{k} w_i + \pi_i$, and we denote $w_i = w_{\mu_i,0}$, $\pi_i = \pi_{\mu_i,0}$, $U_i = w_i + \pi_i$. We have that $J_{q,\lambda}(V) = 4E_{q,\lambda}(U)$, where $q = 5 + \varepsilon$. Observe that we can write

$$E_{q,\lambda}(U) = E_{5,\lambda}(U) + \mathcal{R},$$

where

$$\mathcal{R} \equiv -\frac{1}{6} \int_D \left(e^{(q-5)x} - 1 \right) |V|^6 + 4\pi A_q.$$

A direct computation yields

$$A_q = k(q-5)\left(\frac{1}{6} \int_{-\infty}^{\infty} W^6 \log W \, dx + \frac{1}{36} \int_{-\infty}^{\infty} W^6 \, dx \right) + o(\varepsilon).$$

On the other hand,

$$\mathcal{R} - 4\pi A_q = -\frac{1}{6} \int_D \left[e^{(q-5)x} - 1 \right] V^6 \, dx$$

$$= -\frac{1}{6}(q-5)4\pi \int_D x V^6 \, dx + o(\varepsilon)$$

$$= -\frac{1}{6}(q-5)\left(\sum_{i=1}^{k} \xi_i \right) \int_{-\infty}^{\infty} W^6 \, dx + o(\varepsilon)$$

$$= a_4(q-5) \sum_{j=1}^{k} \log \mu_j + o(\varepsilon).$$

Now we have

$$E_{5,\lambda}(U) = \sum_{j=1}^{k} E_{5,\lambda}(U_j) + \frac{1}{6} \int_D \left[\sum_{i=1}^{k} V_i^6 - \left(\sum_{i=1}^{k} V_i \right)^6 + 6 \sum_{i<j} W_i^5 V_j \right] \tag{4.26}$$

since

$$E_{5,\lambda}(U) - \sum_{j=1}^{k} E_{5,\lambda}(U_j) - \int_D \left[\sum_{i=1}^{k} V_i^6 - \left(\sum_{i=1}^{k} V_i \right)^6 \right]$$

$$= \sum_{i<j} \int_D \left(2\nabla_\Theta V_i \nabla_\Theta V_j + V_i' V_j' + V_i V_j - 2\lambda e^{-4x} V_i V_j \right)$$

$$= \sum_{i<j} \int_D \left(-4\Delta_{S^2} V_i - V_i'' + V_i - 4\lambda e^{-4x} V_i \right) V_j$$

$$= \sum_{i<j} \int_D W_i^5 V_j.$$

To estimate the quantities in (4.26), we consider the numbers

$$\chi_1 = 0, \qquad \chi_l = \frac{1}{2}(\xi_{l-1} + \xi_l), \quad l = 2, \ldots, k, \qquad \chi_{k+1} = +\infty,$$

and decompose

$$E_{5,\lambda}(U) - \sum_{j=1}^{k} E_{5,\lambda}(U_j) = \sum_{1 \leqslant l \leqslant k,\, j > l} \int_{D \cap \{\chi_l < x < \chi_{l+1}\}} V_l^5 V_j + B.$$

A straightforward computation yields $B = o(\varepsilon)$. On the other hand,

$$\sum_{1 \leqslant l \leqslant k,\, j > l} \int_{D \cap \{\chi_l < x < \chi_{l+1}\}} V_l^5 V_j$$

$$= 4\pi \sum_{l=1}^{k} \int_{\chi_l}^{\chi_{l+1}} W_l^5 W_{l+1} + o(\varepsilon)$$

$$= 4\pi \int_{\chi_l}^{\chi_{l+1}} W_l^5 W_{l+1} + o(\varepsilon)$$

$$= 4\pi \int_{\chi_l - \xi_l}^{\chi_{l+1} - \xi_l} W^5(x) W \left(x - (\xi_{l+1} - \xi_l) \right) + o(\varepsilon)$$

$$= 4\pi \sum_{l=1}^{k-1} e^{-|\xi_{l+1} - \xi_l|} (12)^{1/4} \int_{-\infty}^{\infty} e^x W(x)^5 + o(\varepsilon)$$

$$= a_6 \sum_{j=1}^{k-1} \left(\frac{\mu_{j+1}}{\mu_j} \right)^{1/2} + o(\varepsilon).$$

Taking into account the estimate given in Lemma 4.4 and the above estimates, we get (4.25) in the uniform sense. Similar arguments yield that the remainder is as well o(ε) small after a differentiation with respect to the ξ_i's. This concludes the proof. $\square$

REMARK 4.1. Single and multiple bubbling related with supercritical nonlinearity are also present in semilinear elliptic problem with Neumann boundary conditions. As we see below, what determines now the location of the bubble, or of the tower of bubbles in a domain with symmetries, is the presence of a nontrivial critical point situation for the mean curvature of $\partial\Omega$, that we denote by $\mathcal{H}$, with positive value.

Let Ω be a bounded domain in $\mathbb{R}^N$, $N \geq 3$, with smooth boundary $\partial\Omega$ and consider the boundary value problem

$$\begin{cases} -d^2 \Delta u + u = u^p & \text{in } \Omega, \\ u > 0 & \text{in } \Omega, \\ \frac{\partial u}{\partial \nu} = 0 & \text{on } \partial\Omega, \end{cases} \tag{4.27}$$

where $p > 1$ and $d > 0$.

The works [58,69,70] have dealt with precise analysis of least energy solutions to this problem in the subcritical case, $1 < p < \frac{N+2}{N-2}$ namely solutions which minimize the Rayleigh quotient

$$Q(u) = \frac{d^2 \int_\Omega |\nabla u|^2 + \int_\Omega |u|^2}{(\int_\Omega |u|^{p+1})^{2/(p+1)}}, \quad u \in H^1(\Omega) \setminus \{0\}, \tag{4.28}$$

for small d. From those works, it became known that for d sufficiently small, a minimizer u_d of Q has a unique local maximum point x_d which is located on the boundary. Besides, $\mathcal{H}(x_d) \to \max_{x \in \partial\Omega} \mathcal{H}(x)$ and this solution decays exponentially which implies indeed the presence of a very sharp, bounded spike for the solution around x_d.

Concentration phenomena of this type occurs as well in the critical case $p = \frac{N+2}{N-2}$, but with some important differences. Since compactness of the embedding of $H^1(\Omega)$ into $L^{p+1}(\Omega)$ is lost, existence of minimizers of $Q(u)$ becomes nonobvious and in general not true for large d as established in [57]. It is the case however, as shown in [2,86], that such a minimizer does exist if d is sufficiently small. The profile and asymptotic behavior of this least energy solution has been analyzed in [3,68,79]. Again only one local maximum point x_d located around a point of maximum for the mean curvature of $\partial\Omega$ exists. However, unlike the subcritical case now its maximum value $M_d = u_d(x_d) \to +\infty$. Not only that, an important difference with the subcritical case is that now mean curvature is required to be positive at this critical point. In fact, nonnegativity of curvature is actually necessary for existence [4,47,79].

Consider now problem (4.27) when the power p is supercritical, namely $p > \frac{N+2}{N-2}$. Recently in [19] it has been found that if $N \geq 4$, d is left fixed and one considers the exponent p as a parameter approaching the critical exponent from below, then single-bubbling solutions exist in certain cases. In particular, they find existence of single-bubble solutions with maximum points located on the boundary, near critical points of mean curvature with negative value.

Let the parameter d be fixed, with no loss of generality $d = 1$. In [36] we prove that, given a nontrivial critical point situation of the mean curvature of $\partial\Omega$ with positive critical value, a solution exhibiting boundary bubbling around such a point of the following problem

$$\begin{cases} -\Delta u + u = u^{(N+2)/(N-2)+\varepsilon} & \text{in } \Omega, \\ u > 0 & \text{in } \Omega, \\ \frac{\partial u}{\partial v} = 0 & \text{on } \partial\Omega \end{cases} \tag{4.29}$$

exists, as $\varepsilon \to 0$, $\varepsilon > 0$.

Not only this: we are able to construct solutions with just one maximum point for which multiple bubbling is present. For instance if Ω is a ball, there exists a solution whose shape is that of a tower, constituted by superposition of an arbitrary number of single bubbles of different blow-up orders. This phenomenon actually takes place just provided that Ω is symmetric with respect to the first $(N-1)$ variables, and $0 \in \partial\Omega$ is a point with positive mean curvature. Indeed, given $k \geqslant 1$, there exists for all sufficiently small $\varepsilon > 0$ a solution u_ε of (4.29) of the form

$$u_\varepsilon(y) = \alpha_N \sum_{i=1}^{k} \left(\frac{1}{1 + \lambda_i^2 \varepsilon^{-2+(1-i)(4/(N-2))} |y|^2} \right)^{(N-2)/2}$$
$$\times \lambda_i^{(N-2)/2} \varepsilon^{-(N-2)/2-i+1} \left(1 + o(1)\right)$$

for $N \geqslant 4$ and

$$u_\varepsilon(y) = \alpha_3 \sum_{i=1}^{k} \left(\frac{1}{1 + \lambda_i^2 \varepsilon^{2-4i} |\log \varepsilon|^2 |y|^2} \right)^{1/2} \lambda_i^{1/2} \varepsilon^{1/2-i} |\log \varepsilon|^{1/2} \left(1 + o(1)\right)$$

for $N = 3$, where $o(1) \to 0$ uniformly in $\overline{\Omega}$ and λ_i are explicit numbers.

We would like to mention that existence of solutions to problem (4.29) which blow up at an interior point of the domain Ω has been obtained in [80] in the case of dimension $N = 3$ and in [81] for $N \geqslant 4$.

5. Liouville-type equations

5.1. *Proof of Theorem* 1.3

We present the construction of blowing-up families of solutions for problem (1.14) which lifts the nondegeneracy assumption of [9]. More precisely, we consider the role of *nontrivial critical values* of φ_m in existence of solutions of (1.14), which is an example of nontrivial critical point situation (see [31]).

Let $\mathcal{D}$ be an open set in Ω^m compactly contained in $\widetilde{\Omega}^m$ with smooth boundary. We say that φ_m *links in* $\mathcal{D}$ *at critical level* $\mathcal{C}$ *relative to* B *and* B_0 if B and B_0 are closed subsets of $\overline{\mathcal{D}}$ with B connected and $B_0 \subset B$ such that the following conditions hold: Let us set Γ to be the class of all maps $\Phi \in C(B, \mathcal{D})$ with the property that there exists a function $\Psi \in C([0, 1] \times B, \mathcal{D})$ such that

$$\Psi(0, \cdot) = \mathrm{Id}_B, \qquad \Psi(1, \cdot) = \Phi, \qquad \Psi(t, \cdot)|_{B_0} = \mathrm{Id}_{B_0} \quad \text{for all } t \in [0, 1].$$

We assume

$$\sup_{y \in B_0} \varphi_m(y) < C \equiv \inf_{\Phi \in \Gamma} \sup_{y \in B} \varphi_m(\Phi(y)), \tag{5.1}$$

and for all $y \in \partial \mathcal{D}$ such that $\varphi_m(y) = C$, there exists a vector τ_y tangent to $\partial \mathcal{D}$ at y such that

$$\nabla \varphi_m(y) \cdot \tau_y \neq 0. \tag{5.2}$$

Under these conditions a critical point $\bar{y} \in \mathcal{D}$ of φ_m with $\varphi_m(\bar{y}) = C$ exists, as a standard deformation argument involving the negative gradient flow of φ_m shows. Condition (5.1) is a general way of describing a change of topology in the level sets $\{\varphi_m \leqslant c\}$ in $\mathcal{D}$ taking place at $c = C$, while (5.2) prevents intersection of the level set C with the boundary. It is easy to check that the above conditions hold if

$$\inf_{x \in \mathcal{D}} \varphi_m(x) < \inf_{x \in \partial \mathcal{D}} \varphi_m(x) \quad \text{or} \quad \sup_{x \in \mathcal{D}} \varphi_m(x) > \sup_{x \in \partial \mathcal{D}} \varphi_m(x),$$

namely the case of (possibly degenerate) local minimum or maximum points of φ_m. The level C may be taken in these cases respectively as that of the minimum and the maximum of φ_m in $\mathcal{D}$. These hold also if φ_m is C^1-close to a function with a nondegenerate critical point in $\mathcal{D}$. We call C a nontrivial critical level of φ_m in $\mathcal{D}$.

THEOREM 5.1. *Let $m \geqslant 1$ and assume that there is an open set $\mathcal{D}$ compactly contained in $\widetilde{\Omega}^m$ where φ_m has a nontrivial critical level C. Then, there exists a solution u_ε, with*

$$\lim_{\varepsilon \to 0} \varepsilon^2 \int_\Omega e^{u_\varepsilon} = 8m\pi.$$

Moreover, there is an m-tuple $(x_1^\varepsilon, \ldots, x_m^\varepsilon) \in \mathcal{D}$, such that, as $\varepsilon \to 0$,

$$\nabla \varphi_m(x_1^\varepsilon, \ldots, x_m^\varepsilon) \to 0, \qquad \varphi_m(x_1^\varepsilon, \ldots, x_m^\varepsilon) \to C,$$

for which u_ε remains uniformly bounded on $\Omega \setminus \bigcup_{j=1}^m B_\delta(x_i^\varepsilon)$, and

$$\sup_{B_\delta(x_i^\varepsilon)} u_\varepsilon \to +\infty$$

for any $\delta > 0$.

We will see that if Ω is not simply connected, such a set $\mathcal{D}$ actually exists for any $m \geqslant 1$, thus yielding the result of Theorem 1.3. For $m = 1$, a multiplicity result is also available: if Ω has d holes, then there exist at least $d + 1$ solutions u_ε, with

$$\lim_{\varepsilon \to 0} \varepsilon^2 \int_\Omega e^{u_\varepsilon} = 8\pi.$$

If Ω has d holes, namely d bounded components for its complement, then at least $d + 1$ solutions u_ε with

$$\lim_{\varepsilon \to 0} \varepsilon^2 \int_\Omega e^{u_\varepsilon} = 8\pi$$

exist. We observe that $\varphi_1(\xi) = H(\xi)$. Since $H(\xi)$ approaches $+\infty$ as ξ approaches $\partial\Omega$, Ljusternik–Schnirelman theory yields that H has at least $\mathrm{cat}(\Omega) = d + 1$ critical points with critical levels characterized through $d + 1$ min–max quantities. The same property is thus inherited for $F(\xi)$ and the fact is thus established.

We start providing an ansatz for solutions of problem (1.14). The "basic cells" for the construction of an approximate solution of problem (1.14) are the radially symmetric solutions of the problem

$$\begin{cases} \Delta u + e^u = 0 & \text{in } \mathbb{R}^2, \\ u(x) \to -\infty & \text{as } |x| \to \infty, \end{cases} \tag{5.3}$$

which are given by the one-parameter family of functions

$$\omega_\mu(r) = \log \frac{8\mu^2}{(\mu^2 + r^2)^2}, \tag{5.4}$$

where μ is any positive number.

Let m be a positive integer and choose m distinct points in Ω, say $\xi_1, \ldots, \xi_m$. Let μ_j, $j = 1, \ldots, m$, be positive numbers. We observe that the function

$$u_j(x) = \log \frac{8\mu_j^2}{(\mu_j^2 \varepsilon^2 + |x - \xi_j|^2)^2} = \omega_{\mu_j}\left(\frac{|x - \xi_j|}{\varepsilon}\right) + 4\log\frac{1}{\varepsilon}$$

satisfies in entire $\mathbb{R}^2$

$$\Delta u_j + \varepsilon^2 e^{u_j} = 0.$$

We would like to take $\sum_{j=1}^m u_j$ as a first approximation to a solution of the equation. We need to modify it in order to satisfy zero Dirichlet boundary conditions. Let $H_j(x)$ be the solution of

$$\begin{cases} -\Delta H_j(x) = 0 & \text{in } \Omega, \\ H_j(x) = -\omega_{\mu_j}\left(\frac{|x - \xi_j|}{\varepsilon}\right) - 4\log\frac{1}{\varepsilon} & \text{on } \partial\Omega. \end{cases} \tag{5.5}$$

We consider as initial approximation $U = \sum_{i=1}^{m}(u_j + H_j)$, which by definition satisfies the boundary conditions. This approximation is less accurate near ξ_j than u_j alone unless $H_j(\xi_j) + \sum_{i=1, i \neq j}^{m}[H_i(\xi_j) + u_i(\xi_j)] \sim 0$ as $\varepsilon \to 0$. We can achieve this by further adjusting the numbers μ_j. As we will justify below, the good choice of these numbers is

$$\log 8\mu_j^2 = H(\xi_j, \xi_j) + \sum_{l \neq j} G(\xi_l, \xi_j), \tag{5.6}$$

where G and H are Green's function and its regular part as defined in the Introduction. Thus we consider the first approximation

$$U = \sum_{i=1}^{m}(u_i + H_i) = \sum_{i=1}^{m}\left(\omega_i\left(\frac{|x - \xi_i|}{\varepsilon}\right) - \log \varepsilon^4 + H_i\right), \tag{5.7}$$

where $\omega_i = \omega_{\mu_i}$ and with the numbers μ_j defined in (5.6). Let us analyze the asymptotic behavior of H_j as $\varepsilon \to 0$. We observe that for $x \in \partial\Omega$,

$$H_j(x) = -2\log \frac{1}{\mu_j^2 \varepsilon^2 + |x - \xi_j|^2} - \log 8\mu_j^2$$

from where it follows that

$$H_j(x) = H(x, \xi_j) - \log 8\mu_j^2 + O(\mu_j^2 \varepsilon^2), \tag{5.8}$$

uniformly in C^2-sense for x on compact subsets of Ω. Observe also that, away from each ξ_j,

$$w_j = \log 8\mu_j^2 + 4\log \frac{1}{|x - \xi_j|} + O(\mu_j^2 \varepsilon^2)$$

and hence

$$w_j(x) + H_j(x) = G(x, \xi_j) + O(\varepsilon^2), \tag{5.9}$$

where the term $O(\cdot)$ is uniform in C^2-sense on compact subsets of $\overline{\Omega} \setminus \{\xi_j\}$.

A useful observation is that u satisfies equation (1.14) if and only if

$$v(y) = u(\varepsilon y) - 4\log \frac{1}{\varepsilon}$$

satisfies

$$\begin{cases} \Delta v + e^v = 0 & \text{in } \Omega_\varepsilon, \\ u > 0 & \text{in } \Omega_\varepsilon, \\ v = -4\log \frac{1}{\varepsilon} & \text{on } \partial\Omega_\varepsilon, \end{cases} \tag{5.10}$$

where $\Omega_\varepsilon = \varepsilon^{-1}\Omega$. We also write $\xi_i' = \varepsilon^{-1}\xi_i$ and define the initial approximation in expanded variables as $V(y) = U(\varepsilon y) - 4\log\frac{1}{\varepsilon}$. We want to measure how well V solves the above problem. Let us fix a small number $\delta > 0$ and observe that $e^{V(y)} = \varepsilon^4 e^{U(x)}$ with $x = \varepsilon y$, hence we see that

$$k(\varepsilon y)e^{V(y)} = O(\varepsilon^4) \quad \text{if } \left|y - \xi_j'\right| > \frac{\delta}{\varepsilon} \text{ for all } j = 1, \ldots, m. \tag{5.11}$$

Similarly, $\Delta V(y) = \varepsilon^2 \Delta U(x)$ and (5.9) implies

$$\Delta V(y) = O(\varepsilon^4) \quad \text{if } \left|y - \xi_j'\right| > \frac{\delta}{\varepsilon} \text{ for all } j = 1, \ldots, m. \tag{5.12}$$

On the other hand, assume that for certain j, $|y - \xi_j'| < \delta$. Then setting $y = \xi_j' + z$ we get

$$e^{V(y)} = \frac{8\mu_j^2}{(\mu_j^2 + |z|^2)^2}$$

$$\times \exp\left(H_j(\xi_j + \varepsilon z) \right.$$

$$\left. + \sum_{l \neq j} \log\left[\frac{8\mu_l^2}{(\mu_l^2 \varepsilon^2 + |\xi_l - \xi_j + \varepsilon z|^2)^2} \right] + H_l(\xi_j + \varepsilon z) \right).$$

Now, by definition,

$$\log\frac{1}{|\xi_l - \xi_j|^4} + H(\xi_l, \xi_j) = G(\xi_l, \xi_j).$$

Taking into account this relation, the asymptotic expansion (5.8) and the definition of the numbers μ_l in (5.6) we get then that

$$e^{V(y)} = \frac{8\mu_j^2}{(\mu_j^2 + |y - \xi_j'|^2)^2}[1 + O(\varepsilon z) + O(\varepsilon^2)], \quad \left|y - \xi_j'\right| < \frac{\delta}{\varepsilon}. \tag{5.13}$$

We also have in this region

$$\Delta V(y) = \Delta w_{\mu_j}\left(|y - \xi_j'|\right) + O(\varepsilon^4) = -\frac{8\mu_j^2}{(\mu_j^2 + |y - \xi_j'|^2)^2} + O(\varepsilon^4). \tag{5.14}$$

In summary, combining (5.11)–(5.14) we have established the following fact: if we set

$$R = \Delta V(y) + e^{V(y)}, \tag{5.15}$$

then

$$|R(y)| \leqslant C\varepsilon \sum_{j=1}^{m} \frac{1}{1+|y-\xi_j'|^3}.$$

(5.16)

Let us set

$$W(y) = e^{V(y)}.$$

We have

$$W(y) = \sum_{j=1}^{m} \frac{8\mu_j^2}{(\mu_j^2 + |y-\xi_j'|^2)^2} [1 + \theta_\varepsilon(y)],$$

and θ_ε has the property that, for some constant C independent of ε,

$$|\theta_\varepsilon(y)| \leqslant C\varepsilon \sum_{j=1}^{m} [|y-\xi_j'| + 1].$$

In terms of ϕ, problem (5.10) becomes

$$\begin{cases} L(\phi) := \Delta\phi + W\phi = -[R + N(\phi)] & \text{in } \Omega_\varepsilon, \\ \phi = 0 & \text{on } \partial\Omega_\varepsilon, \end{cases}$$

(5.17)

where

$$N(\phi) = W[e^\phi - 1 - \phi].$$

(5.18)

A main step in solving problem (5.17) for small ϕ under a suitable choice of the points ξ_j is that of a solvability theory for the linear operator L. In developing this theory we will take into account the invariance, under translations and dilations, of the problem $\Delta w + e^w = 0$ in $\mathbb{R}^2$.

If we center the system of coordinates at, say ξ_j', by setting $z = y - \xi_j'$, then the operator L formally approaches the linear operator in $\mathbb{R}^2$,

$$L_j(\phi) = \Delta\phi + \frac{8\mu_j^2}{(\mu_j^2 + |z|^2)^2}\phi,$$

namely, equation $\Delta v + e^v = 0$ linearized around the radial solution $v_j(z) = \log \frac{8\mu_j^2}{(\mu_j^2 + |z|^2)^2}$. An important fact is the nondegeneracy of v_j modulo the natural invariance of the equations under translations and dilations, $\zeta \mapsto v_j(z - \zeta)$ and $s \mapsto v_j(sz) - 2\log s$. Thus we set

$$z_{ij}(z) = \frac{\partial}{\partial \zeta_i} v_j(z+\zeta)\Big|_{\zeta=0}, \qquad i=1,2,$$

$$z_{0j}(z) = \frac{\partial}{\partial s}\big[v_j(sz) + 2\log s\big]\Big|_{s=1}.$$

It turns out that the only bounded solutions of $L_j(\phi) = 0$ in $\mathbb{R}^2$ are precisely the linear combinations of the z_{ij}, $i = 0, 1, 2$, see [9] for a proof. Let us denote also $Z_{ij}(y) := z_{ij}(y - \xi'_j)$.

Additionally, let us consider a large but fixed number $R_0 > 0$ and a nonnegative function $\chi(\rho)$ with $\chi(\rho) = 1$ if $\rho < R_0$ and $\chi(\rho) = 0$ if $\rho > R_0 + 1$. We denote

$$\chi_j(y) = \chi\big(|y - \xi'_j|\big).$$

Given h of class $C^{0,\alpha}(\Omega_\varepsilon)$, we consider the linear problem of finding a function ϕ and scalars c_{ij} $i = 1, 2$, $j = 1, \ldots, m$, such that

$$L(\phi) = h + \sum_{i=1}^{2}\sum_{j=1}^{m} c_{ij}\chi_j Z_{ij} \quad \text{in } \Omega_\varepsilon, \tag{5.19}$$

$$\phi = 0 \qquad\qquad\qquad \text{on } \partial\Omega_\varepsilon, \tag{5.20}$$

$$\int_{\Omega_\varepsilon} \chi_j Z_{ij}\phi = 0 \qquad\qquad \text{for all } i = 1, 2,\ j = 1, \ldots, m. \tag{5.21}$$

Our main result for this problem states its bounded solvability, uniform in small ε and points ξ_j uniformly separated from each other and from the boundary. Thus we consider the norms

$$\|\psi\|_\infty = \sup_{y\in\Omega_\varepsilon} |\psi(y)|, \qquad \|\psi\|_* = \sup_{y\in\Omega_\varepsilon} \left(\sum_{j=1}^{m}(1 + |y - \xi'_j|)^{-3} + \varepsilon^2\right)^{-1} |\psi(y)|.$$

PROPOSITION 5.1. *Let $\delta > 0$ be fixed. There exist positive numbers ε_0 and C, such that for any points ξ_j, $j = 1, \ldots, m$, in Ω, with*

$$\mathrm{dist}(\xi_j, \partial\Omega) \geqslant \delta, \qquad |\xi_l - \xi_j| \geqslant \delta \quad \text{for } l \neq j, \tag{5.22}$$

there is a unique solution to problem (5.19)–(5.21) for all $\varepsilon < \varepsilon_0$. Moreover,

$$\|\phi\|_\infty \leqslant C\left(\log\frac{1}{\varepsilon}\right)\|h\|_*. \tag{5.23}$$

Furthermore, the function $\xi' \to \phi$ is C^1 and

$$\|\partial_{\xi'_{hl}} \phi\|_\infty \leqslant C\left(\log\frac{1}{\varepsilon}\right)\|h\|_*. \tag{5.24}$$

We observe that the orthogonality conditions in the problem above are only taken with respect to the elements of the approximate kernel due to translations.

The proof of this result consists of some steps. The first step is to prove uniform a priori estimates for the problem (5.19)–(5.21) when ϕ satisfies additionally orthogonality under dilations. Specifically, we consider the problem

$$L(\phi) = h \qquad\qquad \text{in } \Omega_\varepsilon, \tag{5.25}$$

$$\phi = 0 \qquad\qquad \text{on } \partial\Omega_\varepsilon, \tag{5.26}$$

$$\int_{\Omega_\varepsilon} \chi_j Z_{ij}\phi = 0 \quad \text{for } i = 0, 1, 2,\ j = 1, \ldots, m, \tag{5.27}$$

and prove the following estimate.

LEMMA 5.1. *Let $\delta > 0$ be fixed. There exist positive numbers ε_0 and C, such that for any points ξ_j, $j = 1, \ldots, m$, in Ω, which satisfy relations (5.22), and any solution ϕ to (5.25)–(5.27), one has*

$$\|\phi\|_\infty \leqslant C\|h\|_* \tag{5.28}$$

for all $\varepsilon < \varepsilon_0$.

PROOF. We will carry out the proof of the a priori estimate (5.28) by contradiction. We assume then the existence of sequences $\varepsilon_n \to 0$, points $\xi_j^n \in \Omega$ which satisfy relations (5.22), functions h_n with $\|h_n\|_* \to 0$, ϕ_n with $\|\phi_n\|_\infty = 1$,

$$L(\phi_n) = h_n \qquad\qquad \text{in } \Omega_\varepsilon, \tag{5.29}$$

$$\phi_n = 0 \qquad\qquad \text{on } \partial\Omega_\varepsilon, \tag{5.30}$$

$$\int_{\Omega_\varepsilon} \chi_j Z_{ij}\phi_n = 0 \quad \text{for all } i = 0, 1, 2,\ j = 1, \ldots, m. \tag{5.31}$$

A key step in the proof is the fact that the operator L satisfies maximum principle in Ω_ε outside large balls centered at the points ξ'_j. Consider the function $z_0(r) = \frac{r^2-1}{1+r^2}$, radial solution in $\mathbb{R}^2$ of

$$\Delta z_0 + \frac{8}{(1+r^2)^2} z_0 = 0.$$

Define a comparison function in Ω_ε,

$$Z(y) = \sum_{j=1}^{m} z_0\big(a\big|y - \xi_j'\big|\big), \quad y \in \Omega_\varepsilon.$$

One can prove that if a is taken small and fixed, and $R > 0$ is chosen sufficiently large depending on this a, then we have that $L(Z) < 0$ in $\widetilde{\Omega}_\varepsilon := \Omega_\varepsilon \setminus \bigcup_{j=1}^{m} B(\xi_j', R)$. Since $Z > 0$ in this region, we then conclude that L satisfies maximum principle, namely if $L(\psi) \leqslant 0$ in $\widetilde{\Omega}_\varepsilon$ and $\psi \geqslant 0$ on $\partial\widetilde{\Omega}_\varepsilon$ then $\psi \geqslant 0$ in $\widetilde{\Omega}_\varepsilon$ (see [35] for details).

Let us fix such a number $R > 0$ which we may take larger whenever it is needed. Now, let us consider the "inner norm"

$$\|\phi\|_i = \sup_{\bigcup_{j=1}^{m} B(\xi_j', R)} |\phi|.$$

We make the following claim: There is a constant $C > 0$ such that if $L(\phi) = h$ in Ω_ε then

$$\|\phi\|_\infty \leqslant C\big[\|\phi\|_i + \|h\|_*\big]. \tag{5.32}$$

We will establish this with the use of suitable barriers.

Let M be a large number such that for all j, $\Omega_\varepsilon \subset B(\xi_j', \frac{M}{\varepsilon})$. Consider now the solution of the problem

$$-\Delta\psi_j = \frac{2}{|y - \xi_j'|^3} + 2\varepsilon^2, \quad R < \big|y - \xi_j'\big| < \frac{M}{\varepsilon},$$

$$\psi_j(y) = 0 \quad \text{for } \big|y - \xi_j'\big| = R, \big|y - \xi_j'\big| = \frac{M}{\varepsilon}.$$

A direct computation shows that

$$\psi(r) = \frac{1}{R} - \frac{1}{r} - \varepsilon^2(r - R) - \left[\frac{1}{R} - \frac{1}{r} - \varepsilon^2\left(\frac{M}{\varepsilon} - R\right)\right]\frac{\log(r/R)}{\log(M/(\varepsilon R))},$$

hence these functions have a uniform bound independent of ε as long as $1 < R < \frac{1}{2\varepsilon}$. On the other hand, let us consider the function $Z(y)$ defined above, and let us set

$$\tilde{\phi}(y) = 2\|\phi\|_i Z(y) + \|h\|_* \sum_{j=1}^{m} \psi_j(y).$$

Then, it is easily checked that, choosing R larger if necessary, $L(\tilde{\phi}) \leqslant h$, $\tilde{\phi} \geqslant \phi$ on $\partial\widetilde{\Omega}_\varepsilon$. Hence $\phi \leqslant \tilde{\phi}$ on $\widetilde{\Omega}_\varepsilon$. Similarly, $\phi \geqslant -\tilde{\phi}$ on $\widetilde{\Omega}_\varepsilon$ and the claim follows.

Let us now go back to the contradiction argument. The above claim shows that since $\|\phi_n\|_\infty = 1$, then for some $\kappa > 0$, $\|\phi_n\|_i \geqslant \kappa$. Let us set $\hat{\phi}_n(z) = \phi_n(\xi_j^n + z)$ where the index j is such that $\sup_{|y-\xi_j^n|<R}|\phi_n| \geqslant \kappa$. With no loss of generality we assume that this index j is the same for all n. Elliptic estimates readily imply that $\hat{\phi}_n$ converges uniformly over compacts to a bounded solution $\hat{\phi} \neq 0$ of the problem in $\mathbb{R}^2$

$$\Delta\phi + \frac{8\mu_j^2}{(\mu_j^2 + |z|^2)^2}\phi = 0.$$

This implies that $\hat{\phi}$ is a linear combination of the functions z_{ij}, $i = 0, 1, 2$. However, our assumed orthogonality conditions on ϕ_n pass to the limit and yield $\int \chi(|z|)z_{ij}\hat{\phi} = 0$ and hence necessarily $\hat{\phi} \equiv 0$, a contradiction from which the result of the lemma follows. $\qquad\square$

We want to establish next an a priori estimate for problem (5.25)–(5.27) with the orthogonality conditions $\int \chi_j \phi Z_{0j} = 0$ dropped, namely the problem

$$L(\phi) = h \qquad\qquad \text{in } \Omega_\varepsilon, \tag{5.33}$$

$$\phi = 0 \qquad\qquad \text{on } \partial\Omega_\varepsilon, \tag{5.34}$$

$$\int_{\Omega_\varepsilon} \chi_j Z_{ij}\phi = 0 \quad \text{for } i = 1, 2, \ j = 1, \dots, m. \tag{5.35}$$

LEMMA 5.2. *Let $\delta > 0$ be fixed. There exist positive numbers ε_0 and C, such that for any points ξ_j, $j = 1, \dots, m$, in Ω which satisfy (5.22), and any solution ϕ to problem (5.33)–(5.35), one has*

$$\|\phi\|_\infty \leqslant C\left(\log\frac{1}{\varepsilon}\right)\|h\|_* \tag{5.36}$$

for all $\varepsilon < \varepsilon_0$.

PROOF. Let $R > R_0 + 1$ be a large and fixed number, and let $\hat{z}_{0j}$ be the solution of the problem

$$\Delta\hat{z}_{0j} + \frac{8\mu_j^2}{(\mu_j^2 + |y - \xi_j'|^2)^2}\hat{z}_{0j} = 0,$$

$$\hat{z}_{0j}(y) = z_{0j}(R) \quad \text{for } |y - \xi_j'| = R,$$

$$\hat{z}_{0j}(y) = 0 \qquad \text{for } |y - \xi_j'| = \frac{\delta}{3\varepsilon}.$$

A direct computation shows that this function is explicitly given by

$$\hat{z}_{0j}(y) = z_{0j}(r)\left[1 - \frac{\int_R^r ds/(sz_{0j}^2(s))}{\int_R^{\delta/(3\varepsilon)} ds/(sz_{0j}^2(s))}\right], \quad r = |y - \xi_j'|.$$

Next we consider smooth cut-off functions $\eta_1(r)$ and $\eta_2(r)$ with the following properties: $\eta_1(r) = 1$ for $r < R$, $\eta_1(r) = 0$ for $r > R+1$, $|\eta_1'(r)| \leqslant 2$; $\eta_2(r) = 1$ for $r < \frac{\delta}{4\varepsilon}$, $\eta_2(r) = 0$ for $r > \frac{\delta}{3\varepsilon}$, $|\eta_2'(r)| \leqslant C\varepsilon$, $|\eta''_2(r)| \leqslant C\varepsilon^2$. Then we set

$$\eta_{1j}(y) = \eta_1\big(|y - \xi_j'|\big), \qquad \eta_{2j}(y) = \eta_2\big(|y - \xi_j'|\big) \tag{5.37}$$

and define a test function

$$\tilde{z}_{0j} = \eta_{1j} Z_{0j} + (1 - \eta_{1j})\eta_{2j}\hat{z}_{0j}, \qquad Z_{0j}(y) = z_{0j}\big(|y - \xi_j'|\big).$$

Intuitively, $\tilde{z}_{0j}$ resembles the eigenfunction of the operator L associated to the invariance of L under dilations when L is considered in the whole $\mathbb{R}^2$.

Let ϕ be a solution to (5.33)–(5.35). We will modify ϕ so that the orthogonality conditions with respect to Z_{0j}'s are satisfied. We set

$$\tilde{\phi} = \phi + \sum_{j=1}^{m} d_j \tilde{z}_{0j},$$

where the numbers d_j are defined as

$$d_j \int_{\Omega_\varepsilon} \chi_j |Z_{0j}|^2 + \int_{\Omega_\varepsilon} \chi_j Z_{0j}\phi = 0.$$

Then

$$L(\tilde{\phi}) = h + \sum_{j=1}^{m} d_j L(\tilde{z}_{0j}), \tag{5.38}$$

and $\int_{\Omega_\varepsilon} \chi_j Z_{0i}\tilde{\phi} = 0$ for all i and all j. The previous lemma thus allows us to estimate

$$\|\tilde{\phi}\|_\infty \leqslant C\left[\|h\|_* + \sum_{j=1}^{m} |d_j| \|L(\tilde{z}_{0j})\|_* \right]. \tag{5.39}$$

Testing equation (5.38) against $\tilde{z}_{0l}$ we find

$$\langle \tilde{\phi}, L(\tilde{z}_{0l})\rangle = \langle h, \tilde{z}_{0l}\rangle + d_l\langle L(\tilde{z}_{0l}), \tilde{z}_{0l}\rangle,$$

where $\langle f, g\rangle = \int_{\Omega_\varepsilon} fg$. This relation in combination with (5.39) gives us that

$$d_l\langle L(\tilde{z}_{0l}), \tilde{z}_{0l}\rangle \leqslant C\|h\|_*\big[1 + \|L(\tilde{z}_{0l})\|_*\big] + C \sum_{j=1}^{m} |d_j| \|L(\tilde{z}_{0j})\|_*^2. \tag{5.40}$$

The following estimates hold true (see [35]): if we choose R sufficiently large, then

$$\left\| L(\tilde{z}_{0j}) \right\|_* \leq \frac{C}{\log(1/\varepsilon)} \tag{5.41}$$

and

$$\left\langle L(\tilde{z}_{0l}), \tilde{z}_{0l} \right\rangle \leq -\frac{E}{\log(1/\varepsilon)} \left[1 + O\left(\frac{1}{\log(1/\varepsilon)} \right) \right]. \tag{5.42}$$

Combining relations (5.42) with (5.40) and (5.41) we finally get that

$$|d_j| \leq C \left(\log \frac{1}{\varepsilon} \right) \|h\|_*$$

for all $j = 1, \ldots, m$. We thus conclude from estimate (5.39) that

$$\|\phi\|_\infty \leq C \left(\log \frac{1}{\varepsilon} \right) \|h\|_*.$$

The proof is complete. $\qquad \square$

We are now ready for the proof of Proposition 5.1.

PROOF OF PROPOSITION 5.1. We begin by establishing the validity of the a priori estimate (5.23). The previous lemma yields

$$\|\phi\|_\infty \leq C \left(\log \frac{1}{\varepsilon} \right) \left[\|h\|_* + \sum_{i=1}^{2} \sum_{j=1}^{m} |c_{ij}| \right], \tag{5.43}$$

hence it suffices to estimate the values of the constants $|c_{ij}|$. Let us consider the cut-off function η_{2j} introduced in (5.37). We test equation (5.19) against $Z_{ij}\eta_{2j}$ to find

$$\left\langle L(\phi), \eta_{2j} Z_{ij} \right\rangle = \langle h, \eta_{2j} Z_{ij} \rangle + c_{ij} \int_{\Omega_\varepsilon} \chi_j |Z_{ij}|^2. \tag{5.44}$$

Now

$$\left\langle L(\phi), \eta_{2j} Z_{ij} \right\rangle = \left\langle \phi, L(\eta_{2j} Z_{ij}) \right\rangle.$$

We have

$$L(\eta_{2j} Z_{ij}) = \Delta \eta_{2j} Z_{ij} + 2\nabla \eta_{2j} \nabla Z_{ij} + \varepsilon O\left((1+r)^{-3} \right)$$

with $r = |y - \xi'_j|$. Since $\Delta \eta_{2j} = O(\varepsilon^2)$, $\nabla \eta_{2j} = O(\varepsilon)$, and besides $Z_{ij} = O(r^{-1})$, $\nabla Z_{ij} = O(r^{-2})$, we find

$$L(\eta_{2j} Z_{ij}) = O(\varepsilon^3) + \varepsilon O\big((1+r)^{-3}\big).$$

Thus

$$\big|\langle \phi, L(\eta_{2j} Z_{ij}) \rangle\big| \leqslant C\varepsilon \|\phi\|_\infty.$$

Combining this estimate with (5.44) and (5.43) we obtain

$$|c_{ij}| \leqslant C\left[\|h\|_* + \varepsilon \log \frac{1}{\varepsilon} \sum_{l,k} |c_{lk}|\right]$$

which implies $|c_{ij}| \leqslant C\|h\|_*$. It follows finally from (5.43) that $\|\phi\|_\infty \leqslant C(\log \frac{1}{\varepsilon})\|h\|_*$ and the a priori estimate has been thus proven. It only remains to prove the solvability assertion. To this purpose we consider the space

$$H = \left\{\phi \in H_0^1(\Omega_\varepsilon): \int_{\Omega_\varepsilon} \chi_j Z_{ij} \phi = 0 \text{ for } i = 1, 2, \, j = 1, \ldots, m\right\},$$

endowed with the usual inner product $[\phi, \psi] = \int_{\Omega_\varepsilon} \nabla\phi\nabla\psi$. Problem (5.19)–(5.21) expressed in weak form is equivalent to that of finding a $\phi \in H$, such that

$$[\phi, \psi] = \int_{\Omega_\varepsilon} [-W\phi + h]\psi \, dx \quad \text{for all } \psi \in H.$$

With the aid of Riesz's representation theorem, this equation gets rewritten in H in the operator form $\phi = K(\phi) + \tilde{h}$, for certain $\tilde{h} \in H$, where K is a compact operator in H. Fredholm's alternative guarantees unique solvability of this problem for any h provided that the homogeneous equation $\phi = K(\phi)$ has only the zero solution in H. This last equation is equivalent to (5.19)–(5.21) with $h \equiv 0$. Thus existence of a unique solution follows from the a priori estimate (5.23). We refer the reader to [35] for (5.24). $\qquad\square$

We recall that our goal is to solve problem (5.17). Rather than doing so directly, we shall solve first the intermediate problem

$$L(\phi) = -\big[R + N(\phi)\big] + \sum_{i=1}^{2}\sum_{j=1}^{m} c_{ij} \chi_j Z_{ij} \quad \text{in } \Omega_\varepsilon, \tag{5.45}$$

$$\phi = 0 \qquad \text{on } \partial\Omega_\varepsilon, \tag{5.46}$$

$$\int_{\Omega_\varepsilon} \chi_j Z_{ij} \phi = 0 \quad \text{for all } i = 1, 2, \, j = 1, \ldots, m. \tag{5.47}$$

We assume that the conditions in Proposition 5.1 hold. We have the following result.

LEMMA 5.3. *Under the assumptions of Proposition 5.1 there exist positive numbers C and ε_0, such that problem (5.45)–(5.47) has a unique solution ϕ which satisfies*

$$\|\phi\|_\infty \leqslant C\varepsilon|\log\varepsilon|.$$

Furthermore the map $\xi' \mapsto \phi$ into the space $C(\overline{\Omega}_\varepsilon)$ is C^1 and the derivative $D_{\xi'}\phi$ defines a continuous function of ξ'. Besides, there is a constant $C > 0$ such that

$$\|D_{\xi'}\phi\|_* \leqslant C\varepsilon|\log\varepsilon|^2. \tag{5.48}$$

PROOF. In terms of the operator T defined in Proposition 5.1, problem (5.45)–(5.47) becomes

$$\phi = T\big(-\big(N(\phi) + R\big)\big) \equiv A(\phi). \tag{5.49}$$

For a given number $\gamma > 0$, let us consider the region

$$\mathcal{F}_\gamma \equiv \big\{\phi \in C(\overline{\Omega}_\varepsilon)\colon \|\phi\|_\infty \leqslant \gamma\varepsilon|\log\varepsilon|\big\}.$$

From Proposition 5.1, we get

$$\big\|A(\phi)\big\|_\infty \leqslant C|\log\varepsilon|\big[\big\|N(\phi)\big\|_* + \|R\|_*\big].$$

Estimate (5.16) implies that $\|R\| \leqslant C\varepsilon$. Also, the definition of N in (5.18) immediately yields

$$\big\|N(\phi)\big\|_* \leqslant C\|\phi\|_\infty^2. \tag{5.50}$$

It is also immediate that N satisfies, for $\phi_1, \phi_2 \in \mathcal{F}_\gamma$,

$$\big\|N(\phi_1) - N(\phi_2)\big\|_* \leqslant C\gamma\varepsilon|\log\varepsilon|\|\phi_1 - \phi_2\|_*,$$

where C is independent of γ. Hence we get

$$\big\|A(\phi)\big\|_\infty \leqslant C|\log\varepsilon|\varepsilon\big[\gamma^2\varepsilon|\log\varepsilon|^2 + 1\big],$$
$$\big\|A(\phi_1) - A(\phi_2)\big\|_\infty \leqslant C\gamma\varepsilon|\log\varepsilon|^2\|\phi_1 - \phi_2\|_*.$$

It follows that, for all sufficiently small ε, we get that A is a contraction mapping of $\mathcal{F}_\gamma$, and therefore a unique fixed point of A exists in this region. For the dependence C^1 of ϕ on the variable ξ' and the estimate (5.48), we refer the reader to [35]. $\qquad\square$

After problem (5.45)–(5.47) has been solved, we find a solution to problem (5.17) and hence to the original problem if ξ' is such that

$$c_{ij}(\xi') = 0 \quad \text{for all } i, j. \tag{5.51}$$

This problem is indeed variational: it is equivalent to finding critical points of a function of $\xi = \varepsilon \xi'$. To see that let us consider the energy functional J_ε associated to problem (1.14), namely

$$J_\varepsilon(u) = \frac{1}{2} \int_\Omega |\nabla u|^2 \, dx - \varepsilon^2 \int_\Omega e^u \, dx. \tag{5.52}$$

We define

$$F(\xi) \equiv J_\varepsilon\big(U(\xi) + \tilde{\phi}(\xi)\big), \tag{5.53}$$

where U is the function defined in (5.7) and $\tilde{\phi} = \tilde{\phi}(\xi) = \tilde{\phi}(x, \xi)$ is the function defined on Ω from the relation $\tilde{\phi}(x, \xi) = \phi(\frac{x}{\varepsilon}, \frac{\xi}{\varepsilon})$, with ϕ the solution of problem (5.45)–(5.47) given by Proposition 5.1. Critical points of F correspond to solutions of (5.51) for small ε, as the following result states.

LEMMA 5.4. *Under the assumptions of Proposition 5.1, the functional $F(\xi)$ is of class C^1. Moreover, for all $\varepsilon > 0$ sufficiently small, if $D_\xi F(\xi) = 0$ then ξ satisfies system (5.51).*

PROOF. Define

$$I_\varepsilon(v) = \frac{1}{2} \int_{\Omega_\varepsilon} |\nabla v|^2 \, dy - \int_{\Omega_\varepsilon} e^v \, dy. \tag{5.54}$$

Let us differentiate the function $F(\xi)$ with respect to ξ. Since $J_\varepsilon(U + \tilde{\phi}) = I_\varepsilon(V + \phi)$, we can differentiate directly $I_\varepsilon(V + \phi)$ under the integral sign, so that

$$\partial_{\xi_{kl}} F(\xi) = \varepsilon^{-1} DI_\varepsilon(V + \phi)[\partial_{\xi'_{kl}} V + \partial_{\xi'_{kl}} \phi]$$

$$= \varepsilon^{-1} \sum_{i=1}^{2} \sum_{j=1}^{m} \int_{\Omega_\varepsilon} c_{ij} \chi_j Z_{ij} [\partial_{\xi'_{kl}} V + \partial_{\xi'_{kl}} \phi].$$

From the results of the previous section, this expression defines a continuous function of ξ', and hence of ξ. Let us assume that $D_\xi F(\xi) = 0$. Then

$$\sum_{i=1}^{2} \sum_{j=1}^{m} \int_{\Omega_\varepsilon} c_{ij} \chi_j Z_{ij} [\partial_{\xi'_{kl}} V + \partial_{\xi'_{kl}} \phi] = 0, \quad k = 1, 2, l = 1, \dots, m.$$

We recall that we proved $\|D_{\xi'}\phi\|_\infty \leqslant C\varepsilon |\log \varepsilon|^2$, thus we directly check that as $\varepsilon \to 0$, we have $\partial_{\xi'_{kl}} V + \partial_{\xi'_{kl}} \phi = -[Z_{kl} + o(1)]$ with $o(1)$ small in terms of the L^∞ norm as $\varepsilon \to 0$. We get that $D_\xi F(\xi) = 0$ implies the validity of a system of equations of the form

$$\sum_{i=1}^{m} \sum_{j=1}^{2} c_{ij} \int_{\Omega_\varepsilon} \chi_j Z_{ij} [Z_{kl} + o(1)] = 0, \quad k = 1, 2, l = 1, \dots, m,$$

with o(1) small in the sense of the L^∞ norm as $\varepsilon \to 0$. The above system is diagonal dominant and we thus get $c_{ij} = 0$ for all i, j. This concludes the proof of the lemma. $\qquad \square$

In order to solve for critical points of the function F, a key step is its expected closeness to the function $J_\varepsilon(U)$.

LEMMA 5.5. *The following expansion holds*

$$F(\xi) = J_\varepsilon(U) + \theta_\varepsilon(\xi),$$

where

$$|\theta_\varepsilon| + |\nabla \theta_\varepsilon| \to 0,$$

uniformly on points satisfying the constraints in Proposition 5.1. Furthermore, with the choice (5.6) for the parameters μ_j, the following expansion holds

$$J_\varepsilon(U) = -16m\pi + 8m\pi \log 8 - 16m\pi \log \varepsilon + 32\pi\varphi_m(\xi) + \varepsilon\Theta_\varepsilon(\xi), \qquad (5.55)$$

where the function φ_m is defined by (1.18). Here Θ_ε is a smooth function of $\xi = (\xi_1, \ldots, \xi_m)$, bounded together with its derivatives, as $\varepsilon \to 0$ uniformly on points $\xi_1, \ldots, \xi_m \in \Omega$ that satisfy $\mathrm{dist}(\xi_i, \partial\Omega) > \delta$ and $|\xi_i - \xi_j| > \delta$.

PROOF. Since $I_\varepsilon(V) = J_\varepsilon(U)$, $I_\varepsilon(V + \phi) = J_\varepsilon(U + \tilde{\phi})$, it is enough to show that $\tilde{\theta}_\varepsilon(\xi') = \theta_\varepsilon(\varepsilon\xi')$ satisfies

$$\left|\tilde{\theta}_\varepsilon\right| + \varepsilon^{-1}\left|\nabla_{\xi'}\tilde{\theta}_\varepsilon\right| = o(1).$$

Taking into account $DI_\varepsilon(V + \phi)[\phi] = 0$, a Taylor expansion gives

$$I_\varepsilon(V + \phi) - I_\varepsilon(V)$$

$$= \int_0^1 D^2 I_\varepsilon(V + t\phi)[\phi]^2(1 - t)\,dt$$

$$= \int_0^1 \left(\int_{\Omega_\varepsilon} [N(\phi) + R]\phi + \int_{\Omega_\varepsilon} k(\varepsilon y)e^V\left[1 - e^{t\phi}\right]\phi^2\right)(1 - t)\,dt. \qquad (5.56)$$

Since $\|\phi\|_\infty \leqslant C\varepsilon|\log\varepsilon|$, we get

$$I_\varepsilon(V + \phi) - I_\varepsilon(V) = \tilde{\theta}_\varepsilon = O\left(\varepsilon^2|\log\varepsilon|^3\right).$$

Let us differentiate with respect to ξ'. We use the representation (5.56) and differentiate directly under the integral sign, thus obtaining, for each $k = 1, 2, l = 1, \ldots, m$,

$$
\partial_{\xi'_{kl}} \left[I_\varepsilon(V + \phi) - I_\varepsilon(V) \right]
$$

$$
= \int_0^1 \left(\int_{\Omega_\varepsilon} \partial_{\xi'_{kl}} \left[(N(\phi) + R)\phi \right] + \int_{\Omega_\varepsilon} \partial_{\xi'_{kl}} \left[k(\varepsilon y) e^V [1 - e^{t\phi}] \phi^2 \right] \right) (1 - t) \, \mathrm{d}t.
$$

Using the fact that $\| \partial_{\xi'} \phi \|_* \leqslant C\varepsilon |\log \varepsilon|^2$ and the computations in the proof of Lemma 5.3 we get

$$
\partial_{\xi'_{kl}} \left[I_\varepsilon(V + \phi) - I_\varepsilon(V) \right] = \partial_{\xi'_{kl}} \tilde{\theta}_\varepsilon = O\!\left(\varepsilon^2 |\log \varepsilon|^4 \right).
$$

The continuity in ξ of all these expressions is inherited from that of ϕ and its derivatives in ξ in the L^∞ norm. To obtain (5.55), we just mention that the following asymptotic expansions hold true

$$
\frac{1}{2} \int_\Omega |\nabla U|^2 \, \mathrm{d}x = -8m\pi + \sum_{j=1}^m 16\pi \log \frac{1}{\varepsilon \mu_j}
$$

$$
+ 32\pi \left(\sum_{j=1}^k H(\xi_j, \xi_j) + \sum_{i \neq j} G(\xi_i, \xi_j) \right) + \varepsilon^2 \log \frac{1}{\varepsilon} \Theta_\varepsilon(\xi)
$$

and

$$
\varepsilon^2 \int_\Omega k(x) e^U \, \mathrm{d}x = 8m\pi + \varepsilon \Theta_\varepsilon(\xi).
$$

For the details, see [35]. $\qquad\square$

PROOF OF THEOREM 5.1. Let us consider the set $\mathcal{D}$ as in the statement of the theorem, $\mathcal{C}$ the associated critical value and $\xi \in \mathcal{D}$. According to Lemma 5.4, we have a solution of problem (1.14) if we adjust ξ so that it is a critical point of $F(\xi)$ defined by (5.53). This is equivalent to finding a critical point of

$$
\widetilde{F}(\xi) = F(\xi) + 16m\pi \log \varepsilon.
$$

On the other hand, from Lemma 5.5, we have that for $\xi \in \mathcal{D}$, such that its components satisfy $|\xi_i - \xi_j| \geqslant \delta$,

$$
\alpha \widetilde{F}(\xi) + \beta = \varphi_m(\xi) + \varepsilon \Theta_\varepsilon(\xi),
$$

where Θ_ε and $\nabla_\xi \Theta_\varepsilon$ are uniformly bounded in the considered region as $\varepsilon \to 0$, and $\alpha \neq 0$ and β are universal constants.

Let us observe that if $M > \mathcal{C}$, then assumptions (5.1), (5.2) still hold for the function $\min\{M, \varphi_m(\xi)\}$ as well as for $\min\{M, \varphi_m(\xi) + \varepsilon \Theta_\varepsilon(\xi)\}$. It follows that the function

$\min\{M, \alpha \widetilde{F}(\xi) + \beta\}$ satisfies for all ε small assumptions (5.1), (5.2) in $\mathcal{D}$ and therefore has a critical value $\mathcal{C}_\varepsilon < M$ which is close to $\mathcal{C}$ in this region. If $\xi_\varepsilon \in \mathcal{D}$ is a critical point at this level for $\alpha \widetilde{F}(\xi) + \beta$, then since

$$\alpha \widetilde{F}(\xi_\varepsilon) + \beta \leqslant \mathcal{C}_\varepsilon < M,$$

we have that there exists a $\delta > 0$ such that $|\xi_{\varepsilon,j} - \xi_{\varepsilon,i}| > \delta$, $\mathrm{dist}(\xi_{\varepsilon,j}, \partial\Omega) > 0$. This implies C^1-closeness of $\alpha \widetilde{F}(\xi) + \beta$ and $\varphi_m(\xi)$ at this level, hence $\nabla \varphi_m(\xi_\varepsilon) \to 0$. The function $u_\varepsilon = U(\xi_\varepsilon) + \tilde{\phi}(\xi_\varepsilon)$ is therefore a solution as predicted by the theorem. $\qquad\square$

PROOF OF THEOREM 1.3. According to the result of Theorem 5.1, it is sufficient to establish that given $m \geqslant 1$, φ_m has a nontrivial critical value in some open set $\mathcal{D}$, compactly contained in Ω^m. Our choice of $\mathcal{D}$ is just given by

$$\mathcal{D} = \{y \in \Omega^m \mid \mathrm{dist}(y, \partial\Omega^m) > \delta\},$$

where δ is a small positive number yet to be chosen. We observe that in this set function $\sum_{j=1}^m H(y_j, y_j)$ is bounded and $\sum_{i \neq j} G(y_i, y_j)$ is bounded below. Consequently function $\varphi_m(y)$ is also bounded below in $\mathcal{D}$.

Let Ω_1 be a bounded nonempty component of $\mathbb{R}^2 \setminus \overline{\Omega}$, and consider a closed, smooth Jordan curve γ contained in Ω which encloses Ω_1. We let S to be the image of γ, $B_0 = \emptyset$ and $B = S \times \cdots \times S = S^m$.

Then define

$$\mathcal{C} = \inf_{\Phi \in \Gamma} \sup_{z \in B} \varphi_m(\Phi(z)), \tag{5.57}$$

where $\Phi \in \Gamma$ if and only if $\Phi(z) = \Psi(1, z)$ with $\Psi : [0, 1] \times B \to \mathcal{D}$ continuous and $\Psi(0, z) = z$.

We first need the following lemma.

LEMMA 5.6. *There exists $K > 0$, independent of the small number δ used to define $\mathcal{D}$ such that $\mathcal{C} \geqslant -K$.*

PROOF. We need to prove the existence of $K > 0$ independent of small δ such that if $\Phi \in \Gamma$, then there exists a $\bar{z} \in B$ with

$$\varphi_m(\Phi(\bar{z})) \geqslant -K. \tag{5.58}$$

Let us assume that $0 \in \Omega_1$ and write

$$\Phi(z) = (\Phi_1(z), \ldots, \Phi_m(z)).$$

Identifying the components of the above m-tuple with complex numbers, we shall establish the existence of $\bar{z} \in B$ such that

$$\frac{\Phi_j(\bar{z})}{|\Phi_j(\bar{z})|} = e^{2j\pi i/m} \quad \text{for all } j = 1, \ldots, m. \tag{5.59}$$

Clearly in such a situation, there is a number $\mu > 0$ depending only on m and Ω such that

$$\left| \Phi_j(\bar{z}) - \Phi_l(\bar{z}) \right| \geqslant \mu.$$

This, and the definition of φ_m clearly yields the validity of estimate (5.58) for a number K only dependent of Ω. To prove (5.59), we consider an orientation-preserving homeomorphism $h : S^1 \to S$ and the map $f : T^m \to T^m$ defined as $f(\zeta) = (f_1(\zeta), \ldots, f_m(\zeta))$ with

$$T^m = \underbrace{S^1 \times \cdots \times S^1}_{m}$$

and

$$f_j(\zeta_1, \ldots, \zeta_m) = \frac{\Phi_j(h(\zeta_1), \ldots, h(\zeta_m))}{|\Phi_j(h(\zeta_1), \ldots, h(\zeta_m))|}.$$

We define a homotopy $F : [0, 1] \times T^m \to T^m$ by

$$F_j(t, \zeta) = \frac{\Psi_j(t, h(\zeta_1), \ldots, h(\zeta_m))}{|\Psi_j(t, h(\zeta_1), \ldots, h(\zeta_m))|}.$$

Notice that $F(1, \zeta) = f(\zeta)$ and

$$F(0, \zeta) = \left(\frac{h(\zeta_1)}{|h(\zeta_1)|}, \ldots, \frac{h(\zeta_m)}{|h(\zeta_m)|} \right),$$

which is a homeomorphism of T^m. The existence of $\bar{z}$ such that relation (5.59) holds follows from establishing that f is onto, which we show next.

The torus T^m can be identified with the closed manifold embedded in $\mathbb{R}^{m+1}$ parameterized as

$$\zeta : (\theta_1, \ldots, \theta_m) \in [0, 2\pi)^m \mapsto \left(\rho_1 e^{i\theta_1}, 0_{m-1} \right) + \left(0_1, \rho_2 e^{i\theta_2}, 0_{m-2} \right)$$
$$+ \cdots + \left(0_{m-1}, \rho_m e^{i\theta_m} \right),$$

where $0 < \rho_m < \cdots < \rho_1$ and we have denoted $0_k = \underbrace{(0, \ldots, 0)}_{k}$, $e^{i\theta_j} = (\cos\theta_j, \sin\theta_j)$. We consider as well the solid torus $\widehat{T}^m$ parameterized as

$$(\theta_1, \ldots, \theta_m, \rho) \in [0, 2\pi)^m \times [0, \rho_m] \mapsto \left(\rho_1 e^{i\theta_1}, 0_{m-1} \right) + \left(0_1, \rho_2 e^{i\theta_2}, 0_{m-2} \right)$$
$$+ \cdots + \left(0_{m-1}, \rho e^{i\theta_m} \right).$$

Obviously $\partial \widehat{T}^m = T^m$ in $\mathbb{R}^{m+1}$.

With slight abuse of notation, we consider the map $f : T^m \to T^m$, induced from the original f under the above identification, namely

$$f(\zeta) = \left(\rho_1 f_1(\zeta), 0_{m-1}\right) + \left(0_1, \rho_2 f_2(\zeta), 0_{m-2}\right) + \cdots + \left(0_{m-1}, \rho_m f_m(\zeta)\right).$$

The function f then can be extended continuously to the whole solid torus as $\tilde{f} : \widehat{T}^m \to \mathbb{R}^{m+1}$ defined simply as

$$f(\zeta, \rho) = \left(\rho_1 f_1(\zeta), 0_{m-1}\right) + \left(0_1, \rho_2 f_2(\zeta), 0_{m-2}\right) + \cdots + \left(0_{m-1}, \rho f_m(\zeta)\right).$$

The function $\tilde{f}$ is homotopic to a homeomorphism of $\widehat{T}^m$, along a deformation which applies $\partial \widehat{T}^m$ into itself. Thus if $P \in \mathrm{int}(\widehat{T}^m)$ then $\deg(\tilde{f}, \widehat{T}^m, P) \neq 0$ and hence there exists $Q \in \widehat{T}^m$ such that $\tilde{f}(Q) = P$. Thus if we fix angles $(\theta_1^*, \ldots, \theta_m^*) \in [0, 2\pi)^m$ and $\rho^* \in (0, \rho_m)$ then there exist $\zeta^{**} \in T^m$ and $\rho^{**} \in (0, \rho_m)$ such that

$$\left(\rho_1 f_1(\zeta^{**}), 0_{m-1}\right) + \left(0_1, \rho_2 f_2(\zeta^{**}), 0_{m-2}\right) + \cdots + \left(0_{m-1}, \rho^{**} f_m(\zeta^{**})\right)$$
$$= \left(\rho_1 e^{i\theta_1^*}, 0_{m-1}\right) + \left(0_1, \rho_2 e^{i\theta_2^*}, 0_{m-2}\right) + \cdots + \left(0_{m-1}, \rho^* e^{i\theta_m^*}\right).$$

A direct computation shows then that $f_j(\zeta^{**}) = e^{i\theta_j^*}$ for all j and also $\rho^* = \rho^{**}$. It then follows that f is onto. This concludes the proof. $\qquad\square$

The second step we have to carry out to make Theorem 5.1 applicable is to establish the validity of assumption (5.2). To this end we need to establish a couple of preliminary facts on the half-plane

$$\mathcal{H} = \left\{(x^1, x^2) : x^1 \geq 0\right\}.$$

LEMMA 5.7. *Consider the function of k distinct points on $\mathcal{H}$*

$$\Psi_k(x_1, \ldots, x_k) = -4 \sum_{i \neq j} \log |x_i - x_j|.$$

Let I_+ denote the set of indices i for which $x_i^1 > 0$ and I_0 that for which $x_i^1 = 0$. Then, either

$$\nabla_{x_i} \Psi_k(x_1, \ldots, x_k) \neq 0 \quad \text{for some } i \in I_+$$

or

$$\frac{\partial}{\partial x_{i2}} \Psi_k(x_1, \ldots, x_k) \neq 0 \quad \text{for some } i \in I_0.$$

PROOF. We have that

$$\frac{\partial}{\partial \lambda} \Psi_k(\lambda x_1, \ldots, \lambda x_k)\bigg|_{\lambda=1}$$

$$= \sum_{i \in I_+} \nabla_{x_i} \Psi_k(x_1, \ldots, x_k) x_i + \sum_{i \in I_0} \partial x_{i2} \Psi_k(x_1, \ldots, x_k) x_{i2}.$$

On the other hand,

$$\frac{\partial}{\partial \lambda} \Psi_k(\lambda x_1, \ldots, \lambda x_k)\bigg|_{\lambda=1} = -4 \frac{\partial}{\partial \lambda} \big[k(k-1) \log \lambda \big]\bigg|_{\lambda=1} \neq 0,$$

and the result follows. $\qquad\square$

A second result we need concerns the analogue of the function φ_k, for the half-plane $\mathcal{H}$. Let $x = (x^1, x^2)$, $y = (y^1, y^2)$. Then regular part of Green's function in $\mathcal{H}$ is now given by

$$H(x, y) = -4 \log \frac{1}{|x - \bar{y}|}, \quad \bar{y} = (y^1, -y^2).$$

Then

$$G(x, y) = 4 \log \frac{1}{|x - y|} - 4 \log \frac{1}{|x - \bar{y}|}.$$

Hence the associated function $\bar{\varphi}_k$ is given by

$$\bar{\varphi}_k(x_1, \ldots, x_k) = 4 \sum_{i=1}^{k} \log \frac{1}{|x_i - \bar{x}_i|} + 4 \sum_{i \neq j} \log \frac{|x_i - x_j|}{|x_i - \bar{x}_j|}.$$

With identical proof as the previous lemma we now get the following one.

LEMMA 5.8. *For any k distinct points $x_i \in \mathrm{int}(\mathcal{H})$ we have*

$$\nabla \bar{\varphi}_k(x_1, \ldots, x_k) \neq 0.$$

We will recall here some straightforward to verify facts about the regular part of the Green function $H(x, y) = G(x, y) - 4 \log \frac{1}{|x-y|}$. Let $y \in \Omega$ be a point close to $\partial \Omega$ and let $\bar{y}$ be its uniquely determined reflection with respect to $\partial \Omega$. Set

$$\psi(x, y) = H(x, y) + 4 \log \frac{1}{|x - \bar{y}|}.$$

Then it can be shown that $\psi(x, y)$ is bounded in $\overline{\Omega} \times \overline{\Omega}$ and

$$\left|\nabla_x \psi(x, y)\right| + \left|\nabla_y \psi(x, y)\right| \leqslant C_1. \tag{5.60}$$

Using (5.60) one can derive the following estimates

$$\left|\nabla_x H(x, y)\right| + \left|\nabla_y H(x, y)\right| \leqslant C_1 \min\left\{\frac{1}{|x - y|}, \frac{1}{\mathrm{dist}(y, \partial\Omega)}\right\} + C_2. \tag{5.61}$$

Now we are ready to prove the validity of assumption (5.2) which in this case reads as follows:

LEMMA 5.9. *Given $K > 0$, there exists a $\delta > 0$ such that if $(\xi_1, \ldots, \xi_m) \in \partial\mathcal{D}$ and $|\varphi_m(\xi_1, \ldots, \xi_m)| \leqslant K$, then there is a vector τ, tangent to $\partial\mathcal{D}$ such that*

$$\nabla\varphi_m(\xi_1, \ldots, \xi_m) \cdot \tau \neq 0.$$

PROOF. Let us assume the opposite, namely the existence of a sequence $\delta \to 0$ and of points $\xi = \xi^\delta$ for which $\xi \in \partial\mathcal{D}$ and such that

$$\nabla_{\xi_i} \varphi_m(\xi_1, \ldots, \xi_m) = 0 \quad \text{if } \xi_i \in \Omega_\delta \tag{5.62}$$

and

$$\nabla_{\xi_i} \varphi_m(\xi_1, \ldots, \xi_m) \cdot \tau_i = 0 \quad \text{if } \xi_i \in \partial\Omega_\delta, \tag{5.63}$$

for any vector τ_i tangent to $\partial\Omega_\delta$ at ξ_i, where $\Omega_\delta = \{x \in \Omega : \mathrm{dist}(x, \partial\Omega) > \delta\}$.

From the assumption of the lemma it follows that there is a point $\xi_l \in \partial\Omega_\delta$, such that $H(\xi_l) \to -\infty$ as $\delta \to 0$. Since the value of φ_m remains uniformly bounded, necessarily we must have that at least two points ξ_i and ξ_j that are becoming close. Let $\delta_n = \frac{1}{n}$, $\xi^n = (\xi_1^n, \ldots, \xi_m^n) \in \Omega_{\delta_n}$ be a sequence of points such that (5.62), (5.63) hold, and

$$\rho_n = \inf_{i \neq j} \left|\xi_j^n - \xi_i^n\right| \to 0 \quad \text{as } n \to \infty.$$

Without loss of generality we can assume that $\rho_n = |\xi_1^n - \xi_2^n|$. We define

$$x_j^n = \frac{\xi_1^n - \xi_j^n}{\rho_n}. \tag{5.64}$$

Clearly there exists a k, $2 \leqslant k \leqslant m$, such that

$$\lim_{n \to \infty} \left|x_j^n\right| < \infty, \quad j = 1, \ldots, k, \quad \text{and} \quad \lim_{n \to \infty} \left|x_j^n\right| = \infty, \quad j > k.$$

For $j \leqslant k$ we set

$$\tilde{x}_j = \lim_{n \to \infty} x_j^n.$$

We consider two cases:

(1) either

$$\frac{\text{dist}(\xi_1^n, \partial\Omega_{\delta_n})}{\rho_n} \to \infty;$$

(2) or there exists $c_0 < \infty$ such that for almost all n we have

$$\frac{\text{dist}(\xi_1^n, \partial\Omega_{\delta_n})}{\rho_n} < c_0.$$

CASE 1. It is easy to see that in this case we actually have

$$\frac{\text{dist}(\xi_j^n, \partial\Omega_{\delta_n})}{\rho_n} \to \infty, \quad j = 1, \ldots, k.$$

Furthermore points $\xi_1^n, \ldots, \xi_k^n$ are all interior to Ω_{δ_n} hence (5.62) is satisfied for all partial derivatives $\partial_{\xi_{lj}}$, $j \leqslant k$. Define

$$\tilde{\varphi}_m(x_1, \ldots, x_m) = \varphi_m(\xi_1 + \rho_n x_1, \ldots, \xi_1 + \rho_n x_m).$$

We have for all $l = 1, 2$, $j = 1, \ldots, k$,

$$\partial_{x_{lj}} \tilde{\varphi}_m(x) = \rho_n \partial_{\xi_{lj}} \varphi_m(\xi_1^n + x\rho_n).$$

Then at $\tilde{x} = (\tilde{x}_1, \ldots, \tilde{x}_k, 0, \ldots, 0)$ we have

$$\partial_{x_{lj}} \tilde{\varphi}_m(\tilde{x}) = 0.$$

On the other hand, using (5.61) and letting $\rho_n \to 0$, we get

$$\lim_{n\to\infty} \rho_n \partial_{\xi_{lj}} \varphi_m(\xi_1^n + x\rho_n) = -4 \sum_{i\neq j, i\leqslant k} \partial_{x_{lj}} \log \frac{1}{|\tilde{x}_j - \tilde{x}_i|} = 0.$$

Since this last equality is true for any $j \leqslant k, l = 1, 2$ we arrive at a contradiction with Lemma 5.7 which proves impossibility of the Case 1.

It remains to consider the second case.

CASE 2. In this case there exists a constant C such that

$$\frac{\text{dist}(\xi_j^n, \partial\Omega_{\delta_n})}{\rho_n} \leqslant C, \quad j = 1, \ldots, k.$$

If there points ξ_j^n are all interior to Ω_{δ_n} then after scaling with ρ_n we argue as in Case 1 above to reach a contradiction with Lemma 5.8.

Therefore, if Case 2 is to hold, we assume that for certain $j = j^*$ we have

$$\text{dist}\left(\xi^n_{j*}, \partial\Omega_{\delta_n}\right) = 0.$$

Assume first that there exists a constant C such that $\delta_n \leqslant C\rho_n$. Consider the following sum (summation here is taken with respect to all $i \neq j$)

$$s_n = \sum_{i \neq j} G\left(\xi^n_j, \xi^n_i\right).$$

The leading part of s_n, as $n \to \infty$, comes just from the points that become close as $n \to 0$. We can isolate groups of those points according to the asymptotic form of their mutual distances. For example, we can define

$$\rho^1_n = \inf_{i \neq j, i, j > k} \left|\xi^n_j - \xi^n_i\right|$$

and consider those points whose mutual distances are $O(\rho^1_n)$, and so on. For each group of those points (also those with indices higher than k) the argument given above in the Case 1 applies. This means that not only those points become close to one another but also that their distance to the boundary $\partial\Omega_{\delta_n}$ is comparable with their mutual distance. Applying the asymptotic formula for the Green's function we see that

$$s_n = O(1) \quad \text{as } n \to \infty. \tag{5.65}$$

On the other hand, we have

$$\sum_j H\left(\xi^n_j, \xi^n_j\right) \leqslant H\left(\xi^n_{j*}, \xi^n_{j*}\right) + C \leqslant -4\log\frac{1}{|\xi^n_{j*} - \bar{\xi}^n_{j*}|} + C.$$

Since $|\xi^n_{j*} - \bar{\xi}^n_{j*}| \leqslant 2\delta_n$ (because $\xi^n_{j*} \in \partial\Omega_{\delta_n}$), we have that

$$\sum_j H\left(\xi^n_j, \xi^n_j\right) \to -\infty \quad \text{as } n \to \infty,$$

which together with (5.65) contradicts the fact that $\varphi_m(\xi^n)$ is bounded uniformly in n.

Finally assume that $\rho_n = o(\delta_n)$. In this case after scaling with ρ_n around ξ^n_{j*} and arguing similarly as in the Case 1 we get a contradiction with Lemma 5.7 since those points ξ^n_j that are on $\partial\Omega_{\delta_n}$, after passing to the limit, give rise to points that lie on the same straight line. Thus Case 2 cannot hold. $\qquad\square$

In summary we reached now a contradiction with the assumptions of the lemma. The proof of Theorem 1.3 is complete. $\qquad\square$

REMARK 5.1. Let us mention that in [27,28] we study some elliptic problems in a two-dimensional domain with nonlinear Neumann boundary condition where the nonlinearity is exponential on the boundary. Indeed, in [28] the problem

$$\Delta u = 0 \qquad \text{in } \Omega,$$
$$\frac{\partial u}{\partial \nu} = \varepsilon e^{u} \quad \text{on } \partial\Omega \tag{5.66}$$

was analyzed. In (5.66), Ω is a bounded domain in $\mathbb{R}^2$ with smooth boundary and ε a positive small parameter. We prove that given any domain Ω and any $k \geqslant 1$, for ε sufficiently small, a couple of positive solutions peaking at k points on the boundary of Ω in such a way that $\lambda e^{u} \rightharpoonup 4\pi \sum_{j=1}^{k} \delta_{\xi_j}$ was built up, using as basic cells (after suitable zooming-up) explicit solutions of

$$\Delta v = 0 \qquad \text{in } \mathbb{R}^2_+,$$
$$\frac{\partial v}{\partial \nu} = e^{v} \quad \text{on } \partial\mathbb{R}^2_+,$$

where $\mathbb{R}^2_+$ denotes the upper half-plane $\{(x_1, x_2)\colon x_2 > 0\}$ and ν the unit exterior normal to $\partial\mathbb{R}^2_+$, given by

$$w_{t\mu}(x_1, x_2) = \log \frac{2\mu}{(x_1 - t)^2 + (x_2 + \mu)^2}, \tag{5.67}$$

where $t \in \mathbb{R}$ and $\mu > 0$ are parameters.

These functions are the basic cells to build solutions to a related problem with nonlinearity of exponential type on the boundary, namely

$$\Delta u = 0 \qquad \text{in } \Omega,$$
$$\frac{\partial u}{\partial \nu} = 2\lambda \sinh u \quad \text{on } \partial\Omega. \tag{5.68}$$

In [27] we prove that in any domain Ω and for any $k \geqslant 1$ there are at least two distinct families of solutions to (5.68) which exhibit exactly the qualitative behavior of the explicit solution (5.67) at $2k$ points of the boundary and with alternate signs. See also [89] for a related problem.

5.2. *A related 2-d problem involving nonlinearity with large exponent*

In what is left of this section we deal with the analysis of solutions to another two-dimensional nonlinear elliptic problem, namely the following boundary value problem

$$\begin{cases} \Delta u + u^p = 0 & \text{in } \Omega, \\ u > 0 & \text{in } \Omega, \\ u = 0 & \text{on } \partial\Omega, \end{cases} \tag{5.69}$$

where Ω is a smooth bounded domain in $\mathbb{R}^2$ and p is a large exponent. As we will see, problem (5.69) shares similar patterns with the one discussed above in this section, problem (1.14).

First observe that since $H_0^1(\Omega)$ is compactly embedded in $L^{p+1}(\Omega)$ for any $p > 0$, standard variational methods show that S_p, given by

$$S_p = \inf_{u \in H_0^1(\Omega)\setminus\{0\}} I_p(u),$$

where I_p denotes the Rayleigh quotient

$$I_p(u) = \frac{\int_\Omega |\nabla u|^2}{(\int_\Omega |u|^{p+1})^{2/(p+1)}}, \quad u \in H_0^1(\Omega) \setminus \{0\},$$

is achieved by a positive function u_p which solves problem (5.69). This is known as least energy solution.

In [76,77] the authors show that the least energy solution has L^∞-norm bounded and bounded away from zero uniformly in p, for p large. Furthermore, up to subsequence, the renormalized energy density $p|\nabla u_p|^2$ concentrates as a Dirac delta around a critical point of Robin's function $H(x, x)$ introduced in (1.19).

In [1] and [39] the authors give a further description of the asymptotic behavior of u_p, as $p \to \infty$. Indeed they prove that $\|u_p\|_\infty \to \sqrt{e}$ as $p \to \infty$ and that the limit profile of these solutions, properly translated and normalized, is given by the radially symmetric solution (5.4) to problem (5.3). Hence problem (5.3) is what in literature is known as "limit problem" not only for problem (1.14). Indeed (5.3) is the limit problem associate to (5.69) too.

In the same spirit of the problems discussed above, we can build solutions for problem (5.69) that, up to a suitable normalization, look like a sum of concentrated solutions for the limit profile problem (5.3) centered at several points $\xi_1, \ldots, \xi_m$ as $p \to \infty$. At this point, another analogy with problem (1.14) appears: the function responsible to locate the concentration points $\xi_1, \ldots, \xi_m$ is the same function φ_m defined by (1.18), which is responsible for the location of the concentration point for the Liouville-type problem (1.14).

Our main result, contained in [41], guarantees that, as soon as Ω has a hole, with no restriction on the size of the hole, then problem (5.69) admits a solution with an arbitrary number of point of concentration. This is the result contained in the following theorem.

THEOREM 5.2. *Assume that Ω is not simply connected. Then given any $m \geqslant 1$ there exists $p_m > 0$ such that for any $p \geqslant p_m$ problem (5.69) has a solution u_p with*

$$\lim_{p \to \infty} p \int_\Omega u_p^{p+1} = 8\pi m.$$

As for the Liouville equation (1.14), the previous result can be obtained as a consequence of the following more general theorem.

THEOREM 5.3. *Let $m \geqslant 1$ and assume that there is an open set $\mathcal{D}$ compactly contained in Ω^m where φ_m has a nontrivial critical level C. Then, there exists $p_m > 0$ such that for any $p \geqslant p_m$ problem (5.69) has a solution u_p which concentrates at m different points of Ω, i.e., as p goes to $+\infty$,*

$$pu_p^{p+1} \rightharpoonup 8\pi e \sum_{i=1}^{m} \delta_{\xi_j} \quad \textit{weakly in the sense of measure in } \overline{\Omega} \tag{5.70}$$

for some $\xi \in \mathcal{D}$ such that $\varphi_m(\xi_1, \ldots, \xi_m) = C$ and $\nabla \varphi_m(\xi_1, \ldots, \xi_m) = 0$. More precisely, there is an m-tuple $\xi^p = (\xi_1^p, \ldots, \xi_m^p) \in \mathcal{D}$ converging (up to subsequence) to ξ such that, for any $\delta > 0$, as p goes to $+\infty$,

$$u_p \to 0 \quad \textit{uniformly in } \Omega \setminus \bigcup_{j=1}^{m} B_\delta\big(\xi_i^p\big) \tag{5.71}$$

and

$$\sup_{x \in B_\delta(\xi_i^p)} u_p(x) \to \sqrt{e}. \tag{5.72}$$

The proof of how Theorem 5.3 implies the result contained in Theorem 5.2 is identical as the one to prove Theorem 4.4.

As already mentioned, the case of (possibly degenerate) local maximum or minimum for φ_m is included. This simple fact allows us to obtain an existence result for solutions to problem (5.69) also when Ω is simply connected. Indeed, we can construct simply-connected domains of dumbbell type where a large number of concentrating solutions can be found.

Let h be an integer. By h-dumbbell domain with thin handles we mean the following: let $\Omega_0 = \Omega_1 \cup \cdots \cup \Omega_h$, with $\Omega_1, \ldots, \Omega_h$ smooth bounded domains in $\mathbb{R}^2$ such that $\overline{\Omega}_i \cap \overline{\Omega}_j = \emptyset$ if $i \neq j$. Assume that

$$\Omega_i \subset \big\{(x_1, x_2) \in \mathbb{R}^2 \colon a_i \leqslant x_1 \leqslant b_i\big\}, \quad \Omega_i \cap \{x_2 = 0\} \neq \emptyset,$$

for some $b_i < a_{i+1}$ and $i = 1, \ldots, h$. Let

$$C_\varepsilon = \big\{(x_1, x_2) \in \mathbb{R}^2 \colon |x_2| \leqslant \varepsilon, x_1 \in (a_1, b_h)\big\} \quad \text{for some } \varepsilon > 0.$$

We say that Ω_ε is an h-dumbbell with thin handles if Ω_ε is a smooth simply connected domain such that $\Omega_0 \subset \Omega_\varepsilon \subset \Omega_0 \cup C_\varepsilon$, for some $\varepsilon > 0$.

The following result holds true.

THEOREM 5.4. *There exist $\varepsilon_h > 0$ and $p_h > 0$ such that for any $\varepsilon \in (0, \varepsilon_h)$ and $p \geqslant p_h$ problem (5.69) in Ω_ε has at least $2^h - 1$ families of solutions which concentrate at different points of Ω_ε, according to (5.70)–(5.72), as p goes to $+\infty$. More precisely, for any integer $1 \leqslant m \leqslant h$ there exist $\binom{h}{m}$ families of solutions of (5.69) which concentrate at m different points of Ω_ε.*

The detailed proof of how Theorem 5.3 implies the result contained in Theorem 5.4 can be found in [40].

The proof of all these existence results relies on a Lyapunov–Schmidt procedure, based on a proper choice of the ansatz for the solution we are looking for. Usually, in other related problems of asymptotic analysis, and in particular in all the problems discussed in this chapter, the ansatz for the solution is built as the sum of a main term, which is a solution (properly modified or projected) of the associated limit problem, and a lower-order term, which can be determined by a fixed point argument. In this specific problem, this approach is not enough. Indeed, in order to perform the fix point argument to find the lower-order term in the ansatz (the equivalent version of Lemma 5.3 for problem (5.69)), we need to improve substantially the main term in the ansatz, adding two other terms in the expansion of the solution in order to improve the order of the error from p^{-2} to p^{-4}. This fact is basically due to the fact that, when one write the equation in (5.69) as $L\phi = -R - N(\phi)$, with L a linear operator, R the error for the first approximation to be an actual solution to the problem and $N(\phi)$ a quadratic term in ϕ, one only gets

$$\left\| N(\phi) \right\|_* \leqslant Cp\|\phi\|_\infty^2$$

instead of the expected

$$\left\| N(\phi) \right\|_* \leqslant C\|\phi\|_\infty^2$$

as in (5.50).

By performing a finite-dimensional reduction, we find an actual solution to our problem adjusting points ξ inside Ω to be critical points of a certain function $F(\xi)$ (the equivalent to (5.53) for problem (5.69)). It is quite standard to show that this function $F(\xi)$ is a perturbation of $\varphi_m(\xi)$ in a C^0-sense. On the other hand, it is not at all trivial to show the C^1 closeness between F and φ_m. This difficulty is related to the difference between the exponential decay of the scaling parameters $\delta \sim e^{-p/4}$ and the polynomial decay $\frac{1}{p^4}$ of the error term $\|\Delta U_\xi + U_\xi^p\|_*$ of our approximating function U_ξ. We are able to overcome this difficulty using a Pohozaev-type identity.

For the detailed proof of all these results on problem (5.69) we refer the reader to [41] (see also [42] for some results concerning changing sign solutions).

Acknowledgement

This work has been partly supported by grants Fondecyt 1030840, Fondecyt 1040936 and FONDAP, Chile.

References

[1] Adimurthi and M. Grossi, *Asymptotic estimates for a two-dimensional problem with polynomial nonlinearity*, Proc. Amer. Math. Soc. **132** (4) (2004), 1013–1019.

[2]   Adimurthi and G. Mancini, *The Neumann problem for elliptic equations with critical nonlinearity*, A Tribute in Honour of G. Prodi, Ann. Sc. Norm. Sup. Pisa (1991), 9–25.

[3]   Adimurthi, F. Pacella and S.L. Yadava, *Interaction between the geometry of the boundary and positive solutions of a semilinear Neumann problem with critical nonlinearity*, J. Funct. Anal. **113** (1993), 318–350.

[4]   Adimurthi, F. Pacella and S.L. Yadava, *Characterization of concentration points and L^∞-estimates for solutions of a semilinear Neumann problem involving the critical Sobolev exponent*, Differential Integral Equations **8** (1) (1995), 41–68.

[5]   T. Aubin, *Problèmes isopérimétriques et espaces de Sobolev*, J. Differential Geom. **11** (4) (1976), 573–598.

[6]   A. Bahri, *Critical Points at Infinity in Some Variational Problems*, Pitman Res. Notes Math. Ser., vol. 182, Longman, Harlow (1989).

[7]   A. Bahri and J.M. Coron, *On a nonlinear elliptic equation involving the critical Sobolev exponent: The effect of the topology of the domain*, Comm. Pure Appl. Math. **41** (1988), 255–294.

[8]   A. Bahri, Y. Li and O. Rey, *On a variational problem with lack of compactness: The topological effect of the critical points at infinity*, Calc. Var. Partial Differential Equations **3** (1995), 67–93.

[9]   S. Baraket and F. Pacard, *Construction of singular limits for a semilinear elliptic equation in dimension 2*, Calc. Var. Partial Differential Equations **6** (1) (1998), 1–38.

[10]   D. Bartolucci and G. Tarantello, *The Liouville equation with singular data: A concentration–compactness principle via a local representation formula*, J. Differential Equations **185** (1) (2002), 161–180.

[11]   M. Ben Ayed, K. El Mehdi, M. Grossi and O. Rey, *A nonexistence result of single peaked solutions to a supercritical nonlinear problem*, Commun. Contemp. Math. **5** (2) (2003), 179–195.

[12]   H. Brezis, *Elliptic equations with limiting Sobolev exponent – The impact of topology*, Proceedings 50th Anniv. Courant Inst., Comm. Pure Appl. Math. **39** (1986), 817–839.

[13]   H. Brezis and F. Merle, *Uniform estimates and blow-up behavior for solutions of $-\Delta u = V(x)e^u$ in two dimensions*, Comm. Partial Differential Equations **16** (8/9) (1991), 1223–1253.

[14]   H. Brezis and L. Nirenberg, *Positive solutions of nonlinear elliptic equations involving critical Sobolev exponents*, Comm. Pure Appl. Math. **36** (1983), 437–447.

[15]   H. Brezis and L.A. Peletier, *Asymptotics for elliptic equations involving critical growth*, Partial Differential Equations and the Calculus of Variations, Vol. I, Progr. Nonlinear Differential Equations Appl. (1989), 149–192.

[16]   L.A. Caffarelli, B. Gidas and J. Spruck, *Asymptotic symmetry and local behavior of semilinear elliptic equations with critical Sobolev growth*, Comm. Pure Appl. Math. **42** (3) (1989), 271–297.

[17]   L. Caffarelli and Y.-S. Yang, *Vortex condensation in the Chern–Simons Higgs model: An existence theorem*, Comm. Math. Phys. **168** (2) (1995), 321–336.

[18]   E. Caglioti, P.L. Lions, C. Marchioro and M. Pulvirenti, *A special class of stationary flows for two-dimensional Euler equations: A statistical mechanics description, I; II*, Comm. Math. Phys. **143** (1992), 501–525; **174** (1995), 229–260.

[19]   D. Cao and E.S. Noussair, *The effect of geometry of the domain boundary in an elliptic Neumann problem*, Adv. Differential Equations **6** (8) (2001), 931–958.

[20]   S. Chandrasekhar, *An Introduction to the Study of Stellar Structure*, Dover, New York (1957).

[21]   S.-Y.A. Chang and P. Yang, *Conformal deformation of metrics on S^2*, J. Differential Geom. **27** (1988), 259–296.

[22]   C.-C. Chen and C.-S. Lin, *Topological degree for a mean field equation on Riemann surfaces*, Comm. Pure Appl. Math. **56** (12) (2003), 1667–1727.

[23]   M. Clapp, M. del Pino and M. Musso, *Multiple solutions for a non-homogeneous elliptic equation at the critical exponent*, Proc. Roy. Soc. Edinburgh Sect. A **134** (1) (2004), 69–87.

[24]   J.M. Coron, *Topologie et cas limite des injections de Sobolev*, C. R. Acad. Sci. Paris Ser. I **299** (1984), 209–212.

[25]   M. Crandall and P.H. Rabinowitz, *Some continuation and variational methods for positive solutions of nonlinear elliptic eigenvalue problems*, Arch. Rational Mech. Anal. **58** (3) (1975), 207–218.

[26]   E.N. Dancer, *A note on an equation with critical exponent*, Bull. London Math. Soc. **20** (1988), 600–602.

[27]   J. Dávila, M. del Pino, M. Musso and J. Wei, *Singular limits of a two-dimensional boundary value problem arising in corrosion modelling*, Preprint (2005); Arch. Ration. Mech. Anal., to appear.

[28]   J. Dávila, M. del Pino and M. Musso, *Concentrating solutions in a two-dimensional elliptic problem with exponential Neumann data*, J. Funct. Anal. **227** (2) (2005), 430–490.

[29] M. del Pino, J. Dolbeault and M. Musso, *"Bubble-tower" radial solutions in the slightly supercritical Brezis–Nirenberg problem*, J. Differential Equations **193** (2) (2003), 280–306.

[30] M. del Pino, J. Dolbeault and M. Musso, *Duality in the sub and super critical bubbling in the Brezis–Nirenberg problem for dimension 3*, J. Math. Pures Appl. **83** (12) (2004), 1405–1456.

[31] M. del Pino and P. Felmer, *Semi-classical states for nonlinear Schrödinger equations*, J. Funct. Anal. **149** (1) (1997), 245–265.

[32] M. del Pino, P. Felmer and M. Musso, *Multi-peak solutions for super-critical elliptic problems in domains with small holes*, J. Differential Equations **182** (2) (2002), 511–540.

[33] M. del Pino, P. Felmer and M. Musso, *Two-bubble solutions in the super-critical Bahri–Coron's problem*, Calc. Var. Partial Differential Equations **16** (2) (2003), 113–145.

[34] M. del Pino, P. Felmer and M. Musso, *Multi-bubble solutions for slightly super-critial elliptic problems in domains with symmetries*, Bull. London Math. Soc. **35** (4) (2003), 513–521.

[35] M. del Pino, M. Kowalczyk and M. Musso, *Singular limits in Liouville-type equations*, Calc. Var. Partial Differential Equations **24** (1) (2005), 47–81.

[36] M. del Pino, M. Musso and A. Pistoia, *Super-critical boundary bubbling in a semilinear Neumann problem*, Ann. Inst. H. Poincaré Anal. Non Linéaire **22** (1) (2005), 45–82.

[37] W. Ding, *Positive solutions of $\Delta u + u^{\frac{N+2}{N-2}} = 0$ on contractible domains*, J. Partial Differential Equations **2** (4) (1989), 83–88.

[38] O. Druet, *Elliptic equations with critical Sobolev exponents in dimension 3*, Ann. Inst. H. Poincaré Anal. Non Linéaire **19** (2) (2002), 125–142.

[39] K. El Mehdi and M. Grossi, *Asymptotic estimates and qualitative properties of an elliptic problem in dimension two*, Adv. Nonlinear Stud. **4** (1) (2004), 15–36.

[40] P. Esposito, M. Grossi and A. Pistoia, *On the existence of blowing-up solutions for a mean field equation*, Ann. Inst. H. Poincaré Anal. Non Linéaire **22** (2) (2005), 227–257.

[41] P. Esposito, M. Musso and A. Pistoia, *Concentrating solutions for a planar elliptic problem involving nonlinearities with large exponent*, Preprint (2005); J. Differential Equations, to appear.

[42] P. Esposito, M. Musso and A. Pistoia, *Nodal solutions for a two-dimensional elliptic problem with large exponent in nonlinearity*, Preprint (2005); Proc. London Math. Soc., to appear.

[43] V. Felli and S. Terracini, *Fountain-like solutions for nonlinear equations with critical Hardy potential*, Comm. Contemp. Math. **7** (6) (2005), 867–904.

[44] R.H. Fowler, *Further studies on Emden's and similar differential equations*, Quart. J. Math. **2** (1931), 259–288.

[45] Y. Ge, R. Jing and F. Pacard, *Bubble towers for supercritical semilinear elliptic equations*, J. Funct. Anal. **221** (2) (2005), 251–302.

[46] I.M. Gelfand, *Some problems in the theory of quasilinear equations*, Amer. Math. Soc. Transl. **29** (2) (1963), 295–381.

[47] C. Gui and C.-S. Lin, *Estimates for boundary-bubbling solutions to an elliptic Neumann problem*, J. Reine Angew. Math. **546** (2002), 201–235.

[48] Z.-C. Han, *Asymptotic approach to singular solutions for nonlinear elliptic equations involving critical Sobolev exponent*, Ann. Inst. H. Poincaré Anal. Non Linéaire **8** (2) (1991), 159–174.

[49] N. Hirano, A. Micheletti and A. Pistoia, *Existence of sign changing solutions for some critical problems on $\mathbb{R}^N$*, Commun. Pure Appl. Anal. **4** (1) (2005), 143–164.

[50] R. Jing, *Concentration phenomena for equations with supercritical nonlinearity*, Ph.D. Thesis, University of Paris XII and East China Normal University (2005).

[51] D.D. Joseph and T.S. Lundgren, *Quasilinear problems driven by positive sources*, Arch. Ration. Mech. Anal. **49** (1973), 241–269.

[52] J. Kazdan and F. Warner, *Existence and conformal deformation of metrics with prescribed Gaussian and scalar curvatures*, Ann. of Math. **101** (1975), 317–331.

[53] Y.Y. Li, *On a singularly perturbed elliptic equation*, Adv. Differential Equations **2** (1997), 955–980.

[54] Y.Y. Li, *On a singularly perturbed equation with Neumann boundary condition*, Comm. Partial Differential Equations **23** (3/4) (1998), 487–545.

[55] Y.-Y. Li and I. Shafrir, *Blow-up analysis for solutions of $-\Delta u = V e^u$ in dimension two*, Indiana Univ. Math. J. **43** (4) (1994), 1255–1270.

[56] C.-S. Lin, *Topological degree for mean field equations on S^2*, Duke Math. J. **104** (2000), 501–536.

[57] C.-S. Lin, *Locating the peaks of solutions via the maximum principle, I. The Neumann problem*, Comm. Pure Appl. Math. **54** (2001), 1065–1095.

[58] C.-S. Lin, W.-M. Ni and I. Takagi, *Large amplitude stationary solutions to a chemotaxis system*, J. Differential Equations **72** (1988), 1–27.

[59] J. Liouville, *Sur l' equation aux difference partielles $\frac{d^2 \log \lambda}{du\, dv} \pm \frac{\lambda}{2a^2} = 0$*, C. R. Acad. Sci. Paris **36** (1853), 71–72.

[60] L. Ma and J. Wei, *Convergence for a Liouville equation*, Comment. Math. Helv. **76** (3) (2001), 506–514.

[61] A. Micheletti and A. Pistoia, *On the effect of the domain geometry on the existence of sign changing solutions to elliptic problems with critical and supercritical growth*, Nonlinearity **17** (3) (2004), 851–866.

[62] F. Mignot, F. Murat and J. Puel, *Variation d'un point de retournement par rapport au domaine*, Comm. Partial Differential Equations **4** (11) (1979), 1263–1297.

[63] R. Molle and D. Passaseo, *A finite-dimensional reduction method for slightly supercritical elliptic problems*, Abstr. Appl. Anal. (8) (2004), 683–689.

[64] R. Molle and D. Passaseo, *Positive solutions of slightly supercritical elliptic equations in symmetric domains*, Ann. Inst. H. Poincaré Anal. Non Linéaire **21** (5) (2004), 639–656.

[65] M. Musso and A. Pistoia, *Multispike solutions for a nonlinear elliptic problem involving critical Sobolev exponent*, Indiana Univ. Math. J. **51** (2) (2002), 541–579.

[66] M. Musso and A. Pistoia, *Double blow-up solutions for a Brezis–Nirenberg type problem*, Commun. Contemp. Math. **5** (5) (2003), 1–28.

[67] K. Nagasaki and T. Suzuki, *Asymptotic analysis for two-dimensional elliptic eigenvalue problems with exponentially dominated nonlinearities*, Asymptotic Anal. **3** (2) (1990), 173–188.

[68] W.-M. Ni, X.B. Pan and I. Takagi, *Singular behavior of least-energy solutions of a semilinear Neumann problem involving critical Sobolev exponents*, Duke Math. J. **67** (1) (1992), 1–20.

[69] W.-M. Ni and I. Takagi, *On the shape of least-energy solutions to a semilinear Neumann problem*, Comm. Pure Appl. Math. **44** (1991), 819–851.

[70] W.-M. Ni and I. Takagi, *Locating the peaks of least-energy solutions to a semilinear Neumann problem*, Duke Math. J. **70** (1993), 247–281.

[71] D. Passaseo, *Multiplicity of positive solutions of nonlinear elliptic equations with critical Sobolev exponent in some contractible domains*, Manuscripta Math. **65** (2) (1989), 147–165.

[72] D. Passaseo, *New nonexistence results for elliptic equations with supercritical nonlinearity*, Differential Integral Equations **8** (3) (1995), 577–586.

[73] D. Passaseo, *Nontrivial solutions of elliptic equations with supercritical exponent in contractible domains*, Duke Math. J. **92** (2) (1998), 429–457.

[74] A. Pistoia and O. Rey, *Multiplicity of solutions to the super-critical Bahri–Coron problem in pierced domains*, Preprint.

[75] S. Pohozaev, *Eigenfunctions of the equation $\Delta u + \lambda f(u) = 0$*, Soviet. Math. Dokl. **6** (1965), 1408–1411.

[76] X. Ren and J. Wei, *On a two-dimensional elliptic problem with large exponent in nonlinearity*, Trans. Amer. Math. Soc. **343** (2) (1994), 749–763.

[77] X. Ren and J. Wei, *Single point condensation and least energy solutions*, Proc. Amer. Math. Soc. **124** (1996), 111–120.

[78] O. Rey, *The role of the Green's function in a nonlinear elliptic equation involving the critical Sobolev exponent*, J. Funct. Anal. **89** (1) (1990), 1–52.

[79] O. Rey, *An elliptic Neumann problem with critical nonlinearity in three-dimensional domains*, Comun. Contemp. Math. **1** (1999), 405–449.

[80] O. Rey and J. Wei, *Blowing up solutions for an elliptic Neumann problem with sub- or supercritical nonlinearity, I. $N = 3$*, J. Funct. Anal. **212** (2) (2004), 472–499.

[81] O. Rey and J. Wei, *Blowing up solutions for an elliptic Neumann problem with sub- or supercritical nonlinearity, II. $N \geqslant 4$*, Ann. Inst. H. Poincaré Anal. Non Linéaire **22** (4) (2005), 459–484.

[82] M. Struwe and G. Tarantello, *On multivortex solutions in Chern–Simons gauge theory*, Boll. Unione Mat. Ital. Sez. B Artic. Ric. Mat. **8** (1) (1998), 109–121.

[83] T. Suzuki, *Two-dimensional Emden–Fowler equation with exponential nonlinearity*, Nonlinear Diffusion Equations and Their Equilibrium States, 3 (Gregynog, 1989), Progr. Nonlinear Differential Equations Appl., vol. 7, Birkhäuser, Boston, MA (1992), 493–512.

[84] G. Talenti, *Best constant in Sobolev inequality*, Ann. Mat. Pura Appl. (4) **110** (1976), 353–372.

[85] G. Tarantello, *A quantization property for blow-up solutions of singular Liouville-type equations*, J. Funct. Anal. **219** (2) (2005), 368–399.

[86] X.J. Wang, *Neumann problem of semilinear elliptic equations involving critical Sobolev exponent*, J. Differential Equations **93** (1991), 283–301.

[87] J. Wei, *On the construction of single-peaked solutions to a singularly perturbed semilinear Dirichlet problem*, J. Differential Equations **129** (2) (1996), 315–333.

[88] V.H. Weston, *On the asymptotic solution of a partial differential equation with an exponential nonlinearity*, SIAM J. Math. Anal. **9** (6) (1978), 1030–1053.

[89] D. Ye and F. Zhou, *A generalized two dimensional Emden–Fowler equation with exponential nonlinearity*, Calc. Var. Partial Differential Equations **13** (2) (2001), 141–158.

CHAPTER 4

Singular Elliptic and Parabolic Equations

Jesús Hernández

Departamento de Matemáticas, Universidad Autónoma de Madrid, 28049 Madrid, Spain
E-mail: jesus.hernandez@uam.es

Francisco J. Mancebo

E.T.S.I. Aeronáuticos, Universidad Politécnica de Madrid,
Plaza del Cardenal Cisneros 3, 28040 Madrid, Spain

Contents

1. Introduction 319
2. The spectrum of the linear eigenvalue problem 325
3. Existence theorems in the general case $\alpha > 0$ 333
4. Existence of solutions via sub- and supersolutions 337
5. Uniqueness of positive solutions for sublinear problems 340
6. Regularity and boundary behavior 341
7. Differentiability for some singular nonlinear problems 350
8. The associated parabolic problem: linearized stability 353
9. Stabilization 358
10. Applications 362
    10.1. Power law: $\mathcal{L}u = \lambda K(x)u^q$ 362
    10.2. First combination of two power laws: $\mathcal{L}u + M(x)u^p = \lambda K(x)u^q$ 364
    10.3. Second combination of power laws: $\mathcal{L}u = \lambda K(x)u^q + M(x)u^p$ 367
    10.4. Third combination of power laws: $\mathcal{L}u + M(x)u^p = \lambda K(x)u^q$, with M changing sign 370
11. Variational methods: some multiplicity results 370
12. Results for the radial and the one-dimensional problem 376
13. Free boundary solutions: existence and properties 383
Acknowledgements 388
Appendix A. Proof of Proposition 2.3 389
Appendix B. A strong maximum principle for second-order equations with locally bounded coefficients . 393
References 395

HANDBOOK OF DIFFERENTIAL EQUATIONS
Stationary Partial Differential Equations, volume 3
Edited by M. Chipot and P. Quittner

 J. Hernández and F.J. Mancebo

MSC: 35K65, 35J60, 35B32, 35B35, 35B40, 35B50, 35B65, 35J65, 35P05, 35P30, 35R35, 47J20, 49J40

1. Introduction

Reaction–diffusion equations (and systems) have played an important role in the study of many different phenomena related with applications. These applications include, among many others, population dynamics (Lotka–Volterra systems), chemical reactions, combustion, morphogenesis, nerve impulses (Fitzhugh–Nagumo system), genetics, etc. Very often positive solutions are the only physically meaningful solutions or, at least, the more interesting ones. A very simple, but already interesting model problem is the semilinear parabolic equation,

$$\frac{\partial u}{\partial t} - \Delta u = f(x, u) \quad \text{in } \Omega, \tag{1.1}$$

$$u = 0 \qquad\qquad \text{on } \partial\Omega, \tag{1.2}$$

together with an initial condition. Here Ω is a smooth bounded domain in the space $\mathbb{R}^N$, the ordinary Laplacian is used to model diffusion and the nonlinearity f represents a reaction term in each physical situation. One of the main problems which is considered is the asymptotic (i.e., when time t goes to infinity) behavior of solutions to (1.1) and (1.2). Many different (and difficult to deal with) possibilities are available as, for example, traveling waves, but here we will focus on the situation where the unique positive solution to the parabolic problem (1.1) and (1.2) tends to one of the steady-state positive solutions, i.e., to a solution to the stationary elliptic problem

$$-\Delta u = f(x, u) \quad \text{in } \Omega, \tag{1.3}$$

$$u = 0 \qquad\qquad \text{on } \partial\Omega. \tag{1.4}$$

In general the nonlinear term $f(x, u)$ is smooth and frequently satisfies the condition $f(x, 0) \geqslant 0$. (The so-called nonpositone problems, where $f(x, 0) < 0$ have been also studied recently, but they are less attractive and applicable than the former ones.) Problems with nonlinearities going to infinity when $u > 0$ tends to 0 appear in some applications (see [51,57,58,61], see also [49,89]), like non-Newtonian fluids, chemical heterogeneous catalysts and nonlinear heat equations, and have intrinsic mathematical interest. A model example of a problem of this kind is

$$-\Delta u = \frac{\lambda}{u^\alpha} \quad \text{in } \Omega, \tag{1.5}$$

$$u = 0 \qquad \text{on } \partial\Omega, \tag{1.6}$$

where we have $\alpha > 0$, λ is a real parameter and $\Omega \subset \mathbb{R}^N$ is a smooth bounded domain. The basic results for the linear problem, which are essential for all which follows, are collected in Section 2 but the proofs of some of the main results (which are not relevant for the nonlinear theory) are only included in Appendix A. This involves existence, uniqueness and regularity (including regularity estimates implying the compactness of the associated solution operator) for linear equations with differential operators not necessarily in divergence form and allowing some singularities in the coefficients. An extension of the strong maximum principle to this case is given as well. All these results are then used, together with the Krein–Rutman theorem, in the study of the spectral theory, in particular, the existence of principal eigenvalues (i.e., eigenvalues having a positive eigenfunction) to the linear eigenvalue problem

$$\mathcal{L}U - M(x)U = \lambda \widetilde{N}(x)U \quad \text{in } \Omega, \qquad U = 0 \quad \text{on } \partial\Omega, \tag{1.7}$$

where the singular coefficients M and $\widetilde{N}$ may change sign over Ω. Problems of this type arise when linearizing some nonlinear equations around solutions which are in the interior of the positive cone in the space $C_0^1(\overline{\Omega})$, i.e., functions u such that $u > 0$ in Ω and $\partial u/\partial n < 0$ on $\partial\Omega$. This problem was studied by Bertsch and Rostamian in [16] (see also [6]) in the variational case by using a different approach, namely Hardy–Sobolev inequalities in weighted Sobolev spaces. These results will be useful later when dealing with the differentiability of the Green operator in the interior of the positive cone (Section 7), linearized stability (Section 8) and stabilization (Section 9).

The study of the nonlinear problems begins in Section 3. We will limit ourselves to the case of problem (1.8) and (1.9) in a bounded domain of $\mathbb{R}^N$ ($N \geq 2$), what leaves aside the one-dimensional case, where ordinary differential equation methods can be applied. Something similar happens with the case of unbounded domains (in particular, $\Omega = \mathbb{R}^N$) and nonlinear terms $f(x, u, \nabla u)$ depending on the gradient (see the end of Section 3). Very roughly speaking, there are two main approaches in order to deal with singular problems. The first one is to work in the usual frameworks of Hölder or Sobolev spaces by modifying the known methods if necessary; this is actually the way that most of the authors follow. But it is also very natural to try to "compensate" in some way the singularity by introducing weighted spaces: this has been done by Coclite in [32–34] to study existence by using sub- and supersolutions and by Takáč in [112] for stabilization (see also [9]). First we deal with existence (of positive) solutions and give there the existence results in [38] (see also the work by Stuart [110]), which treat, say, the general situation corresponding to any $\alpha > 0$ in equation (1.5). This is important because, as we will see, most of the theorems in the available references only cover the case $0 < \alpha < 1$. Moreover, they are stated for differential operators in general form. Essentially, two main cases are considered, using in both cases an approximation procedure; first, when $f(x, u)$ is increasing in u, it is shown by using sub- and supersolutions that the approximate problem has a unique solution $u_\varepsilon > 0$ which tends to a positive solution $u > 0$, when ε goes to 0, to the original problem. When this monotonicity condition is not satisfied, a more sophisticated argument is necessary, namely the application to the associated nonlinear eigenvalue problem $\mathcal{L}u = \lambda f(x, u + \varepsilon)$ in $\Omega, u = 0$ on $\partial\Omega$, of a global bifurcation theorem by Rabinowitz providing in this way a continuum of positive solutions C_ε to the approximate problem; then it is shown that, roughly speaking, these continua tend, in some sense, to a continuum C of positive solutions to (1.5), (1.6). Another general existence result valid for $\alpha > 1$ was obtained by del Pino [47] by using a variational argument involving a double approximation procedure.

A different approach to existence is developed in Section 4, where an extension to the singular case of the method of sub- and supersolutions is presented. Instead of solving an approximate problem and then pass to the limit, this direct approach, following [77] (see also [78]), has the advantage of allowing more generality in the coefficients of the differential operators and in the domain of definition of the nonlinearity $f(x, u)$ than in [38] but, on the other side, seems to be restricted to $0 < \alpha < 1$ in the model case (1.5), (1.6). We try to avoid approximation procedures (which were used by many authors, see the remarks below) and work directly trying to get an extension as similar as possible to the general existence theorems between ordered sub- and supersolutions in this framework as, e.g., in [5,90,96,102,107]. We show that this is possible if some new additional conditions are imposed on the sub- and supersolutions; then we get a general existence result (Theorem 4.1) which can be applied directly to different examples (see Section 10). Moreover, it is possible, partially, to get rid of these restrictions on the subsolution by means of the (rather technical) Lemma 4.2. We restrict in this section to this situation where nonlinearities do not depend on the gradient. Some indications on the more general problem $\mathcal{L}u = f(x, u, \nabla u)$ are given and the same has been done for the case $\Omega = \mathbb{R}^N$.

Naturally, many other properties concerning positive solutions are still of interest. A first one is uniqueness: in this respect we state a theorem extending to the singular case a well-known "concavity" argument which is usually credited to Krasnoselskii, and can be found in [36], see also [18]. This corresponds, roughly speaking, to the sublinear case for nonsingular equations.

Concerning the regularity (or smoothness) of positive solutions, is clear that they cannot be in $C^2(\overline{\Omega})$ and it is still possible that they are not in $C^1(\overline{\Omega})$ or even in the Sobolev space $H_0^1(\Omega)$. Several results were obtained in [38] in part as a consequence of a close study of the boundary behavior of solutions. Thinking in the model example (1.5) it was shown that if $\alpha > 1$, then solutions are in the Hölder space $C^{2/(1+\alpha)}(\overline{\Omega})$. Moreover, a necessary and sufficient condition for having solutions being Lipschitz up to the boundary was given and this condition reads $0 < \alpha < 1$ for the model case. Some improvements were obtained in [47,66,88]. Moreover, the latter gave an intriguing necessary and sufficient condition in order that classical solutions in $C^2(\Omega) \cap C(\overline{\Omega})$ are in $H_0^1(\Omega)$: this condition is $\alpha < 3$. A systematic approach of regularity was carried out by Gui and Hua Lin in [69] by a careful study of the boundary behavior using Green's function and comparison arguments. In the paper [77] the regularity was just exhibited by working from the beginning in the corresponding Hölder spaces $C^{1,\delta}(\overline{\Omega})$ for some $0 < \delta < 1$. The implicit function theorem and its generalizations (the Lyapunov–Schmidt reduction, say) and corollaries have been an important tool to show existence of curves of solutions to some nonlinear equations giving at the same time the information about the smoothness of these curves. In the case $f(x, 0) = 0$, i.e., when there is a trivial solution $u = 0$, a necessary condition for a point $(\mu, 0)$ being a bifurcation point is that the implicit function theorem is not applicable at this point or, otherwise stated, that the linearized operator is not an isomorphism. This argument is not meaningful in our context, but it is still possible to apply the implicit function theorem if we restrict ourselves to positive solutions having some additional property, namely to be in the interior of the positive cone. This was proved in [77] and will allow to get results completely similar to the well-known ones in the nonsingular case.

We pass to the parabolic problem in Section 8. Curiously enough, and in a way which is quite different from what happened with reaction–diffusion equations in the nonsingular case, almost all the work has been done here for elliptic problems, with the only exceptions of the first paper by Fulks and Maybee [57] and the recent papers by Dávila and

Montenegro [44,45] and Takáč [112]. The results for the linear problem in Section 2 are instrumental to apply the general machinery of, say, Henry [74] to get local existence (and uniqueness) and the usual dichotomy between blow-up and global existence. These results are, once again, taken from [77] (see also [78]). It is clear that the parabolic problem deserves more attention. Here we give some results extending to this case too the linearized stability theorem telling that the sign of the first eigenvalue to the linearized problem defined in Section 2 gives the (local) stability of the solution. A related interesting problem, the so-called stabilization (i.e., to show that solutions to the parabolic problem converge, when $t \to +\infty$, to one single solution to the stationary problem) has been studied carefully by Takáč [112], this time working in weighted Sobolev spaces and proving that the real analyticity of the corresponding nonlinear operator, which is the main property for proving the result, is satisfied in this framework.

Applications of the above results to existence and other properties of positive solutions to some semilinear elliptic singular problems are given in Section 10. As a rule, we prove existence with sub- and supersolutions using the results in Section 4, which provides a much more general and unified view of results in the literature and also some new ones, and giving at the same time more simple proofs (see details in the remarks). Uniqueness, smoothness of branches (depending on some parameter) of positive solutions and stability are obtained by applying the results in Sections 7 and 8. This has been done in a recent paper [79] (see also [78]), and we give here some examples sketching or skipping most of the proofs.

Systems of reaction–diffusion type have been studied during the last thirty years, especially all which concerns the asymptotic behavior of the solutions. Existence, uniqueness (or multiplicity) of positive solutions and their stability have been studied in different instances, raising many interesting open problems (see, e.g., [96,107]). Until now, it is difficult to say that something similar has happened in the singular case, where there are only a few, and maybe not very genuine, examples. Since from the mathematical point of view it is not difficult, once a satisfactory way to deal with one equation is available, to extend the method of sub- and supersolutions to systems in a way which is completely similar to the nonsingular case ([75,96,102,107]) we do not give here the details (see Remark 10.22).

Variational methods have been one of the main methods to find solutions to semilinear elliptic boundary value problems (see, e.g., the book [109]). In many cases, solutions correspond to critical points to the associated functionals defined in the Sobolev space $H_0^1(\Omega)$; then, if for example the functional is smooth (C^1), satisfies the so-called Palais–Smale condition and has some additional properties, existence of solutions follows for many relevant problems. This approach is not immediately applicable to singular problems. Indeed, even if the functional is well defined (at least for $0 < \alpha < 1$ in (1.5)), it is obvious that it is not Fréchet-differentiable on the whole space and it seems reasonable to take the restriction to the positive cone or some subset of it, since there is a critical point theory for closed convex sets in a Banach space. But even in this case, several problems should be solved. Some approaches have been tried, all of them involving in one way or another Ekeland's principle, see [81,117,122]. We just give in Section 11 an overview of the results, which include existence of at least two weak solutions for some problems, and the methods without going into the (rather complicated) technical points.

In Section 12 we treat in some detail an example borrowed from [76] where the case of a ball is studied by using phase plane arguments for positive radial solutions and sharp multiplicity is obtained in the one-dimensional case.

Section 13 is devoted to the study of nonnegative solutions in the sense of the recent paper [44], where solutions which can annihilate on maybe nonzero measure subsets are studied: this is done only in a particular example, but it seems that most of the arguments can be used in many similar problems. Some questions become more involved, one of them is the regularity of the solutions.

Finally several appendices are added. The first one gives the main results concerning linear elliptic problems with singular data, which are an essential tool for the study of nonlinear problems. The second provides an extended version of the strong maximum principle.

Several related topics are not developed, by one reason or another, in this survey. By suitable changes of variable it is possible to establish some links with the problem of finding positive solutions which "blow up" near to the boundary of the domain to some nonlinear elliptic problems see, e.g., [121] and the references therein. These problems are also sometimes called singular. This connection is probably worth considering with more detail.

Let us define precisely the singular differential operators and the singular problems that will be considered in this survey. We will consider the following problem

$$\mathcal{L}u \equiv -\sum_{i,j=1}^{N} a_{ij}(x)\frac{\partial^2 u}{\partial x_i \,\partial x_j} + \sum_{i=1}^{N} b_i(x)\frac{\partial u}{\partial x_i} = f(x,u) \quad \text{in } \Omega, \tag{1.8}$$

$$u = 0 \quad \text{on } \partial\Omega, \tag{1.9}$$

where the differential operator $\mathcal{L}$ and the function f satisfy the following *assumptions, which hold for some α such that $-1 < \alpha < 1$:*

(H.1) $\Omega \subset \mathbb{R}^N$ is a bounded domain, with a $C^{3,\gamma}$ boundary, for some $\gamma > 0$, if $N > 1$. Note that the distance from $x \in \Omega$ to $\partial\Omega$, $d(x)$, defines a function $d \in C^{3,\gamma}(\overline{\Omega}_1)$, with $\Omega_1 = \{x \in \Omega: d(x) < \rho_1\}$ for some $\rho_1 > 0$.

(H.2) The second-order part of the operator $-\mathcal{L}$ is uniformly, strongly elliptic in Ω. Also, for all $i, j, k = 1, \ldots, N, a_{ij} = a_{ji} \in C^3(\Omega) \cap C(\overline{\Omega})$, $b_i \in C^2(\Omega)$, and there is a constant K such that $|\partial a_{ij}/\partial x_k| + |b_i| < K[1 + d(x)^\alpha]$ and $|\partial^2 a_{ij}/\partial x_i \,\partial x_j| + |\partial b_i/\partial x_j| < K d(x)^{\alpha-1}$ for all $x \in \Omega$.

As a consequence, the functions a_{ij}, $x \to d(x)\,\partial a_{ij}(x)/\partial x^k$ and $x \to d(x)b_i(x)$ are in $C^{0,\delta}(\overline{\Omega})$ whenever $0 < \delta < \min\{\alpha + 1, \gamma\}$.

(H.3) There is an integer $m > 0$ such that $f, \partial^j f/\partial u^j, \partial^j f/\partial u^{j-1}\,\partial x_k \in C(\Omega \times]0, \infty[)$ for all $k = 1, \ldots, N$ and all $j = 1, \ldots, m + 1$. And if $u: \Omega \to \mathbb{R}$ is such that $0 < k_1 d(x) < u(x) < k_2 d(x)$ for all $x \in \Omega$ and some constants k_1 and k_2, then $|f(x, u(x))| < K_0[1 + d(x)^\alpha]$ and $|\partial^j f(x, u(x))/\partial u^j| + \sum_{k=1}^{N} |\partial^j f(x, u(x))/\partial u^{j-1}\,\partial x_k| < K_j d(x)^{\alpha-j}$ for all $x \in \Omega$, all $k = 1, \ldots, N$ and all $j = 1, \ldots, m+1$, where K_j (can depend on k_1 and k_2 but) is independent of u.

For convenience we are allowing (in (H.2)) the coefficients b_i to exhibit an appropriate singularity at the boundary. Also, we are requiring the coefficients of the operator $\mathcal{L}$ to be such that the *adjoint operator* $\mathcal{L}^*$, defined as

$$\mathcal{L}^* u \equiv -\sum \frac{\partial}{\partial x_i}\left(a_{ij}\frac{\partial u}{\partial x_j}\right) - \sum \frac{\partial}{\partial x_i}\left[\left(b_i - \sum \frac{\partial a_{ij}}{\partial x_j}\right)u\right], \qquad (1.10)$$

is such that the equation $\mathcal{L}^* u = f(x, u)$ also satisfies (H.2) and (H.3); $\mathcal{L}^*$ is the formal adjoint of $\mathcal{L}$ with respect to the inner product of $L_2(\Omega)$. Also, *since all sums apply to the values $1, \ldots, N$ of the involved indexes, the limits 1 and N are omitted hereafter in the symbol* $\sum$. Note that assumption (H.3) is satisfied by the usual power-law nonlinearities, $f(x, u) = g(x)u^{\alpha_1}$, whenever $\alpha_1 > -1$ and $g \in C^1(\overline{\Omega})$; or, more generally, when $g \in C^1(\Omega)$ and $|g(x)| < Kd(x)^{\alpha_2}$, for some $K > 0$ and some α_2 such that $|\alpha_1 + \alpha_2| < 1$. Singular elliptic problems of this type where considered, among many others, by Laetsch [86], Cohen and Laetsch [36], Crandall, Rabinowitz and Tartar [38], Brezis and Oswald [20] and Bandle, Pozio and Tesei [12] in bounded domains, and by Spruck [108], Schatzman [103] and Brezis and Kamin [18] in $\mathbb{R}^N$. As a by-product of the results in the chapter, in Section 4.1 we shall extend the existence (of strictly positive, classical solutions, in $C^2(\Omega) \cap C(\overline{\Omega})$) result in [38] to nonlinearities of indefinite sign, which are of interest in, e.g., population dynamics [49,70,71,94]. Note that if the nonlinearity f is singular at $u = 0$ and is negative for $u > 0$ and $x \in \Omega' \neq \emptyset$, then the nonnegative solutions of (1.8), (1.9) can exhibit *free boundaries* between a region $\overline{\Omega}'' \subset \Omega'$ where $u = 0$ ('dead core') and the support of u [49]; these solutions will be excluded, except in Section 13, from the analysis below, where only (strictly) positive solutions will be considered. In Section 4 we shall use the method of sub- and supersolutions, as in [12], where quite weak, not necessarily (strictly) positive solutions were obtained.

The linearization of (1.8), (1.9) around a given positive solution u leads us to consider the linear eigenvalue problem

$$\mathcal{L}U - M(x)U = \lambda U \quad \text{in } \Omega, \qquad U = 0 \quad \text{on } \partial\Omega, \qquad (1.11)$$

where $\mathcal{L}$ is as in (1.8) and $M(x) = f_u(x, u(x))$. Thus the natural assumption on the coefficient M is

(H.3′) $M \in C^1(\Omega)$ and, for all $k = 1, \ldots, N$, $d(x)^{2-\alpha}|\partial M(x)/\partial x_k|$ is bounded in Ω. As a consequence, the function $x \to d(x)^2 M(x)$ is in $C^{0,\delta}(\overline{\Omega})$ whenever $0 < \delta < \min\{\alpha + 1, \gamma\}$, and $d(x)^{1-\alpha}M(x)$ is bounded in Ω.

Note that we are not requiring M to have a constant sign near $\partial\Omega$. In fact, as we shall see in Section 2 (see Lemma 2.1), that sign can be controlled upon a change of variable that affects both the coefficients b_i and the coefficient M itself, with the new coefficients still satisfying (H.2) and (H.3′). This result is of independent interest and appears as surprising at first sight because the sign of M near $\partial\Omega$ plays an important role when applying maximum principles. Similar singular eigenvalue problems in divergence form were considered in [16], where generalized Hardy–Sobolev inequalities [31] (see Remark 2.4) were used to prove that the eigenfunctions are in $C^2(\Omega) \cap H_0^1(\Omega)$. Here we shall prove that the eigenfunctions are also in $C_0^{1,\delta}(\overline{\Omega})$ for all δ such that $0 < \delta < \min\{\alpha + 1, \gamma\}$. *A stronger*

$C^1(\overline{\Omega})$-regularity *(also for the solutions of* (1.8), (1.9) *and of some related linear problems) is necessary in order to apply a straightforward generalization (Appendix B) of the Hopf boundary lemma* [98]. This and our assumption that $\alpha > -1$ in (H.3) and (H.3$'$) will prevent us from using L_p theory [3], Hardy–Sobolev inequalities and embedding theorems [1], which provide $C^1(\overline{\Omega})$-solutions only if $\alpha > -1/N$ (see Remark 2.3). Instead, *we shall use the (integral) reformulation of the various problems through the Green function of the linear problem* $\mathcal{L}u = f$ *in* Ω, $u = 0$ *on* $\partial\Omega$, *and work in* $C_0^1(\overline{\Omega})$ *(or in* $C_0^{1,\delta}(\overline{\Omega})$ *when convenient) but, for the sake of clarity, these problems will be written in differential form in the statements of most results.* Note that the requirement $\alpha > -1$ in assumptions (H.2), (H.3), (H.3$'$) and (H.4) is somewhat optimal when seeking $C^1(\overline{\Omega})$-regularity, as the simplest counterexamples in [16] readily show.

In fact, in Section 2 we shall consider a slightly more general linear eigenvalue problem, namely

$$\mathcal{L}U - M(x)U = \lambda\widetilde{N}(x)U \quad \text{in } \Omega, \qquad U = 0 \quad \text{on } \partial\Omega, \tag{1.12}$$

where $\mathcal{L}$ and M are as defined in (H.2) and (H.3$'$) and $\widetilde{N}$ satisfies

(H.4) $\widetilde{N}$ is strictly positive in Ω, and satisfies assumption (H.3$'$).

2. The spectrum of the linear eigenvalue problem

The object of this section is to analyze the spectrum of the linear eigenvalue problem (1.12). This problem is obtained when linearizing (1.8), (1.9) formally around a positive solution. This is an essential step in order to generalize the method of sub- and supersolutions to singular problems. In this section we generalize many of the spectral properties of the regular nonnecessarily self-adjoint differential operators to singular operators in the sense defined in the Introduction. For a summary of this properties, see the classical book [37] and the extensions in the papers [56,80,91,92], for changing sign coefficients. But before proceeding with the main results we show that the sign of the coefficient M near $\partial\Omega$ can be controlled through a change of variable. This property will allow us to apply the maximum principle to obtain spectral properties of these operators. We follow closely [77] all along this section.

LEMMA 2.1. *Under assumptions* (H.1), (H.2), (H.3$'$) *and* (H.4), *there are two functions* $\varphi^{\pm} \in C^{2,\delta}(\overline{\Omega})$ *for all* δ *such that* $0 < \delta < \delta_0 = \min\{\gamma, \alpha + 1\}$, *such that*

$$\varphi^{\pm} > 0 \quad \text{in } \overline{\Omega}, \tag{2.1}$$

and if $U \in X \equiv C^2(\Omega) \cap C_0^{1,\delta}(\overline{\Omega})$, *with* $0 < \delta < \delta_0$, *is a solution of* (1.12), *then the functions* $U^{\pm} = \varphi^{\pm}U$ *are in X and*

$$\mathcal{L}^{\pm}U^{\pm} \equiv -\sum a_{ij}(x)\frac{\partial^2 U^{\pm}}{\partial x_i\,\partial x_j} + \sum b_i^{\pm}(x)\frac{\partial U^{\pm}}{\partial x_i}$$

$$= M^{\pm}(x)U^{\pm} + \lambda\widetilde{N}(x)U^{\pm} \quad \text{in } \Omega, \tag{2.2}$$

$$U^\pm = 0, \qquad \frac{\partial U^\pm}{\partial n} = \frac{\partial U}{\partial n} \quad \text{on } \partial\Omega, \tag{2.3}$$

where n is the outward unit normal, the coefficients $b_i^\pm$ and $M^\pm$ satisfy assumptions (H.2) and (H.3′), and

$$\pm M^\pm > 0 \quad \text{in a neighborhood of } \partial\Omega. \tag{2.4}$$

PROOF. Let $d(x) = d(x_1, \ldots, x_n)$ be the distance from x to $\partial\Omega$ and let $\psi \in C^3(]0, \infty[) \cap C^0([0, \infty[)$ be a real function such that

$$\delta\psi(\eta) \geqslant 0 \quad \text{if } \eta \geqslant 0, \qquad \delta\psi(\eta) = \eta^{\delta+1} \quad \text{if } 0 \leqslant \eta \leqslant \varepsilon \leqslant \rho_1,$$
$$\psi(\eta) = 0 \quad \text{if } \eta > \rho_1, \tag{2.5}$$

where $\delta \neq 0$ is such that $-1 < \delta < \alpha$, with α and ρ_1 as in assumptions (H.1), (H.2) and (H.3′). The strictly positive constant ε will be selected further.

Now we define the functions $\varphi^\pm$ as $\varphi^\pm(x) = \exp[\mp\psi(d(x))]$. If $U \in X$ is a solution of (1.12) then $U^\pm = \varphi^\pm U$ is such that $U^\pm \in X$ and satisfies (2.2) with

$$b_i^\pm = b_i \mp 2\psi'(d(x)) \sum a_{ij} \frac{\partial d}{\partial x_j},$$

$$M^\pm = M \mp \psi'(d(x)) \sum b_i \frac{\partial d}{\partial x_i}$$
$$+ \sum a_{ij} \left[\frac{\partial d}{\partial x_i} \frac{\partial d}{\partial x_j} \left(\psi'(d(x)) \right)^2 \pm \psi''(d(x)) \right) \pm \frac{\partial^2 d}{\partial x_i \partial x_j} \psi'(d(x)) \right].$$

But according to (2.5),

$$\pm M^\pm d(x)^{1-\delta} \geqslant 0 \quad \text{if } 0 < d(x) < \varepsilon$$

provided that

$$\pm M(x) d(x)^{1-\alpha} d(x)^{\alpha-\delta} - d(x) \sum \frac{b_i}{\delta} \frac{\partial d}{\partial x_i}$$
$$+ \sum a_{ij} \left([1 - (\delta+1)d(x)^{\delta+1}] \frac{\partial d}{\partial x_i} \frac{\partial d}{\partial x_j} + d(x) \frac{\partial^2 d}{\partial x_i \partial x_j} \right) > 0,$$

which holds if $0 < d(x) < \varepsilon$ and ε is sufficiently small, because $|M(x)|d(x)^{1-\alpha}$ is bounded in Ω and the matrix (a_{ij}) is positive definite in Ω, according to assumptions (H.2) and (H.3′). Note that ε is chosen independently of U. Also $b_i^\pm$ and $M^\pm$ satisfy assumptions (H.2) and (H.3′), respectively, and according to (2.5), $U^\pm$ satisfies (2.3). Thus the proof is complete. $\qquad\square$

REMARK 2.2. The result above implies that the point spectrum of (2.2), with Dirichlet boundary data

$$U^\pm = 0 \quad \text{on } \partial\Omega, \tag{2.6}$$

is the same as that of (1.12), provided that the eigenfunctions are in $C^2(\Omega) \cap C_0^1(\overline{\Omega})$ (and this is a natural assumption, as we shall see below). Still, (2.1) and (2.3) imply that

$$U = 0 \ (\text{resp., } U > 0) \quad \text{if and only if} \quad U^\pm = 0 \ \big(\text{resp., } U^\pm > 0\big) \quad \text{in } \Omega,$$

$$\frac{\partial U}{\partial n} = 0 \left(\text{resp., } \frac{\partial U}{\partial n} < 0\right) \quad \text{if and only if}$$

$$\frac{\partial U^\pm}{\partial n} = 0 \left(\text{resp., } \frac{\partial U^\pm}{\partial n} < 0\right) \quad \text{on } \partial\Omega.$$

The first property implies that λ is a principal eigenvalue of (1.12) if and only if it is a principal eigenvalue of (2.2), (2.6). The second property will be quite useful below to apply a Hopf boundary lemma.

In order to analyze the eigenvalue problem (1.12), we first consider the Green operator of the problem

$$\mathcal{L}u = M(x)v \quad \text{in } \Omega, \qquad u = 0 \quad \text{on } \partial\Omega, \tag{2.7}$$

defined as $u = G_1(v)$, with $v \in C_0^{0,1}(\overline{\Omega})$. Since the analysis of this problem is somewhat apart from the remaining part of the paper, it is relegated to Appendix A at the end of the paper, and the result is just stated here; but see Remark 2.4.

PROPOSITION 2.3. *Let Ω, $\mathcal{L}$ and M satisfy assumptions* (H.1), (H.2) *and* (H.3′). *If $v \in C_0^{0,1}(\overline{\Omega})$ then* (2.7) *has a unique solution $u \in C^2(\Omega) \cap C^{1,\delta}(\overline{\Omega})$ for all δ such that $0 < \delta < \delta_0 = \min\{\gamma, \alpha + 1\}$. And there is a constant K, which* (can depend on δ but) *is independent of v, such that*

$$\|u\|_{C^{1,\delta}(\overline{\Omega})} \leqslant K \|v\|_{C^{0,1}(\overline{\Omega})}. \tag{2.8}$$

For the proof see Appendix A.

REMARK 2.4. *If $\alpha > -1/N$ then we can use L_p estimates to show that, under the assumptions in Proposition* 2.3, (2.7) *possesses a unique solution $u \in W_p^2(\Omega)$, for all $p > N$ such that $1 + \alpha p > 0$, and that $\|u\|_{W_p^2(\Omega)} \leqslant K \|v\|_{C^{0,1}(\overline{\Omega})}$, with K independent of v. This* result is readily obtained by first replacing (2.7) by

$$u + G_0\left(\sum b_i \frac{\partial u}{\partial x_i}\right) = G_0\big(M(x)v\big), \tag{2.9}$$

where $G_0 : L_p(\Omega) \to W_p^2(\Omega)$ is the Green operator of the problem $-\sum a_{ij} \frac{\partial^2 u}{\partial x_i \, \partial x_j} = f$ in Ω, $u = 0$ on $\partial \Omega$, and then taking into account that

$$\left\| \sum b_i \frac{\partial u}{\partial x_i} \right\|_{L_p(\Omega)} \leqslant K_1 \|u\|_{C^{0,1}(\overline{\Omega})}, \qquad \left\| M(x)v \right\|_{L_p(\Omega)} \leqslant K_2 \|v\|_{C^{0,1}(\overline{\Omega})} \tag{2.10}$$

with K_1 and K_2 independent of u and v, respectively. The first estimate readily follows from assumption (H.2). The second estimate follows from (H.2) and the inequality

$$|v(x)| \leqslant d(x) \|v\|_{C^{0,1}(\overline{\Omega})} \quad \text{for all } x \in \Omega, \tag{2.11}$$

which holds whenever $v \in C_0^{0,1}(\overline{\Omega})$, as readily obtained when applying the mean value theorem between x and that point of $\partial\Omega$ where the distance $d(x)$ is reached; the second estimate (2.10) can also be obtained via Hardy–Sobolev inequalities [31], but this requires that $\alpha \geqslant 0$ if $p > N$. Now, when using (2.10), the continuity of G_0, the fact that $W_p^2(\Omega)$ is compactly embedded into $C^{0,1}(\overline{\Omega})$, standard maximum principles and standard Riesz theory on compact, linear operators, the result readily follows. Thus if $\alpha > -1/N$ we can proceed with L_p theory and embedding theorems to obtain the results below in a simpler way, which unfortunately is not appropriate to obtain C^1-regularity up to the boundary if $-1 < \alpha \leqslant -1/N$.

The main ingredient to analyze the spectrum of (1.12) when $N > 0$ in Ω is in the following proposition.

PROPOSITION 2.5. *Under assumptions (H.1), (H.2), (H.3$'$) and (H.4), there is a constant k_0 such that if $k > k_0$ and $v \in C_0^{0,1}(\overline{\Omega})$ then the problem*

$$\mathcal{L}u - M(x)u + k\widetilde{N}(x)u = \widetilde{N}(x)v \quad \text{in } \Omega, \qquad u = 0 \quad \text{on } \partial\Omega \tag{2.12}$$

has a unique solution $u \in C^2(\Omega) \cap C^{1,\delta}(\overline{\Omega})$ for all $\delta \in \,]0, \delta_0[\,$, where $\delta_0 = \min\{\gamma, \alpha + 1\}$, and

$$\|u\|_{C^{1,\delta}(\overline{\Omega})} \leqslant K \|v\|_{C^{0,1}(\overline{\Omega})}, \tag{2.13}$$

where K (can depend on k and δ but) is independent of v. If, in addition, $v \geqslant 0$ in Ω and v is not identically zero, then $\partial u/\partial n < 0$ on $\partial\Omega$.

PROOF. We first select k_0 to ensure uniqueness. To this end we rewrite (2.12) as

$$\mathcal{L}^- u^- - M^-(x)u^- + k\widetilde{N}(x)u^- = \widetilde{N}(x)\varphi^- v \quad \text{in } \Omega, \qquad u^- = 0 \quad \text{on } \partial\Omega, \tag{2.14}$$

where $\mathcal{L}^-$, M^- and $\varphi^- > 0$ are as is Lemma 2.1 and $u^- = \varphi^- u$. Since $M^- < 0$ in a neighborhood Ω^1 of $\partial\Omega$ and $\widetilde{N} > 0$ in Ω, $k_0 = \sup\{M^-(x)/\widetilde{N}(x)\colon x \in \Omega \setminus \Omega^1\}$ is well defined, and $k\widetilde{N} - M^- > 0$ in Ω whenever $k > k_0$. Then a standard maximum principle applied to (2.14) ensures uniqueness for that problem, and hence uniqueness for (2.12), if $k > k_0$. Still, the strong maximum principle in Appendix B implies that $\partial u^-/\partial n < 0$ if $v \geqslant 0$ and v is not identically zero. And since $\partial u/\partial n = \partial u^-/\partial n$ on $\partial\Omega$ (see (2.3)) the last statement in Proposition 2.5 also follows if $k > k_0$. Thus only the existence part remains to be proved.

After selecting k_0, we take $k > k_0$, $\delta \in]0, \delta_0[$ and rewrite (2.12) as

$$H(u) \equiv u - G_1(u) + kG_2(u) = G_2(v), \tag{2.15}$$

where $G_1, G_2 \colon C_0^{0,1}(\overline{\Omega}) \to C_0^{1,\delta}(\overline{\Omega})$ are then Green operators ($u = G_1(v)$ and $u = G_2(v)$) of the problems (2.7) and $\mathcal{L}u = \widetilde{N}(x)v$ in Ω, $u = 0$ on $\partial\Omega$, respectively; note that G_1 and G_2 are bounded, according to Proposition 2.3. Since the embedding $i \colon C_0^{1,\delta}(\overline{\Omega}) \to C_0^{0,1}(\overline{\Omega})$ is compact, $\widehat{H} = H \circ i$ is a compact perturbation of the identity in $C_0^{1,\delta}(\overline{\Omega})$, and since $k > k_0$, $\widehat{H}$ is injective. Thus the standard Riesz theory [52] on compact operators applies and $\widehat{H}$ is readily seen to be a linear homeomorphism. Then if $v \in C_0^{0,1}(\overline{\Omega})$, $w = G_2(v) \in C_0^{1,\delta}(\overline{\Omega})$ and (2.12) has a unique solution $u \in C^{1,\delta}(\overline{\Omega})$ such that $\|u\|_{C^{0,\delta}(\overline{\Omega})} \leqslant K'\|G_2(v)\|_{C^{1,\delta}(\overline{\Omega})} \leqslant K\|v\|_{C_0^{0,1}(\overline{\Omega})}$ and the estimate (2.13) follows. Also, as in Proposition 2.3, $u \in C^2(\Omega)$, and the proof is complete. $\qquad\square$

Now we are in a position to analyze the linear eigenvalue problem (1.12). This problem was considered by Bertsch and Rostamian [16] for operators in divergence form via Hardy–Sobolev inequalities, see Remark 2.4. As for the regularity of the eigenfunctions, in [16] it was shown that they are in $C^2(\Omega) \cap H_0^1(\Omega)$ if M and $\widetilde{N}$ satisfy assumption (H.3$'$) above with $\alpha > -1$, and in $C^2(\Omega) \cap C_0^1(\overline{\Omega})$ if $\alpha > 0$. But the corresponding spectrum coincides with that obtained when the eigenfunctions are required to be in $C^2(\Omega) \cap C^1(\overline{\Omega})$, as shown in the following theorem, which also provides a fairly complete characterization of the spectrum for operators in general form.

THEOREM 2.6. *Under assumptions* (H.1), (H.2), (H.3$'$) *and* (H.4), *the spectrum of the linear eigenvalue problem* (1.12), *with* $U \in C^2(\Omega) \cap C^{0,1}(\overline{\Omega})$, *is such that:*

    (i) *It consists at most of an infinite, countable set of eigenvalues which are isolated, and the eigenfunctions are in* $C^{1,\delta}(\overline{\Omega})$ *for all* δ *such that* $0 < \delta < \delta_0 = \min\{\gamma, \alpha + 1\}$.

    (ii) *It contains a unique principal eigenvalue* (*i.e., a real eigenvalue with an associated eigenfunction in the interior of the positive cone of* $C_0^1(\overline{\Omega})$, *namely, such that* $U > 0$ *in* Ω *and* $\partial U/\partial n < 0$ *on* $\partial\Omega$), *which is simple.*

    (iii) *The* (*not necessarily real*) *eigenvalues of* (1.12) *are such that*

$$\mathrm{Re}\,\lambda > \lambda_1 \quad \text{if } \lambda \neq \lambda_1 \quad \text{and} \quad \mathrm{Re}\,\lambda \geqslant c_2 + c_1|\mathrm{Im}\,\lambda|, \tag{2.16}$$

where λ_1 *is the principal eigenvalue of* (1.12) *and the real constants* $c_1 > 0$ *and* c_2 *are independent of* λ.

(iv) *It does not change when the eigenfunctions are only required to be in $C^2(\Omega) \cap H_0^1(\Omega)$, and coincides with the spectrum of the formal adjoint problem* (1.10).

PROOF. We subsequently only prove the statements (i) and (ii); see [77] for (iii) and (iv). Since the operator $\mathcal{L}$ is not necessarily self-adjoint, its spectrum is not necessarily real, and we must work with the complexifications of $\mathcal{L}$ and the various function spaces; this trivial extension will be automatically made below.

(i) If k_0 is as in Proposition 2.5, $k > k_0$ and $0 < \delta < \delta_0$, then the problem (2.12) defines a Green operator $u = G(v)$, with $G : C_0^{0,1}(\overline{\Omega}) \to C_0^{1,\delta}(\overline{\Omega})$ bounded. And if i is the compact embedding of $C_0^{1,\delta}(\overline{\Omega})$ into $C_0^{0,1}(\overline{\Omega})$, then $\widehat{G} = G \circ i : C_0^{1,\delta}(\overline{\Omega}) \to C_0^{1,\delta}(\overline{\Omega})$ is compact. This completes the proof of the second statement in part (i). And the first statement follows by the standard spectral theory for compact operators [52], when taking into account that the eigenvalues of (1.12) and $\widehat{G}$, μ, are related by

$$\mu = \frac{1}{\lambda + k}. \tag{2.17}$$

(ii) Let $\widehat{G}$ be the compact operator defined in the proof of (i). According to Proposition 2.5, $\widehat{G}$ maps the positive cone of $C_0^{1,\delta}(\overline{\Omega})$ into its interior, and the Krein–Rutman theorem [5,40] readily implies that $\widehat{G}$ has a unique principal eigenvalue μ_1, which is simple, strictly positive and such that any other eigenvalue of $\widehat{G}$ satisfies $|\mu| < \mu_1$. And taking into account the relation (2.17) between the spectra of $\widehat{G}$ and (2.4), the statement (ii) readily follows, with $\lambda_1 = 1/\mu_1 - k$ (for any $k > k_0$). Note that, in addition, any other eigenvalue λ of (1.12) satisfies $[|\operatorname{Re}\lambda + k|^2 + |\operatorname{Im}\lambda|^2]^{1/2} = |\lambda + k| = 1/|\mu| > 1/\mu_1 = |\lambda_1 + k|$, and since that inequality holds for all $k > k_0$, we readily obtain

$$\operatorname{Re}\lambda \geqslant \lambda_1 \tag{2.18}$$

and the proof is complete. $\square$

Now we prove that the min–max characterization of the principal eigenvalue introduced in [53] also applies when the coefficients of $\mathcal{L}$ and the function M satisfy (H.2), (H.3′) and (H.4) (Proposition 2.8). The idea of the proof follows that of [14] (see also [53]). The following previous characterization is needed.

LEMMA 2.7. *Under assumptions* (H.1), (H.2), (H.3′) *and* (H.4), *the problem* (1.12) *possesses a strictly positive principal eigenvalue if and only if the operator $\mathcal{L} - M(x)$ satisfies the strong maximum principle, i.e., if $v \in C^2(\Omega) \cap C^1(\overline{\Omega})$ is such that*

$$v \neq 0, \quad \mathcal{L}v - M(x)v \geqslant 0 \quad in \ \Omega, \qquad v \geqslant 0 \quad on \ \partial\Omega, \tag{2.19}$$

then $v > 0$ for all $x \in \Omega$ and $\partial v(x)/\partial n < 0$ (where n is the outward unit normal, as above) for all $x \in \partial\Omega$ such that $v(x) = 0$.

PROOF. If $\mathcal{L} - M(x)$ satisfies a strong maximum principle then the principal eigenvalue of (1.12), λ_1, exists (Theorem 2.6) and is readily seen to be strictly positive. In order to prove the converse we assume without loss of generality (Lemma 2.1) that $M(x) \leqslant 0$ in a neighborhood Ω^1 of $\partial\Omega$. For each function v satisfying (2.19), we consider the function

$$w = v + \varepsilon + \varepsilon k U_1, \tag{2.20}$$

where $\varepsilon > 0$, $U_1 > 0$ is an eigenfunction of (1.12) associated with the principal eigenvalue λ_1 and $k = \sup\{2|M(x)|/[\lambda_1 \widetilde{N}(x) U_1(x)]: x \in \Omega \setminus \Omega^1\}$. Then

$$\mathcal{L}w - M(x)w \geqslant \varepsilon\big[k\lambda_1 \widetilde{N}(x) U_1(x) - M(x)\big] > 0 \quad \text{in } \Omega. \tag{2.21}$$

Moreover, since v is continuous, for each $\varepsilon > 0$ there is a constant $\gamma(\varepsilon) > 0$ such that $w > 0$ in $\Omega_\varepsilon = \{x \in \Omega: d(x) < \gamma(\varepsilon)\}$, and $w > 0$ in $\Omega \setminus \Omega_\varepsilon$, as readily seen upon application of the generalized maximum principle to (2.21) in $\Omega \setminus \Omega_\varepsilon$ (note that U_1 satisfies (2.19), with strict inequalities in $\Omega \setminus \Omega_\varepsilon$). Thus $w > 0$ in Ω for all $\varepsilon > 0$, and by letting $\varepsilon \to 0$, we obtain $v \geqslant 0$ in Ω. Thus, the generalized maximum principle and the strong maximum principle in Appendix B yield $v > 0$ in Ω and $\partial v(x)/\partial n < 0$ if $x \in \partial\Omega$ and $v(x) = 0$, and the proof is complete. $\qquad\square$

PROPOSITION 2.8. *Under assumptions* (H.1), (H.2), (H.3′) *and* (H.4), *the principal eigenvalue of* (1.12) *is given by*

$$\lambda_1 = \sup\left\{\inf\left\{\frac{\mathcal{L}v - M(x)v}{\widetilde{N}(x)v}: x \in \Omega\right\}: v \in \mathcal{P}\right\}, \tag{2.22}$$

where $\mathcal{P}$ *is the set of those functions of* $C^2(\Omega) \cap C_0^1(\overline{\Omega})$ *such that* $v > 0$ *in* Ω.

The following result deals with the dependence of the principal eigenvalue of (1.12) on the coefficient of U in the left-hand side of (1.12). Once again we extend to the singular case results which are well known in the regular problem.

PROPOSITION 2.9. *Let* Ω, $\mathcal{L}$, M *and* $\widetilde{N}$ *satisfy assumptions* (H.1), (H.2), (H.3′) *and* (H.4), *and let* M_1 *be a nonzero function that is nonnegative in* Ω *and satisfies* (H.3′). *For each* $\mu \in \mathbb{R}$, *let* $\lambda = \Lambda(\mu)$ *be the principal eigenvalue of*

$$\big[\mathcal{L} - M(x) + \mu M_1(x)\big]u = \lambda \widetilde{N}(x)u \quad \text{in } \Omega, \qquad u = 0 \quad \text{on } \partial\Omega, \tag{2.23}$$

with $u \in C^2(\Omega) \cap C^1(\overline{\Omega})$. *Then the function* $\Lambda: \mathbb{R} \to \mathbb{R}$ *is analytic, strictly increasing and concave.*

Now we consider the existence of a principal eigenvalue of (1.12) when $\widetilde{N}$ vanishes in a subset of Ω.

THEOREM 2.10. *Let* Ω, $\mathcal{L}$ *and* M *be as in assumptions* (H.1), (H.2), (H.3′), *and let* $\widetilde{N}$ *satisfy* (H.3′) *and be such that* $\widetilde{N} = 0$ *in* C *and* $\widetilde{N} > 0$ *in* $\Omega \setminus C$, *where the closed set* C *is such* *that $\emptyset \neq C \subset \Omega$. Then* (1.12) *possesses a principal eigenvalue if and only if the quantity*

$$\mu_0 = \sup\left\{\inf\left\{\frac{\mathcal{L}v - M(x)v}{v} : x \in \Omega'\right\} : \Omega' \in S, \; v \in \mathcal{P}\right\} \leqslant \infty \tag{2.24}$$

is strictly positive, where S is the set of those open subsets of Ω such that $C \subset \Omega'$ and $\partial \Omega \subset \partial \Omega'$ (that is, Ω' is an open neighborhood of both C and $\partial \Omega$), and $\mathcal{P}$ is the set of those functions $v \in C^2(\Omega) \cap C^1(\overline{\Omega})$ such that $v > 0$ in Ω. Also, if $\mu_0 > 0$ then the principal eigenvalue of (1.12) *is unique and simple.*

REMARK 2.11. Under the assumptions of Theorem 2.10, the existence of a principal eigenvalue is determined by the sign of the quantity μ_0, defined in (2.24). That quantity can be calculated in several cases that have been already considered in the literature [91] for equations with bounded coefficients. Thus we generalize these results to our singular case, see [77] for the details.

Now we consider the principal eigenvalues of (1.12) when the function $\widetilde{N}$ changes sign in Ω. For convenience, we consider the principal eigenvalue μ of the problem

$$\mathcal{L}U - M(x)U - \lambda\widetilde{N}(x)U = \mu U \quad \text{in } \Omega, \qquad U = 0 \quad \text{on } \partial\Omega, \tag{2.25}$$

for varying values of $\lambda \in \mathbb{R}$. Note that μ is also the principal eigenvalue of the adjoint problem

$$\mathcal{L}^*U - M(x)U - \lambda\widetilde{N}(x)U = \mu U \quad \text{in } \Omega, \qquad U = 0 \quad \text{on } \partial\Omega, \tag{2.26}$$

where $\mathcal{L}^*$ is given in (1.10) (Theorem 2.6).

THEOREM 2.12. *Let Ω, $\mathcal{L}$ and M be as in assumptions* (H.1), (H.2) *and* (H.3$'$), *and let $\widetilde{N}$ satisfy assumption* (H.3$'$) *and be such that $\widetilde{N} > 0$ and $\widetilde{N} < 0$ in two nonvoid, open subsets of Ω. Then:*

    (i) *If, for some $\lambda = \lambda_0$, the principal eigenvalue of* (2.25) *is strictly positive then* (1.12) *has exactly two principal eigenvalues, λ_1 and λ_2, which are such that $\lambda_1 < \lambda_0 < \lambda_2$.*

    (ii) *If, for some $\lambda = \lambda_0$, the principal eigenvalue of* (1.11) *is zero then two possibilities arise, depending on the quantity*

$$k_0 = \int_{\Omega} \widetilde{N}(x)U_0(x)U_0^*(x)\,\mathrm{d}x,$$

        where $U_0 > 0$ and $U_0^* > 0$ are eigenfunctions of (2.25) *and* (2.26), *respectively, for $\lambda = \lambda_0$ and $\mu = 0$.*

    (iia) *If $k_0 \neq 0$ then, in addition to λ_0 (which is obviously a principal eigenvalue),* (1.12) *has exactly one principal eigenvalue $\lambda_1 \neq \lambda_0$, and $k_0(\lambda_0 - \lambda_1) > 0$.*

    (iib) *If $k_0 = 0$ then λ_0 is the only principal eigenvalue of* (1.12).

REMARK 2.13. A similar change of variable was used later in [54] for the study of the maximum principle and estimates for the first eigenvalue for some linear differential operators.

REMARK 2.14. The counterexample given in [16] shows that $\alpha = 2$ is the critical negative exponent in order to have a sequence of positive eigenvalues. However, it is still possible to get a finite number of them, see Theorem 3.4 in the same paper. We do not intend to elaborate on this point here.

3. Existence theorems in the general case $\alpha > 0$

We begin by giving the more general existence theorems in which concerns the degree of singularity of the problem, in the sense that they cover the more general case of any $\alpha > 0$ in the model example (1.3), (1.4). These results were proved by Crandall et al. in the paper [38] with two different proofs, a simpler one for monotone nonlinearities and a more involved one for nonmonotone terms. In any case, both use a (different) approximation procedure involving known existence results for the associated nonsingular problems. We consider the problem (1.8), (1.9) where the domain Ω satisfies (H.1), the coefficients satisfy

(H.5) a_{ij}, b_i are in $C^1(\overline{\Omega})$.

The function $f : \overline{\Omega} \times (0, +\infty) \to (0, +\infty)$ satisfies:

(H.6) f is $C^1(\overline{\Omega} \times \mathbb{R})$ and $\lim_{u \to 0+} f(x, u) = +\infty$ when $u \to 0^+$ uniformly in $x \in \Omega$, and

(H.7) $f(x, u)$ is nonincreasing in u for any $x \in \Omega$.

The first result is the following theorem.

THEOREM 3.1. *Assume that assumptions* (H.1), (H.5)–(H.7) *are satisfied. Then there is a unique classical solution $u > 0$ to* (1.8), (1.9) *in* $C^2(\Omega) \cap C(\overline{\Omega})$.

PROOF. We start by considering a naturally associated nonsingular problem. For any $\varepsilon > 0$, let

$$\mathcal{L}u = f(x, u + \varepsilon) \quad \text{in } \Omega, \qquad u = 0 \quad \text{on } \partial\Omega \tag{3.1}$$

be the corresponding problem. Now (3.1) is a nonsingular boundary value problem to which all known techniques may be applied. In particular we can use sub- and supersolutions in the classical version of, e.g., [5]. It is obvious that $u_0 = 0$ is a subsolution for any $\varepsilon > 0$. Indeed, $\mathcal{L}u_0 - f(x, \varepsilon + 0) < 0$. As a supersolution we pick the unique smooth solution $u^0 > 0$ to the linear equation $\mathcal{L}u^0 = f(x, \varepsilon)$ in Ω, $u^0 = 0$ on $\partial\Omega$. Then we have $\mathcal{L}u^0 - f(x, \varepsilon + u^0) = f(x, \varepsilon) - f(x, \varepsilon + u^0) > 0$ by (H.7) and we have a supersolution. This proves existence of a classical positive solution $u_\varepsilon > 0$ for any $\varepsilon > 0$. Moreover, this solution is unique. Indeed, if we have $0 < \varepsilon < \delta$, and $u_\delta > 0$ and $u_\varepsilon > 0$ are the corresponding solutions, then $\mathcal{L}(u_\varepsilon - u_\delta) = f(x, \varepsilon + u_\varepsilon) - f(x, \delta + u_\delta) \geqslant 0$ on A, where $A = \{x \in \Omega : u_\varepsilon(x) < u_\delta(x)\}$, and $u_\varepsilon - u_\delta = 0$ on ∂A, by the monotonicity (H.7) of f. By the maximum principle, $u_\varepsilon - u_\delta \geqslant 0$ on A, a contradiction. Hence $u_\varepsilon < u_\delta$. A completely similar argument shows that $\varepsilon + u_\varepsilon - \delta - u_\delta < 0$. Now we need an auxiliary result from the L_p theory for linear equations. (See [15] and [38], p. 199.)

LEMMA 3.2. *Let* D_0, D *be bounded open domains in* $\mathbb{R}^N$ *with* $\overline{D}_0 \subset D$. *Suppose* $\mathcal{L}$ *is a second-order uniformly elliptic operator with coefficients continuous in* $\overline{D}$ *and* $q > N$. *Then there is a constant* K *such that*

$$\|w\|_{W^{2,q}(D_0)} \leqslant K\big(\|\mathcal{L}w\|_{L^q(D)} + \|w\|_{L^q(D)}\big) \tag{3.2}$$

for all $w \in W^{2,q}(D)$. *The constant* K *depends on* N, q, *the diameter of* D, *the distance from* D_0 *to* ∂D, *the ellipticity constant of* $\mathcal{L}$ *and bounds for the coefficients of* $\mathcal{L}$ *(in* $L^\infty(D)$*) and the moduli of continuity of the coefficients.*

We have obtained $0 < u_\varepsilon - u_\delta < \delta - \varepsilon$. Hence u_ε is a Cauchy sequence and then it converges to a limit u uniformly on $\overline{\Omega}$. Hence in particular, $u > 0$ in Ω and $u = 0$ on $\partial\Omega$. We also have that $\lim_{\varepsilon\to 0^+} f(x, \varepsilon + u_\varepsilon) \to f(x, u)$ uniformly on compact subsets of Ω. Next we pick subsets D_1 and D_2 of Ω such that $\overline{D}_2 \subset D_1 \subset \overline{D}_1 \subset \Omega$. Let $q > N$. Then we get

$$\|u_\varepsilon\|_{W^{2,q}(D_2)} \leqslant K\big(\|\mathcal{L}u_\varepsilon\|_{L^q(D_1)} + \|u_\varepsilon\|_{L^q(D_1)}\big)$$
$$= K\big(\|f(x, \varepsilon + u_\varepsilon)\|_{L^q(D_1)} + \|u_\varepsilon\|_{L^q(D_1)}\big). \tag{3.3}$$

Hence $\{u_\varepsilon\colon 0 < \varepsilon \leqslant \varepsilon_0\}$ is bounded in $W^{2,p}_{\text{loc}}(\Omega)$, and this implies that a subsequence u_ε converges to a limit u weakly in $W^{2,p}_{\text{loc}}(\Omega)$ and by the Sobolev embedding theorem, strongly in $C^{1,\beta}_{\text{loc}}(\Omega)$, where $1 - \beta > N/q$. Now we also have $\lim_{\varepsilon\to 0^+} f(x, \varepsilon + u_\varepsilon) \to f(x, u)$ in $C^1(\Omega)$. This implies that u is in $W^{2,p}_{\text{loc}}(\Omega)$ and passing to the limit we get $\mathcal{L}u = f(x, u)$ in Ω. Since $f(x, u)$ is in $C^1(\Omega)$, the regularity estimates show that u is in $C^2(\Omega)$. Uniqueness is proved in much the same way as before and this ends the proof of the theorem. $\qquad\square$

If only (H.6) is satisfied, but not (H.7), then a more complicated argument should be found. Now we associate to (1.8) the nonlinear eigenvalue problem

$$\mathcal{L}u = \lambda f(x, u) \quad \text{in } \Omega, \qquad u = 0 \quad \text{on } \partial\Omega, \tag{3.4}$$

where the real parameter λ has been introduced. By using a global bifurcation theorem by Rabinowitz and a limit argument the following result can be proved.

THEOREM 3.3. *Assume that assumptions* (H.1), (H.5) *and* (H.6) *are satisfied. Then there is a continuum* C *of positive solutions to* (3.4) *such that* C *is unbounded in the space* $\mathbb{R} \times C_0(\overline{\Omega})$.

PROOF (sketch). Now the singular nonlinear eigenvalue problem (3.4) is replaced by a nonsingular related problem, namely

$$\mathcal{L}u = \lambda f(x, u + \varepsilon) \quad \text{in } \Omega, \qquad u = 0 \quad \text{on } \partial\Omega. \tag{3.5}$$

An application of a global bifurcation theorem by Rabinowitz ([101], Theorem 3.7) gives, for any $\varepsilon > 0$ small enough, the existence of a continuum C_ε of positive solutions to (3.5) which is unbounded in the product space $\mathbb{R} \times C^{1,\delta}(\overline{\Omega})$ and contains the point $(0, 0)$. Next step, i.e., pass to the limit when $\varepsilon \to 0^+$, is much more delicate than in the previous theorem and involves estimates to the solutions in C_ε. Using these estimates and interior $L_p(\Omega)$ estimates we get the existence of a continuum C of positive solutions to (3.4). (The convergence is actually what is called convergence in the sense of Whyburn in general topology.) $\qquad\square$

If more precise information is available on the nonlinearity f, then we can say more on solutions, too. We have the corollary:

COROLLARY 3.4. *Assume that assumptions in Theorem 3.3 are satisfied and that, moreover,*
    (H.8) *there exists $A > 0$ such that $f(x, u) < A$ if $u > 1$ and $x \in \overline{\Omega}$.*
    Then for any $\lambda > 0$, there is at least a classical positive solution to (3.4) and, in particular, a solution to (1.8), (1.9).

REMARK 3.5. Conditions on the differential operator $\mathcal{L}$ and the nonlinearity f can be weakened in different ways. It is also proved in the same paper [38] that if both the coefficients and the nonlinear term are only continuous, then a similar approximation argument allows to exhibit existence of generalized solutions, this time in the sense that they are in the function space $C_0(\Omega) \cap W_{\mathrm{loc}}^{2,q}(\Omega)$. Existence and uniqueness of a solution $u \in W_{\mathrm{loc}}^{2,q} \cap C(\overline{\Omega})$, for some $q > 1$, was proved in [7] for a nonhomogeneous version of the model problem (1.5), (1.6) for less smooth domains and data.

REMARK 3.6. Several alternative approximation procedures have been used by different authors. In [110] Stuart replaced the homogeneous boundary condition $u = 0$ on $\partial\Omega$ by $u = \varphi_\varepsilon$ with $\varphi_\varepsilon > 0$ smooth going to zero with ε. Something similar was done later by Wiegner [115].
    Going in the opposite direction, nonexistence for $\alpha > 1$ in the model case was shown by Zhang [119], Theorem 2.

REMARK 3.7. Asymptotic bifurcation has been used also in order to prove existence theorems. See the work in [9,10,60,61]. The case of the p-Laplacian has been treated in [8].

REMARK 3.8. The singular problem on a ball B,

$$-\Delta u + \frac{1}{u^\alpha} = h(|x|) \quad \text{in } B, \qquad \frac{\partial u}{\partial n} = 0 \quad \text{on } \partial B, \tag{3.6}$$

with $\alpha > 1$ and h radial, was studied in [48] obtaining existence (and also in some cases uniqueness) of radial solutions in terms of the integral $\int_B h$.

The existence of positive solutions for the problem

$$-\Delta u = f(x, u, \nabla u) \quad \text{in } \Omega, \qquad u = 0 \quad \text{on } \partial\Omega \tag{3.7}$$

has been analyzed by Zhang [120], Zhang and Yu [124] and Ghergu and Radulescu [61]. Nevertheless, we will not consider in detail these results. We give a brief summary of the most relevant results obtained by these authors.

The first result we mention is a quite general nonexistence result [120]. More precisely, we have the theorem:

THEOREM 3.9. *Let $\Omega \subset \mathbb{R}^N$ be a bounded domain with a $C^{2+\gamma}$ boundary for some $\gamma \in$ $]0, 1[$. Suppose that $\alpha \geqslant 1$ and f is such that there exists a function $K \in C(]0, +\infty[,$ $]0, +\infty[)$ and a constant C_1 such that $|f(x, u, v)| \leqslant K(u) + C_1|v|^2$ for all $x \in \overline{\Omega}$, $u \in$ $]0, +\infty[$ and $v \in \mathbb{R}^N$. Then the problem*

$$-\Delta u = f(x, u, \nabla u) - \frac{1}{u^\alpha} \quad \text{in } \Omega, \qquad u = 0 \quad \text{on } \partial\Omega \tag{3.8}$$

has no solution in $C^2(\Omega) \cap C(\overline{\Omega})$.

Regarding existence Zhang has proved the following results.

THEOREM 3.10 ([124], Theorems 1.1, 1.2). *Let $\Omega \subset \mathbb{R}^N$ be a bounded domain with a $C^{2+\gamma}$ boundary for some $\gamma \in]0, 1[$, $f(x, u, \nabla u) = u^{-\alpha} + \lambda|\nabla u|^p + \sigma$ for $\sigma \geqslant 0$. If $p = 2$, there is $\bar{\lambda}$ such that there is a unique classical solution for $\lambda < \bar{\lambda}$ and no weak solution if $\lambda \geqslant \bar{\lambda}$. If $0 < p < 2$, there exists $0 < \lambda^* \leqslant +\infty$ such that there is a unique classical solution for $\lambda \in [0, \lambda^*[$ and no classical solution for $\lambda > \lambda^*$. Moreover, if $p \in]1, 2[$ and $\sigma = 0$ (resp., $\sigma > 0$), $\lambda^* = +\infty$ (resp., $\lambda^* < +\infty$).*

The above results have been generalized in [61] (see also [10]). A problem of this kind in one dimension arises in a paper by Gomes and Sprekels [67].

REMARK 3.11. It is possible to try to give a meaning to sign-changing solutions in the one-dimensional situation. This is done by, in some sense, "piecing together" positive and negative solutions on subintervals (see the work in [93] and [13]). But a similar approach seems much harder for nonradial problems in higher dimensions.

REMARK 3.12. Singular problems on the whole space $\mathbb{R}^N$ (or on some unbounded domain) have been also treated in the literature; see, e.g., [87] or [118] and the respective bibliographies. As an example, it was proved in [118] that the problem

$$-\Delta u = p(x)\left(\frac{1}{u^\alpha} + u^s\right) \quad \text{in } \mathbb{R}^N, \qquad \lim_{|x| \to +\infty} u(x) = 0, \tag{3.9}$$

where $0 < \alpha < 1$, $0 < s < 1$, and $p \geqslant 0$ is a smooth radial function, has a radial positive classical solution $u \in C^2_{\text{loc}}(\mathbb{R}^N)$ if and only if p satisfies

$$\int_0^{+\infty} t p(t)\, dt < +\infty. \tag{3.10}$$

Moreover, this radial solution is unique in the sense that if $u(r_0) = v(r_0)$ for some $r_0 > 0$, then $u \equiv v$. Proofs involve mostly ordinary differential equations methods.

Similar problems were considered in [68] for quasilinear problems were the Laplacian was replaced by the p-Laplacian.

4. Existence of solutions via sub- and supersolutions

The method of sub- and supersolutions, widely known for problems with smooth nonlinearities, was used in the proof of Theorem 3.1 as a tool for getting existence of solutions to the (smooth) approximating problems. Then a limiting argument involving classical estimates in the linear theory yields the conclusion. Here we show that if ordered sub- and supersolutions are available for the existence problem, it is possible, under some additional conditions on them, to get an existence theorem rather close in both its statement and its proof to well-known results for smooth nonlinearities.

The results obtained in Section 2 and the strong maximum principle in Appendix B provide the basic ingredients to systematically apply the standard tools mentioned at the beginning of the paper to the analysis of singular equations of the form (1.8), (1.9), under the assumptions (H.1)–(H.3).

Monotone methods can be applied, in quite the same manner as is the regular case, to construct minimal and maximal solutions of (1.8), (1.9) in the positive cone of $C_0^1(\overline{\Omega})$, as seen in the proof of the following theorem.

THEOREM 4.1. *Under assumptions* (H.1)–(H.3), *let us assume that* (1.8), (1.9) *has a sub-solution u_0 and supersolution u^0 such that $u_0, u^0 \in C^2(\Omega) \cap C_0^{1,\delta}(\overline{\Omega})$ for some $\delta > 0$ and*

$$0 < kd(x) < u_0(x) \leqslant u^0(x) \quad \text{for all } x \in \Omega. \tag{4.1}$$

Then (1.8), (1.9) *possesses a minimal and a maximal solution in the interval $[u_0, u^0]$, u_* and u^*, which are such that $u_*, u^* \in C^2(\Omega) \cap C_0^{1,\delta}(\overline{\Omega})$ whenever $0 < \delta < \delta_0 = \min\{\gamma, \alpha + 1\}$ and $u_0 \leqslant u_* \leqslant u^* \leqslant u^0$ in Ω. Also, u_* (resp., u^*) is the $C_0^{1,\delta}(\overline{\Omega})$-limit from below (resp., from above) of a monotone sequence of subsolutions (resp., supersolutions) of* (1.8), (1.9). *Moreover, we have $\partial u^*/\partial n < 0$, $\partial u_*/\partial n < 0$ on $\partial\Omega$.*

PROOF. According to assumption (H.3), $|f_u(x, u(x))| d(x)^{1-\alpha}$ is bounded in Ω under (4.1) if $u : \Omega \to \mathbb{R}$ is such that $u_0 \leqslant u \leqslant u^0$ in Ω. Thus a function $M_1 \in C^1(\Omega)$ exists that satisfies (H.3′) and is such that, for some constants $k > 0$ and $\beta \in {]1 - \alpha, 2[}$,

$$M_1 > 1 \quad \text{in } \Omega \setminus \Omega_1, \qquad M_1(x) = kd(x)^{-\beta} \quad \text{for all } x \in \Omega_1, \tag{4.2}$$

and

$$\frac{f(\cdot, u)}{u} + \left| f_u(\cdot, u) \right| < M_1 \quad \text{in } \Omega \quad \text{if } u_0 \leqslant u \leqslant u^0 \text{ in } \Omega, \tag{4.3}$$

where Ω_1 is as defined in assumption (H.1). Then the mean value theorem and assumption (H.3) imply that the functions $\varphi, \psi : \Omega \to \mathbb{R}$, defined as

$$\varphi(x, u) \equiv M_1(x)u + f(x, u) \equiv M_1(x)\psi(x, u), \tag{4.4}$$

are such that

$$0 < \varphi(\cdot, u) < \varphi(\cdot, v) \quad \text{in } \Omega,$$

$$\psi(\cdot, u) \in C_0^1(\overline{\Omega}), \qquad \left\| \psi(\cdot, u) \right\|_{C^1(\overline{\Omega})} \leqslant K_1 \| u \|_{C^1(\overline{\Omega})} \quad \text{and}$$

$$\left\| \psi(\cdot, v) - \psi(\cdot, u) \right\|_{C^1(\overline{\Omega})} < K_2 \| v - u \|_{C^1(\overline{\Omega})} \quad \text{in } \Omega,$$

$$\text{whenever } u_0 \leqslant u < v \leqslant u^0 \text{ in } \Omega, \tag{4.5}$$

for some constants K_1 and K_2 that are independent of u and v.

Now we consider the sequences $\{u_m\}$, $\{u^m\}$, defined inductively by

$$\mathcal{L}u_m + M_1(x)u_m = \varphi\big(x, u_{m-1}(x)\big) \quad \text{in } \Omega, \qquad u_m = 0 \quad \text{on } \partial\Omega, \tag{4.6}$$

$$\mathcal{L}u^m + M_1(x)u^m = \varphi\big(x, u^{m-1}(x)\big) \quad \text{in } \Omega, \qquad u^m = 0 \quad \text{on } \partial\Omega, \tag{4.7}$$

for $m > 0$, with u_0 and u^0 as defined in Theorem 4.1. Since φ and $\psi = \varphi/M_1$ satisfy (4.5), we only need to apply the maximum principle in Appendix B and Proposition 2.3 to obtain inductively that $\{u_m\}, \{u^m\} \subset C^2(\Omega) \cap C^{1,\delta}(\overline{\Omega})$ whenever $m \geqslant 1$ and $0 < \delta < \delta_0 = \min\{\gamma, \alpha + 1\}$, and as in the regular case, if u is a solution of (1.8), (1.9) such that $u_0 \leqslant u \leqslant u^0$ in Ω, then

$$u_0 \leqslant u_{m-1} \leqslant u_m \leqslant u \leqslant u^m \leqslant u^{m-1} \leqslant u^0 \quad \text{in } \Omega \text{ for all } m > 1. \tag{4.8}$$

Hence the stated result follows if we prove that

$$\{u_m\} \text{ and } \{u^m\} \text{ converge in } C^{1,\delta}(\overline{\Omega}) \text{ whenever } 0 < \delta < \delta_0 = \min\{\gamma, \alpha + 1\} \tag{4.9}$$

(then, according to standard, local, elliptic estimates, the limits must be in $C^2(\Omega)$).

Now, in order to obtain (4.9) for the monotone, bounded sequence $\{u_m\}$ (the other sequence is treated similarly), we first observe that, by the dominated convergence theorem, it converges in $L_q(\Omega)$ for all $q > 1$. Also, according to Proposition 2.3 and properties (4.5), if $p > m > 1$ and δ is as defined in Theorem 4.1, then

$$\| u_p - u_m \|_{C^{1,\delta}(\overline{\Omega})} < K \| u_{p-1} - u_{m-1} \|_{C^1(\overline{\Omega})}, \tag{4.10}$$

with the constant K independent of m and p. In addition we have the interpolation inequality $\|u\|_{C^1(\overline{\Omega})} < \varepsilon \|u\|_{C^{1,\delta}(\overline{\Omega})} + C_{\varepsilon,q} \|u\|_{L_q(\Omega)}$, which holds for all $\varepsilon > 0$ and all $q > N + 2$ (see [3] and [85], p. 80). Thus, if $q > N + 2$ is kept fixed, this inequality (with $\varepsilon > 0$ appropriate) and (4.10) readily yield

$$\|u_p - u_m\|_{C^1(\overline{\Omega})}$$

$$< \frac{1}{4} \|u_{p-1} - u_{m-1}\|_{C^1(\overline{\Omega})} + K_1 \|u_p - u_m\|_{L_q(\Omega)}$$

$$< \frac{\|u_p - u_m\|_{C^1(\overline{\Omega})} + \|u_p - u_{p-1}\|_{C^1(\overline{\Omega})} + \|u_m - u_{m-1}\|_{C^1(\overline{\Omega})}}{4}$$

$$+ K_1 \big(\|u_p - u_m\|_{L_q(\Omega)} \big), \tag{4.11}$$

where K_1 is independent of m and p. Since $\{u_m\}$ converges in L_q, the first inequality in (4.11) (with $p = m + 1$) implies that $\|u_{m+1} - u_m\|_{C^1(\overline{\Omega})} \to 0$ as $m \to \infty$. And, since $\{u_m\}$ is a Cauchy sequence in L_q, (4.11) and (4.10) subsequently imply that $\{u_m\}$ is a Cauchy sequence in $C^1(\overline{\Omega})$ and in $C^{1,\delta}(\overline{\Omega})$. Thus $\{u_m\}$ satisfies (4.9) and the proof is complete. $\qquad \square$

The requirement that $u_0 > kd(x) > 0$ in Ω in Theorem 4.1 is often too strong in applications. For instance, if the unique solution of

$$\mathcal{L}U = M(x) \quad \text{in } \Omega, \qquad U = 0 \quad \text{on } \partial\Omega \tag{4.12}$$

is strictly positive in Ω then (see Section 10) $u_0 = [(1 - \alpha_1)U]^{1/(1-\alpha_1)}$ is a strict subsolution of

$$\mathcal{L}u = M(x)u^{\alpha_1} \quad \text{in } \Omega, \qquad u = 0 \quad \text{on } \partial\Omega. \tag{4.13}$$

But this subsolution does not satisfy the above-mentioned requirement if $U \in C_0^1(\overline{\Omega})$. In order to extend the applicability of Theorem 4.1 to situations like this one, in the following lemma we prove that the above-mentioned requirement in Theorem 4.1 can be weakened.

LEMMA 4.2. *Under the assumptions of Theorem 4.1, let the function f (satisfy (H.3) and) be such that $f(x, u) > -K_1 d(x)^{\alpha_2} u^{\alpha_1}$ and $|f(x, u)| + |f_u(x, u)|u < K_2 d(x)^{\alpha_4} u^{\alpha_3}$ for all $(x, u) \in \Omega \times]0, \infty[$, with $K_1 > 0$, $K_2 > 0$, $|\alpha_1 + \alpha_2| < 1$ and $|\alpha_3 + \alpha_4| < 1$, and let $\tilde{u}_0 \in C^2(\Omega) \cap C_0^{1,\delta}(\overline{\Omega})$, for some $\delta > 0$, be a subsolution of (1.8), (1.9) such that $\tilde{u}_0 > \check{k}d(x)^p$ a.e. in Ω, with $\check{k} > 0$ and $p > 1$. In addition, let us assume that either (a) $\alpha_1 \geqslant 1$, or (b) $0 < (1 - \alpha_1)/(2 + \alpha_2) < \alpha_1 < 1$ and $\alpha_4 + \alpha_3/\alpha_1 > -1$, or (c) $\alpha_1 < (1 - \alpha_1)/(2 + \alpha_2) < 1/p$ and $p\alpha_3 + \alpha_4 > -1$. Then there is a subsolution of (1.8), (1.9), $u_0 \in C^2(\Omega) \cap C_0^{1,\delta}(\overline{\Omega})$ ($\delta > 0$), such that (i) $u_0 > kd(x)$ in Ω for some $k > 0$, and (ii) $u_0 < u$ in Ω, whenever $u \in C^2(\Omega) \cap C_0^{1+\delta}(\overline{\Omega})$ ($\delta > 0$) is a solution of (1.8), (1.9) such that $\tilde{u}_0 < u$ and $0 < k_1 d(x) < u$ in Ω for some $k_1 > 0$.*

For the proof see [77].

REMARK 4.3. Theorem 4.1 is an existence theorem for positive solutions in the interval between ordered sub- and supersolutions, and in this sense it is rather similar to well-known classical results [5,96,102,107]. However, the singularity forces (in our case at least) to impose the additional condition $u^0 = 0$ on $\partial\Omega$, and also that the subsolution satisfies $u_0 > kd(x)$. The boundary condition $u^0 = 0$ on $\partial\Omega$ for the supersolution is also imposed by Zhang [119].

REMARK 4.4. An alternative existence result using also sub- and supersolutions can be found in [106], involving again an approximation replacing the boundary condition $u = g$, with $g \geqslant 0$, by $u = g + \varepsilon_n$, where $\varepsilon_n > 0$ and $\varepsilon_n \to 0$. Conditions on the coefficients in [106] are less general than (H.1)–(H.3). See also the series of papers by Coclite [32–34] working in some kind of $L_p(\Omega)$ weighted spaces (where the weight involves once again the function $d(x)$).

5. Uniqueness of positive solutions for sublinear problems

We obtain a uniqueness result for positive solutions of (1.8), (1.9). The idea in the proof goes back to Krasnoselskii, but we extend to the singular case the proof of Brezis and Kamin [18].

THEOREM 5.1. *Under the assumptions of Theorem* 4.1, *let us also assume that*
   (H.9) $H(x, u) \equiv f(x, u) - uf_u(x, u) > 0$ *in* Ω,
then the problem (1.8), (1.9) *exhibits a unique positive solution.*

PROOF. Note that assumption (H.9) is equivalent to requiring that the function $u \to f(x, u)/u$ be decreasing for all $u > 0$ and all $x \in \Omega$. We prove that if u_1 and u_2 are two solutions to (1.8), (1.9) then $u_1 = u_2$. Since the roles of u_1 and u_2 are interchangeable, we only need to show that

$$1 \in \Lambda \equiv \left\{ t \in [0, 1] : tu_1 \leqslant u_2 \text{ in } \Omega \right\}. \tag{5.1}$$

To this end, we note that Λ contains the interval $[0, \beta]$ for some $\beta > 0$ (recall that u_1 and u_2 are in the positive cone), and assume for contradiction that $t_0 = \sup \Lambda < 1$. Then

$$\mathcal{L}(u_2 - t_0 u_1) = f(x, u_2) - t_0 f(x, u_1) \quad \text{in } \Omega. \tag{5.2}$$

It was shown in the proof of Theorem 4.1 in [77] that there exists a function $M_1(x) \in C^1(\Omega)$ satisfying (H.3$'$) and such that $f(x, u) + M_1(x)u$ is increasing in $u \in [0, \max u_2]$ for any $x \in \Omega$ fixed. Then, adding $M(x)(u_2 - t_0 u_1)$ to both terms of equation (5.2) and invoking assumption (H.9), we have

$$\mathcal{L}(u_2 - t_0 u_1) + M_1(x)(u_2 - t_0 u_1)$$

$$= f(x, u_2) + M_1(x)u_2 - t_0\big(f(x, u_1) + M_1(x)u_1\big)$$

$$\geqslant f(x, t_0 u_1) + t_0 M_1(x) u_1 - t_0 \big(f(x, u_1) + M_1(x) u_1 \big)$$
$$= f(x, t_0 u_1) - t_0 f(x, u_1) \geqslant 0 \quad \text{in } \Omega. \tag{5.3}$$

Moreover,

$$u_2 > t_0 u_1 \quad \text{in } \Omega \quad \text{and} \quad u_2 - t_0 u_1 = 0 \quad \text{on } \partial\Omega, \tag{5.4}$$

where the inequality follows because if $u_2 - t_0 u_1$ were zero at some $x \in \Omega$, then (5.3) would imply that $f(x, t_0 u_1) = t_0 f(x, u_1)$, which is impossible according to assumption (H.9), and the equality follows because $u_1 = u_2 = 0$ on $\partial\Omega$. Then, by the strong maximum principle ([77], Theorem B.2, Appendix B) we obtain that $u_2 - t_0 u_1 > 0$ in Ω and $\partial(u_2 - t_0 u_1)/\partial n < 0$ on $\partial\Omega$, which implies that there is some $\varepsilon > 0$ such that $u_2 - t_0 u_1 > \varepsilon u_1$. But this implies that $t_0 + \varepsilon \in \Lambda$, a contradiction, according to the definition of t_0 just above of (5.2). This completes the proof. $\qquad\square$

REMARK 5.2. Alternative proofs for uniqueness can be given by using the comparison argument in [105] which involves integration by parts ([105], Lemma 2.3); see also for this kind of results [6,18,106,111]. Concerning uniqueness of positive solutions, see also [5,20,21,36,49,75,90,96,102,103,107,108].

6. Regularity and boundary behavior

In the preceding sections we have studied, in particular, existence and uniqueness of positive solutions (more precisely, solutions provided by Theorem 4.1) to the semilinear problem

$$-\Delta u = \frac{p(x)}{u^\alpha} \quad \text{in } \Omega, \qquad u = 0 \quad \text{on } \partial\Omega, \tag{6.1}$$

where $\alpha > 0$ (actually for a larger class of equations) and the function p is smooth (see further).

Concerning the smoothness of these solutions it is easy to see that u does not belong to $C^2(\overline{\Omega})$. Indeed, in such a case, the left-hand side in (6.1) would be in $C(\overline{\Omega})$ whereas the right-hand side blows up close to the boundary $\partial\Omega$ if $p \equiv 1$ or more generally, $p > 0$ at $\partial\Omega$ and near $\partial\Omega$. Thus $C^{1,\beta}(\overline{\Omega})$, for some $0 < \beta < 1$ is the best we can expect, in general. But this is not all we can say on this respect; as we will see further it may also happen that classical solutions (in $C^2(\Omega) \cap C(\overline{\Omega})$) we have exhibited in Theorems 3.1 and 3.3 are not in $C^1(\overline{\Omega})$, even more surprisingly, they are not in the usual Sobolev space $H_0^1(\Omega)$; otherwise stated, there are classical solutions which are not weak solutions.

Solutions obtained by Stuart [110] and Crandall et al. [38] (see again Theorems 3.1 and 3.3) are in $C^2(\Omega) \cap C(\overline{\Omega})$. But the latter were also interested by these topics and they proved in particular that if $\alpha > 1$, then u is Hölder continuous, $u \in C^{2/(\alpha+1)}(\overline{\Omega})$ and also that in the "autonomous" (x-independent) case the solution u behaves as some function of $d(x)$ ([38], Theorem 2.2). Moreover, a characterization of solutions Lipschitz up to the boundary is given in [38], Theorem 2.25, which in our case reads $\alpha < 1$. A result for $\alpha = 1$ was given as well in [38], Theorem 2.7.

The regularity for $\alpha < 1$ was improved by Gomes [66], showing that if $p(x) \leqslant cd(x)^\beta$ with $\alpha - \beta < 1$ not only $|\nabla u|$ is bounded but also $u \in C^1(\overline{\Omega})$. Another contribution in the same direction was obtained by del Pino [47]: if $p \in L_\infty(\Omega)$, $p \geqslant 0$ in Ω and is positive in a subset of positive measure and if, moreover, $|p(x)| \leqslant \varphi(d(x))$ with $\int_0^1 |\varphi(s)/s|^p \, ds < +\infty$ for some $p > 1$, then $|\nabla u|$ is bounded. (If $\varphi(s) = s^{-\beta}$ then the condition reads just $\alpha - \beta < 1$ as above.)

Lazer and McKenna obtained in [88] simpler proofs for some of the above results with several interesting additional results. The first one is the characterization of weak solutions. It is assumed in [88] that $p \in C^\gamma(\overline{\Omega})$ with $0 < \gamma < 1$, and $p > 0$ on $\overline{\Omega}$, a condition stronger than in [47,66] or (H.3).

We need some auxiliary results for the proof of the first main following result (Theorem 6.3). The first one is the lemma:

LEMMA 6.1 ([88], Lemma on p. 726). *We have*

$$\int_\Omega \varphi_1^r \, dx < +\infty \tag{6.2}$$

if and only if $r > -1$.

PROOF. It is a standard argument using the smoothness of Ω to rectify the boundary and a partition of unity. $\qquad\square$

LEMMA 6.2 [88]. *If u is a classical solution to (6.1), then if $\alpha > 1$, $c_1\varphi_1^{2/(1+\alpha)} \leqslant u \leqslant c_2\varphi_1^{2/(1+\alpha)}$ for some constants $c_1, c_2 > 0$.*

PROOF. If we write $w = c\varphi_1^t$ with $t = 2/(1+\alpha)$, then

$$-\Delta w = c^{1+\alpha}\left(t(1-t)|\nabla\varphi_1|^2 + t\lambda_1|\varphi_1|^2\right), \tag{6.3}$$

and since $0 < t < 1$, we can pick $c > 0$ such that

$$-\Delta w - \frac{p(x)}{w^\alpha} \leqslant 0 \quad \text{in } \Omega. \tag{6.4}$$

We claim that $w(x) \leqslant u(x)$. Indeed, if not, there exists $x_0 \in \Omega$ such that $0 < u(x_0) < w(x_0)$ and the continuous function $u - w$ attains its minimum on $\overline{\Omega}$ at x_0. Hence

$$-\Delta(u-w)(x_0) > p(x_0)\left(\frac{-1}{w(x_0)^\alpha} + \frac{1}{u(x_0)^\alpha}\right) > 0, \tag{6.5}$$

a contradiction. The second inequality is proved in a completely similar way. $\qquad\square$

The announced result is the following theorem.

THEOREM 6.3. *Assume that Ω is a $C^{2+\gamma}$ bounded domain in $\mathbb{R}^N$, $p \in C^\gamma(\overline{\Omega})$ with $p > 0$ on $\overline{\Omega}$ and $\gamma > 0$. Then a classical solution (i.e., $C^2(\Omega) \cap C(\overline{\Omega})$) is in $H_0^1(\Omega)$ if and only if $0 < \alpha < 3$.*

PROOF. We take $1 < \alpha < 3$ first. Now it is possible to pick $u_0 = b_1 \varphi_1^t$, where $t = 2/(1+\alpha)$ and $b_1 > 0$ as a subsolution for the approximating problem in the existence proof of [38] or [88], see Theorem 3.1. If v_n denote the iterates for $\varepsilon_n > 0$ going to 0, we get

$$\frac{p(x)v_n}{(v_n + \varepsilon_n)^\alpha} \leqslant \frac{p(x)}{(v_n + \varepsilon_n)^{\alpha-1}} \leqslant \frac{p(x)}{(u_0 + \varepsilon_n)^{\alpha-1}} \leqslant \frac{M}{u_0^{\alpha-1}}, \tag{6.6}$$

where $M = \max\{p(x): x \in \overline{\Omega}\}$. Since $1 < \alpha < 3$, we have $r = 2(1-\alpha)/(1+\alpha) > -1$ and from Lemma 6.1,

$$\int_\Omega u_0^{1-\alpha}\,\mathrm{d}x < +\infty. \tag{6.7}$$

Next, we multiply by v_n the approximating equation

$$-\Delta v_n = \frac{p(x)}{(v_n + \varepsilon_n)^\alpha} \quad \text{in } \Omega, \qquad v_n = 0 \quad \text{on } \partial\Omega \tag{6.8}$$

and integrate by parts on Ω to get

$$\int_\Omega |\nabla v_n|^2\,\mathrm{d}x = \int_\Omega \frac{p(x)v_n}{(v_n + \varepsilon_n)^\alpha}\,\mathrm{d}x \leqslant M \int_\Omega u_0^{1-\alpha} \leqslant C, \tag{6.9}$$

where $C > 0$ is independent of n. Then there is a subsequence v_n converging weakly in $H_0^1(\Omega)$ and a.e. on Ω to a limit v, and $u = v \in H_0^1(\Omega)$.

Let $0 < \alpha < 1$ next. Now $u^0 = c\varphi_1^s$ with s such that $0 < s < 1$ and $s(1+\alpha) < 2$, $c > 0$ is a supersolution for the approximating equation. Taking into account that

$$\frac{p(x)v_n}{(v_n + \varepsilon_n)^\alpha} \leqslant \frac{p(x)}{(v_n + \varepsilon_n)^{\alpha-1}} \leqslant p(x)\left(u^0 + \varepsilon_n\right)^{1-\alpha}, \tag{6.10}$$

again $\|v_n\|_{H_0^1(\Omega)}$ is bounded and the preceding argument still works.

If now $\alpha \geqslant 3$, $u^0 = c\varphi_1^t$ with $t = 2/(1 + \alpha)$ is a supersolution for the approximating problem for some $c > 0$. Now, since $\alpha \geqslant 3$, $t(1 - \alpha) \leqslant -1$ and $u^0 \geqslant u$ by Lemma 6.2, from Lemma 6.1 we obtain

$$\int_\Omega \frac{p(x)}{u^{\alpha-1}}\,\mathrm{d}x \geqslant \int_\Omega \frac{p(x)}{(u^0)^{\alpha-1}}\,\mathrm{d}x = +\infty. \tag{6.11}$$

Suppose that $u \in H_0^1(\Omega)$. Then there exist a sequence $\varphi_n \in C_0^\infty(\Omega)$ converging to u in $H_0^1(\Omega)$. Using that $\varphi_n^+ \in H_0^1(\Omega)$, $\nabla\varphi_n^+ = \nabla\varphi_n$ if $\varphi_n > 0$, $\nabla\varphi_n^+ = 0$ if $\varphi_n < 0$, it follows that φ_n^+ tends to $u \in H_0^1(\Omega)$ and also a.e. along a subsequence, still denoted φ_n^+. Now

$p(x)\varphi_n^+/u^\alpha \geqslant 0$ and, by Fatou's lemma,

$$\lim_{n\to+\infty}\int_\Omega \frac{p(x)\varphi_n^+}{u^\alpha}\,\mathrm{d}x = +\infty. \tag{6.12}$$

Now we multiply (6.1) by φ_n^+ and integrate by parts over Ω to get

$$\int_\Omega -\Delta u\varphi_n^+\,\mathrm{d}x = \int_\Omega \nabla u\nabla\varphi_n^+\,\mathrm{d}x = \int_\Omega \frac{p(x)\varphi_n^+}{(u)^\alpha}\,\mathrm{d}x \tag{6.13}$$

and hence, when $n\to+\infty$,

$$\int_\Omega |\nabla u|^2\,\mathrm{d}x = \lim_{n\to+\infty}\int_\Omega \frac{p(x)\varphi_n^+}{u^\alpha}\,\mathrm{d}x = +\infty, \tag{6.14}$$

a contradiction. $\qquad\square$

A similar comparison argument also provides a simple proof of the fact that if $\alpha > 1$, solutions are not in $C^1(\overline{\Omega})$, a result already contained in [38].

PROPOSITION 6.4. *If $\alpha > 1$ and u is a (classical) solution to (6.1), then $u \notin C^1(\overline{\Omega})$.*

PROOF. As it was shown in Lemma 6.2, $u \geqslant c\varphi_1^t$ for a certain c and $t = 2/(1+\alpha) < 1$. Then we have, for any $\bar{x} \in \partial\Omega$, $\varphi_1(\bar{x}) = 0$ and $\partial\varphi_1/\partial n(\bar{x}) < 0$ and hence for $s > 0$,

$$\frac{u(\bar{x}+sn) - u(\bar{x})}{s} = \frac{u(\bar{x}+sn)}{s} \geqslant c\varphi_1^{t-1}(\bar{x}+sn)\frac{\varphi_1(\bar{x}+sn)}{s} \tag{6.15}$$

which tends to $+\infty$ ($t < 1$). Hence $u \notin C^1(\overline{\Omega})$. $\qquad\square$

The more systematic study concerning boundary behaviour and regularity was carried out by Gui and Hua Lin [69]. They only consider problem (6.1), where now $p \in L_\infty(\Omega)$, $p \geqslant 0$ on Ω and again $\alpha > 0$. However, it is clear that results could be extended to more general operators and nonlinearities and that, in particular, perturbation by smooth nonlinearities is allowed.

We shall consider different situations depending on both the boundary behaviour of $p(x)$ and $\alpha > 0$. Concerning the boundary behaviour we prove first the following theorem.

THEOREM 6.5. *Assume that p is as defined in Theorem 6.3, and that, moreover,*

$$c_1 d(x)^\beta \leqslant p(x) \leqslant c_2 d(x)^\beta \tag{6.16}$$

with $\beta \geqslant 0$ and $c_1, c_2 > 0$. Then if $0 < \alpha - \beta < 1$ we have

$$c_1 d(x) \leqslant u(x) \leqslant c_2 d(x). \tag{6.17}$$

PROOF. The strategy consists in proving first the result for $\Omega = B_1(0)$ and $p(x) = d(x)^\beta$, and then for a general domain. If w is the unique solution to

$$-\Delta w = \frac{d(x)^\beta}{w^\alpha} \quad \text{in } B_1(0), \qquad w = 0 \quad \text{on } \partial B_1(0), \tag{6.18}$$

then w should be a radial function. If ψ_1 is such that

$$-\Delta \psi_1 = \mu_1 d(x)^\beta \psi_1, \quad \psi_1 > 0 \text{ in } B_1(0), \ \psi_1 = 0 \text{ on } \partial B_1(0) \tag{6.19}$$

with principal eigenvalue $\mu_1 > 0$, then $\psi_1(x) \geqslant c_0 d(x)$ by the strong maximum principle. Moreover, if $z = c\psi_1$ we get

$$-\Delta z - \frac{d(x)^\beta}{z^\alpha} = c\frac{d(x)^\beta}{\psi_1^\alpha}\left(\mu_1 \psi_1^{1+\alpha} - c^{-(1+\alpha)}\right) \leqslant 0 \quad \text{in } B_1(0). \tag{6.20}$$

From (6.18)–(6.20), we obtain

$$-\Delta(z - w) + a(x)(z - w) \leqslant 0 \quad \text{in } B_1(0), \qquad z - w = 0 \quad \text{on } \partial B_1(0), \tag{6.21}$$

where $a(x) \geqslant 0$. By the maximum principle,

$$w(x) \geqslant z(x) \geqslant c_3 d(x) \quad \text{for all } x \in \Omega \tag{6.22}$$

for some c_3.

Integrating by parts in (6.18) we get

$$-N\omega_N w'(1) = \int_{B_1(0)} \frac{d(x)^\beta}{w^\alpha}\, dx \leqslant \frac{N\omega_N}{c_3^\alpha} \int_0^1 r^{N-1}(1-r)^{\beta-\alpha}\, dr < +\infty \tag{6.23}$$

since $\alpha - \beta < 1$. Thus $w(x) \leqslant c_4 d(x)$ for some $c_4 > 0$. This ends the proof for $\Omega = B_1(0)$.

A very similar argument works for $\Omega = B_K(0) - \overline{B_1(0)}$ and $q(x) = (|x| - 1)^\beta (K - |x|)^\beta$ and equation

$$-\Delta v = \frac{q(x)}{v^\alpha} \quad \text{in } B_K(0) - \overline{B_1(0)}, \qquad v = 0 \quad \text{on } \partial\left(B_K(0) - \overline{B_1(0)}\right). \tag{6.24}$$

Now we consider the general domain Ω.

From the smoothness of the boundary it follows that there exists a $\delta > 0$ such that, for any $x_0 \in \Omega_\delta = \{x \in \Omega : d(x) \leqslant \delta\}$, $x_0 \in B_\delta(y) \subset \Omega$ together with $d(x_0) + |y - x_0| = d(y) = \delta$ for $y \in \Omega$ and for all $x \in B_\delta(y)$,

$$p(x) \geqslant c_1 \left(\delta - |x - y|\right)^\beta. \tag{6.25}$$

Hence if $z(x) = \bar{c}w((x-y)/\delta)$, with $\bar{c} = (c_1\delta^{\beta+2})^{1/(1+\alpha)}$, we have

$$-\Delta z = \frac{c_1(\delta - |x-y|)^\beta}{z^\alpha} \quad \text{in } B_\delta(y), \qquad z = 0 \quad \text{on } \partial B_\delta(y), \tag{6.26}$$

and then

$$-\Delta(u-z) + c(x)(u-z) \geqslant 0 \quad \text{in } B_\delta(y), \tag{6.27}$$

where $c(x) = c_1(\delta - |x-y|)^\beta(z^{-\alpha} - u^{-\alpha})/(z-u) \geqslant 0$. By the maximum principle,

$$u(x) \geqslant \bar{c}w\left(\frac{x-y}{\delta}\right) \quad \text{in } B_\delta(y) \tag{6.28}$$

and also

$$u(x_0) \geqslant \bar{c}w\left(\frac{x_0 - y}{\delta}\right) \tag{6.29}$$

which together with the above provide the first inequality.

For the second, using again the smoothness of $\partial\Omega$, there exist $R, \delta > 0$ such that, for any $x_0 \in \Omega_\delta$, $\Omega \subset B_R(y) - B_\delta(y)$ for some $y \notin \Omega$ such that

$$d(x_0) + \delta = |x_0 - y| \tag{6.30}$$

and if $x \in \Omega$,

$$p(x) \leqslant c_2(|x-y| - \delta)^\beta. \tag{6.31}$$

Now if $z(x) = cv((x-y)/\delta)$, then for $K = 2R/\delta$

$$-\Delta z - c^{\alpha+1}\delta^{2\beta+2}\frac{(|x-y| - \delta)^\beta(2R - |x-y|)^\beta}{z^\alpha} = 0 \quad \text{in } \Omega, \tag{6.32}$$

and z is a supersolution to (6.1) for c large enough. From the maximum principle it follows that

$$u(x) \leqslant cv\left(\frac{x-y}{\delta}\right) \tag{6.33}$$

and hence $u(x_0) \leqslant cv((x_0 - y)/\delta)$, which ends the proof. $\qquad\square$

THEOREM 6.6. *Assume that p is as defined in Theorem 6.3, that (6.16) is satisfied and that $\alpha - \beta > 1$. Then*

$$c_5 d(x)^{(2+\beta)/(1+\alpha)} \leqslant u(x) \leqslant c_6 d(x)^{(2+\beta)/(1+\alpha)} \tag{6.34}$$

for some $c_5, c_6 > 0$.

PROOF. It is quite similar to the preceding one. If $z(x) = cd(x)^{(2+\beta)/(1+\alpha)}$, then

$$-\Delta z - \frac{d(x)^\beta}{z^\alpha} = \frac{\beta+2}{1+\alpha} cd(x)^{(1+\beta-\alpha)/(1+\alpha)}$$

$$\times \left(\frac{1+\beta-\alpha}{1+\alpha} - \frac{(N-1)d(x)}{|x|} + \frac{1}{c^\alpha} \right) \quad \text{in } B_1(0), \qquad (6.35)$$

$$z = 0 \quad \text{on } \partial B_1(0).$$

Then z is a supersolution of (6.18) for c small enough in $B_1(0) - B_{1/2}(0)$ and $w - z \geqslant 0$ on $\partial B_1(0)$ and from the maximum principle it follows that $w(x) \geqslant c_7 d(x)^{(2+\beta)/(1+\alpha)}$ for c_7 small. The other inequality is proved in a similar way. A similar argument also holds for $\Omega = B_k(0) - \overline{B_1(0)}$ and the problem

$$-\Delta v = \frac{(|x|-1)^\beta (k-|x|)^\beta}{v^\alpha} = 0 \quad \text{in } B_k(0) - \overline{B_1(0)},$$

$$v = 0 \quad \text{on } \partial\big(B_k(0) - \overline{B_1(0)}\big). \tag{6.36}$$

$\square$

Once again a similar but a little more involved argument gives the proof of the theorem:

THEOREM 6.7. *Assume that p is as defined in Theorem 6.3, that (6.16) is satisfied and that $\alpha - \beta = 1$. Then*

$$c_8 d(x)\big(A - \log(d(x))\big)^{1/(1+\alpha)} \leqslant u(x) \leqslant c_9 d(x)\big(A - \log(d(x))\big)^{1/(1+\alpha)}$$

$$\tag{6.37}$$

for some $c_8, c_9, A > 0$.

Now we are ready to state and prove the regularity results. The first one is the following theorem.

THEOREM 6.8. *Assume that p is as defined in Theorem 6.3, that (6.16) is satisfied and that $0 < \alpha - \beta < 1$. Then $u \in C^{1,1+\beta-\alpha}(\overline{\Omega})$.*

PROOF. By using Green's function, we obtain

$$u(x) = \int_\Omega G(x,y) \frac{p(y)}{u^\alpha(y)} \, dy, \tag{6.38}$$

where $G(x,y)$ is obviously the Green function associated to the problem (6.1). Then we also have

$$\nabla u(x) = \int_\Omega G_x(x,y) \frac{p(y)}{u^\alpha(y)} \, dy \tag{6.39}$$

348   *J. Hernández and F.J. Mancebo*

and if $\bar{x}, \bar{\bar{x}} \in \Omega$ and $d(\bar{x}, \bar{\bar{x}}) < \delta$ it is possible to use Lemma 3.2 in [67] to get

$$\left|\nabla u(\bar{x}) - \nabla u(\bar{\bar{x}})\right| \leqslant \int_\Omega \left|G_x(\bar{x}, y) - G_x(\bar{\bar{x}}, y)\right| \frac{p(y)}{u^\alpha(y)}\, dy \tag{6.40}$$

$$= \int_{B_R(\bar{x})} \left|G_x(\bar{x}, y) - G_x(\bar{\bar{x}}, y)\right| \frac{p(y)}{u^\alpha(y)}\, dy$$

$$+ \int_{\Omega - B_R(\bar{x})} \left|G_x(\bar{x}, y) - G_x(\bar{\bar{x}}, y)\right| \frac{p(y)}{u^\alpha(y)}\, dy$$

$$= I + J, \tag{6.41}$$

where $R = (c+4)d(\bar{x}, \bar{\bar{x}})$ and $c > 0$ is given by Lemma 3.2 in [67]. Now we have

$$I \leqslant \int_{B_R(\bar{x})} \left|G_x(\bar{x}, y) - G_x(\bar{\bar{x}}, y)\right| \frac{p(y)}{u^\alpha(y)}\, dy \tag{6.42}$$

$$\leqslant \int_{B_R(\bar{x})} \left|G_x(\bar{x}, y)\right| \frac{p(y)}{u^\alpha(y)}\, dy + \int_{B_{R+d(\bar{x},\bar{\bar{x}})}(\bar{\bar{x}})} \left|G_x(\bar{\bar{x}}, y)\right| \frac{p(y)}{u^\alpha(y)}\, dy \tag{6.43}$$

$$\leqslant c \int_{B_R(\bar{x})} \frac{\min\{|\bar{x} - y|, d(y)\}}{|\bar{x} - y|^N} d(x)^{\beta-\alpha}\, dy$$

$$+ c \int_{B_{R+d(\bar{x},\bar{\bar{x}})}(\bar{\bar{x}})} \frac{\min\{|\bar{\bar{x}} - y|, d(y)\}}{|\bar{\bar{x}} - y|^N} d(x)^{\beta-\alpha}\, dy, \tag{6.44}$$

where here (and in the following) c denotes different positive constants. Now, since $\alpha - \beta > 0$,

$$I \leqslant c \int_{B_R(\bar{x})} |\bar{x} - y|^{N-1}\, dy + c \int_{B_{R+d(\bar{x},\bar{\bar{x}})}(\bar{\bar{x}})} \left|\bar{\bar{x}} - y\right|^{N-1}\, dy. \tag{6.45}$$

On the other hand, we obtain

$$J \leqslant \int_{\Omega - B_R(\bar{x})} \left(\int_0^1 \left|G_{xx}(\xi(t), y)\xi'(t)\right| dt\right) \frac{p(y)}{u^\alpha(y)}\, dy \tag{6.46}$$

$$\leqslant cd(\bar{x}, \bar{\bar{x}}) \int_{\Omega - B_R(\bar{x})} \left(\int_0^1 \frac{\min\{|\xi(t) - y|, d(y)\}}{|\xi(t) - y|^{N+1}}\, dt\right) \frac{p(y)}{u^\alpha(y)}\, dy \tag{6.47}$$

$$\leqslant cd(\bar{x}, \bar{\bar{x}}) \int_{\Omega - B_R(\bar{x})} \frac{\min\{|\bar{x} - y|, d(y)\}}{|\bar{x} - y|^{N+1}} \frac{p(y)}{u^\alpha(y)}\, dy, \tag{6.48}$$

where $\xi(t)$ is the path given in the lemma (which satisfies $|\xi'(t)| \leqslant d(\bar{x}, \bar{\bar{x}})$) and we use that

$$\left|\xi(t)-y\right| \geqslant |\bar{x}-y| - |\bar{x}-\xi(t)| \geqslant (c+4)d(\bar{x},\bar{\bar{x}}) - cd(\bar{x},\bar{\bar{x}}) \geqslant 4d(\bar{x},\bar{\bar{x}})$$

(6.49)

and also

$$\left|\xi(t)-y\right| \leqslant |\bar{x}-y| + |\bar{x}-\xi(t)| \leqslant (2c+4)d(\bar{x},\bar{\bar{x}}) \leqslant \frac{2c+4}{c+4}|\bar{x}-y|, \qquad (6.50)$$

and hence

$$J \leqslant cd(\bar{x},\bar{\bar{x}}) \int_{\Omega-B_R(\bar{x})} \frac{\min\{|\bar{x}-y|, d(y)\}}{|\bar{x}-y|^{N+1}} d^{\beta-\alpha}(y)\, dy \qquad (6.51)$$

by the conclusion of Theorem 6.5.

Now, since $0 < \alpha - \beta < 1$,

$$J \leqslant cd(\bar{x},\bar{\bar{x}}) \int_{\Omega-B_R(\bar{x})} \frac{1}{|\bar{x}-y|^{N+\alpha-\beta}}\, dy$$

(6.52)

$$\leqslant cd(\bar{x},\bar{\bar{x}}) \int_{R}^{+\infty} \frac{1}{r^{N+\alpha-\beta}} r^{N-1}\, dr$$

$$\leqslant cd(\bar{x},\bar{\bar{x}}) R^{\beta-\alpha} = cd(\bar{x},\bar{\bar{x}})^{1+\beta-\alpha}, \qquad (6.53)$$

the conclusion follows. $\qquad\qquad\square$

By using similar arguments one can prove also the theorem:

THEOREM 6.9. *Assume that p is as defined in Theorem 6.3, (6.16) is satisfied and that $\alpha - \beta > 1$. Then $u \in C^{(2+\beta)/(1+\alpha)}(\overline{\Omega})$.*

THEOREM 6.10. *Assume that p is as defined in Theorem 6.3, (6.16) is satisfied and that $\alpha - \beta = 1$. Then $u \in C^{\delta}(\overline{\Omega})$ for any $0 < \delta < 1$.*

REMARK 6.11. Similar regularity results are also obtained in [69] without condition (6.16).

REMARK 6.12. Results concerning the boundary behavior and regularity have been obtained also in other papers; see, e.g., [60,105,115]. In particular, the estimate (6.34) was obtained also in [123] but they do not obtain optimal regularity. It is also proved in the same paper that, under (6.16), $u \in H_0^1(\Omega)$ if and only if $\alpha - 2\beta < 3$. Estimates for some examples were also obtained in the radial case in [95].

7. Differentiability for some singular nonlinear problems

Let us now consider the linearization (in fact, the differentiability) of the semilinear problem (1.8) around a solution $u \in C^2(\Omega) \cap C^1(\overline{\Omega})$ such that

$$u > 0 \quad \text{in } \Omega, \qquad \frac{\partial u}{\partial n} < 0 \quad \text{on } \partial\Omega. \tag{7.1}$$

If $\alpha > -1/N$ then we can treat the problem (1.8), (1.9) in differential form and work in the space $W_p^2(\Omega) \cap W_{1,0}^p(\Omega)$, with $p > N$, which is compactly embedded into $C_0^1(\overline{\Omega})$; but this is not convenient for the general case treated in this paper, $\alpha > -1$ (see Remark 2.4). Instead we shall rewrite (1.8) in integral form, as

$$\mathcal{F}(u) \equiv u - G\big(f(\cdot, u)\big) = 0, \tag{7.2}$$

where $G : C_0^1(\overline{\Omega}) \to C_0^1(\overline{\Omega})$ is the Green operator of

$$\mathcal{L}U = V \quad \text{in } \Omega, \qquad U = 0 \quad \text{on } \partial\Omega \tag{7.3}$$

(defined as $G(V) = U$). Note that, according to Proposition 2.3, (a) G is bounded and can be extended, as a bounded operator, to $C_0^{1,\delta}(\overline{\Omega})$ for all δ such that $0 < \delta < \delta_0 = \min\{\gamma, \alpha + 1\}$; and (b) $u \in C^2(\Omega) \cap C^1(\overline{\Omega})$ satisfies (1.8) if and only if $u \in C_0^1(\overline{\Omega})$ satisfies (7.2).

THEOREM 7.1. *Under assumptions* (H.1)–(H.3), *the operator* $\mathcal{F} : C_0^1(\overline{\Omega}) \to C_0^1(\overline{\Omega})$ *defined in* (7.2) *is of class* C^m *in the interior of the positive cone of* $C_0^1(\overline{\Omega})$ *(that is, the set of those functions of* $C_0^1(\overline{\Omega})$ *that satisfy* (7.1)), *where* $m \geqslant 1$ *is as defined in assumption* (H.3) *and the linear operator* $\mathcal{F}'(u) : C_0^1(\overline{\Omega}) \to C_0^1(\overline{\Omega})$ *is given by*

$$\mathcal{F}'(u)v = v - G\big(f_u(\cdot, u)v\big). \tag{7.4}$$

If $m > 1$ *and* $1 < j \leqslant m$ *then the* j-*linear operator,* $\partial^j \mathcal{F}(u)/\partial u^j \equiv \mathcal{F}^{(j)}(u) : [C_0^1(\overline{\Omega})]^j \to C_0^1(\overline{\Omega})$, *is given by*

$$\mathcal{F}^{(j)}(u)(v_1, \ldots, v_j) = -G\left(\frac{\partial^j f(\cdot, u)}{\partial u^j} v_1 \cdots v_j\right). \tag{7.5}$$

PROOF. The operator $\mathcal{F}$ can be written as

$$\mathcal{F} = I - \mathcal{F}_1 \quad \text{with } I = \text{identity and } \mathcal{F}_1 = G\big(f(\cdot, u)\big). \tag{7.6}$$

Since I is linear and bounded, it is of class C^∞, with its first derivative equal to I and its higher-order derivatives equal to zero. Thus we only need to prove that (a) for $j = 1, \ldots, m$, the jth derivative of $\mathcal{F}_1$ exists and is given by

$$\mathcal{F}_1^{(j)}(u)(v_1, \ldots, v_j) = G\left(\frac{\partial^j f(\cdot, u)}{\partial u^j} v_1 \cdots v_j\right), \tag{7.7}$$

and (b) the mth derivative of $\mathcal{F}_1$ is continuous.

Let us first prove (a) by an induction argument. In order to prove (a) for $j = 1$ we first consider a function $\widetilde{N} \in C^2(\Omega)$ such that $\widetilde{N} \geqslant 1$ in Ω and $\widetilde{N}(x) = d(x)^{\alpha-1}$ for all $x \in \Omega_1$ (with $d(x)$ and Ω_1 as defined in assumption (H.1)). Note that

$$\frac{d(x)^{\alpha-1}}{\widetilde{N}} \quad \text{and} \quad \left|\frac{\partial \widetilde{N}}{\partial x_k}\right| d(x)^{2-\alpha} \quad \text{are uniformly bounded in } \Omega \tag{7.8}$$

for $k = 1, \ldots, N$. Now, take any function u of the positive cone of $C_0^1(\overline{\Omega})$. If $v \in C_0^1(\overline{\Omega})$ is such that $\|v\|_{C_0^1(\overline{\Omega})}$ is sufficiently small then

$$0 < k_1 d(x) < u + \theta v < k_2 d(x) \quad \text{for all } x \in \overline{\Omega} \text{ and all } \theta \in [0, 1], \tag{7.9}$$

where the constants k_1 and k_2 are independent of v. Thus, according to assumption (H.3) and property (7.8), the function

$$W \equiv \frac{f(\cdot, u + v) - f(\cdot, u) - f_u(\cdot, u)v}{\widetilde{N}} \tag{7.10}$$

is such that

$$|W| = \frac{|f_{uu}(x, u + \theta(x)v)|v^2}{\widetilde{N}} \leqslant K_1 \|v\|_{C_0^1(\overline{\Omega})}^2, \tag{7.11}$$

$$\begin{aligned}
|W_{x_k}| &\leqslant \left| f_{ux_k}\left(x, u + \theta(x)v\right) - f_{ux_k}(x, u) \right. \\
&\quad + \left[f_{uu}\left(x, u + \theta(x)v\right) - f_{uu}(x, u) \right] u_{x_k} \left| \frac{|v|}{\widetilde{N}} \right. \\
&\quad + \frac{|f_{uu}(x, u + \theta(x)v)||v v_{x_k}|}{\widetilde{N}} + \frac{|W \widetilde{N}_{x_k}|}{\widetilde{N}^2} \\
&\leqslant K_2 \|v\|_{C_0^1(\overline{\Omega})} \varepsilon\left(\|v\|_{C_0^1(\overline{\Omega})}\right)
\end{aligned} \tag{7.12}$$

for all $k = 1, \ldots, N$ and all $v \in C_0^1(\overline{\Omega})$ such that $\|v\|_{C_0^1(\overline{\Omega})} \leqslant \|u\|_{C_0^1(\overline{\Omega})}$. Here θ and ε stand for functions of the type $\theta: \Omega \to [0, 1]$ and $\varepsilon: \mathbb{R} \to \mathbb{R}$, with $\varepsilon(z) \to 0$ as $z \to 0$, and the constants K_1 and K_2 are independent of v. Thus $\|W\|_{C_0^1(\overline{\Omega})}/\|v\|_{C_0^1(\overline{\Omega})} \leqslant (K_1 + K_2)\varepsilon(\|v\|_{C_0^1(\overline{\Omega})})$ and we only need to take into account the definitions (7.6) and (7.10), and the result in Proposition 2.3 to subsequently obtain

$$\begin{aligned}
&\left\|\mathcal{F}_1(u + v) - \mathcal{F}_1(u) - G\left(f_u(\cdot, u)v\right)\right\|_{C_0^1(\overline{\Omega})} \\
&= \left\|G\left(\widetilde{N}W\right)\right\|_{C_0^1(\overline{\Omega})} \leqslant K\|v\|_{C_0^1(\overline{\Omega})} \varepsilon\left(\|v\|_{C_0^1(\overline{\Omega})}\right)
\end{aligned} \tag{7.13}$$

for all $v \in C_0^1(\overline{\Omega})$ such that $\|v\|_{C_0^1(\overline{\Omega})} \leqslant \|u\|_{C_0^1(\overline{\Omega})}$ and some constant K that is independent of v. This estimate implies that $\mathcal{F}_1'(u)$ exists and is given by (7.7). Thus property (a) above holds for $j = 1$ and the first step of the induction argument is complete.

Let us now assume that property (a) holds for the jth derivative of $\mathcal{F}_1$ and prove that it also holds for its $(j+1)$st derivative. To this end, we take $v_1, \ldots, v_{j+1} \in C_0^1(\overline{\Omega})$ such that $\|v_{j+1}\|_{C_0^1(\overline{\Omega})} \leqslant \|u\|_{C_0^1(\overline{\Omega})}$. Then (7.9) holds with $v = v_{j+1}$ and, as above, according to assumption (H.3), the function

$$W_1 \equiv \frac{\{\partial^j f(\cdot, u + v_{j+1})/\partial u^j - \partial^j f(\cdot, u)/\partial u^j - [\partial^{j+1} f(\cdot, u)/\partial u^{j+1}] v_{j+1}\} v_1 \cdots v_j}{\tilde{N}} \tag{7.14}$$

is seen to be such that

$$\|W_1\|_{C_0^1(\overline{\Omega})} \leqslant K_3 \|v_1\|_{C_0^1(\overline{\Omega})} \cdots \|v_j\|_{C_0^1(\overline{\Omega})} \|v_{j+1}\|_{C_0^1(\overline{\Omega})} \varepsilon\left(\|v_{j+1}\|_{C_0^1(\overline{\Omega})}\right),$$

with K_3 independent of $v_1, \ldots, v_{j+1}$ and ε as above. Thus, as above, we only need to take into account our assumption that $\mathcal{F}_1^{(j)}(u)$ is given by (7.7), the definition (7.14) and the result in Proposition 2.3 to subsequently obtain

$$\left\| [\mathcal{F}_1^{(j)}(u + v_{j+1}) - \mathcal{F}_1^{(j)}(u)](v_1, \ldots, v_j) - G\left(\frac{\partial^{j+1} f(\cdot, u)}{\partial u^{j+1}} v_1 \cdots v_{j+1} \right) \right\|_{C_0^1(\overline{\Omega})}$$

$$= \left\| G(\tilde{N} W_1) \right\|_{C_0^1(\overline{\Omega})}$$

$$\leqslant K \|v_1\|_{C_0^1(\overline{\Omega})} \cdots \|v_j\|_{C_0^1(\overline{\Omega})} \|v_{j+1}\|_{C_0^1(\overline{\Omega})} \varepsilon\left(\|v_{j+1}\|_{C_0^1(\overline{\Omega})}\right), \tag{7.15}$$

with K independent of $v_1, \ldots, v_{j+1}$. This estimate shows that $\mathcal{F}_1^{(j+1)}(u)$ exists and is given by (7.7). Thus the induction argument is complete and property (a) above holds.

Finally, the same argument that led above to (7.15) readily shows that

$$\left\| [\mathcal{F}_1^{(m)}(u + v_{j+1}) - \mathcal{F}_1^{(m)}(u)](v_1, \ldots, v_j) \right\|_{C_0^1(\overline{\Omega})}$$

$$\leqslant K \|v_1\|_{C_0^1(\overline{\Omega})} \cdots \|v_j\|_{C_0^1(\overline{\Omega})} \|v_{j+1}\|_{C_0^1(\overline{\Omega})}, \tag{7.16}$$

with K independent of $v_1, \ldots, v_{j+1}$. According to this estimate, $\mathcal{F}_1^{(m)}$ is continuous. Thus property (b) above also holds, and the proof is complete. $\qquad\square$

If the function f depends also on a parameter, then the same argument in the proof of Theorem 7.1 readily yields the following corollary.

COROLLARY 7.2. *In addition to the assumptions of Theorem 7.1, let us assume that the function f depends on a parameter $\lambda \in \mathbb{R}$ and that, for all $\lambda \in \mathbb{R}$ and all $l = 1, \ldots, r$, the function $\partial^l f/\partial \lambda^l$ satisfies assumption (H.3). If the Green operator G is defined as above, right after (7.2), then the operator $\mathcal{F} : \Omega \times C_0^1(\overline{\Omega}) \times \mathbb{R}$, defined as*

$$\mathcal{F}(u, \lambda) \equiv u - G(f(\cdot, u, \lambda)),$$

is such that, for all $j = 0, \ldots, m$ and all $l = 1, \ldots, r$, the derivative $\partial^{j+l}\mathcal{F}/\partial u^j\,\partial\lambda^l$ exists and is continuous whenever u is in the positive cone of $C_0^1(\overline{\Omega})$ and $\lambda \in \mathbb{R}$. Also, the j-linear operator $\partial^{j+l}\mathcal{F}(u,\lambda)/\partial u^j\,\partial\lambda^l : [C_0^1(\overline{\Omega})]^j \to C_0^1(\overline{\Omega})$ is given by

$$\frac{\partial^{j+l}\mathcal{F}(u,\lambda)}{\partial u^j\,\partial\lambda^l}(v_1,\ldots,v_j) = -G\left(\frac{\partial^{j+l}f(\cdot,u,\lambda)}{\partial u^j\,\partial\lambda^l}v_1\cdots v_j\right).$$

Note that this corollary provides the ingredient to apply implicit-function-like theorems to the problem (1.8), (1.9) when the nonlinearity f is allowed to depend also on a parameter.

REMARK 7.3. These results were obtained in [77] and largely used in [79] to prove the smoothness of branches of positive solutions, see also Section 10. Some related results were given, under rather strong assumptions, this time in weighted spaces, by Aranda and Lami-Dozo in [9]; for an application of the implicit function theorem for a reformulated problem, see [95]. A somewhat similar idea is behind the result on the differentiability of the functional obtained by Zhang in Lemma 1 in [122].

8. The associated parabolic problem: linearized stability

Standard linearized stability results for regular parabolic problems are readily extended to analyze the stability of the solutions of (1.8), (1.9) that are in the positive cone of $C_0^1(\overline{\Omega})$ as steady states of the problem

$$\frac{\partial u}{\partial t} + \mathcal{L}u = f(x,u) \quad \text{in } \Omega, \qquad u = 0 \quad \text{on } \partial\Omega, \tag{8.1}$$

$$u(\cdot,0) = u_0 \quad \text{in } \Omega. \tag{8.2}$$

In fact, if $\alpha > -1/N$ then the operator $\mathcal{L}$ is sectorial in $L_q(\Omega)$ for all $q > N$ and we can apply standard results in the literature [74] to obtain a global existence result on the parabolic problem (8.1), (8.2). Notice that this condition is only necessary because of the singularities depending on α for the coefficients of the linear differential operator that arise in assumption (H.2). In the case of smooth (in particular constant) coefficients the result is valid for any α.

THEOREM 8.1. *In addition to assumptions (H.1)–(H.3), let us assume that $\alpha > -1/N$, let $u_s \in C^2(\Omega) \cap C$ (= the interior of the positive cone of $C_0^1(\overline{\Omega})$) be a solution of (1.8), (1.9), and let $M \equiv f_u(\cdot, u_s)$. If the principal eigenvalue of (1.11) is strictly positive (resp., strictly negative) then u_s is an exponentially stable (resp., unstable) steady state of (8.1) in the Lyapunov sense, with the norm of $C^1(\overline{\Omega})$. Also, if $u_0 \in C$ then the problem (8.1), (8.2) has a unique solution, $t \to u(\cdot,t) \in C$, in a maximal existence interval, $0 \leqslant t < T \leqslant \infty$, and if $T < \infty$ then there is a sequence $\{t_m\}$ such that $t_m \nearrow T$ and either $\max\{u(x,t_m): x \in \Omega\} \to \infty$, or $u(x,t_m) \searrow 0$ for some $x \in \Omega$, or $\min\{\partial u(x,t_m)/\partial n: x \in \partial\Omega\} \to 0$ as $m \to \infty$.*

PROOF. Let q be such that $q > N$ and $1 + \alpha q > 0$, decompose the operator $\mathcal{L}$ as $\mathcal{L} = \mathcal{L}_1 + \mathcal{L}_2$, where $\mathcal{L}_1 u \equiv \sum \partial(a_{ij}\, \partial u/\partial x_i)/\partial x_j$, and consider the operator $\mathcal{L}_1$ in $X = L_q(\Omega)$, with domain $D(\mathcal{L}_1) = W_q^2(\Omega) \cap C \subset C_0^1(\overline{\Omega})$. The self-adjoint operator $\mathcal{L}_1$ is sectoral in X (use the argument in [74], p. 32) and if $(q + N)/(2q) < \beta < 1$ then its fractional power $\mathcal{L}_1^\beta$ is such that $\|u\|_{C^1(\overline{\Omega})} < K\|\mathcal{L}_1^\beta u\|_X$ for all $u \in D(\mathcal{L}_1)$ and some K that is independent of u ([74], Theorem 1.6.1). Also, when using assumption (H.3) and estimates (2.10), and proceeding as in Remark 2.4, the following estimates are obtained

$$\left\| f(\cdot, u) - f(\cdot, u_s) \right\|_X < K_1 \|u - u_s\|_{C_0^1(\overline{\Omega})} < K_2 \left\| \mathcal{L}_1^\beta(u - u_s) \right\|_X,$$

$$\|Mu\|_X + \|\mathcal{L}_2 u\|_X < K_3 \|u\|_{C_0^1(\overline{\Omega})} < K_4 \left\| \mathcal{L}_1^\beta u \right\|_X, \tag{8.3}$$

$$\left\| f(\cdot, u) - f(\cdot, u_s) - M(u - u_s) \right\|_X$$

$$< K_5 \|u - u_s\|^2_{C_0^1(\overline{\Omega})} < K_6 \left\| \mathcal{L}_1^\beta(u - u_s) \right\|^2_X, \tag{8.4}$$

for all $u \in D(\mathcal{L}_1)$ and for all u in a $C^1(\overline{\Omega})$-neighborhood of u_s in $D(\mathcal{L}_1)$, respectively, with the constants $K_1, \ldots, K_6$ independent of u; note that (8.3), (8.4) imply that the operator $u \to \mathcal{L}_1 u - f(\cdot, u)$, of $X^\beta \equiv D(\mathcal{L}_1^\beta) \subset C$ (with the norm $\|u\|_{X^\beta} \equiv \|\mathcal{L}_1^\beta u\|_X$) into X, maps bounded sets into bounded sets and is locally Lipschitzian. Then we only need to apply [74], Theorem 5.1.1, and straightforwardly modify the proofs of [74], Theorems 3.3.3 and 3.3.4, to obtain the stated results and thus to complete the proof. $\square$

Unfortunately the argument above does not apply (and, seemingly, is not straightforwardly extended) if $-1/N \geqslant \alpha > -1$. But still, in this general case, we can use the results in Section 2 to directly derive the following result, which should also yield the linearized stability result in Theorem 8.1 by a well-known argument [102,107], provided that one has a good existence theory for the parabolic problem (8.1), (8.2); the latter has been subsequently analyzed in [57].

REMARK 8.2. Existence and blow-up of positive solutions for a related degenerate parabolic problem were studied by Wiegner in [114,115].

THEOREM 8.3. *In addition to assumptions* (H.1)–(H.3), *let* $u_s \in C^2(\Omega) \cap C$ (= *the positive cone of* $C_0^1(\overline{\Omega})$) *be a solution of* (1.8), (1.9), *and let* $M \equiv f_u(\cdot, u_s)$. *If the principal eigenvalue of* (1.11) *is strictly positive* (*resp., strictly negative*) *then there is a constant* $\varepsilon_0 > 0$ *and a function* $U \in C^2(\Omega) \cap C$ *such that* $u_\varepsilon = u_s + \varepsilon U \in C$ *if* $|\varepsilon| < \varepsilon_0$, *and* u_ε *is a strict subsolution* (*resp., supersolution*) *of* (1.8), (1.9) *if* $-\varepsilon_0 < \varepsilon < 0$, *while* u_ε *is a strict supersolution* (*resp., subsolution*) *of* (1.8), (1.9) *if* $0 < \varepsilon < \varepsilon_0$.

PROOF. Take two functions $u_0, u^0 \in C$ such that $u_s - u_0$ and $u^0 - u_s$ are in C. As in the proof of Theorem 4.1, there is a function $\widetilde{N} \in C^2(\Omega)$ that satisfies (H.3$'$) and is such that

$$1 + \left| d(x) f_{uu}(\cdot, u) \right| < \widetilde{N} \quad \text{in } \Omega \quad \text{if } u_0 \leqslant u \leqslant u^0 \text{ in } \Omega. \tag{8.5}$$

Also, according to Proposition 2.8, if the principal eigenvalue λ_0 of (1.11) is nonzero then the principal eigenvalue λ_1 of (1.12) is such that $\lambda_1\lambda_0 > 0$. In addition, we take an eigenfunction of (1.12) associated with λ_1 such that $U \in \mathcal{C}$, and the constant $\varepsilon_0 > 0$ such that $\varepsilon_0 U < |\lambda_1| d(x)$, $u_0 < u_{-\varepsilon_0}$ and $u_{\varepsilon_0} < u^0$ in Ω, where $u_\varepsilon \equiv u_s + \varepsilon U$ as above. Then we only need to apply the mean value theorem and take into account (8.5) to subsequently obtain

$$\frac{\mathcal{L}u_\varepsilon - f(x, u_\varepsilon)}{\lambda_1 \varepsilon U} = \left\{ \widetilde{N}(x) - \frac{f(x, u_s + \varepsilon U) - f(x, u_s)}{\lambda_1 \varepsilon U} \right\}$$

$$= \widetilde{N}(x) - \frac{f_{uu}(x, u_s + \varepsilon\theta(x)U)\varepsilon U}{\lambda_1} > 1 \quad \text{for all } x \in \Omega$$

and all ε such that $|\varepsilon| < \varepsilon_0$, where $\theta : \Omega \to \mathbb{R}$ stands for a function such that $0 \leqslant \theta \leqslant 1$ in Ω. Since $\lambda_1\lambda_0 > 0$, the stated result follows, and the proof is complete. $\qquad\square$

REMARK 8.4. As pointed out above linearized stability was studied by Bertsch and Rostamian in [16] using Hardy–Sobolev inequality and weighted Sobolev spaces. Some difficult estimates for the convergence rate were obtained as well in this paper. See also the comments on stability in [51].

LEMMA 8.5. *Under assumption* (H.9) *(see Theorem 5.1), let* $u \in C^2(\Omega) \cap C_0^1(\overline{\Omega})$ *be a solution of* (2.1), (2.2) *such that* $u > 0$ *in* Ω *and* $\partial u/\partial n < 0$ *on* $\partial\Omega$. *Then the principal eigenvalue,* μ_1, *of the linearized problem around the solution* u,

$$\mathcal{L}w - f_u(x, u)w = \mu w \quad \text{in } \Omega, \qquad w = 0 \quad \text{on } \partial\Omega, \tag{8.6}$$

is such that $\mu_1 > 0$.

PROOF. According to [77], Theorem 2.6, part (iv), the adjoint eigenvalue problem,

$$\mathcal{L}^*\psi_1 - f_u(x, u)\psi_1 = \mu_1\psi_1 \quad \text{in } \Omega, \qquad \psi_1 = 0 \quad \text{on } \partial\Omega, \tag{8.7}$$

exhibits eigenfunctions $\psi_1 > 0$ associated with the principal eigenvalue μ_1. We now multiply (1.8) by ψ_1, integrate over Ω, integrate by parts and replace (8.7), to obtain

$$\int_\Omega \psi_1 f(x, u) = \int_\Omega \psi_1 \mathcal{L}u = \int_\Omega u\mathcal{L}^*\psi_1 = \int_\Omega \psi_1 u f_u(x, u) + \mu_1 \int_\Omega u\psi_1. \tag{8.8}$$

Hence, we have

$$\int_\Omega \big[f(x, u) - f_u(x, u)u \big]\psi_1 = \mu_1 \int_\Omega u\psi_1 \tag{8.9}$$

and invoking (H.9) we obtain $\mu_1 > 0$, as stated. $\qquad\square$

REMARK 8.6. A similar computation was used by Brown and Hess [21] allowing to get existence and also uniqueness of positive solutions for a nonsingular problem with an indefinite weight. It is shown there that all positive solutions are stable and this yields the results.

Next we consider an alternative and very instrumental approach due to Takáč [112] and consisting in working in the framework of weighted Sobolev spaces, an idea that, as pointed out before, is very natural when dealing with singularities in order to smooth them. This provides also another way to apply the abstract machinery in [74] as in the first part of this section, getting local existence and uniqueness first and proving also the important result an holomorphic semigroup is obtained in a suitable function space.

We begin by defining weighted Lebesgue spaces

$$L_p(\Omega, d^\mu) = \left\{ f \in L_p(\Omega)_{\text{loc}} \colon \int_\Omega |f(x)|^p d(x)^\mu \, dx < +\infty \right\}, \tag{8.10}$$

where $-1 < \mu < +\infty$ and $1 < p < +\infty$, and the associated weighted Sobolev spaces

$$W_p^1(\Omega, d^\mu) = \left\{ f \in W_p^1(\Omega)_{\text{loc}} \colon \int_\Omega (|\nabla f(x)|^p + |f(x)|^p) d(x)^\mu \, dx < +\infty \right\} \tag{8.11}$$

which are Banach spaces with the corresponding norm

$$\|f\|_{W_p^1(\Omega, d^\mu)} = \left(\int_\Omega (|\nabla f(x)|^p + |f(x)|^p) d(x)^\mu \, dx \right)^{1/p}. \tag{8.12}$$

The weighted Sobolev spaces $W_p^k(\Omega, d^\mu)$ are defined in an analogous way for $k \geq 1$ an integer and also in the usual way for fractional exponents (see [84] and [113] for more details and proofs concerning all these matters).

By using Hardy's inequalities it is possible to prove the following embedding theorem.

LEMMA 8.7. *If $\mu \geq 0$, $p > \mu + N$ and $\gamma = 1 - (\mu + N)/p$, we have the compact embedding*

$$W_p^k(\Omega, d^\mu) \to C^{k-1,\gamma}(\overline{\Omega}), \quad k = 1, 2. \tag{8.13}$$

Concerning the traces of functions on the boundary we have the lemma:

LEMMA 8.8. *For $-1 < \mu < p - 1$, the trace mapping*

$$W_p^1(\Omega, d^\mu) \to L_p(\partial\Omega), \tag{8.14}$$

defined in the usual way, is continuous.

Now it is possible to define the space of functions which are zero on the boundary in the sense of traces, namely

$$W_p^1(\Omega, d^\mu)_0 = \left\{ f \in W_p^1(\Omega, d^\mu): f = 0 \text{ on } \partial\Omega \right\} \tag{8.15}$$

which is also a Banach space with the natural norm. Then we also have the following lemma.

LEMMA 8.9. *If $p > N$, $-1 < \mu < p - N$, the embedding*

$$X =: W_p^2(\Omega, d^\mu) \cap W_p^1(\Omega, d^\mu)_0 \to C_0^1(\overline{\Omega}) \tag{8.16}$$

is compact.

It is possible to develop a suitable theory of existence and uniqueness for boundary value problems for linear differential operators in these spaces. More precisely, we deal now with the problem

$$\mathcal{M}(u) = \mathcal{L}u + c(x)u = g(x) \quad \text{in } \Omega, \tag{8.17}$$

$$u = 0 \quad \text{on } \partial\Omega, \tag{8.18}$$

where $\mathcal{L}$ and Ω are defined as above and the following assumptions are satisfied:
- (H.10) The symmetric matrix a_{ij} is uniformly elliptic.
- (H.11) The coefficients $a_{ij} \in C(\overline{\Omega})$ for all $i, j \leqslant N$.
- (H.12) There exists β such that $0 < \beta \leqslant 1$ such that $d(x)^{1-\beta} \partial a_{ij}/\partial x_k$ and $d(x)^{2-\beta} c(x)$ are bounded for any $i, j, k = 1, \ldots, N$.
- (H.13) $d(x)^{1-\beta} b_i(x)$ is bounded in Ω.

Then we obtain the following general result.

THEOREM 8.10. *Suppose that assumptions (H.1), (H.10)–(H.13) are satisfied with $0 < \beta \leqslant 1$, $1 < p < +\infty$ and $(1 - \beta)p - 1 < \mu < p - 1$, then there exists a constant λ_0 such that, for every $\lambda > \lambda_0$ and for every $g \in L_p(\Omega, d^\mu)$, there exists a unique solution $u \in X \equiv W_p^2(\Omega, d^\mu) \cap W_p^1(\Omega, d^\mu)_0$ of the problem (8.17), (8.18).*

Next, we show that the general abstract results of Henry [74] still work in this alternative, and sometimes more convenient framework. For this, it is necessary to show that the linear semigroup associated to the problem (8.17), (8.18) is holomorphic in the weighted Lebesgue space $L_p(\Omega, d^\mu)$ if $1 < p < +\infty$ and $-1 < \mu < p - 1$. The result for $\mu = 0$ can be found in [41], Theorem 1.4.2, for smooth coefficients; see [112] for the general case, where the complex interpolation method of Davies [41] is used.

Now, we consider again the spaces $X_{p,\mu} = W_p^2(\Omega, d^\mu) \cap W_p^1(\Omega, d^\mu)_0$, $Y_{p,\mu} = L_p(\Omega, d^\mu)$ and the linear operator $L_{p,\mu}: X_{p,\mu} \to Y_{p,\mu}$ defined by the existence result (Theorem 8.10). The linear operator $A_{p,\mu}: Y_{p,\mu} \to X_{p,\mu}$ is defined with domain $D(A_{p,\mu}) = X_{p,\mu}$ by $A_{p,\mu}u = L_{p,\mu}u$. We have then the following result.

THEOREM 8.11 (Takáč [112], Theorem B.1). *Suppose that assumptions (H.1), (H.10)– (H.13) are satisfied and that $(1 - \beta)p - 1 < \mu < p - 1$. Then $A_{p,\mu}$ is the infinitesimal generator of a holomorphic semigroup $T_{p,\mu}(t), t \geqslant 0$, on the space $Y_{p,\mu}$. The restriction of $T_{p,\mu}$ to the space $L_\infty(\Omega)$ does not depend on p neither on μ.*

REMARK 8.12. Other interesting questions concerning parabolic problems can be studied. One is existence and properties of periodic solutions in the singular case and the related matter of existence of principal eigenvalues for the periodic–parabolic associated problem, see [65] and the references therein. Traveling waves for problems which are nonsingular in our sense but with non-Lipschitz nonlinearities are considered in the book [89], and the elliptic problems raised there can be studied by the methods of this paper as well.

9. Stabilization

In this section we study again the asymptotic behavior of solutions to the problem (8.1), (8.2), but this time not only in the sense of local stability, as in the preceding paragraph, but trying to get also some interesting information concerning the *global* behavior when t goes to infinity of *every* bounded solution with suitable initial data. This is normally called *stabilization* and has been studied for nonsingular problems by a series of authors (see references in [112]). A fundamental tool for this work has been an inequality due to L. Simon using in an essential way the property of real-analyticity of the nonlinear term in the equation. Simon's very involved arguments have been greatly simplified later and have also been adapted by Takáč [112] to the singular case, by working this time in the rather flexible framework of weighted Sobolev spaces given at the end of the preceding section.

Since the gradient-like structure of the problem is essential for the method applied here, we are limited to differential operators in divergence form, i.e., to "formally self-adjoint" linear operators.

We will consider the following semilinear parabolic boundary value problem

$$\frac{\partial u}{\partial t} - \sum \frac{\partial}{\partial x_i}\left(a_{ij}(x)\frac{\partial u}{\partial x_j}\right) + c(x)u = f(x, u) \quad \text{in } \Omega,$$

$$u = 0 \quad \text{on } \partial\Omega \times [0, +\infty[,$$

$$\tag{9.1}$$

$$u(\cdot, 0) = u_0 \quad \text{in } \overline{\Omega}, \tag{9.2}$$

where Ω still satisfies (H.1) and a_{ij} and c satisfy (H.10)–(H.12). The initial value u_0 is smooth.

Concerning the nonlinearity f we assume that $f : \Omega \times (0, +\infty) \to \mathbb{R}$ is such that $f(x, \cdot)$ is a real analytic function for a.a. $x \in \Omega$ and satisfies following assumption.

(H.13′) There exists $0 < \gamma \leqslant 1$ such that, for any $R > 0$ and any $K > 0$, there exist $r > 0$ and $C > 0$ such that, for a.a. $x \in \Omega$ and for any v such that $Kd(x) \leqslant v \leqslant R$, we have

$$\left|\frac{\partial^k f(x, v)}{\partial v^k}\right| \leqslant Cd(x)^{1-\gamma}k!\left(\frac{r}{d(x)}\right)^k, \quad k = 0, 1, 2, \ldots. \tag{9.3}$$

Condition (H.13$'$) means that the Taylor series $f(x, \cdot): (0, +\infty) \to \mathbb{R}$ converges locally uniformly in $\Omega \times (0, +\infty)$ as will follow from the proof of Theorem 9.1.

Now we associate to the problem (9.1), (9.2) the energy functional

$$E(u) = \int_{\Omega} \left(\frac{1}{2} \sum a_{ij}(x) \frac{\partial u}{\partial x_i} \frac{\partial u}{\partial x_j} + c(x)u^2 - F(x, u) \right) dx, \tag{9.4}$$

where as usual $F(x, u) = \int_0^u f(x, s)\, ds$ for $u > 0$. E is a well-defined functional $E : C \to \mathbb{R}$, where C denotes the interior of the positive cone (in $C_0^1(\overline{\Omega})$).

The following theorem is a key result in what follows.

THEOREM 9.1. *Assume that* $\mu > -1$, $p > 1$ *and* $1 - (\mu + 1)/p < \gamma \leqslant 1$. *If f satisfies* (H.13$'$), *the Nemitskii operator $\mathcal{F}$ associated to $f(x, v)$ is real-analytic mapping from C into $L_p(\Omega, d^\mu)$.*

PROOF. It follows from (9.3) that for $r|h| < d(x)$ the Taylor series can be written as

$$f(x, v + h) - f(x, v) = \sum \frac{1}{n!} f^{(n)}(x, v) h^n. \tag{9.5}$$

We have the estimate

$$\sum \frac{1}{n!} |f^{(n)}(x, v)| |h|^n \leqslant C d(x)^{-1+\gamma} \sum \left(\frac{r|h|}{d(x)} \right)^n$$

$$\leqslant C d(x)^{-1+\gamma} \quad \text{if } 2r|h| \leqslant d(x). \tag{9.6}$$

We fix next $v \in C$ and then pick $R > 0$ and $K > 0$ such that $\|v\|_{L_\infty(\Omega)} \leqslant R$ and $K\|d/v\|_{L_\infty(\Omega)} \leqslant 1$. If $r > 0$ and $C > 0$ are given by (H.13$'$), there exists $\delta > 0$ such that if $h \in C_0^1(\overline{\Omega})$ and $\|h\|_{C_0^1(\overline{\Omega})} \leqslant \delta$, then $2r\|h/d\|_{L_\infty(\Omega)} \leqslant 1$ which implies that (9.3) holds if $\|h\|_{C_0^1(\overline{\Omega})} \leqslant \delta$. Since $d(x)^{-1+\gamma}$ is in $L_p(\Omega, d^\mu)$ if $(-1 + \gamma)p + \mu > -1$ is satisfied, we have proved that the above Taylor series converges absolutely in $L_p(\Omega, d^\mu)$ for $\|h\|_{C_0^1(\overline{\Omega})} \leqslant \delta$. $\qquad\square$

COROLLARY 9.2. *Suppose that* $0 < \gamma < 1$, $p > N$ *and* $(1 - \gamma)p - 1 < \mu < p - N$ *and that* (H.10)–(H.12) *and* (H.13$'$) *are satisfied. Then the mapping*

$$\mathcal{M} : C \cap \left(W_p^2(\Omega, d^\mu) \cap W_p^1(\Omega, d^\mu)_0 \right) \to L_p(\Omega, d^\mu), \tag{9.7}$$

defined by

$$\mathcal{M}(u) = -\sum \frac{\partial}{\partial x_i} \left(a_{ij}(x) \frac{\partial u}{\partial x_j} \right) + c(x)u - f(x, u), \tag{9.8}$$

is real analytic.

REMARK 9.3. Notice that, in principle, E is only analytic with respect to the norm in $C_0^1(\overline{\Omega})$ and may be nondifferentiable in the sense of Fréchet in, e.g., $H_0^1(\Omega)$.

The Fréchet derivative of E is obviously given by (9.8). Hence the set of stationary solutions of the parabolic problem (9.1), (9.2) is given by

$$S = \{\Psi \in C : \mathcal{M}(\Psi) = 0\}. \tag{9.9}$$

Now we are ready to reformulate our problem in terms of the above framework and state and prove our main result.

If $u(x, t)$ is a weak solution to (9.1), (9.2) satisfying, for $k_1, k_2 > 0$,

$$0 < u(x, t) \leqslant k_1 \quad \text{in } \Omega \times (0, +\infty), \tag{9.10}$$

$$-\frac{\partial u}{\partial n}(x, t) \geqslant k_2 \quad \text{in } \partial\Omega \times (0, +\infty), \tag{9.11}$$

then $u : \mathbb{R}^+ \to C_0^1(\overline{\Omega})$ with $u(\cdot, t) \in C_0^1(\overline{\Omega})$ is a trajectory with orbit

$$\mathcal{O}^+(u) = \{u(\cdot, t) : t \geqslant 0\} \tag{9.12}$$

bounded in $L_\infty(\Omega)$. We have $u(\cdot, 0)$ in X and also that

$$\mathcal{O}_{t_0}^+(u) = \{u(\cdot, t + t_0) : t \geqslant 0\} \tag{9.13}$$

is relatively compact in X for any $t_0 > 0$; a fact which follows from Theorems 8.11, 9.1 and Lemma 8.9 by using the results of Henry [74], pp. 54–57. Moreover, by using [74], Theorem 3.5.2, it is proved that $u \in C((0, +\infty), X)$ and $du/dt \in C((0, +\infty), L_p(\Omega, d^\mu))$. Now it is clear that if $p > 2$ and $-1 < \mu < p/2 - 1$ we have the continuous embedding $L_p(\Omega, d^\mu) \to L_2(\Omega)$ and hence the parabolic problem (9.1), (9.2) is equivalent to an abstract initial value problem in the Hilbert space $L_2(\Omega)$, namely

$$\frac{du}{dt} = \mathcal{M}(u), \quad t > 0, \ u(0) = u_0, \ u_0 \in C. \tag{9.14}$$

In particular,

$$\frac{dE}{dt}(u) = \langle \mathcal{M}(u), u' \rangle = -\|u'\|_{L_2(\Omega)}^2 \tag{9.15}$$

and then (9.14) is a gradient system (see [73,74]).

We need for the proof of the main result the following version ([112], Proposition 8.1) of Simon's inequality.

THEOREM 9.4. *Suppose that* (H.1), (H.10)–(H.12) *and* (H.13′) *are satisfied with* $0 < \alpha < 1$. *If* $\varphi \in S$, *there exist* $\Theta > 0$, $\varepsilon > 0$ *and* $0 < \theta < 1/2$ *such that if* $u \in B_\varepsilon(\varphi) \subset X$, *then*

$$\left|E(u) - E(\varphi)\right|^{1-\theta} \leqslant \Theta \|\mathcal{M}u\|_{H^{-1}(\Omega)}. \tag{9.16}$$

The main result is the following theorem.

THEOREM 9.5. *Suppose that* (H.1), (H.10)–(H.12) *and* (H.13′) *are satisfied for* $0 < \alpha < 1$. *If u is a weak solution to* (9.1), (9.2) *such that* $u \in C_0^1(\overline{\Omega})$ *and* (9.10), (9.11) *are satisfied, then there is a stationary solution w to* (9.1), (9.2), $w \in C$ *such that*

$$\lim_{t \to +\infty} \|u(\cdot, t) - w\|_{C_0^1(\overline{\Omega})} = 0. \tag{9.17}$$

PROOF. From the above considerations follows that the ω-limit set

$$\omega(u) = \bigcap_{s \geq 0} \overline{\{u(t) \in X : t \geq s\}}^X \tag{9.18}$$

of the trajectory $u(t)$ is nonempty, connected and compact in X. It follows from (9.10), (9.11) that $\omega(u) \subset \mathcal{S}$ and that, for any $\varphi \in \omega(u)$,

$$\lim_{t \to +\infty} E(u(t)) = E(\varphi). \tag{9.19}$$

Now, from Theorem 9.4, there exist θ, ε and Θ such that (9.3) holds and, moreover, if we pick ε even smaller if necessary, we can get

$$|E(u) - E(\varphi)| \leq 1 \tag{9.20}$$

for every $u \in B_\varepsilon(\varphi)$. The family of open balls $\{B_\varepsilon(\varphi) : \varphi \in \omega(u)\}$ is an open covering for $\omega(u)$ and hence there is a finite subcovering $\{B_{\varepsilon_i}(\varphi_i) : i = 1, \ldots, m\}$. Hence, if we choose Θ and θ such that

$$0 < \theta < \min_{i=1,\ldots,m} \theta_i, \qquad \Theta = \max_{i=1,\ldots,m} \theta_i, \tag{9.21}$$

we obtain

$$|E(u(t)) - E(\varphi)|^{1-\theta} \leq \Theta \|\mathcal{M}u(t)\|_{H^{-1}(\Omega)}. \tag{9.22}$$

Now we integrate (9.15) over $[t, +\infty[$ and get

$$E(u(t)) - E(\varphi) = -\int_t^{+\infty} \frac{dE}{dt}(u(t)) \, dt = \int_t^{+\infty} \|u'(s)\|_{L_2(\Omega)}^2 \, ds \tag{9.23}$$

and this gives

$$\left(\int_t^{+\infty} \|u'(s)\|_{L_2(\Omega)}^2 \, ds \right)^{1-\theta} \leq C \|u'(t)\|_{L_2(\Omega)}^2 \qquad \text{for all } t \geq 0, \tag{9.24}$$

where the constant $C > 0$ does not depend on t. It is enough to apply a result by Feireisl and Simondon [55] to get the convergence of the integral $\int_t^{+\infty} \|u'(s)\|_{L_2(\Omega)}^2 ds$, which implies convergence of the trajectory $\{u(t)\colon t \geqslant 0\}$ in $L_2(\Omega)$ and then in X by compactness. This ends the proof. $\qquad\square$

10. Applications

In this section we apply in a systematic way the general theorems in Sections 4–8 (and Section 9 as well, see Remark 10.3) to a series of nonlinear eigenvalue problems. This list is not exhaustive (see [79] and the references therein) but includes most of the interesting examples; some of them were already considered in the literature and comments are given in the remarks. Concerning the nonlinearities, even if the statements are often more general than in previous work (concerning in particular singular x-dependent coefficients), we do not search for more generality: as a rule, homogeneous nonlinearities may be replaced by nonhomogeneous ones having a similar behavior both at the origin and at infinity; with very slight modifications, the proofs below still work in the general case. The results concerning smoothness of the curve of positive solutions and its stability seem to be new.

We assume all along this section that Ω is a bounded domain in $\mathbb{R}^N$ with the smoothness given by (H.1) and the differential operator $\mathcal{L}$ satisfies assumptions (H.2) where α is as defined in (H.1).

10.1. *Power law:* $\mathcal{L}u = \lambda K(x)u^q$

We consider the problem

$$\mathcal{L}u = \lambda K(x)u^q \quad \text{in } \Omega, \qquad u = 0 \quad \text{on } \partial\Omega, \tag{10.1}$$

which exhibits a unique, asymptotically stable solution for appropriate q, as stated in the following theorem.

THEOREM 10.1. *If $-1 < q < 1$ and $K(x)$ is a function that satisfies $0 < K(x) \leqslant k_1 d(x)^{-\beta}$, where $\beta > 0$ is such that $0 < \beta < 1 + q$, then for any $\lambda > 0$ there exists a unique positive solution $z > 0$ of (10.1), which is such that $z \in C^2(\Omega) \cap C_0^{1,\delta}(\overline{\Omega})$, where $\delta < \delta_1 = \min\{\delta_0, 1 + q - \beta\}$, and δ_0 is as defined in Theorem 4.1. Moreover, z is asymptotically stable and $z = \lambda^{1/(1-q)} z_1$, where z_1 is the solution to (10.1) at $\lambda = 1$.*

PROOF. As a subsolution we try $u_0 \equiv c\varphi_0$, where $\varphi_0 > 0$ is an eigenfunction associated with the principal eigenvalue, $\lambda_0 > 0$, of (Theorem 2.6)

$$\mathcal{L}\varphi_0 = \lambda_0 K(x)\varphi_0 \quad \text{in } \Omega, \qquad \varphi_0 = 0 \quad \text{on } \partial\Omega. \tag{10.2}$$

It follows that

$$\mathcal{L}u_0 - \lambda K(x)u_0^q = \lambda_0 c K(x)\varphi_0 - \lambda K(x)c^q \varphi_0^q$$

$$= \left(c^{1-q}\lambda_0\varphi_0^{1-q} - \lambda\right)K(x)c^q\varphi_0^q,$$

and u_0 will be a subsolution (for $\lambda > 0$) provided that $c^{1-q}\lambda_0\|\varphi_0\|_{L^\infty}^{1-q} \leqslant \lambda$, which holds for $c > 0$ small enough.

In order to find a supersolution, we first define $\varphi_1 > 0$, with $\|\varphi_1\|_{L^\infty} = 1$, as an eigenfunction associated with the principal eigenvalue, λ_1, of (Theorem 2.6)

$$\mathcal{L}\varphi_1 = \lambda_1\varphi_1 \quad \text{in } \Omega, \qquad \varphi_1 = 0 \quad \text{on } \partial\Omega, \tag{10.3}$$

and define $u_1 > 0$ as the unique solution (apply Proposition 2.3 and recall that $0 < \beta - q < 1$) to

$$\mathcal{L}u_1 = K(x)\varphi_1^q \quad \text{in } \Omega, \qquad u_1 = 0 \quad \text{on } \partial\Omega. \tag{10.4}$$

Now we seek a supersolution of the form $u^0 \equiv \mu u_1$, with $\mu = \text{constant}$. It follows that

$$\begin{aligned}
\mathcal{L}u^0 - \lambda K(x)\big(u^0\big)^q &= \mu K(x)\varphi_1^q - \lambda K(x)\mu^q u_1^q \\
&= \mu^q\big(\mu^{1-q}\varphi_1^q - \lambda u_1^q\big)K(x) \quad \text{in } \Omega.
\end{aligned}$$

Thus u^0 is a supersolution if μ is large enough (recall that both u_1 and φ_1 are in the positive cone). Since, in addition, $u^0 > u_0$ in Ω if $\mu > 0$ is large enough, Theorem 4.1 gives existence. Uniqueness and stability follow applying Theorems 5.1 and 4.1, and Lemma 8.5 (with $f(x,u) = K(x)u^q$ and $H(x,u) = (1-q)K(x)u^q > 0$), and Theorem 8.1. The stated dependence of z on λ follows replacing $z = \lambda^{1/(1-q)}z_1$ into (10.1). This completes the proof. $\qquad\qquad\square$

In particular, for $K(x) \equiv 1$ we have the theorem:

THEOREM 10.2. *If* $-1 < q < 1$, *then for any* $\lambda > 0$ *there exists a unique positive solution* z *of*

$$\mathcal{L}z = \lambda z^q \quad \text{in } \Omega, \qquad z = 0 \quad \text{on } \partial\Omega, \tag{10.5}$$

and $z \in C^2(\Omega) \cap C_0^{1,\delta}(\overline{\Omega})$, *where* $\delta < \min\{\delta_0, 1+q\}$. *Moreover,* z *is asymptotically stable and* $z = \lambda^{1/(1-q)}z_1$, *where* z_1 *is the solution to* (10.5) *at* $\lambda = 1$.

REMARK 10.3. Notice that it follows from the stabilization result in the preceding section that the unique positive solution is not only stable but also globally attractive in the sense that solutions with initial data in the interior of the positive cone all converge to this solution. The same argument works whenever there is a unique solution in what follows.

REMARK 10.4. It can be seen, reasoning as in [88], that $u^0 \equiv C\varphi_1^s$ (with φ_1 as defined in (10.3), $0 < s < 1$, $0 < s < 2/(1-q)$ and $-1 < q < 0$) is a supersolution in the usual sense provided that $C > 0$ is large. But $u^0 \notin C^{1,\delta}(\overline{\Omega})$ and we cannot apply Theorem 4.1.

REMARK 10.5. Existence and uniqueness for (10.1) were obtained in [38] and again in [66,88,123] and [47] often for the particular case $\mathcal{L} = -\Delta$ and $K \equiv 1$ using alternative methods and obtaining improvements in the regularity of solutions. See also [104], where the restriction $0 < q < (N+2)/(N-2)$ is required and only the radial case is treated using phase plane arguments.

Theorem 10.1 requires that the function K be nonnegative. If K exhibits both signs, then we can still have existence and uniqueness under additional assumptions, as shown in [77], where the following result was obtained.

THEOREM 10.6 ([77], Theorem 4.4). *Let Ω and $\mathcal{L}$ satisfy assumptions* (H.1) *and* (H.2), *let q be such that $0 < q < 1$, and let $K \in C^1(\Omega)$ be such that* (i) $|K(x)| < K_1 d(x)^\beta$ *for some $K_1 > 0$ and some β such that $-1 < \beta + q < 1$, and* (ii) *the unique solution of*

$$\mathcal{L}e = K(x) \quad \text{in } \Omega, \qquad e = 0 \quad \text{on } \Omega \tag{10.6}$$

satisfies $e > 0$ in Ω and $\partial e/\partial n < 0$ on $\partial \Omega$. Then problem (10.1) *has a unique strictly positive solution for each $\lambda > 0$.*

REMARK 10.7. A careful and detailed study for $\mathcal{L} = -\Delta$ and K smooth and changing sign in Ω was carried out in [12], where existence of nonnegative solutions (possibly with dead cores) was shown by using sub- and supersolutions. But the above result was not included in [12].

REMARK 10.8. The preceding theorem raises some intriguing questions. We have proved existence under the assumption that (10.6) has a smooth positive solution, which is obviously satisfied (by the maximum principle) if $K(x) \geqslant 0$, but also for many K's changing sign in Ω. This condition arises in [39]. A stability result was given in [77], Theorem 4.8.

10.2. *First combination of two power laws:* $\mathcal{L}u + M(x)u^p = \lambda K(x)u^q$

We begin with the case $M \equiv K \equiv 1$, namely

$$\mathcal{L}u + u^p = \lambda u^q \quad \text{in } \Omega, \qquad u = 0 \quad \text{on } \partial\Omega. \tag{10.7}$$

Let us first prove the following a priori estimate.

LEMMA 10.9. *If $u > 0$ is a solution to* (10.7) *with $-1 < q < 1$ and $p > q$, then $0 < u \leqslant \lambda^{1/(p-q)}$ in Ω.*

PROOF. If $u_M = u(x_M) > 0$ is the maximum value of u in Ω, then $x_M \in \Omega$ and invoking the maximum principle, we have $\mathcal{L}u(x_M) = \lambda u_M^q - u_M^p \geqslant 0$, which means that $(u(x) \leqslant) u_M \leqslant \lambda^{1/(p-q)}$ (for all $x \in \Omega$). $\qquad\square$

Using this, we can prove the following theorem.

THEOREM 10.10. *If* $-1 < q < 1$, $p > q$ *and* $\lambda > 0$, *then there is a unique positive solution to* (10.7). *Moreover,* u *is asymptotically stable,* $u \in C_0^{1,\delta}(\overline{\Omega})$, *with* $\delta < \min\{\delta_0, 1 + q\}$, *and the mapping* $\lambda \to u(x, \lambda)$, *from* $]0, \infty[$ *into* $C_0^{1,\delta}(\overline{\Omega})$, *is* C^∞ *and strictly increasing.*

PROOF. As a supersolution, we take $u^0 = z$, the unique solution of (10.5) (Theorem 10.2). And as a subsolution we try $u_0 \equiv c\varphi_1$ with φ_1 given by (10.3) and $c > 0$. It follows that

$$\mathcal{L}u_0 + u_0^p - \lambda u_0^q = \left(\lambda_1 c^{1-q} \varphi_1^{1-q} + c^{p-q} \varphi_1^{p-q} - \lambda\right) c^q \varphi_1^q. \tag{10.8}$$

Thus, u_0 is a subsolution if $c > 0$ small enough. Since, in addition, $u_0 \leqslant u^0$ for sufficiently small $c > 0$, existence follows applying Theorem 4.1.

As for uniqueness and asymptotic stability, we only need to apply Theorem 5.1 and Lemma 8.5, noting that the function $f_\lambda(u) = \lambda u^q - u^p$ satisfies assumption (H.9). This is because (recall that $u \leqslant \lambda^{1/(p-q)}$ in Ω, Lemma 10.9)

$$H_\lambda(x, u) = \lambda(1 - q)u^q - (1 - p)u^p = u^p\left[(1 - q)\lambda u^{q-p} + p - 1\right]$$
$$\geqslant u^p(p - q) > 0 \quad \text{in } \Omega. \tag{10.9}$$

Asymptotic stability and the remaining statements follow applying Theorem 8.1 and Lemma 8.5, taking into account that since f_λ increases with λ, if $\tilde{u}$ and $\hat{u}$ are the solutions at $\lambda = \tilde{\lambda}$ and $\hat{\lambda}$, with $\tilde{\lambda} < \hat{\lambda}$, then $\tilde{u}$ is a strict subsolution of (10.7) at $\lambda = \hat{\lambda}$. Thus the proof is complete. $\qquad\square$

We consider now the opposite relation between p and q.

THEOREM 10.11. *If* $-1 < p < q < 1$, *then there is a positive constant* $\underline{\lambda} > 0$ *such that if* $\lambda > \underline{\lambda}$ *then* (10.7) *possesses a positive solution (which is such that* $u \in C_0^{1,\delta}(\overline{\Omega})$, *with* $\delta < \min\{\delta_0, 1 + p\}$) *and if* $0 < \lambda < \underline{\lambda}$ *then* (10.7) *has no solution.*

PROOF. Let us first see that (10.7) *possesses a solution if* λ *is sufficiently large.* Using the new variable $u = \lambda^{1/(1-q)}v$, (10.7) is rewritten as

$$\mathcal{L}v + \varepsilon v^p = v^q \quad \text{in } \Omega, \qquad v = 0 \quad \text{on } \partial\Omega, \tag{10.10}$$

where $\varepsilon = \lambda^{-(1-p)/(1-q)}$. According to Lemma 8.5 the principal eigenvalue of the problem (10.10) linearized around the unique positive solution of (10.10) for $\varepsilon = 0$ is positive. Then applying the implicit function theorem we obtain that there is a constant $\varepsilon_0 > 0$ such that if $0 \leqslant \varepsilon < \varepsilon_0$ then (10.10) possesses a positive solution, which proves the statement above.

We now see that *the set* $\mathcal{S} = \{\lambda \in]0, +\infty[: (10.7) \text{ possesses a positive solution}\}$ *is an interval.* To this end, we take $\tilde{\lambda}$ and $\hat{\lambda}$ such that $\underline{\lambda} \leqslant \tilde{\lambda} < \hat{\lambda}$, where $\underline{\lambda} = \inf \mathcal{S}$, and prove that if (10.7) possesses a solution $\tilde{u} > 0$ at $\lambda = \tilde{\lambda}$, then (10.7) is also solvable at $\lambda = \hat{\lambda}$. By the argument at the end of the proof of Theorem 10.10, $\tilde{u}$ is a subsolution of (10.7) at $\lambda = \hat{\lambda}$.

In addition, for any $\lambda > 0$, we have the supersolution $u^0 = cz$, where $c \geqslant 1$ and z is the unique solution of (10.5). This is because

$$\mathcal{L}u^0 - \lambda(u^0)^q + (u^0)^p = \left[\lambda(c^{1-q} - 1) + c^{p-q}z^{p-q}\right]c^q z^q.$$

And, obviously, $\tilde{u} < u^0$ if c is sufficiently large. Thus, Theorem 4.1 yields the statement above.

Finally, we show that $\underline{\lambda} > 0$, namely that (10.7) *has no solution if* $\lambda > 0$ *is sufficiently small*. To this end, we note that $u_0 = c\varphi_1$, defined just before (10.8), is such that

$$\mathcal{L}u_0 + u_0^p - \lambda u_0^q > 0$$

if $c > 0$ and λ is smaller than the minimum of the function $t \to \lambda_1 t^{1-q} + t^{p-q}$ in $0 < t < \infty$, which is strictly positive. This means, invoking the strong maximum principle, that for such small $\lambda > 0$, any solution u of (10.7) must be such that $u < c\varphi_1$ for all $c > 0$, which proves nonexistence of positive solutions and completes the proof of the theorem. $\square$

REMARK 10.12. Theorem 10.11 extends [105], Theorem 1.2, [119], Theorem 1, [44], Lemma 4.11, and [124], Theorem 3.3. Zhang also shows that there is no solution in $C^2(\Omega) \cap C(\overline{\Omega})$ if $\alpha \geqslant 1$ (see also [27]). Existence of more than one positive solution is an interesting open problem, see [95] for an interesting result in the radial case and [50] for a regular problem with non-Lipschitz nonlinearities. The linearized stability of the maximal positive solution is proved in [44], Theorem 2.3.

Now we consider the general case

$$\mathcal{L}u + M(x)u^p = \lambda K(x)u^q \quad \text{in } \Omega, \qquad u = 0 \quad \text{on } \partial\Omega, \tag{10.11}$$

where K is as in Theorem 10.1 and $M > 0$ in Ω is such that

(H.14) $|M(x)| \leqslant k_2 d(x)^{-\beta'}$, with $0 < \beta' < 1 + p$.

Let $\delta_2 = \min\{\delta_1, 1 + p - \beta'\}$ where δ_1 is as defined in Theorem 10.1.

Theorems 10.10 and 10.11 are extended in the following one.

THEOREM 10.13. *Let K be as in Theorem* 10.1 *and $M \geqslant 0$ in Ω satisfy assumption* (H.14). *Then*:

(i) *If $-1 < q < 1$, $p > q$, $M < k_4 K d(x)^{q-p}$ and $\lambda > 0$, then there is a positive solution to (10.11), which is such that $u \in C_0^{1,\delta}(\overline{\Omega})$, with $\delta < \delta_2$. If, in addition, either (a) $p \geqslant 1$ and $\lambda > 0$ or (b) $p < 1$ and λ is sufficiently large, then the solution is unique and the map $\lambda \to u(x, \lambda) \to C_0^{1,\delta}(\overline{\Omega})$ is C^∞ and strictly increasing.*

(ii) *If $-1 < p < q < 1$ and $Md(x)^{p-q} > k_3 K$ in Ω, then there is a positive constant $\underline{\lambda} > 0$ such that if $\lambda > \underline{\lambda}$ then (10.11) possesses a positive solution (which is such that $u \in C_0^{1,\delta}(\overline{\Omega})$, with $\delta < \delta_2$) and if $0 < \lambda < \underline{\lambda}$ then (10.11) has no solution.*

For the proof see [79].

REMARK 10.14. The result in part (ii) was proved by Díaz, Morel and Oswald [51] only for $K \in L^\infty(\Omega)$, $M \equiv 1$, $-1 < p < 0$ and $q = 0$ using sub- and supersolutions in a somewhat different way. On the other hand, uniqueness raises interesting and unexpected open problems. For instance, some results and examples provided in [44] and [27] for $M = 1$ suggest that uniqueness/multiplicity may depend very specifically on p, q (and maybe on M and K as well), making it difficult to guess general theorems in this direction.

10.3. *Second combination of power laws:* $\mathcal{L}u = \lambda K(x)u^q + M(x)u^p$

Again, we begin with the case $K(x) \equiv M(x) \equiv 1$, namely

$$\mathcal{L}u = \lambda u^q + u^p \quad \text{in } \Omega, \qquad u = 0 \quad \text{on } \partial\Omega, \tag{10.12}$$

where $-1 < q < 1$ and $-1 < p \neq 1$. We consider first $-1 < p < 1$.

THEOREM 10.15. *If $-1 < q < 1$ and $-1 < p < 1$, then for any $\lambda > 0$ there exists a unique positive solution to (10.12). Moreover, u is asymptotically stable, $u \in C_0^{1,\delta}(\overline{\Omega})$ with $0 < \delta < \min\{\delta_0, 1+p, 1+q\}$, and the mapping $\lambda \to u(x, \lambda)$ from $]0, \infty[$ into $C_0^{1,\delta}(\overline{\Omega})$ is C^∞ and strictly increasing.*

PROOF. If $u_2 = Az$, where z is the unique solution of (10.5), then we have

$$\mathcal{L}u_2 - \lambda u_2^q - u_2^p = \left[\lambda(A^{1-q} - 1) - A^{p-q}z^{p-q}\right]A^q z^q, \tag{10.13}$$

which shows that if $A > 0$ is sufficiently small, then u_2 is a subsolution of (10.12). We look for a supersolution of the form $u^2 \equiv z + \tilde{z} + \rho e$, with $\rho > 0$ and $\tilde{z} > 0$ and $e > 0$ in Ω are uniquely defined by

$$\mathcal{L}\tilde{z} = \tilde{z}^p, \ \mathcal{L}e = 1 \quad \text{in } \Omega, \qquad \tilde{z} = e = 0 \quad \text{on } \partial\Omega. \tag{10.14}$$

It follows that

$$\begin{aligned}
\mathcal{L}u^2 &- \left(u^2\right)^p - \lambda\left(u^2\right)^q \\
&= \lambda z^q + \tilde{z}^p + \rho - \lambda(z + \tilde{z} + \rho e)^q - (z + \tilde{z} + \rho e)^p \\
&> z^q + \tilde{z}^p + \lambda(z + \tilde{z} + \rho e)^{|q|} + (z + \tilde{z} + \rho e)^{|p|} \\
&\quad - \lambda(z + \tilde{z} + \rho e)^q - (z + \tilde{z} + \rho e)^p > 0
\end{aligned} \tag{10.15}$$

provided that $\rho > \lambda(z + \tilde{z} + \rho e)^{|q|} + (z + \tilde{z} + \rho e)^{|p|}$, which holds in Ω if $\rho \geqslant \lambda(\|z + \tilde{z}\|_\infty + \rho\|e\|_\infty)^{|q|} + (\|z + \tilde{z}\|_\infty + \rho\|e\|_\infty)^{|p|}$. Since $0 \leqslant |p| < 1$ and $0 \leqslant |q| < 1$, this latter inequality holds (and thus u^2 is a supersolution) for $\rho > 0$ large enough, when we also have $u_2 < u^2$ in Ω. Invoking Theorem 4.1 we get existence. Uniqueness and asymptotic stability follow applying Theorem 8.1 and Lemma 8.5, noting that $f_\lambda(u) \equiv \lambda u^q + u^p$ is such that $H_\lambda(u) \equiv \lambda(1 - q)u^q + (1 - p)u^p > 0$ (for all $u > 0$ and all $x \in \Omega$) satisfies assumption (H.9). The remaining of the statement follows applying Theorem 8.1, Corollary 7.2 and using the argument at the end of the proof of Theorem 10.10. Thus, the proof is complete. $\qquad\qquad\square$

REMARK 10.16. The result of Theorem 10.15 was proved in [35] for the particular case $\mathcal{L} = -\Delta$ using regularization and sub- and supersolutions. See also Theorem 1.1 in [105] and [30,60,61,111,116].

For $p > 1$ we do not have positive solutions for all $\lambda > 0$ (the case $p = 1$ is treated at the end of this section). More precisely we have the following theorem.

THEOREM 10.17. *If $-1 < q < 1$ and $p > 1$, then there is a constant $\bar{\lambda} > 0$ such that for $0 < \lambda < \bar{\lambda}$ there is a positive solution to (10.12), and if $\lambda > \bar{\lambda}$, then (10.12) has no solution.*

PROOF. Let us first see that (10.12) *has a solution if $\lambda > 0$ is sufficiently small*. As in the proof of Theorem 10.15, (10.13) shows that $u_2 = Az$ is a subsolution of (10.12) if $A > 0$ is sufficiently small. And the result follows applying Theorem 4.1 and noting that if $A > 1$ is fixed and $\lambda > 0$ is sufficiently small then u_2 is also a supersolution, which invoking (10.13) and the expression $z = \lambda^{1/(1-q)}z_1$, with z_1 independent of λ (Corollary 10.2), is equivalent to showing that $\lambda(A^{1-q} - 1) \geq A^{p-q}\|z(\lambda)\|_\infty^{p-q} = A^{p-q}\|z_1\|_\infty^{p-q}\lambda^{(p-q)/(1-q)}$. Since $(p - q)/(1 - q) > 1$, this latter inequality holds provided that $A > 1$ be fixed and $\lambda > 0$ sufficiently small, which completes the proof of the statement above.

We now show that *the set $S = \{\lambda \in]0, +\infty[: (10.12)$ possesses a positive solution$\}$ is an interval*. This statement follows using the argument in the proof of Theorem 10.11, recalling that $u_2 = Az$ is a subsolution if $A > 0$ is sufficiently small, and noting that if $\tilde{\lambda} > \hat{\lambda} > 0$ and (10.12) possesses a solution, $\tilde{u} > 0$, for $\lambda = \tilde{\lambda}$, then $\tilde{u}$ is a supersolution to (10.12) at $\lambda = \hat{\lambda}$.

And as in the proof of Theorem 10.11, it only remains to show that (10.12) *has no solution if λ is sufficiently large*. To this end, we note that if $p > 1$, $q < 1$ and $\lambda_1 > 0$, then there is a constant λ^* such that

$$\lambda_1 t - t^p < \lambda^* t^q \quad \text{for all } t > 0, \tag{10.16}$$

which is equivalent to showing that the function g defined as $g(t) = \lambda^* t^{q-1} + t^{p-1} - \lambda_1$ is strictly positive for all $t > 0$, which follows noting that g is convex and attains its minimum at $t_m = [(1 - q)\lambda^*/(p - 1)]^{1/(p-q)}$, which is such that $t_m \to \infty$ as $\lambda^* \to \infty$. Thus $g(t) \geq g(t_m) = \lambda^* t_m^{q-1} + t_m^{p-1} - \lambda_1 > 0$ if λ^* is large enough and (10.16) follows. Now, let $\lambda_1 > 0$ be the principal eigenvalue of (10.3), let $\varphi_1 > 0$ be an associated eigenfunction and let λ^* be such that (10.16) holds. We have

$$\mathcal{L}\varphi_1 - \lambda\varphi_1^q - \varphi_1^p = \lambda_1\varphi_1 - \lambda\varphi_1^q - \varphi_1^p < (\lambda^* - \lambda)\varphi_1^q < 0 \quad \text{in } \Omega \tag{10.17}$$

if $\lambda > \lambda^*$, which means that φ_1 is a subsolution of (10.12). Since φ_1 is defined up to a constant, $c\varphi_1$ is a subsolution for all $c > 0$, which invoking the strong maximum principle implies nonexistence of positive solutions, and completes the proof. $\qquad\square$

REMARK 10.18. The result in Theorem 10.17 was obtained in [35] for a related problem with $\mathcal{L} = -\Delta$ and a complicated approximation argument involving also sub and supersolutions. See also [6,110]. Existence of a weak solution (in $H_0^1(\Omega)$) was proved in [117] for $\mathcal{L} = -\Delta$. For multiplicity results, see Section 11 and the references therein.

Now we consider the general case

$$\mathcal{L}u = \lambda K(x)u^q + M(x)u^p \quad \text{in } \Omega, \qquad u = 0 \quad \text{on } \partial\Omega. \tag{10.18}$$

Theorems 10.15 and 10.17 are extended in the following one.

THEOREM 10.19. *Let K be as in Theorem 10.1 and let M satisfy assumption* (H.14) *and be such that $Md(x)^{p-q} < k_4 K$ in Ω. Then:*
  (i) *If $-1 < q < 1$, $-1 < p < 1$, $Md(x)^{|p|} < k_5 K$ and $\lambda > 0$, then there is a unique positive solution to* (10.18), *which is asymptotically stable and such that $u \in C_0^{1,\delta}(\overline{\Omega})$ for all λ, with $\delta < \delta_2$. Moreover, the map $\lambda \to u(x,\lambda)$, from $]0,\infty[$ to $C_0^{1,\delta}(\overline{\Omega})$, is C^∞ and strictly increasing.*
  (ii) *If $-1 < q < 1$, $p > 1$ and $M < k_6 K$ in Ω, then there is a positive constant $\bar\lambda > 0$ such that if $0 < \lambda < \bar\lambda$ then* (10.18) *possesses a positive solution (which is such that $u \in C_0^{1,\delta}(\overline{\Omega})$, with $\delta < \delta_2$), and if $\lambda > \bar\lambda$ then* (10.18) *has no solution.*

For the proof see [79].

Now we consider the case $q = 1$ in the general equation (10.18). Since $p \neq 1$, the parameter can be eliminated upon the change of variable $u = \lambda^{1/(p-1)}v$. Thus, we consider the more general case $q = 1$, namely

$$\mathcal{L}u = \lambda K(x)u + M(x)u^p \quad \text{in } \Omega, \qquad u = 0 \quad \text{on } \partial\Omega, \tag{10.19}$$

with $-1 < p < 1$. For this case, we have the following theorem.

THEOREM 10.20. *Let K be as in Theorem 10.1, let M satisfy assumption* (H.14) *and let $\lambda_0 > 0$ be the principal eigenvalue of* (10.2). *Then we have:*
  (i) *If $0 < \lambda < \lambda_0$ and $k_7 M < Kd(x)^{1-p} < k_7^{-1}M$ in Ω, then there is a unique positive solution to* (10.19), *which is asymptotically stable and such that $u \in C_0^{1,\delta}(\overline{\Omega})$ for all λ, with $\delta < \delta_2$. Moreover, the map $\lambda \to u(x,\lambda)$, from $]0,\infty[$ to $C_0^{1,\delta}(\overline{\Omega})$, is C^∞ and strictly increasing.*
  (ii) *If $\lambda > \lambda_0$ then* (10.19) *has no solution.*

For the proof see [79].

10.4. *Third combination of power laws:* $\mathcal{L}u + M(x)u^p = \lambda K(x)u^q$,
   with M changing sign

Now we consider the problem

$$\mathcal{L}u + M(x)u^p = \lambda K(x)u^q \quad \text{in } \Omega, \qquad u = 0 \quad \text{on } \partial\Omega. \tag{10.20}$$

The solutions of the problems encountered in the last two sections can be used as sub- and supersolutions to prove the following theorem.

THEOREM 10.21. *Let K be as in Theorem 10.1 and let M satisfy (H.14) and possibly change sign in Ω. Then we have:*
   (i) *If $-1 < q < p < 1$, $|M|d(x)^{|p|} < k_5 K$ and $|M|d(x)^{p-q} < k_4 K$ in Ω, then (10.20) has at least a solution for all $\lambda > 0$.*
   (ii) *If $-1 < p < q < 1$, $|M|d(x)^{|p|} < k_5 K$ and $k_3 K < |M|d(x)^{p-q}$ in Ω, then there is a constant $\underline{\lambda}$ such that (10.20) has at least a solution for all $\lambda > \underline{\lambda}$.*
   (iii) *If $-1 < q < 1$, $p > 1$ and $|M|d(x)^{p-q} < k_4 K$ in Ω, then there is a constant $\bar{\lambda}$ such that (10.20) has at least a solution whenever $0 < \lambda < \bar{\lambda}$.*

For the proof see [79].

REMARK 10.22. As pointed out in the Introduction, interesting examples of systems arising in applications are more difficult to find in the singular case. (See, however, the papers [26,28,29,59,100].) By using the existence results for linear equations (Proposition 2.3), the maximum principle in Appendix B and the compactness of the Green operator provided by estimate (2.8), a general existence theorem was proved for one equation (Theorem 4.1). Under natural assumptions a completely similar result can be obtained for cooperative systems (see [5,96,102,107]) for the regular case and the noncooperative case can be treated reasoning as in [75].

11. Variational methods: some multiplicity results

Variational methods have been used widely in order to find solutions to large classes of nonlinear elliptic problems (see the book [109] and the references therein). In particular, many results for semilinear elliptic boundary value problems have been obtained by studying the Euler's equation associated to a functional defined on a Banach (or even Hilbert) space and possessing some suitable properties.

Singular problems raise some additional difficulties from this point of view, a fact which is reflected sometimes quite vividly in the literature. However, it is still possible to apply this approach here, as can be seen in the references listed below.

The singularity poses difficulties also when dealing with multiplicity problems. Once a first positive solution of some problems has been obtained, and may be characterized as a local minimum of the associated functional, a more sophisticated variational argument (as, e.g., the mountain pass theorem) is used to get a second positive solution. This method has been used in the paper [6] for a nonsingular problem involving the sum of a concave $(u^q, 0 < q < 1)$ and a convex $(u^p, 1 < p \leqslant (N+2)/(N-2))$ nonlinearities. This again uses a result due to Brezis and Nirenberg [19] showing that for a certain class of problems which does not include the singular case, local minima in the sense of C^1 and H^1 are equivalent. Several alternative ways have been used, often involving Ekeland's principle (see [109]).

Here we begin by considering again the model problem (1.5), (1.6), from the variational point of view. Hence we limit ourselves to operators in divergence form; in fact, to simplify matters, to $\mathcal{L} = -\Delta$. Thus we deal with the problem (1.5), (1.6) where $0 < \alpha < 1$.

The functional associated to (1.5), (1.6) is $\mathcal{I} : H_0^1(\Omega) \to \mathbb{R}$, where

$$\mathcal{I}(u) = \frac{1}{2} \int_\Omega |\nabla u|^2 \, dx - \lambda \int_\Omega F(u) \, dx \tag{11.1}$$

with $F(u) = \int_0^u s^{-\alpha} \, ds = u^{1-\alpha}/(1-\alpha)$. We denote $f(s) = s^{-\alpha}$.

That $\mathcal{I}$ is well defined for any $u \in H_0^1(\Omega)$ is easy to see by using Hölder's inequality and the Sobolev embedding to get

$$\mathcal{I}(u) \geqslant \frac{1}{2} \|u\|_{H_0^1(\Omega)}^2 - \lambda c \|u\|_{H_0^1(\Omega)}^{1-\alpha} \tag{11.2}$$

for some constant $c > 0$, which actually implies that $\mathcal{I}$ is coercive.

Moreover, for any $\varphi, \psi \in H_0^1(\Omega) \cap L_\infty(\Omega)$, the interval $M = [\varphi, \psi] \subset H_0^1(\Omega)$ is closed and convex (thus weakly closed) and $\mathcal{I}$ is sequentially lower semicontinuous on M. Indeed if $u_n \to u$ weakly in $H_0^1(\Omega)$ with $u_n, u \in M$, then passing to a subsequence we can assume that $u_n \to u$ almost everywhere and since $\int_\Omega |F(u_n)| \leqslant C$, we can use Lebesgue's theorem on dominated convergence to get $\int_\Omega |F(u_n)| \to \int_\Omega |F(u)|$, which gives the result.

Now we can provide the version of the method of sub- and supersolutions in this variational framework. For the problem (1.3), (1.4) where $f(x, u) \equiv f(u)$ and $f :]0, +\infty[\to \mathbb{R}$ is a Carathéodory function such that $F : [0, +\infty[\to \mathbb{R}$ is a Carathéodory function, we say that u_0 (resp., u^0) is a weak subsolution (resp., a supersolution) if $u_0, u^0 \in H_0^1(\Omega) \cap L_\infty(\Omega)$ $u_0, u^0 > 0$, and moreover,

$$\int_\Omega \left(\nabla u_0 \nabla \varphi - f(u_0) \varphi \right) dx \leqslant 0 \leqslant \int_\Omega \left(\nabla u^0 \nabla \varphi - f(u^0) \varphi \right) dx \tag{11.3}$$

for any $\varphi \in C_0^\infty(\Omega)$, $\varphi \geqslant 0$ in Ω and, moreover,

$$u_0 = u^0 = 0 \quad \text{on } \partial\Omega. \tag{11.4}$$

The definition of weak solution is similar.

We have the following theorem.

THEOREM 11.1. *Suppose that u_0 (resp., u^0) is a weak subsolution (resp., supersolution) such that $0 < u_0 \leqslant u^0$ on Ω. Then for any $\lambda > 0$ there exists a weak solution $u \in H_0^1(\Omega)$ to (1.3), (1.4) such that $u_0 \leqslant u \leqslant u^0$.*

PROOF. We have seen that $\mathcal{I}$ is coercive and sequentially weakly lower semicontinuous on the closed convex set $M = [u_0, u^0]$ of $H_0^1(\Omega)$. Then it follows from, e.g., [109], Theorem 1.2, that $\mathcal{I}$ attains a local minimum in M, $u \in M$.

Now we can show that u is a weak solution to (1.3), (1.4), the proof is a slight variant of [109], pp. 17–18, and [72], pp. 493–495.

For $\varepsilon > 0$, we define $v_\varepsilon = \min\{u^0, \max\{u_0, u + \varepsilon\varphi\}\}$ with $\varphi \in C_0^\infty(\Omega)$. Then $v_\varepsilon = u + \varepsilon\varphi - \varphi^\varepsilon + \varphi_\varepsilon \in M$, where

$$\varphi^\varepsilon = \max\{0, u + \varepsilon\varphi - u^0\}, \tag{11.5}$$

$$\varphi_\varepsilon = -\min\{0, u + \varepsilon\varphi - u_0\}, \tag{11.6}$$

and hence $\varphi^\varepsilon, \varphi_\varepsilon \in H_0^1(\Omega) \cap L_\infty(\Omega)$. We get

$$0 \leqslant \lim_{t \to 0} \frac{\mathcal{I}(u + t(v_\varepsilon - u)) - \mathcal{I}(u_0)}{t}$$

$$= \int_\Omega \nabla u \nabla (v_\varepsilon - u)\,dx - \lim_{t \to 0} \int_\Omega \left(u + \theta t (v_\varepsilon - u)\right)^{-\alpha}(v_\varepsilon - u)\,dx \tag{11.7}$$

with $0 < t < 1$, $0 < \theta < 1$. Since u_0 is a subsolution, $|v_\varepsilon - u| \in H_0^1(\Omega)$ and we have

$$\int_\Omega u_0^{-\alpha}|v_\varepsilon - u|\,dx < +\infty. \tag{11.8}$$

Moreover, $|(u + \theta t (v_\varepsilon - u))^{-\alpha}(v_\varepsilon - u)| \leqslant u_0^{-\alpha}|v_\varepsilon - u|$ and using again dominated convergence we get

$$\lim_{t \to 0} \int_\Omega \left(u + \theta t (v_\varepsilon - u)\right)^{-\alpha}(v_\varepsilon - u)\,dx = \int_\Omega u^{-\alpha}(v_\varepsilon - u)\,dx. \tag{11.9}$$

Since the limit in the right-hand side exists, the left-hand side limit exists and

$$0 \leqslant \int_\Omega \nabla u \nabla (v_\varepsilon - u)\,dx - \int_\Omega u^{-\alpha}(v_\varepsilon - u)\,dx, \tag{11.10}$$

which implies that, for any $\varphi \in C_0^\infty(\Omega)$, we have

$$\int_\Omega \left(\nabla u \nabla \varphi - \frac{\varphi}{u^\alpha}\right)dx \geqslant \frac{1}{\varepsilon}\left(E^\varepsilon - E_\varepsilon\right), \tag{11.11}$$

with

$$E^\varepsilon = \int_\Omega \left(\nabla u \nabla \varphi^\varepsilon - \frac{\varphi^\varepsilon}{u^\alpha}\right)dx \tag{11.12}$$

and

$$E_\varepsilon = \int_\Omega \left(\nabla u \nabla \varphi_\varepsilon - \frac{\varphi_\varepsilon}{u^\alpha} \right) dx. \tag{11.13}$$

Moreover, since u^0 is a supersolution, we get

$$\frac{1}{\varepsilon} E^\varepsilon = \frac{1}{\varepsilon}\left[\int_\Omega |\nabla(u-u^0)|^2 \, dx + \int_\Omega \nabla(u-u^0)\nabla\varphi \, dx \right]$$
$$+ \frac{1}{\varepsilon} \int_\Omega \left(\frac{1}{u^\alpha} - \frac{1}{(u^0)^\alpha} \right) \varphi^\varepsilon \, dx \tag{11.14}$$
$$\geqslant \int_{\Omega_\varepsilon} \nabla(u-u^0)\nabla\varphi \, dx - \int_\Omega \left| \frac{1}{(u^0)^\alpha} - \frac{1}{u^\alpha} \right| \varphi \, dx = \mathrm{o}(1), \tag{11.15}$$

where $\Omega_\varepsilon = \{ x \in \Omega : \ u(x) + \varepsilon\varphi(x) \geqslant u^0(x) > u(x) \}$. By absolute continuity of Lebesgue integral, $\lim_{\varepsilon \to 0} |\Omega_\varepsilon| = 0$ and analogously,

$$\frac{1}{\varepsilon} E_\varepsilon \leqslant \mathrm{o}(1). \tag{11.16}$$

Hence

$$\int_\Omega \left(\nabla u \nabla \varphi - \frac{\varphi}{u^\alpha} \right) dx \geqslant \mathrm{o}(1) \quad \text{if } \varepsilon \to 0. \tag{11.17}$$

Changing the sign of φ and using the density of $C_0^\infty(\Omega)$ in $H_0^1(\Omega)$ we get

$$\int_\Omega \left(\nabla u \nabla \varphi - \frac{\varphi}{u^\alpha} \right) dx = 0 \tag{11.18}$$

for any $\varphi \in H_0^1(\Omega)$ and u is indeed a weak solution to (1.3), (1.4). Moreover, u is obviously bounded. $\qquad\square$

COROLLARY 11.2. *Assume that u_0 (resp., u^0) is a subsolution (resp., a supersolution) to (1.3), (1.4) in the sense of Section 4. Then for any $\lambda > 0$, there exists a solution $u \in C^2(\Omega) \cap C_0^{1,\delta}(\overline{\Omega})$ to (1.3), (1.4), which is a minimum of $\mathcal{I}$ on $[u_0, u^0]$ (as a subset of $H_0^1(\Omega)$).*

PROOF. It is obvious that assumptions in Theorem 11.1 are satisfied and hence there is $u \in [u_0, u^0]$, $u \in H_0^1(\Omega)$, which is a weak solution to (1.3), (1.4). By the regularity results in Section 2 (Proposition 2.3), u also has the required regularity. $\qquad\square$

REMARK 11.3. It is clear that Theorem 11.1 (and Corollary 11.2) are valid, with the same proof, if we replace $\mathcal{L} = -\Delta$ by a uniformly elliptic operator in divergence form with sufficiently smooth coefficients. In the same vein, we may consider terms of the form $K(x)/u^\alpha$ where $K \geqslant 0$ is bounded, or even singular, in Theorem 11.1.

REMARK 11.4. A rather similar existence result was given by Lair and Shaker [87] for the problem

$$-\Delta u = \frac{K(x)}{u^\alpha} \quad \text{in } \Omega, \qquad u = 0 \quad \text{on } \partial\Omega, \tag{11.19}$$

where $K \geqslant 0$, $K \in L_2(\Omega)$, by using somewhat related arguments. (The problem is actually a little more general.) In the same direction, see the last paragraph of [117].

REMARK 11.5. The fact that both the sub- and the supersolution in Theorem 11.1 are bounded is important for carrying out the proof. Some interesting results in this direction in the same variational framework were obtained in [22] in the nonsingular case.

REMARK 11.6. Other multiplicity results have been obtained by Aranda and Lami-Dozo by using the Lyapunov–Schmidt method [9] and bifurcation at infinity [10].

The rest of the section will be devoted to what could be called the singular version of the problem considered in [6]. We consider the problem

$$-\Delta u = \frac{\lambda}{u^\alpha} + u^p \quad \text{in } \Omega, \qquad u = 0 \quad \text{on } \partial\Omega, \tag{11.20}$$

where Ω is as usual and

$$0 < \alpha < 1 < p \leqslant \frac{N+2}{N-2}. \tag{11.21}$$

If $-1 < \alpha < 0$ this is the nonsingular problem in [6].

Now the associated functional $\mathcal{J} : H_0^1(\Omega) \to \mathbb{R}$ can be written in a similar way as

$$\mathcal{J}(u) = \frac{1}{2} \int_\Omega |\nabla u|^2 \, \mathrm{d}x - \frac{\lambda}{1-\alpha} \int_\Omega |u|^{1-\alpha} \, \mathrm{d}x - \frac{1}{p+1} \int_\Omega |u|^{p+1} \, \mathrm{d}x \tag{11.22}$$

and using again Hölder and Sobolev inequalities we obtain

$$\mathcal{J}(u) \geqslant \frac{1}{2} \|u\|^2_{H_0^1(\Omega)} - C\lambda \|u\|^{1-\alpha}_{H_0^1(\Omega)} - C\|u\|^{p+1}_{H_0^1(\Omega)}. \tag{11.23}$$

Next we state only a multiplicity result for (11.20), namely Theorem 1 in [72]. Since the proof is rather involved from the technical point of view, we only sketch it slightly.

THEOREM 11.7. *Suppose that* (11.21) *is satisfied. Then there exists* $\lambda^* > 0$ *such that for any* $\lambda \in \,]0, \lambda^*[$ *there exists at least two weak solutions* $0 < u_\lambda < v_\lambda$ *of* (11.20), *where* u_λ *is a minimal solution. For* $\lambda = \lambda^*$ *there is at least a weak solution to* (11.20). *Finally, if* $\lambda > \lambda^*$, *there is no weak solution.*

PROOF (sketch). First, the existence of an (actually minimal) weak positive solution can be shown by using again variational arguments in the interval given by ordered sub- and supersolutions as in Theorem 11.1. Here the actual ordered sub- and supersolutions may be chosen as in the proof of Theorem 10.17 above or in any more or less equivalent way. Since the term u^p (with $p > 1$) does not raise now difficulties with respect to (1.5), this part of the proof is very similar to Theorem 11.1. As a byproduct we obtain that the minimal weak solution $u_\lambda > 0$ satisfies $\mathcal{J}(u_\lambda) < 0$, and that $u_\lambda > 0$ is nondecreasing as a function of λ (as in Theorem 10.10).

That there is a $\lambda^* > 0$ such that no weak positive solution exists for $\lambda > \lambda^*$ is proved as in Section 10. Now, for $\lambda = \lambda^*$ existence of a weak positive solution is proved by picking an increasing sequence $\lambda_n \to \lambda$ and showing from energy estimates that the corresponding sequence of u_{λ_n} of solutions is bounded in $H_0^1(\Omega)$. Thus, u_{λ_n} converges weakly to some u_{λ^*} and also pointwise a.e. Using, as usual, dominated convergence (since $u_{\lambda_n} \leqslant u_{\lambda^*}$) we prove that u_{λ^*} is a weak solution.

The existence of a second positive solution is more difficult to obtain in this context. Indeed, since Brezis–Nirenberg [19] result saying that there is always in the interval between ordered sub- and supersolutions a solution which is also a minimum of the associated functional (in the sense of $H_0^1(\Omega)$ and not only of $C_0^1(\overline{\Omega})$) is not available, it is necessary to prove by a more complicated direct method that this is actually the case. This is done in [72] by using an idea of Alama [4].

Now the existence of a second solution is proved by applying Ekeland's principle to the subset $A = \{u \in H_0^1(\Omega)\colon u \geqslant u_\lambda\}$, where u_λ is the above minimal solution. The concluding argument follows the outline of work by Badiale and Tarantello [11] for the critical case of a problem for discontinuous nonlinearities. $\qquad\square$

REMARK 11.8. Apparently, the first multiplicity result for equations (11.19) (with $K(x)$ instead of $K \equiv 1$) was obtained by Yijing et al. [117]. They consider the manifold

$$\Lambda = \left\{ u \in H_0^1(\Omega)\colon \int_\Omega |\nabla u|^2 \, dx - \int_\Omega K(x)|u|^{1-\alpha} \, dx - \lambda \int_\Omega |u|^{p+1} \, dx = 0 \right\} \tag{11.24}$$

which contains the set of weak solutions. Using the splitting of Λ given by

$$\Lambda^+ = \left\{ u \in \Lambda\colon (1+\alpha) \int_\Omega |\nabla u|^2 \, dx - \lambda(p+\alpha) \int_\Omega |u|^{p+1} \, dx > 0 \right\}, \tag{11.25}$$

$$\Lambda_0 = \left\{ u \in \Lambda\colon (1+\alpha) \int_\Omega |\nabla u|^2 \, dx - \lambda(p+\alpha) \int_\Omega |u|^{p+1} \, dx = 0 \right\}, \tag{11.26}$$

$$\Lambda^- = \left\{ u \in \Lambda\colon (1+\alpha) \int_\Omega |\nabla u|^2 \, dx - \lambda(p+\alpha) \int_\Omega |u|^{p+1} \, dx < 0 \right\}, \tag{11.27}$$

and several estimates inspired by Lair and Shaker [87], they are able to show that, for $\lambda > 0$ small enough,

(i) $\Lambda_0 = \{0\}$;

(ii) Λ^- is closed in $H_0^1(\Omega)$;

(iii) if $\|K\|_{L_2(\Omega)}$ is small, for any $u \in H_0^1(\Omega)$, $u \neq 0$, there exists a unique $t(u) > 0$ such that $t(u)u \in \Lambda^-$ (which is nonempty);

(iv) if $u \in \Lambda^-$ then $\mathcal{J}(u) < 0$.

Then an involved variational argument gives the existence of a weak solution $u_1 > 0$ such that $\mathcal{J}(u_1) < 0$. For the second solution they prove first that $\mathcal{J}$ is coercive on Λ^- and solving the minimization problem $\inf_{\Lambda^-} \mathcal{J}$ by using Ekeland's principle gives a weak solution u_2 with $\mathcal{J}(u_2) \geq 0$ and $0 < u_1 < u_2$. No additional regularity is obtained.

REMARK 11.9. The results in Remark 11.8 were improved in some ways in [81]. On one side, they extend the interval of λ's for which existence of two positive solutions is proved. Moreover, they allow $\alpha = 1$ if $p \in \,]1, (N + 2)/(N - 2)[$ and $p = (N + 2)/(N - 2)$ for $\alpha \in \,]0, 1[$. They also employ minimization arguments over different bounded manifolds where $\mathcal{J}$ is bounded below with suitable splittings, then they are able to prove similar multiplicity results. Only some partial regularity is obtained. A related multiplicity result is proved in [2] for a singular critical problem in domains of $\mathbb{R}^2$, allowing $0 < \alpha < 3$ (recall Theorem 6.3). Multiplicity has also been studied by variational methods for some problems concerning the p-Laplacian in [62].

REMARK 11.10. Finally, in a recent paper Zhang [122] gave a much shorter proof of the existence of two positive (weak) solutions. By using results by Chang [24,25] concerning critical point theory and in particular by exploiting the fact that the functional is Gâteaux-differentiable on some closed convex sets, and the properties of invariant sets for the descending flow associated to the functional, he provides this shorter proof and was also able to deal with the case of asymptotically linear nonlinearities. However, an additional condition was imposed on the coefficient $K \in L_2(\Omega)$, $K \geq 0$, namely that $K(x)/\varphi_1^\alpha$ is in $L_q(\Omega)$ with $N/2 < q < 2N/(N - 2)$, where φ_1 is as usual the first eigenfunction.

REMARK 11.11. A much more general setting was given by Canino and de Giovanni in [23]. The main point was to consider generalized solutions to

$$-\Delta u = \frac{1}{u^\alpha} + w \quad \text{in } \Omega, \qquad u = 0 \quad \text{on } \partial\Omega, \tag{11.28}$$

where $w \in L_{1,\mathrm{loc}}(\Omega) \cap H^{-1}(\Omega)$, $\alpha > 0$ only and the boundary condition has a "relaxed" meaning. Notice that any $\alpha > 0$ is allowed here and that Euler's functional is $+\infty$ if $\alpha > 3$ (see [88], Theorem 2). If $w \in C^\gamma(\overline{\Omega})$, the generalized solution is classical, but if $w \in H^{-1}(\Omega)$ only, then the minimum is given by a variational inequality.

12. Results for the radial and the one-dimensional problem

It is well known that phase plane methods, i.e., methods using the theory of ordinary differential equations, can be used to study autonomous problems on balls in which concerns existence of radial solutions. In particular, the case $N = 1$ can be treated in this way. We do not intend to pursue this way here, but only to deal with an interesting example.

In this section we consider the results of [76] about the existence and the exact number of radial solutions for the semilinear boundary value problem

$$-\Delta u = f(u) \quad \text{in } B(0, R), \qquad u = 0 \quad \text{on } \partial B(0, R), \tag{12.1}$$

where $B(0, R)$ is the ball centered at the origin with radius R and $f : \,]0, +\infty[\,\to\,]0, +\infty[$ is locally Lipschitz and such that

$$f(u) \geqslant m > 0 \quad \text{for all } u > 0, \tag{12.2}$$

there exists

$$\lim_{u \to +\infty} \frac{f(u)}{u^p} \in \,]0, +\infty[\quad \text{for some } 1 < p < 2^*, \tag{12.3}$$

where $2^* = (N + 2)/(N - 2)$ if $N \geqslant 3$ and $2^* = \infty$ if $N \leqslant 2$, and there exists

$$\lim_{u \to 0^+} f(u)u^\alpha \in [0, +\infty[\quad \text{for some } 0 < \alpha < 1. \tag{12.4}$$

It is easy to see that the nonlinearity $f(u) = \lambda u^{-\alpha} + u^p$ where

$$0 < \alpha < 1 < p < 2^* \tag{12.5}$$

satisfies (12.2)–(12.4).

We can prove the following existence results for radial solutions.

THEOREM 12.1. *Assume that f satisfies (12.2)–(12.4). Then there exists R^* such that*:
   (i) *If $R < R^*$ there are at least two positive radial solutions to (12.1).*
   (ii) *If $R = R^*$ there is at least one positive radial solution to (12.1).*
   (iii) *If $R > R^*$ there is no positive radial solution to (12.1).*

In order to get exact multiplicity results for $N = 1$ some additional convexity condition seems to be necessary. Here we also assume

$$f \text{ is strictly convex, i.e., } f'' > 0 \text{ in }]0, +\infty[. \tag{12.6}$$

Then we have a sharp multiplicity result.

THEOREM 12.2. *Assume that f satisfies (12.2)–(12.4) and (12.6). Then the results in Theorem 12.1 are sharp if $N = 1$.*

The main tool for obtaining these results is the *time map*.

Let $u(\cdot, c)$ the solution to the initial value problem

$$\left(r^{N-1}u'\right)' + r^{N-1}f(u) = 0, \quad u(0, c) = c, \ u'(0, c) = 0. \tag{12.7}$$

The time map associated to (12.7) is defined as

$$T(c) = \sup\{r > 0 : u(s, c) > 0 \text{ for all } s \in [0, r[\,\} \tag{12.8}$$

and

$$D(T) = \left\{c > 0 : \text{ there exists } R_0 > 0, \ \lim_{r \to R_0} u(r, c) = 0\right\}. \tag{12.9}$$

PROPOSITION 12.3. *Let us assume that*

$$\lim_{u \to +\infty} \frac{f(u)}{u} = +\infty \tag{12.10}$$

and that the solution of (12.7) is decreasing for any $c > 0$. Then for all $\varepsilon > 0$ there exists $c_\varepsilon > 0$ such that, for the solutions of (12.7),

$$u(\varepsilon, c) \leqslant c_\varepsilon \quad \text{for all } c \geqslant c_\varepsilon. \tag{12.11}$$

PROOF. Let ε be an arbitrary positive number. There exists $K_\varepsilon > 0$ such that, for the solution of the linear initial value problem

$$\left(r^{N-1}w'\right)' + r^{N-1}K_\varepsilon w = 0, \quad w(0) > 0, \ w'(0) = 0, \tag{12.12}$$

we have $w(\varepsilon) = 0$ and $w(r) > 0$ in $[0, \varepsilon[$. Since f satisfies (12.10), there exists $c_\varepsilon > 0$ such that

$$f(u) \geqslant K_\varepsilon u \quad \text{for all } u \geqslant c_\varepsilon. \tag{12.13}$$

Assume on the contrary that there is $c \geqslant c_\varepsilon$, for which $u(\varepsilon, c) > c_\varepsilon$. In what follows, we will denote $u(r, c)$ by $u(r)$. Since u is decreasing, $u(r) > c_\varepsilon$ for all $r \in [0, \varepsilon]$. Let us denote by w the solution of (12.12) satisfying the initial condition $w(0) = c$. Since from (12.7), (12.12) and (12.13), it follows that

$$w''(0) > u''(0), \tag{12.14}$$

then for small r we have $w(r) > u(r)$. However, $w(\varepsilon) = 0$ and $u(\varepsilon) > c_\varepsilon$ imply that there exists $r_1 \in]0, \varepsilon[$ such that

$$c_\varepsilon < u(r_1) = w(r_1), \qquad u'(r_1) > w'(r_1), \qquad u(r) < w(r) \quad \text{for all } r \in]0, r_1[.$$

$$\tag{12.15}$$

Let A be the function defined by

$$A(r) = r^{N-1}\big(u'(r)w(r) - u(r)w'(r)\big). \tag{12.16}$$

Then using (12.7), (12.12) and (12.13) we obtain

$$A'(r) = r^{N-1}w(r)\big(K_\varepsilon u(r) - f\big(u(r)\big)\big) < 0. \tag{12.17}$$

From (12.15) and (12.16) we get $A(r_1) > 0$; further, from $A(0) = 0$ and (12.16), there follows $A(r_1) < 0$, which is a contradiction. $\qquad\square$

PROPOSITION 12.4. *Let $N \geqslant 2$ and f satisfy (12.2)–(12.4). Then there exist $u_1 > 0$ and $K > 0$ such that*

$$2NF(u) - (N-2)uf(u) \geqslant Ku^{p+1} \quad \text{for all } u > u_1 \tag{12.18}$$

and there exists $a > 0$ such that

$$2NF(u) - (N-2)uf(u) \geqslant -a \quad \text{for all } u \geqslant 0, \tag{12.19}$$

where $F(u) = \int_0^u f(t)\,dt$.

PROOF. According to the assumptions on f there exist $L > 0$ and $p \in\,]1, 2^*[$ such that $L = \lim_{u\to+\infty} f(u)/u^p$. Let

$$\delta \in\, \Big]0, L\min\Big\{1, \frac{N+2-p(N-2)}{5N-2+p(N-2)}\Big\}\Big[. \tag{12.20}$$

Then there exists $u_0 > 0$ such that

$$\frac{f(u)}{u^p} \in\,]L-\delta, L+\delta[\quad \text{for all } u > u_0. \tag{12.21}$$

Let $u_1 = (L/\delta - 1)^{1/(p+1)}u_0$, then for $u > u_1$ we obtain

$$F(u) \geqslant \int_{u_0}^u f(t)\,dt \geqslant (L-\delta)\int_{u_0}^u t^p\,dt \geqslant (L-2\delta)\frac{u^{p+1}}{p+1} \tag{12.22}$$

and

$$2NF(u) - (N-2)uf(u)$$
$$\geqslant \frac{2N}{p+1}(L-2\delta)u^{p+1} - (N-2)(L+\delta)u^{p+1} \tag{12.23}$$
$$= \Big(\Big(\frac{2N}{p+1} - (N-2)\Big)L - \delta\Big(\frac{4N}{p+1} + N - 2\Big)\Big)u^{p+1}, \tag{12.24}$$

from which, and taking into account (12.20) the first statement follows. The second one follows from the first one and the continuity of the function $2NF(u) - (N-2)uf(u)$ on $[0, +\infty[$. $\qquad\square$

PROPOSITION 12.5. *Let us assume that f satisfies (12.2)–(12.4). For a given $c > 0$ let us define r_c as a solution of the equation $u(r_c, c) = c/2$.*
   Then there exist $M > 0$ and $u_2 > 0$ such that

$$\frac{N}{M}c^{1-p} \leqslant r_c^2 \quad \text{for all } c > 2u_2. \tag{12.25}$$

PROOF. Assumption (12.3) implies that there exist $M > 0$ and $u_2 > 0$ such that

$$f(u) \leqslant Mu^p \quad \text{for all } u > u_2. \tag{12.26}$$

From (12.7) and $u'(r) < 0$ we get

$$-r^{N-1}u'(r) \leqslant \int_0^r s^{N-1} Mu^p(s)\,\mathrm{d}s \leqslant Mc^p\frac{r^N}{N} \tag{12.27}$$

if $u(t) \geqslant u_2$. Dividing this inequality by r^{N-1} and integrating on $[0, t]$ one obtains

$$c - u(t) \leqslant \frac{M}{2N}c^p t^2 \tag{12.28}$$

if $u(t) \geqslant u_2$. If $c > 2u_2$, then for $t = r_c$ we get the desired inequality (12.25). $\qquad\square$

LEMMA 12.6. *Let f satisfy conditions (12.2)–(12.4). Then $\lim_{c\to+\infty} T(c) = 0$.*

PROOF. For $N = 1$ the lemma was proved in [83]. Now we assume $N \geqslant 2$.

STEP 1. Let $\varepsilon > 0$ be an arbitrary positive number. Let us choose c_ε according to Proposition 12.3. We will prove that there exists $d_\varepsilon > c_\varepsilon$ such that

$$\left|u'(r, c)\right| > \frac{c_\varepsilon}{\varepsilon} \quad \text{for all } r \in]\varepsilon, 2\varepsilon[, \ c > d_\varepsilon. \tag{12.29}$$

This inequality implies $T(c) < 2\varepsilon$ for all $c > d_\varepsilon$, since $u(\varepsilon, c) \leqslant c_\varepsilon$ and $u(r, c)$ decreases at least by c_ε on the interval $]\varepsilon, 2\varepsilon[$.

STEP 2. To prove (12.29) we will apply the following general form of the Pohozhaev identity.

$$r^N u'^2(r) + 2r^N F\big(u(r)\big)$$

$$= \int_0^r s^{N-1}\big(2NF\big(u(s)\big) - (N-2)u(s)f\big(u(s)\big)\big)\,\mathrm{d}s$$

$$- (N-2)r^{N-1}u(r)u'(r). \tag{12.30}$$

According to (12.7) and since $u'(r) < 0$,

$$u'^2(r) > r^{-N} \int_0^r s^{N-1}\left(2NF\big(u(s)\big) - (N-2)u(s)f\big(u(s)\big)\right) ds - 2F\big(u(r)\big).$$

(12.31)

Let us choose K and u_1 according to Proposition 12.4, and M and u_2 according to Proposition 12.5. Let

$$\bar{c} = \max\{2c_\varepsilon, 2u_1, 2u_2\}.$$

(12.32)

Let $c > \bar{c}$ and let r_c defined as in Proposition 12.4. From (12.31) we obtain

$$u'^2(r) > r^{-N} \int_0^{r_c} s^{N-1}\left(2NF\big(u(s)\big) - (N-2)u(s)f\big(u(s)\big)\right) ds$$

(12.33)

$$+ r^{-N} \int_{r_c}^r s^{N-1}\left(2NF\big(u(s)\big) - (N-2)u(s)f\big(u(s)\big)\right) ds$$

$$- 2F\big(u(r)\big).$$

(12.34)

STEP 3. Now, we will estimate each term in (12.34). For the first integral in (12.34), since $s < r_c$, we have

$$u(s) > u(r_c) = \frac{c}{2} > u_1.$$

(12.35)

From Proposition 12.4, we get

$$2NF\big(u(s)\big) - (N-2)u(s)f\big(u(s)\big) > Ku^{p+1}(s) > K\left(\frac{c}{2}\right)^{p+1} \quad \text{for all } s < r_c,$$

(12.36)

yielding for $r \in]\varepsilon, 2\varepsilon[$,

$$r^{-N} \int_0^{r_c} s^{N-1}\left(2NF\big(u(s)\big) - (N-2)u(s)f\big(u(s)\big)\right) ds \geq (2\varepsilon)^{-N} K\left(\frac{c}{2}\right)^{p+1} \frac{r_c^N}{N}.$$

(12.37)

Using $c \geq \bar{c} \geq 2u_2$ Proposition 12.5 implies

$$r^{-N} \int_0^{r_c} s^{N-1}\left(2NF\big(u(s)\big) - (N-2)u(s)f\big(u(s)\big)\right) ds$$

$$\geq (2\varepsilon)^{-N} \left(\frac{c}{2}\right)^{p+1} \frac{K}{N} \left(\frac{N}{M}\right)^{N/2} c^{(1-p)N/2}.$$

(12.38)

For the second term in (12.34), we use (12.19)

$$r^{-N}\int_0^{r_c} s^{N-1}\big(2NF\big(u(s)\big)-(N-2)u(s)f\big(u(s)\big)\big)\,ds \geqslant -\frac{a}{r^N}\frac{r^N}{N}=-\frac{a}{N}.$$

(12.39)

For the third term in (12.34), $r\in\,]\varepsilon,2\varepsilon[$ and $c>c_\varepsilon$. Hence

$$F\big(u(r)\big)\leqslant F(c_\varepsilon).$$

(12.40)

STEP 4. Now let us substitute the estimates (12.37), (12.38) and (12.39) into (12.34). Then, for all $r\in\,]\varepsilon,2\varepsilon[$ and $c>\bar c$,

$$u'^2(r,c)>(2\varepsilon)^{-N}\left(\frac{K}{N2^{p+1}}\right)\left(\frac{N}{M}\right)^{N/2}c^{1+p+(1-p)N/2}-\frac{a}{N}-2F(c_\varepsilon).$$

(12.41)

Since $p<2^*$, it is easily seen that (12.41) goes to infinity as c goes to infinity.

Now we are going to analyze the continuity of the time map. For any $\eta>0$ let us introduce the function T_η as follows

$$T_\eta(c)=\min\{r\in\,]0,T(c)[\,:\,u(r,c)=\eta\},$$

(12.42)

where $D(T_\eta)=\,]\eta,+\infty[$, and T_η is continuous. $\square$

LEMMA 12.7. *Let f satisfy (12.2)–(12.4). Then T_η tends uniformly to T on any compact subinterval of $]0,+\infty[$ as η goes to zero.*

PROOF. Let $[a,b]\subset\,]0,+\infty[$ be a compact subinterval. Let $\eta_0<a$, then for all $\eta<\eta_0$, $[a,b]\subset D(T_\eta)$. Let

$$R_0=\min_{[a,b]}T_{\eta_0},$$

(12.43)

then (since $\eta<\eta_0$ implies $T_\eta>T_{\eta_0}$)

$$T_\eta(c)\geqslant R_0\quad\text{for all }c\in[a,b]\text{ and }\eta\in\,]0,\eta_0].$$

(12.44)

Let $c\in[a,b]$, $\eta\in\,]0,\eta_0]$ and $r\in\,]T_\eta(c),T(c)]$. Then using (12.2) and $u'(r)=-r^{1-N}\times\int_0^r s^{N-1}f(u(s))\,ds$,

$$-u'(r,c)\geqslant\frac{m}{N}r\geqslant\frac{m}{N}T_\eta(c)\geqslant\frac{m}{N}R_0,$$

(12.45)

integrating this inequality on $[T_\eta(c), T(c)]$,

$$\eta \geqslant \frac{m}{N} R_0\big(T(c) - T_\eta(c)\big),$$ (12.46)

thus

$$\big|T(c) - T_\eta(c)\big| \leqslant \frac{N\eta}{m R_0}$$ (12.47)

from which the statement in the lemma follows. $\qquad\square$

Now Theorem 12.2 follows from the following properties of the time map:
(1) $D(T) =]0, +\infty[$,
(2) $\lim_{c \to 0} T(c) = \lim_{c \to +\infty} T(c) = 0$,
(3) T is continuous.
When $N = 1$ then Theorem 12.2 is proved and the crucial part of the proof is the property (12.48). Let h be the function defined by $h(r, c) = \partial_c u(r, c)$ where $u(\cdot, c)$ is defined by the initial value problem (12.7). Then

$$\text{for all } c > 0, \text{ the function } h(\cdot, c) \text{ has at most one zero in } \big[0, T(c)\big]. \quad (12.48)$$

The assumption (12.48) is satisfied. Notice that h and v satisfies the same linear equation, from $v(r, c) = -1/(r^{N-1}) \int_0^r s^{N-1} f(v(s, c)) \, ds$ then $v < 0$ in $]0, T(c)[$. Hence Sturm comparison theorem shows that h may have at most one zero in $[0, T(c)]$.

The remaining arguments needed to finish the proof of Theorem 12.2 are not included here since are very technical. Nevertheless, we give a brief idea about them.

First it is proved that the approximate time-map T_η has a unique maximum at some c_0, it is strictly increasing before c_0 and strictly decreasing after c_0. This is done in Lemma 4 and Corollary 3 in [76].

Then it is proved that the time-map T_η cannot be constant on any subinterval of $]0, +\infty[$. This is done in [76], Propositions 4–6 and Lemma 5.

Finally, passing to the limit and using the nondegeneracy of the time-map it is derived that the time-map inherits the properties of the approximated time-map. [76], Corollaries 3 and 4.

REMARK 12.8. Exact multiplicity results for the radial case for a different example were obtained in [95] where a curve of positive solutions "stops" at some point in a solution with zero normal derivative. This result could be related with the work in [50] for a problem with non-Lipschitz continuous nonlinearities.

13. Free boundary solutions: existence and properties

Positive solutions are the more interesting ones for many problems arising in applications and most of the work in the area has been devoted to them. But nonnegative solutions which can annihilate on some subdomains of positive measure may also arise in some situations. These solutions become zero on a subset which is often called *dead core* and has a (more or less regular) boundary which is then called a *free boundary*. This may happen when the nonlinear term f arising in (1.1)–(1.3) is not locally Lipschitz at the origin or, in a somewhat equivalent way, if the usual linear diffusion is replaced by a (slow) nonlinear diffusion $-\Delta u^m$ with $m > 1$. Two typical situations are chemical reactions of order $0 \leqslant p < 1$ and nonlinear diffusion models in population dynamics [70,71,94]. Here the dead core is the subdomain where, respectively, reaction stops (the concentration of reactant is zero) or the population dies. Most results for elliptic problems of this kind are collected in the book by Díaz [49] where many references may be found (see also [12]). Not only existence of these free boundary solutions is studied but also its regularity. Geometrical properties of the free boundary are interesting as well.

These problems have been studied in the singular case by Dávila and Montenegro in a series of papers [42–46]. They actually only consider a specific example but it is clear that it contains most of the interesting features of the general case. The problem is

$$-\Delta u = \chi[u > 0]\left(-\frac{1}{u^\alpha} + \lambda f(x, u)\right) \quad \text{in } \Omega, \qquad u = 0 \quad \text{on } \partial\Omega, \tag{13.1}$$

where Ω is a smooth bounded domain in $\mathbb{R}^N$, $0 < \alpha < 1$ and $f : \Omega \times [0, +\infty[\to [0, +\infty[$ is a smooth ($f_u(x, u)$ is continuous in $\Omega \times]0, +\infty[$) nondecreasing concave sublinear function, in the sense that

$$\lim_{u \to +\infty} \frac{f(x, u)}{u} = 0 \tag{13.2}$$

uniformly in Ω. A typical example is $f(x, u) = u^p$ with $0 < p < 1$. Equation (13.1) has the meaning that the right-hand side is zero for $u = 0$. ($\chi(A)$ denotes, as usual, the characteristic function of the set A.)

A generalized notion of weak solution is instrumental here; it is said that $u \in L_1(\Omega)$, $u \geqslant 0$ is a *weak solution* to (13.1) if

$$g_\lambda(x, u) = \chi[u > 0]\left(-\frac{1}{u^\alpha} + \lambda f(x, u)\right) d(x) \in L_1(\Omega) \tag{13.3}$$

and for any test function $\varphi \in C^2(\overline{\Omega})$ such that $\varphi = 0$ on $\partial\Omega$ we have

$$\int_\Omega u(-\Delta\varphi) = \int_{[u>0]}\left(-\frac{1}{u^\alpha} + \lambda f(x, u)\right)\varphi \, dx. \tag{13.4}$$

Concerning positive solutions of (13.1) it was shown is Section 11 that there exists $\underline{\lambda} > 0$ such that there is no (classical) solution if $0 < \lambda < \underline{\lambda}$ and at least a positive solution in $C^{1,\delta}(\overline{\Omega})$ for any $0 < \delta < \min\{\delta_0, 1 - \alpha\}$ if $f(x, u) = u^p$ (see Theorems 10.11–10.13). For previous work see Remarks 10.12–10.14 and the references therein.

An approximation method which is very similar to the first one used in [38] (see Section 3) allows to obtain existence of weak solutions. As usual its regularity is a more difficult problem due to the presence of the free boundary. Here is the main result in this direction.

THEOREM 13.1 (Theorem 2.1 in [44]). *Under the above assumptions, for any $\lambda > 0$ there exists a unique maximal weak solution u_λ to (13.1). Moreover, there exists $0 < \lambda^* < +\infty$ such that if $\lambda > \lambda^*$ this maximal solution is positive on Ω and $u_\lambda \in C(\overline{\Omega}) \cap C_{\mathrm{loc}}^{1,\beta}(\Omega)$ for any $0 < \beta < 1$. We also have $ad(x) \leqslant u_\lambda \leqslant bd(x)$ for some $a, b > 0$. Finally, if $f \in C^1(\overline{\Omega} \times [0, +\infty[)$, u_λ is a classical solution. If $0 < \lambda \leqslant \lambda^*$ the maximal weak solution u_λ is in $C(\overline{\Omega}) \cap C_{\mathrm{loc}}^{1,(1-\alpha)/(1+\alpha)}(\Omega)$, this regularity is optimal and the null set $\{u_\lambda = 0\}$ has positive measure.*

PROOF (Sketch). First the approximated problem, for $\lambda > 0$ fixed and $\varepsilon > 0$,

$$-\Delta u + \frac{u}{(u+\varepsilon)^{1+\alpha}} = \lambda f(x, u) \quad \text{in } \Omega, \quad u = 0 \quad \text{on } \partial\Omega \tag{13.5}$$

is considered and it is proved, by using sub- and supersolutions for this regular problem, that there is a unique maximal solution u_ε and that u_ε depends increasingly on $\varepsilon > 0$. This implies the existence of a pointwise limit $u = \lim_{\varepsilon \to 0^+} u_\varepsilon$. It is proved, by using dominated convergence and Fatou's lemma, that u is a maximal subsolution to the problem

$$-\Delta u + \chi[u > 0]\frac{1}{u^\alpha} = \lambda f(x, u) \quad \text{in } \Omega, \quad u = 0 \quad \text{on } \partial\Omega. \tag{13.6}$$

Now a rather involved and subtle work is carried out in order to show that $u \in C(\overline{\Omega})$ and also $u \in C^{1,\beta}$ for any $0 < \beta < 1$ restricted to $\{u > 0\}$. Even more, $u \in C_{\mathrm{loc}}^{1,(1-\alpha)/(1+\alpha)}(\Omega)$ and u is a weak solution to (13.1). Finally, the maximal solution u_ε to (13.3) converges uniformly in $\overline{\Omega}$ when ε goes to zero to the maximal solution u to (13.1).

It is possible to show that there is no positive weak solution of (13.1) for $\lambda > 0$ small. Then if we define $\lambda^* = \inf\{\lambda > 0: \text{ there exists a function } u > 0 \text{ a.e. solution of (13.1)}\}$, $0 < \lambda^* < +\infty$ and it is possible to prove that for any $\lambda \geqslant \lambda^*$ there is a positive a.e. weak solution. Moreover, for some $a, b > 0$, we have $ad(x) \leqslant u_\lambda \leqslant bd(x)$ for $\lambda > \lambda^*$; both results are obtained by using sub and supersolutions in the sense of weak solutions as in [17]. $\square$

REMARK 13.2. Weak solutions are actually in $H_0^1(\Omega)$, see also [51]; and not only $\chi[u > 0](d(x)/u^\alpha) \in L_1(\Omega)$ but also $\chi[u > 0](1/u^\alpha) \in L_1(\Omega)$.

REMARK 13.3. The weak solution u_{λ^*} for $\lambda = \lambda^*$ is positive a.e. in Ω. A sufficient condition for $u_{\lambda^*} > 0$ on Ω is given in Theorem 2.4 in [44], the conclusion is actually $u_{\lambda^*} \geqslant cd(x)^{2/(1+\alpha)}$. On the other hand, u_{λ^*} is unique in the class of positive a.e. weak solutions. If $\alpha \geqslant 1$ for any $\lambda > 0$ there is no weak solution positive a.e. in Ω. An example of $u_\lambda \neq 0$ for some $0 < \lambda < \lambda^*$ is given in [43].

REMARK 13.4. If $0 < \lambda < \lambda^*$ the maximal weak solution u_λ may be trivial. If Ω is an interval of $\mathbb{R}$ and $f(x, u) \equiv f(u)$ then the situation is sharp: for any $\lambda > 0$, either $u_\lambda \equiv 0$ or $u_\lambda > 0$ in Ω. An example of $u_\lambda \not\equiv 0$ for some $0 < \lambda < \lambda^*$ is given in [43]. Several examples exhibiting different behavior for the support of the maximal solution u_λ are given in [42].

REMARK 13.5. The question of the possible existence of positive solutions different of the maximal solution $u_\lambda > 0$ was considered by several authors: see again Remark 10.14. In principle, nothing prevents the existence for $\lambda > \lambda^*$ of free boundary solutions.

REMARK 13.6. Concerning the stability of the maximal weak solution u_λ the quantity

$$\Lambda(u) = \inf_{\psi \in C_0^\infty(\Omega)} \frac{\int_\Omega |\nabla \psi|^2 \, dx - \int_\Omega \frac{\partial g_\lambda}{\partial u} \psi^2 \, dx}{\int_\Omega \psi^2 \, dx} \tag{13.7}$$

is well defined for $u \in L_1(\Omega)$, $u > 0$ a.e. in Ω but in general, since $0 < \alpha < 1$, $\Lambda(u) \geqslant -\infty$. Then (Theorem 2.3 in [44]), $\Lambda(u_\lambda) > 0$ if $\lambda > \lambda^*$ and $\Lambda(u_{\lambda^*}) \geqslant 0$. Conversely, if $u > 0$ a.e. in Ω is a weak solution for $\lambda \geqslant \lambda^*$ and $\Lambda(u_\lambda) \geqslant 0$, then $u \equiv u_\lambda$; moreover, the stability of u_λ for $\lambda > \lambda^*$ implies the continuity of the branch u_λ of positive maximal solutions as a mapping into $L_1(\Omega)$. The corresponding result is more delicate to settle for $0 < \lambda < \lambda^*$ due to the presence of the free boundary. See also the observations in [16], p. 397 concerning stability.

REMARK 13.7. It is pointed out in [44] that some arguments used to study the regularity are similar to the ideas used by Phillips [97] in order to get the optimal regularity for minimizers of a different energy functional on a convex set.

A different free boundary problem, but this time on the boundary, was studied in [46]. The equation is now

$$-\Delta u + u = 0 \qquad \text{in } \Omega, \tag{13.8}$$

$$-\frac{\partial u}{\partial n} = \frac{1}{u^\alpha} - \lambda f(x, u) \quad \text{on } \partial\Omega \cap \{u > 0\}, \tag{13.9}$$

where Ω is as above, $0 < \alpha < 1$ and $f : \mathbb{R} \to [0, +\infty[$ is C^1, increasing and satisfies (13.2). Similar results are obtained by using either approximation or variational techniques. In the first case $1/u^\alpha$ is replaced by $u/(u + \varepsilon)^{1+\alpha}$ ($\varepsilon > 0$) and maximal solutions $u_\varepsilon > 0$ (on $\overline{\Omega}$) for the approximated problem converge (uniformly on $\overline{\Omega}$) to a weak solution u (still defined by integration by parts with test functions) to (13.8), (13.9); moreover, $u \in C^{1/(1+\alpha)}(\overline{\Omega})$, and this regularity is optimal. The same regularity is obtained for minimizers (in the Sobolev space $H^1(\Omega)$) of the associated functional. The main results may be collected in the theorem:

THEOREM 13.8. *Under the above assumptions, for any $\lambda > 0$ there is a maximal nonnegative solution u_λ and the mapping $\lambda \to u_\lambda$ is nondecreasing. There exists $0 < \lambda^*$ such that,* *for any $0 < \lambda < \lambda^*$, u_λ is zero on a subset of $\partial\Omega$ with positive measure and for any $\lambda > \lambda^*$, $u_\lambda > 0$ on $\overline{\Omega}$; moreover, $u_{\lambda^*} > 0$ a.e. on $\partial\Omega$. The maximal solution is stable (in a suitable sense) for $\lambda > \lambda^*$.*

REMARK 13.9. If $f(x,u) \equiv f(u)$, then $u_\lambda \equiv 0$ for $\lambda \in {]}0, \lambda^*{[}$ ([46], Proposition 2.1). A sufficient condition for $u_{\lambda^*} > 0$ on $\partial\Omega$ is given in [46], Proposition 1.10.

Maximal positive solutions u_λ to (13.1) were also considered from the variational point of view in [42]. It was proved in this paper that for $\lambda > \lambda^*$, u_λ is a local minimum not only in the sense of C_0^1 but also in the sense of H_0^1. Since the result by Brezis and Nirenberg [19] does not cover this situation, the proof consists in showing first that it is a local minimum in C^1 and then use an associate penalization problem $\mathcal{P}(\varepsilon)$ for $\varepsilon > 0$ in order to prove that it is also a minimum in H^1. The main difficulty is then to get estimates independent of ε for the minimizers u_ε of the penalized problem.

A similar approach may be used in order to deal with the parabolic problem associated to (13.1), namely

$$\frac{\partial u}{\partial t} - \Delta u = \chi[u > 0]\left(-\frac{1}{u^\alpha} + f(u)\right) \quad \text{in } \Omega, \qquad u = 0 \quad \text{on } \partial\Omega, \qquad (13.10)$$

$$u(x, 0) = u_0(x) \quad \text{on } \Omega \qquad (13.11)$$

for some $u_0 > 0$, where now $f(x, u) \equiv f(u)$ is C^2 and satisfies (13.2).

For $T > 0$ and $u_0 \in L_\infty(\Omega)$ with $u_0 \geqslant 0$, it is said that $u \in L_\infty(\Omega \times {]}0, T{[})$, such that $u \geqslant 0$, is a weak solution to (13.10), (13.11) if $g(u) = \chi[u > 0](-1/u^\alpha + f(u)) \in L_1(\Omega \times {]}0, T{[})$ and

$$\int_0^T \left(\int_\Omega \left((\varphi_t + \Delta\varphi)u - \chi[u > 0]g(u)\varphi \right) dx \right) dt + \int_\Omega u_0\varphi(0)\, dx = 0 \qquad (13.12)$$

for any $\varphi \in C^2(\overline{\Omega} \times [0, T])$ such that $\varphi = 0$ on $\partial\Omega \times {]}0, T{[}$ and $\varphi(T) = 0$ on Ω.

The existence proof consists in replacing (13.10) by the approximating problem $\mathcal{P}(\varepsilon)$ obtained replacing $g(u)$ by

$$g_\varepsilon(u) = \frac{u}{(u + \varepsilon)^{1+\alpha}} - f(u) \qquad (13.13)$$

with $\varepsilon > 0$. Now the classical results in [85] can be applied giving a unique solution $u_\varepsilon \in L_\infty(\Omega \times {]}0, T{[})$ to the problem $\mathcal{P}(\varepsilon)$ with $u_\varepsilon > 0$. Going to the limit when $\varepsilon \to 0$ it is possible to prove the following theorem.

THEOREM 13.10 ([44], Theorem 1.1). *Assume that f is as defined just after (13.9) and satisfying (13.2), and that $u_0 \in L_\infty(\Omega \times {]}0, T{[})$ with $u_0 \geqslant 0$. Then the solutions $u_\varepsilon > 0$ to $\mathcal{P}(\varepsilon)$ converge to a limit u when $\varepsilon \to 0$ uniformly on compact subsets of $\Omega \times {]}0, T{[}$. If, moreover, $u_0 \in C(\Omega)$, then $u \in C(\Omega \times {]}0, T{[})$ and u is a weak solution to (13.10), (13.11).*

REMARK 13.11. The same conclusion still holds if $u_0 \in L_\infty(\Omega \times]0, T[)$ and $u_0(x) \geq cd(x)^\gamma$ with $1 < \gamma < 2/(1+\alpha)$. If we only have $u_0 \in L_\infty(\Omega \times]0, T[)$, then (13.10) is satisfied but it is not known if the initial condition is satisfied.

The proof of Theorem 13.10 is very involved and relies on estimates (independent of $\varepsilon > 0$) for both u_ε and its gradient ∇u_ε.

Concerning the asymptotic behavior of weak solutions to (13.10) the following stabilization (cf. Section 9) result can be proved.

THEOREM 13.12 ([44], Proposition 1.5). *Assume that f is as defined in Theorem 3.10 satisfying in particular (13.2), $u_0 \in L_\infty(\Omega)$ and $u_0 \geq 0$. If u is a solution to (13.10), (13.11), then the orbits of u are compact in $C(\Omega)$ (with uniform convergence on compact subsets) and they converge along subsequences to weak solutions to the stationary problem (13.1).*

It follows from Theorems 13.1 and 13.12 that for $\lambda < \lambda^*$ any limit point should be zero somewhere in Ω. Moreover, there is extinction of solutions in the sense that the null set $\{(x, t) \in (\Omega \times]0, +\infty[) : u(x, t) = 0\}$ has positive measure.

It is still possible to prove the following result.

THEOREM 13.13 ([44], Theorem 1.7). *Under the assumptions of Theorem 3.10, if there is a solution to (13.10), (13.11) positive a.e. then there is a solution to (13.1) positive a.e. Conversely, if $w > 0$ a.e. in Ω is a solution to (13.1) and $u_0 \in L_\infty(\Omega)$, with $u_0 \geq w$, $u_0 \not\equiv w$, then the corresponding solution to (13.10), (13.11) is such that*

$$u(x, t) \geq c(t)d(x) \tag{13.14}$$

for some $c:]0, +\infty[\to]0, +\infty[$ continuous and every $t > 0$.

REMARK 13.14. If $\alpha > 1$, there is no positive global classical solution to (13.10), (13.11). On the other side, if $u_0 \in L_\infty(\Omega)$ and $u_0(x) \geq c_1 d(x)^\gamma$, $1 < \gamma < 2/(1+\alpha)$, then the solution u has a similar behavior, i.e., $u(x, t) \geq c_2 d(x)^{2/(1+\alpha)}$ for any $x \in \Omega$, $0 < t < T$.

Only a partial result is already available concerning uniqueness.

THEOREM 13.15. *Under the hypotheses of Theorem 3.10, assume that $u_0 \in L_\infty(\Omega)$ satisfies $u_0(x) \geq cd(x)^\gamma$, $1 < \gamma < 2/(1+\alpha)$. Then there is (at most) a solution to (13.10), (13.11) such that for any $\tau \in]0, T[$ there exists $\bar{c} > 0$ such that $u(x, t) \geq \bar{c}d(x)^\gamma$ for every $t \in]0, \tau[$.*

Acknowledgements

This work was supported by Grants REN 2003-0223-C03 from DGI SGPI (Spain), RTN HPRN-CT-2002-00274 (EC) and MTM2004-03808 from DGI. The authors would like to thank C. Aranda, T. Godoy, E. Lami-Dozo and Junping Shi for sending their preprints and providing useful information, and J.M. Vega for useful discussions and comments.

Appendix A. Proof of Proposition 2.3

The uniqueness part readily follows by a standard maximum principle. The existence part and the estimate (2.8) are obtained by regularizing the coefficients as follows. For each $\varepsilon \in {]0, \rho_1[}$ (with ρ_1 as defined in assumption (H.1)) we consider the problem

$$-\sum a_{ij}(x)\frac{\partial^2 u}{\partial x^i \partial x^j} + \varphi_\varepsilon(x)\sum b_i(x)\frac{\partial u}{\partial x^i} = \varphi_\varepsilon(x)M(x)v \quad \text{in } \Omega,$$

$$u = 0 \quad \text{on } \partial\Omega,$$

(A.1)

where $\varphi_\varepsilon \in C_0^1(\overline{\Omega})$ is defined as $\varphi_\varepsilon(x) = \psi_\varepsilon(d(x))$, with

$$\psi_\varepsilon(\eta) = \frac{(2\varepsilon - \eta)\eta}{\varepsilon^2} \quad \text{if } 0 < \eta < \varepsilon, \qquad \psi_\varepsilon(\eta) = 1 \quad \text{if } \eta \geqslant \varepsilon.$$

(A.2)

Note that, according to assumptions (H.2) and (H.3$'$) and the estimate (2.11), $\varphi_\varepsilon b_i$ and $\varphi_\varepsilon Mv$ are in $C^{0,\delta_0}(\overline{\Omega})$ if $v \in C_0^1(\overline{\Omega})$ and $\varepsilon > 0$. Then (A.1) possesses a unique solution $u \in C^{2,\delta_0}(\overline{\Omega})$. The proof proceeds in three steps.

STEP 1. *For each $\delta \in {]0, \delta_0[}$, there are two constants, $K > 0$ and $\mu \in {]0, \rho_1[}$, independent of v and ε, such that the solution of* (A.1) *satisfies*

$$\|u\|_{C^{1,\delta}(\overline{\Omega}_\mu)} \leqslant K\big[\|v\|_{C^1(\overline{\Omega}_\mu)} + \|u\|_{C^0(\overline{\Omega}_\mu)} + \|u\|_{C^{1,\delta}(\Omega\setminus\Omega_\mu)}\big],$$

(A.3)

where

$$\Omega_\mu = \big\{x \in \Omega : d(x) < \mu\big\}.$$

(A.4)

Since $\partial\Omega$ is of class $C^{3,\gamma}$, as in [85, pp. 95–96], it can be seen that, for each $\mu < \rho_1$, there are two finite families of domains, $\{\Omega_{1\mu}^k\}$ and $\{\Omega_{2\mu}^k\}$ such that for each k:

(a) $\overline{\Omega}_{1\mu}^k \subset \Omega_{2\mu}^k \subset \bigcup_{j=1}^{m_0} \Omega_{1\mu}^k \subset \Omega_\mu, \bigcup_k \Omega_{1\mu}^k = \Omega_\mu$, with m_0 independent of μ and k;

(b) there is a $C^{2,\gamma}$-regular curvilinear, coordinate system in a neighborhood of $\overline{\Omega}_{2\mu}^k$, $\xi = \xi(x)$, such that: (i) $\xi_1(x) = d(x)$, (ii) the domains $\Omega_{1\mu}^k$ and $\Omega_{2\mu}^k$ are given by

$$\Omega_{i\mu}^k = \big\{x \in \mathbb{R}^n : \xi(x) \in \omega_{i\mu}^k\big\},$$

$$\text{with } \omega_{i\mu}^k = \big\{\xi \in \mathbb{R}^n : 0 < \xi_1 < \mu, \ |\xi_2|^2 + \cdots + |\xi_n|^2 < i\mu\big\},$$

(A.5)

for $i = 1$ and 2; and (iii) the $C^{2,\gamma}$-norms of the functions $\xi = \xi(x)$ and $x = x(\xi)$, in $\overline{\Omega}_{2\mu}^k$ and $\overline{\omega}_{2\mu}^k$, respectively, are bounded by a common constant, which is independent of k and μ.

Now, in the new variables, the function

$$U(\xi_1, \ldots, \xi_n) = \int_{\xi_1}^{\mu} u(y, \xi_2, \ldots, \xi_n)\, dy \tag{A.6}$$

is readily seen to satisfy

$$\mathcal{L}_0 U = \mathcal{L}_1 U + \mathcal{L}_2 u + \mathcal{L}_3 u + \mathcal{L}_4 v \quad \text{in a neighborhood of } \bar{\omega}_{2\mu}^k, \tag{A.7}$$

$$\frac{\partial U}{\partial \xi_1} = 0 \quad \text{at } \xi_1 = 0, \qquad U = 0 \quad \text{at } \xi_1 = \mu, \tag{A.8}$$

where

$$\mathcal{L}_0 U = -\sum \tilde{a}_{ij}(\xi) \frac{\partial^2 U}{\partial \xi_i\, \partial \xi_j},$$

$$\mathcal{L}_1 U = \int_{\xi_1}^{\mu} \left[\sum \frac{\partial \tilde{a}_{ij}}{\partial \xi_1} \frac{\partial^2 U}{\partial \xi_i\, \partial \xi_j} \right]_{\xi_1 = y} dy, \tag{A.9}$$

$$\mathcal{L}_2 u = -\int_{\xi_1}^{\mu} \sum \left[\left[\tilde{b}_i^1(\xi) + \psi_\varepsilon(\xi^1)\tilde{b}_i^2(\xi) \right] \frac{\partial u}{\partial \xi_i} \right]_{\xi_1 = y} dy, \tag{A.10}$$

$$\mathcal{L}_3 u = \left[-\tilde{a}_{11} \frac{\partial u}{\partial \xi_1} + 2 \sum \tilde{a}_{1i} \frac{\partial u}{\partial \xi_i} \right]_{\xi_1 = \mu},$$

$$\mathcal{L}_4 v = \int_{\xi_1}^{\mu} \psi_\varepsilon(y) \left[\widetilde{M}(\xi) v(\xi) \right]_{\xi_1 = y} dy. \tag{A.11}$$

Here $\tilde{a}_{ij}$ and $\tilde{b}_i \equiv \tilde{b}_i^1(\xi) + \psi_\varepsilon(\xi^1)\tilde{b}_i^2(\xi)$ denote the coefficients of the operator obtained when using the new variables in the left-hand side of (A.1). Now, according to assumptions (H.2) and (H.3$'$) and property (b), if $0 < \delta < \delta_0 = \min\{\alpha + 1, \gamma\}$ then the functions $\tilde{a}_{ij}$, $\xi \to \xi_1(\partial \tilde{a}_{ij}/\partial \xi_1)$, $\tilde{b}_i^1$, $\xi \to \xi_1 \tilde{b}_i^2$ and $\xi \to (\xi_1)^2 \widetilde{M}$ have $C^{0,\delta}(\bar{\omega}_{2\mu}^k)$-norms that are uniformly bounded by a common constant, which is independent of k and μ. Since, in addition, $0 \leqslant \psi_\varepsilon(\xi^1) \leqslant 1$ in $0 \leqslant \xi^1 \leqslant \mu$ (see (A.2)), we have for all $\delta \in \,]0, \delta_0[$,

$$\|\mathcal{L}_1 U\|_{C^{0,\delta}(\bar{\omega}_{2\mu}^k)}^{(\xi)} \leqslant K \mu^{\delta_0 - \delta} \|U\|_{C^{2,\delta}(\bar{\omega}_{2\mu}^k)}^{(\xi)},$$

$$\|\mathcal{L}_2 u\|_{C^{0,\delta}(\bar{\omega}_{2\mu}^k)}^{(\xi)} \leqslant K \mu^{\delta_0 - \delta} \|u\|_{C^{1,\delta}(\bar{\omega}_{2\mu}^k)}^{(\xi)}, \tag{A.12}$$

$$\|\mathcal{L}_3 u\|_{C^{0,\delta}(\bar{\omega}_{2\mu}^k)}^{(\xi)} \leqslant K \|u\|_{C^{1,\delta}(\Omega \setminus \Omega_\mu)},$$

$$\|\mathcal{L}_4 v\|_{C^{0,\delta}(\bar{\omega}_{2\mu}^k)}^{(\xi)} \leqslant K \mu^{\delta_0 - \delta} \|v\|_{C^{1}(\bar{\omega}_{2\mu}^k)}^{(\xi)}, \tag{A.13}$$

where the constant K is independent of ε, k, μ and v, and the superscript (ξ) indicates that the new coordinates are used in the definition of the norm. The first two estimates follow straightforwardly when taking into account that ($\delta_0 \equiv \min\{\gamma, \alpha + 1\} \leqslant \alpha + 1$ and)

$$\text{if } 0 < \delta < \delta_0 \text{ and } 0 < y_1 < y_2 < \mu \leqslant 1,$$
$$\text{then } 0 < y_2^{\alpha+1} - y_1^{\alpha+1} \leqslant (\alpha + 1)\mu^{\alpha+1-\delta}|y_2 - y_1|^{\delta}. \tag{A.14}$$

The third estimate is a consequence of the facts that the hypersurface $\xi^1 = \mu$ is in $\Omega \setminus \Omega_\mu$, and that, according to property (b), the Hölder norms in the variables x and ξ are equivalent, uniformly on k and μ. The last estimate is obtained when taking into account (A.14) and the inequality

$$\left| v\left(\xi_1, \xi_2^2, \ldots, \xi_n^2\right) - v\left(\xi_1, \xi_2^1, \ldots, \xi_n^1\right) \right|$$
$$\leqslant |2\xi_1|^{1-\delta} \left[\left|\xi_2^2 - \xi_2^1\right|^2 + \cdots + \left|\xi_n^2 - \xi_n^1\right|^2 \right]^{\delta/2} \|v\|_{C^1(\bar\omega_{2\mu}^k)}^{(\xi)}$$

which holds whenever $(\xi_1, \xi_2^1, \ldots, \xi_n^1)$ and $(\xi_1, \xi_2^2, \ldots, \xi_n^2)$ are in $\omega_{2\mu}^k$, and in turn is obtained when taking into account that its left-hand side is bounded above by both

$$2\xi^1 \|v\|_{C^1(\bar\omega_{2\mu}^k)}^{(\xi)} \quad \text{and} \quad \left[\left|\xi_2^2 - \xi_2^1\right|^2 + \cdots + \left|\xi_n^2 - \xi_n^1\right|^2\right]^{1/2} \|v\|_{C^1(\bar\omega_{2\mu}^k)}^{(\xi)},$$

as readily seen when applying the mean value theorem and taking into account that $v = 0$ at $\xi^1 = 0$. Now, if we re-scale ξ as $\xi = \mu\eta$, then in the new variables the domains $\omega_{2\mu}^k$ and ω_μ^k are fixed, and the $C^{0,\delta}$-norms of $\tilde{a}_{ij}$ are bounded above by a common constant, which is independent of k and μ. If we now apply a local Schauder estimate to (A.7) and (A.8) in these new variables, and rewrite this estimate in terms of ξ, we obtain

$$\|U\|_{C^{2,\delta}(\bar\omega_\mu^k)}^{(\xi)} \leqslant K_1 \|\mathcal{L}_1 U + \mathcal{L}_2 u + \mathcal{L}_4 v\|_{C^{0,\delta}(\bar\omega_{2\mu}^k)}^{(\xi)}$$
$$+ \mu^{-\delta} K_2 \|\mathcal{L}_3 u\|_{C^{0,\delta}(\bar\omega_{2\mu}^k)}^{(\xi)} + \mu^{-(2+\delta)} K_3 \|U\|_{C^0(\bar\omega_{2\mu}^k)}^{(\xi)}, \tag{A.15}$$

where the constants K_1, K_2 and K_3 are independent of ε, k, μ and v, and we have taken into account that $\mathcal{L}_1 U + \mathcal{L}_2 u + \mathcal{L}_4 v$ vanishes at $\xi^1 = \mu$. And, when using (A.12)–(A.13) and the fact that (according to property (b)) the Hölder-norms in the variable ξ and x are equivalent, uniformly in k and μ, we have

$$\|U\|_{C^{2,\delta}(\bar\Omega_\mu^k)} \leqslant K_4 \Big[\mu^{\delta_0 - \delta} \big(\|U\|_{C^{2,\delta}(\bar\Omega_{2\mu}^k)} + \|v\|_{C^1(\bar\Omega_{2\mu}^k)} \big)$$
$$+ \mu^{-\delta} \|u\|_{C^{1,\delta}(\Omega \setminus \Omega_\mu)} + \mu^{-1-\delta} \|u\|_{C^0(\bar\Omega_{2\mu}^k)} \Big], \tag{A.16}$$

for some constant K_4 that is independent of ε, k, μ and v. Here we have taken into account that U is independent of the curvilinear coordinate system used in $\Omega_{2\mu}^k$ (U is the integral of u along the normals to $\partial\Omega$, see (A.6)), and that

$$\|u\|_{C^{1,\delta}(\overline{\Omega}_{j\mu}^k)} \leqslant \|U\|_{C^{2,\delta}(\overline{\Omega}_{j\mu}^k)} \quad \text{for } j = 1 \text{ and } 2 \quad \text{and}$$

$$\|U\|_{C^0(\overline{\Omega}_{2\mu}^k)} \leqslant \mu \|u\|_{C^0(\overline{\Omega}_{2\mu}^k)}. \tag{A.17}$$

Now we chose μ such that $K_4 \mu^{\delta_0 - \delta} m_0 < 1/2$, where m_0 is as defined in property (a). Then, if k_1 is that value of k for which the left-hand side of (A.16) is greatest, we have

$$\|U\|_{C^{2,\delta}(\overline{\Omega}_\mu^{k_1})}$$

$$\leqslant \frac{1}{2} \|U\|_{C^{2,\delta}(\overline{\Omega}_\mu^{k_1})}$$

$$+ K_4 \big[\|v\|_{C^1(\overline{\Omega}_{2\mu}^{k_1})} + \mu^{-\delta} \|u\|_{C^{1,\delta}(\Omega \setminus \Omega_\mu)} + \mu^{-(1+\delta)} \|u\|_{C^0(\overline{\Omega}_{2\mu}^{k_1})} \big], \tag{A.18}$$

where we have used the inequality $\|U\|_{C^{2,\delta}(\overline{\Omega}_{2\mu}^{k_1})} \leqslant m_0 \|U\|_{C^{2,\delta}(\overline{\Omega}_\mu^{k_1})}$, which follows from property (a). Thus we only need to use (A.17) and (A.18) and the definition of k_1 to obtain (A.3) and complete the step.

STEP 2. *For each* $\delta \in {]}0, \delta_0{[}$ *there is a constant* K, *independent of* v *and* ε, *such that the solution of* (A.1) *satisfies*

$$\|u\|_{C^{1,\delta}(\overline{\Omega})} \leqslant K \big[\|v\|_{C^1(\overline{\Omega})} + \|u\|_{C^0(\overline{\Omega})} \big]. \tag{A.19}$$

The estimate (A.19) readily follows by first selecting μ as in step 1, and then using (A.3) and the new estimate

$$\|u\|_{C^{2,\delta}(\Omega \setminus \Omega_\mu)} \leqslant K_5 \big[\|v\|_{C^{0,\delta}(\overline{\Omega}^1)} + \|u\|_{C^0(\overline{\Omega}^1)} \big],$$

where $\Omega^1 = \{x \in \Omega : d(x) > \mu/2\}$ and K_5 is independent of v and ε. This latter estimate is just a standard interior Schauder estimate on (A.1) (whose coefficients have $C^1(\overline{\Omega}^1)$-norms that are uniformly bounded in $0 < \varepsilon < \rho_1$).

STEP 3. *If* $v \in C_0^1(\overline{\Omega})$ *then* (2.7) *has a solution* $u \in C^2(\Omega) \cap C^{1,\delta}(\overline{\Omega})$ *for all* $\delta \in {]}0, \delta_0{[}$, *and* (2.8) *holds with* K *independent of* v.

For each $m = 1, 2, \ldots$, let u_m be the solution of (A.1) for $\varepsilon = \rho_1/m$. Let us first see that $\|u_m\|_{C^0(\overline{\Omega})}$ is bounded. To this end, we assume for contradiction that there is a subsequence, also called $\{u_m\}$, such that $\|u_m\|_{C^0(\overline{\Omega})} \to \infty$ as $m \to \infty$. Then $U_m \equiv u_m / \|u_m\|_{C^0(\overline{\Omega})}$ is such that, for all m,

$$-\sum a_{ij} \frac{\partial^2 U_m}{\partial x_i \partial x_j} + \varphi_{\varepsilon_m}(x) \sum b_i \frac{\partial U_m}{\partial x_i} = \varphi_{\varepsilon_m}(x) M(x) \frac{v}{\|u_m\|_{C^0(\overline{\Omega})}} \quad \text{in } \Omega,$$

$$U_m = 0 \quad \text{on } \partial\Omega \quad \text{and} \quad \|U_m\|_{C^0(\overline{\Omega})} = 1, \tag{A.20}$$

where $\varepsilon_m \to 0$ as $m \to \infty$. But the estimate (A.19) applied to (A.20) implies that $\|U_m\|_{C^{1,\delta}(\overline{\Omega})}$ is bounded if $\delta \in \,]0, \delta_0[$ and, since the embedding of $C^{1,\delta}$ into C^1 is compact, there is a subsequence, still called $\{u_m\}$, which converges in $C^1(\overline{\Omega})$ to some U. Now, $U \neq 0$ at some $x \in \Omega$ because $\|U\|_{C^0(\overline{\Omega})} = 1$. Also $U \in C^2(\Omega)$ and satisfies

$$\mathcal{L}U = 0 \quad \text{in } \Omega, \qquad U = 0 \quad \text{on } \partial\Omega, \tag{A.21}$$

as readily obtained when applying interior Schauder estimates to (A.20). But, according to standard maximum principles, (A.21) cannot have nontrivial solutions. Then a contradiction is obtained and the result follows.

Now, since $\|u_m\|_{C^0(\overline{\Omega})}$ is bounded, the estimate (A.19) readily implies that $\|u_m\|_{C^{1,\delta}(\overline{\Omega})}$ is also bounded for each $\delta \in \,]0, \delta_0[$. And, since the embedding of $C^{1,\delta'}$ into $C^{1,\delta}$ is compact whenever $0 < \delta < \delta' < \delta_0 \ (\leqslant 1)$, for each $\delta \in \,]0, \delta_0[$ there is a subsequence, also called $\{u_m\}$, which converges in $C^{1,\delta}(\overline{\Omega})$ to some u $(\in C^{1,\delta}(\overline{\Omega}))$. Also $u \in C^2(\Omega)$ and satisfies (2.7) (thus the existence part of the statement follows) as readily seen when noticing that u_m satisfies (A.1) for $\varepsilon = \varepsilon_m$, with $\varepsilon_m \to 0$ as $m \to \infty$, and applying interior Schauder estimates to this latter equation. And when applying the estimate (A.19) to this latter equation, we obtain

$$\|u\|_{C^{1,\delta}(\overline{\Omega})} \leqslant K\big[\|v\|_{C^1(\overline{\Omega})} + \|u\|_{C^0(\overline{\Omega})}\big], \tag{A.22}$$

where K is independent of v.

Finally, u and v satisfy (2.8), which follows from (A.22) and the estimate

$$\|u\|_{C^0(\overline{\Omega})} \leqslant K\|v\|_{C^1(\overline{\Omega})}, \tag{A.23}$$

with K independent of v. And this latter estimate is readily obtained from (A.22) by a standard contradiction argument, alike to the one already used above (if (A.23) does not hold, then there is a sequence $\{v_m\} \subset C_0^1(\overline{\Omega})$ such that $\|v_m\|_{C^1(\overline{\Omega})} = 1$ for all m, and the corresponding solutions of (2.7) are such that $\|u_m\|_{C^0(\overline{\Omega})} \to \infty$ as $m \to \infty$; but then $U_m = u_m/\|u_m\|_{C^0(\overline{\Omega})}$ possesses a subsequence that converges in $C^2(\Omega) \cap C^1(\overline{\Omega})$ to a nontrivial solution of (A.21), which cannot exist). This completes the step, and the proof of Proposition 2.3.

REMARK A.1. Some related ideas can be found in the paper [63]. For the classical linear theory the reader may consult the books [64] and [15] and the paper [3].

Appendix B. A strong maximum principle for second-order equations with locally bounded coefficients

Here we derive a strong maximum principle for some elliptic and parabolic inequalities with locally bounded coefficients, such as those appearing in this paper. The elliptic case was already considered in [82,99], under essentially the same assumptions made below, but we have been unable to find a proof for the parabolic case in the literature. A very similar version of the Hopf maximum principle was given independently by Takáč [112]. For the sake of brevity we first consider the parabolic case, which contains the elliptic one as a particular case. Of course, the elliptic case could have been directly treated in a similar way.

THEOREM B.1. *Let $u \in C^1(\overline{\Omega} \times [t_0, t_1])$ be such that $u(\cdot, t) \in C^2(\Omega) \cap C^{1,\delta}(\overline{\Omega})$ for all $t \in [t_0, t_1]$ and*

$$N(x)\frac{\partial u}{\partial t} + \mathcal{L}u + M(x)u \leqslant 0 \quad in \ \Omega \times]t_0, t_1[, \tag{B.1}$$

where $0 < \delta < 1$, Ω, $\mathcal{L}$, M and N satisfy assumptions (H.1), (H.2), (H.3$'$) and (H.4), and $M \geqslant 0$ in Ω. Let us assume also that $u \leqslant 0$ in $\Omega \times]t_0, t_1[$, and that $u(x_0, t_1) = 0$. Then the following properties hold:

    (i) *If $x_0 \in \Omega$ then $u = 0$ in $\overline{\Omega} \times [t_0, t_1]$.*

    (ii) *If $x_0 \in \partial\Omega$ and $u < 0$ in $\Omega \times]t_0, t_1[$, then $\partial u/\partial n > 0$ at (x_0, t_1).*

PROOF. Since the coefficients of the linear operator in the left-hand side of (B.1) are locally bounded in Ω and $N > 0$ in Ω, property (i) readily follows when applying the standard strong maximum principle [98].

In order to prove property (ii) assume for contradiction that

$$u < 0 \quad in \ \Omega \times]t_0, t_1[\quad and \quad u = \frac{\partial u}{\partial n} = 0 \quad at \ (x_0, t_1). \tag{B.2}$$

Since, in addition, $u(\cdot, t_1) \in C^{1,\delta}(\overline{\Omega})$, there is a constant $k_2 > 0$ such that

$$|u(x, t_1)| = |u(x, t_1) - u(x_0, t_1)| \leqslant k_2|x - x_0|^{1+\delta} \quad for \ all \ x \in \Omega. \tag{B.3}$$

On the other hand, Ω satisfies the interior sphere condition (because of assumption (H.1)), i.e., there is a hypersphere H, with center at $y_0 \in \Omega$ and radius $\rho_1 > 0$ such that $H \subset \Omega \cup \partial\Omega$ and $H \cap \partial\Omega = \{x_0\}$. Let us consider the function

$$v(x, t) = \left[t - t_1 + \rho_1 - \rho(x)\right]^{1+\delta/2} \quad with \ \rho(x) = |x - y_0|, \tag{B.4}$$

which (when proceeding as in the proof of Lemma 2.1) is seen to satisfy

$$N\frac{\partial v}{\partial t} + \mathcal{L}v + Mv < 0$$

$$in \ A = \left\{(x, t) \in \Omega \times]t_0, t_1]: \ \rho(x) > \rho_2, \ \rho_1 - \rho(x) > t_1 - t \geqslant 0\right\} \tag{B.5}$$

provided that ρ_2 is appropriately close to ρ_1. In that case the function $w_\varepsilon \equiv u + \varepsilon v$ is such that (see (B.1)) $N(x)\partial w_\varepsilon/\partial t + \mathcal{L}w_\varepsilon + M(x)w_\varepsilon < 0$ in A, whenever $\varepsilon > 0$; thus the standard maximum principle [98] implies that the maximum of w_ε in A can be attained neither at an interior point of A nor at $t = t_1$. Thus this maximum must be attained either at $\rho_1 - \rho(x) = t_1 - t$ or at $\rho(x) = \rho_2$; but (a) $w_\varepsilon \equiv u + \varepsilon v = u \leqslant 0$ if $\rho_1 - \rho(x) = t_1 - t \geqslant 0$ and $\varepsilon > 0$, and (b) $w_\varepsilon \equiv u + \varepsilon v < 0$ if $\rho(x) = \rho_2$, $\rho_1 - \rho_2 \geqslant t_1 - t \geqslant 0$ and $\varepsilon > 0$ is appropriately small (see (B.2) and (B.4)). Thus for that value of ε, $w \leqslant 0$ (i.e., $u \leqslant -\varepsilon v$) in A. This property holds, in particular, on the rectilinear segment S of $\Omega \times \{t_1\}$ joining (y_0, t_1) and (x_0, t_1), where $\rho_1 - \rho(x) = d(x)$. Then we have

$$u(x, t_1) \leqslant -\varepsilon d(x)^{1+\delta/2}$$

in $x \in S \cap A \subset \Omega$ (i.e., if $d(x) > 0$ is sufficiently small).

Since $\varepsilon > 0$ and $\delta > 0$, this inequality is in contradiction with (B.3), and the proof is complete. $\qquad\square$

The elliptic case is reduced to the parabolic one as usual, just by noticing that if a function $u = u(x)$ satisfies the elliptic inequality (B.6) below then it also satisfies (B.1), and if that function attains a maximum at $x_0 \in \overline{\Omega}$, then it also attains the maximum at $(x_0, t) \in \overline{\Omega} \times \mathbb{R}$ for all t. Thus the following result follows.

THEOREM B.2. *Let $u \in C^2(\Omega) \cap C^{1,\delta}(\overline{\Omega})$ be such that*

$$\mathcal{L}u + M(x)u \leqslant 0 \quad a.e. \text{ in } \Omega, \tag{B.6}$$

where δ, Ω, $\mathcal{L}$ and M are as in Theorem B.1. Let us assume that $u \leqslant 0$ in $\overline{\Omega}$ and $u(x_0) = 0$ for some $x_0 \in \overline{\Omega}$. Then the following properties hold:
  (i) *If $x_0 \in \Omega$ then $u \equiv 0$ in $\overline{\Omega}$.*
  (ii) *If $x_0 \in \partial\Omega$ and $u < 0$ in Ω then $\partial u/\partial n > 0$ at x_0.*

References

[1] R.A. Adams, *Sobolev Spaces*, Academic Press, Orlando (1975).
[2] Adimurthi and J. Giacomoni, *Multiplicity of positive solutions for a singular and critical elliptic problem in $\mathbb{R}^2$*, to appear.
[3] S. Agmon, A. Douglis and L. Nirenberg, *Estimates near the boundary for solutions of elliptic partial differential equations satisfying general boundary conditions I*, Comm. Pure Appl. Math. **12** (1959), 623–727.
[4] S. Alama, *Semilinear elliptic equations with sublinear indefinite nonlinearities*, Adv. Differential Equations **4** (1999), 813–842.
[5] H. Amann, *Fixed point equations and nonlinear eigenvalue problems in ordered Banach spaces*, SIAM Rev. **18** (1976), 620–709.
[6] A. Ambrosetti, H. Brezis and G. Cerami, *Combined effects of concave and convex nonlinearities in some elliptic problems*, J. Funct. Anal. **122** (1994), 519–543.
[7] C. Aranda and T. Godoy, *On a nonlinear Dirichlet problem with a singularity along the boundary*, Differential Integral Equations **15** (2002), 1313–1324.
[8] C. Aranda and T. Godoy, *Existence and multiplicity of positive solutions for a singular problem associated with the p-Laplacian operator*, Electron. J. Differ. Equ. Conf. **132** (2004), 1–15.
[9] C. Aranda and E. Lami-Dozo, *Multiple solutions to a singular nonlinear Dirichlet problem*, to appear.

[10] C. Aranda and E. Lami-Dozo, *Multiple solutions to a singular Emden–Fowler equation with a convection term*, to appear.

[11] M. Badiale and G. Tarantello, *Existence and multiplicity results for elliptic problems with critical growth and discontinuous nonlinearities*, Nonlinear Anal. **29** (1997), 639–677.

[12] C. Bandle, A. Pozio and A. Tesei, *The asymptotic behavior of the solutions of degenerate parabolic equations*, Trans. Amer. Math. Soc. **303** (1987), 487–501.

[13] V. Benci, A.M. Micheletti and E. Shteto, *On a second order boundary value problem with singular nonlinearity*, Preprint (2004).

[14] H. Berestycki, L. Nirenberg and S.R.S. Varadhan, *The principal eigenvalue and maximum principle for second-order elliptic operators in general domains*, Commun. Pure Appl. Math. **47** (1994), 47–92.

[15] L. Bers, F. John and M. Schechter, *Partial Differential Equations*, Interscience, New York (1964).

[16] M. Bertsch and R. Rostamian, *The principle of linearized stability for a class of degenerate diffusion equations*, J. Differential Equations **57** (1985), 373–405.

[17] H. Brezis, T. Cazenave, Y. Martel and A. Ramiandrisoa, *Blow-up for $u_t - \Delta u = g(u)$ revisited*, Adv. Differential Equations **1** (1996), 73–90.

[18] H. Brezis and S. Kamin, *Sublinear elliptic equations in $\mathbb{R}^n$*, Manuscripta Math. **74** (1992), 87–106.

[19] H. Brezis and L. Nirenberg, *H^1 versus C^1 local minimizers*, C. R. Acad. Sci. Paris Ser. I **317** (1993), 465–472.

[20] H. Brezis and L. Oswald, *Remarks on sublinear elliptic equations*, Nonlinear Anal. **10** (1986), 55–64.

[21] K.J. Brown and P. Hess, *Stability and uniqueness of positive solutions for a semilinear elliptic boundary value problem*, Differential Integral Equations **3** (1990), 201–207.

[22] X. Cabré and P. Majer, *Truncation of nonlinearities in some supercritical elliptic problems*, C. R. Acad. Sci. Paris Ser. I **322** (1996), 1157–1162.

[23] A. Canino and M. de Giovanni, *A variational approach to a class of singular semilinear elliptic equations*, Preprint (2004).

[24] K.C. Chang, *A variant mountain pass lemma*, Sci. China Ser. A **26** (1983), 1241–1255.

[25] K.C. Chang, *Infinite Dimensional Morse Theory and Multiple Solution Problems*, Birkhäuser, Basel (1993).

[26] F. Chen and M. Yao, *Decaying positive solutions to a system of nonlinear elliptic equations*, Preprint (2005).

[27] Y.S. Choi, A.C. Lazer and P.J. McKenna, *Some remarks on a singular elliptic boundary value problem*, Nonlinear Anal. **32** (1998), 305–314.

[28] Y.S. Choi and P.J. McKenna, *A singular Gierer–Meinhardt system of elliptic equations*, Ann. Inst. H. Poincaré Anal. Non Linéaire **17**(4) (2000), 503–522.

[29] Y.S. Choi and P.J. McKenna, *A singular Gierer–Meinhardt system of elliptic equations: The classical case*, Nonlinear Anal. **55** (2003), 521–541.

[30] F. Cirstea, M. Ghergu and V. Radulescu, *Combined effects of asymptotically linear and singular nonlinearities in bifurcation problems of Lane–Emden type*, J. Math. Pures Appl. **84** (2005), 493–508.

[31] P. Clément, D.G. de Figueiredo and E. Mitidieri, *A priori estimates for positive solutions of semilinear elliptic systems via Hardy–Sobolev inequalities*, Nonlinear Partial Differential Equations, Fèz 1994, Pitman Research Notes, vol. 343, Pitman, Boston (1996), 73–91.

[32] M.M. Coclite, *On a singular nonlinear Dirichlet problem II*, Boll. Unione Mat. Ital. Sect. B **5** (1991), 955–975.

[33] M.M. Coclite, *On a singular nonlinear Dirichlet problem III*, Nonlinear Anal. **21** (1993), 547–564.

[34] M.M. Coclite, *On a singular nonlinear Dirichlet problem IV*, Nonlinear Anal. **23** (1994), 925–936.

[35] M.M. Coclite and G. Palmieri, *On a singular nonlinear Dirichlet problem*, Commun. Partial Differential Equations **14** (1989), 1315–1327.

[36] D.S. Cohen and T.W. Laetsch, *Nonlinear boundary value problems suggested by chemical reactor theory*, J. Differential Equations **7** (1970), 217–226.

[37] R. Courant and D. Hilbert, *Methods of Mathematical Physics, Vols. I and II*, Interscience, New York (1962).

[38] M.G. Crandall, P.H. Rabinowitz and L. Tartar, *On a Dirichlet problem with singular nonlinearity*, Commun. Partial Differential Equations **2** (1977), 193–222.

[39] Q. Dai and J. Gu, *Positive solutions for non-homogeneous semilinear elliptic equations with data that changes sign*, Proc. Roy. Soc. Edinburgh **133** (2002), 297–306.

[40] R. Dautray and J.L. Lions, *Mathematical Analysis and Numerical Methods for Science and Technology*, Vol. 3, Springer-Verlag, Berlin (1990).

[41] E.B. Davies, *Heat Kernels and Spectral Theory*, Cambridge University Press (1989).

[42] J. Dávila, *Global regularity for a singular equation and local H^1 minimizers of a nondifferentiable functional*, Commun. Contemp. Math. **6** (2004), 165–193.

[43] J. Dávila and M. Montenegro, *A singular equation with positive and free boundary solutions*, RACSAM **97** (2003), 107–112.

[44] J. Dávila and M. Montenegro, *Positive versus free boundary solutions to a singular equation*, J. Anal. Math. **90** (2003), 303–335.

[45] J. Dávila and M. Montenegro, *Existence and asymptotic behavior for a singular parabolic equation*, Trans. Amer. Math. Soc. **357** (2004), 1801–1828.

[46] J. Dávila and M. Montenegro, *Nonlinear problems with solutions exhibiting a free boundary on the boundary*, Ann. Inst. H. Poincaré Anal. Non Linéaire **22** (2005), 303–330.

[47] M. del Pino, *A global estimate for the gradient in a singular elliptic boundary value problem*, Proc. Roy. Soc. Edinburgh Sect. A **122** (1992), 341–352.

[48] M. del Pino and G.E. Hernández, *Solvability of the Neumann problem in a ball for $-\Delta u + u^{-\nu} = h(|x|)$, $\nu > 1$*, J. Differential Equations **124** (1996), 108–131.

[49] J.I. Díaz, *Nonlinear Partial Differential Equations and Free Boundaries*, Pitman, Boston (1985).

[50] J.I. Díaz and J. Hernández, *Global bifurcation and continua of nonnegative solutions for a quasilinear elliptic problem*, C. R. Acad. Sci. Paris Ser. I **329** (1999), 587–592.

[51] J.I. Díaz, J.M. Morel and L. Oswald, *An elliptic equation with singular nonlinearity*, Commun. Partial Differential Equations **12** (1987), 1333–1344.

[52] J. Dieudonné, *Foundations of Modern Analysis*, Academic Press, New York (1960).

[53] M. Donsker and S.R.S. Varadhan, *On the principal eigenvalue of second-order elliptic differential operators*, Commun. Pure Appl. Math. **29** (1976), 595–621.

[54] A. Fabricant, N. Kutev and T. Rangelov, *Maximum principle for second order elliptic equations in divergence forms*, Preprint, (2004).

[55] E. Feireisl and F. Simondon, *Convergence for semilinear degenerate parabolic equations in several space dimensions*, J. Dynam. Differential Equations, **12** (2000), 647–673.

[56] J. Fleckinger, J. Hernández and F. de Thélin, *Existence of multiple principal eigenvalues for some indefinite linear eigenvalue problems*, Boll. Unione Mat. Ital. **8** (2004), 159–188.

[57] W. Fulks and J.S. Maybee, *A singular nonlinear equation*, Osaka J. Math. **12** (1960), 1–19.

[58] I.M. Gamba and A. Jungel, *Positive solutions to a singular second and third order differential equations for quantum fluids*, Arch. Ration. Mech. Anal. **156** (2001), 183–203.

[59] S. Gaucel and M. Langlais, *Some mathematical problems arising in heterogeneous insular ecological modes*, RACSAM **96** (2002), 389–400.

[60] M. Ghergu and V.D. Radulescu, *Bifurcation and asymptotics for the Lane–Emden equation*, C. R. Acad. Sci. Paris Ser. I **337** (2003), 259–264.

[61] M. Ghergu and V.D. Radulescu, *Multiparameter bifurcation and asymptotics for the singular Lane–Emden–Fowler equation with a convection term*, Proc. Roy. Soc. Edinburgh **135** (2005), 61–84.

[62] J. Giacomoni, I. Schindler and P. Takac, *Sobolev versus Hölder local minimizers and global multiplicity for a singular and quasilinear equation*, Preprint (2006).

[63] D. Gilbarg and L. Hörmander, *Intermediate Schauder estimates*, Arch. Ration. Mech. Anal. **74** (1980), 297–318.

[64] D. Gilbarg and N.S. Trudinger, *Elliptic Partial Differential Equations of Second Order*, Springer-Verlag, Berlin (1983).

[65] T. Godoy, J. Hernández, U. Kaufmann and S. Paczka, *On some singular periodic parabolic problems*, Nonlinear Anal. **62** (2005), 989–995.

[66] S.N. Gomes, *On a singular nonlinear elliptic problem*, SIAM J. Math. Anal. **17** (1986), 1359–1369.

[67] S.N. Gomes and J. Sprekels, *Krasnoselskii's theorem on operators compressing a cone: Application to some singular boundary value problems*, J. Math. Anal. Appl. **153** (1990), 443–459.

[68] J.V. Gonçalves and C.A.P. Santos, *Positive solutions of a class of quasilinear singular equations*, Electron J. Differ. Equ. **56** (2004), 1–15.

[69] C. Gui and F. Hua Lin, *Regularity of an elliptic problem with a singular nonlinearity*, Proc. Roy. Soc. Edinburgh Sect. A **123** (1993), 1021–1029.

[70] W.S.C. Gurney and R.N. Nisbet, *The regulation of inhomogeneous populations*, J. Theor. Biol. **52** (1975), 441–457.

[71] M.E. Gurtin and R.C. MacCamy, *On the diffusion of biological populations*, Math. Biosci. **33** (1977), 35–49.

[72] Y. Haitao, *Multiplicity and asymptotic behavior of positive solutions for a singular semilinear elliptic problem*, J. Differential Equations **189** (2003), 487–512.

[73] J.K. Hale, *Asymptotic Behavior of Dissipative Systems*, Amer. Math. Soc. Providence, RI (1989).

[74] D. Henry, *Geometric Theory of Parabolic Equations*, Lecture Notes in Math., vol. 840, Springer-Verlag, Berlin (1981).

[75] J. Hernández, *Qualitative methods for non-linear diffusion problems*, Nonlinear Diffusion Problems, A. Fasano and M. Primicerio, eds, Lecture Notes in Math., vol. 1224, Springer-Verlag, Berlin (1985), 47–118.

[76] J. Hernández, J. Karátson and P.L. Simon, *Multiplicity for semilinear elliptic equations involving singular nonlinearity*, Nonlinear Anal. (2006).

[77] J. Hernández, F.J. Mancebo and J.M. Vega, *On the linearization of some singular nonlinear elliptic problems and applications*, Ann. Inst. H. Poincaré Anal. Non Linéaire **19** (6) (2002), 777–813.

[78] J. Hernández, F.J. Mancebo and J.M. Vega, *Nonlinear singular elliptic problems: Recent results and open problems*, Progress in Nonlinear Differential Equations and Their Applications, vol. 64, M. Chipot and J. Escher, eds, Birkhäuser, Basel (2005), 227–242.

[79] J. Hernández, F.J. Mancebo and J.M. Vega, *Positive solutions for singular nonlinear elliptic equations*, Preprint (2006).

[80] P. Hess and T. Kato, *On some linear and nonlinear eigenvalue problems with an indefinite weight function*, Commun. Partial Differential Equations **5** (1980), 999–1030.

[81] N. Hirano, C. Saccon and N. Shioji, *Existence of multiple positive solutions for singular elliptic problems with concave and convex nonlinearities*, Adv. Differential Equations **9** (2004), 197–220.

[82] L.I. Kamynin and B.N. Khimchenko, *Development of Aleksandrov's theory of the isotropic strict extremum principle*, Differ. Equ. (English translation) **16** (1980), 181–189.

[83] J. Karátson and P.L. Simon, *Bifurcations of semilinear elliptic equations with convex nonlinearity*, Electron. J. Differ. Equ. **43** (1999), 1–16.

[84] A. Kufner, *Weighted Sobolev Spaces*, Teubner-Verlag, Leipzig (1980).

[85] O.A. Ladyženskaja, V.A. Solonnikov and N.N. Ural'ceva, *Linear and Quasilinear Equations of Parabolic Type*, Amer. Math. Soc., Providence, RI (1968).

[86] T. Laetsch, *Uniqueness of sublinear boundary value problems*, J. Differential Equations **13** (1973), 13–23.

[87] A.V. Lair and A.W. Shaker, *Classical and weak solutions of a singular semilinear elliptic problem*, J. Math. Anal. Appl. **211** (1997), 371–385.

[88] A.C. Lazer and P.J. McKenna, *On a singular nonlinear elliptic boundary value problem*, Proc. Amer. Math. Soc. **111** (1991), 721–730.

[89] J.A. Leach and D.J. Needham, *Matched Asymptotic Expansions in Reaction–Diffusion Theory*, Springer-Verlag, Berlin (2004).

[90] P.L. Lions, *On the existence of positive solutions of semi-linear elliptic equations*, SIAM Rev. **24** (1982), 441–467.

[91] J. López-Gómez, *The maximum principle and the existence of principal eigenvalues for some linear weighted boundary value problems*, J. Differential Equations **127** (1996), 263–294.

[92] A. Manes and A.M. Micheletti, *Un estensione della teoria variazonale classica degli autovalori per operatori ellittici del secondo ordine*, Boll. Unione Mat. Ital. **7** (1973), 285–301.

[93] P.J. McKenna and W. Reichel, *Sign-changing solutions to singular second-order boundary value problems*, Adv. Differential Equations **6** (2001), 441–460.

[94] T. Namba, *Density-dependent dispersal and spatial distribution of a population*, J. Theor. Biol. **86** (1980), 351–363.

[95] T. Ouyang, J. Shi and M. Yao, *Exact multiplicity and bifurcation of solutions of a singular equation*, Preprint (1996).

[96] C.V. Pao, *Nonlinear Parabolic and Elliptic Equations*, Plenum Press, New York (1992).

[97] D. Phillips, *A minimization problem and the regularity of solutions in the presence of a free boundary*, Indiana Univ. Math. J. **32** (1983), 1–17.

[98] M.H. Protter and H.F. Weinberger, *Maximum Principles in Differential Equations*, Springer-Verlag, Berlin (1984).

[99] C. Pucci, *Propietà di massimo e minimo delle soluzioni di equazioni a derivate parziali del secondo ordine di tipo ellittico e parabolico*, Atti Accad. Naz. Lincei Cl. Sci. Fis. Mat. Rend. Lincei **23** (1957), 370–375.

[100] L. Qiu and M. Yao, *Entire positive solutions to a system of nonlinear elliptic equations*, Preprint (2005).

[101] P.H. Rabinowitz, *Some global results for nonlinear eigenvalue problems*, J. Funct. Anal. **7** (1971), 487–513.

[102] D.H. Sattinger, *Monotone methods in nonlinear elliptic and parabolic boundary value problems*, Indiana Univ. Math. J. **21** (1972), 979–1000.

[103] M. Schatzman, *Stationary solutions and asymptotic behavior of a quasilinear degenerate parabolic equation*, Indiana Univ. Math. J. **33** (1984), 1–29.

[104] T. Senba, Y. Ehihara and Y. Furusho, *Dirichlet problem for a semilinear elliptic equation*, Nonlinear Anal. **15** (1990), 299–306.

[105] J. Shi and M. Yao, *On a singular nonlinear semilinear elliptic problem*, Proc. Royal Soc. Edinburgh Sect. A **138** (1998), 1389–1401.

[106] J. Shi and M. Yao, *Positive solutions of elliptic equations with singular nonlinearity*, Electron. J. Differ. Equ. **2005** (04) (2005), 1–11.

[107] J. Smoller, *Shock Waves and Reaction–Diffusion Equations*, Springer-Verlag, Berlin (1983).

[108] J. Spruck, *Uniqueness of a diffusion model of population biology*, Commun. Partial Differential Equations **8** (1983), 53–63.

[109] M. Struwe, *Variational Methods: Applications to Nonlinear Partial Differential Equations and Hamiltonian Systems*, Springer-Verlag, Berlin (1990).

[110] C.A. Stuart, *Existence and approximation of solutions of nonlinear elliptic equations*, Math. Z. **147** (1976), 53–63.

[111] Y. Sun and S. Wu, *Iterative solution for a singular nonlinear elliptic problem*, Appl. Math. Comput. **118** (2001), 53–62.

[112] P. Takáč, *Stabilization of positive solutions for analytic gradient-like systems*, Discrete Contin. Dyn. Syst. **6** (2000), 947–973.

[113] H. Triebel, *Interpolation Theory, Function Spaces, Differential Operators*, North-Holland, Amsterdam (1978).

[114] M. Wiegner, *Blow-up for solutions of some degenerate parabolic equations*, Differential Integral Equations **7** (1994), 1641–1647.

[115] M. Wiegner, *A Degenerate Diffusion Equation with a Nonlinear Source Term*, Nonlinear Anal. **28** (1997), 1977–1995.

[116] S. Wu, *The effect of sublinear term at origin in some elliptic problems*, Proc. Workshop on Morse Theory. Minimax Theory and Their Applications to Nonlinear Differential Equations, International Press, Boston (2003), 257–274.

[117] S. Yijing, W. Shaoping and L. Yiming, *Combined effects of singular and superlinear nonlinearities in some singular boundary value problems*, J. Differential Equations **176** (2001), 511–531.

[118] S. Yijing and L. Shujie, *Structure of ground state solutions of singular semilinear elliptic equations*, Nonlinear Anal. **55** (2003), 399–417.

[119] Z. Zhang, *On a Dirichlet problem with a singular nonlinearity*, J. Math. Anal. Appl. **194** (1995), 103–113.

[120] Z. Zhang, *Nonexistence of positive classical solutions of a singular nonlinear Dirichlet problem with a convection term*, Nonlinear Anal. **27** (1996), 957–961.

[121] Z. Zhang, *Nonlinear elliptic equations with singular boundary conditions*, J. Math. Anal. Appl. **216** (1997), 390–397.

[122] Z. Zhang, *Critical points and positive solutions of singular elliptic boundary value problems*, J. Math. Anal. Appl. **302** (2005), 476–483.

[123] Z. Zhang and J. Cheng, *Existence and optimal estimates of solutions for singular nonlinear Dirichlet problems*, Nonlinear Anal. **57** (2004), 473–484.

[124] Z. Zhang and J. Yu, *On a singular nonlinear Dirichlet problem with a convection term*, SIAM J. Math. Anal. **32** (2000), 916–927.

CHAPTER 5

Schauder-Type Estimates and Applications

Satyanad Kichenassamy

*Laboratoire de Mathématiques, UMR 6056, CNRS and Université de Reims Champagne-Ardenne,
Moulin de la Housse, B.P. 1039, F-51687 Reims cedex 2, France
E-mail: satyanad.kichenassamy@univ-reims.fr*

Contents

1. Introduction 403
   1.1. What are Schauder-type estimates? 403
   1.2. Why do we need Schauder estimates? 404
   1.3. Why so many methods of proof? 405
   1.4. Classification of proofs 406
   1.5. Generalizations and variants 407
   1.6. What process were the Schauder estimates discovered by? 408
   1.7. Outline of the chapter 410
2. Hölder spaces 411
   2.1. First definitions 411
   2.2. Dyadic decomposition 412
   2.3. Weighted norms 413
   2.4. Interpolation inequalities 415
   2.5. Properties of the distance function 416
   2.6. Integral characterization of Hölder continuity 418
3. Interior estimates for the Laplacian 419
   3.1. Direct arguments from potential theory 419
   3.2. $C^{1+\alpha}$ estimates via the maximum principle 423
   3.3. $C^{2+\alpha}$ estimates via Littlewood–Paley theory 425
   3.4. Variational approach 426
   3.5. Other methods 430
4. Perturbation of coefficients 432
   4.1. Basic estimate 432
   4.2. Estimates up to the boundary 434
5. Fuchsian operators on $C^{2+\alpha}$ domains 435
   5.1. First "type (I)" result 436
   5.2. Second "type (I)" result 438

HANDBOOK OF DIFFERENTIAL EQUATIONS
Stationary Partial Differential Equations, volume 3
Edited by M. Chipot and P. Quittner

6. Applications 440
    6.1. Method of continuity 440
    6.2. Basic fixed-point theorems for compact operators 441
    6.3. Fixed-point theory and the Dirichlet problem 445
    6.4. Eigenfunctions and applications 446
    6.5. Method of sub- and supersolutions 448
    6.6. Asymptotics near isolated singularities or at infinity 449
    6.7. Asymptotics for boundary blow-up 451
    6.8. First comparison argument 454
References 461

1. Introduction

The Schauder estimates are among the oldest and most useful tools in the modern theory of elliptic partial differential equations (PDEs). Their influence may be felt in practically all applications of the theory of elliptic boundary-value problems, that is, in fields such as nonlinear diffusion (in biology or environmental sciences), potential theory, field theory or differential geometry and its applications.

Generally speaking, Schauder estimates give Hölder regularity estimates for solutions of elliptic problems with Hölder continuous data; they may be thought of as wide-ranging generalizations of estimates of derivatives of an analytic function in the interior of its domain of analyticity (Cauchy's inequalities) and play a role comparable to that of Cauchy's theory in function theory. They may be viewed as converses to the mean-value theorem: a bound on the solution gives a bound on its derivatives. The estimates generally become false if Hölder continuity is replaced by mere continuity.

Schauder estimates have three aspects, corresponding to three different ways of applying them:

(i) they are regularity results: solutions with minimal regularity must be as regular as data permit;

(ii) they give the boundedness of the inverse of certain elliptic operators;

(iii) they give the compactness of these inverses.

Schauder theory has strongly contributed to the modern idea that solving a PDE is equivalent to obtaining an a priori bound, that is, trying to estimate a solution *before* one has constructed any solution.

We aim in the following pages to give the reader the means to make use of the recent literature on the subject. We assume the reader is familiar with the basic facts of Functional Analysis and elliptic theory (see [11]). For this reason, we give complete proofs of the most commonly used theorems used in actual applications of the estimates; we then survey the main generalizations, with emphasis on recent work. General references on Schauder estimates and their applications include [2,26,35,40,50,53,54,61,64,66,68,72,74].

1.1. *What are Schauder-type estimates?*

It is convenient to distinguish four kinds of estimates: interior, weighted interior, boundary and Fuchsian estimates. Each of them is further divided into second-order and first-order estimates. We begin with the second-order estimates.

The interior Schauder estimate expresses that, if L is a second-order elliptic operator L with Hölder-continuous coefficients,[1] the C^α norm of any second-order derivative of u on the ball of radius r is estimated by the sum of the C^α norm of Lu on the ball of radius $2r$, and the supremum of u on the same ball. It therefore contains the following information:

(i) u is *as smooth as the data allow*: even though Lu is just one particular combination of u and its derivatives of order two or less, the Hölder continuity of Lu implies the same regularity for all second-order derivatives;

---

[1] See Section 2 for the definition of the regularity classes used in this paper. Recall that an operator $L = \sum_{ij} a^{ij} \partial_{ij} + b^i \partial_i + c$ is elliptic if the quadratic form $a^{ij} \xi_i \xi_j$ is positive definite.

(ii) the regularity of u is *local*, in the sense that we require no smoothness assumption on the value of u on the boundary of the ball of radius $2r$;

(iii) the set of all functions u such that $\sup|u|$ and $\|Lu\|_{C^\alpha(|x|<2r)}$ are bounded by a fixed constant M is *relatively compact* in the C^2 topology of the ball of radius r.

The boundary Schauder estimate expresses that if $Lu = f$ on a bounded domain of class $C^{2+\alpha}$ and if u is equal on the boundary to a function of class $C^{2+\alpha}$, then u is of class $C^{2+\alpha}$ up to the boundary.

The scaled, or weighted interior estimates, in their simplest form, express that, if $Lu = f$ is C^α, and f is bounded, then, as one tends to the boundary, (i) ∇u blows up at most like d^{-1}, where d is the distance to the boundary, and $\nabla^2 u$ like d^{-2}; (ii) the expression $|\nabla^2 u(P) - \nabla^2 u(Q)|/|P - Q|^\alpha$ is estimated by $C(\min(d(P), d(Q)))^{-2-\alpha}$.[2] In particular, this estimate does not imply that $d^2 u$ is of class $C^{2+\alpha}$ up to the boundary.

The Fuchsian estimates express that, in the above situation, $d^2 u$ is of class $C^{2+\alpha}$ up to the boundary provided that (i) the a^{ij}/d^2 and b^i/d satisfy a Hölder condition near the boundary and (ii) *either* L satisfies additional sign conditions on the lower-order terms and u is bounded, *or* both u and f satisfy a flatness condition at the boundary. Condition (i) is reminiscent of the scaling behavior of ODE of Fuchs–Frobenius type, hence the terminology.

First-order estimates are similar, with the difference that they give $C^{1+\alpha}$ regularity of the solution if Lu is merely bounded; the conditions on the coefficients of L are also slightly weaker than in the $C^{2+\alpha}$ case. First-order estimates are often as useful as the second-order estimates, and may generalize to nonlinear operators such as the p-Laplacian, for which the $C^{2+\alpha}$ estimates are false.

1.2. *Why do we need Schauder estimates?*

Schauder estimates form the basis of very general existence theorems, because the compactness information they contain makes it possible to apply fixed-point theorems for compact operators (see [11,54,64,66,74]).

Schauder theory has many applications beyond existence theorems; we mention: asymptotic behavior, at infinity or near singularities; properties of eigenfunctions (Riesz–Fredholm theory [11], Krein–Rutman theorem [52]); the method of sub- and supersolutions for nonlinear problems, and bifurcation theory (see [64,68,74]).

Schauder theory has not been rendered obsolete by the more recent developments of Sobolev theory and variational methods for the following reasons:

(i) Schauder theory applies to problems without variational structure;

(ii) it produces existence results without assuming uniqueness;

(iii) it is more convenient than Sobolev theory in the sense that functions in H^k are Hölder continuous only for k greater than the number of space dimensions.

---

[2] Following common practice, we use the "variable constant convention" according to which the same letter C is used to denote constants which may change from line to line. It is consistent as long as (i) the context makes clear on what quantities the constants depend and (ii) one is not interested in the value of the constant, but only in its existence. The convention may have been influenced by an observation by Schauder to the effect that the best constants in Schauder estimates are not well understood.

Of course, modern studies of nonlinear problems often use Sobolev or de Giorgi–Nash theory to obtain a modicum of regularity, and improve it using the Schauder estimates.

This survey is by no means an exhaustive report on regularity theory; in particular, the de Giorgi–Nash theorem on Hölder estimates for operators with bounded measurable coefficients, and the literature it gave rise to, is not discussed. Special results on particular equations such as the Monge–Ampère equation, or the Laplace equation on polyhedral domains, are only briefly discussed. Regularity estimates for parabolic problems, including probabilistic methods for diffusion processes [76], fall outside the scope of this volume devoted to stationary problems, although many of the techniques are similar to those in the stationary case.

1.3. *Why so many methods of proof?*

The wide variety of methods for the derivation of Schauder estimates may be understood as follows: all proofs require the following ingredients:
- an estimate for a model problem (the Laplacian on the unit ball of $\mathbb{R}^n$, a half-space or a half-ball),
- a scaling argument which transfers estimates to balls of radius R,
- a linear change of coordinates which yields the result for constant-coefficient operators,
- a passage to continuous coefficients.

The second step is often formulated in terms of weighted norms involving the distance to the boundary; the third is achieved using Korn's device, which consists in comparing the given operator with the operator with coefficients "frozen" at one point; the fourth is streamlined by the use of interpolation inequalities for (weighted) Hölder norms. In fact, the first three steps follow from the invariance properties of the Laplace operator; the variety of proofs essentially comes from the different ways to exploit these properties of the Laplace operator.

The period 1882–1934 has seen the emergence of *derivations* of estimates of which the definitive form was only gradually discovered. In this first stage, one notices a tendency to try and replace the estimation of Green's function by a direct estimation of the solution. As a result, solution methods based on the construction of integral operators which provide (approximate) inverses developed separately from regularity theory for nonlinear problems, and eventually gave rise to pseudodifferential analysis.

The subsequent period, say from 1934 to 1964, was devoted to a streamlining and elucidation of the methods, and culminated in the generalization to systems [1] of Schauder's estimates. At the end of this period, estimates on Green's function had been evacuated from the variable-coefficient case. They would re-appear indirectly with the introduction of pseudodifferential inverses of elliptic operators, but pseudodifferential techniques with symbols of limited regularity are still not very well understood [77].

Once the estimates had been discovered, it became possible to look for *verifications*: efficient ways to prove that the estimates hold once they have been proved by other methods. This search has brought about a change in perspective, triggered by the needs of new applications: once the passage from constant to variable coefficients had been streamlined, it became clear that potential theory was still needed to prove the estimates in the case of the Laplacian. In other words, all of the refinements of Schauder theory were ultimately based on the direct proof of the estimates for the Laplacian. And the various proofs in this case ultimately make use of the invariance of the Laplacian under translation, rotation and scaling.

Now, a problem in which a singularity occurs at a specific point in space cannot be translation invariant. The first step in handling such problems would be to consider scaled Schauder estimates in balls which become smaller as one approaches the singularity; these "blow-up" methods lead naturally to weighted Schauder estimates.[3] This approach does not yield optimal regularity. From 1990 onwards, the author showed that the correct regularity, first for hyperbolic problems, and more recently, for elliptic problems, may be understood by reducing the problem to a local model of the typical form

$$d^2 \Delta u + \lambda d \nabla d \cdot \nabla u + \mu u = f(d, P), \tag{1}$$

where d is the distance to the singularity locus, λ and μ are usually constants.[4] This leads to the Fuchsian estimates mentioned above, which form part of a systematic technique for finding the asymptotic behavior of solutions (see [47,48,50] and Sections 5 and 6.9).

Note that Fuchsian operators had been studied for their own sake from the 1970s onwards, but the results obtained at this time were slightly weaker than those required for application to nonlinear problems. Of course, the idea of Fuchsian reduction is not to be found in earlier work.

1.4. *Classification of proofs*

The various approaches to the estimates differ in their treatment of the model case, and the characterization of Hölder continuity they use. The modern theory is dominated by the fact that interior H^k estimates for harmonic functions are now considered more or less obvious (see the beginning of the proof of Theorem 16).

A first proof is based on the direct estimation of the Newtonian potential [27,35,41,42, 63,69–71]. A second proof is based on the search for comparison functions, and therefore uses only the maximum principle [9,10]. A third proof rests on the dyadic decomposition of the Fourier transform of u [75]. A variant may be based on a characterization of Hölder continuity by mollification generalizing a property of the Poisson kernel [79]. A fourth proof rests on an integral characterization of Hölder continuity [20,19,32,58,60]. The regularity problem for minimal surfaces has led to a fifth approach: consider scaled versions of the graph of u corresponding to smaller and smaller scales and characterize their limit by a Liouville theorem [73]. A sixth proof consists in rescaling $u - P$ where P is a second-degree polynomial [15,17,39].

---

[3] The question of behavior at infinity is of a similar nature, because infinity may be replaced by an isolated singularity by inversion.

[4] In some cases, it is convenient to allow them to be operators.

1.5. *Generalizations and variants*

The most important cases to which the second-order estimates on bounded smooth domains may be generalized are: higher-order equations of Agmon–Douglis–Nirenberg type, for which it is possible to find a fundamental solution for a model problem with constant coefficients [1] and equations on unbounded domains [22,65]. Scaling interior estimates yields several, nonequivalent results [22,34,47,78]. The first-order estimates may hold under weaker conditions on the coefficients [16,24,25]. It is also possible to obtain estimates in cases when the right-hand side is only Hölder with respect to some of the variables [30]. Slightly stronger results hold in two dimensions ([35], Chapter 12). A simple example in which the model problem is quite difficult is the case of the Laplace equation on a polyhedral domain.[5]

Higher-order estimates may be obtained in the obvious manner, by differentiating the equation, provided the nonlinearities are smooth. The Schauder estimates are actually true for certain fully nonlinear equations with nonsmooth nonlinearities [4,15,35].

There are cases in which the model case is not a linear, constant-coefficient problem: for instance,

(i) the p-Laplacian – also invariant under a similar group – has the property that solutions with right-hand side zero are not necessarily of class C^2 (see [29,44]);

(ii) Fuchsian operators also admit nonsmooth solutions with smooth data [38,47,48];

(iii) subelliptic operators, such as those related to Carnot groups, are not close to the Laplacian either [21];

(iv) even the Laplacian on polyhedral domains presents new features not found in the regularity theory in smooth domains. All this led to a very recent surge of activity on very simple models. Since the simplest nontrivial model beyond the Laplacian is the Fuchsian case, we briefly explain how such problems arise naturally.

When trying to generalize Schauder estimates to problems with boundary degeneracy, we saw that the local model is not the Laplacian any more: it is a problem with quadratic degeneracy of special form; it is *scale invariant* but not *translation invariant*. Let us mention a few further contexts where such PDEs arise: Axisymmetric potential theory leads to problems with singular coefficients such as the (elliptic) Euler–Poisson–Darboux equation

$$u_{rr} + \frac{\lambda}{r} u_r + u_{zz} = -4\pi\rho, \tag{2}$$

where λ is a constant. Many authors, especially Alexander Weinstein (and Hadamard) stressed long ago the usefulness of this equation and noted its remarkable behavior under transformations of the form $u \mapsto r^\gamma u(r, z)$. It may be treated within the framework of the general theory of degenerate elliptic PDEs (Fichera), writing it in the form

$$r u_{rr} + \lambda u_r + r u_{zz} = -4\pi r\rho.$$

In this form, the problem is reminiscent of Legendre's equation, which also admits a linear degeneracy (at ± 1). Motivated by the search for a higher-dimensional generalization of the

---

[5] Separation of variables shows that the smoothness of harmonic functions on a wedge-like domain, with Dirichlet conditions, depends on the opening angle of the wedge.

expansion into Legendre functions to several variables, a general theory of the Dirichlet problem for elliptic equations with linear degeneracy on the boundary was developed in the 1960s and 1970s. The prototype of such problems is

$$d \Delta u + \lambda \nabla d \cdot \nabla u = f(P) \quad \text{in } \Omega, \tag{3}$$

where d is a smooth function of $P \in \Omega$, equivalent to the distance to $\partial\Omega$ near the boundary. An analogue of Schauder estimates may be derived by an explicit computation of Green's function if $\lambda > 0$ [38]; the essential step is the analysis of a model problem on a half-plane, by Laplace transform in the normal variable. This method does not seem to generalize to the case of quadratic degeneracy.

These considerations took a new meaning when, in the 1990s, one realized that nonlinear problems give rise, by a systematic process of reduction (see [49,50]), to problems modeled upon the general Fuchsian-type problem

$$d^2 \Delta u + \lambda d \operatorname{grad} d \cdot \operatorname{grad} u + \mu u = f(P) \quad \text{in } \Omega. \tag{4}$$

Because of the quadratic degeneracy, the Laplace transform is not helpful. Nevertheless, it is possible to analyze indirectly this model problem (see [47,48,50] and Sections 5 and 6.7).

1.6. *What process were the Schauder estimates discovered by?*

Many steps in the derivation of the Schauder estimates become clearer if one recalls the historical development which led from potential theory to the Schauder estimates. For this reason, we give a historical sketch, starting from Poisson (1813).

1.6.1. *Does Poisson's equation hold?* Consider the Newtonian potential in three dimensions:

$$V(P) = \int_{\mathbb{R}^3} \frac{\rho(Q)}{|P - Q|} \, dQ, \tag{5}$$

where $P \in \mathbb{R}^3$ and $|P - Q|$ is the distance from P to Q and integrals are extended over $\mathbb{R}^3$. This integral represents, up to a constant factor, the gravitational potential generated by the mass density $\rho(Q)$, if $\rho \geqslant 0$. If ρ takes positive and negative values, it may be interpreted in terms of an electrostatic potential. If the density is bounded and has limited support, V is defined by a convergent integral, and so is the corresponding force field proportional to the gradient of $-V$, formally given by

$$-\nabla V(P) = \int \frac{Q - P}{|P - Q|^3} \rho(Q) \, dQ.$$

If P lies outside the support of ρ, the integral may be differentiated again, to yield Laplace's equation

$$\Delta V = 0,$$

where $\Delta = \sum_{i=1}^{3} \partial_i^2$. However, if $\rho(P) \neq 0$, differentiation of the force field yields a divergent integral, because $1/|P - Q|^3$ is not integrable. Poisson (in 1813) showed that, if ρ is constant in the neighborhood of P, V nevertheless satisfies Poisson's equation at the point P

$$-\Delta V = 4\pi\rho. \tag{6}$$

Indeed, one may split the density into two parts: a constant density in a ball around P, and a density which vanishes in a neighborhood of P. The first part yields a potential which may be computed exactly: it is quadratic near P; the second yields a solution of Laplace's equation. Gauss [31] then proved that Poisson's equation is valid if the density is continuously differentiable. After investigations by Riemann, Dirichlet and Clausius, Hölder (in 1882) [41] proved that the second derivatives of the potential are continuous, and that Poisson's equation holds, under the *Hölder condition of order* α,

$$\big|\rho(P) - \rho(Q)\big| \leqslant C|P - Q|^\alpha \tag{7}$$

for some $\alpha \in (0, 1)$. In fact, the second derivatives of V also satisfy a Hölder condition. Furthermore, if ρ is merely continuous, the first-order derivatives of V satisfy a Hölder condition for any α. The argument was streamlined by Neumann [62].[6] This substantiates Poisson's idea that the potential should be well approximated by a quadratic function near every point where ρ is well approximated by a constant.

1.6.2. *Emergence of the Dirichlet problem.* At the same time it became clear that the Newtonian potential is merely one among all possible solutions of Poisson's equation; in fact, solutions may be parameterized by their values on the boundary of sufficiently smooth bounded domains $\Omega \subset \mathbb{R}^3$: this leads us to *Dirichlet problem*

$$\begin{cases} -\Delta V = 4\pi\rho & \text{in } \Omega, \\ V = g & \text{on } \partial\Omega. \end{cases} \tag{8}$$

It seemed at first sight that the Dirichlet problem should have a unique solution on the grounds that it should represent the equilibrium potential in Ω when the potential is prescribed on the boundary and continuous. Dirichlet and Riemann worked on the assumption that V could be obtained by minimizing the *Dirichlet integral*

$$E[u, \Omega] = \int_\Omega \big|\nabla u(Q)\big|^2 \, \mathrm{d}Q \tag{9}$$

among all sufficiently regular u which agree with g on $\partial\Omega$. Weierstrass pointed out that such an argument may fail for certain variational principles, and it was only with the advent of Hilbert spaces that a justification of this method could be made, for smooth domains.

---

[6]The continuity of ρ is not sufficient to ensure that V is twice continuously differentiable. Necessary and sufficient conditions for the existence of second derivatives were studied by Petrini.

But then, if we find a function V which admits integrable first-order derivatives and minimizes Dirichlet's integral, how do we know that it has second-order derivatives and that it solves Poisson's equation? There is a second difficulty: the Dirichlet problem may have no continuous solution if the boundary presents a sharp inward spike ("Lebesgue spine"). Even for $\rho = 0$, the Poincaré *balayage* method, reformulated and simplified by Perron into the *method of sub- and supersolutions*, only proves that, for continuous g, there is a unique solution which is continuous up to the boundary if $\partial \Omega$ is well behaved[7] but does not prove that the solution is smooth if the data (Ω, ρ and g) are smooth.

The corresponding issues for equations with variable coefficients and nonlinearities also led to the need for regularity estimates: the Calculus of Variations and Conformal Mapping led to nonlinear elliptic equations such as the equation of minimal surfaces and Liouville's equation ($\Delta u = e^{u}$) in two variables. Picard emphasized the advantages of iterative methods for PDEs. Now, if one wishes to solve iteratively an equation of the form

$$\Delta u = f(u)$$

to fix ideas, one should define a sequence of functions obtained by solving the Poisson equations

$$\Delta u_n = f(u_{n-1})$$

with $n = 1, 2, \ldots$. In view of the above results, it seems appropriate to work in a space of functions the second derivatives of which satisfy a Hölder condition. The first results in this direction seem to be due to Bernstein (see [12]). The *continuity method* may be viewed as a outgrowth of these efforts. But the iterative approach only allows one to reach problems close to Poisson's equation. Other approaches, based on the reduction to an integral equation on the boundary, required detailed estimates on the Green function for operators with variable coefficients. In the course of this development, estimates for second derivatives of solutions of PDEs with variable coefficients in n variables were obtained (Korn, Hopf, Giraud, Kellogg, Schauder and others, see [14,36,37,42,43,55]).

Schauder's approach is different: it reduces the problem to a new fixed-point theorem: the Leray–Schauder theorem for compact operators; the compactness is provided by estimates of second derivatives in Hölder spaces. Schauder's proof bypasses the construction of Green's function for variable-coefficient operators, and opens the door to the solution of wide classes of nonlinear equations.

1.7. *Outline of the chapter*

Section 2 collects several characterizations of Hölder spaces, and gives the main interpolation results which enable the passage from constant to variable coefficients.

Section 3 illustrates the main proof techniques on the case of the interior estimates for the Laplacian.

---

[7]For instance, it is sufficient that $\partial \Omega$ satisfy an exterior sphere condition. A necessary and sufficient condition is due to Wiener.

Section 4 deals with the passage from the model case (Laplacian on a ball) to variable coefficients and general domains.

Section 5 gives the main general-purpose Fuchsian estimates.

Section 6 collects the most important applications; self-contained proofs of the major topological tools are also included.

2. Hölder spaces

2.1. *First definitions*

Let $\Omega \subset \mathbb{R}^n$ be a domain (i.e., an open and connected set).

DEFINITION 1. A function u is Hölder-continuous at the point P of Ω, with exponent $\alpha \in (0, 1)$, if

$$[u]_{\alpha,\Omega,P} := \sup_{Q \in \Omega, Q \neq P} \frac{|u(P) - u(Q)|}{|P - Q|^\alpha} < \infty.$$

It is Hölder-continuous over Ω, or *of class* $C^\alpha(\Omega)$ if it satisfies this condition for every $P \in \Omega$. We write $[u]_{\alpha,\Omega} := \sup_P [u]_{\alpha,\Omega}$.

It is of class $C^\alpha(\Omega)$ if

$$\|u\|_{C^\alpha(\Omega)} := \sup_\Omega |u| + [u]_{\alpha,\Omega}.$$

Functions of class C^α are in particular uniformly continuous. If $\partial\Omega$ is smooth, one can extend u by continuity to a continuous function on $\overline{\Omega}$; for this reason, it is sometimes convenient to write $C^\alpha(\overline{\Omega})$ for $C^\alpha(\Omega)$ in this case, to emphasize that u is continuous up to the boundary.

It is easy to check that

$$[uv]_{\alpha,\Omega} \leqslant \|u\|_{C^\alpha(\Omega)} \|v\|_{C^\alpha(\Omega)}.$$

Higher-order Hölder spaces $C^{k+\alpha}(\Omega)$ are defined in the natural way: first, write $|\nabla^k u|$ for the sum of the absolute values of the derivatives of u of order k, and define $[\nabla^k u]_{\alpha,\Omega}$ similarly. Let

$$\|u\|_{C^k(\Omega)} := \max_{0 \leqslant j \leqslant k} \sup_\Omega |\nabla^j u|$$

and

$$\|u\|_{C^{k+\alpha}(\Omega)} := \|u\|_{C^k(\Omega)} + [\nabla^k u]_{\alpha,\Omega}.$$

In all these norms, the reference domain Ω will be omitted whenever it is clear from the context.

2.2. *Dyadic decomposition*

The Hölder spaces defined above are all Banach spaces, but smooth functions are not dense in them: even in one dimension, if (f_m) is a sequence of smooth functions and $f \in C^\alpha(\mathbb{R})$ is such that $\|f - f_m\|_{C^\alpha(\mathbb{R})} \to 0$, one proves easily that for any P and any $\varepsilon > 0$, there is a neighborhood of P on which $|f(P) - f(Q)| \leqslant \varepsilon|P - Q|^\alpha$. In other words, $\lim_{Q \to P} |f(P) - f(Q)||P - Q|^{-\alpha} = 0$. Any function f which does not satisfy this property cannot be approximated by smooth functions in the C^α norm.

Nevertheless, there is a systematic way to decompose Hölder-continuous functions on $\mathbb{R}^n$ into a uniformly convergent sum of smooth functions: define the Fourier transform of u by

$$\hat{u}(\xi) = \int_{\mathbb{R}^n} e^{-ix \cdot \xi} u(x) \, dx$$

and consider $\varphi \in C_0^\infty(\mathbb{R})$ such that $0 \leqslant \varphi \leqslant 1$, $\varphi = 1$ for $|x| \leqslant 1$, $\varphi = 0$ for $|x| \geqslant 0$. Define

$$\hat{u}_0 = \varphi(|\xi|)\hat{u}(\xi), \qquad \hat{u}_j = \left[\varphi\left(2^{-j}|\xi|\right) - \varphi\left(2^{-(j-1)}|\xi|\right)\right]\hat{u}(\xi) \quad \text{for } j \geqslant 1.$$

We let $\hat{v}_j = \hat{u}_0 + \cdots + \hat{u}_j$.

DEFINITION 2. The decomposition

$$u = \sum_{j \geqslant 0} u_j$$

is the Littlewood–Paley (LP), or dyadic decomposition of u [75].

By Fourier inversion, we have

$$u_j = \psi_j * u \quad \text{with } \psi_j(x) = 2^{jn}\psi\left(2^j x\right),$$

where $\psi(x) = (2\pi)^{-n} \int_{\mathbb{R}^n} [\varphi(|\xi|/2) - \varphi(|\xi|)] \exp(ix \cdot \xi) \, d\xi$. Note that $\hat{\psi}$ vanishes near the origin; in particular, $\hat{\psi}_j(0) = \int_{\mathbb{R}^n} \psi_j \, dx = 0$.

THEOREM 3. *Let* $0 < \alpha < 1$.
1. (*Bernstein's inequality.*) *There is a constant* C *such that, for any* k, $\sup_x (|\nabla^k u_j| + |\nabla^k v_j| \leqslant C2^{jk} \sup_x |u(x)|$.
2. *If* $u \in C^\alpha(\mathbb{R}^n)$, *there is a constant* C *independent of* j *such that* $\sup_x |u_j(x)| \leqslant C2^{-j\alpha}\|u\|_{C^\alpha}$.
3. *Conversely, if the above inequality holds for every* $j \geqslant 1$, *then* $u \in C^\alpha(\mathbb{R}^n)$.

PROOF. (1) On the one hand, we have $|u_j(x)| \leqslant \|\psi\|_{L^1} \sup |u|$ and $|v_j(x)| \leqslant \|\phi\|_{L^1} \times \sup |u|$. On the other hand, if a is a multiindex of length k,

$$\left| \nabla^a u_j(x) \right| = \left| \int u(y) 2^{jk} \nabla^a \psi \left[2^j (x-y) \right] 2^{jn} \, dy \right|$$

$$= C 2^{jk} \sup |u|.$$

The result follows.

(2) Since $\int \psi(y) \, dy = 0$, u_j may be written, for $j \geqslant 1$,

$$u_j(x) = \int \left[u(x-y) - u(x) \right] 2^{jn} \psi \left(2^j y \right) dy$$

$$= \int \left[u \left(x - \frac{z}{2^j} \right) - u(x) \right] \psi(z) \, dz.$$

If $u \in C^\alpha$, it follows that

$$\left| u_j(x) \right| \leqslant 2^{-j\alpha} [u]_\alpha \int |z|^\alpha |\psi(z)| \, dz.$$

(3) Conversely, if the u_j are of order $2^{-j\alpha}$, the series $u_0 + u_1 + \cdots$ converges uniformly. Call its sum u; it is readily seen that the u_j do give its LP decomposition. We may apply (1) to $u_{j-1} + u_j + u_{j+1}$ and obtain

$$\sup_x \left| \nabla u_j(x) \right| \leqslant C 2^{j(1-\alpha)}.$$

Writing $u = v_{j-1} + w_j$, where $w_j = u_j + u_{j+1} + \cdots$, we find that

$$\left| u(x) - u(y) \right| \leqslant \sum_{j>k} |x-y| \sup |\nabla u_j| + 2 \sup |w_j|$$

$$\leqslant C|x-y| \left(1 + \cdots + 2^{(j-1)(1-\alpha)} \right) + C 2^{-j\alpha}$$

$$\leqslant C \left[2^{-j\alpha} + |x-y| 2^{j(1-\alpha)} \right].$$

Choose j such that $2^{-j} \leqslant |x-y| \leqslant 2^{-(j-1)}$. A bound on $[u]_\alpha$ follows. $\qquad\square$

2.3. *Weighted norms*

Several of the results we shall prove estimate the Hölder norm of a function u on a ball of radius R in terms of bounds on the ball of radius $2R$ with the same center. In order to exploit these inequalities in a systematic fashion, it is useful to define Hölder norms weighted by the distance to the boundary.

Let $\Omega \neq \mathbb{R}^n$ and let $d(P)$ denote the distance from P to $\partial\Omega$, and

$$d_{P,Q} = \min\big(d(P), d(Q)\big).$$

Let also δ be a smooth function in all of Ω which is equivalent to d for d sufficiently small.[8] Define, for $k = 0, 1, \dots,$

$$\|u\|_{k,\Omega}^{\#} = \sum_{j=0}^{k} \sup_{\Omega} d^j \big|\nabla^j u\big|$$

and

$$\|u\|_{k+\alpha,\Omega}^{\#} = \sum_{j=0}^{k} \big\|\delta^j u\big\|_{C^{j+\alpha}(\Omega)}.$$

The spaces corresponding to these norms are called $C_{\#}^{k}(\Omega)$, $C_{\#}^{k+\alpha}(\Omega)$. The space $C_{*}^{k+\alpha}(\Omega)$ has the norm

$$\|u\|_{k+\alpha,\Omega}^{*} = \|u\|_{k,\Omega}^{*} + [u]_{k+\alpha,\Omega},$$

where

$$\|u\|_{k,\Omega}^{*} = \sum_{j=0}^{k} [u]_{j,\Omega}^{*},$$

with $[u]_{k,\Omega}^{*} = \sup_{\Omega} d^k |\nabla^k u|$ and

$$[u]_{k+\alpha,\Omega}^{*} = \sup_{P,Q\in\Omega} d_{P,Q}^{k+\alpha} \frac{|\nabla^k u(P) - \nabla^k u(Q)|}{|P - Q|^{\alpha}}.$$

We also need the further definitions:

$$[u]_{\alpha,\Omega}^{(\sigma)} = \sup_{P,Q\in\Omega} d_{P,Q}^{\alpha+\sigma} \frac{|u(P) - u(Q)|}{|P - Q|^{\alpha}}, \qquad \|u\|_{\alpha,\Omega}^{(\sigma)} = \sup_{\Omega} |d^\sigma u| + [u]_{\alpha,\Omega}^{(\sigma)}.$$

As before, the mention of Ω will be omitted whenever possible.

---

[8] Such a function is easy to construct if Ω is bounded and smooth. Note that even in this case, d is smooth only near the boundary; see Section 2.5.

2.4. *Interpolation inequalities*

THEOREM 4. *For any $\varepsilon > 0$, there is a constant C_ε such that*

$$[u]_1^* \leqslant \varepsilon [u]_2^* + C_\varepsilon \sup |u|,$$

$$[u]_1^* \leqslant \varepsilon [u]_{1+\alpha}^* + C_\varepsilon \sup |u|,$$

$$[u]_2^* \leqslant \varepsilon [u]_{2+\alpha}^* + C_\varepsilon [u]_1^*,$$

$$[u]_{1+\alpha}^* \leqslant \varepsilon [u]_2^* + C_\varepsilon \sup |u|.$$

PROOF. Recall the elementary inequality, for C^2 functions of one variable $t \in [a, b]$,[9]

$$\sup |f'| \leqslant \frac{2}{b-a} \sup |f| + (b-a) \sup |f''|.$$

Fix $\theta \in (0, \frac{1}{2})$ and $P \in \Omega$. Let $r = \theta d(P)$. If $Q \in B_r(P)$ and $Z \in \partial\Omega$, we have

$$|Z - Q| \geqslant |Z - P| - |P - Q| \geqslant d(P)(1 - \theta) \geqslant \frac{1}{2} d(P) \geqslant r \geqslant |P - Q|.$$

It follows in particular that $d(Q) \geqslant d(P)(1 - \theta) \geqslant \frac{1}{2} d(P)$, hence

$$d_{P,Q} \geqslant \frac{1}{2} d(P).$$

Applying the elementary inequality to u restricted to the segment $[P, P + re_i]$,[10] where e_i is the ith basis vector, we find

$$|\partial_i u(P)| \leqslant \frac{2}{r} \sup_{B_r} |u| + r \sup_{B_r} |\partial_{ii} u|.$$

It follows that

$$\sup_{B_r} |\partial_{ii} u| \leqslant \sup d(Q)^{-2} \sup d(Q)^2 |\partial_{ii} u| \leqslant \frac{[u]_2^*}{d(P)^2 (1 - \theta)^2}.$$

Therefore

$$[u]_1^* = \sup_{B_r} |d(Q) \, \partial_i u(Q)| \leqslant \frac{2}{\theta} \sup |u| + \frac{\theta}{(1 - \theta)^2} [u]_2^*.$$

If we choose θ so that $\theta(1 - \theta)^{-2} \leqslant \varepsilon$, we arrive at the first of the desired inequalities.

---

[9]For the proof, write $f'(t) = f'(s) + \int_s^t f''(\tau) \, d\tau$, where s satisfies $f'(s) = (f(b) - f(a))/(b - a)$.

[10]By the choice of r, this segment lies entirely within Ω.

For the second inequality, we note that, using again the mean-value theorem, there is on the segment $[P, P + re_i]$ some $\widetilde{P}$ such that $|\partial_i u(\widetilde{P})| \leqslant (2/r)\sup_{B_r}|u|$. It follows that

$$
\left|\partial_i u(P)\right| \leqslant \left|\partial_i u(\widetilde{P})\right| + \left|\partial_i u(P) - \partial_i u(\widetilde{P})\right|
$$

$$
\leqslant \frac{2}{r}\sup_{\Omega}|u|
$$

$$
+ \left(\sup_{Q\in B_r(P)} d_{P,Q}^{-1-\alpha}\right)|P - \widetilde{P}|^{\alpha} \sup_{Q\in B_r(P)} d_{P,Q}^{1+\alpha}\frac{|\nabla u(P) - \nabla u(Q)|}{|P - Q|^{\alpha}}
$$

$$
\leqslant \frac{2}{r}\sup_{\Omega}|u| + \left(\frac{2}{d(P)}\right)^{1+\alpha}\left(\theta d(P)\right)^{\alpha}[u]_{1+\alpha}^{*}.
$$

Multiplying through by $d(P) = r/\theta$, we find the second inequality.

A similar argument gives the third and fourth inequalities. $\qquad\square$

2.5. *Properties of the distance function*

We prove a few properties of the function $d(x) = \text{dist}(x, \partial\Omega)$, when Ω is bounded with boundary of class $C^{2+\alpha}$. Without smoothness assumption on the boundary, all we can say is that d is Lipschitz; indeed, since the boundary is compact, there is, for every x a $z \in \partial\Omega$ such that $d(x) = |x - z|$. If y is any other point in Ω, we have $d(y) \leqslant |y - z| \leqslant |y - x| + |x - z| = |y - x| + d(x)$. It follows that $|d(x) - d(y)| \leqslant |x - y|$. For more regular $\partial\Omega$, we have the following results.

THEOREM 5. *If $\partial\Omega$ is bounded of class $C^{2+\alpha}$,*

(1) there is a $\delta > 0$ such that every point such that $d(x) < \delta$ has a unique nearest point on the boundary;

(2) in this domain, d is of class $C^{2+\alpha}$; furthermore, $|\nabla d| = 1$ and

$$
-\Delta d = \sum_{j}\frac{\kappa_j}{1 - \kappa_j d},
$$

where $\kappa_1, \ldots, \kappa_{n-1}$ are the principal curvatures of $\partial\Omega$. In particular, $-\Delta d/(n - 1)$ is equal to the mean curvature of the boundary.

PROOF. We work near the origin, which we may take on $\partial\Omega$. Our proofs will give local information near the origin, which can be made global by a standard compactness argument.

Choose the coordinate axes so that Ω is locally represented $\{x_n > h(x')\}$, where $x' = (x_1, \ldots, x_n)$ and h is of class $C^{2+\alpha}$ with $h(0) = 0$ and $\nabla h(0) = 0$. We may also assume that the axes are rotated so that the Hessian $(\partial_{ij}h(0))$ is diagonal. Its eigenvalues are, by definition, the principal curvatures $\kappa_1, \ldots, \kappa_{n-1}$ of the boundary. Their average is, again by definition, the mean curvature of the boundary.

At any boundary point, the vector with components

$$(\nu_i) = \frac{-\partial_1 h, \ldots, -\partial_{n-1} h, 1}{\sqrt{1 + |\nabla h|^2}}$$

is the *inward* normal to $\partial\Omega$ at that point. One checks $\partial_j \nu_i(0) = -\partial_{ij} h(0) = \kappa_j \delta_{ij}$ for i and j less than n. Thus, ν is of class C^1. For any $T > 0$ and $y \in \mathbb{R}^{n-1}$, both small, consider the point $x(Y, T) = (Y, h(Y)) + T\nu(Y)$; this represents the point obtained by traveling the distance T into Ω, starting from the boundary point $(Y, h(Y))$, and traveling along the normal. We write

$$\Phi : (Y, T) \mapsto x(Y, T).$$

We want to prove that all points in a neighborhood of the boundary are obtained by this process, in a unique manner: in other words, $(Y, h(Y))$ is the unique closest point from $x(Y, T)$ on the boundary, provided that T is positive and small. It suffices to argue for $Y = 0$; in that case, since h is C^2, it is bounded below by an expression of the form $a|Y|^2$, which implies that for T sufficiently small, the sphere of radius T about $x(Y, T)$ contains no point of the boundary except the origin.[11] We may now consider the new coordinate system (Y, T) thus defined. We compute, for $Y = 0$, but T not necessarily zero,

$$\frac{\partial x_i}{\partial Y_j} = \delta_{ij}(1 - \kappa_j T)$$

for i and $j < n$, while

$$\frac{\partial x_n}{\partial Y_j} = \frac{\partial x_i}{\partial T} = 0, \qquad \frac{\partial x_n}{\partial T} = 1.$$

The inverse function theorem shows that, near the origin, the map Φ and its inverse are of class C^1, and that the Jacobian of Φ^{-1} is, for $Y = 0$,

$$\frac{\partial(Y, T)}{\partial x} = \mathrm{diag}\left(\frac{1}{1 - \kappa_1 T}, \ldots, \frac{1}{1 - \kappa_{n-1} T}, 1\right).$$

In fact, Φ^{-1} is of class $C^{1+\alpha}$. Indeed, Φ has this regularity, and the differential of Φ^{-1} is given by $[\Phi' \circ \Phi^{-1}]^{-1}$, and the map $A \mapsto A^{-1}$ on invertible matrices is a smooth map. Since $\nu(Y)$, which is equal to the gradient of d, is a $C^{1+\alpha}$ function of Y, we see that it is also a $C^{1+\alpha}$ function of the x coordinates. It follows that d is of class $C^{2+\alpha}$. The computation of the second derivatives of d is now a consequence of the computation of the first-order derivatives of ν.

It follows from this discussion that $T = d$ near the boundary, and that $|\nabla d| = 1$; in fact, $\nabla d = \nu$. $\qquad\square$

---

[11] Indeed, the equation of this sphere is $x_n = T - \sqrt{T^2 - |Y|^2}$, which, by inspection, is bounded below by $a|Y|^2$ for $2aT < 1$.

2.6. *Integral characterization of Hölder continuity*

Let Ω be a bounded domain. Write $\Omega(x, r)$ for $\Omega \cap B(x, r)$. We assume that the measure of $\Omega(x, r)$ is at least Ar^n for some positive constant A, if $x \in \Omega$ and $r \leqslant 1$. This condition is easily verified if Ω has a smooth boundary. Define the average of u:

$$u_{x,r} = \left| \Omega(x, r) \right|^{-1} \int_{\Omega(x,r)} u \, dx.$$

THEOREM 6. *The space $C^\alpha(\Omega)$ coincides with the space of (classes of) measurable functions which satisfy*

$$\int_{\Omega(x,r)} \left| u(y) - u_{x,r} \right|^2 dy \leqslant Cr^{n+2\alpha}$$

for $0 < r < \operatorname{diam} \Omega$. The smallest constant C, denoted by $\|u\|_{\mathcal{L}^{2,n+2\alpha}}$ is equivalent to the $C^\alpha(\Omega)$ norm.

REMARK 1. If one defines $\mathcal{L}^{p,\lambda}$ by the property $\int_{\Omega(x,r)} |u(y) - u_{x,r}|^p \, dy \leqslant Cr^\lambda$, with $n < \lambda < n + p$, one obtains a characterization of the space $C^{(\lambda-n)/p}$.

PROOF OF THEOREM 6. The integral estimate is clearly true for Hölder continuous functions. Let us therefore focus on the converse. We first prove that u is uniformly approximated by its averages, and then derive a modulus of continuity for u.

If $x_0 \in \Omega$ and $0 < \rho < r \leqslant 1$, we have

$$A\rho^n |u_{x_0,\rho} - u_{x_0,r}|^2 \leqslant \int_{\Omega(x_0,\rho)} |u_{x_0,\rho} - u_{x_0,r}|^2 \, dx$$

$$\leqslant 2\left(\int_{\Omega(x_0,\rho)} |u - u_{x_0,\rho}|^2 \, dx + \int_{\Omega(x_0,r)} |u - u_{x_0,r}|^2 \, dx \right)$$

$$\leqslant C\left(r^\lambda + \rho^\lambda \right).$$

Letting $r_j = r2^{-j}$ and $u_j = u_{x_0,\rho_j}$ for $j \geqslant 0$, we find

$$|u_{j+1} - u_j| \leqslant C2^{j(n-\lambda)/2} r^{(\lambda-n)/2} = C2^{-j\alpha} r^\alpha.$$

For almost every x_0, the Lebesgue differentiation theorem ensures that $u_j \to u(x_0)$ as $j \to \infty$. It follows that

$$|u(x_0) - u_{x_0,r}| \leqslant \sum_j |u_{j+1} - u_j| \leqslant Cr^\alpha.$$

Since $u_{x,r}$ is continuous in x and converges uniformly as $r \to 0$, it follows that u may be identified, after modification on a null set, with a continuous function.

To estimate its modulus of continuity, we need the following result.

LEMMA 7. *Let $u \in \mathcal{L}^{2,n+2\alpha}$, x, y two points in Ω and $r = |x - y|$; we have*

$$|u_{x,r} - u_{y,r}| \leqslant Cr^\alpha.$$

PROOF. We may assume $r = |x - y| \leqslant 1$. If $z \in B_r(x)$, we have $|z - y| \leqslant r + |x - y| \leqslant 2r$. Therefore $\Omega(y, 2r) \supset \Omega(x, r)$. It follows that $\Omega(x, 2r) \cap \Omega(y, 2r) \supset \Omega(x, r)$ has measure Ar^n at least. We therefore have

$$\left|\Omega(x, 2r) \cap \Omega(y, 2r)\right|\left|u_{x,2r} - u_{y,2r}\right|$$

$$\leqslant \int_{\Omega(x,2r)} \left|u(z) - u_{x,2r}\right| dz + \int_{\Omega(y,2r)} \left|u(z) - u_{y,2r}\right| dz$$

$$\leqslant \left[\int_{\Omega(x,2r)} \left|u(z) - u_{x,2r}\right|^2 dz\right]^{1/2} \left|\Omega(x, 2r)\right|^{1/2}$$

$$+ \left[\int_{\Omega(y,2r)} \left|u(z) - u_{y,2r}\right|^2 dz\right]^{1/2} \left|\Omega(y, 2r)\right|^{1/2}$$

$$\leqslant Cr^{\alpha+n/2}r^{n/2}.$$

It follows that

$$|u_{x,2r} - u_{y,2r}| \leqslant CA^{-1}r^\alpha, \qquad \qquad \Box$$

To conclude the proof of the theorem, it suffices to estimate $|u(x) - u(y)|$ by $|u(x) - u_{x,r}| + |u_{x,r} - u_{y,r}| + |u_{y,r} - u(y)| \leqslant 2Cr^\alpha + |u_{x,r} - u_{y,r}|$. $\qquad \Box$

3. Interior estimates for the Laplacian

3.1. *Direct arguments from potential theory*

Let $n \geqslant 2$, and let $B_R(P)$ denote the open ball of radius R about P. Mention of the point P is omitted whenever this does not create confusion. The volume of B_R is $\omega_n R^n$ and its surface $n\omega_n R^{n-1}$. The Newtonian potential in n dimensions is

$$g(P, Q) = \frac{|P - Q|^{2-n}}{(2 - n)n\omega_n} \quad \text{for } n \geqslant 3$$

and

$$\frac{1}{2\pi} \ln|P - Q| \quad \text{for } n = 2.$$

It is helpful to note that:

1. The derivatives of g of order $k \geqslant 1$ with respect to P are $O(|P - Q|^{2-n-k})$.

2. The average of each of these second derivatives over the sphere $\{Q\colon |P-Q|=\text{const}\}$ vanishes.[12]

Next, consider, for $f \in L^1 \cap L^\infty(\mathbb{R}^n)$, the integral

$$u(P) = \int_{\mathbb{R}^n} g(P, Q) f(Q)\, dQ.$$

We wish to estimate u and its derivatives in terms of bounds on f. Because of the behavior of g as $P \to Q$, g and its first derivatives are locally integrable, but its second derivative is not.

It is easy to see that, if the point P lies outside the support of f, u is smooth near P and satisfies $\Delta u = 0$. For this reason, it suffices to study the case in which the density f is supported in a neighborhood of P.

We prove in the next three theorems: (i) a pointwise bound on u and its first-order derivatives; (ii) a representation of the second-order derivatives which involves only locally integrable functions; (iii) a direct estimation of $\nabla^2 u(P) - \nabla^2 u(Q)$ using this representation.

THEOREM 8. *If f vanishes outside $B_R(0)$, we have*

$$\sup_{B_R}\big(|u| + |\nabla u|\big) \leqslant CR^2 \sup f,$$

and ∇u is given by formally differentiating the integral defining u.

PROOF. Consider a cut-off function $\varphi_\varepsilon(P, Q) := \varphi(|P - Q|/\varepsilon)$, where $\varphi(t)$ is smooth, takes its values between 0 and 1, vanishes for $t \leqslant 1$ and equals 1 for $t \geqslant 2$. Considering the functions

$$u_\varepsilon(P) = \int g(P, Q)\varphi_\varepsilon(P, Q) f(Q)\, dQ,$$

which are smooth, it is easy to see that the $\partial_i u_\varepsilon$ converge uniformly, as $\varepsilon \downarrow 0$, to $\int \partial_i g(P, Q) f(Q)\, dQ$. Similarly, u_ε converges to u. Therefore, u is continuously differentiable. Using the growth properties of g and its derivatives, we may estimate $\partial_i u(P)$ by

$$C \int_{B_{2R}(P)} C|P - Q|^{1-n} \sup |f|\, dQ,$$

because $B_R(0) \subset B_{2R}(P)$. Taking polar coordinates centered at P, the result follows. $\square$

The case of second derivatives is more delicate, since the second derivatives of g are not locally integrable. We know since Poisson that the integral defining u is smooth near

---

[12] To check this, it is useful to note that the average of x_i^2/r^2 over the unit sphere $\{r = 1\}$ is equal to $1/n$, and similarly, using symmetry, the average of $(x_i - y_i)(x_j - y_j)/|x - y|^2$ over the set $\{|x - y| = \text{const}\}$ vanishes for $i \neq j$.

P if f is constant in a neighborhood of P. This suggests a reduction to the case in which f vanishes at P. We therefore first prove, for such f, a representation of the second-order derivatives which circumvents the fact that the second-order derivatives of g are not integrable.

THEOREM 9. *If f has support in a bounded neighborhood Ω of the origin, with smooth boundary, and if $f \in C^\alpha(\mathbb{R}^n)$ for some $\alpha \in (0, 1)$, then all second-order derivatives of u exist, and are equal to*

$$w_{ij} := \int_\Omega \partial_{ij} g(P, Q)\big[f(Q) - f(P)\big]\,\mathrm{d}Q - f(P) \int_{\partial\Omega} \partial_i g\, n_j\,\mathrm{d}s(Q),$$

where derivatives of g are taken with respect to its first argument and n_j are the components of the outward normal to $\partial\Omega$.

PROOF. To establish the existence of second derivatives, we consider

$$v_{i\varepsilon}(P) = \int \partial_i g(P, Q)\varphi_\varepsilon(P, Q)f(Q)\,\mathrm{d}Q,$$

which converges pointwise to $\partial_i u(P)$; in fact, since $1 - \varphi_\varepsilon$ is supported by a ball of radius 2ε, a direct computation yields $|u_i - v_{i\varepsilon}|(P) = \mathrm{O}(\varepsilon \sup |f|)$. Now, writing $P = (x_i)$ and $Q = (y_i)$, we have

$$\partial_j v_{i\varepsilon}(P) = \int_\Omega \partial_{x_j}(\varphi_\varepsilon \partial_{x_i} g)(P, Q)\big[f(Q) - f(P)\big]\,\mathrm{d}Q$$

$$+ f(P) \int_\Omega (\varphi_\varepsilon \partial_{x_i} g)(P, Q)\,\mathrm{d}Q.$$

Now, since φ_ε and g only depend on $|P - Q|$, we may replace $\partial/\partial x_j$ by $-\partial/\partial y_j$ and integrate by parts. This yields

$$\partial_j v_{i\varepsilon}(P) = \int_\Omega \partial_{x_j}(\varphi_\varepsilon \partial_{x_i} g)(P, Q)\big[f(Q) - f(P)\big]\,\mathrm{d}Q$$

$$- f(P) \int_{\partial\Omega} \varphi_\varepsilon \partial_{x_i} g(P, Q)n_j(Q)\,\mathrm{d}s(Q).$$

We may now estimate the difference $\partial_j v_{i\varepsilon} - w_{ij}$ using the same method as for the first-order derivatives. It follows that $\partial_{ij} u = w_{ij}$. $\qquad\square$

We now give the main estimate for second-order derivatives.

THEOREM 10. *Let*

$$u(P) = \int_{B_{2R}(0)} g(P, Q)f(Q)\,\mathrm{d}Q,$$

where $f \in C^\alpha(B_{2R})$, with $0 < \alpha < 1$. Then

$$\sup_{B_R} |\nabla^2 u| + [\nabla^2 u]_{\alpha, B_R} \leqslant C\left(\sup_{B_{2R}} |f| + R^\alpha [f]_{\alpha, B_{2R}}\right). \tag{10}$$

PROOF. We wish to estimate the regularity of $\partial_{ij} u$; we therefore study $|\partial_{ij} u(P) - \partial_{ij} u(P')|$ for P, P' in $B_R(0)$, where the second derivatives are given by the expressions in the previous theorem. The main step is to decompose the first integrand in the resulting expression for $w_{ij}(P) - w_{ij}(P')$ into

$$[f(Q) - f(P')][\partial_{ij} g(P, Q) - \partial_{ij} g(P', Q)] + [f(P') - f(P)]\partial_{ij} g(P, Q).$$

We therefore need to estimate the following quantities:

    (I) $f(P)[\partial_i g(P, Q) - \partial_i g(P', Q)]$ for $Q \in \partial B_{2R}$,

    (II) $[f(P) - f(P')]\partial_i g(P', Q)$ for $Q \in \partial B_{2R}$,

    (III) $[f(P') - f(P)]\partial_{ij} g(P, Q)$ for $Q \in B_{2R}$,

    (IV) $[f(Q) - f(P')][\partial_{ij} g(P, Q) - \partial_{ij} g(P', Q)]$ for $Q \in B_{2R}$.

The first boundary term (I) is easy to estimate using the mean-value theorem:

$$\left| \partial_i g(P, Q) - \partial_i g(P', Q) \right| \leqslant |P - P'| \sup_{\xi \in [P, P']} \left| \nabla \partial_i g(\xi, Q) \right|.$$

Now, since $Q \in \partial B_{2R}$, and $\xi \in B_R$, we have $|\xi - Q| \geqslant 2R - R = R$, hence the supremum in the above formula is bounded by a multiple of R^{-n}. Integrating we get a contribution $O(|P - P'|/R)$, which is a fortiori $O(|P - P'|^\alpha / R^\alpha)$.

Expression (II) is $O(|P - P'|^\alpha)$ since f is of class C^α.

To estimate (III) and (IV), let $r_0 = |P - P'|$ and M be the midpoint of $[P, P']$. We distinguish two cases:

    (i) When $|Q - M| > r_0$, the distance from Q to any point on the segment $[P, P']$ is comparable to $|Q - M|$; this will enable a direct estimation of (IV) using the mean-value theorem, and of (III) by integration by parts.

    (ii) On the set on which $|Q - M| \leqslant r_0$, we may directly estimate the sum of (III) and (IV); the smallness of the region of integration compensates the singularity of the derivatives of g.

We begin with the first case: consider first the integral of (III) over the set

$$A := \left\{ Q \in B_{2R} : |Q - M| > r_0 \right\}.$$

Its boundary is included in $\partial B_{2R}(0) \cup \partial B_{r_0}(M)$. Integrating by parts and using the fact that, on this set, $|P - Q|$ is bounded below by $\min(R, r_0/2)$, we find that (III) $= O(|P - P'|^\alpha)$. For the term (IV), integrated over the same set, we estimate $\partial_{ij} g(P, Q) - \partial_{ij} g(P', Q)$ by $C|P - P'||\xi - Q|^{-n-1}$, for some $\xi \in [P, P']$. Using the Hölder continuity of f, the integral of (IV) is estimated by

$$C r_0 \frac{|Q - P'|^\alpha}{|Q - \xi|^{n+1}}.$$

Its integral over A is estimated by its integral over

$$A' := \{Q: |Q - M| > r_0\}.$$

On A',

$$|Q - P'| \leqslant |Q - M| + |M - P'| = |Q - M| + \frac{1}{2}r_0 \leqslant \frac{3}{2}|Q - M|.$$

On the other hand,

$$|Q - \xi| \geqslant |Q - M| - |M - \xi| \geqslant |Q - M| - \frac{1}{2}r_0 \geqslant \frac{1}{2}|Q - M|.$$

Combining the two pieces of information, we find

$$\int_{A'} Cr_0 \frac{|Q - P'|^\alpha}{|Q - \xi|^{n+1}} \, dQ \leqslant Cr_0 \int_{A'} |Q - M|^{\alpha - n - 1} \, dQ$$

$$= Cr_0 \int_{r_0}^{\infty} r^{\alpha - 2} \, dr = C|P - P'|^\alpha.$$

This completes the analysis of the integrals of (III) and (IV) over A.

It remains to consider (III) and (IV) over the part of B_{2R} on which $|Q - M| \leqslant r_0$. In this case, $|P - Q| \leqslant |P - M| + |M - Q| \leqslant \frac{3}{2}r_0$, and similarly for $|P' - Q|$. We therefore estimate directly the sum of (III) and (IV), namely

$$[f(Q) - f(P)]\partial_{ij} g(P, Q) - [f(Q) - f(P')]\partial_{ij} g(P', Q),$$

by

$$C[f]_{\alpha, B_{2R}} \int_{|Q - M| < r_0} \left(|Q - P|^{\alpha - n} + |Q - P'|^{\alpha - n}\right) dQ$$

$$\leqslant C[f]_{\alpha, B_{2R}} \int_0^{3r_0/2} |Q - P|^{\alpha - 1} \, d|Q - P| \leqslant Cr_0^\alpha.$$

Since $r_0 = |P - P'|$, this completes the proof. $\qquad\square$

3.2. $C^{1+\alpha}$ estimates via the maximum principle

We give two estimates for function such that Δu is bounded. The result is essentially optimal, and relies only on the maximum principle. The choice of comparison functions is motivated by numerical approximations for second-order derivatives; in this sense, the argument may be compared with Nirenberg's "method of translations" for the proof of

L^2-type estimates. Second-order estimates may also be derived by this method, but the choice of comparison functions is much more complicated.

We begin with the C^1 estimate.

THEOREM 11. *If $\Delta u = f$ on $K = \{|x_i| < s$ for $i = 1, \ldots, n\}$, then*

$$|\partial_n u(0)| \leqslant \frac{n}{s} \sup_K |u| + \frac{d}{2} \sup_K |f|.$$

PROOF. Let $M = \sup_K |u|$, $N = \sup_K |f|$,

$$v(x', x_n) = \frac{1}{2}\left[u(x', x_n) - u(x', -x_n)\right]$$

and

$$w(x', x_n) = \frac{M}{s^2}\left[|x'|^2 + x_n\left(ns - (n-1)x_n\right)\right] + \frac{1}{2}Nx_n(s - x_n).$$

Applying the maximum principle to $w \pm v$ on $K \cap \{0 < x_n < s\}$ we obtain

$$\frac{1}{2x_n}\left|u(x', x_n) - u(x', -x_n)\right| \leqslant \frac{M}{s^2}\left(ns - (n-1)x_n\right) + \frac{N}{2}(s - x_n).$$

Letting $x_n \to 0$, one finds the desired inequality. $\qquad\square$

We now turn to the continuity of the gradient of u. The result implies interior $C^{1+\alpha}$ regularity for every $\alpha < 1$.

THEOREM 12. *Let $\mu = \sup_K |\nabla u|$. There is a constant k such that*

$$\frac{1}{2}\left|\partial_i u(0, x_n) - \partial_i u(0, -x_n)\right| \leqslant \mu \frac{x_n}{s} + kx_n \ln \frac{x_n}{s}$$

for $|x_n| \leqslant s/4$.

PROOF. It suffices to prove the result for $i = n$ and $i = n - 1$.

For the case $i = n$, we consider the function of $n + 1$ variables (x', y, z) defined by

$$\phi(x', y, z) = \frac{1}{4}\left[u(x', y + z) - u(x', y - z) - u(x', -y + z) + u(x', -y - z)\right],$$

and the operator $L = \sum_{i<n} \partial_i^2 + \frac{1}{2}(\partial_y^2 + \partial_z^2)$, so that one checks $|L\phi| \leqslant N$. We then compare ϕ with

$$W = \frac{4M}{s}yz + kyz \ln \frac{2s}{y + z},$$

where

$$k = \frac{4}{3}\left(N + \frac{8M}{s^2}(n-1)\right),$$

on the set $K' = \{|x_i| < \frac{s}{2} \text{ if } i \leqslant n-1; 0 < y, z < \frac{s}{4}\}$. Since

$$LW = \frac{8M}{s^2}(n-1) + k\left[-1 + \frac{yz}{(y+z)^2}\right] \leqslant \frac{8M}{s^2}(n-1) - \frac{3}{4}k = -N$$

on K', the maximum principle yields

$$|\phi(0, y, z)| \leqslant y\left[\frac{4\mu}{s} + k\ln\frac{2s}{y+z}\right].$$

Letting $z \to 0$ gives the first inequality in the theorem.

For the case $i = n-1$, we work with functions of n variables $(\tilde{x}, y, z)$, where $\tilde{x} = (x_1, \ldots, x_{n-2})$, on the set $K'' = \{|x_i| < \frac{s}{2} \text{ if } i \leqslant n-2; 0 < y, z < \frac{s}{2}\}$, and the auxiliary functions

$$\psi(\tilde{x}, y, z) = \frac{1}{4}\left[u(\tilde{x}, y, z) - u(\tilde{x}, y, -z) - u(\tilde{x}, -y, z) + u(\tilde{x}, -y, -z)\right]$$

and

$$\widetilde{W} = \frac{4L}{s^2}|\tilde{x}|^2 + yz\left[\frac{4\mu}{s} + \tilde{k}\ln\frac{2s}{y+z}\right],$$

where $\tilde{k} = (2/3)[N + (8M/s^2)(n-2)]$. One finds $|\Delta\psi| \leqslant N$ and $-\Delta\widetilde{W} \geqslant N$ on K'', and the maximum principle yields the desired result as before. $\qquad\square$

3.3. $C^{2+\alpha}$ estimates via Littlewood–Paley theory

LP decomposition provides a simple proof of the basic interior estimate for the constant-coefficient case. This is essentially due to the fact that the Fourier transform is rotation-invariant and has a simple scaling behavior. The argument is however tailored to isotropic situations. We give the argument in its simplest form, with no aim at generality.

THEOREM 13. *Let $\rho \in C^\alpha(\mathbb{R}^n)$ be such that $\hat{\rho} = O(|\xi|^5)$ near $\xi = 0$. Then there is a $u \in C^{2+\alpha}(\mathbb{R}^n)$ such that $-\Delta u = \rho$.*

REMARK 2. The condition $\hat{\rho} = O(|\xi|^5)$ means that the first few moments of ρ vanish; it may be achieved by subtracting from ρ a smooth potential with prescribed multipolar moments.

PROOF OF THEOREM 13. Consider the LP decomposition $\rho_0 + \rho_1 + \cdots$ of ρ. Define u_j by $\hat{u}_j = \hat{\rho}_j/|\xi|^{-2}$. Then $u = u_0 + u_1 + \cdots$ is well defined, and the flatness assumption ensures that u_0 and its first two derivatives are bounded. In particular, u_0 is of class $C^{2+\alpha}$. Consider the Fourier transform of $-\partial_{kl}u$

$$\xi_k \xi_l \hat{u}_j(\xi) = \frac{\xi_k \xi_l}{|\xi|^2} \hat{\rho}(\xi) \hat{\psi}\left(2^j |\xi|\right) = \hat{\rho}(\xi) \frac{(2^j \xi_k)(2^j \xi_l) \hat{\psi}(2^j |\xi|)}{(2^j |\xi|)^2}.$$

Recall that $\hat{\psi}$ is flat at the origin, so that there is no singularity for $\xi = 0$. Applying point (2) of Theorem 3 to the decomposition of ρ in which ψ would be replaced by ψ', with $\hat{\psi}'(\xi) = \xi_k \xi_l |\xi|^{-2} \hat{\psi}(|\xi|)$, we find $\sup |(\partial_{kl}u)_j| = O(2^{-(2+\alpha)j})$. From the characterization of $C^{2+\alpha}$ spaces, the result follows. $\square$

3.4. *Variational approach*

We turn to a different approach, based on the integral characterization of Hölder spaces. The techniques involved have many other applications beyond the one discussed here; in particular, they allow a "direct approach" to regularity theory for minimizers of coercive functionals, without having to consider the Euler equation. We begin with a simple result.

LEMMA 14. *For any* $u \in H^1(B_R(x_0))$ *and any* $r < 1$, $\int_{B_r} |u - c|^2 \, dx$ *is minimum when* $c = u_{x_0,r}$ *and Poincaré's inequality*

$$\int_{B_r(x_0)} |u - u_{x_0,r}|^2 \, dx \leqslant C \int_{B_r(x_0)} |\nabla u|^2 \, dx$$

holds. If we assume in addition that u is harmonic and $0 < a < 1$, we have

$$\int_{B_{ar}(x_0)} |\nabla u|^2 \leqslant c(a) r^{-2} \int_{ar < |x-x_0| < r} |u - u_{x_0,r}|^2 \, dx.$$

The right-hand side is in particular estimated by $c(a) r^{-2} \int_{B_r(x_0)} |u|^2$.

PROOF. Poincaré's inequality is classical, see, e.g., [11]. Let $x_0 = 0$ and choose a smooth, nonnegative function η supported by B_r, equal to 1 for $|x| \leqslant ar$, such that $|\nabla \eta| \leqslant C/r$.[13] Multiply the Laplace equation by $(u - m)\eta^2$, where m is any constant. We find, integrating by parts,

$$\int_{B_r} \left[\eta^2 |\nabla u|^2 + 2(u - m)\nabla \eta \cdot \nabla u\right] dx = 0,$$

---

[13] It suffices to find such a function η_0 for the case $r = 1$ and then let $\eta(x) = \eta_0(x/r)$.

hence, estimating $\nabla\eta$ by C/r and using Hölder's inequality, we find

$$\int_{\eta=1} |\nabla u|^2 \, dx \leqslant \int \eta^2 |\nabla u|^2 \, dx \leqslant \frac{C}{r^2} \int_{\nabla\eta\neq 0} |u - m|^2 \, dx.$$

Taking m to be the average of u on B_r, we obtain in particular the desired inequality. The inequality corresponding to $m = 0$ is also occasionally useful. $\qquad\square$

Since the derivatives of a harmonic function are themselves harmonic, this result implies that higher-order derivatives are locally square-integrable; the Sobolev inequality then shows easily that any harmonic function is smooth. We now turn to a more precise estimate which enables one to compare a harmonic function and its spherical mean.

THEOREM 15. *Let u be harmonic in the ball of radius R_0 about $x_0 \in \mathbb{R}^n$. If $0 < r < R < R_0$, we have*

$$\sup_{B_{R/2}(x_0)} |u|^2 \leqslant C R^{-n} \int_{B_R(x_0)} |u|^2 \, dx, \tag{11}$$

$$\int_{B_r(x_0)} |u|^2 \, dx \leqslant C \left(\frac{r}{R}\right)^n \int_{B_R(x_0)} |u|^2 \, dx, \tag{12}$$

$$\int_{B_r(x_0)} |u - u_{x_0,r}|^2 \, dx \leqslant C \left(\frac{r}{R}\right)^{n+2} \int_{B_R(x_0)} |u - u_{x_0,R}|^2 \, dx, \tag{13}$$

where C is independent of u, r, R and R_0.

PROOF. We may take $x_0 = 0$. It suffices to prove the first inequality for $R = 1$ and scale variables. If k is an integer, we find, applying the preceding lemma repeatedly,

$$\int_{B_{1/2}} |\nabla^k u|^2 \, dx \leqslant c(k) \int_{B_1} u^2 \, dx.$$

The result follows by the Sobolev inequality.

Regarding the last two inequalities, it suffices to prove them for $r \leqslant R/2$ since they are obvious for $r \geqslant R$. In that case, we have, using the first inequality,

$$\int_{B_r} |u|^2 \, dx \leqslant \omega_n r^n \sup_{B_{R/2}} |u|^2 \leqslant C \left(\frac{r}{R}\right)^n \int_{B_R} |u|^2 \, dx$$

as desired. Similarly, since the derivatives of u are also harmonic, Poincaré's inequality yields

$$\int_{B_r} |u - u_{x_0,r}|^2 \, dx \leqslant C\rho^2 \int_{B_r} |\nabla u|^2 \, dx \leqslant C\rho^2 \left(\frac{r}{R}\right)^n \int_{B_{3R/4}(x_0)} |\nabla u|^2 \, dx.$$

We conclude using Lemma 14. $\square$

We turn to the estimation of second derivatives of the solutions of Poisson's equation $-\Delta v = f$. It is equivalent to seek an estimate for the *first* derivatives of solutions of $-\Delta u = \partial_k f$. It turns out to be convenient to consider more generally the problem

$$\Delta u + \sum_k \partial_k f^k = 0, \tag{14}$$

where $u \in H^1(B_R)$ and the f^k are of class C^α. Recall that function of class C^α correspond to the class $\mathcal{L}^{2,\lambda}$ for $\lambda = n + 2\alpha$. This suggests the following theorem.

THEOREM 16. *Assume that* $\mathbf{f} := (f_1, \ldots, f_n) \in \mathcal{L}^{2,\lambda}$ *with* $0 \leqslant \lambda < n + 2$, *and that* $u \in H^1(B_R)$ *solves* (14), *then* ∇u *is locally of class* $\mathcal{L}^{2,\lambda}$. *In particular, if the* f^k *are locally* C^α, *with* $0 < \alpha < 1$, *so is* ∇u.

PROOF. We must analyze the behavior of the integrals

$$F(r) := \int_{B_r(x_0)} \left| \nabla u - (\nabla u)_{x_0, r} \right|^2 dx,$$

defined for given $x_0 \in B_R$ and $r < R - |x_0|$, as $r \to 0$. We first prove the estimate

$$F(\rho) \leqslant A \left(\frac{\rho}{r} \right)^{n+2} F(r) + B r^\lambda \tag{15}$$

for $\rho < r$. We then deduce from it an estimate of the form $F(r) = O(r^\lambda)$, from which the result follows.

Consider the solution of $\Delta v = 0$ in $B_r(x_0)$ such that $u - v \in H_0^1(B_r(x_0))$. From Theorem 15, we have the inequality

$$\int_{B_\rho} \left| \nabla v - (\nabla v)_{x_0, \rho} \right|^2 dx \leqslant C \left(\frac{\rho}{r} \right)^{n+2} \int_{B_r} \left| \nabla v - (\nabla v)_{x_0, \rho} \right|^2 dx.$$

For any $w \in H_0^1(B_r(x_0))$, we have, writing $\partial_k f^k = \partial_k (f^k - f^k_{x_0, r})$,

$$\int_{B_r} \nabla (u - v) \cdot \nabla w \, dx = - \int (\mathbf{f} - \mathbf{f}_{x_0, r}) \cdot \nabla w \, dx.$$

Taking $w = u - v$, we find

$$\int_{B_r} \left| \nabla (u - v) \right|^2 \leqslant \int_{B_r} |\mathbf{f} - \mathbf{f}_{x_0, r}|^2 \, dx.$$

From now on, we omit the mention of the point x_0 in averages. Since $\nabla u - (\nabla u)_\rho = [\nabla v - (\nabla v)_\rho] + [\nabla w - (\nabla w)_\rho]$ and

$$\int_{B_\rho} |\nabla w - (\nabla w)_\rho|^2 \, dx \leqslant \int_{B_\rho} |\nabla w|^2 \, dx \leqslant \int_{B_r} |\nabla w|^2 \, dx,$$

we find

$$\int_{B_\rho} |\nabla u - (\nabla u)_\rho|^2 \, dx \leqslant 2 \int_{B_\rho} |\nabla v - (\nabla v)_\rho|^2 \, dx + 2 \int_{B_r} |\nabla w|^2 \, dx$$

$$\leqslant C \left(\frac{\rho}{r} \right)^{n+2} \int_{B_r} |\nabla v - (\nabla v)_r|^2 \, dx + C \int_{B_r} |\mathbf{f} - \mathbf{f}_r|^2 \, dx.$$

The second term is $O(r^\lambda)$ thanks to the hypothesis on $\mathbf{f}$. We now estimate the first term in terms of $F(r)$. To this end, we need the following result.

LEMMA 17. $\int_{B_r} |\nabla v - (\nabla v)_r|^2 \, dx \leqslant \int_{B_r} |\nabla u - (\nabla u)_r|^2 \, dx.$

PROOF. Since $u - v$ is in H_0^1, we have

$$\int_{B_r} |\nabla v|^2 \, dx \leqslant \int_{B_r} |\nabla u|^2 \, dx$$

and, as soon as $\mathbf{g}$ is constant,

$$\int_{B_r} \nabla(v - u) \cdot \mathbf{g} \, dx = 0.$$

It follows that

$$\int_{B_r} |\nabla v - (\nabla v)_r|^2 \, dx - \int_{B_r} |\nabla u - (\nabla u)_r|^2 \, dx$$

$$= \int_{B_r} \nabla(v - u) \cdot \left(\nabla(v + u) - (\nabla(v + u))_r \right) dx$$

$$= \int_{B_r} \left(|\nabla v|^2 - |\nabla u|^2 \right) dx \leqslant 0. \qquad \Box$$

The proof of inequality (15) is now complete. To conclude the proof of the theorem, we argue as follows: Fix $\gamma \in (\lambda, n + 2)$ and choose $t \in (0, 1)$ such that

$$2At^{n+2} \leqslant t^\gamma.$$

Fix j such that $t^{j+1}r < \rho \leqslant t^j r$. We find, since F is nondecreasing,

$$
\begin{aligned}
F(\rho) \leqslant F\left(t^j\right) &\leqslant t^\gamma F\left(t^{j-1}r\right) + B\left(t^{j-1}r\right)^\lambda \\
&\leqslant t^\gamma\left[t^\gamma F\left(t^{j-2}r\right) + B\left(t^{j-2}r\right)^\lambda\right] + B\left(t^{j-1}r\right)^\lambda \\
&= t^{2\gamma} F\left(t^{j-2}r\right) + Br^\lambda t^{(j-1)\lambda}\left[1 + t^{\gamma-\lambda}\right] \\
&\leqslant \cdots \leqslant t^{j\gamma} F(r) + Br^\lambda \frac{t^{(j-1)\lambda}}{1 - t^{\gamma-\lambda}} \\
&\leqslant t^{-\gamma}\left(\frac{\rho}{r}\right)^\gamma F(r) + \frac{Bt^{-2\lambda}}{1 - t^{\gamma-\lambda}}\left(\frac{\rho}{r}\right)^\lambda.
\end{aligned}
$$

Since $\gamma > \lambda$, this implies $F(\rho) = O(\rho^\lambda)$ as desired. This gives the Hölder regularity of the gradient of u. $\qquad\square$

REMARK 3. For more general problems, it is useful to note that the last part of the argument also applies in the more general situation in which F is a nonnegative, nondecreasing function satisfying

$$
F(\rho) \leqslant A\left[\left(\frac{\rho}{r}\right)^a + \varepsilon\right]F(r) + Br^b
$$

for $0 < \rho < r \leqslant R$, with $a > b$. Taking t as before, we find that if $\varepsilon \leqslant t^a$, then F satisfies an inequality of the form

$$
F(\rho) \leqslant c(a, b, A)\left[\left(\frac{\rho}{r}\right)^b F(R) + B\rho^b\right].
$$

REMARK 4. A similar argument may be applies to nondivergence operators with Hölder-continuous coefficients, using the previous remark. As expected, the argument consists in writing the operator as the sum of a constant-coefficient operator and an operator with small coefficients.

3.5. *Other methods*

We briefly outline two other approaches.

3.5.1. *A regularization method.*   For any standard mollifier $\rho_\varepsilon(x) = \varepsilon^{-n}\rho(x/\varepsilon)$, consider, for any $u(x)$,

$$
u_\varepsilon(x) = \rho_\varepsilon * u = \int \rho(z)u(x - \varepsilon z)\, dz.
$$

It is easy to see that $u(x) - u_\varepsilon(x) = \mathrm{O}(\varepsilon^\alpha)$ if $u \in C^\alpha(\mathbb{R}^n)$, and that, similarly, the derivatives of u_ε with respect to ε or x are $\mathrm{O}(\varepsilon^{\alpha-1})$. Conversely, if these derivatives are $\leqslant M\varepsilon^{\alpha-1}$, it turns out that one may estimate the Hölder constant of u: first of all,

$$\left|u(x) - u(x_\varepsilon)\right| \leqslant \varepsilon \int_0^1 \left|\frac{\partial u}{\partial \varepsilon}\right|(x, \varepsilon\sigma)\,\mathrm{d}\sigma \leqslant M\varepsilon^\alpha \int_0^1 \frac{\mathrm{d}\sigma}{\sigma^{1-\alpha}} = \frac{M\varepsilon^\alpha}{\alpha}.$$

We now find the estimate

$$\left|u(x) - u(y)\right| = \left|u(x) - u(x_\varepsilon)\right| + \left|u(x_\varepsilon) - u(y_\varepsilon)\right| + \left|u(y_\varepsilon) - u(y)\right|$$

$$= \frac{2}{\alpha} M\varepsilon^\alpha + |x - y|\left|\nabla_x u_\varepsilon(z)\right|$$

for some $z \in [x, y]$. One then estimates $|\nabla_x u_\varepsilon(z)|$ by $M\varepsilon^{\alpha-1}$ and takes $\varepsilon = |x - y|$. One also proves that mixed derivatives of u_ε with respect to x and ε are controlled by the Hölder norm of second derivatives of u with respect to x.

One then considers equation $-\Delta u = f(x) = f(x_0) + g(x)$, where $|g| \leqslant R^\alpha[f]_{\alpha, B_R(x_0)}$. We have $-\Delta u_\varepsilon = g_\varepsilon$. If ∇^2 represents any second-order derivative, in x and ε, we find $-\Delta\nabla^2 u_\varepsilon = \nabla^2 g_\varepsilon$. Applying the interior C^1 estimate Theorem 11, and letting $R = N\varepsilon$, where N is to be chosen later, one gets

$$\varepsilon^{1-\alpha}\left|\nabla\partial_{ij}u(x_0, \varepsilon)\right| \leqslant C\left\{N^{\alpha-1}\left[\nabla_x^2 u\right]_\alpha + N^{\alpha+1}R^{-\alpha} \sup_{B_{(1+N)\varepsilon}(x_0)} |g|\right\},$$

where one has estimated $\sup_{B_R(x_0)} |\nabla^2 g|$ by $\varepsilon^{-2}\sup_{B_{R+\varepsilon}(x_0)} |g|$. This quantity is itself estimated by $[f]_\alpha(R + \varepsilon)^\alpha$. Taking N so large that $CN^{\alpha-1} < 1/2$, we find an estimate of $[\nabla_x^2 u]_\alpha$, as desired.

3.5.2. *Blow-up method.* We sketch the idea of the proof of the interior estimate for the Laplacian; a similar idea, with somewhat more complicated details, applies to other situations.

Assume there is no estimate of the form $[\nabla^2 u]_\alpha \leqslant C[\Delta u]_\alpha$ for functions of class $C^{2+\alpha}(\mathbb{R}^n)$. In that case, there must be some sequence u_k such that

$$\left[\nabla^2 u_k\right]_\alpha = 1 > 2k[\Delta u_k]_\alpha.$$

We may therefore find indices i and j, and sequences x_k, a_k of vectors in $\mathbb{R}^n$ such that

$$1 \geqslant \frac{1}{|a_k|^\alpha}\left|\partial_{ij}u_k(x_k) - u_k(x_k + a_k)\right| \geqslant \frac{1}{2} \geqslant k[\Delta u_k]_\alpha.$$

Subtracting an affine function, we may assume that u_k and its first-order derivatives vanish at the point x_k. Subtracting a quadratic function, we may also assume that Δu_k vanishes at x_k. Performing a (k-dependent) rotation of axes, we may also assume that

$a_k = (h_k, 0, \ldots, 0)$. Considering $v_k(y) = h_k^{-(2+\alpha)} u_k(x_k + h_k y)$ we see that $[\Delta v_k]_\alpha = [\Delta u_k]_\alpha \to 0$, while v_k and its first-order derivatives vanish at the origin and grows at most like $|y|^{2+\alpha}$ at infinity. In addition, we have

$$\left| \partial_{ij} v_k(0) - v_k(e_1) \right| \geq \frac{1}{2}, \tag{16}$$

where $e_1 = (1, 0, \ldots, 0)$. After extraction of a subsequence, we are left with a harmonic function which grows at most like $|y|^{2+\alpha}$, and satisfies equation (16). A variant of the Liouville property ensures that v is quadratic, which contradicts (16).

4. Perturbation of coefficients

4.1. *Basic estimate*

Working on a relatively compact subset Ω' of Ω, we may assume that $[u]^*_{2+\alpha}$; since the constants in the various inequalities will not depend on the choice of Ω, the full result will follow.

Consider $x_0 \in \Omega$ and let $r = \theta d(x_0)$ with $\theta \leq 1/2$. Let $L_0 = \sum_{ij} a^{ij}(x_0) \partial_{ij}$ (the "tangential operator", with coefficients "frozen" at x_0). We define

$$F := L_0 u = \sum_{ij} \left(a^{ij}(x_0) - a^{ij}(x) \right) \partial_{ij} u - \sum b^i \partial_i u - cu + f.$$

We apply the constant-coefficient interior estimates on the ball $B_r(x_0)$. Let $y_0 \neq x_0$ such that $d(y_0) \geq d(x_0)$.

If $|x_0 - y_0| < r/2$, we have

$$\left(\frac{r}{2} \right)^{2+\alpha} [\nabla^2 u]_{\alpha, x_0, y_0} \leq C \left(\sup_{B_r} |u| + \sup_{B_r} |r^2 F| + \sup_{B_r \times B_r} r^{2+\alpha} \frac{|F(x) - F(y)|}{|x - y|^\alpha} \right).$$

Therefore

$$d(x_0)^{2+\alpha} [\nabla^2 u]_{\alpha, x_0, y_0} \leq C\theta^{-2-\alpha} \left(\sup |u| + \|F\|^{(2)}_{\alpha, B_r} \right). \tag{17}$$

If $|x_0 - y_0| \geq r/2$, we have

$$d(x_0)^{2+\alpha} [\nabla^2 u]_{\alpha, x_0, y_0} \leq 2[u]^*_2 \frac{d(x_0)^\alpha}{|x_0 - y_0|^\alpha} \leq 2[u]^*_2 \left(\frac{2}{\theta} \right)^\alpha. \tag{18}$$

The issue is therefore the estimation of $\|F\|^{(2)}_{\alpha, B_r}$ in terms of norms of u and its derivatives over Ω.

For clarity, we begin with three lemmas.

LEMMA 18. $\|uv\|_{\alpha,\Omega}^{(s+t)} \leqslant \|u\|_{\alpha,\Omega}^{(s)} \|v\|_{\alpha,\Omega}^{(t)}.$

PROOF. Direct verification. $\qquad\square$

LEMMA 19. *If* $r = \theta d(x, \partial\Omega)$, *with* $0 < \theta \leqslant \frac{1}{2}$ *(so that* $B_r(x) \subset \Omega$*), we have*

$$\|\nabla^2 u\|_{\alpha,B_r}^{(2)} \leqslant 8\big[\theta^2 \|\nabla^2 u\|_{2,\Omega}^* + \theta^{2+\alpha} [u]_{2+\alpha,\Omega}^*\big], \tag{19}$$

$$\|f\|_{\alpha,B_r}^{(2)} \leqslant 8\theta^2 \|f\|_{\alpha,\Omega}^{(2)}. \tag{20}$$

PROOF. We need to estimate, for $y \in B_r(x)$, $d(y, \partial B_r(x))$ and d_{x,y,B_r} in terms of the corresponding distances relative to Ω. On the one hand, $d(y, \partial B_r) \leqslant r - |x - y| \leqslant r = \theta d(x)$. On the other hand, if $z \in B_r(x)$ and $d(y, \partial B_r(x)) \leqslant d(z, \partial B_r(x))$, we have $d_{y,z,B_r} \leqslant \theta d(x)$ and also $d(y) \geqslant d(y, \partial B_r(x)) \geqslant (1 - \theta) d(x)$; it follows that $d(x) \leqslant (1 - \theta)^{-1} d_{x,y,\Omega}$. Therefore

$$d(y, \partial B_r) \leqslant \theta d(x)$$

and

$$d_{y,z,B_r} \leqslant \frac{\theta}{1 - \theta} d_{y,z,\Omega}.$$

The two desired inequalities follow. $\qquad\square$

LEMMA 20. *If* $x \in B_r(x_0)$ *with* $r = \theta d(x_0)$, *with* $0 < \theta \leqslant 1/2$, *we have*

$$\big\|a(x) - a(x_0)\big\|_{\alpha,B_r}^{(0)} \leqslant C\theta^\alpha [a]_{\alpha,\Omega}^*.$$

PROOF. If $d(x) \leqslant d(y)$ and $|x - y| \leqslant r = \theta d(x_0)$ with $\theta \leqslant 1$,

$$\big|a(x) - a(y)\big| \leqslant d(x)^\alpha \frac{|a(x) - a(y)|}{|x - y|^\alpha}\left(\frac{|x - y|}{d(x)}\right)^\alpha \leqslant C\theta^\alpha [a]_\alpha^{(0)}$$

since $(1 - \theta) d(x_0) \leqslant d(x) \leqslant (1 + \theta) d(x_0)$. Therefore, estimating $|a(x) - a(x_0)|$ by $r^\alpha [a]_{\alpha,\Omega}^*$, we find the announced inequality. $\qquad\square$

We now resume the proof of the estimate of $[\nabla^2 u]_\alpha$: first,

$$\big\|(a(x) - a(x_0))\nabla^2 u(x)\big\|_{\alpha,B_r}^{(2)}$$

$$\leqslant \big\|a(x) - a(x_0)\big\|_{\alpha,B_r}^{(0)} \big\|\nabla^2 u\big\|_{\alpha,B_r}^{(2)}$$

$$\leqslant C\theta^{2+\alpha} \|a\|_{\alpha,\Omega}^{(0)} \big(\big\|\nabla^2 u(x)\big\|_{\alpha,\Omega}^* + \theta^\alpha [u]_{2+\alpha,\Omega}^*\big).$$

Similarly,

$$\left\| b \nabla u(x) \right\|_{\alpha, B_r}^{(2)} \leqslant 8\theta^2 \|b\nabla u\|_{\alpha,\Omega}^{(2)}$$

$$\leqslant 8\theta^2 \|b\|_{\alpha,\Omega}^{(1)} \|\nabla u\|_{\alpha,\Omega}^{(1)}$$

$$\leqslant C\theta^2 \|b\|_{\alpha,\Omega}^{(1)} \left\{ \theta^{2\alpha} [u]_{2+\alpha,\Omega}^* + \sup |u| \right\}.$$

Finally,

$$\|cu\|_{\alpha,B_r}^{(2)} \leqslant 8\theta^2 \|cu\|_{\alpha,\Omega}^{(2)} \leqslant 8\theta^2 \|c\|_{\alpha,\Omega}^{(2)} \|u\|_{\alpha,\Omega}^{(0)} \leqslant 8\theta^2 \left\{ \theta^{2\alpha} [u]_{2+\alpha,\Omega}^* + \sup |u| \right\}.$$

It follows that

$$\|F\|_{\alpha,B_r}^{(2)} \leqslant C\theta^{2+2\alpha} [u]_{2+\alpha,\Omega}^* + c(\theta) \left(\sup |u| + \|f\|_{\alpha,\Omega}^{(2)} \right).$$

Therefore, using this inequality in (17) and (18), we find

$$d(x_0)^{2+\alpha} [u]_{2+\alpha,\Omega}^* \leqslant C\theta^\alpha [u]_{2+\alpha,\Omega}^* + c'(\theta) \left(\sup |u| + \|f\|_{\alpha,\Omega}^{(2)} \right).$$

The desired estimate on $[u]_{2+\alpha,\Omega}^*$ follows.

4.2. *Estimates up to the boundary*

The potential-theoretic argument extends easily to the case of Poisson's equation on the half-ball for the following reason: if we apply the formula for the second-order derivatives of the Newtonian potential (Theorem 9) to the case in which Ω is the half-ball $B_R \cap \{x_n > 0\}$, we find that the contribution to the boundary integral of the part of the boundary on which $x_n = 0$ vanishes if $j < n$, because the component n_j of the outward normal then vanishes. The subsequent argument therefore goes through without change, and yields the Hölder continuity up to the boundary of all second-order derivatives of u except $\partial_{x_n}^2 u$; but the latter is given in terms of the former using Poisson's equation. We therefore obtain the $C^{2+\alpha}$ estimates up to the boundary for the Newtonian potential of a density f of class C^α in the half-ball.

To obtain regularity up to $x_n = 0$ for the solution of the Dirichlet problem on the half-ball, we use *Schwarz' reflection principle.*

LEMMA 21. *Let f be of class C^α in the closed half-ball. If u is of class C^2 on the open half-ball of radius R, is continuous on the closed ball, satisfies $\Delta u = f$ in the half-ball, and vanishes for $x_n = 0$, it may be extended to the entire ball as a solution of an equation of the form $\Delta u = f_1$. In particular, u is of class $C^{2+\alpha}$ on any compact subset of the closed half-ball which does not meet the spherical part of its boundary.*

PROOF. Write $x = (x', x_n)$, and extend f to an even function f_1 on the ball. Using the inequality $a^\alpha + b^\alpha \leqslant 2(a+b)^\alpha$, we see that f_1 is of class C^α. Now, the Newtonian potential of f_1 does not satisfy the Dirichlet boundary condition. We therefore consider

$$W(x) := \int_{B_R \cap \{x_n > 0\}} \big[g(x-y) - g(x-\tilde{y})\big] f(y) \, dy,$$

where $\tilde{y} = (y', -y_n)$ is the reflection of y across $\{x_n = 0\}$. It is easy to see that $\Delta W = 0$ in the half-ball, and that $W = 0$ for $x_n = 0$. It is also of class $C^{2+\alpha}$ by the variant of Theorem 9 already indicated. We now consider $V := u - W$, which is harmonic in the half-ball, and vanishes for $x_n = 0$. Extend V to an *odd* function of x_n on the entire ball. Consider the solution of the Dirichlet problem on the ball with boundary data equal to V. This problem has a unique solution V^* by the Poincaré–Perron method, which is independent of Schauder theory. Since $-V^*(x', -x_n)$ solves the same problem, we find that V^* must be odd with respect to x_n. Therefore V^* is also the solution of the Dirichlet problem on the half-ball, with boundary value given by V on the spherical part of the boundary, and value zero on the flat part of the boundary (where $x_n = 0$). Therefore V^* must be equal to V on the half-ball, and therefore on the ball as well. This proves that $V = V^*$ has the required regularity up to $x_n = 0$, as desired. $\qquad\qquad\square$

The perturbation from constant to variable coefficients then proceeds by a variant of the argument used for the interior estimates [1,27,35].

5. Fuchsian operators on $C^{2+\alpha}$ domains

We now consider operators satisfying an asymptotic scale invariance condition near the boundary. These operators arise naturally as local models near singularities through the process of Fuchsian reduction [50]. We develop the basic estimates for such operators without condition on the sign of the lower-order terms. A typical example of the more precise theorems one obtains under such conditions is given in Theorem 43. We distinguish two types of Fuchsian operators.

An operator A is said to be *of type* (I) (on a given domain Ω) if it can be written

$$A = \partial_i \big(d^2 a^{ij} \, \partial_j\big) + db^i \, \partial_i + c,$$

with (a^{ij}) uniformly elliptic and of class C^α, and b^i, c bounded.

REMARK 5. One can also allow terms of the type $\partial_i (b'^i u)$ in Au, if b'^i is of class C^α, but this refinement will not be needed here.

An operator is said to be *of type* (II) if it can be written

$$A = d^2 a^{ij} \, \partial_{ij} + db^i \, \partial_i + c,$$

with (a^{ij}) uniformly elliptic and a^{ij}, b^i, c of class C^α.

REMARK 6. One checks directly that types (I) and (II) are invariant under changes of coordinates of class $C^{2+\alpha}$. In particular, to check that an operator is of type (I) or (II), we may work indifferently in coordinates x or (T, Y) defined in Section 2.5. All proofs will be performed in the (T, Y) coordinates; an operator is of type (II) precisely if it has the above form with d replaced by T, and the coefficients a^{ij}, b^i, c are of class C^{α} as functions of T and Y; a similar statement holds for type (I).

The basic results for type (I) operators are the following theorem, in which Ω' denotes a neighborhood of the boundary.

THEOREM 22. *If $Ag = f$, where f and g are bounded and A is of type* (I) *on Ω', then $d\nabla g$ is bounded and dg and $d^2\nabla g$ belong to $C^{\alpha}(\Omega' \cup \partial\Omega)$.*

THEOREM 23. *If $Ag = df$, where f and g are bounded, $g = O(d^{\alpha})$ and A is of type* (I) *on Ω', then $g \in C^{\alpha}(\Omega' \cup \partial\Omega)$ and $dg \in C^{1+\alpha}(\Omega' \cup \partial\Omega)$.*

These two results are proved in the next subsection. The main result for type (II) operators is the following theorem.

THEOREM 24. *If $Ag = df$, where $f \in C^{\alpha}(\Omega' \cup \partial\Omega)$, $g = O(d^{\alpha})$ and A is of type* (II) *on Ω', then d^2g belongs to $C^{2+\alpha}(\Omega' \cup \partial\Omega)$.*

PROOF. The assumptions ensure that $a^{ij}\,\partial_{ij}(d^2 f)$ is Hölder-continuous and that f is bounded; $d^2 f$ therefore solves a Dirichlet problem to which the Schauder estimates apply near $\partial\Omega$. Therefore $d^2 f$ is of class $C^{2+\alpha}$ up to the boundary. Since we already know that $f \in C^{\alpha}(\overline{\Omega}_\delta)$ and df is of class $C^{1+\alpha}(\overline{\Omega}_\delta)$, we have indeed f of class $C_\sharp^{2+\alpha}(\overline{\Omega}_{\delta'})$ for $\delta' < \delta$. $\qquad\square$

Let $\rho > 0$ and $t \leqslant 1/2$. Throughout the proofs, we shall use the sets

$$Q = \left\{ (T, Y)\colon 0 \leqslant T \leqslant 2 \text{ and } |y| \leqslant 3\rho \right\},$$

$$Q_1 = \left\{ (T, Y)\colon \frac{1}{4} \leqslant T \leqslant 2 \text{ and } |y| \leqslant 2\rho \right\},$$

$$Q_2 = \left\{ (T, Y)\colon \frac{1}{2} \leqslant T \leqslant 1 \text{ and } |y| \leqslant \frac{\rho}{2} \right\},$$

$$Q_3 = \left\{ (T, Y)\colon 0 \leqslant T \leqslant \frac{1}{2} \text{ and } |y| \leqslant \frac{\rho}{2} \right\}.$$

We may assume, by scaling coordinates, that $Q \subset \Omega'$. It suffices to prove the announced regularity on Q_3.

5.1. *First "type (I)" result*

We prove Theorem 22.

Let $Af = g$, with A, f, g satisfying the assumptions of the theorem over Q, and let y_0 be such that $|y_0| \leqslant \rho$.

For $0 < \varepsilon \leqslant 1$ and $(T, Y) \in Q_1$, let

$$f_\varepsilon(T, Y) = f(\varepsilon T, y_0 + \varepsilon Y)$$

and similarly for g and other functions. We have $f_\varepsilon = (Ag)_\varepsilon = A_\varepsilon f_\varepsilon$, where

$$A_\varepsilon = \partial_i \left(T^2 a_\varepsilon^{ij} \partial_j\right) + T b_\varepsilon^i \partial_i + c_\varepsilon$$

is also of type (I), with coefficient norms independent of ε and y_0, and is uniformly elliptic in Q_1.

Interior estimates give

$$\|g_\varepsilon\|_{C^{1+\alpha}(Q_2)} \leqslant M_1 := C_1\left(\|f_\varepsilon\|_{L^\infty(Q_1)} + \|g_\varepsilon\|_{L^\infty(Q_1)}\right). \tag{21}$$

The assumptions of the theorem imply that M_1 is independent of ε and y_0.

We, therefore, find

$$\left|\varepsilon \nabla g(\varepsilon T, y_0 + \varepsilon Y)\right| \leqslant M_1, \tag{22}$$

$$\varepsilon\left|\nabla g(\varepsilon T, y_0 + \varepsilon Y) - \nabla g\left(\varepsilon T', y_0\right)\right| \leqslant M_1\left(\left|T - T'\right| + |Y|\right)^\alpha \tag{23}$$

if $\frac{1}{2} \leqslant T, T' \leqslant 1$ and $|Y| \leqslant \rho/2$. It follows, in particular, taking $Y = 0$, $\varepsilon = t \leqslant 1$, $T = 1$, and recalling that $|y_0| \leqslant \rho$, that

$$\left|t \nabla g(t, y)\right| \leqslant M_1 \quad \text{if } |y| \leqslant \rho, t \leqslant 1. \tag{24}$$

This proves the first statement in the theorem.

Taking $\varepsilon = 2t \leqslant 1$, $T = 1/2$ and letting $y = y_0 + \varepsilon Y$, $t' = \varepsilon T'$,

$$2t\left|\nabla g(t, y) - \nabla g\left(t', y_0\right)\right| \leqslant M_1\left(\left|t - t'\right| + |y - y_0|\right)^\alpha (2t)^{-\alpha}$$

for $|y - y_0| \leqslant \rho t$ and $t \leqslant t' \leqslant 2t \leqslant 1$.

Let us prove that

$$\left|t^2 \nabla g(t, y) - t'^2 \nabla g\left(t', y_0\right)\right| \leqslant M_2\left(\left|t - t'\right| + |y - y_0|\right)^\alpha \tag{25}$$

for $|y|, |y_0| \leqslant \rho$ and $0 \leqslant t \leqslant t' \leqslant \frac{1}{2}$, which will prove

$$t^2 \nabla g \in C^\alpha(Q_3).$$

It suffices to prove this estimate in the two cases: (i) $t = t'$ and (ii) $y = y_0$; the result then follows from the triangle inequality. We distinguish three cases.

1. If $t = t'$, we need only consider the case $|y - y_0| \geqslant \rho t$. We then find

$$t^2 \big| \nabla g(t, y) - \nabla g(t, y_0) \big| \leqslant 2M_1 t \leqslant 2M_1 \frac{|y - y_0|}{\rho}.$$

2. If $y = y_0$ and $t \leqslant t' \leqslant 2t \leqslant 1$, we have $t + t' \leqslant 2t'$, hence

$$\begin{aligned}
\big| t^2 \nabla g(t, y_0) - t'^2 \nabla g(t', y_0) \big| & \\
&\leqslant t^2 \big| \nabla g(t, y_0) - \nabla g(t', y_0) \big| + |t - t'|(t + t') \big| \nabla g(t', y_0) \big| \\
&\leqslant M_1 2^{-1-\alpha} t^{1-\alpha} |t - t'|^\alpha + 2M_1 |t - t'| \\
&\leqslant M_2 |t - t'|^\alpha.
\end{aligned}$$

3. If $y = y_0$ and $2t \leqslant t' \leqslant \frac{1}{2}$, we have $t + t' \leqslant 3(t' - t)$, and

$$\begin{aligned}
\big| t^2 \nabla g(t, y_0) - t'^2 \nabla g(t', y_0) \big| &\leqslant M_1(t + t') \\
&\leqslant 3M_1 |t - t'|.
\end{aligned}$$

This proves estimate (25).

On the other hand, since g and $T\nabla g$ are bounded over Q_3,

$$Tg \in \mathrm{Lip}(Q_3) \subset C^\alpha(Q_3).$$

This completes the proof of Theorem 22.

5.2. *Second "type (I)" result*

We prove Theorem 23.

The argument is similar, except that M_1 is now replaced by $M_3 \varepsilon^\alpha$, with M_3 independent of ε and y_0. It follows that

$$\big| t \nabla g(t, y) \big| \leqslant M_3 t^\alpha \quad \text{if } |y| \leqslant \rho, t \leqslant 1. \tag{26}$$

Taking $\varepsilon = 2t \leqslant 1$, $T = 1/2$, letting $y = y_0 + \varepsilon Y$, $t' = \varepsilon T'$ and noting that $\varepsilon^\alpha(|T - T'| + |Y|)^\alpha = (|t - t'| + |y - y_0|)^\alpha$, we find

$$2t \big| \nabla g(t, y) - \nabla g(t', y_0) \big| \leqslant M_3 \big(|t - t'| + |y - y_0| \big)^\alpha$$

for $|y - y_0| \leqslant \rho t$ and $t \leqslant t' \leqslant 2t \leqslant 1$. Let us prove that

$$\big| t \nabla g(t, y) - t' \nabla g(t', y_0) \big| \leqslant M_4 \big(|t - t'| + |y - y_0| \big)^\alpha \tag{27}$$

for $|y|$, $|y_0| \leqslant \rho$ and $0 \leqslant t \leqslant t' \leqslant \frac{1}{2}$, which will prove

$$T \nabla g \in C^\alpha(Q_3).$$

We again distinguish three cases.

1. If $t = t'$, $|y - y_0| \geqslant \rho t$, we find

$$t \big| \nabla g(t, y) - \nabla g(t, y_0) \big| \leqslant 2M_3 t^\alpha \leqslant 2M_3 \left(\frac{|y - y_0|}{\rho} \right)^\alpha.$$

2. If $y = y_0$ and $t \leqslant t' \leqslant 2t \leqslant 1$, we have $|t - t'| \leqslant t \leqslant t'$, hence

$$\big| t \nabla g(t, y_0) - t' \nabla g(t', y_0) \big|$$
$$\leqslant \frac{1}{2} M_3 |t - t'|^\alpha + |t - t'| \big| \nabla g(t', y_0) \big|$$
$$\leqslant M_3 |t - t'|^\alpha \left(\frac{1}{2} + t'^{1-\alpha} t'^{\alpha-1} \right) \leqslant 2M_3 |t - t'|^\alpha.$$

3. If $y = y_0$ and $2t \leqslant t' \leqslant 1/2$, we have $t \leqslant t' \leqslant 3(t' - t)$, and

$$\big| t \nabla g(t, y_0) - t' \nabla g(t', y_0) \big| \leqslant M_3 \big(t^\alpha + t'^\alpha \big)$$
$$\leqslant 2M_3 \big(3|t - t'| \big)^\alpha.$$

Estimate (27) therefore holds.

The same type of argument shows that

$$g \in C^\alpha(Q_3).$$

In fact, we have, with again $\varepsilon = 2t$, $\|g_\varepsilon\|_{C^\alpha(Q_2)} \leqslant M_5 \varepsilon^\alpha$, where M_5 depends on the right-hand side and the uniform bound assumed on f. This implies

$$\big| g(t, y) - g(t', y_0) \big| \leqslant M_5 \big(|t - t'| + |y - y_0| \big)^\alpha,$$

if $t \leqslant t' \leqslant 2t \leqslant 1$ and $|y - y_0| \leqslant \rho t$. The assumptions of the theorem yield, in particular,

$$\big| g(t, y) \big| \leqslant M_5 t^\alpha$$

for $t \leqslant \frac{1}{2}$ and $|y| \leqslant \rho$.

If $\rho t \leqslant |y - y_0| \leqslant \rho$ and $t \leqslant \frac{1}{2}$, we have

$$\big| g(t, y) - g(t, y_0) \big| \leqslant 2M_5 t^\alpha \leqslant 2M_5 \left(\frac{|y - y_0|}{\rho} \right)^\alpha.$$

If $2t \leqslant t' \leqslant \frac{1}{2}$ and $y = y_0$,

$$\left| g(t, y_0) - g(t', y_0) \right| \leqslant M_5 \left(t^\alpha + t'^\alpha \right) \leqslant 2M_5 \left(3|t - t'| \right)^\alpha .$$

If $t \leqslant t' \leqslant 2t \leqslant 1/2$, we already have

$$\left| g(t, y_0) - g(t', y_0) \right| \leqslant M_5 |t - t'|^\alpha .$$

The Hölder continuity of g follows.

Combining these pieces of information, we conclude that

$$g \in C_{\#}^{1+\alpha}(Q_3).$$

6. Applications

6.1. *Method of continuity*

The principle of the method of continuity consists in solving a problem (P) by embedding it into a one-parameter family (P_t) of problems, such that (P_0) admits a unique solution, and (P_1) coincides with problem (P). One then proves that the set of parameter values for which (P_t) admits a unique solution is both open and closed in $[0, 1]$. The openness usually follows from the implicit function theorem in Hölder spaces, and the closedness from Ascoli's theorem; thus, both steps are made possible by Schauder estimates.

We give an example in which a simplified procedure based on the contraction mapping principle suffices.

THEOREM 25. *Let L be an elliptic operator with C^α coefficients and $c \leqslant 0$, in a bounded domain Ω of class $C^{2+\alpha}$. Then, for any $g \in C^{2+\alpha}(\overline{\Omega})$, $Lu = f$ admits a solution in $C^{2+\alpha}(\overline{\Omega})$ which is equal to g on $\partial \Omega$.*

PROOF. Considering $u - g$, we may restrict our attention to the case $g = 0$. We let $L_t u = t L u + (1 - t) \Delta u$ and consider the problem (P_t) which consists in solving $L_t u = f$ with Dirichlet conditions. L_t is a bounded operator from $C^{2+\alpha}(\overline{\Omega}) \cup \{u = 0 \text{ on } \partial \Omega\}$ to $C^\alpha(\overline{\Omega})$. We know that L_0 is invertible, and we wish to invert L_1. By the maximum principle, the assumption $c \leqslant 0$ ensures that any solution of (P_t) satisfies $\sup_x |u(x)| \leqslant C \sup_x |f(x)|$, with a constant C independent of t. Therefore, if T is the set of t such that L_t is invertible, the Schauder estimates show that L_t^{-1} is bounded, and that its norm admits a bound m independent of t. This fact makes the rest of the proof simpler: let $t \in T$; for any s, the equation $L_t u = f$ is equivalent to $u = L_t^{-1} f + M(t, s)u$, where

$$M(s, t)u = (s - t)L_t^{-1}(L_0 - L_1)u.$$

If $|t - s| < \delta := [m(\|L_0\| + \|L_1\|)]^{-1}$, $M(t, s)$ is a contraction, and (P_s) is uniquely solvable. Covering $[0, 1]$ by a finite number of open intervals of length δ, we find that L_t is invertible for every t. The result follows. $\qquad\square$

For a typical example of the application of the method of continuity, see [2], Theorem 7.14.

6.2. *Basic fixed-point theorems for compact operators*

We prove several versions of the Schauder fixed-point theorem. The first ingredient in the proofs is the Brouwer fixed-point theorem:

THEOREM 26. *A continuous mapping $g : B \to B$, where B is the closed unit ball in $\mathbb{R}^n$, has at least one fixed point.*

PROOF. We begin with the case of smooth g. Assume that g has no fixed point. Let $\tilde{x} = x + a(x - g(x))$, where a is the largest root of the (quadratic) equation $|\tilde{x}|^2 = 1$. The point $\tilde{x}$ is on the intersection of the segment $[x, g(x)]$ with the unit sphere, and is chosen so that x lies between $\tilde{x}$ and $g(x)$. The map from B to its boundary defined by $x \mapsto \tilde{x}$ is well defined and smooth; in fact,

$$0 = |\tilde{x}|^2 - 1 = |x - g(x)|^2 a^2 + 2(x, x - g(x))a + |x|^2 - 1,$$

where $(\cdot, \cdot)$ denotes the usual dot product. The discriminant of this quadratic is $4[(x, x - g(x))^2 + (1 - |x|^2)|x - g(x)|^2]$, which is nonnegative, and vanishes only if $|x| = 1$ and $(x, g(x)) = 1$. Since $g(x)$ has norm one at most, we are in the limiting case of the Cauchy–Schwarz inequality, and we must have $g(x) = x$, which contradicts the hypothesis. Therefore, our quadratic equation has two distinct real roots, obviously smooth.

For $|x| = 1$, we find that $a = 0$ since $(x, x - g(x)) \geq 0$.

Define $f : \mathbb{R} \times B \to \mathbb{R}^n$ by

$$f(t, x_1, \ldots, x_n) = x + ta(x)(x - g(x)).$$

We find by inspection that (i) if $|x| = 1$, $f(t, x) = x$ and $\partial_t f(t, x) = 0$; (ii) $f(0, x) = x$ for every x in B; (iii) $|f(1, x)| = 1$ for every x in B (by construction of a).

Write x_0 for t, and define the determinants

$$D_i = \det\left(f_{x_0}, \ldots, \hat{f}_{x_i}, \ldots, f_{x_n}\right),$$

where i runs from 0 to n; a hat indicates that the corresponding vector is omitted and the subscripts denote derivatives. Define further

$$I(t) = \int_B D_0(t, x)\, dx.$$

We have $I(0) = 1$ since $f(0, x) = x$. For $t = 1$, since f lies on the boundary of the unit sphere, $f_{x_1}, \ldots, f_{x_n}$ are all tangent to the sphere, and are linearly dependent; therefore, $I(1) = 0$.

We prove that $I(t)$ is constant, which will generate a contradiction to the hypothesis that g has no fixed point. We need the following lemma.

LEMMA 27. $\sum_{i=0}^{n}(-1)^i \partial_{x_i} D_i = 0$.

PROOF. We have, for every i,

$$\partial_{x_i} D_i = \sum_{j<i}(-1)^j C_{ij} + \sum_{j>i}(-1)^{j-1} C_{ij},$$

where

$$C_{ij} = \det\left(f_{x_i x_j}, f_{x_0}, \ldots, \hat{f}_{x_i}, \ldots, \hat{f}_{x_j}, \ldots, f_{x_n}\right) = C_{ji}.$$

Therefore $\sum_{i=0}^{n}(-1)^i \partial_{x_i} D_i = \sum_{i,j=0}^{n}(-1)^{i+j} C_{ij}\sigma_{ij}$, where $\sigma_{ij} = 1$ for $j < i$, -1 for $j > i$, and zero for $i = j$. Since $(-1)^{i+j}C_{ij}$ is symmetric in i and j, and σ_{ij} is antisymmetric, the result follows. $\qquad\square$

Now, for $i > 0$, D_i vanishes on the boundary of B because $\partial_t f = 0$ there. If n_i is the ith component of the outward normal to B, we find

$$\int_B \partial_{x_i} D_i \, dx = \int_{\partial B} n_i D_i \, ds = 0.$$

(This may be proved without using Stokes' theorem, by integrating with respect to the x_i variable keeping the others fixed.) Using the lemma, we find

$$\frac{dI(t)}{dt} = \int_B \partial_t D_0 \, dx = \sum_{i>0} \pm\partial_{x_i} D_i \, dx = 0.$$

This completes the proof in the smooth case.

Finally, we extend the result to the case of continuous g. By the Stone–Weierstrass theorem, there is a sequence of polynomial (vector-valued) mappings p_n such that $|g - p_n| \leqslant \varepsilon_n \to 0$ uniformly over B. Since $p_n/(1 + \varepsilon_n)$ maps B to itself, there is a y_n such that $p_n(y_n) = (1 + \varepsilon_n)y_n$. Extracting a subsequence, we may assume y_n has a limit y. It follows that $g(y) = y$. $\qquad\square$

The Brouwer fixed-point theorem may be extended as the following theorem.

THEOREM 28. *Let K be the closed convex hull of a set of N vectors $x_1, \ldots, x_N$ in n-dimensional space. A continuous map from K to itself has a fixed point.*

PROOF. Let $\bar{x} = \frac{1}{N}\sum_k x_k$. Decreasing n if necessary, and relabeling the x_k, we may assume that the $(x_k - \bar{x})_{k \leqslant n}$ generate $\mathbb{R}^n$. We prove that K is homeomorphic to the unit ball, so that the result follows from the Brouwer fixed point theorem. First, $\bar{x}$ is interior to K, because, $\bar{x} + \sum_{k \leqslant n} \varepsilon_k (x_k - \bar{x})$ is a convex combination of the x_k if the ε_k are small enough. Let ε be such that $B_\varepsilon(\bar{x}) \subset \operatorname{int} K$. Let, for any unit vector y, $s(y) = \sup\{s\colon \bar{x} + sy \in L\}$. It is well defined and bounded; also, $s(y) \geqslant \varepsilon$. We need the following lemma.

LEMMA 29. *$s(y)$ is continuous.*

PROOF. If $y_m \to y$ and $s(y_m) \to s$ as $m \to \infty$, with $\bar{x} + s(y_m)y_m \in K$ for all m, we find $\bar{x} + sy \in K$, hence $s \leqslant s(y)$. If $s' < s(y)$, define $t = s'/s(y) \in [0, 1]$, $(1 - t)B_\varepsilon(\bar{x}) + ts(y)y$ is included in K (which is convex), and is a neighborhood of $\bar{x} + s'y$. This implies that $\bar{x} + s'y_m \in K$ for m sufficiently large; it follows that $s(y_m) \geqslant s'$ for m large. Therefore, $s \geqslant s(y)$. $\qquad\square$

We now construct the required homeomorphism from B to K by letting $x \mapsto xs(x/|x|)$, which inverse $x \mapsto x/s(x/|x|)$. We just proved that these maps are continuous at all points other than 0; the continuity at the origin follows from the fact that s and $1/s$ are bounded. $\qquad\square$

We now turn to fixed-point theorems in infinite dimensions.

THEOREM 30. *If K is a compact convex subset of a Banach space E, and $T : K \to K$ is continuity, then T admits a fixed point.*

PROOF. For any integer p, there is an integer $N = N(p)$ and points $x_1, \ldots, x_N$ in K such that $K \subset B(x_1, 1/p) \cap \cdots \cap B(x_N, 1/p)$. Let $B_k = B(x_k, 1/p)$. Consider the closed convex hull K_p of $x_1, \ldots, x_N$ which is a convex set which lies in some finite-dimensional subspace of E; it is a subset of K. The map

$$F_p : x \mapsto \frac{\sum_k x_k d(x, K \setminus B_k)}{\sum_k d(x, K \setminus B_k)}$$

is well defined and continuous (the denominator does not vanish because the B_k cover K). Since any term on the numerator contributes to the sum only if $|x - x_k| \leqslant 1/p$, we have $\|F_p(x) - x\|_E \leqslant 1/p$.

The map $F_p \circ T$ therefore admits a fixed point y_p: $F_p(T(y_p)) = y_p$. We may extract a subsequence $y_{p'}$ which tends to $y \in K$. We have $T(y_{p'}) \to T(y)$, and $\|F_{p'}(T(y_{p'})) - T(y_{p'})\|_E \to 0$. It follows that $T(y) = y$. $\qquad\square$

THEOREM 31. *If K is a closed convex subset of a Banach space E, and $T : K \to K$ is continuous, then if $T(K)$ has compact closure, T admits a fixed point.*

PROOF. One approach would consist in working in the closure of the convex hull of $T(K)$; this requires first proving that this set is compact. A more direct argument is to apply the same method of proof as in the previous theorem, with the difference that K is replaced by the closure of $T(K)$ in the definition of F_p. The map $F_p \circ T$ is continuous on the closed convex hull of $x_1, \ldots, x_N$, and therefore has a fixed point y_p as before. We may extract a subsequence $y_{p'}$ such that $Ty_{p'}$ tends to some z in the closure of $T(K)$. Since $\|F_{p'}(T(y_{p'})) - T(y_{p'})\|_E \to 0$, $y_{p'}$ also tends to z. It follows that $Tz = z$. $\qquad\square$

A useful variant is the following theorem.

THEOREM 32. *Let F be a continuous mapping from the closed unit ball in a Banach space E, with values in E and with precompact image. If $\|x\|_E = 1$ implies $\|T(x)\|_E < 1$, then T has a fixed point.*

PROOF. It suffices to consider the mapping

$$S : x \mapsto \frac{T(x)}{\max(1, \|T(x)\|_E)},$$

which is continuous with precompact image from the unit ball to itself. It therefore possesses a fixed point y. If $\|T(y)\|_E \geqslant 1$, we find that $y = T(y)/\|T(y)\|_E$ has norm 1; the assumption now yields $\|T(y)\|_E < 1$: a contradiction. Therefore $\|T(y)\|_E < 1$ and $T(y) = y$. $\qquad\square$

The next theorem asserts the existence of a fixed point as soon as we have an *a priori* bound. Let E denote a Banach space. Recall that a compact operator is an operator which maps bounded sets to relatively compact sets.

THEOREM 33. *Let $S : E \to E$ be compact, and assume that there is an $r > 0$ such that if u solves $u = \sigma S(u)$ for some $\sigma \in [0, 1]$, $\|u\|_E < r$. Then S admits a fixed point in the ball of radius r in E.*

PROOF. Let $T(u) = S(u)$ if $\|S(u)\|_E \leqslant r$ and $T(u) = rS(u)/\|S(u)\|_E$ otherwise. Then the previous theorem applies and yields a fixed point u for T. If $\|S(u)\|_E \geqslant r$, $\|T(u)\|_E = r$ and $u = T(u) = \sigma S(u)$, with $\sigma = r/\|S(u)\|_E \in [0, 1]$. Therefore, $\|u\|_E < r$. Since $u = T(u)$, we find $\|T(u)\|_E < r$, which is impossible. Therefore, $\|S(u)\|_E < r$ and $u = T(u) = S(u)$. $\qquad\square$

We note two useful variants.

THEOREM 34. *Let $T : \mathbb{R} \times E \to E$ be compact, and satisfy $T(0, u) = 0$ for every $u \in E$. Let $C_\pm$ denote the connected component of $(0, 0)$ in the set*

$$\big\{ (\lambda, u) \in \mathbb{R} \times E : u = T(\lambda, u) \text{ and } \pm \lambda \geqslant 0 \big\}.$$

Then C_+ and C_- are both unbounded.

For this result see [54,66].

THEOREM 35. *Let $T : [0, 1] \times E \to E$ be compact, and satisfy $T(0, u) = 0$ for every $u \in E$. Assume that the relation $u = T(\sigma, u)$ implies $\|u\|_E < r$. Then equation $T(x, 1) = x$ has a solution.*

PROOF. Changing the norm on E, we may assume that $r = 1$.

Let $\varepsilon > 0$, and consider the mapping F_ε defined by

$$F_\varepsilon(x) = T\left(\frac{x}{\|x\|_E}, \frac{1 - \|x\|_E}{\varepsilon}\right) \quad \text{if } 1 - \varepsilon \leqslant \|x\|_E \leqslant 1$$

and

$$F_\varepsilon(x) = T\left(\frac{x}{1 - \varepsilon}, 1\right) \quad \text{if } \|x\|_E \leqslant 1 - \varepsilon,$$

which is continuous with precompact image. Note that

$$F_\varepsilon(x) = T\left(\frac{x}{\max(1 - \varepsilon, \|x\|_E)}, \min\left(1, \frac{1 - \|x\|_E}{\varepsilon}\right)\right).$$

If $\|x\|_E = 1$, $F_\varepsilon(x) = 0$. Theorem 31 applies and yields x_ε in the (open) unit ball such that $F_\varepsilon(x_\varepsilon) = x_\varepsilon$. For any integer $k \geqslant 1$, let $y_p = x_{1/p}$ and $\sigma_p = \min(p(1 - \|y_p\|_E), 1)$. Since the image of T is precompact and the σ_p are bounded, we may extract a subsequence such that $(x_{p'}, \sigma_{p'})$ tends to a point $(x_\infty, \sigma_\infty) \in E \times [0, 1]$.

If $\sigma_\infty < 1$, all $\sigma_{p'}$ are less than 1 for large p', which means that $1 - \|y_{p'}\|_E \geqslant 1/p'$. It follows that $\|x_\infty\| = 1$. The relation $x_\infty = T(x_\infty, \sigma_\infty)$ now implies that $\|x_\infty\| < 1$: a contradiction.

Therefore, $\sigma_\infty = 1$. From the second expression for F_ε, it follows, by passing to the limit, that $x_\infty = T(x_\infty, 1)$, so that $x \mapsto T(x, 1)$ has a fixed point. $\qquad\square$

6.3. *Fixed-point theory and the Dirichlet problem*

We now apply the abstract theorems we just proved.

We begin with an application of Theorem 33. Let α and β denote two numbers in $(0, 1)$. Consider the nonlinear operator

$$A : u \mapsto \sum_{ij} a^{ij}(x, u, \nabla u)\, \partial_{ij} u + b(x, u, \nabla u),$$

where a^{ij} and b are of class C^α in their arguments say, globally, to fix ideas.[14] Let g be a function of class $C^{2+\alpha}(\overline{\Omega})$. We wish to solve $Au = 0$ in Ω, with $u = g$ on the boundary.

To A, we associate linear operators A_v, parameterized by a function v,

$$A_v : u \mapsto \sum_{ij} a^{ij}(x, v, \nabla v)\, \partial_{ij} u + b(x, v, \nabla v),$$

---

[14]In many cases, the argument below automatically yields a priori bounds for u and its derivatives, so that one may truncate the nonlinearities for large values of their arguments.

and an operator T defined for $v \in C^{1+\beta}(\overline{\Omega})$, by $T(v) = u$, where u is the solution of the Dirichlet problem for equation

$$A_v u = 0$$

in Ω, with $u = g$ on the boundary. Since $b(x, v, \nabla v)$ is easily seen to be of class $C^{\alpha\beta}$, the Schauder estimates ensure that u thus defined belongs to $C^{2+\alpha\beta}(\overline{\Omega})$. Note that $u = \sigma T(u)$ means that $\sum_{ij} a^{ij}(x, u, \nabla v)\,\partial_{ij}u + \sigma b(x, u, \nabla u)$ in Ω, and $u = \sigma g$ on the boundary.

THEOREM 36. *If there is $\beta \in (0, 1)$ such that solutions in $C^{2+\alpha\beta}$ of equation $A(u) = 0$ in Ω, with $u = \sigma g$ on the boundary admit an a priori bound of the form $\|u\|_{C^{1+\beta}} \leqslant M$, with M independent of u and $\sigma \in [0, 1]$, then equation $A(u) = 0$ admits at least one solution with $u = g$ on the boundary.*

PROOF. Operator T maps bounded sets of $C^{1+\beta}$ to bounded sets of $C^{2+\alpha\beta}$, which, by Ascoli's theorem, are relatively compact in $C^{1+\beta}$. If $v_n \to v$ in $C^{1+\beta}$, the functions $u_n = T(v_n)$ are bounded in $C^{2+\alpha\beta}$ by Schauder estimates, and therefore, admit a convergent subsequence $u_{n'} \to u$ in the C^2 topology, and a fortiori in $C^{1+\beta}$. Since

$$\sum_{ij} a^{ij}(x, v_n, \nabla v_n)\,\partial_{ij}u_n + b(x, v_n, \nabla v_n) = 0,$$

it follows that $A_v(u) = 0$. Therefore T is continuous and compact. The result now follows from Theorem 33. $\qquad\square$

We now turn to an application of Theorem 35, which arises naturally if we wish σ to enter in the definition of A_v, which gives some flexibility in the perturbation argument. We simply define $u = T(v, \sigma)$ by solving

$$\sum_{ij} a^{ij}(x, v, \nabla v, \sigma)\,\partial_{ij}u + b(x, v, \nabla v, \sigma) = 0,$$

with $u = \sigma g$ on the boundary. Here again, the existence of an a priori $C^{1+\beta}$ bound enables one to conclude that $T(v, 1)$ has a fixed point.

6.4. *Eigenfunctions and applications*

Since the inverse of the Laplacian (with Dirichlet boundary condition) is compact, Riesz–Fredholm theory (see [11]) ensures that the Laplacian admits a sequence of real eigenvalues of finite multiplicity, tending to $+\infty$. The Fredholm alternative holds: $\Delta u + \lambda u = f$ is solvable if and only if f is orthogonal to the eigenspace corresponding to the eigenvalue λ.

We mention two important techniques related to Schauder theory: bifurcation from a simple eigenvalue (see [67,74]) and the Krein–Rutman theorem (see [52,66,74]).

6.4.1. *Bifurcation from a simple eigenvalue.* Consider, to fix ideas, the problem

$$-\Delta u + \lambda u = u^2 \quad \text{on } \Omega,$$

with Dirichlet boundary conditions. Assume we have an eigenfunction ϕ_0, for the simple eigenvalue λ_0,

$$-\Delta \phi_0 + \lambda \phi_0 = 0,$$

with $\phi_0 = 0$ on the boundary. Let $Q[u] = \int_\Omega u\phi_0 \, dx$ and $P[u] = u - \phi_0 Q[u]$. We seek a family $(\mu(\varepsilon), v(\varepsilon))$ such that our nonlinear problem admits the solutions (λ, u), where

$$u = \varepsilon \phi_0 + \varepsilon^2 v(\varepsilon); \qquad \lambda = \lambda_0 + \varepsilon \mu(\varepsilon).$$

In other words, we seek a curve of solutions which is tangent to the eigenspace for the eigenvalue λ_0. If $\varepsilon \mu$ is small, it is easy to see that $-\Delta + \lambda$ is invertible on the orthogonal complement of this eigenspace. Projecting on the orthogonal complement of ϕ_0, we find

$$v = (-\Delta + \lambda)^{-1} P\big[(\phi_0 + \varepsilon v)^2\big],$$

which may be solved for v as a function of μ, by the implicit function theorem. This gives a map $v = \Psi[\varepsilon, \mu]$. Projecting the equation on ϕ_0 now yields an equation for $\mu(\varepsilon)$

$$\mu(\varepsilon) = Q\big[(\phi_0 + \varepsilon \Psi[\varepsilon, \mu])^2\big],$$

which may be solved for $\mu(\varepsilon)$, again by an implicit function theorem. We find $\mu(\varepsilon) = Q[\phi_0^2] + O(\varepsilon)$. For variants of this argument, see, e.g., [45], Chapter 5.

6.4.2. *Krein–Rutman theorem.* We wish to generalize to infinite dimensions a classical property of matrices with nonnegative entries.

We first need a variant of Theorem 34, which follows from it using an extension theorem due to Dugundji (see [28,66,74]).

THEOREM 37. *Let K be a closed convex cone with vertex 0, and let $T : \mathbb{R}^+ \times K \to K$ be compact, and assume $T(0, u) = 0$ for every u. Then the connected component of $(0, 0)$ in the set of all solutions (λ, u) of $u = T(\lambda, u)$ is unbounded.*

As a consequence, we derive the "compression of a cone" theorem.

THEOREM 38. *Let K be a closed convex cone with vertex 0 and nonempty interior, with the property*

$$K \cap (-K) = \{0\}.$$

Let L denote a bounded linear operator on E which maps $K \setminus \{0\}$ to the interior of K. Then there is a unit vector in K and a positive real μ such that $Lx_0 = \mu$.

REMARK 7. A typical application: let $E = C^{1+\alpha}(\Omega)$, with Ω bounded and smooth, take for L the inverse of an elliptic operator, such as $-\Delta + c(x)$, with $c \geqslant 0$, and for K the closure of $\{u \in E : u > 0 \text{ in } \Omega \text{ and } \partial u/\partial n < 0 \text{ on } \partial\Omega\}$, where $\partial/\partial n$ denotes the outward normal derivative. As usual, the compactness is ensured by the Schauder estimates. The fact that L is a "compression" of the cone K, i.e., sends $K \setminus \{0\}$ to the interior of K, follows from the Hopf maximum principle. Note that the conclusion $x_0 \in K$ gives directly the information that the first eigenfunction is positive throughout Ω.

PROOF OF THEOREM 38. In this proof only, we write $u \geqslant v$ when $u - v \in K$. Fix $u \in K \setminus \{0\}$; in particular, Lu, which is interior to K, cannot be equal to 0. There is a positive M such that $Lu \geqslant u/M$, for otherwise, we would have $Lu - u/M \notin K$ for all $M > 0$, and letting $M \to \infty$, we would find $Lu \notin \text{int } K$.

For any $\varepsilon > 0$, consider the compact operator defined by $T_\varepsilon(\lambda, x) = \lambda L(x + \varepsilon u)$. Let C_ε be the connected component of $(0, 0)$ in $\mathbb{R}_+ \times K$ of the set of solutions of $x = T_\varepsilon(\lambda, x)$; we know that it is unbounded. For such a solution, we have, since $x \in K$, $x = \lambda Lx + \lambda\varepsilon u \geqslant \lambda\varepsilon u$. Since K is invariant under L, we find $Lx \geqslant \lambda\varepsilon Lu \geqslant \lambda\varepsilon u/M$. We also have $x \geqslant \lambda Lx$; therefore, $x \geqslant \lambda^2\varepsilon u/M$ and $Lx \geqslant (\lambda/M)^2\varepsilon u$. By induction, we find $Lx \geqslant (\lambda/M)^n\varepsilon u$ for every $n \geqslant 1$. If $\lambda > M$, we find, letting $n \to \infty$, that $\varepsilon u \leqslant 0$, which means $u \in -K$. Since $u \in K$ and $u \neq 0$, this is impossible. Therefore, C_ε lies in $[0, M] \times K$. Since C_ε is unbounded and contains $(0, 0)$, there is, for every $\varepsilon > 0$, a unit vector $x_\varepsilon \in K$ such that

$$x_\varepsilon = \lambda_\varepsilon L(x_\varepsilon + \varepsilon u) \quad \text{and} \quad 0 \leqslant \lambda_\varepsilon \leqslant M.$$

Since L is compact, there is a sequence $\varepsilon_n \to 0$ and a $(\mu, x_0) \in [0, M] \times K$ such that $x_{\varepsilon_n} \to x_0$ and $\lambda_{\varepsilon_n} \to \mu$. It follows that $x_0 = \mu L x_0$ and $\|x_0\|_E = 1$. Since $x_0 \neq 0$, we must have $\mu > 0$ and also $x_0 \in \text{int } K$. This completes the proof. $\qquad\square$

6.5. *Method of sub- and supersolutions*

Consider the problem

$$-\Delta u = f(u) \tag{28}$$

with Dirichlet boundary conditions on a smooth bounded domain Ω, and f smooth, such that f and df/du are both bounded.[15] We assume that we are given two ordered *sub- and supersolutions* v and w: v and w are of class $C^2(\overline{\Omega})$, vanish on $\partial\Omega$ and satisfy, over Ω,

$$v \leqslant w, \qquad -\Delta v \leqslant f(v), \qquad -\Delta w \geqslant f(w).$$

We then have the following theorem.

---

[15] The boundedness condition is not as restrictive as it seems: for instance, if u represents a concentration, it must lie between 0 and 1, and f may be redefined outside $[0, 1]$ so that it is bounded.

THEOREM 39. *Problem* (28) *admits two solutions* $\underline{u}$ *and* $\bar{u}$ *such that*

$$v \leqslant \underline{u} \leqslant \bar{u} \leqslant w.$$

In addition, if u is any solution of (28) *which lies between v and w, then necessarily* $\underline{u} \leqslant u \leqslant \bar{u}$.

REMARK 8. For more results of this kind see, e.g., [67,68].

PROOF OF THEOREM 39. Choose a constant m such that $g(u) := f(u) + mu$ is strictly increasing. Define inductively two sequences $(v_j)_{j \geqslant 0}$ and $(w_j)_{j \geqslant 0}$ by the relations: $v_0 = v$, $w_0 = w$,

$$-\Delta v_j + mv_j = g(v_{j-1}), \qquad -\Delta w_j + mw_j = g(w_{j-1}) \quad \text{for } j \geqslant 1,$$

and $v_j = w_j = 0$ on $\partial \Omega$. We have $(-\Delta + m)(v_1 - v_0) \geqslant g(v_0) - g(v_0) = 0$, which implies $v_1 \geqslant v_0$ by the maximum principle.[16] Since $(-\Delta + m)(v_{j+1} - v_j) = g(v_j) - g(v_{j-1})$, we find by induction $(-\Delta + m)(v_{j+1} - v_j) \geqslant 0$, hence $v_{j+1} - v_j \geqslant 0$. Therefore, the sequence (v_j) is nondecreasing. Similarly, (w_j) is nonincreasing. In addition, $(-\Delta + m)(w_0 - v_0) = g(w_0) - g(v_0) \geqslant 0$ and $(-\Delta + m)(w_j - v_j) = g(w_{j-1}) - g(v_{j-1})$ for $j \geqslant 1$. It follows that $w_0 \geqslant v_0$ and, by induction, $w_j \geqslant v_j$. We conclude that $\underline{u} := \lim_{j \to \infty} v_j$ and $\bar{u} := \lim_{j \to \infty} w_j$ exist and satisfy

$$v_0 \leqslant v_1 \leqslant \cdots \leqslant \underline{u} \leqslant \bar{u} \leqslant \cdots \leqslant w_1 \leqslant w_0.$$

By construction, the v_j are bounded. Therefore, $(-\Delta + m)v_j$ is bounded independently of j. Consider now any ball B_r such that $\overline{B}_{2r} \subset \Omega$, and fix $\alpha \in (0, 1)$. The interior $C^{1+\alpha}$ Schauder estimates ensure first that the v_j are, for $j \geqslant 1$, bounded in $C^1(B_{3r/2})$, independently of j. This implies in particular a C^α bound on $g(v_j)$. The $C^{2+\alpha}$ Schauder estimates now ensure that the v_j are bounded in $C^2(B_r)$ for $j \geqslant 1$, and that their second derivatives are equicontinuous. It follows that one may extract a subsequence $v_{j'}$ which converges to $\underline{u}$ in $C^2(B_r)$. It follows that $(-\Delta + m)\underline{u} = f(\underline{u}) + m\underline{u}$; so that $\underline{u}$ solves (28). A similar argument applies to $\bar{u}$. Finally, if u is a solution such that $v_0 \leqslant u \leqslant w_0$, we have $(-\Delta + m)(v_0 - u) \leqslant g(v_0) - g(u)$ and $(-\Delta + m)(v_j - u) = g(v_{j-1}) - g(u)$ for $j \geqslant 1$. It follows, by induction, that $v_j \leqslant u$ for all j. Similarly, $w_j \geqslant u$ for all j. Passing to the limit, we find $\underline{u} \leqslant u \leqslant \bar{u}$. $\qquad\square$

6.6. *Asymptotics near isolated singularities or at infinity*

We give three simple examples where Schauder estimates help understand the behavior of solutions at infinity or at isolated singularities.

---

[16]See, e.g., [11] for a simple proof.

6.6.1. *Liouville property.* Regularity theory gives a simple proof of the Liouville property for scale-invariant equations. Consider for instance the p-Laplace equation $A_p u := \operatorname{div}(|\nabla u|^{p-2}\nabla u) = 0$, where $p > 1$. We have [44] an interior C^1 estimate of the form

$$\|u\|_{C^1(B_1)} \leqslant C \sup_{B_2} |u|.$$

Applying it to $u(Rx)$, we find, since $\nabla(u(Rx)) = R(\nabla u)(Rx)$,

$$\sup_{B_R} |\nabla u| \leqslant \frac{C}{R} \sup_{B_{2R}} |u|.$$

Letting $R \to \infty$, it follows immediately that any solution which is bounded on all of $\mathbb{R}^n$ is constant. A more subtle result of this type is: any nonnegative solution on $\mathbb{R}^n \setminus \{0\}$ is necessarily constant (see [51], p. 602).

6.6.2. *Asymptotics at infinity.* If u solves $Lu = f$ on an exterior domain $\{|x| > \rho\}$, where the coefficients of L tend to constants at infinity, one may hope to apply the above scaling argument on balls $B_R(x_R)$, where, say, $|x_R| \geqslant 2R \to \infty$. In this way, it is possible to obtain weighted estimates at infinity, which are useful in solving the constraints equations in General Relativity [22] or in asymptotics for solutions of the Ginzburg–Landau equation [65].

6.6.3. *Asymptotics near isolated singularities.* The $C^{1+\alpha}$ Schauder-type estimates for the p-Laplace equation $A_p u = 0$ may be used to determine the behavior at the origin of positive solutions in a punctured neighborhood of the origin. For instance, if $n \geqslant 2$ and $p < n$ and

$$\mu(r) = \frac{p-1}{n-p}(n\omega_n)^{-1/(p-1)} r^{(p-n)/(p-1)},$$

respectively $\mu(r) = (n\omega_n)^{-1/(n-1)} \ln(1/r)$ for $p = n$, then any solution which is bounded above and below by positive multiples of $\mu(|x|)$ must in fact be of the form $\gamma\mu(|x|) + O(1)$ for some constant γ. In fact,

$$-A_p u = \gamma |\gamma|^{p-2}\delta_0,$$

in the sense of distributions, where δ_0 is the Dirac distribution at the origin.

Regularity estimates enter the argument as follows: to consider the family of functions $u_r(y) = u(ry)/\mu(r)$, which, by Schauder-type $C^{1+\alpha}$ estimates, satisfies a compactness condition on annular domains. Letting $r \to 0$ along a suitable sequence, we find that u_r tends to a solution v of $A_p v = 0$ outside the origin, and we may arrange so that $v(y)/\mu(|y|)$ has an interior maximum γ. At such a maximum, the gradient of v is proportional to the gradient of μ and thus does not vanish, so that the equation is in fact uniformly elliptic near the point of maximum; this makes it possible to conclude that v/μ

is in fact constant, using the strong maximum principle (as pointed out in [35], p. 263, the difference $w = u - \gamma\mu$ solves a linear elliptic equation). See [44,51] for details and further results. For $p = n$, one can see that $u - \gamma\mu$ has a limit at the origin; this fact has found recent applications [3,23]. For similar results for semilinear equations, see [18,33].

6.7. *Asymptotics for boundary blow-up*

We give a typical application of Fuchsian reduction to elliptic problems [48,49]. The proof structure hinges on general properties of the Fuchsian reduction process and is therefore liable of application to many other situations.

6.7.1. *Main result and structure of proof.* Let $\Omega \subset \mathbb{R}^n$, $n \geqslant 3$, be a bounded domain of class $C^{2+\alpha}$, where $0 < \alpha < 1$. Consider the Loewner–Nirenberg equation in the form

$$-\Delta u + n(n-2)u^{(n+2)/(n-2)} = 0. \tag{29}$$

It is known [5–7,56,57] that this equation admits a maximal solution u_Ω, which is positive and smooth inside Ω; it is the limit of the increasing sequence $(u_m)_{m \geqslant 1}$ of solutions of (29) which are equal to m on the boundary. It arises in many contexts [5,56]. We note for later reference the monotonicity property: if $\Omega \subset \Omega'$, then any classical solution in Ω' restricts to a classical solution in Ω, so that

$$u_{\Omega'} \leqslant u_\Omega; \tag{30}$$

it follows easily from the maximality of u_Ω. The *hyperbolic radius* of Ω is the function

$$v_\Omega := u_\Omega^{-2/(n-2)};$$

it vanishes on $\partial\Omega$. Let $d(x)$ denote the distance of x to $\partial\Omega$. It is of class $C^{2+\alpha}$ near $\partial\Omega$. We prove the following theorem.

THEOREM 40. *If Ω is of class $C^{2+\alpha}$, then $v_\Omega \in C^{2+\alpha}(\overline{\Omega})$, and*

$$v_\Omega(x) = 2d(x) - d(x)^2\big[H(x) + o(1)\big]$$

as $d(x) \to 0$, where $H(x)$ is the mean curvature at the point of $\partial\Omega$ closest to x.

This result is optimal, since H is of class C^α on the boundary. It follows from Theorem 40 that v_Ω is a *classical solution* of

$$v_\Omega \Delta v_\Omega = \frac{n}{2}\big(|\nabla v_\Omega|^2 - 4\big),$$

even though u_Ω cannot be interpreted as a weak solution of (29), insofar as $u_\Omega^{(n+2)/(n-2)} \sim (2d)^{-1-n/2} \notin L^1(\Omega)$.

We now give an idea of the proof.

We begin by performing a Fuchsian reduction, that is, we introduce the degenerate equation solved by a renormalized unknown, which governs the higher-order asymptotics of the solution; in this case, a convenient renormalized unknown is

$$w := \frac{v_\Omega - 2d}{d^2}.$$

It follows from general arguments, see the overview in [49,50], that the equation for w has a very special structure: the coefficient of the derivatives of order k is divisible by d^k for $k = 0$, 1 and 2, and the nonlinear terms all contain a factor of d. Such an equation is said to be *Fuchsian*.

In the present case, one finds

$$\frac{2v^{n/2}}{n-2}\left\{-\Delta u_\Omega + n(n-2)u_\Omega^{(n+2)/(n-2)}\right\} = Lw + 2\Delta d - M_w(w), \tag{31}$$

where

$$L := d^2\Delta + (4-n)\,d\nabla d \cdot \nabla + (2 - 2n),$$

and M_w is a linear operator with w-dependent coefficients, defined by

$$M_w(f) := \frac{nd^2}{2(2+dw)}[2f\nabla d \cdot \nabla w + d\nabla w \cdot \nabla f] - 2df\Delta d.$$

The proof now consists in a careful bootstrap argument in which better and better information on w results in better and better properties of the degenerate linear operator $L - M_w$. A key step is the inversion of the analogue of L in the half-space, which plays the role of the Laplacian in the usual Schauder theory.

Equation (31) needs only to be studied in the neighborhood of the boundary. Let us therefore introduce $C^{2+\alpha}$ thin domains $\Omega_\delta = \{0 < d < \delta\}$, such that $d \in C^{2+\alpha}(\overline{\Omega}_\delta)$, and $\partial\Omega_\delta = \partial\Omega \cup \Gamma$ consists of two hypersurfaces of class $C^{2+\alpha}$.

Recall that

$$\|u\|_{C_\#^{k+\alpha}(\overline{\Omega}_\delta)} := \sum_{j=0}^{k} \|d^j u\|_{C^{j+\alpha}(\overline{\Omega}_\delta)}.$$

The proof proceeds in five steps, corresponding to five theorems: first, a comparison argument combined with Schauder estimates gives the following theorem.

THEOREM 41. *w and $d^2\nabla w$ are bounded near $\partial\Omega$.*

Theorem 41 ensures that $L - M_w$ is of type (I). Theorem 22 then implies that $d\nabla w$ is bounded near the boundary; going back to the definition of M_w, we find $M_w(w) = O(d)$; this yields the next theorem.

THEOREM 42. *$d\nabla w$ and $M_w(w)/d$ are bounded near $\partial\Omega$.*

At this stage, we have $Lw + 2\Delta w = O(d)$. In order to use Theorem 22, we need to subtract from w a function w_0 such that $Lw_0 + 2\Delta = 0$ with controlled boundary behavior, and $w - w_0 = O(d)$; the function w_0 is constructed in the following theorem.

THEOREM 43. *If δ is sufficiently small, there is a $w_0 \in C_\#^{2+\alpha}(\overline{\Omega}_\delta)$ such that*

$$Lw_0 + 2\Delta d = 0 \tag{32}$$

in Ω_δ. Furthermore,

$$w_0|_{\partial\Omega} = -H, \tag{33}$$

where $H = -(\Delta d)/(n-1)$ is the mean curvature of the boundary.

Incidentally, we see how the curvature of the boundary enters into the asymptotics. We now use a comparison function of the form $w_0 + Ad$, where A is a constant, to bound $w - w_0$.

THEOREM 44. *Near the boundary,*

$$\tilde{w} := w - w_0 = O(d).$$

At this stage, we know that

$$L\tilde{w} = O(d) \quad \text{and} \quad \tilde{w} = O(d)$$

near $\partial\Omega$. Theorem 23 yields that $\tilde{w}$ is in $C_\#^{1+\alpha}(\overline{\Omega}_\delta)$ for δ small enough. It follows that $M_w(w) \in C^\alpha(\overline{\Omega}_\delta)$. We may now use Theorem 24 to conclude that $d^2 w$ is of class $C^{2+\alpha}$ near the boundary. Since $\tilde{w} = O(d)$, $w|_{\partial\Omega}$ is equal to $-H$. This completes the proof of Theorem 40.

We write henceforth u and v for u_Ω and v_Ω, respectively. The rest of this section is devoted to the proofs of the above theorems.

It remains to prove Theorems 41, 43 and 44.

Theorem 41 is proved in Section 6.8.4 by a comparison argument combined with regularity estimates, as in Section 6.6.3.

Theorem 43 is proved in three steps: first, one decomposes L into a sum $L_0 + L_1$ in a coordinate system adapted to the boundary, where L_0 is the analogue of L in a half-space in the new coordinates (Section 6.8.1); next, one solves $Lf = g + O(d^\alpha)$ in this coordinate system for any function of class C^α – such as $-2\Delta d$ – by inverting a model operator closely related to L_0 (Section 6.8.2); finally, we patch the results to obtain a function w_0 such that $Lw_0 = g$ (Section 6.8.3).

Theorem 44 is proved in Section 6.8.4 by a second comparison argument.

6.8. *First comparison argument*

Since $\partial\Omega$ is $C^{2+\alpha}$, it satisfies a uniform interior and exterior sphere condition, and there is a positive r_0 such that any $P \in \Omega$ such that $d(P) \leqslant r_0$ admits a unique nearest point Q on the boundary, and such that there are two points C and C' on the line determined by P and Q, such that

$$B_{r_0}(C) \subset \Omega \subset \mathbb{R}^n \setminus B_{r_0}(C'),$$

these two balls being tangent to $\partial\Omega$ at Q. We now define two functions u_i and u_e. Let

$$u_i(M) = \left(r_0 - \frac{CM^2}{r_0}\right)^{1-n/2} \quad \text{and} \quad u_e(M) = \left(\frac{C'M^2}{r_0} - r_0\right)^{1-n/2}.$$

u_i and u_e are solutions of equation (29) in $B_{r_0}(C)$ and $\mathbb{R} \setminus B_{r_0}(C')$, respectively.

If we replace r_0 by $r_0 - \varepsilon$ in the definition of u_e, we obtain a classical solution of (29) in Ω, which is therefore dominated by u_Ω. It follows that

$$u_e \leqslant u_\Omega \quad \text{in } \Omega.$$

The monotonicity property (30) yields

$$u_\Omega \leqslant u_i \quad \text{in } B_{r_0}(C).$$

In particular, the inequality

$$u_e(M) \leqslant u_\Omega(M) \leqslant u_i(M)$$

holds if M lies on the semiopen segment $[P, Q)$. Since Q is then also the point of the boundary closest to M, we have $QM = d(M)$, $CM = r_0 - d$ and $C'M = r_0 + d$; it follows that

$$\left(2d + \frac{d^2}{r_0}\right)^{1-n/2} \leqslant u_\Omega(M) \leqslant \left(2d - \frac{d^2}{r_0}\right)^{1-n/2}.$$

Since $u_\Omega = (2d + d^2 w)^{1-n/2}$, it follows that

$$|w| \leqslant \frac{1}{r_0} \quad \text{if } d \leqslant r_0.$$

Next, consider $P \in \Omega$ such that $d(P) = 2\sigma$, with $3\sigma < r_0$. For x in the closed unit ball $\overline{B}_1$, let

$$P_\sigma := P + \sigma x, \qquad u_\sigma(x) := \sigma^{(n-2)/2} u(P_\sigma).$$

One checks that u_σ is a classical solution of (29) in $\overline{B}_1$. Since $d \mapsto 2d \pm \frac{1}{r_0}d^2$ is increasing for $d < r_0$, and $d(P_\sigma)$ varies between σ and 3σ if x varies in $\overline{B}_1$, we have

$$\left(6 + \frac{9\sigma}{r_0}\right)^{1-n/2} \leqslant u_\sigma(M) \leqslant \left(2 - \frac{\sigma}{r_0}\right)^{1-n/2}.$$

This provides a uniform bound for u_σ on B_1. Applying interior regularity estimates as in [7,44], we find that ∇u_σ is uniformly bounded for $x = 0$. Recalling that $\sigma = \frac{1}{2}d(P)$, we find that

$$d^{n/2-1}u \text{ and } d^{n/2}\nabla u \text{ are bounded near } \partial\Omega.$$

It follows that $u^{-n/(n-2)} = O(d^{n/2})$, and since $d^2w = -2d + u^{-2/(n-2)}$, we have

$$d^2\nabla w = -2(1 + dw)\nabla d - \frac{2}{n-2}u^{-n/(n-2)}\nabla u,$$

hence $d^2\nabla w$ is bounded near $\partial\Omega$. This completes the proof of Theorem 41.

6.8.1. *Decomposition of L in adapted coordinates.* Since $\partial\Omega$ is compact, there is a positive r_0 such that in any ball of radius r_0 centered at a point of $\partial\Omega$, one may introduce a coordinate system (Y, T) in which $T = d$ is the last coordinate. The formulae of Section 2.5 apply. It will be convenient to assume that the domain of this coordinate system contains a set of the form

$$0 < T < \theta \quad \text{and} \quad |Y_j| < \theta \quad \text{for } j \leqslant n - 1.$$

Let $\partial_j = \partial_{x_j}$, and write d_n and d_j for $\partial d/\partial x_n$ and $\partial d/\partial x_j$, respectively. Primes denote derivatives with respect to the Y variables: $\partial'_j = \partial_{Y_j}$, $\nabla' = \nabla_Y$, $\Delta' = \sum_{j<n}\partial'^2_j$, etc. We write $\widetilde{\nabla}d = (d_1, \dots, d_{n-1})$. Recall that $|\nabla d| = 1$. We let throughout

$$D = T\,\partial_T.$$

The transformation formulae are

$$T = d(x_1, \dots, x_n), \qquad Y_j = x_j \quad \text{for } j < n,$$
$$\partial_n = d_n\,\partial_T, \qquad \partial_j = d_j\,\partial_T + \partial'_j.$$

We recall that $\Delta d = (1 - n)H$, where H is the mean curvature of $\partial\Omega$. We further have

$$d\nabla d \cdot \nabla w = \left(D + T\widetilde{\nabla}d \cdot \nabla'\right)w,$$
$$|\nabla w|^2 = w_T^2 + |\nabla'w|^2 + 2w_T\widetilde{\nabla}d \cdot \nabla'w,$$
$$\Delta w = w_{TT} + \Delta'w + 2\widetilde{\nabla}d \cdot \nabla'w_T + w_T\Delta d.$$

It follows that

$$Lw = L_0 w + L_1 w,$$

where

$$L_0 w = (D+2)(D+1-n)w + T^2 \Delta' w$$

and

$$L_1 w = (4-n)\widetilde{\nabla}d \cdot \nabla'(Tw) + 2T\widetilde{\nabla}d \cdot \nabla'(Dw) + T(Dw)\Delta d.$$

6.8.2. *Solution of $Lf = k + \mathrm{O}(d^\alpha)$.* We now solve approximately equation $Lf = k$ by solving exactly a model problem, related to the operator L_0.

Let C^α_{per} denote the space of functions $k(Y,T) \in C^\alpha (0 \leqslant T \leqslant \theta)$ which satisfy $k(Y_j + 2\theta, T) = k(Y_j, T)$ for $1 \leqslant j \leqslant n-1$. We prove the following theorem.

THEOREM 45. *Let $\theta > 0$, and $k(Y,T)$ of class C^α_{per}. Then there is a function f such that*
  (1) $L_0 f = k + \mathrm{O}(d^\alpha)$,
  (2) f is of class $C^{2+\alpha}_{\#} (0 \leqslant T \leqslant \theta)$,
  (3) $f(Y,0) = k(Y,0)/(2-2n)$ and
  (4) $L_1 f = \mathrm{O}(d^\alpha)$.

PROOF. Let

$$L'_0 = (D+2)(D-1) + T^2 \Delta' = L_0 + (n-2)(D+2).$$

We first solve the equation $L'_0 f_0 = k$.

LEMMA 46. *There is a bounded linear operator*

$$G : C^\alpha_{\mathrm{per}} \to C^{2+\alpha}_{\#} (0 \leqslant T \leqslant \theta)$$

such that $f_0 := G[k]$ verifies
  (1) $L'_0 f_0 = k$,
  (2) f_0 is of class $C^{2+\alpha}_{\#} (0 \leqslant T \leqslant \theta)$,
  (3) $f_0(Y,0) + k(Y,0)/2 = 0$, $Df_0(Y,0) = 0$ and
  (4) $L_1 f_0 = \mathrm{O}(d^\alpha)$.

PROOF. One first constructs $\tilde{k}$ such that $(D-1)\tilde{k} = -k$, and $\tilde{k}$ and $D\tilde{k}$ are both C^α up to $T = 0$. One may take

$$\tilde{k} = \int_1^\infty F_1[k](Y, T\sigma)\frac{d\sigma}{\sigma^2},$$

where F_1 is an extension operator, so that $F_1[k] = k$ for $T \leqslant \theta$.

One checks that $\tilde{k} = k$ for $T = 0$.

One then solves $(\partial_{TT} + \Delta')h + \tilde{k} = 0$ with periodic boundary conditions, of period 2θ, in each of the Y_j, and $h(Y, 0) = h_T(Y, \theta) = 0$; this yields

$$h \text{ is of class } C^{2+\alpha} (0 \leqslant T \leqslant \theta)$$

by the Schauder estimates. In particular, h_T is continuous up to $T = 0$, and $Dh = 0$ for $T = 0$ and $T = \theta$.

Since $h = 0$ for $T = 0$, we also have $\Delta' h = 0$ for $T = 0$. The equation for h therefore gives

$$h_{TT} = -\tilde{k} = -k \quad \text{for } T = 0.$$

In addition,

$$(\partial_{TT} + \Delta')Dh = D(\partial_{TT} + \Delta')h + 2h_{TT} = k - \tilde{k} + 2h_{TT},$$

which is C^α. Since, on the other hand, Dh is of class C^1 and $Dh = 0$ for $T = 0$ and $T = \theta$, we conclude, using again the Schauder estimates, that

$$Dh \text{ is of class } C^{2+\alpha} (0 \leqslant T \leqslant \theta).$$

We now define f_0 by

$$f_0 := T^{-2}(D - 1)h = \partial_T \left(\frac{h}{T} \right) = \int_0^1 \sigma h_{TT}(Y, T\sigma) \, d\sigma. \tag{34}$$

Since f_0 is itself uniquely determined by h, itself defined in terms of k we define a map G by

$$f_0 = G[k].$$

A direct computation yields $L'_0 f_0 = k$:

$$\begin{aligned}
L'_0 f_0 &= (D + 2)(D - 1)T^{-2}(D - 1)h + (D - 1)\Delta' h \\
&= T^{-2}D(D - 3)(D - 1)h + (D - 1)\{-T^{-2}D(D - 1)h - \tilde{k}\} \\
&= T^{-2}D(D - 1)(D - 3)h - T^{-2}(D - 3)D(D - 1)h - (D - 1)\tilde{k} \\
&= k.
\end{aligned}$$

Let us now consider the regularity of f_0 up to $\partial\Omega$, and the values of f_0 and its derivatives on $\partial\Omega$.

Consider $g_0 := T^2 f_0$. Since $g_0 = (D-1)h \in C^{2+\alpha}(0 \leqslant T \leqslant \theta)$ and vanishes for $T = 0$, we have $g_0 = \int_0^1 g_{0T}(Y, T\sigma)T\,d\sigma$. It follows that

$$T f_0(Y, T) = \int_0^1 g_{0T}(Y, T\sigma)\,d\sigma \in C^{1+\alpha}(0 \leqslant T \leqslant \theta).$$

Since, on the other hand, $G[k] = \int_0^1 \sigma h_{TT}(Y, T\sigma)\,d\sigma$, we find $f_0 \in C^{\alpha}(0 \leqslant T \leqslant \theta)$, and

$$f_0(Y, 0) = \frac{1}{2}h_{TT}(Y, 0) = -\frac{1}{2}k(Y, 0).$$

We therefore have

$$f_0 \text{ is of class } C_{\#}^{2+\alpha}(0 \leqslant T \leqslant \theta).$$

Since

$$(D+2)f_0 = T^{-2}D(D-1)h = h_{TT},$$

we find $D f_0(Y, 0) = h_{TT}(Y, 0) - 2 f_0(Y, 0) = 0$. By differentiation with respect to the Y variables, we obtain that $\widetilde{\nabla}d \cdot \nabla'(T f_0)$ is of class C^{α} and vanishes for $T = 0$. The same is true of $T(D f_0)\Delta d$. Similarly,

$$2T\widetilde{\nabla}d \cdot \nabla' D f_0 = 2\widetilde{\nabla}d \cdot \nabla'\big[\partial_T\big(T^2 f_0\big) - 2T f_0\big]$$

is of class C^{α}, and vanishes for $T = 0$ because this is already the case for $T D f_0$. It follows that $L_1 f_0$ is a C^{α} function which vanishes for $T = 0$; it is therefore $O(d^{\alpha})$ as desired. $\quad\square$

We are now ready to prove Theorem 45. Let a be a constant, and $f = G[ak]$. We therefore have $L_0' f = ak$ and, for $T = 0$, $f = -\frac{1}{2}ak$. Since $L_1 f \in C^{\alpha}$, and $L_1 f$ and Df both vanish for $T = 0$, it follows that, for $T = 0$,

$$Lf - k = \big(L_0' - (n-2)(D+2) + L_1\big)f - k = \big[a + (n-2)a - 1\big]k.$$

Taking $a = 1/(n-1)$, we find that f has the announced properties. $\quad\square$

6.8.3. *Solution of $L w_0 = g$.*   Let us now consider a function g of class $C^{\alpha}(\overline{\Omega}_\delta)$.

Recall that there is a positive $r_0 < \delta$ such that any ball of radius r_0, centered at a point of the boundary, is contained in a domain in which we have a system of coordinates of the type (Y, T). Let us cover (a neighborhood of) $\partial\Omega$ by a finite number of balls $(V_\lambda)_{\lambda \in \Lambda}$ of radius $r_1 < r_0$ and centers on $\partial\Omega$, and consider the balls $(U_\lambda)_{\lambda \in \Lambda}$ of radius r_0 with the same centers. Thus, we may assume that every U_λ is associated with a coordinate system (Y_λ, T_λ) of the type considered in Section 2.5; taking r_1 smaller if necessary, we may also assume that $\overline{V}_\lambda \subset Q_\lambda \subset U_\lambda$, where Q_λ has the form

$$Q_\lambda := \big\{(Y_{\lambda,1,\ldots}, Y_{\lambda,n-1}, T_\lambda)\colon 0 \leqslant Y_{\lambda,j} \leqslant \theta \text{ for every } j, \text{ and } 0 < T_\lambda < \theta\big\}.$$

Consider a smooth partition of unity (φ_λ) and smooth functions (Φ_λ), such that

(1) $\sum_{\lambda \in \Lambda} \varphi_\lambda = 1$ near $\partial\Omega$;
(2) $\operatorname{supp} \varphi_\lambda \subset V_\lambda$;
(3) $\sup \Phi_\lambda \subset U_\lambda \cap \{T < \theta\}$;
(4) $\Phi_\lambda = 1$ on V_λ.

In particular, $\Phi_\lambda \varphi_\lambda = \varphi_\lambda$.

The function $g\varphi_\lambda$ is of class $C^\alpha(\overline{Q}_\lambda)$; it may be extended by successive reflections to an element of C^α_{per}, with period 2θ in the Y_λ variables; this extension will be denoted by the same symbol for simplicity.

Let us apply Theorem 45, and consider, for every λ, the function $w_\lambda := G[g\varphi_\lambda/(n-1)]$. We have

$$ L w_\lambda = g\varphi_\lambda + R_\lambda, $$

in $U_\lambda \cap \{T < \theta\}$, where R_λ is Hölder continuous for $T \leqslant \theta$ and vanishes on $\partial\Omega$; as a consequence, $R_\lambda = \mathrm{O}(d^\alpha)$.

The function $\Phi_\lambda w_\lambda$ is compactly supported in U_λ, and may be extended, by zero, to all of Ω; it is of class $C^{2+\alpha}_{\#}(\overline{\Omega})$. We may therefore consider

$$ w_1 := \sum_{\lambda \in \Lambda} \Phi_\lambda w_\lambda, $$

which is supported near $\partial\Omega$. Now, near $\partial\Omega$,

$$ \sum_\lambda L(\Phi_\lambda w_\lambda) = \sum_\lambda \Phi_\lambda L(w_\lambda) + 2d^2 \nabla\Phi_\lambda \cdot \nabla w_\lambda $$

$$ + d^2 w_\lambda \Delta\Phi_\lambda + (4-n) w_\lambda d\nabla d \cdot \nabla\Phi_\lambda $$

$$ = \sum_\lambda g\Phi_\lambda \varphi_\lambda + R'_\lambda = g + f, $$

where $f = \sum_\lambda R'_\lambda$ has the same properties as R_λ. It therefore suffices to solve $L w_2 = f$ when f is a Hölder continuous function which vanishes on the boundary.

LEMMA 47. *For any $f \in C^\alpha(\overline{\Omega})$, there is, for δ small enough, an element $w_2 \in C^{2+\alpha}_{\#}(\overline{\Omega}_\delta)$ such that*

$$ L w_2 = f \quad \text{and} \quad w_2 = \mathrm{O}(d^\alpha) \qquad \text{near } \partial\Omega. $$

PROOF. Consider the solution w_ε of the Dirichlet problem $L w_\varepsilon = f$ on a domain of the form $\{\varepsilon < d(x) < \delta\}$, with zero boundary data. As before, δ is taken small enough to ensure that $d \in C^{2+\alpha}(\overline{\Omega}_\delta)$. Schauder theory gives $w_\varepsilon \in C^{2+\alpha}(\{\varepsilon \leqslant d(x) \leqslant \delta\})$. By assumption, $|f| \leqslant ad^\alpha$ for some constant a. Let $A > (\alpha+2)(n-1-\alpha)$. Since

$$ -L(d^\alpha) = d^\alpha\big[(\alpha+2)(n-1-\alpha) - \alpha d\Delta d\big], $$

$Ad(x)^\alpha$ is a supersolution if δ is small, and the maximum principle gives us a uniform bound on w_ε/d^α. By interior regularity, we obtain that, for a sequence $\varepsilon_n \to 0$, the w_{ε_n} converge in C^2, in every compact away from the boundary, to a solution w_2 of $Lw_2 = f$ with $w_2 = \mathrm{O}(d^\alpha)$. Since the right-hand side f is also $\mathrm{O}(d^\alpha)$, we obtain, by the "type (I)" Theorem 23, that w_2 of class $C_\#^{1+\alpha}(\overline{\Omega}_\delta)$. Theorem 24 now ensures that w_2 is in fact of class $C_\#^{2+\alpha}(\overline{\Omega}_\delta)$. $\qquad\square$

It now suffices to take $g = -2\Delta d$ and let

$$w_0 = w_1 - w_2.$$

By construction, $Lw_0 + 2\Delta d = 0$ near the boundary, and w_0 is of class $C_\#^{2+\alpha}(\overline{\Omega}_\delta)$ if δ is small. In addition, we know from Theorem 45 that $w_1|_{\partial\Omega} = (2\Delta d)/(2n - 2)$, which is equal to $-H$ on $\partial\Omega$. Lemma 47 gives us $w_2 = \mathrm{O}(d^\alpha)$. We conclude that $w_0|_{\partial\Omega} = -H$ on the boundary.

This completes the proof of Theorem 43.

6.8.4. *Second comparison argument.*	At this stage, we have the following information, where $\Omega_\delta = \{x\colon 0 < d(x) < \delta\}$, for δ small enough:
  (1) w and $d\nabla w$ are bounded near $\partial\Omega$;
  (2) $w = w_0 + \tilde{w}$, where $L\tilde{w} = M_w(w) = \mathrm{O}(d)$, and
  (3) w_0 is of class $C_\#^{2+\alpha}(\overline{\Omega}_\delta)$ for δ small enough.
We wish to estimate $\tilde{w}$. Write $|M_w(w)| \leqslant cd$, where c is constant.
  For any constant $A > 0$, define

$$w_A := w_0 + Ad.$$

Since $L(d) = 3(2 - n)d + d^2\Delta d$, we have

$$L(w_A - w) = L(Ad - \tilde{w}) \leqslant Ad\big[3(2 - n) + d\Delta d\big] + cd.$$

Choose δ so that, say, $2(2 - n) - d\Delta d \leqslant 0$ for $d \leqslant \delta$. Then, choose A so large that (i) $w_0 + A\delta \geqslant w$ for $d = \delta$, and (ii) $(2 - n)A + c \leqslant 0$. We then have

$$L(w_A - w) \leqslant 0 \quad \text{in } \Omega_\delta \quad \text{and} \quad w_A - w \geqslant 0 \quad \text{for } d = \delta.$$

Next, choose δ and a constant B such that $nB + (2 + Bd)\Delta d \geqslant 0$ on Ω_δ. We have, by direct computation,

$$L\big(d^{-2} + Bd^{-1}\big) = -(nB + 2\Delta d)d^{-1} - B\Delta d \leqslant 0$$

on Ω_δ. Therefore, for any $\varepsilon > 0$, $z_\varepsilon := \varepsilon[d^{-2} + Bd^{-1}] + w_A - w$ satisfies $Lz_\varepsilon \leqslant 0$, and the maximum principle ensures that z_ε has no negative minimum in Ω_δ. Now, z_ε tends to $+\infty$

as $d \to 0$. Therefore, z_ε is bounded below by the least value of its negative part restricted to $d = \delta$. In other words, for $d \leqslant \delta$, we have, since $w_A - w \geqslant 0$ for $d = \delta$,

$$w_A - w + \varepsilon\left[d^{-2} + Bd^{-1}\right] \geqslant \varepsilon \min\left(\delta^{-2} + B\delta^{-1}, 0\right).$$

Letting $\varepsilon \to 0$, we obtain $w_A - w \geqslant 0$ in Ω_δ.

Similarly, for suitable δ and A, $w - w_{-A} \geqslant 0$ in Ω_δ.

We now know that w lies between $w_0 + Ad$ and $w_0 - Ad$ near $\partial\Omega$, hence $|w - w_0| = O(d)$.

References

[1] S. Agmon, A. Douglis and L. Nirenberg, *Estimates near the boundary for solutions of elliptic partial differential equations satisfying general boundary conditions, I & II*, Commun. Pure Appl. Math. **12** (1959), 623–7271; **17** (1964), 35–92.

[2] T. Aubin, *Nonlinear Analysis on Manifolds. Monge-Ampère Equations*, Springer-Verlag, Berlin (1982).

[3] Z. Balogh, I. Holopainen and J.T. Tyson, *Singular solutions, homogeneous norms and quasiconformal mappings in Carnot groups*, Math. Ann. **324** (2002), 159–186.

[4] C. Bandle and M. Essén, *On the solution of quasilinear elliptic problems with boundary blow-up*, Symposia Math. **35** (1994), 93–111.

[5] C. Bandle and M. Flucher, *Harmonic radius and concentration of energy; hyperbolic radius and Liouville's equations $\Delta U = e^U$ and $\Delta U = U^{(n+2)/(n-2)}$*, SIAM Rev. **38** (1996), 191–238.

[6] C. Bandle and M. Marcus, *Asymptotic behavior of solutions and their derivatives, for semilinear elliptic problems with blowup on the boundary*, Ann. Inst. H. Poincaré Anal. Non Linéaire **12** (1995), 155–171.

[7] C. Bandle and M. Marcus, *On second-order effects in the boundary behavior of large solutions of semilinear elliptic problems*, Differential Integral Equations **11** (1998), 23–34.

[8] L. Bers, *Local behavior of solutions of general linear elliptic equations*, Commun. Pure Appl. Math. **8** (1955), 473–496.

[9] A. Brandt, *Estimates for difference quotients of solutions of Poisson type difference equations*, Math. Comp. **20** (1966), 473–499.

[10] A. Brandt, *Interior estimates for second-order elliptic differential (or finite-difference) equations via the maximum principle*, Israel J. Math. **7** (1969), 95–121, 254–262.

[11] H. Brezis, *Analyse fonctionnelle: Théorie et applications*, Masson, Paris (1983).

[12] H. Brezis and F. Browder, *Partial differential equations in the 20th century*, Adv. Math. **135** (1998), 76–144.

[13] H. Brezis and L.C. Evans, *A variational inequality approach to the Bellman–Dirichlet equation for two elliptic operators*, Arch. Ration. Mech. Anal. **71** (1979), 1–13.

[14] R. Caccioppoli, *Sulle equazioni ellittiche non lineari a derivate parziali*, Rend. Accad. Naz. Lincei **18** (1933), 103–106; *Sulle equazioni ellittiche non lineari a derivate parziali con n variabli independenti*, Rend. Accad. Naz. Lincei **19** (1934), 83–89.

[15] X. Cabré and L.A. Caffarelli, *Interior $C^{2+\alpha}$ regularity theory for a class of nonconvex fully nonlinear elliptic equations*, J. Math. Pures Appl. **82** (2003), 573–612.

[16] L.A. Caffarelli, *Interior a priori estimates for solutions of fully non linear equations*, Ann. Math. **130** (1989), 189–213.

[17] L.A. Caffarelli and X. Cabré, *Fully Nonlinear Elliptic Equations*, Amer. Math. Soc. Colloq. Publ., vol. 43, Amer. Math. Soc., Providence, RI (1995).

[18] L.A. Caffarelli, B. Gidas and J. Spruck, *Asymptotic symmetry and local behavior of semilinear elliptic equations with critical Sobolev growth*, Commun. Pure Appl. Math. **42** (1989), 271–297.

[19] S. Campanato, *Proprietà di Hölderianità di alcune classi di funzioni*, Ann. Sc. Norm. Sup. Pisa **17** (1963), 175–188; *Proprietà di une famiglia di spazi funzionali*, Ann. Sc. Norm. Sup. Pisa **18** (1964), 137–160.

[20] S. Campanato, *Equazioni ellittiche des secondo ordine e spazi $\mathcal{L}^{2,\lambda}$*, Ann. Mat. Pura Appl. **69** (1965), 321–380.

[21] L. Capogna and Q. Han, *Pointwise Schauder estimates for subelliptic operators in Carnot groups*, Proc. AMS-IMS-SIMA Harmonic Analysis Conference, Mt. Holyoke College, 2001, W. Beckner et al., eds, Contemporary Mathematics, vol. 32, Amer. Math. Soc., Providence, RI (2003), pp. 45–69.

[22] A. Chaljub-Simon and Y. Choquet-Bruhat, *Problèmes elliptiques du second ordre sur une variété euclidienne à l'infini*, Ann. Fac. Sci. Toulouse **1** (1979), 9–25; A. Chaljub-Simon, Rend. Circ. Mat. Palermo Ser. II **30** (1981), 300–310; Y. Choquet-Bruhat, S. Deser, Ann. Phys. **81** (1973), 165.

[23] A. Colesanti and P. Cuoghi, *The Brunn–Minkowski inequality for the n-dimensional logarithmic capacity of convex bodies*, Potential Anal. **22** (2005), 289–304.

[24] H.O. Cordes, *Vereinfachter Beweis der Existenz einer Apriori-Hölderkonstante*, Math. Ann. **138** (1959), 155–178.

[25] H.O. Cordes, *Zero order a priori estimates for solutions of elliptic differential equations*, Proc. Symp. Pure Math. **4** (1961), 157–178.

[26] R. Courant and D. Hilbert, *Methods of Mathematical Physics*, Vol. 2, Wiley, New York (1962).

[27] A. Douglis and L. Nirenberg, *Interior estimates for elliptic systems of partial differential equations*, Commun. Pure Appl. Math. **8** (1955), 503–538.

[28] J. Dugundji, *An extension of Tietze's theorem*, Pacific J. Math. **1** (1951), 353–367 (see Theorem 4-1 and also J.T. Schwartz's notes, Theorem 3.1).

[29] L.C. Evans, *A new proof of local $C^{1+\alpha}$ regularity for solutions of certain elliptic p.d.e.*, J. Differential Equation **45** (1982), 356–373.

[30] P. Fife, *Schauder estimates under incomplete Hölder continuity assumptions*, Pacific J. Math. **13** (1963), 511–550.

[31] C.F. Gauss, *Allgemeine Lehrsätze in Beziehung auf die im verkehrten Verhältnisse des Quadrats des Entfernung wirkenden Anziehung- und Abstossungs-kräfte*, Werke, Bd. 5, Georg Olms Verlag, Hildesheim (1973), pp. 197–244. Reprint of original edition published by the Köninglichen Gesellschaft der Wissenschaften zu Götingen, Götingen (1863–1874).

[32] M. Giaquinta, *Multiple Integrals in the Calculus of Variations and Nonlinear Elliptic Systems*, Ann. Math. Studies, vol. 105, Princeton Univ. Press, Princeton, NJ (1983). See also, *Introduction to Regularity Theory for Nonlinear Elliptic Systems*, Lectures in Mathematics, ETH Zürich, Birkhäuser, Basel (1993).

[33] B. Gidas and J. Spruck, *Global and local behavior of positive solutions of nonlinear elliptic equations*, Commun. Pure and Appl. Math. **34** (1981), 525–598.

[34] D. Gilbarg and L. Hörmander, *Intermediate Schauder estimates*, Arch. Ration. Mech. Anal. **74** (1980), 297–318.

[35] D. Gilbarg and N.S. Trudinger, *Elliptic Partial Differential Equations of Second Order*, Springer-Verlag, New York (1983).

[36] G. Giraud, *Sur le problème de Dirichlet généralisé; équations non linéaires à m variables*, Ann. Sci. École Norm. Sup. (4) **43** (1926), 1–128.

[37] G. Giraud, *Sur le problème de Dirichlet généralisé (Deuxième mémoire)*, Ann. Sci. École Norm. Sup. (4) **46** (1929), 131–245.

[38] C. Goulaouic and N. Shimakura, *Régularité Hölderienne de certains problèmes aux limites elliptiques dégénérés*, Ann. Sc. Norm. Sup. Pisa Ser. 4 **10** (1983), 79–108.

[39] Q. Han, *Schauder estimates for elliptic operators with applications to nodal sets*, J. Geom. Anal. **10** (2000), 455–480.

[40] Q. Han and F.H. Lin, *Elliptic Partial Differential Equations*, Courant Inst. Lect. Notes, vol. 1, Courant Inst. Math. Sci./Amer. Math. Soc., Providence, RI (1997).

[41] O. Hölder, *Beiträge zur Potentialtheorie*, Doctoral Dissertation, Tübingen, Stuttgart (1882).

[42] E. Hopf, *Über den Funktionalen, insbesondere den analytischen Charakter, der Lösingen elliptischer Differentialgleichungen zweiter Ordnung*, Math. Z. **34** (1931), 194–233.

[43] O.D. Kellogg, *On the derivatives of harmonic functions on the boundary*, TAMS **33** (1931), 486–510.

[44] S. Kichenassamy, *Quasilinear problems with singularities*, Manuscripta Math. **57** (1987), 281–313. (This paper and [51] are part of the author's thesis under the direction of H. Brezis.)

[45] S. Kichenassamy, *Nonlinear Wave Equations*, Dekker, New York (1996).

[46] S. Kichenassamy, *Régularité du rayon hyperbolique*, C. R. Acad. Sci. Paris, Sér. I **3381** (2004), 13–18.

[47] S. Kichenassamy, *Boundary blow-up and degenerate equations*, J. Funct. Anal. **215** (2004), 271–289.

[48] S. Kichenassamy, *Boundary behavior in the Loewner–Nirenberg problem*, J. Funct. Anal. **222** (2005), 98–113.

[49] S. Kichenassamy, *Recent progress on boundary blow-up*, Elliptic and Parabolic Problems: A Special Tribute to the Work of Haïm Brezis, C. Bandle et al., eds, Prog. Nonlinear Differential Equations Appl., vol. 63, Birkhäuser, Boston, MA (2005), pp. 329–341.

[50] S. Kichenassamy, *Fuchsian Reduction: Lasers, Cosmology, Combustion, and Geometry*, Progress in Non Linear Differential Equations and Their Applications, Birkhäuser, Boston, MA, to appear.

[51] S. Kichenassamy and L. Véron, *Singular solutions of the p-Laplace equation*, Math. Ann. **275** (1986), 599–615.

[52] M.G. Krein and M.A. Rutman, *Linear operators leaving invariant a cone in a Banach space*, Amer. Math. Soc. Transl. Ser. I **10** (1950), 199–325; transl. of Uspekhi Mat. Nauk (N.S.) **3** (1948), 2–95.

[53] O. Ladyzhenskaya and N. Ural'tseva, *Linear and Quasilinear Elliptic Equations*, Academic Press, New York (1968).

[54] J. Leray and J. Schauder, *Topologie et équations fonctionnelles*, Ann. Sci. École Norm. Sup. (3) **51** (1934), 45–78.

[55] L. Lichtenstein, *Neure Entwicklungen der Theorie partieller Differentialgleichungen zweiter Ordnung*, Encykl. Math. Wiss., Bd. II.C, Heft 8 (1924), 1277–1334; complements in: Math. Z. **20** (1924), 194–212.

[56] C. Loewner and L. Nirenberg, *Partial differential equations invariant under conformal or projective transformations*, Contributions to Analysis, L. Ahlfors et al., eds, Academic Press, New York (1974), pp. 245–272.

[57] M. Marcus and L. Véron, *Uniqueness and asymptotic behavior of solutions with boundary blow-up for a class of nonlinear elliptic equations*, Ann. Inst. H. Poincaré Anal. Non Linéaire **14** (1997), 237–274.

[58] N.G. Meyers, *Mean oscillation over cubes and Hölder continuity*, Proc. Amer. Math. Soc. **15** (1964), 717–721.

[59] C. Miranda, *Partial Differential Equations of Elliptic Type*, Springer-Verlag, Berlin (1970).

[60] C.B. Morrey, *Second-order elliptic systems of differential equations*, Contributions to the Theory of Partial Differential Equations, Ann. Math. Stud., vol. 33, Princeton Univ. Press, Princeton, NJ (1954), pp. 101–159.

[61] C.B. Morrey, *Multiple Integrals in the Calculus of Variations*, Springer-Verlag, New York (1966).

[62] C. Neumann, *Über einige Fundamentalsätze der Potentialtheorie*, Berichte über die Verhandlungen des königlichen sachsischen Gesellschaft der Wissenschaften zu Leipzig–Mathematisch-Physische Klasse **42** (1890), 327–340.

[63] L. Nirenberg, *On non linear partial differential equations and Hölder continuity*, Commun. Pure Appl. Math. **6** (1953), 103–156.

[64] L. Nirenberg, *Topics in Nonlinear Functional Analysis*, New York University, Courant Inst. Math. Sci. (1973–1974). Reprinted as Courant Inst. Lect. Notes, vol. 6, Courant Inst. Math. Sci./Amer. Math. Soc., Providence, RI (2001).

[65] F. Pacard and T. Rivière, *Linear and Nonlinear Vortices: The Ginzburg–Landau Model*, Birkhäuser, Boston (2000).

[66] P. Rabinowitz, *Méthodes topologiques et problèmes aux limites non lináires*, Notes rédigées par H. Berestycki, Paris (1984).

[67] D. Sattinger, *Monotone methods in nonlinear elliptic and parabolic equations*, Indiana Univ. Math. J. **21** (1972), 979–1000.

[68] D. Sattinger, *Topics in Stability Bifurcation Theory*, Lecture Notes in Math., vol. 309, Springer-Verlag, Berlin (1973).

[69] J. Schauder, *Potentialtheoretische Untersuchungen*, Math. Z. **33** (1931), 602–640; see also his remark on this paper: Math. Z. **35** (1932), 536–538.

[70] J. Schauder, *Über lineare elliptische Differentialgleichungen zweiter Ordnung*, Math. Z. **38** (1934), 257–282.

[71] J. Schauder, *Numerische Abschätzungen in elliptischen linearen Differentialgleichungen*, Studia Math. **5** (1934), 34–42.

[72] N. Shimakura, *Partial Differential Equations of Elliptic Type*, Transl. Math. Monogr., vol. 99, Amer. Math. Soc., Providence, RI (1992); transl. and rev. by the author from the 1978 Japanese edition.

[73] L. Simon, *Schauder estimates by scaling*, Calc. Var., Partial Differential Equations **5** (1997), 391–407.

[74] J. Smoller, *Shock Waves and Reaction–Diffusion Equations*, Springer-Verlag, Berlin (1983).

[75] E.M. Stein, *Harmonic Analysis*, Princeton Univ. Press, Princeton, NJ (1993).
[76] D. Stroock and S.R.S. Varadhan, *Multidimensional Diffusion Processes*, Springer-Verlag, Berlin (1979).
[77] M. Taylor, *Pseudodifferential Operators and Nonlinear P.D.E.*, Birkhäuser, Boston (1991).
[78] G. Troianello, Estimates of the Caccioppoli–Schauder type in weighted function spaces, Trans. Amer. Math. Soc. **334** (1992), 551–573.
[79] N. Trudinger, *A new approach to Schauder estimates for linear elliptic equations*, Proc. CMA, vol. 14, Canberra (1986), pp. 52–59.

CHAPTER 6

The Dam Problem

A. Lyaghfouri

*Mathematical Sciences Department, King Fahd University of Petroleum and Minerals,
Dhahran 31261, Saudi Arabia
E-mail: lyaghfo@kfupm.edu.sa*

Contents

Introduction 467
  The variational inequalities approach by Baiocchi 467
  The weak formulation of Alt and Brezis–Kinderlehrer–Stampacchia 468
  Outline of the chapter 469
  Notation 470
1. A unified formulation of the dam problem 471
  1.1. Formulation of the problem 471
  1.2. Existence of a solution 475
  1.3. Regularity and monotonicity of the solutions 486
2. The dam problem with Dirichlet boundary condition 491
  2.1. Some properties of the solutions 491
  2.2. Continuity of the free boundary 495
  2.3. Existence and uniqueness of minimal and maximal solutions 498
  2.4. Reservoirs-connected solution 507
  2.5. Uniqueness of the reservoirs-connected solution 510
3. The dam problem with leaky boundary condition 516
  3.1. Properties of the solutions 516
  3.2. Continuity of the free boundary 526
  3.3. Existence and uniqueness of minimal and maximal solutions 533
  3.4. Reservoirs-connected solution 544
  3.5. Uniqueness of the reservoirs-connected solution 545
Acknowledgement 551
References 551

HANDBOOK OF DIFFERENTIAL EQUATIONS
Stationary Partial Differential Equations, volume 3
Edited by M. Chipot and P. Quittner

Introduction

The steady-state dam problem consists of studying the filtration of a fluid (say water) through a porous medium Ω assuming an equilibrium has been reached. Then we look for the saturated region S (see Figure 1) and the fluid pressure p inside Ω. We are also concerned with the regularity of the interface that separates wet and dry regions, called free boundary.

The variational inequalities approach by Baiocchi

The study of this problem goes back to the early seventies with the pioneering work of Baiocchi [6,7], who is probably the first one to solve this problem in the case of a rectangular dam. By introducing the transformation $u(x_1, x_2) = \int_0^H p(x_1, t)\,dt$, where H is the height of the dam, he showed that u is the unique solution of the variational inequality

$$\begin{cases} \text{find } u \in \mathbb{K}_1 \text{ such that} \\ \int_\Omega \nabla u \cdot \nabla(\zeta - u) \geqslant -\int_\Omega (\zeta - u) \quad \forall \zeta \in \mathbb{K}_1, \end{cases}$$

where $\mathbb{K}_1 = \{\zeta \in H^1(\Omega)/\zeta = g \text{ on } \partial\Omega, \zeta \geqslant 0 \text{ in } \Omega\}$ and g is a given Lipschitz continuous function. He also proved that the free boundary is an analytic curve $x_2 = \Phi(x_1)$. In [8] and [9], he generalized these results to dams with horizontal bottoms.

For heterogeneous dams, the first results where also established in the rectangular case via the theory of variational inequalities. Indeed using the following generalized Baiocchi's transformation $u(x_1, x_2) = \int_{x_2}^H k_2(s) p(x_1, s)\,ds$, allowed authors to handle the case of a matrix permeability of the form $k(x_1, x_2) I_2$, where I_2 is the 2×2 unit matrix and

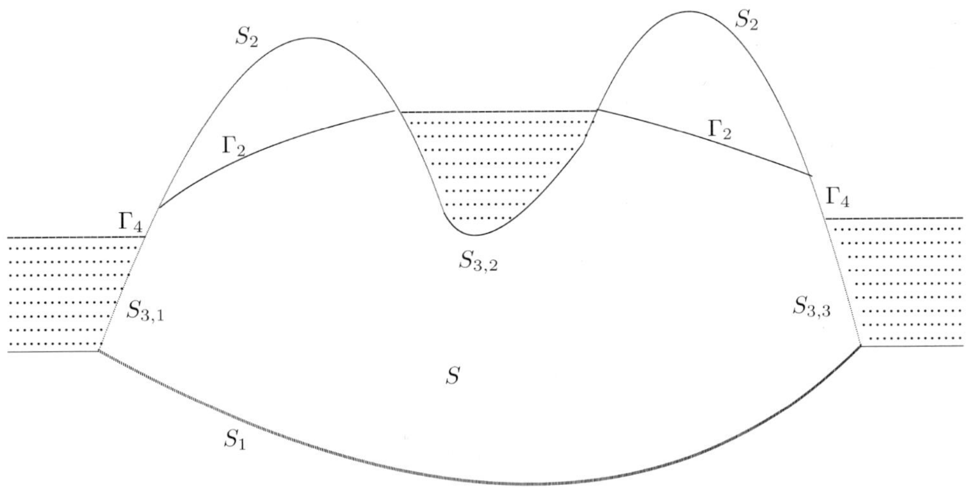

Fig. 1.

$k(x_1, x_2) = k_1(x_1)k_2(x_2)$. Hence, Benci proved in [14] that u is the unique solution of the following variational inequality

$$\begin{cases} \text{find } u \in \mathbb{K}_2 \text{ such that} \\ \int_\Omega \frac{k_1(x_1)}{k_2(x_2)} \nabla u \cdot \nabla(\zeta - u) \geqslant - \int_\Omega k_1(x_1)(\zeta - u) \quad \forall \zeta \in \mathbb{K}_2, \end{cases}$$

where $\mathbb{K}_2 = \{\zeta \in H^1(\Omega) / \zeta = h \text{ on } \partial\Omega, \zeta \geqslant 0 \text{ in } \Omega\}$ and h is a given function. He also proved that the free boundary is a curve $x_1 = \Psi(x_2)$. When $k_2'(x_2) \leqslant 0$, he proved that it is a curve of a continuous decreasing function $\Phi(x_1)$. Baiocchi and Friedman [11] extended these results assuming only that $\ln(\int_{x_2}^H k_2(t)\,dt)$ is concave. In [17] Caffarelli and Friedman proved that the free boundary is a curve $x_1 = \Psi(x_2)$ provided that $k(x_1, x_2) = k_2(x_2)$ is a nonincreasing step function.

Note that Baiocchi's transformation was used locally by Alt in [1] to prove the regularity of the free boundary.

The weak formulation of Alt and Brezis–Kinderlehrer–Stampacchia

Given that the variational approach is not possible for dams with general geometry, Brezis, Kinderlehrer and Stampacchia in [16] and independently Alt in [2] introduced the following formulation:

$$\begin{cases} \text{find } (p, \chi) \in H^1(\Omega) \times L^\infty(\Omega) \text{ such that} \\ \text{(i) } p \geqslant 0, 0 \leqslant \chi \leqslant 1, p(1 - \chi) = 0 \text{ a.e. in } \Omega, \\ \text{(ii) } p = \varphi \text{ on } S_2 \cup S_3, \\ \text{(iii) } \int_\Omega (\nabla p + \chi e) \cdot \nabla \xi \, dx \leqslant 0, e = (0, 1), \\ \qquad \forall \xi \in H^1(\Omega), \xi = 0 \text{ on } S_3, \xi \geqslant 0 \text{ on } S_2, \end{cases} \quad (P_1)$$

where χ is a bounded function characterizing the wet part of the dam, S_1 is the impervious part of the dam, S_2 is the part in contact with air and S_3 is the part in contact with the reservoirs, φ represents the exterior pressure.

Existence of a solution was proved by approaching χ with an approximation of the Heaviside graph. Regarding the free boundary, Alt proved in [1], that it is an analytic surface $x_n = \Phi(x_1, \ldots, x_{n-1})$ when Ω is a Lipschitz domain of $\mathbb{R}^n$ with $n \geqslant 2$. The uniqueness of the so-called S_3-connected or reservoirs-connected solution was proved by Carrillo and Chipot in [18] and also by Alt and Gilardi in [3].

In [20] Carrillo and Lyaghfouri considered this problem, assuming the flow governed by the following nonlinear Darcy law [28] $|v|^{m-1}v = -\nabla(p + x_2)$, $m > 0$. The authors formulated the problem in terms of the hydrostatic head $u = p + x_2$ instead of the pressure, which led to the following problem:

$$\begin{cases} \text{find } (u, g) \in W^{1,q}(\Omega) \times L^\infty(\Omega) \text{ such that} \\ \text{(i) } u \geqslant x_2, 0 \leqslant g \leqslant 1, g(u - x_2) = 0 \text{ a.e. in } \Omega, \\ \text{(ii) } u = \varphi + x_2 \text{ on } S_2 \cup S_3, \\ \text{(iii) } \int_\Omega (|\nabla u|^{q-2}\nabla u - ge) \cdot \nabla \xi \, dx \leqslant 0, \\ \qquad \forall \xi \in W^{1,q}(\Omega), \xi = 0 \text{ on } S_3, \xi \geqslant 0 \text{ on } S_2. \end{cases} \quad (P_2)$$

Then they showed the existence of a solution, proved the continuity of the free boundary $x_2 = \Phi(x_1)$ and the uniqueness of the reservoirs-connected solution in dimension 2. For dimension $n \geqslant \max(2, q)$, they proved the existence and uniqueness of a minimal solution.

The general heterogeneous dam with general geometry was formulated first in [2] by Alt. In [27] and [37] the authors showed that for a permeability matrix $a(x) = k(x_1, x_2)I_2$ with $\frac{\partial k}{\partial x_2} \geqslant 0$ in $\mathcal{D}'(\Omega)$, the free boundary is a continuous curve $x_2 = \Phi(x_1)$ and the reservoirs-connected solution is unique. These results were generalized by Lyaghfouri [33] to the case where

$$a(x) = \begin{pmatrix} a_{11}(x) & 0 \\ a_{21}(x) & a_{22}(x) \end{pmatrix} \quad \text{and} \quad \frac{\partial a_{22}}{\partial x_2} \geqslant 0 \quad \text{in } \mathcal{D}'(\Omega).$$

The model with leaky boundary condition, i.e., when the flow through the reservoirs' bottoms is equal to a function of the difference between exterior and interior pressures,

$$v \cdot \nu = -\beta(x, \varphi - p)$$

was considered first in [9] in the rectangular case. The general situation was considered in [19,36] and in [23] via the following formulation:

$$\begin{cases} \text{find } (p, \chi) \in H^1(\Omega) \times L^\infty(\Omega) \text{ such that} \\ \text{(i) } p \geqslant 0, 0 \leqslant \chi \leqslant 1, p(1 - \chi) = 0 \text{ a.e. in } \Omega, \\ \text{(ii) } p = \varphi \text{ on } S_2, \\ \text{(iii) } \int_\Omega a(x)(\nabla p + \chi e) \cdot \nabla \xi \, dx \leqslant \int_{S_3} \beta(x, \varphi - p)\xi \, d\sigma(x) \\ \qquad \forall \xi \in H^1(\Omega), \xi \geqslant 0 \text{ on } S_2. \end{cases} \qquad (P_3)$$

The continuity of the free boundary and the uniqueness of the reservoirs-connected solution were established in [23] for the two-dimensional and homogeneous case. In [35] the continuity of the free boundary was extended to heterogeneous media with linear or nonlinear Darcy's laws together with various uniqueness results including the situation corresponding to the linear Darcy law with a diagonal permeability matrix.

The main difference between the model corresponding to Dirichlet condition and the one with leaky condition is the fact that the region below the reservoirs is always saturated in the first model while it is not necessarily the case for the second one. However we know [23, 35], that if $\beta(x, \varphi) \geqslant a_{22}\nu_2$ on a connected subset T of S_3, where ν_2 is the second entry of the outward unit normal vector ν to $\partial\Omega$, then the region below T is completely saturated provided its lower boundary is impervious.

Differentiability of the free boundary is still an open problem for the heterogeneous case even in dimension 2 except when the permeability does not depend on the variable x_2 [23]. In dimension $n \geqslant 3$, the continuity of the free boundary is also an open problem except for the homogeneous case with Dirichlet conditions.

Outline of the chapter

In this study, we shall be concerned mainly with a two-dimensional heterogeneous dam with general geometry, assuming the flow governed by a nonlinear Darcy's law. We refer to [21] or [26] for the homogeneous case with flow obeying to the linear Darcy law. For the variational inequalities approach, we refer to [10]. The chapter is organized as follows: in Section 1 we derive the weak formulations of the dam problem with Dirichlet boundary condition and with leaky boundary condition, respectively. Then we prove the existence of a solution (u, g, γ) to a unified formulation of the problem. Moreover, we show that u is uniformly bounded and locally Hölder continuous in Ω. Under additional assumptions on the permeability, we also establish a monotonicity result for g, which with the continuity of u allows us to define the free boundary as an $x_1 x_2$-graph. In Section 2 we address the case of Dirichlet condition. We prove the continuity of the free boundary and show that g is the characteristic function of the dry part of the dam. Then we prove the uniqueness of the reservoirs-connected solution. In Section 3 we consider the leaky condition case. We give a sufficient condition for saturation under the reservoirs and prove the continuity of the free boundary. As a consequence, we obtain the expression of g which unlike the previous case is not equal to the characteristic function of the dry part. Finally, the uniqueness of the reservoirs-connected solution is established in two situations.

Notation

$$x = (x_1, x_2) \quad - \quad \text{a typical point in } \mathbb{R}^2 \text{ and } |x| = \sqrt{x_1^2 + x_2^2}$$

$\overline{S}, \mathrm{Int}(S)$ and ∂S — respectively the closure, the interior and the boundary of the set S

$\chi(S)$ — the characteristic function of S

$\Omega \setminus S = \{x \in \Omega \mid x \notin S\}$

$|S|$ — the Lebesgue measure of S in $\mathbb{R}^2$

$B_r(x)$ — the open ball with center x and radius r

If u — a real-valued function, then $u^+ = \max(u, 0)$, $u^- = (-u)^+$, $u_{x_i} = \frac{\partial u}{\partial x_i}$ and $\nabla u = (u_{x_1}, u_{x_2})$

$C^{0,\alpha}_{\mathrm{loc}}(\Omega)$ — the set of all functions $u : \Omega \to \mathbb{R}$ that are locally Hölder continuous in Ω

$C^{0,\alpha}_{\mathrm{loc}}(\Omega \cup S)$ — the set of all functions $u : \Omega \cup S \to \mathbb{R}$ that possess an extension to $C^{0,\alpha}_{\mathrm{loc}}(\Omega')$ for some open set Ω' which contains $\Omega \cup S$, where $S \subset \partial \Omega$

$C^{1,\alpha}_{\mathrm{loc}}(\Omega \cup S)$ — the set of all functions $u : \Omega \cup S \to \mathbb{R}$ whose first partial derivatives are in $C^{0,\alpha}_{\mathrm{loc}}(\Omega \cup S)$, where $S \subset \partial \Omega$

$C^1(\Omega)$ — the set of all continuously differentiable functions $u : \Omega \to \mathbb{R}$

$C^1(\overline{\Omega})$ — the set of all functions $u : \overline{\Omega} \to \mathbb{R}$ that possess an extension to $C^1(\mathbb{R}^2)$

$\mathcal{D}(\Omega)$ — the set of all indefinitely continuously differentiable functions $u : \Omega \to \mathbb{R}$ with compact support

$\mathcal{D}'(\Omega)$ — the set of distributions

$L^p(\Omega)$ — the set of all Lebesgue measurable functions $u : \Omega \to \mathbb{R}$ such that $|u|_{p,\Omega} < \infty$, where $|u|_{p,\Omega} = (\int_\Omega |u(x)|^p \, dx)^{1/p}$ if $1 \leqslant p < \infty$ and $|u|_{\infty,\Omega} = \mathrm{ess\,sup}_\Omega |u|$

$\mathbb{L}^p(\Omega) = (L^p(\Omega))^2$

$L^{p'}(\Omega)$ — the dual space of $L^p(\Omega)$, where $1/p + 1/p' = 1$

$W^{1,p}(\Omega) = \{u : \Omega \to \mathbb{R} \mid u, u_{x_1}, u_{x_2} \in L^p(\Omega)\}$

$|u_{1,p}|$ – the norm of $W^{1,p}(\Omega)$, $|u|_{1,p} = (|u|^p_{p,\Omega} + |u_{x_1}|^p_{p,\Omega} + |u_{x_2}|^p_{p,\Omega})^{1/p}$
$H^1(\Omega) = W^{1,2}(\Omega)$.

1. A unified formulation of the dam problem

1.1. *Formulation of the problem*

A porous medium that we denote by Ω is supplied by several reservoirs of a fluid which infiltrates through Ω. We assume that Ω is a bounded locally Lipschitz domain of $\mathbb{R}^2$ with boundary $\partial\Omega = S_1 \cup S_2 \cup S_3$, where S_1 is the impervious part, S_2 is the part in contact with air and $S_3 = \bigcup_{i=1}^{N} S_{3,i}$ with $S_{3,i}$ the part in contact with the bottom of the ith reservoir. We assume that the flow in Ω has reached a steady state and we look for the fluid pressure p and the saturated region S of the porous medium. The boundary of S is divided into four parts (see Figure 1):

- $\Gamma_1 \subset S_1$ – the impervious part,
- $\Gamma_2 \subset \Omega$ – the free boundary,
- $\Gamma_3 \subset S_3$ – the part covered by fluid,
- $\Gamma_4 \subset S_2$ – the part where the fluid flows out of Ω.

In the saturated region, the flow is governed by the following nonlinear Darcy law:

$$v = -\mathcal{A}\big(x, \nabla(p + x_2)\big) = -\mathcal{A}(x, \nabla u), \tag{1.1}$$

where v is the fluid velocity, $u = p + x_2$ is the hydrostatic head and $\mathcal{A} : \Omega \times \mathbb{R}^2 \to \mathbb{R}^2$ is a mapping that satisfies the following assumptions for some constants $q > 1$ and $0 < \lambda \leqslant M < \infty$:

$$\begin{cases}
\text{(i) } x \mapsto \mathcal{A}(x, \xi) \text{ is measurable } \forall \xi \in \mathbb{R}^2, \\
\text{(ii) } \xi \mapsto \mathcal{A}(x, \xi) \text{ is continuous for a.e. } x \in \Omega, \\
\text{(iii) for all } \xi \in \mathbb{R}^2 \text{ and for a.e. } x \in \Omega, \\
\quad \mathcal{A}(x, \xi) \cdot \xi \geqslant \lambda|\xi|^q \text{ and } |\mathcal{A}(x, \xi)| \leqslant M|\xi|^{q-1}, \\
\text{(iv) for all } \xi, \zeta \in \mathbb{R}^2 \text{ and for a.e. } x \in \Omega, \\
\quad \big(\mathcal{A}(x, \xi) - \mathcal{A}(x, \zeta)\big) \cdot (\xi - \zeta) \geqslant 0.
\end{cases} \tag{1.2}$$

A first example of such an operator $\mathcal{A}$ corresponds to the classical Darcy law [13],

$$\mathcal{A}(x, \xi) = a(x)\xi,$$

where $a(x)$ is the permeability matrix of the medium.

Another example corresponds to the nonlinear Darcy law [28],

$$\mathcal{A}(x, \xi) = \big|a(x)\xi \cdot \xi\big|^{(q-2)/2} a(x)\xi.$$

In the following, we shall derive weak formulations of the problem with Dirichlet condition on $S_2 \cup S_3$ and leaky condition on S_3, respectively. Then we give a unified formulation that covers both formulations. First due to the incompressibility of the fluid one has

$$\operatorname{div}(v) = 0 \quad \text{in } S.$$

Thus for each sufficiently smooth function ξ and assuming v and S smooth enough, one has

$$0 = \int_S \operatorname{div}(v) \cdot \xi \, dx = -\int_S v \cdot \nabla\xi \, dx + \int_{\partial S} v \cdot \nu \xi \, d\sigma(x),$$

where ν is the outward unit normal vector to ∂S. This reads by (1.1)

$$\int_S \mathcal{A}(x, \nabla u) \cdot \nabla\xi \, dx = \int_{\partial S} -v \cdot \nu \xi \, d\sigma(x).$$

Assuming that the exterior atmospheric pressure is normalized to 0 and extending p by 0 to $\Omega \setminus S$, we obtain

$$\int_\Omega \big(\mathcal{A}(x, \nabla u) - \chi(\Omega \setminus S)\mathcal{A}(x, e)\big) \cdot \nabla\xi \, dx = \int_{\partial S} -v \cdot \nu \xi \, d\sigma(x). \tag{1.3}$$

Note that we are looking for a $p \geqslant 0$ or equivalently,

$$u \geqslant x_2 \quad \text{in } \Omega.$$

Thus $\chi(\Omega \setminus S)$ is a function, that we denote by g, such that

$$0 \leqslant g \leqslant 1, \quad g \cdot (u - x_2) = 0 \quad \text{in } \Omega.$$

We assume that no fluid can flow through the part of ∂S that is contained in Ω, i.e., the free boundary. This leads to

$$v \cdot \nu = 0 \quad \text{on } \partial S \cap \Omega.$$

It follows from (1.3) that for $\Gamma = \partial\Omega$,

$$\int_\Omega \big(\mathcal{A}(x, \nabla u) - g\mathcal{A}(x, e)\big) \cdot \nabla\xi \, dx = \int_{\partial S \cap \Gamma} -v \cdot \nu \xi \, d\sigma(x). \tag{1.4}$$

Now if we assume that Γ_1 is impervious and since there is overflow on Γ_4, we obtain for $\xi \geqslant 0$ on S_2

$$\int_\Omega \big(\mathcal{A}(x, \nabla u) - g\mathcal{A}(x, e)\big) \cdot \nabla\xi \, dx \leqslant \int_{\Gamma_3} -v \cdot \nu \xi \, d\sigma(x). \tag{1.5}$$

Finally, we denote by φ the exterior pressure on $S_2 \cup S_3$ which is equal to the atmospheric pressure on S_2, and is equal to the fluid pressure on S_3. A first model for the flow through the reservoirs bottoms' consists on assuming the continuity of the pressure, i.e., that we have

$$u = \psi = \varphi + x_2 \quad \text{on } S_3. \tag{1.6}$$

Now assuming that $\xi = 0$ on S_3, we get the following weak formulation (see, for example, [2,16] and [20]):

$$\begin{cases} \text{find } (u, g) \in W^{1,q}(\Omega) \times L^\infty(\Omega) \text{ such that} \\ \text{(i) } u \geqslant x_2, 0 \leqslant g \leqslant 1, g(u - x_2) = 0 \text{ a.e. in } \Omega, \\ \text{(ii) } u = \psi \text{ on } S_2 \cup S_3, \\ \text{(iii) } \int_\Omega \big(\mathcal{A}(x, \nabla u) - g\mathcal{A}(x, e) \big) \cdot \nabla \xi \, dx \leqslant 0 \\ \qquad \forall \xi \in W^{1,q}(\Omega), \xi = 0 \text{ on } S_3, \xi \geqslant 0 \text{ on } S_2. \end{cases} \tag{P_D}$$

A second model for the flow through S_3 consists on prescribing the flux instead of the pressure, i.e.,

$$-v \cdot v = \beta(x, \varphi - p) \quad \text{on } S_3, \tag{1.7}$$

where $\beta(x, v)$ is a function that is nondecreasing with respect to v. In this case, we obtain from (1.5) and (1.7), for $\xi \geqslant 0$ on S_2 (see [19,22] and [23]),

$$\begin{cases} \text{find } (u, g) \in W^{1,q}(\Omega) \times L^\infty(\Omega) \text{ such that} \\ \text{(i) } u \geqslant x_2, 0 \leqslant g \leqslant 1, g(u - x_2) = 0 \text{ a.e. in } \Omega, \\ \text{(ii) } u = \psi \text{ on } S_2, \\ \text{(iii) } \int_\Omega \big(\mathcal{A}(x, \nabla u) - g\mathcal{A}(x, e) \big) \cdot \nabla \xi \, dx \leqslant \int_{S_3} \beta(x, \psi - u)\xi \, d\sigma(x) \\ \qquad \forall \xi \in W^{1,q}(\Omega), \xi \geqslant 0 \text{ on } S_2. \end{cases} \tag{P_L}$$

Now we would like to replace the above boundary conditions by the following unified boundary condition

$$-v \cdot v \in \mathcal{B}(x, \varphi - p) \quad \text{on } \Gamma, \tag{1.8}$$

where for a.e. $x \in \Gamma$, $\mathcal{B}(x, \cdot)$ is a multivalued monotone function.

Note that if $\mathcal{B}$ is given by

$$\mathcal{B}(x, \cdot) = \begin{cases} \mathbb{R} \times \{0\} & \text{for a.e. } x \in S_1, \\ \{0\} \times \mathbb{R} & \text{for a.e. } x \in S_2 \cup S_3, \end{cases} \tag{1.9}$$

we obtain

$$v \cdot v = 0 \quad \text{on } S_1 \quad \text{and} \quad p = \varphi \quad \text{on } S_2 \cup S_3,$$

which corresponds to (P_D).

If for a.e. $x \in S_3$, $\beta(x, \cdot)$ is a continuous nondecreasing function, and $\mathcal{B}$ is given by

$$\mathcal{B}(x, \cdot) = \begin{cases} \mathbb{R} \times \{0\} & \text{for a.e. } x \in S_1, \\ \{0\} \times \mathbb{R} & \text{for a.e. } x \in S_2, \\ \beta(x, \cdot) & \text{for a.e. } x \in S_3, \end{cases} \tag{1.10}$$

we obtain

$$v \cdot \nu = 0 \quad \text{on } S_1, \quad p = \varphi \quad \text{on } S_2$$

and

$$v \cdot \nu = -\beta(x, \varphi - p) \quad \text{on } S_3,$$

which corresponds to $(\mathrm{P_L})$.

For $\mathcal{B}$, we assume that

$$\text{for a.e. } x \in \Gamma, \quad \mathcal{B}(x, \cdot) \text{ is a maximal monotone graph of } \mathbb{R}^2, \tag{1.11}$$

$$\text{for a.e. } x \in \Gamma, \quad 0 \in \mathcal{B}(x, 0), \tag{1.12}$$

$$\text{for a.e. } x \in \Gamma, \quad D(\mathcal{B}(x, \cdot)) = [a(x), b(x)],$$

$$-\infty \leqslant a(x) \leqslant 0 \leqslant b(x) \leqslant +\infty. \tag{1.13}$$

Taking into account the assumptions (1.11)–(1.13), there exist for a.e. $x \in \Gamma$, two maximal monotone graphs $\mathcal{B}_1(x, \cdot)$ and $\mathcal{B}_2(x, \cdot)$ in $\mathbb{R}^2$ such that

$$\begin{cases} D(\mathcal{B}_2(x, \cdot)) = \mathbb{R}, \\ \mathcal{B}_2(x, \cdot) = \mathcal{B}(x, \cdot) & \text{in } (a(x), b(x)), \\ \mathcal{B}_2(x, s) = \{(s, \mathcal{B}^0(x, a(x)))\} & \forall s \leqslant a(x) \text{ if } a(x) > -\infty, \\ \mathcal{B}_2(x, s) = \{(s, \mathcal{B}^0(x, b(x)))\} & \forall s \geqslant b(x) \text{ if } b(x) < \infty, \\ D(\mathcal{B}_1(x, \cdot)) = D(\mathcal{B}(x, \cdot)), \\ \mathcal{B}_1(x, \cdot) \equiv 0 & \text{in } (a(x), b(x)), \\ \mathcal{B} = \mathcal{B}_1 + \mathcal{B}_2 & \text{in } [a(x), b(x)], \end{cases} \tag{1.14}$$

where for a.e. $x \in \Gamma$, $\mathcal{B}^0(x, \cdot)$ is the minimal section of $\mathcal{B}(x, \cdot)$, i.e., for each $s \in D(\mathcal{B}(x, \cdot))$, $|\mathcal{B}^0(x, s)| = \min_{\gamma \in \mathcal{B}(x, s)} |\gamma|$.

Moreover, we assume that

$$\forall R > 0, \exists C_R > 0: \text{ for a.e. } x \in \Gamma, \forall s \in [-R, R] \cap [a(x), b(x)],$$

$$\mathcal{B}_2(x, s) \subset \{s\} \times [-C_R, C_R]. \tag{1.15}$$

Taking into account (1.5) and (1.8), we obtain the following unified weak formulation (see [5] and [34]):

$$
\begin{cases}
\text{find } (u, g, \gamma) \in W^{1,q}(\Omega) \times L^{\infty}(\Omega) \times L^{q'}(\Gamma) \text{ such that} \\
\text{(i) } \psi(x) - u(x) \in D\big(\mathcal{B}(x, \cdot)\big) \text{ for a.e. } x \in \Gamma, \\
\text{(ii) } u \geqslant x_2, 0 \leqslant g \leqslant 1, g(u - x_2) = 0 \text{ a.e. in } \Omega, \\
\text{(iii) } \gamma(x) \in \mathcal{B}\big(x, \psi(x) - u(x)\big) \text{ for a.e. } x \in \Gamma \\
\qquad \text{and } \gamma(x) \leqslant 0 \text{ for a.e. } x \in \Gamma \text{ such that } \psi(x) = x_2, \\
\text{(iv) } \int_{\Omega} \big(\mathcal{A}(x, \nabla u) - g\mathcal{A}(x, e)\big) \cdot \nabla(\xi - u) \, dx \geqslant \int_{\Gamma} \gamma \cdot (\xi - u) \, d\sigma(x) \\
\qquad \forall \xi \in \mathbb{K} = \big\{ \xi \in W^{1,q}(\Omega) \mid a(x) \leqslant \psi(x) - \xi(x) \leqslant c(x) \text{ for a.e. } x \in \Gamma \big\},
\end{cases}
\tag{P_U}
$$

where for a.e. $x \in \Gamma$, $c(x) = b(x)$ if $\varphi(x) > 0$ and $c(x) = +\infty$ if $\varphi(x) = 0$.

1.2. *Existence of a solution*

In this subsection, we establish the existence of a solution of the problem (P_U).

THEOREM 1.1. *Assume that φ is a nonnegative Lipschitz continuous function, $\mathcal{A}$ satisfies (1.2) and $\mathcal{B}$ satisfies (1.11)–(1.15). Then there exists a solution (u, g, γ) to the problem (P_U).*

For $\varepsilon > 0$, we introduce the following approximated problem:

$$
\begin{cases}
\text{find } u_{\varepsilon} \in V = \big\{ \xi \in W^{1,q}(\Omega) \mid \xi_{|\Gamma} \in L^{q'}(\Gamma) \big\} \text{ such that} \\
\int_{\Omega} \varepsilon\big(|u_{\varepsilon}|^{q-2} u_{\varepsilon} - |x_2|^{q-2} x_2\big) \cdot \xi \\
\qquad + \big(\mathcal{A}(x, \nabla u_{\varepsilon}) - G_{\varepsilon}(u_{\varepsilon})\mathcal{A}(x, e)\big) \cdot \nabla \xi \, dx \\
\qquad + \int_{\Gamma} \varepsilon\big(|u_{\varepsilon}|^{q'-2} u_{\varepsilon} - |x_2|^{q'-2} x_2\big) \cdot \xi \, d\sigma(x) \\
\quad = \int_{\Gamma} \big(\mathcal{B}_1^{\varepsilon}(x, \psi - u_{\varepsilon}) + \mathcal{B}_2^{\varepsilon}(x, \psi - u_{\varepsilon})\big) \cdot \xi \, d\sigma(x) \quad \forall \xi \in V,
\end{cases}
\tag{P_{ε}}
$$

where $G_{\varepsilon} : L^q(\Omega) \to L^{\infty}(\Omega)$ is defined for a.e. $x \in \Omega$ by $G_{\varepsilon}(v(x)) = 1 - H_{\varepsilon}(v(x) - x_2)$ and $H_{\varepsilon}(s) = \min(\frac{s^+}{\varepsilon}, 1)$. $\mathcal{B}_i^{\varepsilon}$, $i = 1, 2$, denotes the Yoshida approximation of $\mathcal{B}_i$. Note that $\mathcal{B}_i^{\varepsilon}$ is nondecreasing and uniformly Lipschitz continuous with respect to the second variable (see [12,15]), with Lipschitz constant equal to $1/\varepsilon$. Taking into account (1.12), we deduce that $\mathcal{B}_i^{\varepsilon}(x, 0) = 0$ for a.e. $x \in \Gamma$. By the monotonicity of $\mathcal{B}_i^{\varepsilon}(x, \cdot)$, this leads to

$$
\mathcal{B}_i^{\varepsilon}(x, u) \cdot u \geqslant 0 \quad \text{for a.e. } x \in \Gamma, \forall u \in \mathbb{R}.
\tag{1.16}
$$

We shall equip V with the norm $\|u\| = |u|_{1,q} + |u|_{q',\Gamma}$.

REMARK 1.1. The approximation of (P_U) by a similar problem to (P_{ε}) was first introduced in [2] and [16] to solve the problem with Dirichlet condition and linear Darcy's law. It was extended in [19] to address the problem for leaky condition. Finally it was used in the form given here in [34] and in [5] for $\mathcal{A}(x, \xi) = \xi$.

We first establish the existence of a solution to (P_ε).

THEOREM 1.2. *Assume that φ is a nonnegative Lipschitz continuous function, that $\mathcal{A}$ satisfies (1.2) and $\mathcal{B}$ satisfies (1.11)–(1.15). Then there exists a solution u_ε to (P_ε).*

PROOF. The proof is based on the Schauder fixed point theorem. First we consider for $u \in V$ the operator $Au : V \to \mathbb{R}$ defined by

$$\xi \to \langle Au, \xi \rangle = \int_\Omega \varepsilon |u|^{q-2} u \cdot \xi + \mathcal{A}(x, \nabla u) \cdot \nabla \xi \, dx$$

$$+ \int_\Gamma \varepsilon |u|^{q'-2} u \cdot \xi \, d\sigma(x)$$

$$- \int_\Gamma \left(\mathcal{B}_1^\varepsilon(x, \psi - u) + \mathcal{B}_2^\varepsilon(x, \psi - u) \right) \cdot \xi \, d\sigma(x).$$

Clearly A defines a continuous operator from V to V'. Moreover, A is strictly monotone due to the monotonicity of $\mathcal{A}(x, \cdot)$, $\mathcal{B}_i^\varepsilon(x, \cdot)$ and the function $u \to |u|^{r-2} u$ for $r > 1$.

We claim that A is also coercive. Indeed, we have for each $u \in V$,

$$\langle Au, u \rangle = \int_\Omega \varepsilon |u|^q + \mathcal{A}(x, \nabla u) \cdot \nabla u \, dx$$

$$+ \int_\Gamma \varepsilon |u|^{q'} \, d\sigma(x)$$

$$- \sum_{i=1}^2 \int_\Gamma \mathcal{B}_i^\varepsilon(x, \psi - u) \cdot u \, d\sigma(x). \tag{1.17}$$

Using (1.16) and the Lipschitz continuity of $\mathcal{B}_i^\varepsilon(x, \cdot)$, we have

$$\mathcal{B}_i^\varepsilon(x, \psi - u) \cdot u \leqslant \mathcal{B}_i^\varepsilon(x, \psi - u) \cdot \psi \leqslant \frac{1}{\varepsilon} |\psi - u| \cdot |\psi| \leqslant \frac{1}{\varepsilon} \left(|\psi|^2 + |u| \cdot |\psi| \right)$$

which leads by Hölder's inequality, for some positive constants c_0, c_1, to

$$\int_\Gamma \mathcal{B}_i^\varepsilon(x, \psi - u) \cdot u \, d\sigma(x) \leqslant c_0 + c_1 |u|_{q', \Gamma}. \tag{1.18}$$

Using (1.2)(iii), (1.17) and (1.18), we get for a positive constant c_2

$$\langle Au, u \rangle \geqslant c_2 \left(|u|_{1,q}^q + |u|_{q', \Gamma}^{q'} \right) - 2 c_1 |u|_{q', \Gamma} - 2 c_0 \quad \forall u \in V. \tag{1.19}$$

Since $q, q' > 1$, we obtain $\lim_{\|u\| \to +\infty} \langle Au, u \rangle / \|u\| = +\infty$.

Now for $v \in L^q(\Omega)$, we consider the mapping $F_v : V \to \mathbb{R}$ defined by

$$F_v(\xi) = \int_\Omega \varepsilon |x_2|^{q-2} x_2 \cdot \xi \, dx + \int_\Omega G_\varepsilon(v) \mathcal{A}(x, e) \cdot \nabla \xi \, dx$$

$$+ \int_\Gamma \varepsilon |x_2|^{q'-2} x_2 \cdot \xi \, d\sigma(x).$$

Given that F_v is a continuous linear form on V, and A is strictly monotone, continuous and coercive, there exists, for each $v \in L^q(\Omega)$, a unique solution u_ε [31] for the variational problem

$$\begin{cases} u_\varepsilon \in V, \\ \langle Au_\varepsilon, w \rangle = \langle F_v, w \rangle \quad \forall w \in V. \end{cases} \tag{1.20}$$

This defines a mapping $\mathcal{F}_\varepsilon : L^q(\Omega) \to V$, $v \to u_\varepsilon$, which satisfies

$$\exists R_\varepsilon > 0 \mid \|\mathcal{F}_\varepsilon(v)\| \leqslant R_\varepsilon \ \forall v \in L^q(\Omega), \tag{1.21}$$

$$\mathcal{F}_\varepsilon(\overline{B}_{R_\varepsilon}) \subset \overline{B}_{R_\varepsilon}, \tag{1.22}$$

$$\mathcal{F}_\varepsilon : L^q(\Omega) \to L^q(\Omega) \text{ is continuous}, \tag{1.23}$$

where $\overline{B}_{R_\varepsilon}$ is the closed ball of $L^q(\Omega)$ of center 0 and radius R_ε.

Indeed, using u_ε as a test function for (1.20) and taking into account (1.19), we get

$$c_2 \big(|u_\varepsilon|_{1,q}^q + |u_\varepsilon|_{q',\Gamma}^{q'} \big) - 2c_1 |u_\varepsilon|_{q',\Gamma} - 2c_0 \leqslant c_3 \big(|u_\varepsilon|_{1,q} + |u_\varepsilon|_{q',\Gamma} \big)$$

which can be written

$$|u_\varepsilon|_{1,q}^q + |u_\varepsilon|_{q',\Gamma}^{q'} \leqslant c_3' \big(|u_\varepsilon|_{1,q} + |u_\varepsilon|_{q',\Gamma} \big) + c_0'. \tag{1.24}$$

We discuss three cases
- $|u_\varepsilon|_{1,q} \leqslant 1$: In this case, we deduce from (1.24) that $|u_\varepsilon|_{q',\Gamma}^{q'} \leqslant c_3' |u_\varepsilon|_{q',\Gamma} + c_0' + c_3'$, which leads to $|u_\varepsilon|_{q',\Gamma} \leqslant c_4$ for some constant c_4. Thus $|u_\varepsilon|_{1,q} + |u_\varepsilon|_{q',\Gamma} \leqslant c_4 + 1$.
- $|u_\varepsilon|_{q',\Gamma} \leqslant 1$: In this case, we deduce from (1.24) that $|u_\varepsilon|_{1,q}^q \leqslant c_3' |u_\varepsilon|_{1,q} + c_0' + c_3'$, which leads to $|u_\varepsilon|_{1,q} \leqslant c_5$ for some constant c_5. Thus $|u_\varepsilon|_{1,q} + |u_\varepsilon|_{q',\Gamma} \leqslant c_5 + 1$.
- $|u_\varepsilon|_{1,q}, |u_\varepsilon|_{q',\Gamma} > 1$: Let $r = \min(q, q')$. Then we have

$$\big(|u_\varepsilon|_{1,q} + |u_\varepsilon|_{q',\Gamma} \big)^r \leqslant 2^{r-1} \big(|u_\varepsilon|_{1,q}^r + |u_\varepsilon|_{q',\Gamma}^r \big)$$

$$\leqslant 2^{r-1} \big(|u_\varepsilon|_{1,q}^q + |u_\varepsilon|_{q',\Gamma}^{q'} \big)$$

$$\leqslant c_3' 2^{r-1} \big(|u_\varepsilon|_{1,q} + |u_\varepsilon|_{q',\Gamma} \big) + c_0' 2^{r-1}$$

which leads to $|u|_{1,q} + |u_\varepsilon|_{q',\Gamma} \leqslant c_6$ for some constant c_6.

Finally, we have proved (1.21). Now (1.22) is a consequence of (1.21) since $|u_\varepsilon|_{q,\Omega} \leqslant \|u_\varepsilon\|$.

To prove (1.23), let $(v_k)_k$ be a sequence of $L^q(\Omega)$ which converges to v in $L^q(\Omega)$. We denote $\mathcal{F}_\varepsilon(v_k)$ by u_ε^k. Since $u_\varepsilon^k - u_\varepsilon$ is a suitable test function for (1.20), we obtain by subtracting the equations written for u_ε^k and u_ε, respectively,

$$\int_\Omega \left(\mathcal{A}(x, \nabla u_\varepsilon^k) - \mathcal{A}(x, \nabla u_\varepsilon)\right) \cdot \nabla\left(u_\varepsilon^k - u_\varepsilon\right) dx$$

$$+ \int_\Omega \varepsilon\left(|u_\varepsilon^k|^{q-2}u_\varepsilon^k - |u_\varepsilon|^{q-2}u_\varepsilon\right) \cdot \left(u_\varepsilon^k - u_\varepsilon\right) dx$$

$$+ \int_\Gamma \varepsilon\left(|u_\varepsilon^k|^{q'-2}u_\varepsilon^k - |u_\varepsilon|^{q'-2}u_\varepsilon\right) \cdot \left(u_\varepsilon^k - u_\varepsilon\right) d\sigma(x)$$

$$= \int_\Omega \left(G_\varepsilon(v_k) - G_\varepsilon(v)\right)\mathcal{A}(x, e) \cdot \nabla\left(u_\varepsilon^k - u_\varepsilon\right) dx$$

$$+ \sum_{i=1}^{2} \int_\Gamma \left(\mathcal{B}_i^\varepsilon(x, \psi - u_\varepsilon^k) - \mathcal{B}_i^\varepsilon(x, \psi - u_\varepsilon)\right) \cdot \left(u_\varepsilon^k - u_\varepsilon\right) d\sigma(x)$$

which leads by the monotonicity of $\mathcal{A}(x, \cdot)$, $s \to |s|^{q'-2}s$, and $\mathcal{B}_i^\varepsilon(x, \cdot)$ to

$$\int_\Omega \varepsilon\left(|u_\varepsilon^k|^{q-2}u_\varepsilon^k - |u_\varepsilon|^{q-2}u_\varepsilon\right) \cdot \left(u_\varepsilon^k - u_\varepsilon\right) dx$$

$$\leqslant \int_\Omega \left(G_\varepsilon(v_k) - G_\varepsilon(v)\right)\mathcal{A}(x, e) \cdot \nabla\left(u_\varepsilon^k - u_\varepsilon\right) dx. \tag{1.25}$$

Using (1.2)(iii), the fact that G_ε is Lipschitz continuous with Lipschitz constant equal to $1/\varepsilon$ and the Hölder inequality, we obtain since $q' > 1$ and $|G_\varepsilon(v_k) - G_\varepsilon(v)| \leqslant 1$,

$$\int_\Omega \left(G_\varepsilon(v_k) - G_\varepsilon(v)\right)\mathcal{A}(x, e) \cdot \nabla\left(u_\varepsilon^k - u_\varepsilon\right) dx$$

$$\leqslant M\left(\int_\Omega |G_\varepsilon(v_k) - G_\varepsilon(v)|^{q'} dx\right)^{1/q'} \cdot \left(\int_\Omega |\nabla(u_\varepsilon^k - u_\varepsilon)|^q dx\right)^{1/q}$$

$$\leqslant M\|u_\varepsilon^k - u_\varepsilon\|\left(\int_\Omega \frac{1}{\varepsilon}|v_k - v| dx\right)^{1/q'}$$

$$\leqslant \frac{M|\Omega|^{1/q'^2}}{\varepsilon^{1/q'}}\|u_\varepsilon^k - u_\varepsilon\|\left(\int_\Omega |v_k - v|^q dx\right)^{1/(qq')}. \tag{1.26}$$

Combining (1.21), (1.25) and (1.26), we obtain for some constant $C(\varepsilon)$,

$$\int_\Omega \left(|u_\varepsilon^k|^{q-2}u_\varepsilon^k - |u_\varepsilon|^{q-2}u_\varepsilon\right) \cdot \left(u_\varepsilon^k - u_\varepsilon\right) dx \leqslant C(\varepsilon)|v_k - v|_{q,\Omega}^{1/q'}$$

which leads to

$$\lim_{k \to \infty} \int_{\Omega} \left(|u_{\varepsilon}^k|^{q-2} u_{\varepsilon}^k - |u_{\varepsilon}|^{q-2} u_{\varepsilon} \right) \cdot \left(u_{\varepsilon}^k - u_{\varepsilon} \right) dx = 0. \tag{1.27}$$

Using the following inequalities for some $\mu > 0$:

(i) if $q \geqslant 2, \forall x, y \in \mathbb{R}^2, \quad \mu |x - y|^q \leqslant \left(|x|^{q-2} x - |y|^{q-2} y \right) \cdot (x - y),$

(ii) if $1 < q < 2, \forall x, y \in \mathbb{R}^2,$

$$\mu |x - y|^2 \leqslant \left(|x| + |y| \right)^{2-q} \cdot \left(|x|^{q-2} x - |y|^{q-2} y \right) \cdot (x - y),$$

we easily obtain from (1.27) that $u_{\varepsilon}^k \to u_{\varepsilon}$ in $L^q(\Omega)$. Hence the continuity of $\mathcal{F}_{\varepsilon}$ is established. In particular, $\mathcal{F}_{\varepsilon} : \overline{B}(0, R_{\varepsilon}) \to \overline{B}(0, R_{\varepsilon})$ is continuous. Moreover, by (1.21), $\mathcal{F}_{\varepsilon}(\overline{B}(0, R_{\varepsilon}))$ is relatively compact in $L^q(\Omega)$. Thus we can apply the Schauder fixed point theorem to obtain a fixed point for $\mathcal{F}_{\varepsilon}$ which is a solution of $(\mathrm{P}_{\varepsilon})$. $\qquad\square$

Now we have the following estimates.

LEMMA 1.1. *Let H and ε_0 be two positive constants such that $H > \varepsilon_0 + \max_{x \in \overline{\Omega}} \psi(x)$. Then we have for all $\varepsilon \in (0, \varepsilon_0)$,*

$$x_2 \leqslant u_{\varepsilon} \leqslant H \quad \textit{a.e. in } \Omega. \tag{1.28}$$

PROOF. (i) Using $(u_{\varepsilon} - x_2)^-$ as a test function for $(\mathrm{P}_{\varepsilon})$ and taking into account that $G_{\varepsilon}(u_{\varepsilon}) = 1$ a.e. in $[u_{\varepsilon} \leqslant x_2]$, we obtain

$$\int_{\Omega} \mathcal{A}(x, \nabla u_{\varepsilon}) \cdot \nabla (u_{\varepsilon} - x_2)^- dx + \int_{\Omega} \varepsilon \left(|u_{\varepsilon}|^{q-2} u_{\varepsilon} - |x_2|^{q-2} x_2 \right) \cdot (u_{\varepsilon} - x_2)^- dx$$

$$+ \int_{\Gamma} \varepsilon \left(|u_{\varepsilon}|^{q'-2} u_{\varepsilon} - |x_2|^{q'-2} x_2 \right) \cdot (u_{\varepsilon} - x_2)^- d\sigma(x)$$

$$= \int_{\Omega} \mathcal{A}(x, e) \cdot \nabla (u_{\varepsilon} - x_2)^- dx$$

$$+ \sum_{i=1}^{2} \int_{\Gamma} \mathcal{B}_i^{\varepsilon}(x, \psi - u_{\varepsilon}) \cdot (u_{\varepsilon} - x_2)^- d\sigma(x). \tag{1.29}$$

Using the fact that $\psi \geqslant x_2$, the monotonicity of $\mathcal{B}_i^{\varepsilon}(x, \cdot)$ and (1.16), one has for $i = 1, 2$,

$$\int_{\Gamma} \mathcal{B}_i^{\varepsilon}(x, \psi - u_{\varepsilon}) \cdot (u_{\varepsilon} - x_2)^- d\sigma(x)$$

$$\geqslant \int_{\Gamma \cap [u_{\varepsilon} \leqslant x_2]} \mathcal{B}_i^{\varepsilon}(x, x_2 - u_{\varepsilon}) \cdot (x_2 - u_{\varepsilon}) d\sigma(x) \geqslant 0. \tag{1.30}$$

Combining (1.29) and (1.30), we get

$$\int_{\Omega\cap[u_\varepsilon\leqslant x_2]} \big(\mathcal{A}(x,\nabla u_\varepsilon) - \mathcal{A}(x,\nabla x_2)\big)\cdot\nabla(u_\varepsilon - x_2)\,dx$$

$$+ \int_{\Omega\cap[u_\varepsilon\leqslant x_2]} \varepsilon\big(|u_\varepsilon|^{q-2}u_\varepsilon - |x_2|^{q-2}x_2\big)\cdot(u_\varepsilon - x_2)\,dx$$

$$+ \int_{\Gamma\cap[u_\varepsilon\leqslant x_2]} \varepsilon\big(|u_\varepsilon|^{q'-2}u_\varepsilon - |x_2|^{q'-2}x_2\big)\cdot(u_\varepsilon - x_2)\,d\sigma(x) \leqslant 0. \tag{1.31}$$

Using the monotonicity of $\mathcal{A}(x,\cdot)$ and $u\to|u|^{r-2}u$ for $r=q,q'$, and (1.31), we obtain $u_\varepsilon \geqslant x_2$ a.e. in Ω.

(ii) Note that for $\varepsilon\in(0,\varepsilon_0)$ and for $u_\varepsilon(x)\geqslant H$, one has $u_\varepsilon(x)\geqslant H\geqslant\varepsilon_0+\psi\geqslant\varepsilon+x_2$, and therefore $G_\varepsilon(u_\varepsilon(x))=0$. It follows that

$$G_\varepsilon\big(u_\varepsilon(x)\big)\mathcal{A}(x,e)\cdot\nabla(u_\varepsilon - H)^+ = 0 \quad\text{for a.e. } x\in\Omega. \tag{1.32}$$

Using (1.16) and the monotonicity of $\mathcal{B}_i^\varepsilon(x,\cdot)$, one has

$$\mathcal{B}_i^\varepsilon(x,\psi - u_\varepsilon)(u_\varepsilon - H)^+ \leqslant \mathcal{B}_i^\varepsilon(x, H - u_\varepsilon)(u_\varepsilon - H)^+$$

$$\leqslant 0 \quad\text{for a.e. } x\in\Gamma. \tag{1.33}$$

Using $(u_\varepsilon - H)^+$ as a test function for (P_ε) and taking into account (1.32) and (1.33), we obtain

$$\int_\Omega \mathcal{A}(x,\nabla u_\varepsilon)\cdot\nabla(u_\varepsilon - H)^+\,dx + \int_\Omega \varepsilon\big(|u_\varepsilon|^{q-2}u_\varepsilon - |x_2|^{q-2}x_2\big)\cdot(u_\varepsilon - H)^+\,dx$$

$$+ \int_\Gamma \varepsilon\big(|u_\varepsilon|^{q'-2}u_\varepsilon - |x_2|^{q'-2}x_2\big)\cdot(u_\varepsilon - H)^+\,d\sigma(x) \leqslant 0$$

which can be written since $H\geqslant x_2$ for all $x\in\Omega$,

$$\int_\Omega \mathcal{A}\big(x,\nabla(u_\varepsilon - H)\big)\cdot\nabla(u_\varepsilon - H)^+\,dx$$

$$+ \int_\Omega \varepsilon\big(|u_\varepsilon|^{q-2}u_\varepsilon - |H|^{q-2}H\big)\cdot(u_\varepsilon - H)^+\,dx$$

$$+ \int_\Gamma \varepsilon\big(|u_\varepsilon|^{q'-2}u_\varepsilon - |H|^{q'-2}H\big)\cdot(u_\varepsilon - H)^+\,d\sigma(x) \leqslant 0.$$

Using the monotonicity of $\mathcal{A}(x,\cdot)$ and $u\to|u|^{q'-2}u$, we get

$$\int_\Omega \varepsilon\big(|u_\varepsilon|^{q-2}u_\varepsilon - |H|^{q-2}H\big)\cdot(u_\varepsilon - H)^+\,dx \leqslant 0.$$

This clearly leads to $u_\varepsilon \leqslant H$ a.e. in Ω. $\qquad\qquad\square$

Here we give an estimate for ∇u_ε.

LEMMA 1.2. *Under the assumptions of Lemma 1.1, we have for some positive constant C independent of ε*

$$\forall \varepsilon \in (0, \varepsilon_0), \quad \int_\Omega |\nabla u_\varepsilon|^q \, dx \leqslant C. \tag{1.34}$$

PROOF. Using $u_\varepsilon - \psi$ as a test function for (P_ε) and taking into account (1.16), we obtain

$$\int_\Omega \mathcal{A}(x, \nabla u_\varepsilon) \cdot \nabla(u_\varepsilon - \psi) \, dx + \int_\Omega \varepsilon \left(|u_\varepsilon|^{q-2} u_\varepsilon - |x_2|^{q-2} x_2 \right) \cdot (u_\varepsilon - \psi) \, dx$$

$$+ \int_\Gamma \varepsilon \left(|u_\varepsilon|^{q'-2} u_\varepsilon - |x_2|^{q'-2} x_2 \right) \cdot (u_\varepsilon - \psi) \, d\sigma(x)$$

$$\leqslant \int_\Omega G_\varepsilon(u_\varepsilon) \mathcal{A}(x, e) \cdot \nabla(u_\varepsilon - \psi) \, dx$$

which can be written, for a positive constant C_1 independent of ε, by using (1.2), (1.28) and the fact that $|G_\varepsilon(u_\varepsilon)| \leqslant 1$,

$$\int_\Omega |\nabla u_\varepsilon|^q \, dx \leqslant C_1 + C_1 \int_\Omega |\nabla u_\varepsilon|^{q-1} \, dx + \int_\Omega |\nabla u_\varepsilon| \, dx.$$

Using Young's inequality $ab \leqslant a^q/q + b^{q'}/q'$ for appropriate a and b, one gets for another positive constant independent of ε still denoted by C_1,

$$\int_\Omega |\nabla u_\varepsilon|^q \, dx \leqslant C_1 + \frac{1}{2} \int_\Omega |\nabla u_\varepsilon|^q \, dx$$

which is (1.34) with $C = 2C_1$. $\qquad\square$

PROOF OF THEOREM 1.1. The proof consists in passing to the limit as $\varepsilon \to 0$ in (P_ε). First we have $0 \leqslant G_\varepsilon(u_\varepsilon) \leqslant 1$, u_ε is bounded in $W^{1,q}(\Omega)$ by (1.28) and (1.34). Also by (1.2) and (1.34), $\mathcal{A}(x, \nabla u_\varepsilon)$ is bounded in $\mathbb{L}^{q'}(\Omega)$. It follows by the Relich–Kondrachov theorem, the complete continuity of the trace operator, and the reflexivity of the Lebesgue space $L^r(\Omega)$ for $r > 1$, that there exists a subsequence (u_{ε_k}) of (u_ε), functions $u \in W^{1,q}(\Omega)$, $g \in L^{q'}(\Omega)$, and $\mathcal{A}_0 \in \mathbb{L}^{q'}(\Omega)$ such that

$$G_{\varepsilon_k}(u_{\varepsilon_k}) \rightharpoonup g \qquad \text{in } L^{q'}(\Omega), \tag{1.35}$$

$$u_{\varepsilon_k} \rightharpoonup u \qquad \text{in } W^{1,q}(\Omega), \tag{1.36}$$

$$u_{\varepsilon_k} \to u \qquad \text{in } L^q(\Omega) \text{ and a.e. in } \Omega, \tag{1.37}$$

$$u_{\varepsilon_k} \to u \qquad \text{in } L^q(\Gamma) \text{ and a.e. in } \Gamma, \tag{1.38}$$

$$\mathcal{A}(x, \nabla u_{\varepsilon_k}) \rightharpoonup \mathcal{A}_0 \quad \text{in } \mathbb{L}^{q'}(\Omega). \tag{1.39}$$

482   A. Lyaghfouri

Moreover, by (1.28) and the monotonicity of $\mathcal{B}_2^{\varepsilon_k}(x, \cdot)$, we have for some positive constant H_1 and for a.e. $x \in \Gamma$,

$$\mathcal{B}_2^{\varepsilon_k}(x, -H_1) \leqslant \mathcal{B}_2^{\varepsilon_k}(x, \psi - H) \leqslant \mathcal{B}_2^{\varepsilon_k}(x, \psi - u_{\varepsilon_k}),$$
$$\mathcal{B}_2^{\varepsilon_k}(x, \psi - u_{\varepsilon_k}) \leqslant \mathcal{B}_2^{\varepsilon_k}(x, \psi - x_2) = \mathcal{B}_2^{\varepsilon_k}(x, \varphi) \leqslant \mathcal{B}_2^{\varepsilon_k}(x, H_1).$$

Using (1.15), we deduce that we have for some positive constant C_{H_1},

$$\left|\mathcal{B}_2^{\varepsilon_k}(x, \psi - u_{\varepsilon_k})\right| \leqslant \max\left(\left|\mathcal{B}_2^0(x, -H_1)\right|, \left|\mathcal{B}_2^0(x, H_1)\right|\right) \leqslant C_{H_1}$$

from which follows that $(\mathcal{B}_2^{\varepsilon_k}(x, \psi - u_{\varepsilon_k}))$ is bounded in $L^\infty(\Gamma)$. Therefore, there exists a subsequence of (u_{ε_k}) still denoted by (u_{ε_k}) and an element γ of $L^{q'}(\Gamma)$ such that

$$\mathcal{B}_2^{\varepsilon_k}(x, \psi - u_{\varepsilon_k}) \rightharpoonup \gamma \quad \text{in } L^{q'}(\Gamma). \tag{1.40}$$

We shall prove that (u, g, γ) is a solution of $(\mathrm{P_U})$.

Since the sets $\{v \in W^{1,q}(\Omega) \mid v \geqslant x_2 \text{ a.e. in } \Omega\}$ and $\{v \in L^{q'}(\Omega) \mid 0 \leqslant v \leqslant 1 \text{ a.e. in } \Omega\}$ are weakly closed in $W^{1,q}(\Omega)$ and $L^{q'}(\Omega)$ respectively, and contain respectively u_{ε_k} and $G_{\varepsilon_k}(u_{\varepsilon_k})$, we obtain

$$u \geqslant x_2 \quad \text{and} \quad 0 \leqslant g \leqslant 1 \text{ a.e. in } \Omega. \tag{1.41}$$

Moreover, we have for all $s \geqslant 0$, $0 \leqslant (1 - \min(\frac{s}{\varepsilon}, 1))s = (1 - \frac{s}{\varepsilon})s\,\chi([0 \leqslant s \leqslant \varepsilon]) \leqslant \varepsilon$. Therefore

$$0 \leqslant \int_\Omega G_{\varepsilon_k}(u_{\varepsilon_k}) \cdot (u_{\varepsilon_k} - x_2)\,dx \leqslant \varepsilon_k |\Omega|$$

which leads by letting $\varepsilon_k \to 0$ and taking into account (1.35) and (1.37), to

$$\int_\Omega g \cdot (u - x_2)\,dx = 0.$$

By (1.41), we obtain

$$g \cdot (u - x_2) = 0 \quad \text{a.e. in } \Omega. \tag{1.42}$$

Now since we have for a.e. $x \in \Gamma$ such that $\psi = x_2$, $\mathcal{B}_2^{\varepsilon_k}(x, \psi - u_{\varepsilon_k}) = \mathcal{B}_2^{\varepsilon_k}(x, x_2 - u_{\varepsilon_k}) \leqslant 0$, we deduce that

$$\gamma(x) \leqslant 0 \quad \text{for a.e. } x \in \Gamma \text{ such that } \psi = x_2. \tag{1.43}$$

Moreover, using (1.38) and (1.40), we obtain ([12], Lemma 1.3, p. 42)

$$\gamma(x) \in \mathcal{B}_2\big(x, \psi(x) - u(x)\big) \quad \text{for a.e. } x \in \Gamma. \tag{1.44}$$

Using $\psi - u_{\varepsilon_k}$ as a test function for $(\mathrm{P}_{\varepsilon_k})$ and taking into account (1.16), (1.28) and (1.34), we get for some constant C independent of ε_k

$$0 \leqslant \int_\Gamma \mathcal{B}_1^{\varepsilon_k}(x, \psi - u_{\varepsilon_k}) \cdot (\psi - u_{\varepsilon_k}) \, d\sigma(x) \leqslant C$$

which can be written, since for a.e. $x \in \Gamma$ and for all $u \in D(\mathcal{B})$

$$\mathcal{B}_1^{\varepsilon_k}(x, u) = \frac{1}{\varepsilon_k}\left((u - b)^+ - (a - u)^+\right)$$

$$0 \leqslant \int_\Gamma \left((\psi - u_{\varepsilon_k} - b)^+ - (a - \psi + u_{\varepsilon_k})^+\right) \cdot (\psi - u_{\varepsilon_k}) \, d\sigma(x) \leqslant \varepsilon_k C.$$

Letting $\varepsilon_k \to 0$, we obtain

$$\int_\Gamma \left((\psi - u - b)^+ - (a - \psi + u)^+\right) \cdot (\psi - u) \, d\sigma(x) = 0.$$

Since $a \leqslant 0 \leqslant b$ a.e. in Γ, one has $(\psi - u - b)^+ \cdot (\psi - u) \geqslant 0$ and $-(a - \psi + u)^+ \cdot (\psi - u) \geqslant 0$ for a.e. $x \in \Gamma$. It follows that $((\psi - u - b)^+ - (a - \psi + u)^+) \cdot (\psi - u) = 0$ for a.e. $x \in \Gamma$, which leads to $a \leqslant \psi - u \leqslant b$ a.e. in Γ. Hence $\psi - u \in D(\mathcal{B}(x, \cdot))$ for a.e. $x \in \Gamma$. We deduce then from (1.44) and the definition of $\mathcal{B}_2$ that $\gamma(x) \in \mathcal{B}(x, \psi(x) - u(x))$ for a.e. $x \in \Gamma$.

So far, we have proved (P_U)(i), (ii) and (iii). It remains to show (P_U)(iv). First remark that because of (1.28), any element of $W^{1,q}(\Omega)$ is a test function for $(\mathrm{P}_{\varepsilon_k})$. Next let $\xi \in \mathbb{K}$ and note that

$$\mathcal{B}_1^{\varepsilon_k}(x, \psi - u_{\varepsilon_k}) \cdot (\xi - u_{\varepsilon_k}) \geqslant 0 \quad \text{a.e. in } \Gamma. \tag{1.45}$$

Indeed one has first

$$-(a - \psi + u_{\varepsilon_k})^+ \cdot (\xi - u_{\varepsilon_k}) = (a - \psi + u_{\varepsilon_k})^+ \cdot (u_{\varepsilon_k} - \xi)$$
$$\geqslant (a - \psi + u_{\varepsilon_k})^+ \cdot (u_{\varepsilon_k} - \psi + a)$$
$$\geqslant 0 \quad \text{a.e. in } \Gamma.$$

To show that $(\psi - u_{\varepsilon_k} - b)^+ \cdot (\xi - u_{\varepsilon_k}) \geqslant 0$ a.e. in Γ, we consider two cases:

- $\psi > x_2$: In this case $c = b$ and $\psi - \xi \leqslant b$. Therefore,

$$(\psi - u_{\varepsilon_k} - b)^+ \cdot (\xi - u_{\varepsilon_k}) \geqslant (\psi - u_{\varepsilon_k} - b)^+ \cdot (\psi - b - u_{\varepsilon_k}) \geqslant 0 \quad \text{a.e. in } \Gamma.$$

- $\psi = x_2$: In this case $(\psi - u_{\varepsilon_k} - b)^+ = (x_2 - u_{\varepsilon_k} - b)^+ = 0$ since $u_{\varepsilon_k} - x_2 \geqslant 0$ and $b \geqslant 0$ a.e. in Γ.

Taking $\xi - u_{\varepsilon_k}$ as a test function for (P_{ε_k}) and using (1.45), we get

$$\int_\Omega \left(A(x, \nabla u_{\varepsilon_k}) - G_{\varepsilon_k}(u_{\varepsilon_k}) A(x, e) \right) \cdot \nabla(\xi - u_{\varepsilon_k}) \, dx$$

$$+ \int_\Omega \varepsilon_k \left(|u_{\varepsilon_k}|^{q-2} u_{\varepsilon_k} - |x_2|^{q-2} x_2 \right) \cdot (\xi - u_{\varepsilon_k}) \, dx$$

$$+ \int_\Gamma \varepsilon \left(|u_{\varepsilon_k}|^{q'-2} u_{\varepsilon_k} - |x_2|^{q'-2} x_2 \right) \cdot (\xi - u_{\varepsilon_k}) \, d\sigma(x)$$

$$\geqslant \int_\Gamma \mathcal{B}_2^{\varepsilon_k}(x, \psi - u_{\varepsilon_k}) \cdot (\xi - u_{\varepsilon_k}) \, d\sigma(x). \tag{1.46}$$

It remains to verify $(P_U)(iv)$ by passing to the limit in (1.46), which we will be able to do after proving the following lemma.

LEMMA 1.3. *We have*

$$\int_\Omega A(x, \nabla u) \cdot \nabla \xi \, dx = \int_\Omega A_0(x) \cdot \nabla \xi \, dx \quad \forall \xi \in W^{1,q}(\Omega), \tag{1.47}$$

$$\lim_{k \to \infty} \int_\Omega A(x, \nabla u_{\varepsilon_k}) \cdot \nabla u_{\varepsilon_k} \, dx = \int_\Omega A(x, \nabla u) \cdot \nabla u \, dx. \tag{1.48}$$

PROOF. Choosing $\xi = u$ in (1.46) and taking into account that u_{ε_k} is uniformly bounded, one gets

$$\int_\Omega A(x, \nabla u_{\varepsilon_k}) \cdot \nabla u_{\varepsilon_k} \, dx$$

$$\leqslant \int_\Omega A(x, \nabla u_{\varepsilon_k}) \cdot \nabla u \, dx$$

$$+ \int_\Omega G_{\varepsilon_k}(u_{\varepsilon_k}) A(x, e) \cdot \nabla(u_{\varepsilon_k} - u) \, dx$$

$$+ \int_\Gamma \mathcal{B}_2^{\varepsilon_k}(x, \psi - u_{\varepsilon_k}) \cdot (u_{\varepsilon_k} - u) \, d\sigma(x) + C\varepsilon_k. \tag{1.49}$$

Using (1.38) and (1.40), we obtain

$$\lim_{k \to \infty} \int_\Gamma \mathcal{B}_2^{\varepsilon_k}(x, \psi - u_{\varepsilon_k}) \cdot (u_{\varepsilon_k} - u) \, d\sigma(x) = 0. \tag{1.50}$$

By (1.39), we have

$$\lim_{k \to \infty} \int_\Omega A(x, \nabla u_{\varepsilon_k}) \cdot \nabla u \, dx = \int_\Omega A_0(x) \cdot \nabla u \, dx. \tag{1.51}$$

Note that

$$\int_\Omega G_{\varepsilon_k}(u_{\varepsilon_k})\mathcal{A}(x,e)\cdot\nabla(u_{\varepsilon_k}-u)\,\mathrm{d}x$$

$$=\int_\Omega G_{\varepsilon_k}(u_{\varepsilon_k})\mathcal{A}(x,e)\cdot\nabla(u_{\varepsilon_k}-x_2)\,\mathrm{d}x$$

$$-\int_\Omega G_{\varepsilon_k}(u_{\varepsilon_k})\mathcal{A}(x,e)\cdot\nabla(u-x_2)\,\mathrm{d}x. \tag{1.52}$$

Using (1.35) and taking into account (1.42), we have

$$\lim_{k\to\infty}\int_\Omega G_{\varepsilon_k}(u_{\varepsilon_k})\mathcal{A}(x,e)\cdot\nabla(u-x_2)\,\mathrm{d}x$$

$$=\int_\Omega g\mathcal{A}(x,e)\cdot\nabla(u-x_2)\,\mathrm{d}x$$

$$=0. \tag{1.53}$$

For the second integral in the right-hand side of (1.52), we rewrite it as

$$\int_\Omega G_{\varepsilon_k}(u_{\varepsilon_k})\mathcal{A}(x,e)\cdot\nabla(u_{\varepsilon_k}-x_2)\,\mathrm{d}x=\int_\Omega \mathcal{A}(x,e)\cdot\nabla v_k\,\mathrm{d}x,$$

where $v_k=\int_0^{u_{\varepsilon_k}-x_2}(1-H_{\varepsilon_k}(s))\,\mathrm{d}s$.

Since we have $\nabla v_k=(1-H_{\varepsilon_k}(u_{\varepsilon_k}-x_2))\cdot\nabla(u_{\varepsilon_k}-x_2)$, $|v_k(x)|\leqslant\varepsilon_k$ for a.e. $x\in\Omega$, and $|v_k|_{1,q}$ is bounded in $W^{1,q}(\Omega)$, we deduce that $v_k\rightharpoonup 0$ weakly in $W^{1,q}(\Omega)$. Therefore

$$\lim_{k\to\infty}\int_\Omega G_{\varepsilon_k}(u_{\varepsilon_k})\mathcal{A}(x,e)\cdot\nabla(u_{\varepsilon_k}-x_2)\,\mathrm{d}x=0. \tag{1.54}$$

Using (1.52)–(1.54) we get

$$\lim_{k\to\infty}\int_\Omega G_{\varepsilon_k}(u_{\varepsilon_k})\mathcal{A}(x,e)\cdot\nabla(u_{\varepsilon_k}-u)\,\mathrm{d}x=0. \tag{1.55}$$

Combining (1.49)–(1.51) and (1.55), we get

$$\limsup_{k\to\infty}\int_\Omega \mathcal{A}(x,\nabla u_{\varepsilon_k})\cdot\nabla u_{\varepsilon_k}\,\mathrm{d}x\leqslant\int_\Omega \mathcal{A}_0(x)\cdot\nabla u\,\mathrm{d}x. \tag{1.56}$$

Let now $v\in W^{1,q}(\Omega)$. By (1.2), we have

$$\int_\Omega\big(\mathcal{A}(x,\nabla u_{\varepsilon_k})-\mathcal{A}(x,\nabla v)\big)\cdot\nabla(u_{\varepsilon_k}-v)\,\mathrm{d}x\geqslant 0\quad\forall k$$

or

$$\int_{\Omega} \mathcal{A}(x, \nabla u_{\varepsilon_k}) \cdot \nabla u_{\varepsilon_k} \, dx - \int_{\Omega} \mathcal{A}(x, \nabla u_{\varepsilon_k}) \cdot \nabla v \, dx$$

$$- \int_{\Omega} \mathcal{A}(x, \nabla v) \cdot \nabla (u_{\varepsilon_k} - v) \, dx \geqslant 0 \quad \forall k. \tag{1.57}$$

Passing to the limsup in (1.57) and taking into account (1.36), (1.39) and (1.56), we obtain

$$\int_{\Omega} \mathcal{A}_0(x) \cdot \nabla u \, dx - \int_{\Omega} \mathcal{A}_0(x) \cdot \nabla v \, dx - \int_{\Omega} \mathcal{A}(x, \nabla v) \cdot \nabla (u - v) \, dx \geqslant 0$$

or

$$\int_{\Omega} \mathcal{A}_0(x) \cdot \nabla (u - v) \, dx \geqslant \int_{\Omega} \mathcal{A}(x, \nabla v) \cdot \nabla (u - v) \, dx. \tag{1.58}$$

Choosing $v = u \pm t\xi$, with $t \in [0, 1]$ and $\xi \in W^{1,q}(\Omega)$, in (1.58), we obtain

$$\int_{\Omega} \mathcal{A}_0(x) \cdot \nabla \xi \, dx = \int_{\Omega} \mathcal{A}(x, \nabla u \pm t \nabla \xi) \cdot \nabla \xi \, dx.$$

Letting $t \to 0$ and using (1.2)(ii) and the Lebesgue theorem, we obtain (1.47).
    Using $\xi = u$ in (1.47), and taking into account (1.56), we obtain

$$\limsup_{k \to \infty} \int_{\Omega} \mathcal{A}(x, \nabla u_{\varepsilon_k}) \cdot \nabla u_{\varepsilon_k} \, dx \leqslant \int_{\Omega} \mathcal{A}(x, \nabla u) \cdot \nabla u \, dx. \tag{1.59}$$

Now rewriting (1.57) for $v = u$, and passing to the liminf, we obtain

$$\liminf_{k \to \infty} \int_{\Omega} \mathcal{A}(x, \nabla u_{\varepsilon_k}) \cdot \nabla u_{\varepsilon_k} \, dx \geqslant \int_{\Omega} \mathcal{A}_0(x) \cdot \nabla u \, dx$$

$$= \int_{\Omega} \mathcal{A}(x, \nabla u) \cdot \nabla u \, dx. \tag{1.60}$$

Combining (1.59) and (1.60), we get (1.48). $\qquad \square$

1.3. *Regularity and monotonicity of the solutions*

Throughout this section, we shall denote a solution of (P_U) by (u, g, γ). We show that u is bounded and locally Hölder continuous in Ω. Under suitable assumptions on $\mathcal{A}$, we also give a monotonicity property for g which together with the continuity of u, allows to define the free boundary as an $x_1 x_2$-graph.

PROPOSITION 1.1. *We have for some positive constant h_0*

$$u \leqslant h_0 \quad \text{a.e. in } \Omega. \tag{1.61}$$

PROOF. Let h be such that $h > \max_{x \in \bar{\Omega}} \psi(x)$. Note that $\xi = u - (u-h)^+$ is a test function for $(\mathrm{P_U})$. Indeed

$$\text{if } u \leqslant h, \text{ then } \psi - \xi = \psi - u + (u-h)^+ = \psi - u \in [a, b] \subset [a, c],$$

$$\text{if } u > h, \text{ then } \psi - \xi = \psi - h \geqslant \psi - u \geqslant a \text{ and } \psi - h \leqslant 0 \leqslant c.$$

So we have

$$\int_\Omega \left(\mathcal{A}(x, \nabla u) - g \mathcal{A}(x, e) \right) \cdot \nabla (u-h)^+ \, dx \leqslant \int_\Gamma \gamma \cdot (u-h)^+ \, d\sigma(x). \tag{1.62}$$

Since $h \geqslant \psi(x) \geqslant x_2$, we have by $(\mathrm{P_U})$(ii) $g \cdot \nabla (u-h)^+ = 0$ a.e. in Ω. For the same reason, we get by (1.12), $(\mathrm{P_U})$(iii) and the monotonicity of $\mathcal{B}(x, \cdot)$ that $\gamma \cdot (u-h)^+ \leqslant 0$ a.e. in Γ.

It follows from (1.62) and (1.2)(iii) that

$$\int_\Omega \lambda \left| \nabla (u-h)^+ \right|^q \, dx \leqslant 0$$

which leads to $\nabla (u-h)^+ = 0$ a.e. in Ω. Therefore $(u-h)^+ = C$ for some positive constant C. Hence $u \leqslant h + C = h_0$ a.e. in Ω. $\qquad\square$

PROPOSITION 1.2. *We have*

$$\mathrm{div}\left(\mathcal{A}(x, \nabla u) - g \mathcal{A}(x, e) \right) = 0 \quad \text{in } \mathcal{D}'(\Omega). \tag{1.63}$$

If $\mathrm{div}\left(\mathcal{A}(x, e) \right) \geqslant 0$, *then* $\mathrm{div}\left(\mathcal{A}(x, \nabla u) \right) = \mathrm{div}\left(g \mathcal{A}(x, e) \right) \geqslant 0 \quad \text{in } \mathcal{D}'(\Omega).$

$$\tag{1.64}$$

PROOF. (i) (1.63) follows immediately by taking $u \pm \xi$ as a test function for $(\mathrm{P_U})$, where $\xi \in \mathcal{D}(\Omega)$.

(ii) Let $\xi \in \mathcal{D}(\Omega)$, $\xi \geqslant 0$ and $\varepsilon > 0$. Using $u \pm \min(\frac{u-x_2}{\varepsilon}, 1)\xi$ as test functions for $(\mathrm{P_U})$, and taking into account $(\mathrm{P_U})$(ii), we get

$$\int_\Omega \mathcal{A}(x, \nabla u) \cdot \nabla \left(\min\left(\frac{u - x_2}{\varepsilon}, 1 \right) \xi \right) \, dx = 0. \tag{1.65}$$

Since $\mathrm{div}(\mathcal{A}(x, e)) \geqslant 0$, we have

$$\int_\Omega \mathcal{A}(x, e) \cdot \nabla \left(\left(1 - \min\left(\frac{u - x_2}{\varepsilon}, 1 \right) \right) \xi \right) \, dx \leqslant 0. \tag{1.66}$$

Adding (1.65) and (1.66), and using the fact that $\nabla x_2 = e$, we obtain

$$\int_\Omega \min\left(\frac{u - x_2}{\varepsilon}, 1\right)\left(\mathcal{A}(x, \nabla u) - \mathcal{A}(x, \nabla x_2)\right) \cdot \nabla \xi \, dx$$

$$+ \frac{1}{\varepsilon} \int_{[u - x_2 \leqslant \varepsilon]} \xi\left(\mathcal{A}(x, \nabla u) - \mathcal{A}(x, \nabla x_2)\right) \cdot (\nabla u - \nabla x_2) \, dx$$

$$\leqslant - \int_\Omega \mathcal{A}(x, e) \cdot \nabla \xi. \tag{1.67}$$

Letting $\varepsilon \to 0$ in (1.67) and using the monotonicity of $\mathcal{A}(x, \cdot)$, we get

$$\int_\Omega \mathcal{A}(x, \nabla u) \cdot \nabla \xi \, dx \leqslant 0.$$

$\square$

PROPOSITION 1.3. *We have* $u \in C_{\mathrm{loc}}^{0,\alpha}(\Omega)$ *for some* $\alpha \in (0, 1)$, *and the set* $[u > x_2]$ *is open.*

PROOF. This is a consequence of (1.61) and (1.63) [29]. $\square$

REMARK 1.2. (i) Assume that

$$\mathcal{A}(x, e) = k(x)e \quad \text{for a.e. } x \in \Omega \text{ with } k : \Omega \to \mathbb{R} \quad \text{and}$$

$$\frac{\partial k}{\partial x_2} \geqslant 0 \quad \text{in } \mathcal{D}'(\Omega). \tag{1.68}$$

Then we obtain from (1.64) that gk is nondecreasing in x_2.

(ii) We also deduce from (1.63) and $(\mathrm{P_U})$(ii) that $\mathrm{div}(\mathcal{A}(x, \nabla u)) = 0$ in $\mathcal{D}'([u > x_2])$, i.e., u is $\mathcal{A}$-harmonic in $[u > x_2]$. Therefore if there exist nonnegative constants κ, σ and positive constants α_0, α_1 with $\sigma \leqslant 1$ and $\alpha_1 \geqslant \alpha_0$ such that, for all $x, y \in \Omega, \zeta, \xi \in \mathbb{R}^2$,

$$\sum_{i,j} \frac{\partial \mathcal{A}^i}{\partial \zeta_j}(x, \zeta)\xi_i \xi_j \geqslant \alpha_0\left(\kappa + |\zeta|^{q-2}\right)|\xi|^2, \tag{1.69}$$

$$\left|\frac{\partial \mathcal{A}^i}{\partial \zeta_j}(x, \zeta)\right| \leqslant \alpha_1\left(\kappa + |\zeta|^{q-2}\right), \tag{1.70}$$

$$\left|\mathcal{A}(x, \zeta) - \mathcal{A}(y, \zeta)\right| \leqslant \alpha_1\left(1 + |\zeta|^{q-1}\right)\left(|x - y|^\sigma\right), \tag{1.71}$$

then we have [25], $u \in C_{\mathrm{loc}}^{1,\delta}([u > x_2])$ for some $\delta \in (0, 1)$.

Assumptions (1.69)–(1.71) are satisfied, for example, if $\mathcal{A}(x, \zeta) = |a(x)\zeta \cdot \zeta|^{(q-2)/2} \times a(x)\zeta$ with $a(x)$ a bounded 2×2 matrix satisfying $a \in C_{\mathrm{loc}}^{0,\sigma}(\Omega)$ for some $\sigma \in (0, 1)$. In particular if $q = 2$, we have $\mathcal{A}(x, \zeta) = a(x)\zeta$ and (1.69)–(1.71) are satisfied obviously.

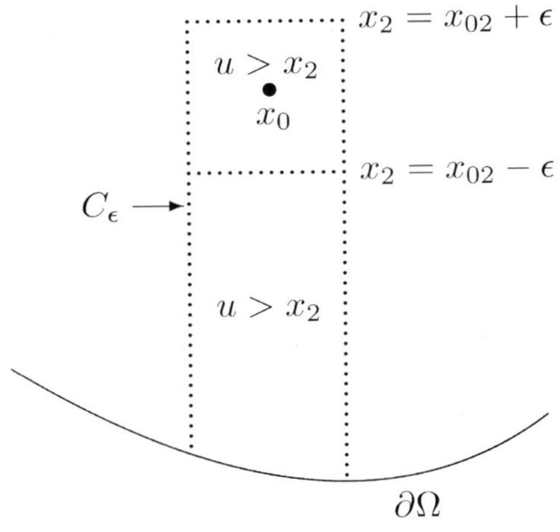

Fig. 2.

In the rest of the chapter, we shall assume that (1.69)–(1.71) are satisfied except when $\mathcal{A}(x, \zeta) = a(x)\zeta$. Moreover, we assume that Ω is vertically convex, i.e.,

$$\forall (x_1, x_2), (x_1, x_2') \in \Omega, \quad \{x_1\} \times [x_2, x_2'] \subset \Omega.$$

We also define the functions s_- and s_+ for $x_1 \in \pi_{x_1}(\Omega)$ by

$$s_-(x_1) = \inf\{x_2 : (x_1, x_2) \in \Omega\}, \qquad s_+(x_1) = \sup\{x_2 : (x_1, x_2) \in \Omega\}$$

and assume that s_- (resp. s_+) is continuous except on a finite set S_- (resp. S_+). π_{x_1} is the orthogonal projection on the x_1-axis.

Then we have the following theorem.

THEOREM 1.3. *If* $x_0 = (x_{01}, x_{02}) \in [u > x_2]$, *then there exists* $\varepsilon > 0$ *such that (see Figure 2):*

$$u(x_1, x_2) > x_2 \quad \forall (x_1, x_2) \in C_\varepsilon = \{(x_1, x_2) \in \Omega \mid |x_1 - x_{01}| < \varepsilon, x_2 < x_{02} + \varepsilon\}.$$

We need the following strong comparison principle proved in [24].

LEMMA 1.4. *Let D be a domain of $\mathbb{R}^2$ and let $u_1, u_2 \in C^1(D)$ such that*

$$\begin{cases} \operatorname{div}(\mathcal{A}(x, \nabla u_1)) \geqslant \operatorname{div}(\mathcal{A}(x, \nabla u_2)) & \text{in } \mathcal{D}'(D), \\ u_1 \leqslant u_2 & \text{in } D \quad \text{and} \quad S = \{x \in D \mid \nabla u_1(x) = \nabla u_2(x) = 0\} = \emptyset. \end{cases}$$

Then we have

$$\text{either} \quad u_1 \equiv u_2 \quad \text{in } D \quad \text{or} \quad u_1 < u_2 \quad \text{in } D.$$

PROOF OF THEOREM 1.3. **By the continuity of u, there exists $\varepsilon > 0$ such that**

$$Q_\varepsilon = \{(x_1, x_2) \in \Omega \mid |x_1 - x_{01}| < \varepsilon, |x_2 - x_{02}| < \varepsilon\} \subset [u > x_2].$$

From (P_U)(ii), we deduce that $g = 0$ and then $gk = 0$ a.e. in Q_ε. But since gk is nondecreasing in x_2 and $gk \geqslant 0$ a.e. in Ω, we obtain $gk = 0$ a.e. in C_ε. Using (1.63) we obtain

$$\mathrm{div}\big(\mathcal{A}(x, \nabla u)\big) = 0 \quad \text{in } \mathcal{D}'(C_\varepsilon).$$

Since $u \geqslant x_2$ in C_ε, $\mathrm{div}(\mathcal{A}(x, \nabla x_2)) = \frac{\partial k}{\partial x_2} \geqslant 0$ in $\mathcal{D}'(C_\varepsilon)$, $u > x_2$ in Q_ε and $\nabla x_2 = e \neq 0$, we deduce from Lemma 1.4 that $u(x) > x_2$ $\forall x \in C_\varepsilon$. $\qquad\square$

COROLLARY 1.1. *If $u(x_{01}, x_{02}) = x_{02}$, then $u(x_{01}, x_2) = x_2$ $\forall x_2 \in [x_{02}, s_+(x_{01}))$.*

REMARK 1.3. Theorem 1.3 means that if a point (x_{01}, x_{02}) is wet, then so are all the points below. This is due to gravity and most likely to the fact that by (1.68), an important component of the permeability is nondecreasing with respect to x_2.

Now we are able to define a function Φ that represents the free boundary:

$$\forall x_1 \in \pi_{x_1}(\Omega),$$

$$\Phi(x_1) = \begin{cases} \sup\{x_2 \colon (x_1, x_2) \in [u > x_2]\} & \text{if this set is not empty,} \\ s_-(x_1) & \text{otherwise.} \end{cases}$$

Then we have the following proposition.

PROPOSITION 1.4. *Φ is lower semicontinuous (l.s.c.) on $\pi_{x_1}(\Omega)$ except perhaps on S_-. Moreover,*

$$[u > x_2] = \big[x_2 < \Phi(x_1)\big]. \tag{1.72}$$

PROOF. Let $x_{01} \in \pi_{x_1}(\Omega) \setminus S_-$. Since s_- is continuous on $\pi_{x_1}(\Omega) \setminus S_-$ and $\Phi(x_1) \geqslant s_-(x_1)$, it is clear that Φ is l.s.c. at x_{01} if $\Phi(x_{01}) = s_-(x_{01})$.

Now assume that $\Phi(x_{01}) > s_-(x_{01})$ and let $\varepsilon > 0$ small enough. There exists x_{02} such that $x_0 = (x_{01}, x_{02}) \in [u > x_2]$ and $\Phi(x_{01}) > x_{02} > \Phi(x_{01}) - \varepsilon > s_-(x_{01})$. By continuity of u, there exists $\eta > 0$ small enough such that $u(x) > x_2$ in $B_\eta(x_0) \subset \Omega$. Using Corollary 1.1, we obtain $u(x) > x_2$ in $((x_{01} - \eta, x_{01} + \eta) \times (-\infty, x_{02})) \cap \Omega$. This leads to $\Phi(x_1) \geqslant x_{02} > \Phi(x_{01}) - \varepsilon$ for all $x_1 \in (x_{01} - \eta, x_{01} + \eta)$.

Let $(x_{01}, x_{02}) \in [u > x_2]$. By Theorem 1.3, there exists $\varepsilon > 0$ small enough such that $u(x_{01}, x_2) > x_2$ for all $x_2 \in (s_-(x_{01}), x_{02} + \varepsilon)$. In particular, $\Phi(x_{01}) \geqslant x_{02} + \varepsilon > x_{02}$ and $(x_{01}, x_{02}) \in [x_2 < \Phi(x_1)]$.

Conversely, let $(x_{01}, x_{02}) \in [x_2 < \Phi(x_1)]$. If $u(x_{01}, x_{02}) = x_{02}$, then by Corollary 1.1 we would have $u(x_{01}, x_2) = x_2$ for all $x_2 \in [x_{02}, s_+(x_{01}))$ and therefore $\Phi(x_{01}) \leqslant x_{02}$, which contradicts the assumption. $\qquad\square$

2. The dam problem with Dirichlet boundary condition

In this section, we assume that $\mathcal{B}$ is given by (1.9). In this case we obtain the problem

$$
\begin{cases}
\text{find } (u, g) \in W^{1,q}(\Omega) \times L^{\infty}(\Omega) \text{ such that} \\
\quad \text{(i) } u = \psi \text{ on } S_2 \cup S_3, \\
\quad \text{(ii) } u \geqslant x_2, 0 \leqslant g \leqslant 1, g(u - x_2) = 0 \text{ a.e. in } \Omega, \\
\quad \text{(iii) } \int_{\Omega} \big(\mathcal{A}(x, \nabla u) - g\mathcal{A}(x, e)\big) \cdot \nabla \xi \, dx \leqslant 0 \\
\qquad\quad \forall \xi \in W^{1,q}(\Omega) \text{ such that } \xi = 0 \text{ on } S_3, \text{ and } \xi \geqslant 0 \text{ on } S_2.
\end{cases}
\tag{P_D}
$$

From now on, we assume that $\varphi = 0$ on S_2 and $\varphi > 0$ on S_3.

2.1. *Some properties of the solutions*

Throughout this subsection, we shall denote a solution of (P_D) by (u, g). First we have the following regularity result.

PROPOSITION 2.1. $u \in C^{0,\alpha}_{\text{loc}}(\Omega \cup S_2 \cup S_3)$ *for some* $\alpha \in (0, 1)$.

PROOF. This is a consequence of (1.61), (1.63) and (P_D)(i) (see [29]). $\qquad\square$

COROLLARY 2.1. *The dam is saturated below* S_3, *i.e., we have*

$$
u(x_1, x_2) > x_2 \quad \forall (x_1, x_2) \in \Omega, x_1 \in \pi_{x_1}(S_3).
$$

PROOF. Let $x_0 = (x_{01}, x_{02}) \in S_{3,i}$ for some $i \in \{1, \ldots, N\}$. Since $u(x_{01}, x_{02}) = \psi(x_{01}, x_{02}) > x_{02}$, we deduce from Proposition 2.1 that for some $\varepsilon > 0$ small enough one has $u(x) > x_2$ in $B_\varepsilon(x_0) \cap \Omega$. Using Theorem 1.3, we deduce that $u(x) > x_2$ below $B_\varepsilon(x_0) \cap \Omega$. $\qquad\square$

REMARK 2.1. By Corollary 2.1, we have

$$
\forall x_1 \in \text{Int}\big(\pi_{x_1}(S_3)\big), \quad \Phi(x_1) = s_+(x_1).
$$

It follows that Φ is continuous on $\text{Int}(\pi_{x_1}(S_3)) \setminus S_+$.

The following theorem will be used several times in this section.

THEOREM 2.1. *Let* C_h *be a connected component of* $[u > x_2] \cap [x_2 > h]$ *and* $Z_h = \Omega \cap (\pi_{x_1}(C_h) \times (h, +\infty))$. *Assume that* $\overline{Z}_h \cap S_3 = \emptyset$. *Then we have*

$$
\int_{Z_h} \big(\mathcal{A}(x, \nabla u) - g\mathcal{A}(x, e)\big) \cdot e \, dx \leqslant 0.
\tag{2.1}
$$

PROOF. Let $(a_1, a_2) = \pi_{x_1}(C_h)$ and let for $\delta > 0$ small enough, $\alpha_\delta \in \mathcal{D}((a_1, a_2))$ be a function such that $0 \leqslant \alpha_\delta(x_1) \leqslant 1$ and $\alpha_\delta = 1$ in $(a_1 + \delta, a_2 - \delta)$.

First we have

$$\int_{Z_h} \big(\mathcal{A}(x, \nabla u) - g\mathcal{A}(x, e)\big) \cdot e \, dx$$

$$= \int_{Z_h} \big(\mathcal{A}(x, \nabla u) - g\mathcal{A}(x, e)\big) \cdot \nabla\big(\alpha_\delta(x_2 - h)\big) \, dx$$

$$+ \int_{Z_h} \big(\mathcal{A}(x, \nabla u) - g\mathcal{A}(x, e)\big) \cdot \nabla\big((1 - \alpha_\delta)(x_2 - h)\big) \, dx. \tag{2.2}$$

Since $\overline{Z}_h \cap S_3 = \emptyset$, $\chi(Z_h)\alpha_\delta(x_2 - h)$ is a test function for (P$_D$) and we have

$$\int_{Z_h} \big(\mathcal{A}(x, \nabla u) - g\mathcal{A}(x, e)\big) \cdot \nabla\big(\alpha_\delta(x_2 - h)\big) \, dx \leqslant 0. \tag{2.3}$$

Set $\zeta_\delta = (1 - \alpha_\delta)(x_2 - h)$ and remark that for $\varepsilon > 0$, $\pm\chi(Z_h) \cdot ((u - x_2)/\varepsilon \wedge \zeta_\delta)$ are test functions for (P$_D$). So we have by taking into account (P$_D$)(ii)

$$\int_{Z_h} \mathcal{A}(x, \nabla u) \cdot \nabla\left(\frac{u - x_2}{\varepsilon} \wedge \zeta_\delta\right) dx = 0. \tag{2.4}$$

Using the monotonicity of $\mathcal{A}$, we get from (2.4)

$$\int_{Z_h \cap [u - x_2 \geqslant \varepsilon\zeta_\delta]} \big(\mathcal{A}(x, \nabla u) - \mathcal{A}(x, e)\big) \cdot \nabla\zeta_\delta \, dx$$

$$\leqslant - \int_{Z_h} \mathcal{A}(x, e) \cdot \nabla\left(\frac{u - x_2}{\varepsilon} \wedge \zeta_\delta\right) dx. \tag{2.5}$$

Note that

$$\int_{Z_h} \chi\big([u > x_2]\big)\mathcal{A}(x, e) \cdot \nabla\zeta_\delta \, dx$$

$$= \int_{Z_h} \chi\big([u > x_2]\big)\mathcal{A}(x, e) \cdot \nabla\left(\zeta_\delta - \frac{u - x_2}{\varepsilon}\right)^+ dx$$

$$+ \int_{Z_h} \mathcal{A}(x, e) \cdot \nabla\left(\frac{u - x_2}{\varepsilon} \wedge \zeta_\delta\right) dx. \tag{2.6}$$

Since k is nondecreasing in x_2, we deduce by the second mean value theorem, that for a.e.

$x_1 \in (a_1, a_2)$, there exists $h^*(x_1) \in [h, \Phi(x_1)]$ such that

$$\int_{Z_h} \chi([u > x_2]) \mathcal{A}(x, e) \cdot \nabla \left(\zeta_\delta - \frac{u - x_2}{\varepsilon} \right)^+ dx$$

$$= \int_{a_1}^{a_2} \left(\int_h^{\Phi(x_1)} k(x) \left(\zeta_\delta - \frac{u - x_2}{\varepsilon} \right)_{x_2}^+ dx_2 \right) dx_1$$

$$= \int_{a_1}^{a_2} k(x_1, \Phi(x_1)_-) \left(\int_{h^*(x_1)}^{\Phi(x_1)} \left(\zeta_\delta - \frac{u - x_2}{\varepsilon} \right)_{x_2}^+ (x_1, x_2) \, dx_2 \right) dx_1$$

$$\leqslant \int_{a_1}^{a_2} k(x_1, \Phi(x_1)_-) \zeta_\delta(x_1, \Phi(x_1)) \, dx_1, \tag{2.7}$$

where for a.e. $x_1 \in (a_1, a_2)$, $k(x_1, \Phi(x_1)_-)$ is the left limit of $k(x_1, \cdot)$ at $\Phi(x_1)$.

Adding (2.5), (2.6) and using (2.7), we get

$$\int_{Z_h \cap [u - x_2 \geqslant \varepsilon \zeta_\delta]} \left(\mathcal{A}(x, \nabla u) - \mathcal{A}(x, e) \right) \cdot \nabla \zeta_\delta \, dx$$

$$+ \int_{Z_h} \chi([u > x_2]) \mathcal{A}(x, e) \cdot \nabla \zeta_\delta \, dx$$

$$\leqslant \int_{a_1}^{a_2} k(x_1, \Phi(x_1)_-) \zeta_\delta(x_1, \Phi(x_1)) \, dx_1$$

which leads by letting $\varepsilon \to 0$ to

$$\int_{Z_h} \left(\mathcal{A}(x, \nabla u) - \chi([u = x_2]) \mathcal{A}(x, e) \right) \cdot \nabla \zeta_\delta \, dx$$

$$\leqslant \int_{a_1}^{a_2} k(x_1, \Phi(x_1)_-) \zeta_\delta(x_1, \Phi(x_1)) \, dx_1. \tag{2.8}$$

Using (2.2), (2.3) and (2.8) we obtain

$$\int_{Z_h} \left(\mathcal{A}(x, \nabla u) - g \mathcal{A}(x, e) \right) \cdot e \, dx \leqslant \int_{Z_h} \left(\chi([u = x_2]) - g \right) k(x) (1 - \alpha_\delta)$$

$$+ \int_{a_1}^{a_2} k(x_1, \Phi(x_1)_-) \zeta_\delta(x_1, \Phi(x_1)) \, dx_1.$$

Letting δ go to 0, we get (2.1). $\qquad \square$

REMARK 2.2. Let $Z_h = ((a_1, a_2) \times (h, +\infty)) \cap \Omega$. If $\overline{Z}_h \cap S_3 = \emptyset$ and for $i = 1, 2$, we have $u(a_i, x_2) = x_2 \ \forall x_2 \geqslant h$, then inequality (2.1) holds for the domain Z_h also.

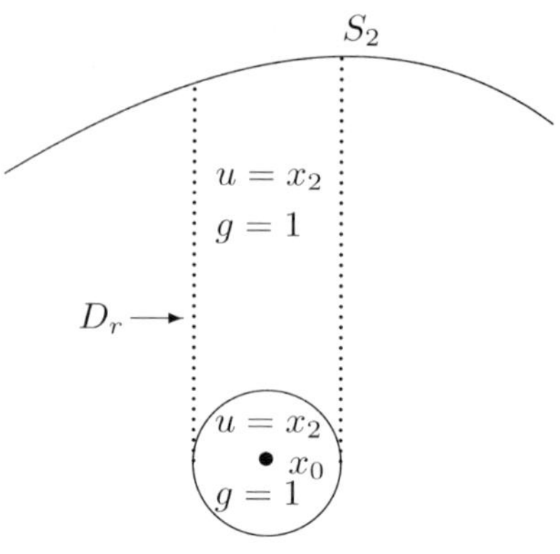

Fig. 3.

From now on, we assume that there is no impervious part above Ω. Then we have the following theorem.

THEOREM 2.2. *Let $x_0 = (x_{01}, x_{02}) \in \Omega$ and $B_r = B_r(x_0) \subset \Omega$. If $u = x_2$ in B_r, then we have (see Figure 3)*

$$g = 1 \quad \text{a.e. in } D_r = \{(x_1, x_2) \in \Omega \mid |x_1 - x_{01}| < r \text{ and } x_{02} < x_2\} \cup B_r. \quad (2.9)$$

PROOF. It is clear by Theorem 1.3 that we have $u = x_2$ in D_r, and therefore $\pi_{x_1}(B_r) \subset \pi_{x_1}(S_2)$. Applying Theorem 2.1 with domains $Z_h \subset D_r$ as in Remark 2.2, we obtain

$$0 \leqslant \int_{Z_h} k(x)(1-g)\, dx = \int_{Z_h} \big(\mathcal{A}(x, \nabla u) - g\mathcal{A}(x, e)\big) \cdot e\, dx \leqslant 0.$$

This leads to $g = 1$ a.e. in Z_h. Thus $g = 1$ a.e. in D_r.   □

Now we prove a nonoscillation result.

THEOREM 2.3. *Let $x_0 = (x_{01}, x_{02}) \in \Omega$ such that $B_r = B_r(x_0) \subset \Omega$. Then the following situations (see Figure 4) are impossible*:

$$\text{(i)} \quad \begin{cases} u(x_1, x_2) = x_2 & \forall(x_1, x_2) \in B_r \cap [x_1 = x_{01}], \\ u(x_1, x_2) > x_2 & \forall(x_1, x_2) \in B_r, \; x_1 \neq x_{01}; \end{cases}$$

$$\text{(ii)} \quad \begin{cases} u(x_1, x_2) > x_2 & \forall(x_1, x_2) \in B_r \cap [x_1 < x_{01}], \\ u(x_1, x_2) = x_2 & \forall(x_1, x_2) \in B_r \cap [x_1 \geqslant x_{01}]; \end{cases}$$

$$\text{(iii)} \quad \begin{cases} u(x_1, x_2) = x_2 & \forall(x_1, x_2) \in B_r \cap [x_1 \leqslant x_{01}], \\ u(x_1, x_2) > x_2 & \forall(x_1, x_2) \in B_r \cap [x_1 > x_{01}]. \end{cases}$$

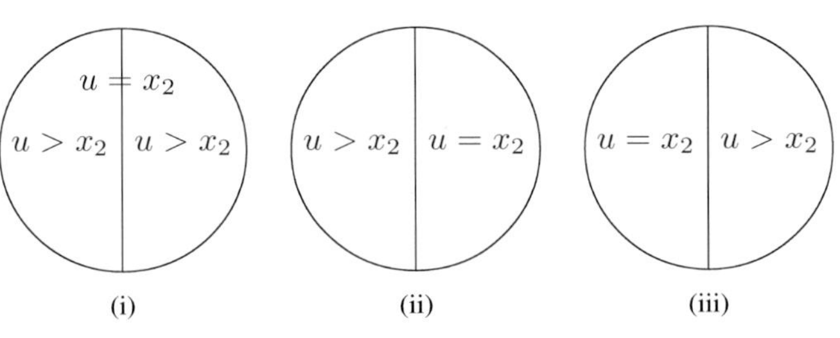

Fig. 4.

PROOF. (i) From the assumption and $(\mathrm{P_D})$(ii), we have $g = 0$ a.e. in B_r and by (1.63) this leads to $\mathrm{div}(\mathcal{A}(x, \nabla u)) = 0 \leqslant \mathrm{div}(\mathcal{A}(x, \nabla x_2))$ in $\mathcal{D}'(B_r)$. Since neither $u > x_2$ in B_r nor $u \equiv x_2$ in B_r, we get a contradiction with the strong maximum principle (Lemma 1.4).

(ii) From the assumption, $(\mathrm{P_D})$(ii), and (2.9), we have $g\mathcal{A}(x, e) = \chi(B_r \cap [x_1 > x_{01}]) \times k(x)e$. Then by using (1.63) and (1.68), we obtain in $\mathcal{D}'(B_r)$

$$\mathrm{div}(\mathcal{A}(x, \nabla u)) = \mathrm{div}(\chi(B_r \cap [x_1 > x_{01}])k(x)e) \leqslant \mathrm{div}(k(x)e)$$
$$= \mathrm{div}(\mathcal{A}(x, \nabla x_2)).$$

Hence we get a contradiction with Lemma 1.4, as in the previous case.

(iii) Similar to (ii). $\qquad\square$

2.2. Continuity of the free boundary

For the rest of this section, we assume that $\mathcal{A}$ is strictly monotone, i.e.,

$$(\mathcal{A}(x, \xi) - \mathcal{A}(x, \zeta)) \cdot (\xi - \zeta) > 0 \quad \forall \xi, \zeta \in \mathbb{R}^2, \xi \neq \zeta \text{ a.e. } x \in \Omega. \tag{2.10}$$

The main result of this section is the continuity of the function Φ.

THEOREM 2.4. Φ *is continuous at each point* $x_{01} \in \mathrm{Int}(\pi_{x_1}(S_2))$ *such that* $(x_{01}, \Phi(x_{01})) \in \Omega$.

PROOF. Let $x_{01} \in \mathrm{Int}(\pi_{x_1}(S_2))$ such that $x_0 = (x_{01}, \Phi(x_{01})) \in \Omega$ and let $\varepsilon > 0$. Using the continuity of u, there exists a ball $B_{\varepsilon'}(x_0)$, $0 < \varepsilon' < \varepsilon$, such that

$$\pi_{x_1}(B_{\varepsilon'}(x_0)) \subset S_2 \quad \text{and} \quad u(x) \leqslant x_2 + \varepsilon \quad \forall x \in B_{\varepsilon'}(x_0). \tag{2.11}$$

By Theorem 2.3, we have, for example,

$$\exists \underline{x} = (\underline{x}_1, \underline{x}_2) \in B_{\varepsilon'}(x_0) \quad \text{such that} \quad \underline{x}_1 < x_{01} \quad \text{and} \quad u(\underline{x}) = \underline{x}_2. \tag{2.12}$$

Then we set (see Figure 5) $h = \max(\underline{x}_2, \Phi(x_{01}))$, $Z = ((\underline{x}_1, x_{01}) \times (h, +\infty)) \cap \Omega$, $v = (\varepsilon + h - x_2)^+ + x_2$ and $\xi = (u - v)^+$. Using (2.11) and (2.12), the fact that $u(x_0) = x_{02}$, and Corollary 1.1, it is clear that $\xi = 0$ on ∂Z and therefore $\pm \xi$ are test

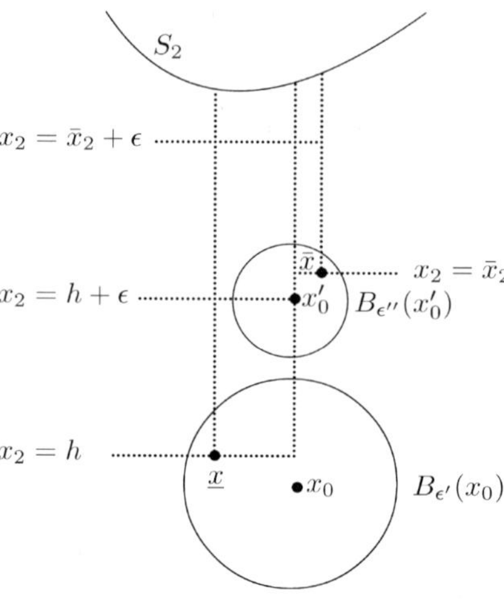

Fig. 5.

functions for $(\mathrm{P_D})$. So we have

$$\int_Z \big(\mathcal{A}(x, \nabla u) - g\mathcal{A}(x, e)\big) \cdot \nabla(u - v)^+ \, dx = 0. \tag{2.13}$$

A simple calculation shows that

$$\int_Z \big(\mathcal{A}(x, \nabla v) - \chi\big([v = x_2]\big)\mathcal{A}(x, e)\big) \cdot \nabla(u - v)^+ \, dx = 0. \tag{2.14}$$

Since by (2.11) $\partial Z \cap S_3 = \emptyset$, we have by Theorem 2.1 and Remark 2.2

$$\int_{Z \cap [v = x_2]} \big(\mathcal{A}(x, \nabla u) - g\mathcal{A}(x, e)\big) \cdot e \, dx \leqslant 0. \tag{2.15}$$

Subtracting (2.14) from (2.13) and adding (2.15) to the result, we obtain

$$\int_{Z \cap [v > x_2]} \big(\mathcal{A}(x, \nabla u) - \mathcal{A}(x, \nabla v)\big) \cdot \nabla(u - v)^+ \, dx$$

$$+ \int_{Z \cap [v = x_2]} \mathcal{A}(x, \nabla u) \cdot \nabla u - g\mathcal{A}(x, e) \cdot e \, dx \leqslant 0. \tag{2.16}$$

Note that

$$\int_{Z \cap [v = x_2]} \mathcal{A}(x, \nabla u) \cdot \nabla u - g\mathcal{A}(x, e) \cdot e \, dx$$

$$= \int_{Z \cap [u > v = x_2]} \mathcal{A}(x, \nabla u) \cdot \nabla u + \int_{Z \cap [u = v = x_2]} (1 - g)k(x) \, dx \geqslant 0.$$

It follows then from (2.16) that

$$\int_{Z\cap[v>x_2]} \big(\mathcal{A}(x,\nabla u) - \mathcal{A}(x,\nabla v)\big)\cdot \nabla(u-v)^+ \, dx \leqslant 0$$

which leads by (2.10) to $\nabla(u-v)^+ = 0$ a.e. in $Z\cap[v>x_2]$. By (2.11), we deduce that $(u-v)^+ = 0$ in $Z\cap[v>x_2]$. In particular, we obtain $u(x_1, h+\varepsilon) = h+\varepsilon \ \forall x_1 \in (\underline{x}_1, x_{01})$ which leads by Corollary 1.1 to $u = x_2$ in $Z\cap[x_2 \geqslant h+\varepsilon]$.

Let $x_0' = (x_{01}, h+\varepsilon)$. Since $u(x_0') = h+\varepsilon$, we deduce from the continuity of u that there exists a ball $B_{\varepsilon''}(x_0')$, $0 < \varepsilon'' < \varepsilon'$, such that $u(x) \leqslant x_2 + \varepsilon$ for all $x \in B_{\varepsilon''}(x_0')$. Taking into account this result and Theorem 2.3, there exists $\bar{x} = (\bar{x}_1, \bar{x}_2) \in B_{\varepsilon''}(x_0')$ such that: $x_{01} < \bar{x}_1$, $h+\varepsilon \leqslant \bar{x}_2$ and $u(\bar{x}) = \bar{x}_2$. Set $Z' = ((x_{01}, \bar{x}_1) \times (\bar{x}_2, +\infty)) \cap \Omega$, $w = (\varepsilon + \bar{x}_2 - x_2)^+ + x_2$ and $\xi = (u-w)^+$. Then one can argue as in the previous step, to conclude that $(u-w)^+ = 0$ in $Z'\cap[w>x_2]$ and then $u = x_2$ in $Z'\cap[x_2 \geqslant \bar{x}_2 + \varepsilon]$.

Finally, we have proved that $u = x_2$ in Z'', where $Z'' = ((\underline{x}_1, \bar{x}_1) \times (\bar{x}_2 + \varepsilon, +\infty)) \cap \Omega$. This leads to $\Phi(x_1) \leqslant \bar{x}_2 + \varepsilon \leqslant \Phi(x_{01}) + 3\varepsilon \ \forall x_1 \in (\underline{x}_1, \bar{x}_1)$. Hence Φ is upper semicontinuous at x_{01}. Taking into account Proposition 1.4, we obtain the continuity of Φ at x_{01} and the theorem is proved. $\qquad\square$

REMARK 2.3. For each $x_1 \in \pi_{x_1}(S_2 \cap S_3)$ such that $\{x_1\} \times (s_-(x_1), s_+(x_1)) \subset \Omega$, we have $\Phi(x_1) = s_+(x_1)$. Indeed otherwise we have $\Phi(x_{01}) < s_+(x_{01})$ for some point $x_{01} \in \pi_{x_1}(S_2 \cap S_3)$, with $\{x_{01}\} \times (s_-(x_{01}), s_+(x_{01})) \subset \Omega$. For clarity, we assume that the connected component of $\overline{S}_2$ (resp. $\overline{S}_3$) which contains $(x_{01}, s_+(x_{01}))$ is located to the left (resp. right) of the line $x_1 = x_{01}$. So there exists $\eta > 0$ such that $(x_{01} - \eta, x_{01}) \subset \text{Int}(\pi_{x_1}(S_2))$ and $(x_{01}, x_{01} + \eta) \subset \text{Int}(\pi_{x_1}(S_3))$. Now from Corollary 2.1, we have $u > x_2$ in $Z_+ = ((x_{01}, x_{01} + \eta) \times (\Phi(x_{01}), +\infty)) \cap \Omega$. Arguing as in the proof of Theorem 2.4, one can show that for some $\varepsilon > 0$ small enough we have $u = x_2$ in $Z_- = ((x_{01} - \varepsilon, x_{01}) \times (\Phi(x_{01}) + \varepsilon, +\infty)) \cap \Omega$. Thus we get a contradiction with Theorem 2.3(iii).

REMARK 2.4. Φ is continuous at each point $x_{01} \in \text{Int}(\pi_{x_1}(S_2)) \setminus S_-$ such that $\Phi(x_{01}) = s_-(x_{01})$. Indeed in this case, one has for each $\varepsilon > 0$ small enough, $(x_{01}, \Phi(x_{01}) + \varepsilon) \in \Omega$ and $u(x_{01}, \Phi(x_{01}) + \varepsilon) = \Phi(x_{01}) + \varepsilon$. Therefore one can adapt the proof of Theorem 2.4 to get $u = x_2$ in $((x_{01} - \varepsilon', x_{01} + \varepsilon') \times (\Phi(x_{01}) + 3\varepsilon, +\infty)) \cap \Omega$ (for some $\varepsilon' > 0$) which means the upper semicontinuity of Φ at x_{01}.

As a consequence of the continuity of the function Φ, we obtain the expression of g.

COROLLARY 2.2. *We have*

$$g = \chi\big([u=x_2]\big). \tag{2.17}$$

PROOF. First by (1.72) and (P_D)(ii), we have

$$g = 0 \quad \text{a.e. in } \big[x_2 < \Phi(x_1)\big]. \tag{2.18}$$

Now let $x_0 = (x_{01}, x_{02}) \in [x_2 > \Phi(x_1)]$. We have necessarily $s_+(x_{01}) > \Phi(x_{01})$. Moreover, from Remarks 2.1 and 2.3, we deduce that $x_{01} \in \text{Int}(\pi_{x_1}(S_2))$. By Theorem 2.4 and Remark 2.4, Φ is continuous in $\text{Int}(\pi_{x_1}(S_2)) \setminus S_-$. Assume that $x_{01} \notin S_-$. By continuity, there exists a ball $B_r(x_0)$ such that $B_r(x_0) \subset [x_2 > \Phi(x_1)]$. From (2.9), we have $g = 1$ a.e. in $B_r(x_0)$. It follows that

$$g = 1 \quad \text{a.e. in } \big[x_2 > \Phi(x_1)\big]. \tag{2.19}$$

Finally because Φ is continuous except on a finite set, the set $[x_2 = \Phi(x_1)]$ is of Lebesgue's measure zero. Thus we get, by (2.18) and (2.19),

$$g = \chi\big(\big[x_2 > \Phi(x_1)\big]\big)$$

which is (2.17). $\qquad\qquad\qquad\qquad\qquad\qquad\qquad\qquad\qquad\qquad\qquad\qquad\qquad\qquad\square$

2.3. *Existence and uniqueness of minimal and maximal solutions*

In this subsection, we show the existence and uniqueness of two solutions which minimize (resp. maximize) a functional. Moreover, one is minimal and the other one is maximal in the usual sense among all solutions. First we establish a key result.

THEOREM 2.5. *Let (u_1, g_1) and (u_2, g_2) be two solutions of* (P_D).
   Set $u_m = \min(u_1, u_2)$, $u_M = \max(u_1, u_2)$, $g_m = \min(g_1, g_2)$, and $g_M = \max(g_1, g_2)$. Then we have, for $i = 1, 2$ and for all $\zeta \in W^{1,q}(\Omega)$,

$$\text{(i)} \quad \int_\Omega \big((\mathcal{A}(x, \nabla u_i) - \mathcal{A}(x, \nabla u_m)) - (g_i - g_M)\mathcal{A}(x, e)\big) \cdot \nabla \zeta \, dx = 0,$$

$$\text{(ii)} \quad \int_\Omega \big((\mathcal{A}(x, \nabla u_i) - \mathcal{A}(x, \nabla u_M)) - (g_i - g_m)\mathcal{A}(x, e)\big) \cdot \nabla \zeta \, dx = 0.$$

PROOF. (i) Let $\zeta \in C^1(\overline{\Omega})$, $\zeta \geqslant 0$. For $\delta, \varepsilon > 0$, we consider $\alpha_\delta(x) = (1 - d(x, A_m)/\delta)^+$ and $\xi = \min(\alpha_\delta \zeta, \frac{u_i - u_m}{\varepsilon})$, where $A_m = [u_m > x_2]$. We have

$$\int_\Omega \big((\mathcal{A}(x, \nabla u_i) - \mathcal{A}(x, \nabla u_m)) - (g_i - g_M)\mathcal{A}(x, e)\big) \cdot \nabla \zeta \, dx$$

$$= \int_\Omega \big((\mathcal{A}(x, \nabla u_i) - \mathcal{A}(x, \nabla u_m)) - (g_i - g_M)\mathcal{A}(x, e)\big) \cdot \nabla(\alpha_\delta \zeta) \, dx$$

$$+ \int_\Omega \big((\mathcal{A}(x, \nabla u_i) - \mathcal{A}(x, \nabla u_m)) - (g_i - g_M)\mathcal{A}(x, e)\big) \cdot \nabla\big((1 - \alpha_\delta)\zeta\big) \, dx. \tag{2.20}$$

Since $(1 - \alpha_\delta)\zeta$ is a test function for (P_D), we have

$$\int_\Omega \big(\mathcal{A}(x, \nabla u_i) - g_i \mathcal{A}(x, e)\big) \cdot \nabla\big((1 - \alpha_\delta)\zeta\big) \, dx \leqslant 0. \tag{2.21}$$

Given that $(1 - \alpha_\delta)\zeta = 0$ on A_m and by (2.17) $g_M = 1$ a.e. in $[u_m = x_2]$, we obtain

$$\int_\Omega \left(\mathcal{A}(x, \nabla u_m) - g_M \mathcal{A}(x, e)\right) \cdot \nabla\left((1 - \alpha_\delta)\zeta\right) dx$$

$$= \int_{[u_m = x_2]} k(x)(1 - g_M)\left((1 - \alpha_\delta)\zeta\right)_{x_2} dx = 0. \tag{2.22}$$

Subtracting (2.22) from (2.21), we get

$$\int_\Omega \left((\mathcal{A}(x, \nabla u_i) - \mathcal{A}(x, \nabla u_m)) - (g_i - g_M)\mathcal{A}(x, e)\right) \cdot \nabla\left((1 - \alpha_\delta)\zeta\right) dx \leqslant 0. \tag{2.23}$$

Now clearly $\pm\xi$ are test functions for (P_D). So we have for $i, j = 1, 2$ with $i \neq j$,

$$\int_\Omega \left((\mathcal{A}(x, \nabla u_i) - \mathcal{A}(x, \nabla u_j)) - (g_i - g_j)\mathcal{A}(x, e)\right) \cdot \nabla\xi \, dx = 0. \tag{2.24}$$

Since $\xi = 0$ on the set $[u_i = u_m]$, we have to integrate only on the set $[u_i - u_m > 0]$ where $u_m = u_j$. So (2.24) becomes

$$\int_\Omega \left((\mathcal{A}(x, \nabla u_i) - \mathcal{A}(x, \nabla u_m)) - (g_i - g_M)\mathcal{A}(x, e)\right) \cdot \nabla\xi \, dx = 0$$

which can be written by the monotonicity of $\mathcal{A}$

$$\int_{[u_i - u_m \geqslant \varepsilon\zeta]} \left(\mathcal{A}(x, \nabla u_i) - \mathcal{A}(x, \nabla u_m)\right) \cdot \nabla(\alpha_\delta\zeta) \, dx$$

$$- \int_\Omega k(x)(g_i - g_M) \cdot (\alpha_\delta\zeta)_{x_2} \, dx$$

$$\leqslant \int_\Omega k(x)(g_M - g_i)\left(\alpha_\delta\zeta - \frac{u_i - u_m}{\varepsilon}\right)^+_{x_2} dx. \tag{2.25}$$

Using (2.17) we have

$$\int_\Omega k(x)(g_M - g_i)\left(\alpha_\delta\zeta - \frac{u_i - u_m}{\varepsilon}\right)^+_{x_2} dx$$

$$= \int_{[u_i > u_m = x_2]} k(x)\left(\alpha_\delta\zeta - \frac{u_i - u_m}{\varepsilon}\right)^+_{x_2} dx$$

$$= \int_{D_i} dx_1 \int_{\Phi_m(x_1)}^{\Phi_i(x_1)} k(x)\left(\alpha_\delta\zeta - \frac{u_i - u_m}{\varepsilon}\right)^+_{x_2} dx_2 \tag{2.26}$$

with

$$D_i = \left\{ x_1 \in \pi_{x_1}(\Omega) \mid \Phi_m(x_1) < \Phi_i(x_1) \right\}, \quad i = 1, 2,$$

and

$$\Phi_m = \min(\Phi_1, \Phi_2).$$

Since k is nondecreasing in x_2, we deduce by the second mean value theorem that for a.e. $x_1 \in D_i$, there exists $\Phi_*(x_1) \in [\Phi_m(x_1), \Phi_i(x_1)]$ such that

$$\int_{\Phi_m(x_1)}^{\Phi_i(x_1)} k(x) \left(\alpha_\delta \zeta - \frac{u_i - u_m}{\varepsilon} \right)^+_{x_2} dx_2$$
$$= k\big(x_1, \Phi_i(x_1)_-\big) \int_{\Phi_*(x_1)}^{\Phi_i(x_1)} \left(\alpha_\delta \zeta - \frac{u_i - u_m}{\varepsilon} \right)^+_{x_2} dx_2. \tag{2.27}$$

Then by (2.25)–(2.27), we get

$$\int_{[u_i - u_m \geqslant \varepsilon\zeta]} \big(\mathcal{A}(x, \nabla u_i) - \mathcal{A}(x, \nabla u_m)\big) \cdot \nabla(\alpha_\delta \zeta) \, dx$$
$$- \int_\Omega k(x)(g_i - g_M) \cdot (\alpha_\delta \zeta)_{x_2} \, dx$$
$$\leqslant \int_{D_i} k\big(x_1, \Phi_i(x_1)_-\big)(\alpha_\delta \zeta)\big(x_1, \Phi_i(x_1)\big) \, dx_1$$

which leads by letting ε go to zero to

$$\int_\Omega \big(\mathcal{A}(x, \nabla u_i) - \mathcal{A}(x, \nabla u_m) - (g_i - g_M)\mathcal{A}(x, e)\big) \cdot \nabla(\alpha_\delta \zeta) \, dx$$
$$\leqslant \int_{D_i} k\big(x_1, \Phi_i(x_1)_-\big)(\alpha_\delta \zeta)\big(x_1, \Phi_i(x_1)\big) \, dx_1$$
$$\leqslant \int_{D_i} (k\alpha_\delta \zeta)\big(x_1, \Phi_i(x_1)\big) \, dx_1. \tag{2.28}$$

Using (2.20), (2.23) and (2.28), we get

$$\int_\Omega \big(\mathcal{A}(x, \nabla u_i) - \mathcal{A}(x, \nabla u_m) - (g_i - g_M)\mathcal{A}(x, e)\big) \cdot \nabla \zeta \, dx$$
$$\leqslant \int_{D_i} (k\alpha_\delta \zeta)\big(x_1, \Phi_i(x_1)\big) \, dx_1. \tag{2.29}$$

Now since for each $x_{01} \in D_i$, we have $(x_{01}, \Phi_i(x_{01})) \notin \overline{A}_m$, we deduce that $\alpha_\delta(x_{01}, \Phi_i(x_{01}))$ converges to 0 when δ goes to 0. Using the Lebesgue theorem we obtain, by letting $\delta \to 0$ in (2.29),

$$\int_\Omega \big(\mathcal{A}(x, \nabla u_i) - \mathcal{A}(x, \nabla u_m) - (g_i - g_M)\mathcal{A}(x, e)\big) \cdot \nabla \zeta \, dx \leqslant 0.$$

Remarking that the last inequality holds also for $M - \zeta$ with $M = \max_{\overline{\Omega}} \zeta$, we get

$$\int_\Omega \big(\mathcal{A}(x, \nabla u_i) - \mathcal{A}(x, \nabla u_m) - (g_i - g_M)\mathcal{A}(x, e)\big) \cdot \nabla \zeta \, dx = 0$$

$$\forall \zeta \in C^1(\overline{\Omega}), \zeta \geqslant 0.$$

Since $C^1(\overline{\Omega})$ is dense in $W^{1,q}(\Omega)$ and since each function $\zeta \in W^{1,q}(\Omega)$ can be written as $\zeta = \zeta^+ - \zeta^-$, we get

$$\int_\Omega \big(\mathcal{A}(x, \nabla u_i) - \mathcal{A}(x, \nabla u_m) - (g_i - g_M)\mathcal{A}(x, e)\big) \cdot \nabla \zeta \, dx = 0$$

$$\forall \zeta \in W^{1,q}(\Omega).$$

(ii) It is enough to establish the result for $i = 1$ since it is similar for $i = 2$. We have for $\zeta \in W^{1,q}(\Omega)$,

$$\int_\Omega \big(\mathcal{A}(x, \nabla u_1) - \mathcal{A}(x, \nabla u_M) - (g_1 - g_m)\mathcal{A}(x, e)\big) \cdot \nabla \zeta \, dx$$

$$= \int_{[u_2 \geqslant u_1]} \big(\mathcal{A}(x, \nabla u_1) - \mathcal{A}(x, \nabla u_2) - (g_1 - g_2)\mathcal{A}(x, e)\big) \cdot \nabla \zeta \, dx$$

$$= -\int_{[u_2 \geqslant u_1]} \big(\mathcal{A}(x, \nabla u_2) - \mathcal{A}(x, \nabla u_1) - (g_2 - g_1)\mathcal{A}(x, e)\big) \cdot \nabla \zeta \, dx$$

$$= -\int_\Omega \big(\mathcal{A}(x, \nabla u_2) - \mathcal{A}(x, \nabla u_m) - (g_2 - g_M)\mathcal{A}(x, e)\big) \cdot \nabla \zeta \, dx = 0. \qquad \square$$

As a consequence of Theorem 2.5, we obtain the following corollary.

COROLLARY 2.3. *Let (u_1, g_1) and (u_2, g_2) be two solutions of* (P_D). *Then* $(\min(u_1, u_2),$ $\max(g_1, g_2))$ *and* $(\max(u_1, u_2), \min(g_1, g_2))$ *are also solutions of* (P_D).

PROOF. We will only prove that $(\min(u_1, u_2), \max(g_1, g_2))$ is a solution for (P_D). The proof that $(\max(u_1, u_2), \min(g_1, g_2))$ is a solution, is similar.

Indeed, first it is clear that we have $(u_m, g_M) \in W^{1,q}(\Omega) \times L^\infty(\Omega)$, $u_m = \min(u_1, u_2) \geqslant x_2$ and $0 \leqslant g_M = \max(g_1, g_2) \leqslant 1$ a.e. in Ω. Moreover, if $u_m > x_2$ then $u_1 > x_2$ and

$u_2 > x_2$, which leads to $g_1 = g_2 = 0$ and therefore $g_M = 0$. Since $u_1 = u_2 = \psi$ on $S_2 \cup S_3$, we have $u_m = \psi$ on $S_2 \cup S_3$.

Finally, let $\zeta \in W^{1,q}(\Omega)$ such that $\zeta = 0$ on S_3 and $\zeta \geqslant 0$ on S_2. Then we have, by Theorem 2.5 and since (u_i, g_i) is a solution of (P$_D$),

$$\int_\Omega \big(\mathcal{A}(x, \nabla u_m) - g_M \mathcal{A}(x, e)\big) \cdot \nabla \zeta \, dx$$

$$= \int_\Omega \big(\mathcal{A}(x, \nabla u_i) - g_i \mathcal{A}(x, e)\big) \cdot \nabla \zeta \, dx \leqslant 0. \qquad \square$$

Consider now the set of all solutions of (P$_D$)

$$\mathcal{S}_D = \big\{(u, g) \in W^{1,q}(\Omega) \times L^\infty(\Omega) \mid (u, g) \text{ is a solution of (P}_D) \big\}.$$

We define the following mapping $\mathcal{I}_D$ on $\mathcal{S}_D$ by

$$\forall (u, g) \in \mathcal{S}_D, \quad \mathcal{I}_D(u, g) = \int_\Omega \mathcal{A}(x, \nabla u) \cdot \nabla u \, dx - \frac{1}{q} \int_\Omega g \mathcal{A}(x, e) \cdot e \, dx.$$

The main result of this section is the following theorem.

THEOREM 2.6. *There exist a unique minimal solution (u_m, g_M) and a unique maximal solution (u_M, g_m) in $\mathcal{S}_D$ in the following sense:*

$$\mathcal{I}_D(u_m, g_M) = \min_{(u,g) \in \mathcal{S}_D} \mathcal{I}_D(u, g), \qquad \mathcal{I}_D(u_M, g_m) = \max_{(u,g) \in \mathcal{S}_D} \mathcal{I}_D(u, g)$$

$$\forall (u, g) \in \mathcal{S}_D,\ u_m \leqslant u \leqslant u_M, g_m \leqslant g \leqslant g_M \text{ in } \Omega.$$

We first prove a monotonicity result for $\mathcal{I}_D$.

LEMMA 2.1. *$\mathcal{I}_D$ is strictly monotone, i.e., for each $(u_1, g_1), (u_2, g_2) \in \mathcal{S}_D$,*

(i)　$u_1 \leqslant u_2, g_2 \leqslant g_1$ *in* Ω　$\implies$　$\mathcal{I}_D(u_1, g_1) \leqslant \mathcal{I}_D(u_2, g_2);$

(ii)　$u_1 \leqslant u_2, g_2 \leqslant g_1$ *in* Ω　*and*　$u_1 \neq u_2$　$\implies$　$\mathcal{I}_D(u_1, g_1) < \mathcal{I}_D(u_2, g_2).$

PROOF. (i) Let $(u_1, g_1), (u_2, g_2) \in \mathcal{S}_D$ such that $u_1 \leqslant u_2$ and $g_2 \leqslant g_1$ a.e. in Ω. Then we have

$$\mathcal{I}_D(u_1, g_1) - \mathcal{I}_D(u_2, g_2)$$

$$= \int_\Omega \big(\mathcal{A}(x, \nabla u_1) \cdot \nabla u_1 - g_1 \mathcal{A}(x, e) \cdot e\big) \, dx$$

$$- \int_\Omega \big(\mathcal{A}(x, \nabla u_2) \cdot \nabla u_2 - g_2 \mathcal{A}(x, e) \cdot e\big) \, dx$$

$$+ \frac{1}{q'} \int_\Omega (g_1 - g_2) \mathcal{A}(x, e) \cdot e \, dx$$

$$= \int_\Omega \big(\mathcal{A}(x, \nabla u_1) - g_1 \mathcal{A}(x, e)\big) \cdot \nabla u_1 \, dx$$

$$- \int_\Omega \big(\mathcal{A}(x, \nabla u_2) - g_2 \mathcal{A}(x, e)\big) \cdot \nabla u_2 \, dx + \frac{1}{q'} \int_\Omega (g_1 - g_2) \mathcal{A}(x, e) \cdot e \, dx$$

$$= \int_\Omega \big(\mathcal{A}(x, \nabla u_1) - g_1 \mathcal{A}(x, e)\big) \cdot \nabla (u_1 - \psi) \, dx$$

$$- \int_\Omega \big(\mathcal{A}(x, \nabla u_2) - g_2 \mathcal{A}(x, e)\big) \cdot \nabla (u_2 - \psi) \, dx$$

$$+ \frac{1}{q'} \int_\Omega (g_1 - g_2) \mathcal{A}(x, e) \cdot e \, dx$$

$$+ \int_\Omega \big(\big(\mathcal{A}(x, \nabla u_1) - \mathcal{A}(x, \nabla u_2)\big) - (g_1 - g_2) \mathcal{A}(x, e)\big) \cdot \nabla \psi \, dx.$$

Using Theorem 2.5(i) and the fact that $u_i - \psi$, $i = 1, 2$, are test functions for $(\mathrm{P_D})$, we get

$$\mathcal{I}_\mathrm{D}(u_1, g_1) - \mathcal{I}_\mathrm{D}(u_2, g_2) = \frac{1}{q'} \int_\Omega (g_1 - g_2) k(x) \, dx \leqslant 0$$

which proves (i).

(ii) Assume that $u_1 \leqslant u_2$ and $g_2 \leqslant g_1$ a.e. in Ω and $\mathcal{I}_\mathrm{D}(u_1, g_1) = \mathcal{I}_\mathrm{D}(u_2, g_2)$. Then $\int_\Omega (g_1 - g_2) k(x) \, dx = 0$ and therefore $g_1 = g_2$ a.e. in Ω. Using Theorem 2.5(i) for $\xi = u_1 - u_2$, we get

$$\int_\Omega \big(\mathcal{A}(x, \nabla u_1) - \mathcal{A}(x, \nabla u_2)\big) \cdot \nabla (u_1 - u_2) \, dx = 0$$

which leads by (2.10) to $\nabla(u_1 - u_2) = 0$ a.e. in Ω. But since $u_1 - u_2 = 0$ on $S_2 \cup S_3$, we obtain $u_1 = u_2$ and (ii) is proved. $\qquad\square$

PROOF OF THEOREM 2.6. First remark that for each $(u, g) \in \mathcal{S}_\mathrm{D}$, we have

$$\mathcal{I}_\mathrm{D}(u_1, g_1) \geqslant -\frac{1}{q} \int_\Omega k(x) g \, dx \geqslant -\frac{|\Omega|}{q} M.$$

We deduce that there exists a minimizing sequence $(u_k, g_k)_{k \in \mathbb{N}}$ for $\mathcal{I}_\mathrm{D}$ i.e.

$$\forall k \in \mathbb{N}, \quad (u_k, g_k) \in \mathcal{S}_\mathrm{D} \quad \text{and} \quad \lim_{k \to +\infty} \mathcal{I}_\mathrm{D}(u_k, g_k) = m = \inf_{(u,g) \in \mathcal{S}_\mathrm{D}} \mathcal{I}_\mathrm{D}(u, g).$$

$$(2.30)$$

Now we define another sequence $(v_k, f_k)_{k \in \mathbb{N}}$ by

$$
\begin{cases}
(v_0, f_0) = (u_0, g_0), \\
(v_{k+1}, f_{k+1}) = \big(\min(u_{k+1}, v_k), \max(g_{k+1}, f_k)\big) \quad \forall k \in \mathbb{N}.
\end{cases}
$$

By Corollary 2.3 and Lemma 2.1, it is clear that for all $k \in \mathbb{N}$, $(v_k, f_k) \in \mathcal{S}_{\mathrm{D}}$ and we have

$$
\begin{cases}
\forall k \in \mathbb{N}, \quad v_k \leqslant u_k, \quad\ \ g_k \leqslant f_k, \\
\forall k \in \mathbb{N}, \quad m \leqslant \mathcal{I}_{\mathrm{D}}(v_k, f_k) \leqslant \mathcal{I}_{\mathrm{D}}(u_k, g_k).
\end{cases}
$$

This clearly leads to

$$
m = \lim_{k \to +\infty} \mathcal{I}_{\mathrm{D}}(v_k, f_k). \tag{2.31}
$$

From the definition of $(v_k, f_k)_{k \in \mathbb{N}}$, we deduce that

$$
\forall k \in \mathbb{N}, \quad v_{k+1} \leqslant v_k \quad \text{and} \quad f_k \leqslant f_{k+1} \quad \text{a.e. in } \Omega.
$$

But since v_k and f_k are uniformly bounded ($x_2 \leqslant v_k \leqslant h_0$ by Proposition 1.1) and $0 \leqslant f_k \leqslant 1$ a.e. in Ω, we obtain by Beppo Levi's theorem that there exists $(v, f) \in L^q(\Omega) \times L^{q'}(\Omega)$ such that

$$
\begin{cases}
v_k \to v \quad \text{in } L^q(\Omega) \text{ and a.e. in } \Omega, \\
f_k \to f \quad \text{in } L^{q'}(\Omega) \text{ and a.e. in } \Omega.
\end{cases} \tag{2.32}
$$

Now since $\pm(v_k - \psi)$ are test functions for $(\mathrm{P_D})$, we obtain, by using (1.2)(iii), $(\mathrm{P_D})$(ii) and the Hölder inequality,

$$
\int_\Omega \mathcal{A}(x, \nabla v_k) \cdot \nabla v_k \, dx = \int_\Omega \mathcal{A}(x, \nabla v_k) \cdot \nabla \psi \, dx - \int_\Omega g \mathcal{A}(x, e) \cdot \nabla \varphi \, dx
$$

$$
\leqslant C \left(\int_\Omega |\nabla v_k|^q \, dx \right)^{1/q'} + C,
$$

where C is some positive constant. By (1.2)(iii), we deduce that $(v_k)_k$ is bounded in $W^{1,q}(\Omega)$. So we have up to a subsequence

$$
v_{k_p} \rightharpoonup v \quad \text{in } W^{1,q}(\Omega), \tag{2.33}
$$

$$
v_{k_p} \to v \quad \text{in } L^q(S_2 \cup S_3) \text{ and a.e. in } S_2 \cup S_3. \tag{2.34}
$$

From the continuity of the trace operator and (2.34), we have $v = \psi$ on $S_2 \cup S_3$. From (2.32), we obtain that

$$
v \geqslant x_2, 0 \leqslant f \leqslant 1, \quad f(v - x_2) = 0 \quad \text{a.e. in } \Omega.
$$

We would like to prove that $(v, f) \in \mathcal{S}_\mathrm{D}$. It suffices to verify that it satisfies $(\mathrm{P_D})(\mathrm{iii})$.

From the fact that (v_k) is bounded in $W^{1,q}(\Omega)$, we deduce that up to a subsequence still denoted by (v_{k_p}), one has

$$\mathcal{A}(x, \nabla v_{k_p}) \rightharpoonup \mathcal{A}_0 \quad \text{in } \mathbb{L}^{q'}(\Omega). \tag{2.35}$$

Let $p, s \in \mathbb{N}$ such that $p \leqslant s$. We have, by Theorem 2.5(i),

$$\int_\Omega \left\{ \left(\mathcal{A}(x, \nabla v_{k_p}) - \mathcal{A}(x, \nabla v_{k_s}) \right) - (f_{k_p} - f_{k_s})\mathcal{A}(x, e) \right\} \cdot \nabla v_{k_p} \, dx = 0$$

from which we deduce, by letting respectively $s \to +\infty$ and $p \to +\infty$, and using (2.33) and (2.35),

$$\lim_{p \to +\infty} \int_\Omega \mathcal{A}(x, \nabla v_{k_p}) \cdot \nabla v_{k_p} \, dx = \int_\Omega \mathcal{A}_0 \cdot \nabla v \, dx. \tag{2.36}$$

Using the monotonicity of $\mathcal{A}$, (2.33), (2.35) and (2.36), we easily obtain

$$\mathcal{A}(x, \nabla v_{k_p}) \rightharpoonup \mathcal{A}(x, \nabla v) \quad \text{in } \mathbb{L}^{q'}(\Omega). \tag{2.37}$$

Finally, let $\xi \in W^{1,q}(\Omega)$ such that $\xi \geqslant 0$ on S_2 and $\xi = 0$ on S_3. For any $p \in \mathbb{N}$, we have

$$\int_\Omega \left(\mathcal{A}(x, \nabla v_{k_p}) - f_{k_p}\mathcal{A}(x, e) \right) \cdot \nabla \xi \, dx \leqslant 0. \tag{2.38}$$

Using (2.32) and (2.37) we get, by letting $p \to +\infty$ in (2.38),

$$\int_\Omega \left(\mathcal{A}(x, \nabla v) - f\mathcal{A}(x, e) \right) \cdot \nabla \xi \, dx \leqslant 0.$$

Thus (v, f) is a solution of $(\mathrm{P_D})$.

Now, using (2.31)–(2.32) and (2.35)–(2.37), we obtain

$$
\begin{aligned}
m &= \lim_{p \to +\infty} \mathcal{I}_\mathrm{D}(v_{k_p}, f_{k_p}) \\
&= \lim_{p \to +\infty} \int_\Omega \mathcal{A}(x, \nabla v_{k_p}) \cdot \nabla v_{k_p} \, dx - \frac{1}{q} \int_\Omega f_{k_p}\mathcal{A}(x, e) \cdot e \, dx \\
&= \int_\Omega \mathcal{A}(x, \nabla v) \cdot \nabla v \, dx - \frac{1}{q} \int_\Omega f\mathcal{A}(x, e) \cdot e \, dx \\
&= \mathcal{I}_\mathrm{D}(v, f).
\end{aligned}
$$

Let $(u, g) \in \mathcal{S}_\mathrm{D}$. Since $(\min(u, v), \max(g, f)) \in \mathcal{S}_\mathrm{D}$, we deduce that $\mathcal{I}_\mathrm{D}(v, f) = \mathcal{I}_\mathrm{D}(\min(u, v), \max(g, f))$ which leads by Lemma 2.1 to $(v, f) = (\min(u, v), \max(g, f))$, i.e., $v \leqslant u$ and $g \leqslant f$ a.e. in Ω. The uniqueness of (v, f) is clear. This achieves the proof of the first part of Theorem 2.6.

Let us prove the second part of the theorem. First remark that

$$\mathcal{I}_{\mathrm{D}}(u, g) = \int_{\Omega} \mathcal{A}(x, \nabla u) \cdot \nabla u \, dx - \frac{1}{q} \int_{\Omega} gk(x) \, dx$$

$$\leqslant \int_{\Omega} \mathcal{A}(x, \nabla u) \cdot \nabla u \, dx \leqslant M \int_{\Omega} |\nabla u|^q \, dx.$$

Moreover,

$$\int_{\Omega} \big(\mathcal{A}(x, \nabla u) - g\mathcal{A}(x, e)\big) \cdot \nabla(u - \psi) \, dx = 0$$

which leads by (1.2)(iii) to

$$\lambda \int_{\Omega} |\nabla u|^q \, dx \leqslant \int_{\Omega} \mathcal{A}(x, \nabla u) \cdot \nabla u \, dx$$

$$= \int_{\Omega} \mathcal{A}(x, \nabla u) \cdot \nabla \psi \, dx - \int_{\Omega} g\mathcal{A}(x, e) \cdot \nabla \varphi \, dx$$

$$\leqslant M \left(\int_{\Omega} |\nabla u|^q \, dx \right)^{1/q'} \cdot \left(\int_{\Omega} |\nabla \psi|^q \, dx \right)^{1/q} + c$$

and then

$$\int_{\Omega} |\nabla u|^q \, dx \leqslant C \quad \text{for some positive constant } C.$$

Thus $\mathcal{I}_{\mathrm{D}}(u, g)$ is bounded for all $(u, g) \in \mathcal{S}_{\mathrm{D}}$. Let then $(u_k, g_k)_{k \in \mathbb{N}}$ be a sequence in $\mathcal{S}_{\mathrm{D}}$ such that

$$\lim_{k \to +\infty} \mathcal{I}_{\mathrm{D}}(u_k, g_k) = \sup_{(u,g) \in \mathcal{S}_{\mathrm{D}}} \mathcal{I}_{\mathrm{D}}(u, g).$$

We consider the following sequence $(w_k, h_k)_{k \in \mathbb{N}}$ defined by

$$\begin{cases} (w_0, h_0) = (u_0, g_0), \\ (w_{k+1}, h_{k+1}) = \big(\max(u_{k+1}, w_k), \min(g_{k+1}, h_k)\big) \quad \forall k \in \mathbb{N}. \end{cases}$$

By Corollary 2.3, we have $(w_k, h_k) \in \mathcal{S}_{\mathrm{D}}$ for each $k \in \mathbb{N}$. By Lemma 2.1, we obtain

$$\forall k \in \mathbb{N}, \quad \mathcal{I}_{\mathrm{D}}(u_k, g_k) \leqslant \mathcal{I}_{\mathrm{D}}(w_k, h_k) \leqslant \sup_{(u,g) \in \mathcal{S}_{\mathrm{D}}} \mathcal{I}_{\mathrm{D}}(u, g).$$

Therefore,

$$\sup_{(u,g)\in\mathcal{S}_{\mathrm{D}}} \mathcal{I}_{\mathrm{D}}(u,g) = \lim_{k\to+\infty} \mathcal{I}_{\mathrm{D}}(w_k,h_k). \tag{2.39}$$

We have also

$$\forall k \in \mathbb{N}, \quad w_k \leqslant w_{k+1} \quad \text{and} \quad h_{k+1} \leqslant h_k \quad \text{a.e. in } \Omega.$$

Using the monotonicity of $(w_k,h_k)_{k\in\mathbb{N}}$, (2.39) and arguing as above, we prove that for a subsequence $(w_{k_p},h_{k_p})_{p\in\mathbb{N}}$, we have

$$w_{k_p} \rightharpoonup w \quad \text{in } W^{1,q}(\Omega),$$

$$w_{k_p} \to w \quad \text{in } L^q(\Omega) \text{ and a.e. in } \Omega,$$

$$w_{k_p} \to w \quad \text{in } L^q(S_3) \text{ and a.e. in } S_3,$$

$$\mathcal{A}(x,\nabla w_{k_p}) \rightharpoonup \mathcal{A}(x,\nabla w) \quad \text{in } \mathbb{L}^{q'}(\Omega),$$

$$h_{k_p} \to h \quad \text{in } L^{q'}(\Omega) \text{ and a.e. in } \Omega.$$

Thus we obtain that (w,h) is a solution of (P_{D}) which satisfies $\mathcal{I}_{\mathrm{D}}(w,h) = \sup_{(u,g)\in\mathcal{S}_{\mathrm{D}}} \mathcal{I}_{\mathrm{D}}(u,g)$. We also prove, as in the case of minimal solution, that for all $(u,g) \in \mathcal{S}_{\mathrm{D}}$: $u \leqslant w, h \leqslant g$ a.e. in Ω. $\qquad\square$

2.4. *Reservoirs-connected solution*

Assume that we are in the situation of Figure 6 with C and C' denoting the regions shown in the figure. Then it is not difficult to verify that

$$(u,g) = \begin{cases} (h,0) & \text{in } C, \\ (x_2,1) & \text{otherwise}, \end{cases}$$

and

$$(u,g) = \begin{cases} (h,0) & \text{in } C, \\ (h',0) & \text{in } C', \\ (x_2,1) & \text{elsewhere}, \end{cases}$$

are solutions of (P_{D}). Moreover we can obtain more solutions just by replacing h' by any $0 < k < h'$. This example is an extension of an example given in [21] in the case of linear Darcy's law. It shows that in general the solution of the problem (P_{D}) is not unique. The first solution in the previous example is such that the only connected component of $[u > x_2]$ is connected to the unique reservoir. It seems that it is the only solution that is relevant from the physical point of view. This type of solution was introduced in [18] under the name of S_3-connected solution. Here we call it reservoirs-connected solution. Hence we have the definition:

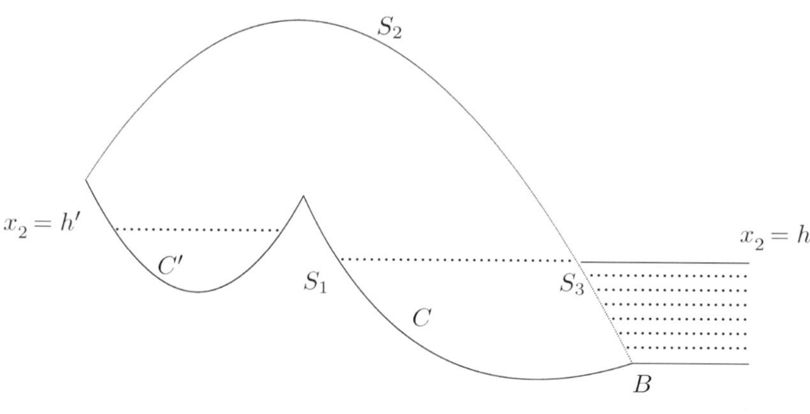

Fig. 6.

DEFINITION 2.1. A solution (u, g) of (P_D) is called a reservoirs-connected solution if for each connected component C of $[u > x_2]$, we have $\overline{C} \cap S_3 \neq \emptyset$.

REMARK 2.5. If C_i is the connected component of $[u > x_2]$ that contains $S_{3,i}$ on its boundary, then by continuity and thanks to Remark 2.1, C_i contains the strip of Ω below $S_{3,i}$.

The following theorem characterizes the connected components of $[u > x_2]$ which are not related to S_3.

THEOREM 2.7. *Let (u, g) be a solution of (P_D) and C a connected component of $[u > x_2]$ such that $\overline{C} \cap S_3 = \emptyset$. If we set $h_c = \sup\{x_2 \mid (x_1, x_2) \in C\}$, then we have*

$$\begin{cases} C = \{(x_1, x_2) \in \Omega \mid x_1 \in \pi_{x_1}(C), x_2 < h_c\}, \\ u = x_2 + (h_c - x_2)^+ \cdot \chi(C), \qquad g = 1 - \chi(C) \quad in \ Z = \Omega \cap \left(\pi_{x_1}(C) \times \mathbb{R}\right). \end{cases}$$

PROOF. By assumption and Theorem 1.3, we have $\pi_{x_1}(C) \subset \pi_{x_1}(S_2)$. Then $\pm \chi(Z)(u - x_2) = \pm \chi(C)(u - x_2)$ are test functions of (P_D) and we have

$$\int_Z \left(\mathcal{A}(x, \nabla u) - g\mathcal{A}(x, e)\right) \cdot \nabla(u - x_2) = 0. \tag{2.40}$$

Applying Theorem 2.1 to Z we obtain

$$\int_Z \left(\mathcal{A}(x, \nabla u) - g\mathcal{A}(x, e)\right) \cdot e \leqslant 0. \tag{2.41}$$

Adding (2.40) and (2.41) we get

$$\int_Z \left(\mathcal{A}(x, \nabla u) \cdot \nabla u - kg\right) \leqslant 0$$

which leads by (1.2)(iii) to

$$\int_{Z\cap[u>x_2]} \lambda|\nabla u|^q + \int_{Z\cap[u=x_2]} k(1-g) \leqslant 0.$$

It follows that $\nabla u = 0$ a.e. in $Z \cap [u > x_2] = C$ and $g = 1$ a.e. in $Z \cap [u = x_2] = Z \setminus C$. Hence we obtain $u = x_2 + (h_c - x_2)^+ \cdot \chi(C)$ and $g = 1 - \chi(C)$ a.e. in Z. $\qquad\square$

The result of Theorem 2.7 leads to the following definition (see [18] and also [20] for an extension).

DEFINITION 2.2. We call a pool in Ω a pair of functions defined in Ω by $(p, \chi) = ((h - x_2)^+, 1)\chi(C)$, where C is a connected component of $\Omega \cap [x_2 < h]$.

REMARK 2.6. Thanks to this definition, Theorem 2.7 becomes:
For each solution (u, g) of $(\mathrm{P_D})$ and each connected component C of $[u > x_2]$ such that $\overline{C} \cap S_3 = \emptyset$, $(u - x_2, 1 - g)$ agrees with a pool in the strip $\Omega \cap (\pi_{x_1}(C) \times \mathbb{R})$.

Now we have the following theorem.

THEOREM 2.8. *Each solution (u, g) of $(\mathrm{P_D})$ can be written as*

$$u = u_r + \sum_{i\in I} p_i \quad and \quad g = g_r - \sum_{i\in I} \chi_i,$$

where (u_r, g_r) is a reservoirs-connected solution and (p_i, χ_i) are pools.

PROOF. Let $(C_i)_{i\in I}$ be the family of all connected components of $[u > x_2]$ such that $C_i \cap S_3 = \emptyset$. Set

$$(u', g') = (u, g) - \sum_{i\in I}\big(\chi(C_i)(u - x_2), -\chi(C_i)\big).$$

Since each connected component C of $[u' > x_2]$ is such that $C \cap S_3 \neq \emptyset$, it follows that if (u', g') is a solution of $(\mathrm{P_D})$, it will be a reservoirs-connected solution. Let us then verify that (u', g') is a solution of $(\mathrm{P_D})$.
   (i) $(u', g') \in W^{1,q}(\Omega) \times L^\infty(\Omega)$. Since we have

$$(u - x_2)\chi(C_i) \in W^{1,q}(\Omega) \quad and$$

$$\nabla\big(\chi(C_i)(u - x_2)\big) = \chi(C_i)\nabla(u - x_2) \quad \forall i \in I \quad [18],$$

we deduce that

$$\nabla u' = \nabla u - \sum_{i\in I} \chi(C_i)\nabla(u - x_2) \in L^q(\Omega).$$

Moreover, since $0 \leqslant g' \leqslant 1$ a.e. in Ω, we have $g' \in L^{\infty}(\Omega)$.

(ii) Clearly we have $u' \geqslant x_2$ and $g'(u' - x_2) = 0$ a.e. in Ω.

(iii) Let $\xi \in W^{1,q}(\Omega)$ such that $\xi \geqslant 0$ on S_2 and $\xi = 0$ on S_3. Using Theorem 2.7, we have

$$\int_{\Omega} \left(\mathcal{A}(x, \nabla u') - g' \mathcal{A}(x, e) \right) \cdot \nabla \xi = \int_{\Omega} \left(\mathcal{A}(x, \nabla u) - g \mathcal{A}(x, e) \right) \cdot \nabla \xi \leqslant 0.$$

Thus the theorem is proved. $\qquad\square$

We deduce immediately from Theorem 2.8 the following corollary.

COROLLARY 2.4. *The minimal solution (u_m, g_M) is a reservoirs-connected solution.*

2.5. *Uniqueness of the reservoirs-connected solution*

In this subsection, we address the question of uniqueness of the reservoirs-connected solution.

2.5.1. *The case of linear Darcy's law.* Here we assume that

$$\mathcal{A}(x, \xi) = a(x) \cdot \xi \quad \forall \xi \in \mathbb{R}^2, \text{ a.e. } x \in \Omega,$$

$$a(x) = \left(a_{ij}(x) \right) \text{ is a } 2 \times 2 \text{ matrix,} \tag{2.42}$$

$$\exists \lambda, M > 0: \quad \lambda |\xi|^2 \leqslant a(x)\xi \cdot \xi \leqslant M |\xi|^2 \quad \forall \xi \in \mathbb{R}^2, \text{ a.e. } x \in \Omega, \tag{2.43}$$

$$a_{12}(x) = 0 \quad \text{for a.e. } x \in \Omega, \tag{2.44}$$

$$\frac{\partial a_{22}}{\partial x_2} \geqslant 0 \quad \text{in } \mathcal{D}'(\Omega). \tag{2.45}$$

THEOREM 2.9. *Assume that (2.42)–(2.45) are satisfied. Then there is one and only one reservoirs-connected solution.*

PROOF. Let (u, g) be a reservoirs-connected solution of (P_D). Let C_i (resp. $C_{m,i}$) be the connected component of $[u > x_2]$ (resp. $[u_m > x_2]$) such that $\overline{C}_i \cap S_{3,i} \neq \emptyset$ (resp. $\overline{C}_{m,i} \cap S_{3,i} \neq \emptyset$). Using Remark 2.1, we know that C_i and $C_{m,i}$ contain $S_{3,i}$ on their boundaries as well as the strip below it.

Consider $\zeta \in \mathcal{D}(C_{m,i} \cup B_r(x_i))$, where $x_i \in S_{3,i}$ and r is a small positive number. Then from Theorem 2.5(i), we have since $g = g_M = 0$ in $C_{m,i}$,

$$\int_{C_{m,i}} a(x) \nabla (u - u_m) \cdot \nabla \zeta \, dx = 0. \tag{2.46}$$

Define w in $\Omega_i = C_{m,i} \cup B_r(x_i)$ by $w = \chi(C_{m,i})(u - u_m)$. Since $u = u_m = \psi$ on S_3, it is clear that $w \in H^1(\Omega_i)$ and if one extends a by I_2 into $B_r(x_i) \setminus C_{m,i}$, we obtain from (2.46)

$$\int_{\Omega_i} a(x)\nabla w \cdot \nabla \zeta \, dx = 0 \quad \forall \zeta \in \mathcal{D}(\Omega_i). \tag{2.47}$$

Using the strict ellipticity of a, the fact that $w \geqslant 0$ in Ω_i, $w = 0$ in $B_r(x_i) \setminus C_{m,i}$ and the strong maximum principle, we get from (2.47) that $w = 0$ in Ω_i which leads to $u = u_m$ in $C_{m,i}$. Now we prove that $C_i = C_{m,i}$. Indeed since $C_{m,i}$ is a nonempty open set in the connected set C_i, it suffices to prove that $C_{m,i}$ is also closed relative to C_i. Indeed let $(x_k)_k$ be a sequence of points in $C_{m,i}$ which converges to an element x in C_i. By continuity of u and u_m, we obtain $u(x) = u_m(x)$. Since $u(x) > x_2$, we obtain $u_m(x) > x_2$ which means that $x \in C_{m,i}$.

Hence $u = u_m$ in Ω and from Corollary 2.2, we get $g = g_M$ in Ω. $\qquad\square$

2.5.2. *The case of a nonlinear Darcy's law.* In this subsection, we prove the uniqueness of the reservoirs-connected solution for a Darcy's law corresponding to

$$\mathcal{A}(x, \xi) = \left| a(x)\xi \cdot \xi \right|^{(q-2)/2} a(x)\xi,$$

$$q > 1, q \neq 2 \text{ and } a(x) = \left(a_{ij}(x) \right) \text{ is a } 2 \times 2 \text{ matrix}. \tag{2.48}$$

Moreover, we assume that

$$a(x) \text{ is symmetric and belongs to } C^{0,1}(\Omega), \tag{2.49}$$

$$\exists \lambda, M > 0: \quad \lambda |\xi|^2 \leqslant a(x)\xi \cdot \xi \leqslant M|\xi|^2 \quad \forall \xi \in \mathbb{R}^2, \text{ a.e. } x \in \Omega, \tag{2.50}$$

$$\text{for each } i \in \{1, \ldots, N\}, \quad \varphi = h_i - x_2 \quad \text{on } S_{3,i}, \tag{2.51}$$

$$\exists r_i > 0, \exists x_i \in S_{3,i}, \exists \alpha_i \in (0, 1): \quad S_{3,i}^{r_i} = S_{3,i} \cap B_r(x_i) \text{ is } C^{1,\alpha_i}, \tag{2.52}$$

$$a_{12}(x) = 0 \quad \text{for a.e. } x \in \Omega, \tag{2.53}$$

$$\frac{\partial a_{22}^{q/2}}{\partial x_2} \geqslant 0 \quad \text{in } \mathcal{D}'(\Omega). \tag{2.54}$$

Then we have the following theorem.

THEOREM 2.10. *Assume that* (2.48)–(2.54) *are satisfied. Then there is one and only one reservoirs-connected solution.*

To prove Theorem 2.10, we need three lemmas. We shall denote by (u, g) a reservoirs-connected solution of (P_D) and for each $i \in \{1, \ldots, N\}$, we shall denote by C_i (resp. $C_{m,i}$) the connected component of $[u > x_2]$ (resp. $[u_m > x_2]$) which contains $S_{3,i}$ on its boundary.

LEMMA 2.2. *For each $i \in \{1, \ldots, N\}$, we have the following alternatives:*
   (i) *either* $\exists x_i' \in S_{3,i}^{r_i}, \exists r_i' \in (0, r_i): \forall x \in \overline{B(x_i', r_i')} \cap \Omega, \nabla u(x) \neq 0,$

512 A. Lyaghfouri

(ii) *or $u = h_i$ in C_i.*

PROOF. First note that u satisfies

$$\begin{cases} \operatorname{div}\big(\mathcal{A}(x, \nabla u)\big) = 0 & \text{in } B(x_i, r_i) \cap \Omega, \\ u = \psi = h_i & \text{on } B(x_i, r_i) \cap \partial\Omega. \end{cases}$$

We deduce that for all $r \in (0, r_i)$, we have $u \in C^{1,\alpha_i}(\overline{B(x_i, r) \cap \Omega})$ [30]. So either (i) is true or we must have $\nabla u(x) = 0 \ \forall x \in S_{3,i}^{r_i}$. Assume that we are in the second case and set

$$\begin{cases} w_0(x) = u(x) - h_i & \text{for } x \in B(x_i, r_i) \cap \overline{\Omega}, \\ w_0(x) = 0 & \text{for } x \in B(x_i, r_i) \setminus \Omega. \end{cases}$$

Since $u - h_i = 0$ on $S_{3,i}^{r_i}$, we have $w_0 \in W^{1,q}(B(x_i, r_i))$. Moreover, because $u \in C^{1,\alpha_i}(B(x_i, r_i) \cap \overline{\Omega})$ and $\nabla u = 0$ on $S_{3,i}^{r_i}$, we deduce that $\operatorname{div}(\mathcal{A}(x, \nabla w_0)) = 0$ in $\mathcal{D}'(B(x_i, r_i))$.

Now since $\nabla w_0 = 0$ in $B(x_i, r_i) \setminus \Omega$ and since the zeros of the gradient of a nonconstant $\mathcal{A}$-harmonic function, under the conditions (2.48)–(2.50) are isolated [4], we conclude that $w_0 = 0$ in $B(x_i, r_i) \cap \Omega$, i.e., $u = h_i$ in $B(x_i, r_i) \cap \Omega$.

Arguing as before, $u - h_i$ is an $\mathcal{A}$-harmonic function in C_i such that $u - h_i = 0$ and $\nabla(u - h_i) = 0$ in $B(x_i, r_i) \cap C_i$. We conclude that $u - h_i = 0$ in C_i. $\square$

LEMMA 2.3. *If u and u_m are not both constant in C_i and $C_{m,i}$, respectively, then there exists $x_i' \in B(x_i, r_i) \cap S_{3,i}$, $r_i' \in (0, r_i)$, $0 < \lambda_0, \lambda_1 < +\infty$ such that*

$$\forall x \in \overline{B(x_i', r_i') \cap \Omega}, \quad \lambda_0 \leqslant \lambda(x) \leqslant \lambda_1, \tag{2.55}$$

where $\lambda(x) = \int_0^1 |\nabla w_t(x)|^{q-2} \, dt$ and $w_t = tu + (1 - t)u_m$.

PROOF. We will consider only the case where u is not constant in C_i. So the situation (i) of Lemma 2.2 holds. Since u and u_m are of class C^1 in $\overline{B(x_i, r_i) \cap \Omega}$, there exists x_i', $r_i' \in (0, r_i)$, $c_i, c_i' > 0$ such that

$$c_i \leqslant |\nabla u(x)| \leqslant c_i' \quad \forall x \in K_i = \overline{B(x_i', r_i') \cap \Omega}, \tag{2.56}$$

$$|\nabla u_m(x)| \leqslant c_m \quad \forall x \in K_i. \tag{2.57}$$

We distinguish two cases.

Case 1: $q > 2$. Using (2.56) and (2.57) we obtain

$$\lambda(x) = \int_0^1 |\nabla w_t(x)| \, dt \leqslant (c_m + c_i')^{q-2} = \lambda_1 \quad \forall x \in K_i. \tag{2.58}$$

Now clearly $\lambda(x)$ is continuous on K_i. Let us denote by λ_0 the minimum value of $\lambda(x)$ on K_i. There exists $x_* \in K_i$ such that $\lambda_0 = \lambda(x_*)$. We claim that $\lambda_0 > 0$. Indeed otherwise we will have, since $q > 2$, $\nabla w_t(x_*) = 0$ for all $t \in [0, 1]$. This leads to $\nabla u(x_*) = \nabla u_m(x_*) = 0$ which is impossible.

Case 2: $1 < q < 2$. Using (2.56) and (2.57) we obtain

$$\lambda(x) = \int_0^1 \left| \nabla w_t(x) \right| \, dt \geqslant \left(c_m + c_i' \right)^{q-2} = \lambda_0 \quad \forall x \in K_i.$$

We would like to show that $\lambda(x) \leqslant \lambda_1 < \infty$ in K_i.

If $\nabla w_t(x)$ does not vanish for each $(t, x) \in [0, 1] \times K_i$, then $|\nabla w_t(x)|^{q-2}$ is continuous in $[0, 1] \times K_i$ and therefore $\lambda(x)$ is continuous in K_i. If we denote by λ_1 the maximum value of $\lambda(x)$ on K_i, then we have $\lambda(x) \leqslant \lambda_1 < \infty$ in K_i.

If $\nabla w_t(x) = 0$ for some $(t, x) \in [0, 1] \times K_i$, then $t \in [0, 1)$. Otherwise we will have $\nabla u(x) = 0$. Moreover, if there exist two values $t_1 \neq t_2 \in [0, 1]$ such that $\nabla w_{t_i}(x) = 0$, then $\nabla u(x) = \nabla u_m(x) = 0$. Therefore for each $x \in K_i$, there exists at most one value $t(x) \in [0, 1)$ such that $\nabla w_{t(x)}(x) = 0$. In this case, we have $\nabla u_m(x) = -t(x)/(1 - t(x)) \times \nabla u(x)$ and

$$\lambda(x) = \frac{|\nabla u(x)|^{q-2}}{(1 - t(x))^{q-2}} \int_0^1 \left| t - t(x) \right|^{q-2} \, dt$$

$$= \frac{(1 - t(x))^{2-q}}{q - 1} \left[(1 - t(x))^{q-1} + t^{q-1}(x) \right] |\nabla u(x)|^{q-2} \leqslant \frac{2 c_i^{q-2}}{q - 1}. \qquad \square$$

LEMMA 2.4. *For each $i \in \{1, \ldots, N\}$, there exist $x_i' \in B(x_i, r_i) \cap S_{3,i}$, $r_i' \in (0, r_i)$ such that*

$$u = u_m \quad in \ B\left(x_i', r_i' \right) \cap \Omega.$$

PROOF. If u is constant in C_i and u_m is constant in $C_{m,i}$, then the result is trivial by Lemma 2.2.

In the following, we assume that either u is not constant in C_i or u_m is not constant in $C_{m,i}$.

By Lemma 2.3, we know that there exist $x_i' \in S_{3,i}^{r_i}$, $r_i' \in (0, r_i)$ and $\lambda_0, \lambda_1 > 0$ such that

$$\forall x \in \overline{B\left(x_i', r_i' \right) \cap \Omega}, \quad \lambda_0 \leqslant \lambda(x) \leqslant \lambda_1 < +\infty. \tag{2.59}$$

Since $B(x_i', r_i') \cap \Omega \subset C_i \cap C_{m,i}$ and $g = g_M = 0$ a.e. in $C_i \cap C_{m,i}$, we obtain from Theorem 2.5(i)

$$\int_{B(x_i',r_i')\cap\Omega} \left(\mathcal{A}(x, \nabla u) - \mathcal{A}(x, \nabla u_m) \right) \cdot \nabla \zeta \, dx = 0 \quad \forall \zeta \in \mathcal{D}\left(B\left(x_i', r_i' \right) \right). \tag{2.60}$$

Note that for each $x \in B(x'_i, r'_i) \cap \Omega$, we have

$$\mathcal{A}(x, \nabla u) - \mathcal{A}(x, \nabla u_m) = \int_0^1 \frac{\mathrm{d}}{\mathrm{d}t} \mathcal{A}\big(x, \nabla w_t(x)\big)\, \mathrm{d}t. \tag{2.61}$$

Writing $w = u - u_m$, $A(x) = (A_{ij}(x))$, with $A_{ij}(x) = \int_0^1 \frac{\partial}{\partial \zeta_j} \mathcal{A}_i(x, \nabla w_t(x))\, \mathrm{d}t$, we deduce form (2.60) and (2.61) that

$$\int_{B(x'_i, r'_i) \cap \Omega} A(x)(\nabla w) \cdot \nabla \zeta \, \mathrm{d}x = 0 \quad \forall \zeta \in \mathcal{D}\big(B(x'_i, r'_i)\big). \tag{2.62}$$

If we denote $\frac{\partial}{\partial \zeta_j} \mathcal{A}_i(x, \nabla w_t(x))$ by $A_{ij}(t, x)$ and set $A(t, x) = (A_{ij}(t, x))$, then a simple calculation shows that

$$A_{ij}(t, x) = a_{ij}\big|a(x)\nabla w_t \cdot \nabla w_t\big|^{(q-2)/2}$$

$$+ (q-2)\left(\sum_{k=1}^2 a_{ik} w_{tx_k}\right)\left(\sum_{k=1}^2 a_{jk} w_{tx_k}\right)\big|a(x)\nabla w_t \cdot \nabla w_t\big|^{(q-4)/2}$$

$$= \big|a(x)\nabla w_t \cdot \nabla w_t\big|^{(q-2)/2}$$

$$\times \left(a_{ij} + (q-2)\frac{(\sum_{k=1}^2 a_{ik} w_{tx_k})(\sum_{k=1}^2 a_{jk} w_{tx_k})}{|a(x)\nabla w_t \cdot \nabla w_t|}\right).$$

This means that

$$A(t, x) = \big|a(x)\nabla w_t \cdot \nabla w_t\big|^{(q-2)/2}\left(a(x) + (q-2)\frac{(a(x)\nabla w_t) \otimes (a(x)\nabla w_t)}{|a(x)\nabla w_t \cdot \nabla w_t|}\right),$$

where for $h = (h_1, h_2), k = (k_1, k_2) \in \mathbb{R}^2$, $h \otimes k$ denotes the matrix $(h_i k_j)$. Moreover, since a is symmetric, we can write

$$A(t, x) = \big|a(x)\nabla w_t \cdot \nabla w_t\big|^{(q-2)/2}\sqrt{a(x)}$$

$$\times \left(I_2 + (q-2)\frac{(\sqrt{a(x)}\nabla w_t) \otimes (\sqrt{a(x)}\nabla w_t)}{|a(x)\nabla w_t \cdot \nabla w_t|}\right)\sqrt{a(x)},$$

where $\sqrt{a(x)}$ is the symmetric definite positive matrix satisfying $\sqrt{a(x)}\sqrt{a(x)} = a(x)$.

Let $y = (y_1, y_2) \in \mathbb{R}^2$. We have

$$A(t, x) \cdot y \cdot y = \big|a(x)\nabla w_t \cdot \nabla w_t\big|^{(q-2)/2} B(t, x) \cdot \big(\sqrt{a(x)}y\big) \cdot \big(\sqrt{a(x)}y\big) \tag{2.63}$$

with

$$B(t, x) = I_2 + (q-2)\frac{(\sqrt{a(x)}\nabla w_t) \otimes (\sqrt{a(x)}\nabla w_t)}{|a(x)\nabla w_t \cdot \nabla w_t|}.$$

We claim that

$$\min(1, q-1)|y|^2 \leqslant B(t,x)\cdot y\cdot y \leqslant \max(1, q-1)|y|^2. \tag{2.64}$$

Indeed if $\sqrt{a(x)} = (b_{ij})$, then (2.64) is an immediate consequence of

$$0 \leqslant \left(\sqrt{a(x)}\nabla w_t\right) \otimes \left(\sqrt{a(x)}\nabla w_t\right)\cdot y\cdot y = \left(y_1\sum_{k=1}^{2}b_{1k}w_{tx_k} + y_2\sum_{k=1}^{2}b_{2k}w_{tx_k}\right)^2$$

$$\leqslant |y|^2\left|\sqrt{a(x)}\nabla w_t\right|^2.$$

It follows from (2.63) and (2.64) that $\forall t \in [0,1]$, $\forall x \in \overline{B(x_i', r_i')} \cap \Omega$,

$$\min(1, q-1)\lambda\left|a(x)\nabla w_t \cdot \nabla w_t\right|^{(q-2)/2}|y|^2$$

$$\leqslant A(t,x)\cdot y\cdot y$$

$$\leqslant \max(1, q-1)M\left|a(x)\nabla w_t \cdot \nabla w_t\right|^{(q-2)/2}|y|^2$$

which leads for some positive constants C_1, C_2 depending only on q, λ and M, to

$$C_1\lambda(x)|y|^2 \leqslant A(x)\cdot y\cdot y \leqslant C_2\lambda(x)|y|^2 \quad \forall x \in \overline{B(x_i', r_i')} \cap \Omega.$$

Using (2.59) we obtain, for some other positive constants $\tilde{\lambda}_0$, $\tilde{\lambda}_1$ depending only on q, λ, M, λ_0 and λ_1,

$$\tilde{\lambda}_0|y|^2 \leqslant A(x)\cdot y\cdot y \leqslant \tilde{\lambda}_1|y|^2 \quad \forall x \in \overline{B(x_i', r_i')} \cap \Omega. \tag{2.65}$$

Finally, we extend w by 0 to $B(x_i', r_i') \setminus \Omega$. Since $w = 0$ on $B(x_i', r_i') \cap S_{3,i}$, the obtained function belongs to $W^{1,q}(B(x_i', r_i'))$. Moreover, we extend $A(x)$ by $\tilde{\lambda}_0 I_2$ to $B(x_i', r_i') \setminus \Omega$. Thanks to (2.65), the obtained matrix remains bounded and strictly elliptic in $B(x_i', r_i')$. Now thanks to (2.62), w satisfies $\operatorname{div}(A(x)\nabla w) = 0$ in $W^{-1,q'}(B(x_i', r_i'))$, $w \geqslant 0$, and $w = 0$ in $B(x_i', r_i') \setminus \Omega$. We conclude by the strong maximum principle that $w = 0$ in $B(x_i', r_i')$, which means that $u = u_m$ in $B(x_i', r_i') \cap \Omega$. $\qquad\square$

PROOF OF THEOREM 2.10. First note that since $u_m \leqslant u$ in Ω, we have $C_{m,i} \subset C_i$. Moreover, u and u_m are $\mathcal{A}$-harmonic in $C_{m,i}$ and by Lemma 2.4, $u = u_m$ in $B(x_i', r_i') \cap \Omega$. It follows from [4], Theorem 4.1, that $u = u_m$ in $C_{m,i}$.

As in the linear case, one can prove that $C_i = C_{m,i}$. Thus $u = u_m$ in $C_{m,i} = C_i$ $\forall i \in \{1, \ldots, N\}$ and $u = u_m$ in Ω. From Corollary 2.2, we deduce that $g = g_M$ in Ω. $\qquad\square$

3. The dam problem with leaky boundary condition

In this section, we assume that $\mathcal{B}$ is given by (1.10) and we study the problem:

$$
\begin{cases}
\text{find } (u, g) \in W^{1,q}(\Omega) \times L^\infty(\Omega) \text{ such that} \\
\text{(i) } u = \psi \text{ on } S_2, \\
\text{(ii) } u \geqslant x_2, 0 \leqslant g \leqslant 1, g(u - x_2) = 0 \text{ a.e. in } \Omega, \\
\text{(iii) } \int_\Omega \big(\mathcal{A}(x, \nabla u) - g\mathcal{A}(x, e)\big) \cdot \nabla\xi \, dx \leqslant \int_{S_3} \beta(x, \psi - u)\xi \, d\sigma(x) \\
\qquad \forall \xi \in W^{1,q}(\Omega) \text{ such that } \xi \geqslant 0 \text{ on } S_2.
\end{cases}
\tag{P_L}
$$

The main difference between the model we are considering here and the one we studied in Section 2, is the fact that the region below a reservoir is not necessarily completely saturated if the flux through the bottom is not strong enough, which makes it possible to have a free boundary there. Moreover, as we shall prove it, the function g is not a characteristic function of the dry part of the dam.

For $\beta : S_3 \times \mathbb{R} \to \mathbb{R}$, we assume that

$$\beta(0) = 0$$

$$\text{for a.e. } x \in S_3, \quad \beta(x, \cdot) \text{ is nondecreasing}$$

$$\forall s \in \mathbb{R}, \exists C_s > 0: \quad \text{for a.e. } x \in S_3, \quad \big|\beta(x, s)\big| \leqslant C_s.$$

3.1. *Properties of the solutions*

Throughout this subsection, we shall denote a solution of (P_L) by (u, g). First we give a regularity result.

PROPOSITION 3.1. *$u \in C_{\mathrm{loc}}^{0,\alpha}(\Omega \cup S_2)$ for some $\alpha \in (0, 1)$.*

PROOF. This is a consequence of (1.61), (1.63) and (P_L)(i) (see [29]). $\qquad\qquad\square$

From now on, we assume that the function $x \mapsto \beta(x, \varphi(x))$ extends to S_2 so that

$$\beta\big(x, \varphi(x)\big) = 0 \quad \text{a.e. } x \in S_2.$$

The following theorem will play the same role as Theorem 2.1 of the previous section.

THEOREM 3.1. *Let C_h be a connected component of $[u > x_2] \cap [x_2 > h]$ and $Z_h = \Omega \cap (\pi_{x_1}(C_h) \times (h, +\infty))$. Then we have for each nonnegative function $f \in W^{1,q}(Z_h)$ depending only on x_2,*

$$
\int_{Z_h} \big(\mathcal{A}(x, \nabla u) - g\mathcal{A}(x, e)\big) \cdot f e \, dx \leqslant \int_{Z_h} \gamma \mathcal{A}(x, e) \cdot f e \, dx
\tag{3.1}
$$

where $\gamma(x) = \beta(x, \varphi)/(k(x)v_2) = \beta((x_1, s_+(x_1)), \varphi(x_1, s_+(x_1)))/(k(x)v_2(x_1, s_+(x_1)))$.

PROOF. Let $(a_1, a_2) = \pi_{x_1}(C_h)$ and let for $\delta > 0$ small enough, $\alpha_\delta \in \mathcal{D}((a_1, a_2))$ be a function such that $0 \leqslant \alpha_\delta(x_1) \leqslant 1$ and $\alpha_\delta = 1$ in $(a_1 + \delta, a_2 - \delta)$. First we have

$$\int_{Z_h} \left(\mathcal{A}(x, \nabla u) - g\mathcal{A}(x, e) \right) \cdot f(x_2) e \, dx$$

$$= \int_{Z_h} \left(\mathcal{A}(x, \nabla u) - g\mathcal{A}(x, e) \right) \cdot \nabla \left(\alpha_\delta \int_h^{x_2} f(s) \, ds \right) dx$$

$$+ \int_{Z_h} \left(\mathcal{A}(x, \nabla u) - g\mathcal{A}(x, e) \right) \cdot \nabla \left((1 - \alpha_\delta) \int_h^{x_2} f(s) \, ds \right) dx. \tag{3.2}$$

Since $\chi(Z_h)\alpha_\delta \int_h^{x_2} f(s) \, ds$ is a test function for (P$_L$), we have with $S_3^{Z_h} = S_3 \cap \partial Z_h$

$$\int_{Z_h} \left(\mathcal{A}(x, \nabla u) - g\mathcal{A}(x, e) \right) \cdot \nabla \left(\alpha_\delta \int_h^{x_2} f(s) \, ds \right) dx$$

$$\leqslant \int_{S_3^{Z_h}} \beta(x, \psi - u) \cdot \left(\alpha_\delta \int_h^{s_+(x_1)} f(s) \, ds \right) d\sigma(x). \tag{3.3}$$

Set $\zeta_\delta = (1 - \alpha_\delta) \int_h^{x_2} f(s) \, ds$ and remark that for $\varepsilon > 0$, $\pm \chi(Z_h) \cdot ((u - x_2)/\varepsilon \wedge \zeta_\delta)$ are test functions for (P$_L$). So we have, by taking into account (P$_L$)(ii),

$$\int_{Z_h} \mathcal{A}(x, \nabla u) \cdot \nabla \left(\frac{u - x_2}{\varepsilon} \wedge \zeta_\delta \right) dx = \int_{S_3^{Z_h}} \beta(x, \psi - u) \cdot \left(\frac{u - x_2}{\varepsilon} \wedge \zeta_\delta \right) d\sigma(x)$$

which leads by the monotonicity of $\mathcal{A}$ to

$$\int_{Z_h \cap [u - x_2 \geqslant \varepsilon \zeta_\delta]} \left(\mathcal{A}(x, \nabla u) - \mathcal{A}(x, e) \right) \cdot \nabla \zeta_\delta \, dx$$

$$\leqslant \int_{S_3^{Z_h}} \beta(x, \psi - u) \cdot \left(\frac{u - x_2}{\varepsilon} \wedge \zeta_\delta \right) d\sigma(x)$$

$$- \int_{Z_h} \mathcal{A}(x, e) \cdot \nabla \left(\frac{u - x_2}{\varepsilon} \wedge \zeta_\delta \right) dx. \tag{3.4}$$

Note that

$$\int_{Z_h} \chi([u > x_2]) \mathcal{A}(x, e) \cdot \nabla \zeta_\delta \, dx$$

$$= \int_{Z_h} \chi([u > x_2]) \mathcal{A}(x, e) \cdot \nabla \left(\zeta_\delta - \frac{u - x_2}{\varepsilon} \right)^+ dx$$

$$+ \int_{Z_h} \mathcal{A}(x, e) \cdot \nabla \left(\frac{u - x_2}{\varepsilon} \wedge \zeta_\delta \right) dx. \tag{3.5}$$

518   *A. Lyaghfouri*

Since k is nondecreasing in x_2, we deduce by using the second mean value theorem, that for a.e. $x_1 \in (a_1, a_2)$, there exists $h^*(x_1) \in [h, \Phi(x_1)]$ such that

$$\int_{Z_h} \chi\left([u > x_2]\right) A(x, e) \cdot \nabla\left(\zeta_\delta - \frac{u - x_2}{\varepsilon}\right)^+ dx$$

$$= \int_{a_1}^{a_2} \left(\int_h^{\Phi(x_1)} k(x)\left(\zeta_\delta - \frac{u - x_2}{\varepsilon}\right)^+_{x_2} dx_2\right) dx_1$$

$$= \int_{a_1}^{a_2} k\left(x_1, \Phi(x_1)_-\right)\left(\int_{h^*(x_1)}^{\Phi(x_1)} \left(\zeta_\delta - \frac{u - x_2}{\varepsilon}\right)^+_{x_2} (x_1, x_2)\, dx_2\right) dx_1$$

$$\leqslant \int_{a_1}^{a_2} k\left(x_1, \Phi(x_1)_-\right) \zeta_\delta\left(x_1, \Phi(x_1)\right) dx_1. \tag{3.6}$$

Adding (3.4) and (3.5) and taking into account (3.6), we get

$$\int_{Z_h \cap [u - x_2 \geqslant \varepsilon \zeta_\delta]} \left(A(x, \nabla u) - A(x, e)\right) \cdot \nabla \zeta_\delta\, dx$$

$$+ \int_{Z_h} \chi\left([u > x_2]\right) A(x, e) \cdot \nabla \zeta_\delta\, dx$$

$$\leqslant \int_{S_3^{Z_h}} \beta(x, \psi - u) \cdot \left(\frac{u - x_2}{\varepsilon} \wedge \zeta_\delta\right) d\sigma(x)$$

$$+ \int_{a_1}^{a_2} k\left(x_1, \Phi(x_1)_-\right) \zeta_\delta\left(x_1, \Phi(x_1)\right) dx_1$$

which leads by letting $\varepsilon \to 0$ to

$$\int_{Z_h} \left(A(x, \nabla u) - \chi\left([u = x_2]\right) A(x, e)\right) \cdot \nabla \zeta_\delta\, dx$$

$$\leqslant \int_{S_3^{Z_h} \cap [u > x_2]} \beta(x, \psi - u) \zeta_\delta\, d\sigma(x)$$

$$+ \int_{a_1}^{a_2} k\left(x_1, \Phi(x_1)_-\right) \zeta_\delta\left(x_1, \Phi(x_1)\right) dx_1. \tag{3.7}$$

Using (3.2), (3.3) and (3.7), we obtain

$$\int_{Z_h} \left(A(x, \nabla u) - g A(x, e)\right) \cdot f(x_2) e\, dx$$

$$\leqslant \int_{S_3^{Z_h}} \beta(x, \psi - u)\left(\int_h^{s_+(x_1)} f(s)\, ds\right) d\sigma(x)$$

$$+ \int_{Z_h} \big(\chi\big([u = x_2]\big) - g\big)k(x)(1 - \alpha_\delta)f(x_2)$$

$$+ \int_{a_1}^{a_2} k\big(x_1, \Phi(x_1)_-\big)\zeta_\delta\big(x_1, \Phi(x_1)\big)\,\mathrm{d}x_1. \tag{3.8}$$

Letting δ go to 0 in (3.8), we get

$$\int_{Z_h} \big(\mathcal{A}(x, \nabla u) - g\mathcal{A}(x, e)\big) \cdot fe\,\mathrm{d}x$$

$$\leqslant \int_{S_3^{Z_h}} \beta(x, \psi - u)\left(\int_h^{s_+(x_1)} f(s)\,\mathrm{d}s\right)\mathrm{d}\sigma(x)$$

$$\leqslant \int_{S_3^{Z_h}} \beta(x, \varphi)\left(\int_h^{s_+(x_1)} f(s)\,\mathrm{d}s\right)\mathrm{d}\sigma(x)$$

$$= \int_{Z_h} \gamma\mathcal{A}(x, e) \cdot fe\,\mathrm{d}x$$

and the lemma is proved. $\qquad\square$

REMARK 3.1. If $Z_h = ((a_1, a_2) \times (h, +\infty)) \cap \Omega$ and for $i = 1, 2$, we have $u(a_i, x_2) = x_2$ $\forall x_2 \geqslant h$, then the inequality (3.1) holds for the domain Z_h also.

As a consequence of Theorem 3.1, we obtain a characterization of the unsaturated region.

THEOREM 3.2. *Let $x_0 = (x_{01}, x_{02})$ such that $B_r = B_r(x_0) \subset \Omega$. If $B_r \subset \Omega$ and $u = x_2$ in B_r, then we have*

$$g = 1 - \gamma \quad \text{a.e. in } D_r = \big\{(x_1, x_2) \in \Omega \mid |x_1 - x_{01}| < r \text{ and } x_{02} < x_2\big\} \cup B_r.$$

$$\tag{3.9}$$

PROOF. By Corollary 1.1, we have $u = x_2$ in D_r. Moreover, it is clear that it is enough to prove (3.9) in the following two cases: $\pi_{x_1}(B_r) \subset \pi_{x_1}(S_2)$ or $\pi_{x_1}(B_r) \subset \pi_{x_1}(S_3)$.

(i) $\pi_{x_1}(B_r) \subset \pi_{x_1}(S_2)$: Applying Theorem 3.1 with $f = 1$ and domains $Z_h \subset D_r$ of type as in Remark 3.1, we obtain

$$0 \leqslant \int_{Z_h} k(x)(1 - g)\,\mathrm{d}x = \int_{Z_h} \big(\mathcal{A}(x, \nabla u) - g\mathcal{A}(x, e)\big) \cdot e\,\mathrm{d}x \leqslant 0$$

from which we deduce that $g = 1$ a.e. in Z_h. Thus $g = 1$ a.e. in D_r.

(ii) $\pi_{x_1}(B_r) \subset \pi_{x_1}(S_3)$: Let $\xi \in W^{1,q}(D_r)$ such that $\xi = 0$ on $\partial D_r \cap \Omega$. Then $\pm\chi(D_r)\xi$ are test functions for $(\mathrm{P_L})$ and we have

$$\int_{D_r} k(x)(1 - g)\xi_{x_2}\,\mathrm{d}x = \int_{\partial D_r \cap S_3} \beta(x, \varphi)\xi\,\mathrm{d}\sigma(x). \tag{3.10}$$

Using (1.63) we get in the distributional sense $(k(x)(1-g))_{x_2} = 0$ in D_r, which leads, by applying Green's formula in (3.10), to $k(x)(1-g)v_2 = \beta(x,\varphi)$ a.e. on $\partial D_r \cap S_3$. Hence we get

$$g(x) = 1 - \frac{\beta(x,\varphi)}{k(x)v_2} = 1 - \gamma(x) \quad \text{a.e. in } D_r. \qquad \square$$

Now we have a nonoscillation result.

THEOREM 3.3. *Let $x_0 = (x_{01}, x_{02}) \in \Omega$ such that $B_r = B_r(x_0) \subset \Omega$. Then the following situations are impossible*

(i) $\begin{cases} u(x_1, x_2) = x_2 & \forall(x_1, x_2) \in B_r \cap [x_1 = x_{01}], \\ u(x_1, x_2) > x_2 & \forall(x_1, x_2) \in B_r, x_1 \neq x_{01}; \end{cases}$

(ii) $\begin{cases} u(x_1, x_2) > x_2 & \forall(x_1, x_2) \in B_r \cap [x_1 < x_{01}], \\ u(x_1, x_2) = x_2 & \forall(x_1, x_2) \in B_r \cap [x_1 \geqslant x_{01}]; \end{cases}$

(iii) $\begin{cases} u(x_1, x_2) = x_2 & \forall(x_1, x_2) \in B_r \cap [x_1 \leqslant x_{01}], \\ u(x_1, x_2) > x_2 & \forall(x_1, x_2) \in B_r \cap [x_1 > x_{01}]. \end{cases}$

PROOF. (i) By the assumption and (P)(ii), we have $g = 0$ a.e. in B_r. By (1.63), this leads to $\operatorname{div}(\mathcal{A}(x, \nabla u)) = 0 \leqslant \operatorname{div}(\mathcal{A}(x, \nabla x_2))$ in $\mathcal{D}'(B_r)$. Using the strong maximum principle (Lemma 1.4) applied for $u_1 = x_2$ and $u_2 = u$, we deduce that either $u > x_2$ in B_r or $u \equiv x_2$ in B_r. This contradicts the assumption.

(ii) Let $\xi \in \mathcal{D}(B_r)$, $\xi \geqslant 0$. Setting $B_r^+ = B_r \cap [x_1 > x_{01}]$ we have, by (3.9),

$$\int_{B_r} \mathcal{A}(x, \nabla u) \cdot \nabla \xi \, dx = \int_{B_r} g \mathcal{A}(x, e) \cdot \nabla \xi \, dx$$

$$= \int_{B_r} k(x) g \xi_{x_2} \, dx$$

$$= \int_{B_r^+} k(x) \left(1 - \frac{\beta((x_1, s_+(x_1)), \varphi(x_1, s_+(x_1)))}{k(x)v_2(x_1, s_+(x_1))} \right) \xi_{x_2} \, dx$$

$$= \int_{B_r^+} k \xi_{x_2} \, dx$$

$$= \int_{B_r} k \xi_{x_2} \, dx - \int_{B_r^-} k \xi_{x_2} \, dx$$

$$\geqslant \int_{B_r} k \xi_{x_2} \, dx.$$

It follows that $\operatorname{div}(\mathcal{A}(x, \nabla u)) \leqslant \operatorname{div}(\mathcal{A}(x, \nabla x_2))$ in $\mathcal{D}'(B_r)$. Applying Lemma 1.4, we get a contradiction with the assumption (ii).

(iii) Similar to (ii). $\qquad \square$

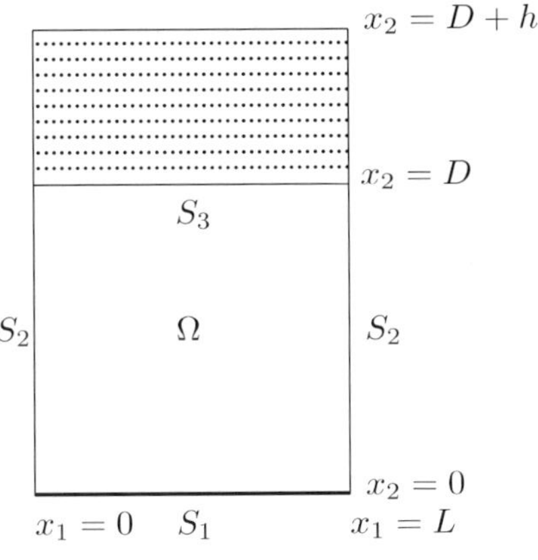

Fig. 7.

Assuming that the dam is a rectangular domain supplied by a unique reservoir located on its top (see Figure 7) and that $\mathcal{A}(x, \xi) = \xi$, Carrillo and Chipot showed in [19] that the dam is unsaturated above the line $x_2 = L/2\sqrt{\beta/(1 - \beta)}$ provided that $\beta = \beta(h) \leqslant 4D^2/(4D^2 + L^2)$.

In [32] the author showed that if $\beta(h) \geqslant 1$, then the dam is completely saturated. The following theorem is an extension of that simple result which gives a sufficient condition to get total saturation in a more general framework.

THEOREM 3.4. *We have*

$$0 \leqslant g \leqslant \mu(x) = 1 - \min(1, \gamma) \quad a.e. \ in \ \Omega. \tag{3.11}$$

We first prove a lemma.

LEMMA 3.1. *Let* $i \in \{1, \ldots, N\}$ *and* $Z = ((a_1, a_2) \times (h, +\infty)) \cap \Omega$ *with* $\pi_{x_1}(Z) \subset \pi_{x_1}(S_{3,i})$ *and* $(a_1, a_2) \times \{h\} \subset \Omega$ *(see Figure 8). Then we have*

$$\int_Z k(x)(g - \mu)^+ \xi_{x_2} \, dx \leqslant 0 \quad \forall \xi \in H^1(Z), \xi \geqslant 0 \ and \ \xi = 0 \ on \ \partial Z \cap \Omega. \tag{3.12}$$

PROOF. Let ξ as in the lemma. For $\varepsilon > 0$, $\pm(H_\varepsilon(u - x_2) - 1)\xi$ are test functions for (P_L) and then we have, with $S_3^Z = S_3 \cap \overline{Z}$,

$$\int_Z \big(\mathcal{A}(x, \nabla u) - g\mathcal{A}(x, e)\big) \cdot \nabla\big((H_\varepsilon(u - x_2) - 1)\xi\big) \, dx$$

$$= \int_{S_3^Z} \beta(x, \psi - u)\big(H_\varepsilon(u - x_2) - 1\big)\xi \, d\sigma(x)$$

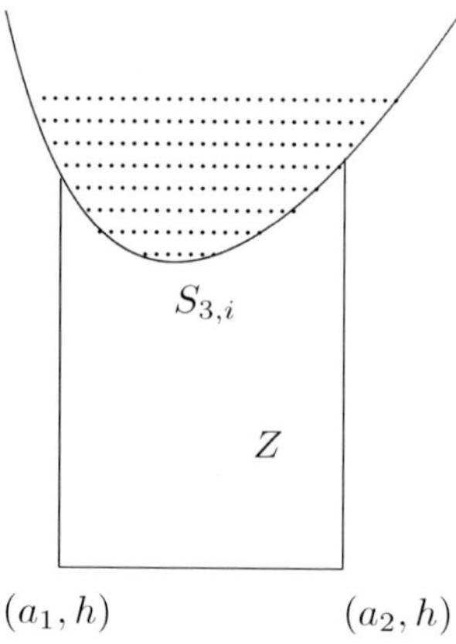

Fig. 8.

from which we deduce that

$$
\int_Z gk\xi_{x_2}\,dx = \int_Z g\mathcal{A}(x, e)\cdot\nabla\xi\,dx
$$

$$
= \int_Z \big(1 - H_\varepsilon(u - x_2)\big)\mathcal{A}(x, \nabla u)\cdot\nabla\xi\,dx
$$

$$
- \int_Z H_\varepsilon'(u - x_2)\xi\,\mathcal{A}(x, \nabla u)\cdot(\nabla u - \nabla x_2)\,dx
$$

$$
+ \int_Z g\mathcal{A}(x, e)\cdot\nabla\big(H_\varepsilon(u - x_2)\xi\big)
$$

$$
+ \int_{S_3^Z} \beta(x, \psi - u)\big(H_\varepsilon(u - x_2) - 1\big)\xi\,d\sigma(x)
$$

$$
\leqslant \int_Z \big(1 - H_\varepsilon(u - x_2)\big)\mathcal{A}(x, \nabla u)\cdot\nabla\xi\,dx
$$

$$
- \int_Z H_\varepsilon'(u - x_2)\xi\,\mathcal{A}(x, e)\cdot\nabla(u - x_2)\,dx
$$

$$
+ \int_{S_3^Z} \beta(x, \psi - u)\big(H_\varepsilon(u - x_2) - 1\big)\xi\,d\sigma(x)
$$

$$
= \int_Z \big(1 - H_\varepsilon(u - x_2)\big)\big(\mathcal{A}(x, \nabla u) - \mathcal{A}(x, e)\big)\cdot\nabla\xi\,dx
$$

$$
+ \int_Z \big(H_\varepsilon(u - x_2) - 1\big)\xi k_{x_2}\,dx - \int_{S_3^Z} k\nu_2\big(H_\varepsilon(u - x_2) - 1\big)\xi
$$

$$
+ \int_{S_3^Z} \beta(x, \psi - u)\big(H_\varepsilon(u - x_2) - 1\big)\xi\,d\sigma(x). \tag{3.13}
$$

Now set $\alpha(x) = \min(1, \gamma(x))$. Then $\mu(x) = 1 - \alpha(x)$ and since $(\alpha k)_{x_2} \geqslant 0$, we have

$$
\int_Z -\mu k \xi_{x_2} \, dx
$$

$$
= \int_Z k_{x_2} \xi \, dx - \int_{S_3^Z} k v_2 \xi \, d\sigma(x) + \int_{S_3^Z} \alpha k v_2 \xi \, d\sigma(x) - \int_Z \xi (\alpha k)_{x_2} \, dx
$$

$$
\leqslant \int_Z k_{x_2} \xi \, dx + \int_{S_3^Z} (\alpha(x) - 1) k v_2 \xi \, d\sigma(x). \tag{3.14}
$$

Adding (3.13) and (3.14) we get

$$
\int_Z (g - \mu) k \xi_{x_2} \, dx
$$

$$
\leqslant \int_Z \big(1 - H_\varepsilon(u - x_2)\big)\big(\mathcal{A}(x, \nabla u) - \mathcal{A}(x, e)\big) \cdot \nabla \xi \, dx
$$

$$
+ \int_Z H_\varepsilon(u - x_2) k_{x_2} \xi \, dx
$$

$$
+ \int_{S_3^Z} \big[\big(1 - H_\varepsilon(u - x_2)\big)\big(k v_2 - \beta(x, \psi - u)\big) + (\alpha - 1) k v_2\big] \xi \, d\sigma(x).
$$

$$
\tag{3.15}
$$

The last integral in (3.15) can be written as

$$
\int_{S_3^Z} \big(1 - H_\varepsilon(u - x_2)\big)\big(\beta(x, \varphi) - \beta(x, \psi - u)\big) \xi \, d\sigma(x)
$$

$$
+ \int_{S_3^Z} \left[\big(1 - H_\varepsilon(u - x_2)\big)\left(1 - \frac{\beta(x, \varphi)}{k v_2}\right) + (\alpha - 1)\right] k v_2 \xi \, d\sigma(x)
$$

$$
\leqslant \int_{S_3^Z} \big(1 - H_\varepsilon(u - x_2)\big)\big(\beta(x, \varphi) - \beta(x, \psi - u)\big) \xi \, d\sigma(x)
$$

$$
= J_1^\varepsilon. \tag{3.16}
$$

We get from (3.15) and (3.16),

$$
\int_Z (g - \mu) k \xi_{x_2} \, dx \leqslant \int_Z \big(1 - H_\varepsilon(u - x_2)\big)\big(\mathcal{A}(x, \nabla u) - \mathcal{A}(x, e)\big) \cdot \nabla \xi \, dx
$$

$$
+ \int_Z H_\varepsilon(u - x_2) k_{x_2} \xi \, dx + J_1^\varepsilon.
$$

524 A. Lyaghfouri

Moreover, we have

$$\lim_{\varepsilon \to 0} J_1^\varepsilon = \int_{S_3^Z} \chi[u = x_2]\big(\beta(x, \varphi) - \beta(x, \psi - x_2)\big)\xi \, d\sigma(x) = 0. \tag{3.17}$$

Therefore we get, by letting $\varepsilon \to 0$ and taking into account (3.17),

$$\int_Z (g - \mu)k\xi_{x_2} \, dx \leqslant \int_Z \chi\big([u > x_2]\big)k_{x_2}\xi \, dx$$

$$\forall \xi \in H^1(Z), \xi = 0 \text{ on } \partial Z \cap \Omega \text{ and } \xi \geqslant 0. \tag{3.18}$$

In what follows, we extend the functions k, $(g - \mu)$ and $\chi([u > x_2])k_{x_2}$ by 0 and we denote the extensions respectively by k, $(g - \mu)$ and θ. Note that (3.18) holds in particular for functions with compact support in $Z \cup S_3^Z$. Let then $\xi \in C^{0,1}(\mathbb{R}^2)$, $\xi \geqslant 0$ with $\xi/\overline{Z}$ having a compact support in $Z \cup S_3^Z$. Set $\varepsilon_0 = d(\text{supp}(\xi/\overline{Z}), \partial Z \cap \Omega)$ and let $\varepsilon \in (0, \varepsilon_0/2)$. For each $y \in B_\varepsilon(0)$, the function $x \to \xi(x + y)$ is nonnegative, belongs to $H^1(Z)$ and has a compact support in $Z \cup S_3^Z$. Therefore it can be used in (3.18) to get

$$\int_{\mathbb{R}^2} \big(g(x) - \mu(x)\big)k(x)\xi_{x_2}(x + y) \, dx \leqslant \int_{\mathbb{R}^2} \theta(x)\xi(x + y) \, dx$$

from which we deduce that

$$\int_{\mathbb{R}^2} \rho_\varepsilon \bigg(\int_{\mathbb{R}^2} (\bar{g} - \bar{\mu})\xi_{x_2}(x + y) \, dx\bigg) dy \leqslant \int_{\mathbb{R}^2} \rho_\varepsilon(y) \bigg(\int_{\mathbb{R}^2} \theta(x)\xi(x + y) \, dx\bigg) dy,$$

where $\bar{g} = gk$, $\bar{\mu} = \mu k$ and ρ_ε is a smooth function satisfying $\rho_\varepsilon \geqslant 0$, $\text{supp}\,\rho_\varepsilon \subset B_\varepsilon(0)$ and $\int_{\mathbb{R}^2} \rho_\varepsilon = 1$. Writing $f_\varepsilon = \rho_\varepsilon * f$ for a function f, we get

$$\int_{\mathbb{R}^2} (\bar{g}_\varepsilon - \bar{\mu}_\varepsilon)\xi_{x_2} \, dx \leqslant \int_{\mathbb{R}^2} \theta_\varepsilon \xi \, dx \quad \forall \xi \in C^{0,1}(\mathbb{R}^2), \xi \geqslant 0, \text{supp}\,\frac{\xi}{\overline{Z}} \subset Z \cup S_3^Z.$$

In particular, we obtain, for the function $\xi = \min(1, \frac{(\bar{g}_\varepsilon - \bar{\mu}_\varepsilon)^+}{\delta})\zeta$, with $\delta > 0$, $\zeta \in H^1(\mathbb{R}^2)$, $\zeta \geqslant 0$, $\text{supp}(\zeta/\overline{Z}) \subset Z \cup S_3^Z$,

$$\int_{\mathbb{R}^2} (\bar{g}_\varepsilon - \bar{\mu}_\varepsilon)\bigg(\min\bigg(1, \frac{(\bar{g}_\varepsilon - \bar{\mu}_\varepsilon)^+}{\delta}\bigg)\zeta\bigg)_{x_2} dx$$

$$\leqslant \int_{\mathbb{R}^2} \theta_\varepsilon \min\bigg(1, \frac{(\bar{g}_\varepsilon - \bar{\mu}_\varepsilon)^+}{\delta}\bigg)\zeta \, dx.$$

As $\delta \to 0$, we get

$$* \quad \int_{\mathbb{R}^2} \min\bigg(1, \frac{(\bar{g}_\varepsilon - \bar{\mu}_\varepsilon)^+}{\delta}\bigg)(\bar{g}_\varepsilon - \bar{\mu}_\varepsilon)\zeta_{x_2} \to \int_{\mathbb{R}^2} (\bar{g}_\varepsilon - \bar{\mu}_\varepsilon)^+ \zeta_{x_2},$$

$$* \quad \int_{\mathbb{R}^2} (\bar{g}_\varepsilon - \bar{\mu}_\varepsilon)\zeta\left(\min\left(1, \frac{(\bar{g}_\varepsilon - \bar{\mu}_\varepsilon)^+}{\delta}\right)\right)_{x_2}$$

$$= \int_{\mathbb{R}^2 \cap [0 < \bar{g}_\varepsilon - \bar{\mu}_\varepsilon < \delta]} \frac{\zeta}{\delta}(\bar{g}_\varepsilon - \bar{\mu}_\varepsilon)(\bar{g}_\varepsilon - \bar{\mu}_\varepsilon)_{x_2}$$

$$= \int_{\mathbb{R}^2} \frac{\zeta}{2\delta}\left(\left(\min\left(\delta, (\bar{g}_\varepsilon - \bar{\mu}_\varepsilon)^+\right)\right)^2\right)_{x_2}$$

$$= -\frac{1}{2\delta} \int_{\mathbb{R}^2} \left(\min\left(\delta, (\bar{g}_\varepsilon - \bar{\mu}_\varepsilon)^+\right)\right)^2 \zeta_{x_2} \to 0,$$

$$* \quad \int_{\mathbb{R}^2} \theta_\varepsilon \min\left(1, \frac{(\bar{g}_\varepsilon - \bar{\mu}_\varepsilon)^+}{\delta}\right)\zeta \leqslant \int_{\mathbb{R}^2} \theta_\varepsilon \zeta.$$

It follows that

$$\int_{\mathbb{R}^2} (\bar{g}_\varepsilon - \bar{\mu}_\varepsilon)^+ \zeta_{x_2} \leqslant \int_{\mathbb{R}^2} \theta_\varepsilon \zeta$$

which leads by letting $\varepsilon \to 0$, to

$$\int_Z (g - \mu)^+ k \zeta_{x_2} \leqslant \int_Z \chi\left([u > x_2]\right)\zeta k_{x_2}.$$

Now it is not difficult to extend this inequality to functions as in the lemma. Hence if we write it for $\zeta = (1 - H_\varepsilon(u - x_2))\xi$, with $\xi \in H^1(Z)$, $\xi = 0$ on $\partial Z \cap \Omega$, $\xi \geqslant 0$, we obtain since $g(u - x_2) = 0$,

$$\int_Z (g - \mu)^+ k \xi_{x_2} \leqslant \int_Z \chi\left([u > x_2]\right)\left(1 - H_\varepsilon(u - x_2)\right)k_{x_2}\xi.$$

Letting $\varepsilon \to 0$, the lemma follows. $\qquad \square$

PROOF OF THEOREM 3.4. We give the proof only below S_3, since below S_2, one has $\mu = 1 \geqslant g$ a.e. So let $i \in \{1, \ldots, N\}$ and let Z be a domain below $S_{3,i}$ such that $Z = ((a_1, a_2) \times (h, +\infty)) \cap \Omega$ and such that $(a_1, a_2) \times \{h\} \subset \Omega$. Using Lemma 3.1 for $\xi = \kappa(x_1) \cdot (x_2 - h)$ with $\kappa(x_1) = (x_1 - a_1)(a_2 - x_1)$, we obtain

$$\int_Z k(x)(g - \mu)^+ \kappa(x_1)\,dx = 0$$

which leads to $(g - \mu)^+ = 0$ a.e. in Z and $g \leqslant \mu$ a.e. in Z. Hence the theorem is proved. $\qquad \square$

COROLLARY 3.1. *Let T be a domain contained in $S_3 \setminus S_+$. If $\beta(x, \varphi) \geqslant k(x)v_2$ a.e. on T, then the subset Z_T of Ω located below T is totally saturated.*

PROOF. We deduce from Theorem 3.4 and the assumption, that $g = 0$ a.e. in Z_T. From (1.63) we then get $\mathrm{div}(\mathcal{A}(x, \nabla u)) = 0$ in $\mathcal{D}'(Z_T)$. This leads by Lemma 1.4 to $u = x_2$ in Z_T or $u > x_2$ in Z_T. Suppose that $u = x_2$ in Z_T and let $\xi \in W^{1,q}(Z_T)$ with $\xi = 0$ on $\partial Z_T \cap \Omega$. Using $\pm\xi$ as test functions for (P$_L$) and the fact that $u = x_2$ and $g = 0$ in Z_T, we obtain

$$\int_{Z_T} k\xi_{x_2} = \int_T \beta(x, \varphi)\xi \, d\sigma(x)$$

which leads to

$$\int_{T \cup (\partial Z_T \cap S_1)} k\xi v_2 \, d\sigma(x) = \int_T \beta(x, \varphi)\xi \, d\sigma(x).$$

It follows that $\beta(x, \varphi) = 0$ on T and we get a contradiction with $\beta(x, \varphi) \geqslant k(x)v_2 > 0$. Hence $u > x_2$ in Z_T. $\square$

3.2. Continuity of the free boundary

From now on, we assume that $\mathcal{A}$ is strictly monotone in the sense of (2.10). We first give a theorem dealing with the continuity of the free boundary below S_2.

THEOREM 3.5. Φ is continuous at each point $x_{01} \in \mathrm{Int}(\pi_{x_1}(S_2))$ such that $(x_{01}, \Phi(x_{01})) \in \Omega$ (resp. $x_{01} \notin S_-$ and $\Phi(x_{01}) = s_-(x_{01})$).

PROOF. Since for test functions vanishing on S_3, the problem (P$_L$) behaves exactly as the problem (P$_D$), the proof is exactly the same as the one of Theorem 2.4 by taking into account Remark 2.4. $\square$

In the following theorem, we prove the continuity of the free boundary below S_3. For this purpose, we need the following assumptions which will be assumed in the sequel:

$$\mathcal{A}(x, te) = t^{q-1} \mathcal{A}(x, e) \text{ for all } x \text{ below } S_3, \forall t > 0; \tag{3.19}$$

$$\gamma \text{ is continuous at each } x \text{ below } S_3 \text{ such that } \gamma(x) < 1; \tag{3.20}$$

$$\exists \mu_0 > 0 \text{ such that } k_{x_2}(x) \leqslant \mu_0 \text{ for all } x \text{ below } S_3. \tag{3.21}$$

Then we have the following theorem.

THEOREM 3.6. Φ is continuous at each $x_{01} \in \mathrm{Int}(\pi_{x_1}(S_3))$ such that $x_0 = (x_{01}, \Phi(x_{01})) \in \Omega$ and $\gamma(x_0) \in [0, 1)$.

PROOF. Without loss of generality, we may assume that $\mu_0 \geqslant 1$. Since $\gamma(x_0) < 1$, γ is continuous at x_0 and k is nondecreasing with respect to x_2, there exists $\gamma_+ \in (0, 1)$ and a ball $B(x_0, \varepsilon_0) \subset \Omega$ such that

$$\gamma(x) \leqslant \gamma_+ < 1 \quad \forall x \in Z = \left(\pi_{x_1}(B(x_0, \varepsilon_0)) \times (\Phi(x_{01}), +\infty)\right) \cap \Omega.$$

Since $\Phi(x_{01}) < s_+(x_{01})$ and $\gamma_+ < 1$, there exists $\delta > 0$ small enough such that

$$\Phi(x_{01}) < s_+(x_{01}) - \frac{\delta}{2} \quad \text{and} \quad \gamma(x) \leqslant \gamma_+ < \exp\left(-\frac{\delta\mu_0}{\lambda}\right) \quad \forall x \in Z. \qquad (3.22)$$

Because $0 \leqslant \gamma(x_0) < 1$, we have necessarily s_+ continuous at x_{01}. So there exists $\varepsilon'_0 \in (0, \varepsilon_0)$ small enough such that

$$s_+(x_{01}) - \frac{\delta}{2} < s_+(x_1) < s_+(x_{01}) + \frac{\delta}{2} \quad \forall x_1 \in \left[x_{01} - \varepsilon'_0, x_{01} + \varepsilon'_0\right].$$

Setting $h_0 = s_+(x_{01}) - \delta/2$, we obtain

$$\Phi(x_{01}) < h_0 < s_+(x_1) < h_0 + \delta \quad \forall x_1 \in \left[x_{01} - \varepsilon'_0, x_{01} + \varepsilon'_0\right]. \qquad (3.23)$$

Let now $\varepsilon > 0$ small enough such that

$$\varepsilon < \varepsilon'_0 \quad \text{and} \quad h_0 + 3\varepsilon < s_+(x_1) \quad \forall x_1 \in \left[x_{01} - \varepsilon'_0, x_{01} + \varepsilon'_0\right]. \qquad (3.24)$$

Since $u(x_{01}, h_0) = h_0$ and u is continuous at (x_{01}, h_0), there exists $\varepsilon''_0 \in (0, \varepsilon)$ such that

$$u(x) \leqslant x_2 + \int_0^\varepsilon \left(1 - \exp(-sv_0)\right) ds \quad \forall x \in B_{\varepsilon''_0}(x_{01}, h_0), \qquad (3.25)$$

where $v_0 = \mu_0/(\lambda(q-1))$.

Using Theorem 3.3, we are in one of the following situations:

$$\begin{cases} \text{(i) } \exists \underline{x}_0 = (\underline{x}_{01}, \underline{x}_{02}) \in B_{\varepsilon''_0}(x_{01}, h_0) \\ \qquad \text{such that} \quad \underline{x}_{01} < x_{01} \quad \text{and} \quad u(\underline{x}_0) = \underline{x}_{02}, \\ \text{(ii) } \exists \bar{x}_0 = (\bar{x}_{01}, \bar{x}_{02}) \in B_{\varepsilon''_0}(x_{01}, h_0) \\ \qquad \text{such that} \quad \bar{x}_{01} > x_{01} \quad \text{and} \quad u(\bar{x}_0) = \bar{x}_{02}. \end{cases}$$

Assume for example that (i) holds and set $h'_0 = \max(\underline{x}_{02}, h_0)$, $Z_0 = ((\underline{x}_{01}, x_{01}) \times (h'_0, +\infty)) \cap \Omega$, $g_0(t) = 1 - \exp(-v_0(t - h'_0))$, and

$$w_0(x) = \begin{cases} x_2 + \int_{x_2}^{h'_0 + \varepsilon} g_0(t)\, dt & \text{if } h'_0 \leqslant x_2 \leqslant h'_0 + \varepsilon, \\ x_2 & \text{if } x_2 > h'_0 + \varepsilon. \end{cases}$$

Note that

$$w_0\left(x_1, h'_0\right) = h'_0 + \int_{h'_0}^{h'_0 + \varepsilon} \left(1 - \exp\left(-v_0(t - h'_0)\right)\right) dt$$

$$= h'_0 + \int_0^\varepsilon \left(1 - \exp(-sv_0)\right) ds.$$

This leads by (3.25) to

$$u(x_1, h_0') \leqslant w_0(x_1, h_0') \quad \forall x_1 \in [\underline{x}_{01}, x_{01}]. \tag{3.26}$$

Given that $u(x_0) = x_{02}$ and $u(\underline{x}_0) = \underline{x}_{02}$, we obtain by Corollary 1.1 that

$$u(x_{01}, x_2) = u(\underline{x}_{01}, x_2) = x_2 \quad \forall x_2 \geqslant h_0'. \tag{3.27}$$

Using (3.26) and (3.27) we obtain $(u - w_0)^+ = 0$ on $\partial Z_0 \cap \Omega$. It follows that $\pm \chi(Z_0)(u - w_0)^+$ are test functions for $(\mathrm{P_L})$ and we have

$$\int_{Z_0} \big(\mathcal{A}(x, \nabla u) - g\mathcal{A}(x, e)\big) \cdot \nabla(u - w_0)^+ \, dx$$
$$= \int_{S_3^{Z_0}} \beta(x, \psi - u) \cdot (u - w_0)^+ \, d\sigma(x). \tag{3.28}$$

Moreover, one has for $Z_0^+ = ((\underline{x}_{01}, x_{01}) \times (h_0', h_0' + \varepsilon)) \cap \Omega$,

$$\int_{Z_0} \big(\mathcal{A}(x, \nabla w_0) - \chi\big([w_0 = x_2]\big)\mathcal{A}(x, e)\big) \cdot \nabla(u - w_0)^+ \, dx$$
$$= \int_{Z_0^+} \big(1 - g_0(x_2)\big)^{q-1} k(x) \cdot (u - w_0)_{x_2}^+ \, dx. \tag{3.29}$$

Remarking that

$$\int_{Z_0} \big(1 - g_0(x_2)\big)^{q-1} k(x) \cdot (u - w_0)_{x_2}^+ \, dx$$
$$= -\int_{Z_0} \big((1 - g_0(x_2))^{q-1} k(x)\big)_{x_2} \cdot (u - w_0)^+ \, dx$$
$$+ \int_{S_3^{Z_0}} \big(1 - g_0(x_2)\big)^{q-1} k(x) \cdot (u - w_0)^+ v_2 \, d\sigma(x)$$

we obtain, for $Z_0^0 = ((\underline{x}_{01}, x_{01}) \times (h_0' + \varepsilon, +\infty)) \cap \Omega$,

$$\int_{Z_0^+} \big(1 - g_0(x_2)\big)^{q-1} k(x) \cdot (u - w_0)_{x_2}^+ \, dx$$
$$= -\int_{Z_0^0} \big(1 - g_0(x_2)\big)^{q-1} k(x) \cdot (u - x_2)_{x_2}^+ \, dx$$
$$- \int_{Z_0} \big((1 - g_0(x_2))^{q-1} k(x)\big)_{x_2} \cdot (u - w_0)^+ \, dx$$
$$+ \int_{S_3^{Z_0}} \big(1 - g_0(x_2)\big)^{q-1} k(x) \cdot (u - w_0)^+ v_2 \, d\sigma(x). \tag{3.30}$$

Subtracting (3.29) from (3.28) and taking into account (3.30), we get

$$\int_{Z_0^+} \big(\mathcal{A}(x,\nabla u) - \mathcal{A}(x,\nabla w_0)\big)\cdot \nabla(u-w_0)^+\,dx$$

$$+ \int_{Z_0^0}\big(\mathcal{A}(x,\nabla u) - g\mathcal{A}(x,e)\big)\cdot\nabla(u-x_2)\,dx$$

$$= \int_{Z_0^0}\big(1-g_0(x_2)\big)^{q-1}k(x)\cdot(u-x_2)_{x_2}\,dx$$

$$+ \int_{Z_0}\big((1-g_0(x_2))^{q-1}k(x)\big)_{x_2}\cdot(u-w_0)^+\,dx$$

$$+ \int_{S_3^{Z_0}}\big(\beta(x,\psi-u) - (1-g_0(x_2))^{q-1}k(x)v_2\big)\cdot(u-w_0)^+\,d\sigma(x).$$

$$(3.31)$$

Note that by (3.19),

$$\int_{Z_0^0}\big(\mathcal{A}(x,\nabla u) - g\mathcal{A}(x,e)\big)\cdot\nabla(u-x_2)\,dx$$

$$= \int_{Z_0^0}\big(\mathcal{A}(x,\nabla u) - g\mathcal{A}(x,e)\big)\cdot\big(\nabla u - (1-g_0)e\big)\,dx$$

$$- \int_{Z_0^0}\big(\mathcal{A}(x,\nabla u) - g\mathcal{A}(x,e)\big)\cdot g_0 e\,dx$$

$$= \int_{Z_0^0}\big(\mathcal{A}(x,\nabla u) - \mathcal{A}(x,(1-g_0)e)\big)\cdot\big(\nabla u - (1-g_0)e\big)\,dx$$

$$+ \int_{Z_0^0}\big((1-g_0)^{q-1} - g\big)k(x)\big((u-x_2)_{x_2} + g_0\big)\,dx$$

$$- \int_{Z_0^0}\big(\mathcal{A}(x,\nabla u) - g\mathcal{A}(x,e)\big)\cdot g_0 e\,dx$$

$$= \int_{Z_0^0}\big(\mathcal{A}(x,\nabla u) - \mathcal{A}(x,(1-g_0)e)\big)\cdot\big(\nabla u - (1-g_0)e\big)\,dx$$

$$+ \int_{Z_0^0}(1-g_0)^{q-1}k(x)(u-x_2)_{x_2}\,dx + \int_{Z_0^0}\big((1-g_0)^{q-1} - g\big)k(x)g_0\,dx$$

$$- \int_{Z_0^0}\big(\mathcal{A}(x,\nabla u) - g\mathcal{A}(x,e)\big)\cdot g_0 e\,dx.$$

$$(3.32)$$

530   *A. Lyaghfouri*

Using (3.31) and (3.32) we obtain

$$
\int_{Z_0^+} \big(\mathcal{A}(x, \nabla u) - \mathcal{A}(x, \nabla w_0)\big) \cdot \nabla (u - w_0)^+ \, dx
$$

$$
+ \int_{Z_0^0} \big(\mathcal{A}(x, \nabla u) - \mathcal{A}(x, (1 - g_0)e)\big) \cdot \big(\nabla u - (1 - g_0)e\big) \, dx
$$

$$
+ \int_{Z_0^0} \big((1 - g_0)^{q-1} - g\big) k(x) g_0 \, dx
$$

$$
\leqslant \int_{Z_0^0} \big(\mathcal{A}(x, \nabla u) - g\mathcal{A}(x, e)\big) \cdot g_0 e \, dx
$$

$$
+ \int_{Z_0} \big((1 - g_0(x_2))^{q-1} k(x)\big)_{x_2} \cdot (u - w_0)^+ \, dx
$$

$$
+ \int_{S_3^{Z_0}} k(x) v_2 \big(\gamma - (1 - g_0(x_2))^{q-1}\big) \cdot (u - x_2) \, d\sigma(x). \tag{3.33}
$$

By (3.1), we have, for $f = g_0$ and $Z_h = Z_0^0$,

$$
\int_{Z_0^0} \big(\mathcal{A}(x, \nabla u) - g\mathcal{A}(x, e)\big) \cdot g_0 e \, dx \leqslant \int_{Z_0^0} \gamma k g_0 \, dx. \tag{3.34}
$$

Moreover, by (3.23) and since $h_0 \leqslant h_0'$, one has for all $x_1 \in [x_{01} - \varepsilon_0'', x_{01} + \varepsilon_0'']$, $s_+(x_1) - h_0' \leqslant \delta$ which leads by (3.22) to

$$
\gamma \leqslant \gamma_+ < \exp\left(-\frac{\mu_0}{\lambda}\big(s_+(x_1) - h_0'\big)\right) = \exp\big(-\nu_0(q - 1)\big(s_+(x_1) - h_0'\big)\big)
$$

$$
= \big(1 - g_0(s_+(x_1))\big)^{q-1},
$$

so that

$$
\int_{S_3^{Z_0}} k(x) v_2 \big(\gamma - (1 - g_0(x_2))^{q-1}\big) \cdot (u - x_2) \, d\sigma(x) \leqslant 0. \tag{3.35}
$$

Now since $\lambda k_{x_2} \leqslant k \mu_0$ a.e. in Z_0, we have

$$
\int_{Z_0} \big((1 - g_0(x_2))^{q-1} k(x)\big)_{x_2} \cdot (u - w_0)^+ \, dx
$$

$$
= \int_{Z_0} \big((1 - g_0)^{q-1} k_{x_2} + \big((1 - g_0)^{q-1}\big)_{x_2} k\big) \cdot (u - w_0)^+ \, dx
$$

$$
= \int_{Z_0} (1 - g_0)^{q-1} \left(k_{x_2} - \frac{\mu_0 k}{\lambda}\right) \cdot (u - w_0)^+ \, dx \leqslant 0. \tag{3.36}
$$

Then (3.33) becomes by (3.34)–(3.36),

$$\int_{Z_0^+} \big(\mathcal{A}(x, \nabla u) - \mathcal{A}(x, \nabla w_0)\big) \cdot \nabla (u - w_0)^+ \, dx$$

$$+ \int_{Z_0^0 \cap [u > x_2]} \big((1 - g_0)^{q-1} - \gamma\big) k g_0 \, dx + \int_{Z_0^0 \cap [u = x_2]} (1 - g - \gamma) k g_0 \, dx$$

$$+ \int_{Z_0^0 \cap [u > x_2]} \big(\mathcal{A}(x, \nabla u) - \mathcal{A}\big(x, (1 - g_0)e\big)\big) \cdot \big(\nabla u - (1 - g_0)e\big) \, dx$$

$$\leqslant 0. \tag{3.37}$$

By (3.22) and (3.23), we have

$$\gamma(x) \leqslant \gamma_+ < \exp\left(\frac{-\delta \mu_0}{\lambda}\right)$$

$$\leqslant \exp\big(-\nu_0(x_2 - h_0')\big) = \big(1 - g_0(x_2)\big)^{q-1} \quad \forall x \in Z_0^0.$$

Indeed for $x \in Z_0^0$, we have $h_0 \leqslant h_0' < x_2 < h_0 + \delta$, so that $x_2 - h_0' \leqslant \delta$. Since $\mu_0 \geqslant 0$, we obtain

$$\exp\left(-\frac{\delta \mu_0}{\lambda}\right) \leqslant \exp\left(-\frac{\mu_0}{\lambda}(x_2 - h_0')\right) = \big(1 - g_0(x_2)\big)^{q-1}.$$

So

$$\int_{Z_0^0 \cap [u > x_2]} \big((1 - g_0)^{q-1} - \gamma\big) k g_0 \, dx \geqslant 0. \tag{3.38}$$

Using (3.37), (3.38) and (3.11) we get

$$\int_{Z_0^+} \big(\mathcal{A}(x, \nabla u) - \mathcal{A}(x, \nabla w_0)\big) \cdot \nabla (u - w_0)^+ \, dx \leqslant 0.$$

This leads by (2.10) to $\nabla(u - w_0)^+ = 0$ a.e. in Z_0^+ and then by (3.26) to $u \leqslant w_0$ in Z_0^+ which gives in particular $u(x_1, h_0' + \varepsilon) = h_0' + \varepsilon \; \forall x_1 \in (\underline{x}_{01}, x_{01})$. By Corollary 1.1, we obtain $u(x) = x_2 \; \forall x \in Z_0^0$.

Using Theorem 3.3 once again we deduce that there exists $\bar{x}_0 = (\bar{x}_{01}, \bar{x}_{02}) \in \Omega$ such that $x_{01} < \bar{x}_{01} < x_{01} + \varepsilon_0'', h_0' + \varepsilon < \bar{x}_{02} < h_0' + 2\varepsilon$ and $u(\bar{x}_0) = \bar{x}_{02}$. Arguing as above we prove that

$$u(x) = x_2 \quad \forall x \in \big((x_{01}, \bar{x}_{01}) \times (h_0' + 2\varepsilon, +\infty)\big).$$

Hence,

$$u(x) = x_2 \quad \forall x \in \left((\underline{x}_{01}, \bar{x}_{01}) \times (h_0' + 2\varepsilon, +\infty) \right).$$

Now we consider a subdivision $(k_i)_{0 \leqslant i \leqslant N}$ of the interval $[\Phi(x_{01}) + 2\varepsilon, h_0' + 2\varepsilon]$ with $k_i - k_{i+1} = (h_0' - \Phi(x_{01}))/N < \varepsilon/2$. Repeating the above step we can prove successively that $u = x_2$ in $Z_1 \cap [x_2 \geqslant k_1]$, $Z_2 \cap [x_2 \geqslant k_2], \dots, Z_N \cap [x_2 \geqslant k_N]$, where Z_i are domains of type Z_0 such that $\pi_{x_1}(Z_{i+1}) \subset \pi_{x_1}(Z_i)$. Thus we obtain

$$\Phi(x_1) \leqslant \Phi(x_{01}) + 2\varepsilon \quad \forall x_1 \in (\underline{x}_{1N}, \bar{x}_{1N})$$

and the upper semicontinuity of Φ at x_{01} follows. Taking into account Proposition 1.4 the theorem is proved. $\qquad \square$

REMARK 3.2. Φ is continuous at each point $x_{01} \in \mathrm{Int}(\pi_{x_1}(S_3)) \setminus S_-$ such that $\Phi(x_{01}) = s_-(x_{01})$ and $\gamma(x_{01}, x_2) \in [0, 1)$ for all $x_2 \in (s_-(x_{01}), s_+(x_{01}))$. Indeed in this case, one has for each $\varepsilon > 0$ small enough, $x_0' = (x_{01}, \Phi(x_{01}) + \varepsilon) \in \Omega$, $u(x_0') = \Phi(x_{01}) + \varepsilon$ and γ is continuous at x_0'. Therefore one can adapt the proof of Theorem 3.6 to get $u = x_2$ in $((x_{01} - \varepsilon', x_{01} + \varepsilon') \times (\Phi(x_{01}) + 3\varepsilon + \infty)) \cap \Omega$ for some $\varepsilon' > 0$, which means the upper semicontinuity and thus the continuity of Φ at x_{01}.

As a consequence of the continuity of Φ, we obtain the expression of g which is not a characteristic function of the dry region.

COROLLARY 3.2. *Under the assumptions of Theorem 3.6, we have*

$$g = (1 - \gamma)^+ \cdot \chi([u = x_2]). \tag{3.39}$$

PROOF. First note that since $\gamma = 0$ below S_2, the formula (3.39) can be obtained below S_2 in the same way we obtained (2.17).

Next by (3.11), we have $g = 0$ a.e. in $\Omega \cap [\gamma \geqslant 1]$ and then

$$g = g\chi([\gamma < 1]). \tag{3.40}$$

By Proposition 1.4, we have $[u > x_2] = [x_2 < \Phi(x_1)]$. So

$$g = 0 \quad \text{a.e. in } [x_2 < \Phi(x_1)]. \tag{3.41}$$

Let $x_0 = (x_{01}, x_{02}) \in [x_2 > \Phi(x_1)]$ such that x_0 is located below S_3 and $\gamma(x_0) < 1$. By continuity of γ, there exists a ball $B_r(x_0)$ contained in $[\gamma < 1]$ such that $x_{02} - r > \Phi(x_{01})$. Since γ is nonincreasing with respect to x_2, we have $Z = ((x_{01} - \frac{r}{2}, x_{01} + \frac{r}{2}) \times (x_{02} - \frac{r}{2}, +\infty)) \cap \Omega \subset [\gamma < 1]$. Moreover, $u(x_{01}, x_2) = x_2 \ \forall x_2 \geqslant x_{02} - r/2$. Then one can adapt the proof of Theorem 3.6 to show that $u(x_1, x_2) = x_2 \ \forall (x_1, x_2) \in Z' = ((x_{01} - r', x_{01} + r') \times$

$(x_{02} - r', +\infty)) \cap \Omega$ for some $r' \in (0, r/2)$. By (3.9), we obtain $g = 1 - \gamma$ a.e. in Z'. We deduce that

$$g = 1 - \gamma \quad \text{a.e. in } [x_2 > \Phi(x_1)] \cap [\gamma < 1].$$

Now the set $[x_2 = \Phi(x_1)] \cap [\gamma < 1]$ being of measure zero (since Φ is continuous at any point x_1 such that $(x_1, \Phi(x_1)) \in \Omega$ and $\gamma(x_1, \Phi(x_1)) < 1$), we get, by (3.40) and (3.41),

$$g = (1 - \gamma)\chi\left([x_2 \geqslant \Phi(x_1)]\right) \cdot \chi\left([\gamma < 1]\right)$$

which is (3.39). $\qquad\square$

3.3. *Existence and uniqueness of minimal and maximal solutions*

In this subsection, we show the existence and uniqueness of two solutions which minimize (resp. maximize) a functional. Moreover, one is minimal and the other one is maximal in the usual sense among all solutions. From now on, we assume that

$$\beta(x, u) > 0 \quad \forall u > 0, \text{ a.e. } x \in S_3. \tag{3.42}$$

We first prove a result similar to Theorem 2.5.

THEOREM 3.7. *Let (u_1, g_1) and (u_2, g_2) be two solutions of $(\mathrm{P_L})$. Set $u_m = \min(u_1, u_2)$, $u_M = \max(u_1, u_2)$, $g_m = \min(g_1, g_2)$ and $g_M = \max(g_1, g_2)$. Then we have, for $i = 1, 2$ and for all $\zeta \in W^{1,q}(\Omega)$,*

$$\text{(i)} \int_\Omega \left(\left(A(x, \nabla u_i) - A(x, \nabla u_m)\right) - (g_i - g_M)A(x, e)\right) \cdot \nabla \zeta \, dx = 0,$$

$$\text{(ii)} \int_\Omega \left(\left(A(x, \nabla u_i) - A(x, \nabla u_M)\right) - (g_i - g_m)A(x, e)\right) \cdot \nabla \zeta \, dx = 0,$$

$$\text{(iii)} \ \beta(x, \psi - u_1) = \beta(x, \psi - u_2) \quad \text{a.e. } x \in S_3.$$

PROOF. (i) Let $\zeta \in C^1(\overline{\Omega})$, $\zeta \geqslant 0$. For $\delta, \varepsilon > 0$, we consider $\alpha_\delta(x) = (1 - d(x, A_m)/\delta)^+$ where $A_m = [u_m > x_2] \cup [\gamma \geqslant 1]$ and $\xi = \min(\alpha_\delta\zeta, \frac{u_i - u_m}{\varepsilon})$. We have

$$\int_\Omega \left(\left(A(x, \nabla u_i) - A(x, \nabla u_m)\right) - (g_i - g_M)A(x, e)\right) \cdot \nabla \zeta \, dx$$

$$= \int_\Omega \left(\left(A(x, \nabla u_i) - A(x, \nabla u_m)\right) - (g_i - g_M)A(x, e)\right) \cdot \nabla(\alpha_\delta\zeta) \, dx$$

$$+ \int_\Omega \left(\left(A(x, \nabla u_i) - A(x, \nabla u_m)\right) - (g_i - g_M)A(x, e)\right) \cdot \nabla\left((1 - \alpha_\delta)\zeta\right) \, dx. \tag{3.43}$$

Since $(1 - \alpha_\delta)\zeta$ is a test function for (P_L), we have

$$\int_\Omega \big(\mathcal{A}(x, \nabla u_i) - g_i \mathcal{A}(x, e)\big) \cdot \nabla\big((1 - \alpha_\delta)\zeta\big)\, dx$$

$$\leq \int_{S_3} \beta(x, \psi - u_i)\big((1 - \alpha_\delta)\zeta\big)\, d\sigma(x). \tag{3.44}$$

Since $(1 - \alpha_\delta)\zeta = 0$ on A_m, we have

$$\int_\Omega \big(\mathcal{A}(x, \nabla u_m) - g_M \mathcal{A}(x, e)\big) \cdot \nabla\big((1 - \alpha_\delta)\zeta\big)\, dx$$

$$= \int_{[u_m = x_2] \cap [\gamma < 1]} k(x)(1 - g_M)\big((1 - \alpha_\delta)\zeta\big)_{x_2}\, dx$$

$$= \int_{[u_m = x_2] \cap [\gamma < 1]} \left(\frac{\beta(x, \varphi)}{v_2}(1 - \alpha_\delta)\zeta\right)_{x_2}\, dx$$

$$= \int_\Omega \left(\frac{\beta(x, \varphi)}{v_2}(1 - \alpha_\delta)\zeta\right)_{x_2}\, dx$$

$$= \int_{\partial\Omega} \frac{\beta(x, \varphi)}{v_2}(1 - \alpha_\delta)\zeta n_2\, d\sigma(x)$$

$$= \int_{S_3} \beta(x, \varphi)(1 - \alpha_\delta)\zeta\, d\sigma(x) + \int_{S_1'} \frac{\beta(x, \varphi)}{v_2}(1 - \alpha_\delta)\zeta n_2\, d\sigma(x), \tag{3.45}$$

where $S_1' = \bigcup_{i=1}^N S_{1,i}$, $S_{1,i} = \{(x_1, s_-(x_1)) \mid x_1 \in \pi_{x_1}(S_{3,i})\}$, and n_2 denotes the second entry of the outward unit normal vector n to S_1'.

We claim that

$$\int_{S_1'} \frac{\beta(x, \varphi)}{v_2}(1 - \alpha_\delta)\zeta n_2\, d\sigma(x) = 0.$$

Indeed, first remark that

$$\int_{S_1'} \frac{\beta(x, \varphi)}{v_2}(1 - \alpha_\delta)\zeta n_2\, d\sigma(x) = \int_{S_1' \setminus \overline{A}_m} \frac{\beta(x, \varphi)}{v_2}(1 - \alpha_\delta)\zeta n_2\, d\sigma(x).$$

Next let $x_0 = (x_{01}, s_-(x_{01})) \in S_1' \setminus \overline{A}_m$ such that $\beta((x_{01}, s_+(x_{01})), \varphi(x_{01}, s_+(x_{01}))) > 0$. Without loss of generality, we assume that s_- and s_+ are continuous at x_{01}.

Since $x_0 \notin \overline{A}_m$, there exists $\varepsilon_0 > 0$ such that $B_{\varepsilon_0}(x_0) \cap A_m = \emptyset$ which means that $B_{\varepsilon_0}(x_0) \cap \Omega \subset [u_m = x_2] \cap [\gamma < 1]$. Note that we can choose ε_0 small enough to ensure that s_- and s_+ are continuous in $(x_{01} - \varepsilon_0, x_{01} + \varepsilon_0)$. Moreover, using the continuity of γ in $[\gamma < 1]$, one can also assume that $\beta((x_1, s_+(x_1)), \varphi(x_1, s_+(x_1))) > 0$ in $(x_{01} - \varepsilon_0, x_{01} + \varepsilon_0)$.

By Corollary 1.1 and the fact that γ is nonincreasing with respect to x_2, we get for some $\varepsilon \in (0, \varepsilon_0)$

$$u_m = x_2 \quad \text{and} \quad \gamma < 1 \quad \text{in } D_\varepsilon = \big((x_{01} - \varepsilon, x_{01} + \varepsilon) \times \mathbb{R}\big) \cap \Omega.$$

By Theorem 3.6, we know that Φ_1 and Φ_2 are continuous in $(x_{01} - \varepsilon, x_{01} + \varepsilon)$.

Now we distinguish three cases:

* $\forall x_1 \in (x_{01} - \varepsilon, x_{01} + \varepsilon)$, $\Phi_1(x_1) = s_-(x_1)$.

In this case, we have $u_1(x) = x_2 \ \forall x \in D_\varepsilon$. Let $\xi \in W^{1,q}(\Omega)$ such that $\xi = 0$ on $\partial D_\varepsilon \cap \Omega$. Since $\pm \chi(D_\varepsilon)\xi$ are test functions for $(\mathrm{P_L})$, we have

$$\int_{D_\varepsilon} (1 - g_1)k\xi_{x_2} \, dx = \int_{\partial D_\varepsilon \cap S_3} \beta(x, \varphi)\xi \, d\sigma(x).$$

Using Corollary 3.2, we get

$$\int_{D_\varepsilon} (1 - g_1)k\xi_{x_2} \, dx = \int_{D_\varepsilon} \frac{\beta(x, \varphi)}{v_2} \xi_{x_2} \, dx$$

$$= \int_{\partial D_\varepsilon \cap S_3} \beta(x, \varphi)\xi \, d\sigma(x) + \int_{\partial D_\varepsilon \cap S_1'} \frac{\beta(x, \varphi)}{v_2} n_2 \xi \, d\sigma(x).$$

This leads to

$$\int_{\partial D_\varepsilon \cap S_1'} \frac{\beta(x, \varphi)}{v_2} n_2 \xi \, d\sigma(x) = 0 \quad \forall \xi \in W^{1,q}(\Omega) \text{ such that } \xi = 0 \text{ on } \partial D_\varepsilon \cap \Omega.$$

We obtain $\beta(x, \varphi) = 0$ on $\partial D_\varepsilon \cap S_3$ which contradicts (3.42).

* $\forall x_1 \in (x_{01} - \varepsilon, x_{01} + \varepsilon)$, $\Phi_1(x_1) = s_+(x_1)$.

In this case, we have $u_1(x) > x_2 \ \forall x \in D_\varepsilon$ and then $u_2(x) = u_m(x) = x_2 \ \forall x \in D_\varepsilon$. This leads to $\Phi_2(x_1) = s_-(x_1) \ \forall x_1 \in (x_{01} - \varepsilon, x_{01} + \varepsilon)$. Then we get a contradiction as in the previous case.

* $\exists x_{01}' \in (x_{01} - \varepsilon, x_{01} + \varepsilon)$, $s_-(x_{01}') < \Phi_1(x_{01}') < s_+(x_{01}')$.

By continuity there exists a small $\delta' \in (0, \varepsilon - |x_{01} - x_{01}'|)$ such that $x_{01} - \varepsilon < x_{01}' - \delta' < x_{01}' + \delta' < x_{01} + \varepsilon$ and $s_-(x_1) < \Phi_1(x_1) < s_+(x_1) \ \forall x_1 \in (x_{01}' - \delta', x_{01}' + \delta')$.

Since $u_m = x_2$ in D_ε, this leads to $\Phi_2(x_1) = s_-(x_1) \ \forall x_1 \in (x_{01}' - \delta', x_{01}' + \delta')$. Again we obtain $\beta(x, \varphi) = 0$ on $\partial D_{\delta'} \cap S_3$, with $D_{\delta'} = ((x_{01}' - \delta', x_{01}' + \delta') \times \mathbb{R}) \cap \Omega$ which is impossible.

We conclude that $S_1' \setminus \overline{A_m} = \emptyset$ and therefore

$$\int_{S_1'} \frac{\beta(x, \varphi)}{v_2} (1 - \alpha_\delta) \zeta n_2 \, d\sigma(x) = 0.$$

Subtracting then (3.45) from (3.44) we get by the monotonicity of β,

$$\int_{\Omega} \big((\mathcal{A}(x, \nabla u_i) - \mathcal{A}(x, \nabla u_m)) - (g_i - g_M)\mathcal{A}(x, e) \big) \cdot \nabla\big((1 - \alpha_\delta)\zeta\big)\, dx$$

$$\leqslant \int_{S_3} \big(\beta(x, \psi - u_i) - \beta(x, \psi - x_2)\big)(1 - \alpha_\delta)\zeta\, d\sigma(x) \leqslant 0. \tag{3.46}$$

Now clearly $\pm\xi$ are test functions for $(\mathrm{P_L})$. So we have, for $i, j = 1, 2$ with $i \neq j$,

$$\int_{\Omega} \big((\mathcal{A}(x, \nabla u_i) - \mathcal{A}(x, \nabla u_j)) - (g_i - g_j)\mathcal{A}(x, e) \big) \cdot \nabla\xi\, dx$$

$$= \int_{S_3} \big(\beta(x, \psi - u_i) - \beta(x, \psi - u_j)\big) \cdot \xi\, d\sigma(x). \tag{3.47}$$

Given that we integrate only on the set $[u_i - u_m > 0]$ where $u_m = u_j$, (3.47) becomes by the monotonicity of β,

$$\int_{\Omega} \big((\mathcal{A}(x, \nabla u_i) - \mathcal{A}(x, \nabla u_m)) - (g_i - g_M)\mathcal{A}(x, e) \big) \cdot \nabla\xi\, dx \leqslant 0$$

which can be written by the monotonicity of $\mathcal{A}$,

$$\int_{[u_i - u_m \geqslant \varepsilon\zeta]} \big(\mathcal{A}(x, \nabla u_i) - \mathcal{A}(x, \nabla u_m)\big) \cdot \nabla(\alpha_\delta\zeta)\, dx$$

$$- \int_{\Omega} k(x)(g_i - g_M) \cdot (\alpha_\delta\zeta)_{x_2}\, dx$$

$$\leqslant \int_{\Omega} k(x)(g_M - g_i)\left(\alpha_\delta\zeta - \frac{u_i - u_m}{\varepsilon} \right)^+_{x_2} dx. \tag{3.48}$$

Using (3.39) we have

$$\int_{\Omega} k(x)(g_M - g_i)\left(\alpha_\delta\zeta - \frac{u_i - u_m}{\varepsilon} \right)^+_{x_2} dx$$

$$= \int_{[u_i > u_m = x_2] \cap [\gamma < 1]} \left(k(x) - \frac{\beta(x, \varphi)}{\nu_2} \right)\left(\alpha_\delta\zeta - \frac{u_i - u_m}{\varepsilon} \right)^+_{x_2} dx$$

$$= \int_{[u_i > u_m = x_2]} k(x)(1 - \gamma)^+\left(\alpha_\delta\zeta - \frac{u_i - u_m}{\varepsilon} \right)^+_{x_2} dx$$

$$= \int_{D_i} dx_1 \int_{\Phi_m(x_1)}^{\Phi_i(x_1)} k(x)(1 - \gamma)^+\left(\alpha_\delta\zeta - \frac{u_i - u_m}{\varepsilon} \right)^+_{x_2} dx_2. \tag{3.49}$$

with $D_i = \{x_1 \in \pi_{x_1}(\Omega) \mid \Phi_m(x_1) < \Phi_i(x_1)\}$, $i = 1, 2$ and $\Phi_m = \min(\Phi_1, \Phi_2)$.

Since for a.e. $x_1 \in D_i$, $k(1-\gamma)^+(x_1, \cdot)$ is nondecreasing with respect to x_2, we deduce, by the second mean-value theorem,

$$\int_{\Phi_m(x_1)}^{\Phi_i(x_1)} k(1-\gamma)^+ \left(\alpha_\delta\zeta - \frac{u_i - u_m}{\varepsilon}\right)_{x_2}^+ dx_2$$

$$= \left(k(1-\gamma)^+\right)\left(x_1, \Phi_i(x_1)_-\right) \int_{\Phi_*(x_1)}^{\Phi_i(x_1)} \left(\alpha_\delta\zeta - \frac{u_i - u_m}{\varepsilon}\right)_{x_2}^+ dx_2$$

$$\leqslant \left(k(1-\gamma)^+\right)\left(x_1, \Phi_i(x_1)_-\right)(\alpha_\delta\zeta)\left(x_1, \Phi_i(x_1)\right) \tag{3.50}$$

with $\Phi_*(x_1) \in [\Phi_m(x_1), \Phi_i(x_1)]$ and $(k(1-\gamma)^+)(x_1, \Phi_i(x_1)_-)$ is the left limit of $(k(1-\gamma)^+)(x_1, \cdot)$ at $\Phi_i(x_1)$.

Taking into account (3.48)–(3.50) we get

$$\int_{[u_i - u_m \geqslant \varepsilon\zeta]} \left(\mathcal{A}(x, \nabla u_i) - \mathcal{A}(x, \nabla u_m)\right) \cdot \nabla(\alpha_\delta\zeta)\, dx$$

$$- \int_\Omega k(x)(g_i - g_M) \cdot (\alpha_\delta\zeta)_{x_2}\, dx$$

$$\leqslant \int_{D_i} \left(k(1-\gamma)^+\right)\left(x_1, \Phi_i(x_1)_-\right)(\alpha_\delta\zeta)\left(x_1, \Phi_i(x_1)\right) dx_1$$

which leads by letting ε go to zero to

$$\int_\Omega \left(\mathcal{A}(x, \nabla u_i) - \mathcal{A}(x, \nabla u_m) - (g_i - g_M)\mathcal{A}(x, e)\right) \cdot \nabla(\alpha_\delta\zeta)\, dx$$

$$\leqslant \int_{D_i} \left(k(1-\gamma)^+\right)\left(x_1, \Phi_i(x_1)_-\right)(\alpha_\delta\zeta)\left(x_1, \Phi_i(x_1)\right) dx_1$$

$$\leqslant \int_{D_i} \left(k(1-\gamma)^+\alpha_\delta\zeta\right)\left(x_1, \Phi_i(x_1)\right) dx_1. \tag{3.51}$$

Using (3.43), (3.46) and (3.51) we get

$$\int_\Omega \left(\mathcal{A}(x, \nabla u_i) - \mathcal{A}(x, \nabla u_m) - (g_i - g_M)\mathcal{A}(x, e)\right) \cdot \nabla\zeta\, dx$$

$$\leqslant \int_{D_i} \left(k(1-\gamma)^+\alpha_\delta\zeta\right)\left(x_1, \Phi_i(x_1)\right) dx_1. \tag{3.52}$$

Now let $x_{01} \in D_i$ such that $\gamma(x_{01}, \Phi_i(x_{01})) < 1$. By continuity there exists a small $\eta > 0$ such that $\gamma(x) < 1 \ \forall x \in B_\eta(x_{01}, \Phi_i(x_{01}))$. Moreover, one can assume that $\Phi_m(x_{01}) < \Phi_i(x_{01}) - \eta/2$. Since $\gamma(x_{01}, \Phi_i(x_{01}) - \eta/2) < 1$ and $u_m(x_{01}, \Phi_i(x_{01}) - \eta/2) = \Phi_i(x_{01}) - \eta/2$, one can argue as in the proof of Theorem 3.6 to get, for some $\eta' \in (0, \eta)$,

$$u_m(x) = x_2 \quad \forall x \in \left((x_{01} - \eta', x_{01} + \eta') \times (\Phi_i(x_{01}) - \eta', +\infty)\right) \cap \Omega.$$

It follows that $(x_{01}, \Phi_i(x_{01})) \notin \overline{A}_m$. Thus $\alpha_\delta(x_{01}, \Phi_i(x_{01}))$ converges to 0 when δ goes to 0. Using the Lebesgue theorem we obtain

$$\lim_{\delta \to 0} \int_{D_i} \left(k(1-\gamma)^+ \alpha_\delta \zeta\right)(x_1, \Phi_i(x_1)) \, dx_1 = 0$$

which leads to

$$\int_\Omega \left(\mathcal{A}(x, \nabla u_i) - \mathcal{A}(x, \nabla u_m) - (g_i - g_M)\mathcal{A}(x, e)\right) \cdot \nabla \zeta \, dx \leqslant 0.$$

At this step, we argue as at the end of the proof of Theorem 2.5 to get

$$\int_\Omega \left(\mathcal{A}(x, \nabla u_i) - \mathcal{A}(x, \nabla u_m) - (g_i - g_M)\mathcal{A}(x, e)\right) \cdot \nabla \zeta \, dx = 0$$

$$\forall \zeta \in W^{1,q}(\Omega).$$

(ii) We argue as for the proof of (ii) of Theorem 2.5.

(iii) Let $\xi \in W^{1,q}(\Omega)$, $\xi = 0$ on S_2. Using (i) for $i = 1, 2$, we obtain

$$\int_\Omega \left(\mathcal{A}(x, \nabla u_1) - \mathcal{A}(x, \nabla u_2) - (g_1 - g_2)\mathcal{A}(x, e)\right) \cdot \nabla \xi \, dx = 0.$$

But since $\pm\xi$ are test functions for $(\mathrm{P_L})$, we deduce that

$$\int_{S_3} \left(\beta(x, \psi - u_2) - \beta(x, \psi - u_1)\right)\xi \, d\sigma(x) = 0 \quad \forall \xi \in W^{1,q}(\Omega), \xi = 0 \text{ on } S_2$$

which gives $\beta(x, \psi - u_1) = \beta(x, \psi - u_2)$ a.e. on S_3. $\qquad\square$

As a consequence of Theorem 3.7, we deduce by arguing as for the proof of Corollary 2.3:

COROLLARY 3.3. *Let (u_1, g_1) and (u_2, g_2) be two solutions of $(\mathrm{P_L})$. Then $(\min(u_1, u_2), \max(g_1, g_2))$ and $(\max(u_1, u_2), \min(g_1, g_2))$ are also solutions of $(\mathrm{P_L})$.*

We consider the set of all solutions of $(\mathrm{P_L})$:

$$\mathcal{S}_\mathrm{L} = \left\{(u, g) \in W^{1,q}(\Omega) \times L^\infty(\Omega) \mid (u, g) \text{ is a solution of } (\mathrm{P_L})\right\}.$$

Then we define the following functional $\mathcal{I}_\mathrm{L}$ on $\mathcal{S}_\mathrm{L}$ by

$$\forall (u, g) \in \mathcal{S}_\mathrm{L}, \quad \mathcal{I}_\mathrm{L}(u, g) = \int_\Omega \mathcal{A}(x, \nabla u) \cdot \nabla u \, dx - \frac{1}{q} \int_\Omega g \mathcal{A}(x, e) \cdot e \, dx$$

$$+ \int_{S_3} \beta(x, \psi - u)(\psi - u) \, d\sigma(x). \qquad (3.53)$$

We assume that

$$\text{for a.e. } x \in S_3 \quad \beta(x, \cdot) \text{ is a continuous function.} \tag{3.54}$$

Here is the main result of this section.

THEOREM 3.8. *There exist a unique minimal solution* (u_m, g_M) *and a unique maximal solution* (u_M, g_m) *in* $\mathcal{S}_L$ *in the following sense*

$$\mathcal{I}_L(u_m, g_M) = \min_{(u,g)\in\mathcal{S}_L} \mathcal{I}_L(u, g), \qquad \mathcal{I}_L(u_M, g_m) = \max_{(u,g)\in\mathcal{S}_L} \mathcal{I}_L(u, g)$$

$$\forall (u, g) \in \mathcal{S}_L, u_m \leqslant u \leqslant u_M, g_m \leqslant g \leqslant g_M \quad \text{in } \Omega.$$

We first prove a monotonicity result.

LEMMA 3.2. $\mathcal{I}_L$ *is strictly monotone, i.e.,* $\forall (u_1, g_1), (u_2, g_2) \in \mathcal{S}_L$

(i) $u_1 \leqslant u_2, g_2 \leqslant g_1$ *in* $\Omega \quad \Longrightarrow \quad \mathcal{I}_L(u_1, g_1) \leqslant \mathcal{I}_L(u_2, g_2),$

(ii) $u_1 \leqslant u_2, g_2 \leqslant g_1$ *in* $\Omega \quad and \quad u_1 \neq u_2 \quad \Longrightarrow \quad \mathcal{I}_L(u_1, g_1) < \mathcal{I}_L(u_2, g_2).$

PROOF. (i) Let $(u_1, g_1), (u_2, g_2) \in \mathcal{S}_L$ satisfying $u_1 \leqslant u_2$ and $g_2 \leqslant g_1$ a.e. in Ω. Then we have, since by Theorem 3.7(iii) $\beta(x, \psi - u_1) = \beta(x, \psi - u_2)$ a.e. in S_3.

$$\mathcal{I}_L(u_1, g_1) - \mathcal{I}_L(u_2, g_2)$$

$$= \int_\Omega \left(\mathcal{A}(x, \nabla u_1) \cdot \nabla u_1 - g_1 \mathcal{A}(x, e) \cdot e \right) dx - \int_{S_3} \beta(x, \psi - u_1) u_1 \, d\sigma(x)$$

$$- \int_\Omega \left(\mathcal{A}(x, \nabla u_2) \cdot \nabla u_2 - g_2 \mathcal{A}(x, e) \cdot e \right) dx + \int_{S_3} \beta(x, \psi - u_2) u_2 \, d\sigma(x)$$

$$+ \frac{1}{q'} \int_\Omega (g_1 - g_2) \mathcal{A}(x, e) \cdot e \, dx$$

$$= \int_\Omega \left(\mathcal{A}(x, \nabla u_1) - g_1 \mathcal{A}(x, e) \right) \cdot \nabla (u_1 - x_2) \, dx$$

$$- \int_{S_3} \beta(x, \psi - u_1)(u_1 - x_2) \, d\sigma(x)$$

$$- \int_\Omega \left(\mathcal{A}(x, \nabla u_2) - g_2 \mathcal{A}(x, e) \right) \cdot \nabla (u_2 - x_2) \, dx$$

$$+ \int_{S_3} \beta(x, \psi - u_2)(u_2 - x_2) \, d\sigma(x) + \frac{1}{q'} \int_\Omega (g_1 - g_2) \mathcal{A}(x, e) \cdot e \, dx$$

$$+ \int_\Omega \left((\mathcal{A}(x, \nabla u_1) - \mathcal{A}(x, \nabla u_2)) - (g_1 - g_2) \mathcal{A}(x, e) \right) \cdot \nabla x_2 \, dx.$$

By Theorem 3.7(i) and since $\pm(u_i - x_2)$, $i = 1, 2$, are test functions for ($\mathrm{P_L}$), we get

$$\mathcal{I}_{\mathrm{L}}(u_1, g_1) - \mathcal{I}_{\mathrm{L}}(u_2, g_2) = \frac{1}{q'} \int_{\Omega} (g_1 - g_2) k(x) \, dx \leqslant 0.$$

(ii) Assume that $u_1 \leqslant u_2$, $g_2 \leqslant g_1$ in Ω and $\mathcal{I}_{\mathrm{L}}(u_1, g_1) = \mathcal{I}_{\mathrm{L}}(u_2, g_2)$. From the proof of (i), we get $\int_{\Omega} (g_1 - g_2) k(x) \, dx = 0$ which leads to $g_1 = g_2$ a.e. in Ω. Using Theorem 3.7(i) for $\xi = u_1 - u_2$, we obtain

$$\int_{\Omega} \big(\mathcal{A}(x, \nabla u_1) - \mathcal{A}(x, \nabla u_2)\big) \cdot \nabla(u_1 - u_2) \, dx = 0$$

which leads by (2.10) to $\nabla(u_1 - u_2) = 0$ a.e. in Ω. But since $u_1 - u_2 = 0$ on S_2, we deduce that $u_1 = u_2$. $\qquad\square$

PROOF OF THEOREM 3.8. First remark that for each $(u, g) \in \mathcal{S}_{\mathrm{L}}$, we have by (1.2) and the monotonicity of $\beta(x, \cdot)$

$$\mathcal{I}_{\mathrm{L}}(u_1, g_1) \geqslant -\frac{1}{q} \int_{\Omega} k(x) g \, dx \geqslant -\frac{|\Omega|}{q} M$$

from which we deduce that there exists a minimizing sequence $(u_k, g_k)_{k \in \mathbb{N}}$ for $\mathcal{I}_{\mathrm{L}}$ i.e.

$$\forall k \in \mathbb{N}, \quad (u_k, g_k) \in \mathcal{S}_{\mathrm{L}} \quad \text{and} \quad \lim_{k \to +\infty} \mathcal{I}_{\mathrm{L}}(u_k, g_k) = m = \inf_{(u,g) \in \mathcal{S}_{\mathrm{L}}} \mathcal{I}_{\mathrm{L}}(u, g).$$

$$(3.55)$$

As in the Dirichlet case, we define the sequence $(v_k, f_k)_{k \in \mathbb{N}}$ by

$$\begin{cases} (v_0, f_0) = (u_0, g_0), \\ (v_{k+1}, f_{k+1}) = \big(\min(u_{k+1}, v_k), \max(g_{k+1}, f_k)\big) & \forall k \in \mathbb{N}. \end{cases}$$

Using Corollary 3.3 and Lemma 3.2 and arguing as in the proof of Theorem 2.6, one can verify that for each $k \in \mathbb{N}$, $(v_k, f_k) \in \mathcal{S}_{\mathrm{L}}$ and that we have, for some $(v, f) \in L^q(\Omega) \times L^{q'}(\Omega)$,

$$m = \lim_{k \to +\infty} \mathcal{I}_{\mathrm{L}}(v_k, f_k), \tag{3.56}$$

$$v_k \to v \text{ in } L^q(\Omega) \text{ and a.e. in } \Omega, \tag{3.57}$$

$$f_k \to f \text{ in } L^{q'}(\Omega) \text{ and a.e. in } \Omega. \tag{3.58}$$

Now since $\pm(v_k - x_2)$ are test functions for ($\mathrm{P_L}$), we have, by ($\mathrm{P_L}$)(ii), Proposition 1.1

and (1.2)(iii),

$$\lambda \int_{\Omega} |\nabla v_k|^q \, dx$$

$$\leqslant \int_{\Omega} \mathcal{A}(x, \nabla v_k) \cdot \nabla v_k \, dx$$

$$= \int_{S_3} \beta(x, \psi - v_k)(v_k - x_2) \, d\sigma(x) + \int_{\Omega} \mathcal{A}(x, \nabla v_k) \cdot \nabla x_2 \, dx$$

$$\leqslant \int_{S_3} (h_0 - x_2)\beta(x, \psi - x_2) \, d\sigma(x) + M|\Omega|^{1/q} \left(\int_{\Omega} |\nabla v_k|^q \, dx \right)^{1/q'}.$$

We deduce that $(v_k)_k$ is bounded in $W^{1,q}(\Omega)$. So we have up to a subsequence

$$v_{k_p} \rightharpoonup v \quad \text{in } W^{1,q}(\Omega), \tag{3.59}$$

$$v_{k_p} \rightarrow v \quad \text{in } L^q(S_2 \cup S_3) \text{ and a.e. in } S_2 \cup S_3. \tag{3.60}$$

By (3.60), we have $v = 0$ on S_2. By (3.57) and (3.58), we deduce that

$$v \geqslant x_2, 0 \leqslant f \leqslant 1, \quad f(v - x_2) = 0 \quad \text{a.e. in } \Omega.$$

Hence to prove that $(v, f) \in \mathcal{S}_{\mathrm{L}}$, it remains to prove that it satisfies $(\mathrm{P_L})$(iii).

From the fact that (v_k) is bounded in $W^{1,q}(\Omega)$, we deduce that up to a subsequence still denoted by (v_{k_p}), one has

$$\mathcal{A}(x, \nabla v_{k_p}) \rightharpoonup \mathcal{A}_0 \quad \text{in } \mathbb{L}^{q'}(\Omega). \tag{3.61}$$

Let $p, s \in \mathbb{N}$ such that $p \leqslant s$. We have, by Theorem 3.7,

$$\int_{\Omega} \left\{ \left(\mathcal{A}(x, \nabla v_{k_p}) - \mathcal{A}(x, \nabla v_{k_s}) \right) - (f_{k_p} - f_{k_s})\mathcal{A}(x, e) \right\} \cdot \nabla v_{k_p} \, dx = 0$$

from which we deduce, by letting first $s \rightarrow +\infty$ and then $p \rightarrow +\infty$,

$$\lim_{p \rightarrow +\infty} \int_{\Omega} \mathcal{A}(x, \nabla v_{k_p}) \cdot \nabla v_{k_p} \, dx = \int_{\Omega} \mathcal{A}_0 \cdot \nabla v \, dx. \tag{3.62}$$

Using the monotonicity of $\mathcal{A}$, (3.59), (3.61) and (3.62), we easily obtain

$$\mathcal{A}(x, \nabla v_{k_p}) \rightharpoonup \mathcal{A}(x, \nabla v) \quad \text{in } \mathbb{L}^{q'}(\Omega). \tag{3.63}$$

Finally, let $\xi \in W^{1,q}(\Omega)$ such that $\xi \geqslant 0$ on S_2. For each $p \in \mathbb{N}$, we have

$$\int_{\Omega} \left(\mathcal{A}(x, \nabla v_{k_p}) - f_{k_p} \mathcal{A}(x, e) \right) \cdot \nabla \xi \, dx \leqslant \int_{S_3} \beta(x, \psi - v_{k_p}) \xi \, d\sigma(x). \tag{3.64}$$

542   *A. Lyaghfouri*

Using (3.54), (3.58), (3.60) and (3.63) we get, by letting $p \to +\infty$ in (3.64),

$$\int_\Omega \big(\mathcal{A}(x, \nabla v) - f\mathcal{A}(x, e)\big) \cdot \nabla \xi \, dx \leqslant \int_{S_3} \beta(x, \psi - v)\xi \, d\sigma(x).$$

Thus (v, f) is a solution of $(\mathrm{P_L})$.

Now, using (3.56), (3.58) and (3.60)–(3.63), we obtain

$$
\begin{aligned}
m &= \lim_{p \to +\infty} \mathcal{I}_\mathrm{L}(v_{k_p}, f_{k_p}) \\[2mm]
&= \lim_{p \to +\infty} \int_\Omega \mathcal{A}(x, \nabla v_{k_p}) \cdot \nabla v_{k_p} \, dx - \frac{1}{q} \int_\Omega f_{k_p} \mathcal{A}(x, e) \cdot e \, dx \\[2mm]
&\quad + \int_{S_3} \beta(x, \psi - v_{k_p})(\psi - v_{k_p}) \, d\sigma(x) \\[2mm]
&= \int_\Omega \mathcal{A}(x, \nabla v) \cdot \nabla v \, dx - \frac{1}{q} \int_\Omega f \mathcal{A}(x, e) \cdot e \, dx \\[2mm]
&\quad + \int_{S_3} \beta(x, \psi - v)(\psi - v) \, d\sigma(x) \\[2mm]
&= \mathcal{I}_\mathrm{L}(v, f).
\end{aligned}
$$

Let $(u, g) \in \mathcal{S}_\mathrm{L}$. Since by Corollary 3.3, $(\min(u, v), \max(g, f)) \in \mathcal{S}_\mathrm{L}$, we deduce by Lemma 3.2(i) that

$$m \leqslant \mathcal{I}_\mathrm{L}\big(\min(u, v), \max(g, f)\big) \leqslant \mathcal{I}_\mathrm{L}(v, f) = m$$

which leads by Lemma 3.2(ii) to $(v, f) = (\min(u, v), \max(g, f))$, i.e., $v \leqslant u$ and $g \leqslant f$ a.e. in Ω. The uniqueness of (v, f) is then clear. This achieves the proof of the first part of Theorem 3.8.

Let us prove the second part of the theorem. First we have, by (1.2)(iii) and the monotonicity of $\beta(x, \cdot)$,

$$
\begin{aligned}
\mathcal{I}_\mathrm{L}&(u, g) \\[2mm]
&= \int_\Omega \mathcal{A}(x, \nabla u) \cdot \nabla u \, dx - \frac{1}{q} \int_\Omega gk(x) \, dx + \int_{S_3} \beta(x, \psi - u)(\psi - u) \, d\sigma(x) \\[2mm]
&\leqslant \int_\Omega \mathcal{A}(x, \nabla u) \cdot \nabla u \, dx + \int_{S_3} \beta(x, \varphi)\varphi \, d\sigma(x) \\[2mm]
&\leqslant M \int_\Omega |\nabla u|^q \, dx + \int_{S_3} \beta(x, \varphi)\varphi \, d\sigma(x).
\end{aligned}
$$

Moreover, since $\pm(u - x_2)$ are test functions for $(\mathrm{P_L})$, we have

$$\int_\Omega \big(\mathcal{A}(x, \nabla u) - g\mathcal{A}(x, e)\big) \cdot \nabla(u - x_2) \, dx = \int_{S_3} \beta(x, \psi - u)(u - x_2) \, d\sigma(x).$$

This leads by (1.2)(iii) to

$$\lambda \int_{\Omega} |\nabla u|^q \, dx \leqslant \int_{\Omega} \mathcal{A}(x, \nabla u) \cdot \nabla u \, dx$$

$$= \int_{\Omega} \mathcal{A}(x, \nabla u) \cdot e \, dx + \int_{S_3} \beta(x, \psi - u)(u - x_2) \, d\sigma(x)$$

$$\leqslant M |\Omega|^{1/q} \left(\int_{\Omega} |\nabla u|^q \, dx \right)^{1/q'} + \int_{S_3} \beta(x, \varphi)(h_0 - x_2) \, d\sigma(x)$$

and

$$\int_{\Omega} |\nabla u|^q \, dx \leqslant C \quad \text{for some positive constant } C.$$

Thus $\mathcal{I}_{\mathrm{L}}(u, g)$ is bounded for all $(u, g) \in \mathcal{S}_{\mathrm{L}}$. Let then $(u_k, g_k)_{k \in \mathbb{N}}$ be a sequence of solutions of $(\mathrm{P}_{\mathrm{L}})$ such that

$$\lim_{k \to +\infty} \mathcal{I}_{\mathrm{L}}(u_k, g_k) = \sup_{(u,g) \in \mathcal{S}_{\mathrm{L}}} \mathcal{I}_{\mathrm{L}}(u, g).$$

We consider the following sequence $(w_k, h_k)_{k \in \mathbb{N}}$ defined by

$$\begin{cases} (w_0, h_0) = (u_0, g_0), \\ (w_{k+1}, h_{k+1}) = \big(\max(u_{k+1}, w_k), \min(g_{k+1}, h_k)\big) \quad \forall k \in \mathbb{N}. \end{cases}$$

Then we have, by Corollary 3.3 and Lemma 3.2(i),

$$\forall k \in \mathbb{N}, \quad (w_k, h_k) \in \mathcal{S}_{\mathrm{L}} \quad \text{and} \quad \mathcal{I}_{\mathrm{L}}(u_k, g_k) \leqslant \mathcal{I}_{\mathrm{L}}(w_k, h_k) \leqslant \sup_{(u,g) \in \mathcal{S}_{\mathrm{L}}} \mathcal{I}_{\mathrm{L}}(u, g).$$

It follows that

$$\sup_{(u,g) \in \mathcal{S}_{\mathrm{L}}} \mathcal{I}_{\mathrm{L}}(u, g) = \lim_{k \to +\infty} \mathcal{I}_{\mathrm{L}}(w_k, h_k). \tag{3.65}$$

Using the monotonicity of $(w_k, h_k)_{k \in \mathbb{N}}$, (3.65) and arguing as above, we prove that, for a subsequence $(w_{k_p}, h_{k_p})_{p \in \mathbb{N}}$, we have

$$w_{k_p} \rightharpoonup w \quad \text{in } W^{1,q}(\Omega),$$

$$w_{k_p} \to w \quad \text{in } L^q(\Omega) \quad \text{and a.e. in } \Omega,$$

$$w_{k_p} \to w \quad \text{in } L^q(S_2 \cup S_3) \text{ and a.e. in } S_2 \cup S_3,$$

$$\mathcal{A}(x, \nabla w_{k_p}) \rightharpoonup \mathcal{A}(x, \nabla w) \quad \text{in } \mathbb{L}^{q'}(\Omega),$$

$$h_k \to h \quad \text{in } L^{q'}(\Omega) \text{ and a.e. in } \Omega.$$

Thus it becomes easy to verify that (w, h) is a solution of $(\mathrm{P_L})$ which satisfies $\mathcal{I}_\mathrm{L}(w, h) = \sup_{(u,g)\in \mathcal{S}_\mathrm{L}} \mathcal{I}_\mathrm{L}(u, g)$. We can then prove, as in the case of minimal solution, that for any $(u, g) \in \mathcal{S}_\mathrm{L}$: $u \leqslant w$, $h \leqslant g$ a.e. in Ω. $\qquad\square$

3.4. Reservoirs-connected solution

In this subsection we first extend the definition of the reservoirs-connected solution. Then we prove the uniqueness of this solution in three situations.

THEOREM 3.9. *Let (u, g) be a solution of $(\mathrm{P_L})$. For each $i \in \{1, \ldots, N\}$ and each interval $I \subset \pi_{x_1}(S_{3,i})$, one cannot have $u \equiv x_2$ in $(I \times \mathbb{R}) \cap \Omega$.*

PROOF. Assume that $u \equiv x_2$ in $Z = (I \times \mathbb{R}) \cap \Omega$. By (3.9), one has $g = 1 - \gamma$ a.e. in Z. Let now $\xi \in W^{1,q}(\Omega)$ such that $\xi = 0$ on $\partial Z \cap \Omega$. Then $\pm\xi$ are test functions for $(\mathrm{P_L})$ and we have

$$\int_Z \gamma \mathcal{A}(x, e)\nabla\xi \, dx = \int_{S_3^Z} \beta(x, \varphi)\xi \, d\sigma(x)$$

or

$$\int_Z \frac{\beta(x, \varphi)}{v_2}\xi_{x_2} \, dx = \int_{S_3^Z} \beta(x, \varphi)\xi \, d\sigma(x)$$

which leads to

$$\int_{\partial Z \cap S_1} \frac{\beta(x, \varphi)}{v_2}\xi n_2 \, d\sigma(x) = 0,$$

where n_2 is the second entry of the outward unit normal vector n to S_1. Hence we obtain $\beta(x, \varphi) = 0$ on $\partial Z \cap S_1$, which contradicts (3.42). $\qquad\square$

In the following we give similar results as in Section 2.4.

DEFINITION 3.1. A solution (u, g) of $(\mathrm{P_L})$ is called a reservoirs-connected solution if for each connected component C of $[u > x_2]$, we have $\overline{\pi_{x_1}(C)} \cap \pi_{x_1}(S_3) \neq \emptyset$.

REMARK 3.3. Suppose that we have $\beta(x, \varphi) \geqslant k(x)v_2$ a.e. in some open connected subset T of $S_{3,i}$ for some $i \in \{1, \ldots, N\}$. If C is a connected component of $[u > x_2]$ such that $\pi_{x_1}(C) \cap \pi_{x_1}(T) \neq \emptyset$, then by Corollary 3.1, C contains the strip of Ω below T and T on its boundary.

THEOREM 3.10. *Let (u, g) be a solution of $(\mathrm{P_L})$ and C a connected component of $[u > x_2]$ such that $\overline{\pi_{x_1}(C)} \cap \pi_{x_1}(S_3) = \emptyset$. If we set $h_\mathrm{c} = \sup\{x_2 \mid (x_1, x_2) \in C\}$, then we have*

$$\begin{cases} C = \{(x_1, x_2) \in \Omega \mid x_1 \in \pi_{x_1}(C), x_2 < h_\mathrm{c}\}, \\ u = x_2 + (h_\mathrm{c} - x_2)^+ \cdot \chi(C), \qquad g = 1 - \chi(C) \quad in \ Z = \big(\pi_{x_1}(C) \times \mathbb{R}\big) \cap \Omega. \end{cases}$$

For the proof see that of Theorem 2.7.

REMARK 3.4. Thanks to Definition 2.2, Theorem 3.10 becomes:

For each solution (u, g) of (P_L) and each connected component C of $[u > x_2]$ such that $\overline{\pi_{x_1}(C)} \cap \pi_{x_1}(S_3) = \emptyset$, $(u - x_2, 1 - g)$ agrees with a pool in the strip $\Omega \cap (\pi_{x_1}(C) \times \mathbb{R})$.

We get by adapting the proof of Theorem 2.8.

THEOREM 3.11. *Any solution (u, g) of (P_L) can be written as*

$$u = u_r + \sum_{i \in I} p_i \quad and \quad g = g_r - \sum_{i \in I} \chi_i$$

where (u_r, g_r) is a reservoirs-connected solution and (p_i, χ_i) are pools.

It follows from Theorem 3.11.

COROLLARY 3.4. *The minimal solution (u_m, g_M) is a reservoirs-connected solution.*

3.5. *Uniqueness of the reservoirs-connected solution*

In this subsection, we establish the uniqueness of the reservoirs-connected solution in three situations.

3.5.1. *The case of linear Darcy's law with a diagonal permeability matrix.* Here we assume that

$$\mathcal{A}(x, \xi) = a(x) \cdot \xi \quad \forall \xi \in \mathbb{R}^2, \text{ a.e. } x \in \Omega,$$

$$a(x) = \big(a_{ij}(x)\big) \text{ is a } 2 \times 2 \text{ matrix}, \tag{3.66}$$

$$\exists \lambda, M > 0: \quad \lambda |\xi|^2 \leqslant a(x)\xi \cdot \xi \leqslant M |\xi|^2 \quad \forall \xi \in \mathbb{R}^2, \text{ a.e. } x \in \Omega, \tag{3.67}$$

$$a_{12}(x) = a_{21}(x) = 0 \quad \text{for a.e. } x \in \Omega, \tag{3.68}$$

$$\frac{\partial a_{22}}{\partial x_2} \geqslant 0 \quad \text{in } \mathcal{D}'(\Omega), \tag{3.69}$$

$$\text{for a.e. } x \in S_3, \quad \beta(x, \cdot) \text{ is an increasing function.} \tag{3.70}$$

Then we have the following theorem.

THEOREM 3.12. *Assume that (3.66)–(3.70) are satisfied. Then there is one and only one reservoirs-connected solution.*

We first prove a lemma.

LEMMA 3.3. *Let (u, g) be a solution of* $(\mathrm{P_L})$. *Then we have*

$$\nabla(u - u_m) = (g - g_M)e \quad \text{a.e. in } \Omega. \tag{3.71}$$

PROOF. Set $p = u - x_2$, $\chi = 1 - g$, $p_m = u_m - x_2$ and $\chi_m = 1 - g_M$. Then from Theorem 3.7(i), we have, by taking $\zeta = p - p_m$ and $\zeta = x_2$, respectively,

$$\int_\Omega a(x)\big(\nabla(p - p_m) + (\chi - \chi_m)e\big) \cdot \nabla(p - p_m)\,dx = 0, \tag{3.72}$$

$$\int_\Omega a(x)\big(\nabla(p - p_m) + (\chi - \chi_m)e\big) \cdot e\,dx = 0. \tag{3.73}$$

Using the fact that $0 \leqslant \chi - \chi_m \leqslant 1$ and $\chi\nabla(p - p_m) = \nabla(p - p_m)$ a.e. in Ω, (3.73) becomes

$$\int_\Omega a(x)\big(\nabla(p - p_m)\big) \cdot \chi e\,dx + \int_\Omega a(x)\big((\chi - \chi_m)e\big) \cdot (\chi - \chi_m)e\,dx \leqslant 0. \tag{3.74}$$

Since $\pm p_m$ and $\pm p$ are test functions for $(\mathrm{P_L})$, we have

$$\int_\Omega a(x)(\nabla p + \chi e) \cdot \nabla p_m\,dx = \int_{S_3} \beta(x, \varphi - p)p_m\,d\sigma(x), \tag{3.75}$$

$$\int_\Omega a(x)(\nabla p_m + \chi_m e) \cdot \nabla p\,dx = \int_{S_3} \beta(x, \varphi - p_m)p\,d\sigma(x). \tag{3.76}$$

Subtracting (3.76) from (3.75) and taking into account the fact that $p = p_m$ a.e. in S_3 which is a consequence of Theorem 3.7(iii) and (3.70), we obtain

$$\int_\Omega a(x)(\chi e) \cdot \nabla p_m\,dx - \int_\Omega a(x)(\chi_m e) \cdot \nabla p\,dx$$
$$= \int_\Omega a(x)(\nabla p_m) \cdot \nabla p\,dx - \int_\Omega a(x)(\nabla p) \cdot \nabla p_m\,dx = 0 \tag{3.77}$$

since by (3.68), a is a symmetric matrix.

Now because $\chi\nabla p_m = \chi_m\nabla p_m$ a.e. in Ω, (3.77) becomes

$$\int_\Omega a(x)(-\chi_m e) \cdot \nabla(p - p_m)\,dx = 0$$

or

$$\int_\Omega a(x)\nabla(p - p_m) \cdot (-\chi_m e)\,dx = 0. \tag{3.78}$$

Then, if we add (3.74) and (3.78), we get

$$\int_{\Omega} a(x)\big(\nabla(p - p_m) + (\chi - \chi_m)e\big) \cdot (\chi - \chi_m)e\, dx \leqslant 0. \tag{3.79}$$

Finally, adding (3.72) and (3.79), we obtain

$$\int_{\Omega} a(x)\big(\nabla(p - p_m) + (\chi - \chi_m)e\big) \cdot \big(\nabla(p - p_m) + (\chi - \chi_m)e\big)\, dx \leqslant 0$$

which leads by (3.67) to

$$\nabla(p - p_m) = -(\chi - \chi_m)e \quad \text{a.e. in } \Omega$$

and (3.71) is proved. $\qquad\qquad\square$

PROOF OF THEOREM 3.12. Let (u, g) be a reservoirs-connected solution of $(\mathrm{P_L})$. Let $C_{m,i}$ be a connected component of $[u_m > x_2]$ such that $\pi_{x_1}(C_{m,i}) \cap \pi_{x_1}(S_{3,i}) \neq \emptyset$ and let C_i be the connected component of $[u > x_2]$ which contains $C_{m,i}$.

First we deduce from (3.71) that we have, for some nonnegative constant c_i,

$$u - u_m = c_i \qquad \text{in } C_{m,i}. \tag{3.80}$$

We shall prove that $u = u_m$ in $C_{m,i}$. To do this, we distinguish two cases:

(i) $\overline{C_{m,i}} \cap (S_2 \cup S_3) \neq \emptyset$: Note that $w = u - u_m$ satisfies

$$\begin{cases} \operatorname{div}\big(a(x)(\nabla w)\big) = \operatorname{div}\big((g - g_M)a(x)(e)\big) & \text{in } \mathcal{D}'(\Omega), \\ w = 0 & \text{on } S_2 \cup S_3. \end{cases}$$

So $w \in C^{0,\alpha}(\Omega \cup S_2 \cup S_3)$ and from (3.80), we obtain $w = 0$ in $C_{m,i}$.

(ii) $\overline{C_{m,i}} \cap (S_2 \cup S_3) = \emptyset$: Again we distinguish two cases:

- $\partial C_i \cap \partial C_{m,i} \cap \Omega \neq \emptyset$: Since $u - u_m = 0$ on $\partial C_i \cap \partial C_{m,i} \cap \Omega$, we deduce that $u - u_m = c_i = 0$ in $C_{m,i}$ and therefore $u = u_m$ in $C_{m,i}$.
- $\partial C_i \cap \partial C_{m,i} \cap \Omega = \emptyset$: Since $\partial C_i \cap \partial C_{m,i} \cap \Omega = \emptyset$, we have $C_i \setminus C_{m,i} \neq \emptyset$ and $\partial C_{m,i} \cap \Omega \subset C_i$. Again we deduce from (3.71) that, for some constant c_i',

$$u = c_i' \quad \text{in } C_i \setminus C_{m,i}. \tag{3.81}$$

Using (3.80), (3.81) and the continuity of u, we get

$$\Phi_m(x_1) = c_i' - c_i = k_i \quad \forall x_1 \in \pi_{x_1}(C_{m,i}). \tag{3.82}$$

Since $\partial C_{m,i} \cap (S_2 \cup S_3) = \emptyset$, we deduce that $\pm\xi = \pm(u_m - k_i)\chi(C_{m,i})$ are suitable test functions for $(\mathrm{P_L})$ and then we have

$$\int_{\Omega} a(x)(\nabla u_m - g_M e) \cdot \nabla \xi\, dx = 0$$

which can be written

$$\int_{C_{m,i}} a(x)\nabla u_m \cdot \nabla u_m \, dx = 0.$$

Using (3.67), we deduce that $\nabla u_m = 0$ a.e. in $C_{m,i}$ and

$$u_m = k_i \quad \text{in } C_{m,i}. \tag{3.83}$$

Now let $x_0 = (x_{01}, \Phi_m(x_{01})) = (x_{01}, k_i) \in \partial C_{m,i} \cap \Omega$ and let $r > 0$ small enough. Let $\zeta \in \mathcal{D}(B_r(x_0))$. Since $\pm\zeta$ are test functions for (P_L), we deduce, by taking into account (3.82) and (3.83),

$$\int_{B_r(x_0)} a(x)(\nabla u_m - g_M e) \cdot \nabla\zeta \, dx = 0$$

or

$$\int_{B_r(x_0)\cap[x_2=k_i]} \frac{\beta(x,\varphi)}{v_2}\zeta \, dx_1 = 0 \quad \forall \zeta \in \mathcal{D}(B_r(x_0))$$

which leads to $\beta(\cdot,\varphi) = 0$ a.e. in $\pi_{x_1}(B_r(x_0))$. But this contradicts (3.42).

Finally, we have proved that $u = u_m$ in $C_{m,i}$ for all $i \in \{1,\ldots,N\}$. As in Section 2.5, one can prove that $u = u_m$ in Ω and by Corollary 3.2, we get $g = g_M$ in Ω. $\qquad\square$

REMARK 3.5. Theorem 3.12 remains valid if we assume that

$$\beta(x,\varphi) \geqslant a_{22}(x)v_2 \quad \text{for a.e. } x \text{ below } S_3, \tag{3.84}$$

and if we replace (3.68) and (3.70) respectively by

$$a_{12}(x) = 0 \quad \text{for a.e. } x \in \Omega, \tag{3.85}$$

and

for each $i \in \{1,\ldots,N\}$, there exists a nonempty domain $S'_{3,i} \subset S_{3,i}$:

for a.e. $x \in S'_{3,i}, u \to \beta(x,u)$ is an increasing function. $\tag{3.86}$

Indeed by (3.84) and Corollary 3.1, we know that, for each $i \in \{1,\ldots,N\}$, the strip Z_i below $S_{3,i}$ is completely saturated. Now let (u, g) be a reservoirs-connected solution of (P_L) and let C_i (resp. $C_{m,i}$) be the connected component of $[u > x_2]$ (resp. $[u_m > x_2]$) that contains Z_i. We have $g = g_M = 0$ a.e. in Z_i. Moreover, by Theorem 3.7(iii) and (3.86), we have also $u = u_m$ a.e. in $S'_{3,i}$. Then one can adapt the proof of Theorem 2.9 to show that $u = u_m$ in $C_i = C_{m,i}$.

3.5.2. *The case of a nonlinear Darcy's law.* In this section, we prove the uniqueness of the reservoirs-connected solution for a Darcy's law corresponding to

$$\mathcal{A}(x,\xi) = \left|a(x)\xi \cdot \xi\right|^{(q-2)/2} a(x)\xi,$$

$$q > 1, q \neq 2 \text{ and } a(x) = (a_{ij}) \text{ is a } 2 \times 2 \text{ matrix}, \tag{3.87}$$

subject to the assumptions:

$$a(x) \text{ is symmetric and belongs to } C^{0,1}(\Omega), \tag{3.88}$$

$$\exists \lambda, M > 0: \quad \lambda|\xi|^2 \leqslant a(x)\xi \cdot \xi \leqslant M|\xi|^2 \quad \forall \xi \in \mathbb{R}^2, \text{ a.e. } x \in \Omega, \tag{3.89}$$

$$a_{12}(x) = 0 \quad \text{for a.e. } x \in \Omega, \tag{3.90}$$

$$\frac{\partial a_{22}^{q/2}}{\partial x_2} \geqslant 0 \quad \text{in } \mathcal{D}'(\Omega), \tag{3.91}$$

$$\forall i \in \{1, \ldots, N\}, \quad \varphi = h_i - x_2 \quad \text{on } S_{3,i}, \tag{3.92}$$

$$\forall i \in \{1, \ldots, N\}, \exists r_i > 0, \exists x_i \in S_{3,i}, \exists \alpha_i \in (0,1):$$

$$S_{3,i}^{r_i} = S_{3,i} \cap B_r(x_i) \text{ is } C^{1,\alpha_i}, \tag{3.93}$$

$$\forall i \in \{1, \ldots, N\}, \exists \kappa_i \in (0,1): \quad \beta \in C^{0,\kappa_i}\left(S_{3,i}^{r_i} \times \mathbb{R}\right), \tag{3.94}$$

$$\beta(x,\varphi) \geqslant a_{22}^{q/2}(x)\nu_2 \quad \text{for a.e. } x \text{ below } S_3, \tag{3.95}$$

$$\forall x \in S_{3,i}^{r_i}, \quad u \mapsto \beta(x,u) \text{ is an increasing function.} \tag{3.96}$$

Then we have the following theorem.

THEOREM 3.13. *Under the assumptions* (3.87)–(3.96), *there is one and only one reservoirs-connected solution of* (P_L).

To prove Theorem 3.13, we need three lemmas. We shall denote by (u, g) a reservoirs-connected solution of (P_L). By (3.95) and Corollary 3.1, we know that for all $i \in \{1, \ldots, N\}$, the strip Z_i below $S_{3,i}$ is completely saturated. For each $i \in \{1, \ldots, N\}$, we denote by C_i (resp. $C_{m,i}$) the connected component of $[u > x_2]$ (resp. $[u_m > x_2]$) which contains Z_i.

LEMMA 3.4. *For each $i \in \{1, \ldots, N\}$, we have the following alternatives*:
    (i) *either* $\exists x_i' \in B(x_i, r_i) \cap S_{3,i}, \exists r_i' \in (0, r_i): \forall x \in \overline{B(x_i', r_i')} \cap \Omega, \nabla u(x) \neq 0$,
    (ii) *or* $u = h_i$ *in* C_i.

PROOF. First note that since $g = 0$ a.e. in Z_i, we have by (1.63) $\operatorname{div}(\mathcal{A}(x, \nabla u)) = 0$ in $B(x_i, r_i) \cap \Omega$. Taking into account the assumptions (3.87)–(3.89) and (3.92)–(3.94), we have (see [30]) $u \in C^{1,\alpha_i}(\overline{B(x_i,r) \cap \Omega})$ for all $r \in (0, r_i)$. So either (i) is true or we must have $\nabla u(x) = 0 \;\forall x \in B(x_i, r_i) \cap S_{3,i}$ which leads to

$$\beta(x, \psi - u) = \left| a(x) \nabla u \cdot \nabla u \right|^{(q-2)/2} a(x) \nabla u \cdot v = 0 \quad \text{in } S_{3,i}^{r_i}.$$

Using (3.96) we obtain $u = \psi = h_i$ in $S_{3,i}^{r_i}$.

Now set

$$\begin{cases} w_0(x) = u(x) - h_i & \forall x \in B(x_i, r_i) \cap \Omega, \\ w_0(x) = 0 & \forall x \in B(x_i, r_i) \setminus \Omega. \end{cases}$$

Since $w_0 = 0$ and $\nabla w_0 = 0$ on $S_{3,i}^{r_i}$, it is clear that $w_0 \in C^1(B(x_i, r_i))$ and that $\operatorname{div}(\mathcal{A}(x, \nabla w_0)) = 0$ in $\mathcal{D}'(B(x_i, r_i))$. The rest of the proof follows the proof of Lemma 2.2.
$\square$

LEMMA 3.5. *If u and u_m are not both constant in C_i and $C_{m,i}$, respectively, then there exist $x_i' \in B(x_i, r_i) \cap S_{3,i}$, $r_i' \in (0, r_i)$, $0 < \lambda_0, \lambda_1 < +\infty$ such that*

$$\forall x \in \overline{B(x_i', r_i') \cap \Omega}, \quad \lambda_0 \leqslant \lambda(x) \leqslant \lambda_1,$$

where $\lambda(x)$ is as in Lemma 2.3.

PROOF. One can adapt the proof of Lemma 2.3.
$\square$

LEMMA 3.6. *For each $i \in \{1, \ldots, N\}$, there exists $x_i' \in B(x_i, r_i) \cap S_{3,i}$, $r_i' \in (0, r_i)$ such that*

$$u = u_m \quad \text{in } B(x_i', r_i') \cap \Omega.$$

PROOF. If u is constant in C_i and u_m is constant in $C_{m,i}$, we have by Lemma 3.4 $u = h_i$ in C_i and $u_m = h_i$ in $C_{m,i}$ and Lemma 3.6 follows in this case.

In the following we assume that either u is not constant in C_i or u_m is not constant in $C_{m,i}$.

By Lemma 3.5, we know that there exist $x_i' \in S_{3,i}^{r_i}$ and $r_i' \in (0, r_i)$, $\lambda_0, \lambda_1 > 0$ such that

$$\forall x \in \overline{B(x_i', r_i') \cap \Omega}, \quad \lambda_0 \leqslant \lambda(x) \leqslant \lambda_1 < +\infty.$$

Since $B(x_i', r_i') \cap \Omega \subset C_i \cap C_{m,i}$ and $g = g_M = 0$ a.e. in $C_i \cap C_{m,i}$, we deduce from Theorem 3.7(i) that

$$\int_{B(x_i', r_i') \cap \Omega} \left(\mathcal{A}(x, \nabla u) - \mathcal{A}(x, \nabla u_m) \right) \cdot \nabla \zeta \, dx = 0 \quad \forall \zeta \in \mathcal{D}(B(x_i', r_i')).$$

The rest of the proof follows the proof of Lemma 2.4.
$\square$

PROOF OF THEOREM 3.13. Using Lemma 3.6 and arguing as in the proof of Theorem 2.10, we prove that for each $i \in \{1, \ldots, N\}$, $u = u_m$ in $C_{m,i}$ and that $C_{m,i} = C_i$. Hence $u = u_m$ in Ω and by Corollary 3.2, we deduce that $g = g_M$ in Ω. $\square$

REMARK 3.6. The case $q = 2$ was discussed in Remark 3.5. In particular, we neither need the $C^{0,1}$ regularity of the matrix $a(x)$ nor the $C^{0,\alpha}$ regularity of any part of $S_{3,i}$.

As a special case of Theorem 3.13, we obtain the uniqueness of the solution for a rectangular dam supplied by two reservoirs.

The uniqueness of the reservoirs-connected solution in general is still an open problem for $q \neq 2$ even for $\mathcal{A}(x, \xi) = |\xi|^{q-2}\xi$.

Acknowledgement

The author is grateful for the financial support and facilities provided by KFUPM.

References

[1] H.W. Alt, *The fluid flow through porous media. Regularity of the free surface*, Manuscripta Math. **21** (1977), 255–272.

[2] H.W. Alt, *Strömungen durch inhomogene poröse Medien mit freiem Rand*, J. Reine Angew. Math. **305** (1979), 89–115.

[3] H.W. Alt and G. Gilardi, *The behavior of the free boundary for the dam problem*, Ann. Sc. Norm. Sup. Pisa Cl. Sci. (4) **9** (1981), 571–626.

[4] G. Alessandrini and M. Sigalotti, *Geometric properties of solutions to the anisotropic p-Laplace equation in dimension two*, Ann. Acad. Sci. Fenn. Ser. A I Math. **26** (2001), 249–266.

[5] A. Alonso and J. Carrillo, *A unified formulation for the boundary conditions in some convection–diffusion problem*, Elliptic and Parabolic Problems, Pont-à-Mousson (1994), 51–63; Pitman Res. Notes Math. Ser., vol. 325, Longman, Harlow (1995).

[6] C. Baiocchi, *Sur un problème à frontière libre traduisant le filtrage de liquides à travers des milieux poreux*, C. R. Acad. Sci. Paris Ser. A **273** (1971), 1215–1217.

[7] C. Baiocchi, *Su un problema di frontiera libera connesso a questioni di idraulica*, Ann. Mat. Pura Appl. **92** (1972), 107–127.

[8] C. Baiocchi, *Free boundary problems in the theory of fluid flow through porous media*, Proceedings of the International Congress of Mathematicians (Vancouver, 1974), Vol. 2, Canad. Math. Congress, Montreal (1975), 237–243.

[9] C. Baiocchi, *Free boundary problems in fluid flows through porous media and variational inequalities*, Free Boundary Problems – Proceedings of a Seminar Held in Pavia, September–October, 1979, Vol. 1, Roma (1980), 175–191.

[10] C. Baiocchi and A. Capelo, *Variational and Quazivariational Inequalities. Applications to Free Boundary Problems*, Wiley, New York (1984); transl. from Italian by Lakshmi Jayakar.

[11] C. Baiocchi and A. Friedman, *A filtration problem in a porous medium with variable permeability*, Ann. Mat. Pura Appl. **114** (1977), 377–393.

[12] V. Barbu, *Nonlinear Semigroups and Differential Equations in Banach Spaces*, Noordhoff, Leiden (1976); transl. from Romanian, Editura Academiei Republicii Socialiste România, Bucharest.

[13] J. Bear and A. Verruijt, *Modeling Groundwater Flow and Pollution*, Reidel, Holland (1992).

[14] V. Benci, *On a filtration through a porous medium*, Ann. Mat. Pura Appl. **100** (1974), 191–209.

[15] H. Brézis, *Opérateurs maximaux monotones et semi-groupes de contractions dans les espaces de Hilbert*, North-Holland Math. Stud., vol. 5. Notas de Matemática (50). North-Holland, Amsterdam–London/Elsevier, New York (1973).

552   *A. Lyaghfouri*

[16] H. Brézis, D. Kinderlehrer and G. Stampacchia, *Sur une nouvelle formulation du problème de l'écoulement à travers une digue*, C. R. Acad. Sci. Paris Ser. A **287** (1978), 711–714.

[17] L.A. Caffarelli and A. Friedman, *The dam problem with two layers*, Arch. Ration. Mech. Anal. **68** (1978), 125–154.

[18] J. Carrillo and M. Chipot, *On the dam problem*, J. Differential Equations **45** (1982), 234–271.

[19] J. Carrillo and M. Chipot, *The dam problem with leaky boundary conditions*, Appl. Math. Optim. **28** (1993), 57–85.

[20] J. Carrillo and A. Lyaghfouri, *The dam problem for nonlinear Darcy's laws and Dirichlet boundary conditions*, Ann. Sc. Norm. Sup. Pisa Cl. Sci. (4) **26** (1998), 453–505.

[21] M. Chipot, *Variational Inequalities and Flow in Porous Media*, Springer-Verlag, New York (1984).

[22] M. Chipot and A. Lyaghfouri, *The dam problem with nonlinear Darcy's law and leaky boundary conditions*, Math. Methods Appl. Sci. **20** (1997), 1045–1068.

[23] M. Chipot and A. Lyaghfouri, *The dam problem with linear Darcy's law and leaky boundary conditions*, Adv. Differential Equations **3** (1998), 1–50.

[24] L. Damascelli, *Comparison theorems for some quasilinear degenerate elliptic operators and applications to symmetry and monotonicity results*, Ann. Inst. H. Poincaré Anal. Non Linéaire **15** (1998), 493–516.

[25] E. DiBenedetto, C^1 *local regularity of weak solutions of degenerate elliptic equations*, Nonlinear Anal. **7** (1983), 827–850.

[26] A. Friedman, *Variational Principles and Free-Boundary Problems*, Krieger, Malabar, FL (1988).

[27] A. Friedman and S.-Y. Huang, *The inhomogeneous dam problem with discontinuous permeability*, Ann. Sc. Norm. Sup. Pisa Cl. Sci. (4) **14** (1987), 49–77.

[28] R.A. Greenkorn, *Flow Phenomena in Porous Media: Fundamental and Applications in Petroleum, Water and Food Production*, Dekker, New York (1983).

[29] Le Dung, *On a class of singular quasilinear elliptic equations with general structure and distribution data*, Nonlinear Anal. **28** (1997), 1879–1902.

[30] G.M. Lieberman, *Boundary regularity for solutions of degenerate elliptic equations*, Nonlinear Anal. **12** (1988), 1203–1219.

[31] J.L. Lions, *Quelques méthodes de résolution de problèmes aux limites non linéaires*, Dunod Gauthier–Villars, Paris (1969).

[32] A. Lyaghfouri, *Sur quelques problèmes d'écoulement dans les milieux poreux*, Ph.D. Thesis, Metz, France (1994).

[33] A. Lyaghfouri, *The inhomogeneous dam problem with linear Darcy's law and Dirichlet boundary conditions*, Math. Models Methods Appl. Sci. **8** (1996), 1051–1077.

[34] A. Lyaghfouri, *A unified formulation for the dam problem*, Riv. Mat. Univ. Parma (6) **1** (1998), 113–148.

[35] A. Lyaghfouri, *A free boundary problem for a fluid flow in a heterogeneous porous medium*, Ann. Univ. Ferrara Sez. VII (N.S.) **IL** (2003), 209–262.

[36] J.F. Rodrigues, *On the dam problem with leaky boundary condition*, Portugal. Math. **39** (1980), 399–411.

[37] R. Stavre and B. Vernescu, *Incompressible fluid flow through a nonhomogeneous and anistropic dam*, Nonlinear Anal. **9** (1985), 799–810.

Nonlinear Eigenvalue Problems for Higher-Order Model Equations

L.A. Peletier

Mathematical Institute, Leiden University, PB 9512, 2300 RA Leiden, The Netherlands
E-mail: peletier@math.leidenuniv.nl

Contents

1. Introduction 555
2. Decreasing nonlinearity $f(u)$ 560
   2.1. Existence of a periodic solution 561
   2.2. Uniqueness 565
   2.3. Asymptotics 569
3. A superlinear bifurcation problem 573
   3.1. Preliminaries 574
   3.2. Multibump periodic solutions 576
   3.3. Branches of periodic solutions 583
4. A sublinear bifurcation problem 585
5. A multiscale analysis 593
   5.1. Multibump solutions 594
   5.2. Branches of multibump periodic solutions 599
Acknowledgements 603
References 603

HANDBOOK OF DIFFERENTIAL EQUATIONS
Stationary Partial Differential Equations, volume 3
Edited by M. Chipot and P. Quittner

1. Introduction

In studying complex spatio-temporal pattern formation, new, often higher order, model equations have recently been proposed. When looking for special solutions such as stationary solutions or traveling wave solutions of these equations there are many instances in which one is led to an equation which can be reduced to one of the form

$$\frac{\mathrm{d}^4 u}{\mathrm{d}x^4} + q \frac{\mathrm{d}^2 u}{\mathrm{d}x^2} + f(u) = 0. \tag{1.1}$$

Here $q \in \mathbb{R}$ is a parameter and f a given, usually nonlinear, function. In this chapter we study the existence and qualitative properties of *bounded* solutions of equation (1.1) in unbounded and bounded domains.

Equation (1.1) belongs to a class of *Pattern Forming Equations* [15,16] in which the parameter q is seen as a measure of what has been called the *pattern forming tendency*. It determines the relative strength of the second- and the fourth-order derivative, and as we shall see, it so plays an important role in determining the qualitative properties of solutions of (1.1). Specifically, we shall find that in general, as q increases, the complexity of the patterns increases.

As to the nonlinearity f, we shall assume throughout that it is a smooth function defined on the whole of $\mathbb{R}$, which vanishes at the origin, i.e., $f(0) = 0$. Therefore, equation (1.1) will always have the trivial solution $u = 0$. Typical functions we shall consider in this chapter are

$$f(s) = -s \pm |s|^{p-1}s \quad \text{and} \quad f(s) = s \pm |s|^{p-1}s, \quad p > 1. \tag{1.2}$$

Examples of equations which lead to (1.1) are the *Swift–Hohenberg* (SH) *equation*:

$$\frac{\partial u}{\partial t} = \kappa u - \left(1 + \frac{\partial^2}{\partial x^2}\right)^2 u - u^3, \quad \kappa \in \mathbb{R}, \tag{1.3}$$

proposed in 1977 by Swift and Hohenberg [36] in studies of Rayleigh–Bénard convection [11,22], and the *Extended Fisher–Kolmogorov* (EFK) *equation*:

$$\frac{\partial u}{\partial t} = -\gamma \frac{\partial^4 u}{\partial x^4} + \frac{\partial^2 u}{\partial x^2} + u - u^3, \quad \gamma > 0, \tag{1.4}$$

proposed in 1988 by Dee and van Saarloos [12] as a higher-order model equation for systems which are bi-stable. Stationary solutions of these equations yield equation (1.1) after an appropriate transformation. Traveling wave solutions, $u(x, t) = v(x - ct)$ $(c \in \mathbb{R})$ of the *Beam equation* [35]:

$$\frac{\partial^2 u}{\partial t^2} + P \frac{\partial^4 u}{\partial x^4} + Q(u) = 0, \quad P \in \mathbb{R}, \tag{1.5}$$

                                         L.A. Peletier

also yield equation (1.1), as do solutions of the *Nonlinear Schrödinger equation* [2]:

$$i\frac{\partial u}{\partial x} + \frac{\partial^2 u}{\partial t^2} - \frac{\partial^4 u}{\partial t^4} + |u|^2 u = 0, \tag{1.6}$$

with a sinusoidal spatial profile, i.e., solutions of the form $u(x,t) = v(t)e^{ikx}$, where $v(t) \in \mathbb{R}$ and $k \in \mathbb{R}$. Finally, we mention an important equation in this field of study, which arises in the theory of water waves,

$$\frac{\mathrm{d}^4 u}{\mathrm{d}x^4} + P\frac{\mathrm{d}^2 u}{\mathrm{d}x^2} + u - u^2 = 0, \quad P \in \mathbb{R}, \tag{1.7}$$

and is already in the form of equation (1.1).

Equation (1.1) is endowed with two properties, which make it more accessible than most fourth-order equations.

(i) Equation (1.1) fits into a variational structure and is the Euler–Lagrange equation of the functional

$$J(u) = \int_\Omega \left\{ \frac{1}{2}(u'')^2 - \frac{q}{2}(u')^2 + F(u) \right\} \mathrm{d}x, \quad \text{where } F(s) = \int_0^s f(t)\,\mathrm{d}t \tag{1.8}$$

and Ω is an appropriate interval.

(ii) Equation (1.1) can be integrated once to yield

$$\mathcal{E}(u) \stackrel{\text{def}}{=} u'u''' - \frac{1}{2}(u'')^2 + \frac{q}{2}(u')^2 + F(u) = E, \tag{1.9}$$

where E is a constant, which is often referred to as the *energy*.

In recent years equation (1.1) has been found to possess a rich structure and, for certain values of q and E, a multitude of qualitatively different solutions. We mention the work of Buffoni, Champneys and Toland [6,7,38] who utilize the Hamiltonian structure of (1.1) and that of Peletier and Troy [26,33,32,43] who devised a method of topological shooting, combined with continuation arguments. We also mention the work of Kalies and van der Vorst [17,18], that of M.A. Peletier [27] and that of Chaparova, Peletier and Tersian [9, 10,37], whose work is based on the variational structure of (1.1), and that of van den Berg who applied the maximum principle to this *fourth*-order equation [39–41]. For an extensive bibliography of papers up to 2001, we refer to the monograph [32] of that year.

As with second-order equations, the solution set of equation (1.1) depends crucially on the choice of function f. In the studies mentioned above the functions

$$f(s) = -s + s^3 \quad \text{and} \quad f(s) = s - s^2, \tag{1.10}$$

have received a great deal of attention. The cubic function (cf. [28–30]) arises in bi-stable systems and the quadratic equation arises in water wave theory (cf. (1.7)). Here we should also mention the two related functions (one is a smooth version of the other)

$$f(s) = (s+1)_+ - 1 \quad \text{and} \quad f(s) = e^s - 1, \tag{1.11}$$

where $s_+ = \max\{s, 0\}$, which were introduced by McKenna and Walter [23,24] in connection with the study of traveling waves in suspension bridges (see also [32] and [31]).

In this chapter we sketch some of the results that have been obtained about the solution structure of equation (1.1) for different functions f, present new results about nonlinearities $f(s)$ which are decreasing for large s, and establish a series of asymptotic results for large values of the parameter q.

About decreasing nonlinearities we prove the following theorem.

THEOREM 1.1. *Let $f \in C^1(\mathbb{R})$ in (1.1) have the following properties*:

$$f'(s) < 0 \quad \text{for } s \in \mathbb{R} \quad \text{and} \quad f(0) = 0, \tag{1.12}$$

and let q be an arbitrary constant in $\mathbb{R}$. Then, for every $E \geqslant 0$, there exists no periodic solution of (1.1) and for every $E < 0$ there exists precisely one symmetric periodic solution $u(x)$ on $\mathbb{R}$. This solution is symmetric with respect to each of its critical points.

Plainly, the solutions described in Theorem 1.1 have a very simple structure. For more complex structures, $f(s)$ will have to be increasing for some values of $s \in \mathbb{R}$. As a first model nonlinearity we shall discuss the function

$$f(s) = s + |s|^{p-1}s, \quad \text{where } p > 1, \tag{1.13}$$

which arises in the context of the Swift–Hohenberg equation (1.3) when $0 < \alpha < 1$.

The following nonexistence theorem was proved in [32].

THEOREM 1.2. *Let f in (1.1) be given by (1.13). Then, if $q \leqslant 0$, there exist no periodic solutions of (1.1).*

For $q > 0$, the situation becomes more interesting and a great variety of qualitatively different periodic solutions emerge as q increases. Specifically, at the values

$$q_{n,m} = \frac{n}{m} + \frac{m}{n}, \quad n = 1, 2, \ldots, m = 1, 2, \ldots, n \geqslant m, \tag{1.14}$$

branches of periodic solutions with zero energy ($E = 0$) bifurcate *supercritically* from the trivial solution and extend al the way to $q = \infty$. In Figure 1(a) we show branches bifurcating from $q_{1,1}$, $q_{3,1}$, $q_{5,1}$ and $q_{7,1}$.

The qualitative properties of solutions on these branches are characterized by the numbers n and m:

(a) The number of monotone segments, or *laps*, between points of symmetry, i.e., points where $u' = 0$ and $u''' = 0$, is given by n in (1.14).

(b) The number of zeros between points of symmetry is given by m.

In Figure 2 we show solutions on branches which bifurcate from $q_{1,1}$, $q_{3,1}$ and $q_{5,1}$, which clearly exhibit the properties (a) and (b).

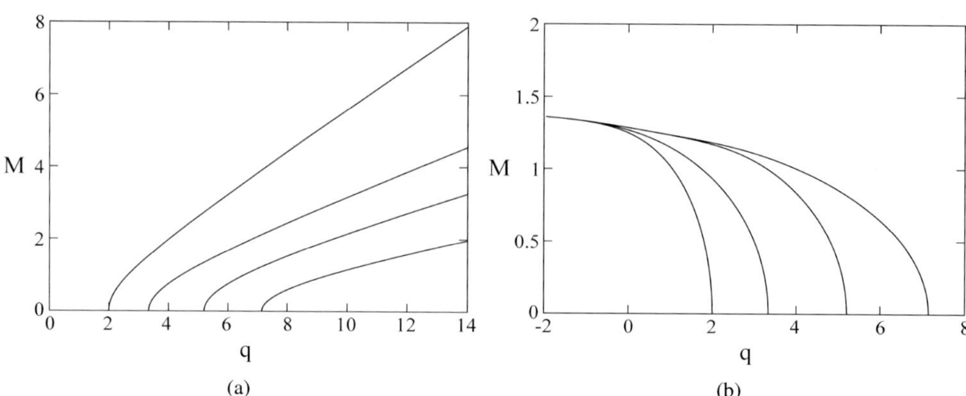

Fig. 1. Branches of one-lap, three-lap, 5-lap and 7-lap solutions of equation (1.1) in the (q, M)-plane, where $M = \sup\{|u(x)|:\ x \in \mathbb{R}\}$. (a) Branches for $f(s) = s + s^3$; (b) branches for $f(s) = s - s^3$.

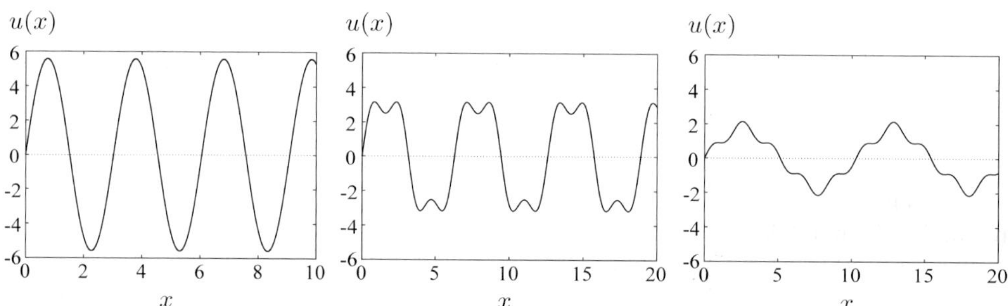

Fig. 2. Periodic solutions of (1.1) with $f(s) = s + s^3$ corresponding to $n = 1$, $n = 3$ and $n = 5$ and $m = 1$.

As a second model equation we discuss equation (1.1) when f is given by

$$f(s) = s - |s|^{p-1}s, \quad p > 1, \tag{1.15}$$

which arises in the context of the nonlinear Schrödinger equation (1.6). When linearizing about the trivial solution, we obtain the same linear equation as with (1.13) and so periodic solutions bifurcate from $u = 0$ at the same critical values $q_{n,m}$ of q. However, in this case the branches all bifurcate sublinearly and extend back to $q = -2$. In Figure 1(b) we show branches bifurcating from $q_{1,1}$, $q_{3,1}$, $q_{5,1}$ and $q_{7,1}$, and we see that they all bend back and converge to one point as $q \to -2^-$. The solutions on these branches have the same qualitative properties as those shown in Figure 2.

As q increases, the two length scales become evident and it is possible to use multiscale asymptotic methods to obtain accurate expressions for the different periodic solutions. Introducing the scaled variables,

$$x^* = q^{-1/2}x \quad \text{and} \quad u^*(x^*) = u(x), \tag{1.16}$$

we can write equation (1.1) as

$$\varepsilon^2 u^{\text{iv}} + u'' + f(u) = 0, \qquad \varepsilon = q^{-1}, \tag{1.17}$$

where we have omitted the asterisks again. In this section we choose

$$f(s) = s + g(s), \qquad g(s) = \mathrm{o}(s) \text{ as } s \to 0,$$

and we assume that g is an odd function. We shall show that for $E = \mathrm{O}(1)$ and large values of q, any periodic solution $u(x)$ can be approximated by an expression of the form

$$u(x) = B_0(x) \pm \varepsilon \rho(\varepsilon, \alpha) \cos \frac{x}{\varepsilon} + \mathrm{O}(\varepsilon^2) \quad \text{as } \varepsilon \to 0, \tag{1.18}$$

where $B_0(x)$ is a solution of the reduced equation

$$u'' + u + g(u) = 0, \tag{1.19}$$

and $\rho(\varepsilon, \alpha)$ is a known constant which depends on ε and on $\alpha = u(0)$. Thus, we see that $u(x)$ can be viewed as a *Baseline solution B_0* with superimposed on it a small amplitude *high frequency oscillation*. In the presentation of this results we shall make frequent use of recent results of Kuske and Peletier [21]. The asymptotic expression (1.18) for periodic solutions makes it possible to obtain accurate approximations of bifurcation curves in the region in which q is large.

The main objective of this chapter is to present methods which have been successful in analyzing fourth-order equations such as equation (1.1) and to indicate basic elements of the structure of solution sets of their often complex solution sets. For the sake of transparency we have focused on equation (1.1) with just a few typical nonlinearities $f(u)$. However, the methods in this chapter are applicable to a wide class of equations. As further examples we mention the equation

$$u^{\text{iv}} + h(u)u'' - \frac{1}{2} h'(u)(u')^2 + f(u) = 0, \tag{1.20}$$

which has been studied in [8,19] and more recently in [4,5] when $f(u) = u^3 - u$, and the system

$$\begin{cases} w'' + w(v - 1) = 0, \\ v'' - \alpha v + \frac{1}{2} w^2 = 0. \end{cases} \tag{1.21}$$

This system arises in the study of coupled nonlinear Schrödinger equations [46] and has been studied in [44,46] and [45]

The plan of the chapter is the following. We begin in Section 2 with a detailed analysis of periodic solutions for decreasing nonlinearities f: their existence, uniqueness, estimates for their period, and their asymptotic behavior for large values of q and E. Then, in Section 3 we turn to an analysis of the solution set when f is given by (1.13), and in Section 4 we discuss equation (1.1) when f is given by (1.15). Finally, in Section 5 we present the multianalysis of periodic solutions when q is large and $E = O(1)$.

The solution graphs in this chapter have been made with XPPAUT [14], the bifurcation graphs with AUTO [13] and the numerical comparison in Section 5 with Matlab.

2. Decreasing nonlinearity $f(u)$

In this section we study periodic solutions of the basic equation

$$u^{\mathrm{iv}} + qu'' + f(u) = 0 \tag{1.1}$$

when the function f is strictly decreasing,

$$f'(s) < 0 \qquad \text{for } s \in \mathbb{R} \quad \text{and} \quad f(0) = 0. \tag{2.1}$$

We shall show that, for any $E < 0$ and any $q \in \mathbb{R}$, there exists a unique periodic solution of equation (1.1), and we study its dependence on the eigenvalue q and the energy E, and its asymptotic behavior as $q \to \pm\infty$ and $E \to -\infty$.

We begin with a few preliminary observations. Let $u(x)$ be a periodic solution of equation (1.1). Then there exist points where $u' = 0$, and at those critical points the energy identity (1.9) becomes

$$E = -\frac{1}{2}(u'')^2 + F(u). \tag{2.2}$$

From (2.1) we conclude that $sf(s) < 0$ for $s \in \mathbb{R} \setminus \{0\}$ so that

$$F(s) = \int_0^s f(t)\,dt < 0 \quad \text{for } s \in \mathbb{R} \setminus \{0\}.$$

This implies the following nonexistence result.

LEMMA 2.1. *If u is a nontrivial periodic solution of equation* (1.1), *then its energy must be negative*: $\mathcal{E}(u) = E < 0$.

Next, we derive a priori upper and lower bounds for any periodic solution $u(x)$ of equation (1.1) in terms of its energy E. Let

$$M_+ = \max\{u(x) : x \in \mathbb{R}\} \quad \text{and} \quad M_- = \min\{u(x) : x \in \mathbb{R}\}. \tag{2.3}$$

Plainly, there exist points at which $u = M_+$ and where $u = M_-$. At these points we have $u' = 0$, and we conclude from the energy identity (1.9) that

$$F(M_\pm) \geqslant -|E|.$$

Hence, since $M_- \leqslant u(x) \leqslant M_+$ on $\mathbb{R}$ and $F'' = f' < 0$ on $\mathbb{R}$, we conclude that

$$F\big(u(x)\big) \geqslant -|E| \quad \text{for } x \in \mathbb{R}. \tag{2.4}$$

Let us denote the roots of the equation

$$F(s) = -|E| \tag{2.5}$$

by $c_-(E)$ and $c_+(E)$, where $c_- < 0 < c_+$. Then (2.4) yields the following a priori bounds.

LEMMA 2.2. *Let $u(x)$ be a periodic solution of equation* (1.1) *with energy $E < 0$. Then*

$$c_-(E) \leqslant u(x) \leqslant c_+(E) \quad \text{for } x \in \mathbb{R}. \tag{2.6}$$

In the next subsection we shall show that for every $E < 0$ there exists a periodic solution of equation (1.1).

2.1. *Existence of a periodic solution*

We construct a periodic solution which is symmetric with respect to *all* its critical points, i.e., at each critical point $u''' = 0$ as well. As in [32] we call such points, *points of symmetry*. The number of monotone segments between two nearest points of symmetry will be called the number of *laps*. Thus, in this section we construct *one-lap* periodic solutions.

Without loss of generality we shall place the origin at a local *minimum*, so that

$$u''(0) \geqslant 0. \tag{2.7}$$

If the first positive critical point – a local maximum – is located at $x = \xi$, then integration of equation (1.1) over $(0, \xi)$ shows that

$$\int_{u(0)}^{u(\xi)} f(s)\,\mathrm{d}s = 0,$$

which implies that

$$u(0) < 0 \quad \text{and} \quad u(\xi) > 0. \tag{2.8}$$

The main result of this subsection is the following existence theorem.

THEOREM 2.1. *For each $q \in \mathbb{R}$ and each $E < 0$, there exists a periodic solution $u(x)$ of equation* (1.1) *with energy E, which is symmetric with respect to each of its critical points.*

We prove Theorem 2.1 by means of the topological shooting argument that was first developed in [30,28,29] (see also [32]). Since the solution is assumed to be symmetric with respect to the origin we shall be considering the initial value problem

$$\begin{cases} u^{\mathrm{iv}} + qu'' + f(u) = 0, & x > 0, \\ (u, u', u'', u''')(0) = (\alpha, 0, \beta, 0), \end{cases} \tag{2.9}$$

where α and β are constants which need to be determined. When we compute the energy E at the origin, where we have put a local minimum, we obtain the following relation between α, β and E:

$$E = -\frac{1}{2}\beta^2 + F(\alpha). \tag{2.10}$$

This relation enables us to express β in terms of α and E,

$$\beta = \pm\sqrt{2\{|E| + F(\alpha)\}}. \tag{2.11}$$

When we choose $\alpha \in [c_-, c_+]$, then $F(\alpha) + |E| \geqslant 0$, so that the expression for β is well defined. In view of the assumption (2.7), we choose

$$u(0) = \alpha < 0 \quad \text{and} \quad u''(0) = \beta \geqslant 0. \tag{2.12}$$

Plainly, problem (2.9) has a unique local solution; we denote it by $u = u(x, \alpha)$.

The basic idea of the shooting argument is to find a value of α such that at the first positive critical point ξ, not only $u' = 0$, but also $u''' = 0$. The solution is then symmetric with respect to ξ. Since by construction it is also symmetric with respect to the origin, the solution can be continued over the whole real line as a periodic solution with period 2ξ.

Thus, we write

$$\xi(\alpha) = \sup\{x > 0 : u'(\cdot, \alpha) > 0 \text{ on } (0, x)\}.$$

LEMMA 2.3. *There exists a constant $\delta > 0$ such that if $c_- < \alpha < c_- + \delta$, then*

$$u'\big(\xi(\alpha), \alpha\big) = 0 \quad \text{and} \quad u'''\big(\xi(\alpha), \alpha\big) < 0. \tag{2.13}$$

PROOF. Note that if $\alpha = c_-$, then

$$u''(0) = 0, \qquad u'''(0) = 0, \qquad u^{\mathrm{iv}}(0) = -f(c_-) < 0.$$

Hence, in a right-neighborhood of the origin we have

$$u''' < 0 \quad \text{and} \quad u'' < 0.$$

If we now increase α slightly, we find that u' changes sign in a right-neighborhood of the origin, so that $\xi(\alpha)$ exists, and that

$$\xi(\alpha) \to 0 \quad \text{and} \quad u\big(\xi(\alpha), \alpha\big) \to c_- \quad \text{as } \alpha \searrow c_-.$$

Moreover, integration of the equation shows that $u'''(\xi(\alpha), \alpha) < 0$ for α close to c_-, as asserted. $\qquad\square$

Next, we seek a value of α for which $u'''(\xi(\alpha), \alpha) > 0$. Note that if $u(\xi) < c_+$, then by the energy identity,

$$u''(\xi) = -2\sqrt{2\big\{|E| + F\big(u(\xi)\big)\big\}} < 0.$$

Therefore, by the Implicit Function Theorem, if $u(\xi(\tilde{\alpha}), \tilde{\alpha}) \in (c_-, c_+)$ for some $\tilde{\alpha} > c_-$, then $\xi(\alpha)$ depends continuously on α in a neighborhood of $\tilde{\alpha}$.

Define the set

$$\mathcal{A} = \big\{\alpha > c_- : u\big(\xi(\alpha'), \alpha'\big) < c_+ \text{ for } c_- < \alpha' < \alpha\big\}.$$

Plainly, $\mathcal{A}$ is an interval of the form (c_-, α^*).

LEMMA 2.4. (a) $\xi(\alpha)$ *is finite for every* $\alpha \in \mathcal{A}$, *and*

$\quad$ (b) $\quad \xi(\alpha^*) < \infty \quad and \quad u\big(\xi(\alpha^*), \alpha^*\big) = c_+;$

$\quad$ (c) $\quad u''\big(\xi(\alpha^*), \alpha^*\big) = 0 \quad and \quad u'''\big(\xi(\alpha^*), \alpha^*\big) > 0.$

PROOF. (a) Suppose that $\xi = \infty$ for some $\alpha \in \mathcal{A}$. Then

$$u'(x) > 0 \quad \text{and} \quad u(x) \leqslant c_+ \quad \text{for all } x > 0.$$

This means that

$$\lim_{x \to \infty} u(x) \text{ exists} \overset{\text{def}}{=} \ell.$$

Since $u = 0$ is the only constant solution of the differential equation, it follows that $\ell = 0$. However, integration of the equation over $\mathbb{R}^+$ shows that $\ell > 0$, a contradiction.

$\quad$ (b) By definition, $u(\xi(\alpha^*), \alpha^*) \geqslant c_+$. However, by the energy identity (1.9), u' cannot vanish if $u > c_+$. Hence $u(\xi(\alpha^*), \alpha^*) = c_+$, as asserted. That $\xi(\alpha^*) < \infty$ follows as in the proof of part (a).

$\quad$ (c) Let $\alpha = \alpha^*$. Then since $u(\xi) = c_+$, the energy identity implies that $u''(\xi) = 0$. Since $u' > 0$ in a left-neighborhood of ξ, it follows that $u'''(\xi) \geqslant 0$. Suppose that $u'''(\xi) = 0$. Then, because $u^{\mathrm{iv}}(\xi) = -f(c_+) > 0$, it follows that $u > c_+$ in left-neighborhood of ξ, a contradiction. $\qquad\square$

PROOF OF THEOREM 2.1. We are now ready to prove that $u'''(\xi(\alpha), \alpha)$ has a zero on $(c_-, \alpha^*]$. For convenience we write

$$\varphi(\alpha) \stackrel{\text{def}}{=} u'''(\xi(\alpha), \alpha). \tag{2.14}$$

Since $\xi(\alpha)$ and $u(x, \alpha)$ depend continuously on α, so does $\varphi(\alpha)$. In Lemma 2.3 we have shown that $\varphi(\alpha) < 0$ for α close to c_-, and in Lemma 2.4 that $\varphi(\alpha^*) > 0$. Therefore, there exists an $\alpha_0 \in (c_-, \alpha^*)$ such that $\varphi(\alpha_0) = 0$. Plainly, by symmetry arguments, we see that the solution $u(x, \alpha_0)$ is a one-lap periodic solution on $\mathbb{R}$ with period 2ξ. Since for the energy E we had chosen an arbitrary negative number, this completes the proof of Theorem 2.1.   □

Let $u(x)$ be a one-lap periodic solution which is symmetric with respect to its critical points, and let its period be denoted by $2L$. In the following lemma we establish an upper bound for the half-period when $q > 0$.

LEMMA 2.5. *Let $u(x)$ be a periodic solution of equation (1.1) which is symmetric with respect to its critical points. If $q \geqslant 0$, its half-period L satisfies the inequality*

$$L < \frac{\pi}{\sqrt{q}}.$$

PROOF. We shift $u(x)$ so that its minimum lies at the origin. Then $u' > 0$ on $(0, L)$. We differentiate equation (1.1), multiply by $\sin(\pi x/L)$ and integrate over $(0, L)$. This yields, after repeated integrations by part,

$$\int_0^L \sin\left(\frac{\pi x}{L}\right) u'(x) \left\{ \frac{\pi^2}{L^2}\left(\frac{\pi^2}{L^2} - q\right) + f'(u(x)) \right\} dx = 0. \tag{2.15}$$

Since $f' < 0$ on $\mathbb{R}$ and $u' > 0$ on $(0, L)$, this implies that

$$\frac{\pi^2}{L^2} > q \quad \text{or} \quad L < \frac{\pi}{\sqrt{q}}. \qquad \square$$

REMARK. Lemma 2.5 shows that as $q \to \infty$, the period shrinks to zero at a rate of $O(q^{-1/2})$. More precise estimates will be given in Section 2.3.

REMARK. Lemma 2.5 can be further refined. Let

$$m = \min\{|f'(s)| : c_- \leqslant s \leqslant c_+\}.$$

Then $f'(s) < -m$, and we deduce from (2.15) that

$$\int_0^L \sin\left(\frac{\pi x}{L}\right) u'(x) \left\{ \frac{\pi^2}{L^2}\left(\frac{\pi^2}{L^2} - q\right) - m \right\} dx > 0. \tag{2.16}$$

This means that

$$\frac{\pi^2}{L^2}\left(\frac{\pi^2}{L^2} - q\right) - m > 0,$$

and hence that

$$L < \frac{\pi\sqrt{2}}{\sqrt{q + \sqrt{q^2 + 4m}}}. \tag{2.17}$$

2.2. *Uniqueness*

In this subsection we show that the periodic solution obtained in Theorem 2.1 is *unique*. The proof is based on ideas due to van den Berg [42].

THEOREM 2.2. *Suppose that the nonlinearity $f(u)$ has the properties* (2.1). *Then for any $q \in \mathbb{R}$ and any $E < 0$, there exists at most one bounded solution of equation* (1.1).

PROOF. We deal with the cases $q \leqslant 0$ and $q > 0$ separately.

Case 1: $q \leqslant 0$. Suppose that u_1 and u_2 are periodic solutions of equation (1.1). Without loss of generality we may assume that u_1 and u_2 have local minima at the origin, and that $u_2(0) \geqslant u_1(0)$. By the energy identity, (2.2) evaluated at critical points, this implies that $u_2''(0) \geqslant u_1''(0)$. Finally, by possibly applying the transformation $x \to -x$, we can ensure that $u_2'''(0) \geqslant u_1'''(0)$. Thus, writing $v = u_2 - u_1$ we obtain

$$v \geqslant 0, \qquad v' = 0, \qquad v'' \geqslant 0, \qquad v''' \geqslant 0 \quad \text{at } x = 0, \tag{2.18}$$

where, by uniqueness, at least one of the inequalities must be strict, and

$$v^{\mathrm{iv}} = |q|v'' - f'(\theta)v, \qquad \theta \in (u_1, u_2). \tag{2.19}$$

It is evident from (2.18) and (2.19) that $v > 0$ and $v' > 0$ in a right-neighborhood of the origin. Let

$$y = \sup\{x > 0 \colon v' > 0 \text{ on } (0, x)\}.$$

Since $v(x)$ is bounded, it follows that $y < \infty$ and $v'(x) \to 0$ as $x \to y$. But, integration of (2.19) over $(0, x)$ yields

$$v'''(x) = v'''(0) + |q|v'(x) - \int_0^x f'(\theta)v \, \mathrm{d}t > 0 \quad \text{for } 0 < x < y.$$

Since $v''(0) > 0$, this implies that $v''(x) > 0$ on $(0, y)$, and in particular, $v'(x)$ cannot tend to 0 as $x \to y$, a contradiction.

Case 2: $q > 0$. Let u_1 be a periodic solution as constructed in Theorem 2.1 with minimum at the origin and maximum at $x = L$, and let u_2 be a bounded solution of equation (1.1), also with a local minimum at the origin. Without loss of generality we may assume that one of the following two sets of inequalities hold:

$$u_2 \geqslant u_1 \quad \text{and} \quad u_2''' \geqslant u_1''' \quad \text{at } x = 0, \tag{A}$$

$$u_2 \leqslant u_1 \quad \text{and} \quad u_2''' \leqslant u_1''' \quad \text{at } x = 0. \tag{B}$$

By uniqueness, in (A) as well as in (B), one of the two inequalities must be strict.

Let ξ be the first zero of u_2' on $\mathbb{R}^+$. Then we shall prove the following inequalities.

(i) If (A) holds, then

$$u_2(\xi) - u_1(L) > u_2(0) - u_1(0) \quad \text{and} \quad u_2'''(\xi) > 0. \tag{2.20}$$

(ii) If (B) holds, then

$$u_2(\xi) - u_1(L) < u_2(0) - u_1(0) \quad \text{and} \quad u_2'''(\xi) < 0. \tag{2.21}$$

By applying (2.20) and (2.21) alternatively, we conclude that the sequence of maxima of u_2 increases and that its sequence of minima decreases. Since u_2 is bounded, these extrema tend to limits, so that u_2 tends to a periodic solution $\bar{u}$, whose maxima and minima lie, respectively, above and below those of u_1. Repeating the argument with $\bar{u}$ instead of u_2 we are led to a contradiction.

Suppose that (A) holds. Let $v = u_2 - u_1$. Then

$$v \geqslant 0, \qquad v' = 0, \qquad v'' \geqslant 0, \qquad v''' \geqslant 0 \quad \text{at } x = 0.$$

Also, define $w = v'' + qv$, and let

$$y = \sup\{x > 0 : v > v(0) \text{ on } (0, x)\}.$$

Then

$$w'' = v^{\text{iv}} + qv'' = -\{f(u_2) - f(u_1)\} > 0 \quad \text{on } (0, y),$$

and since $w' = 0$ and $w \geqslant 0$ at $x = 0$,

$$w' > 0 \quad \text{and} \quad w > 0 \quad \text{on } (0, y]. \tag{2.22}$$

We claim that

$$y > \frac{\pi}{\sqrt{q}} \quad \text{and} \quad v' > 0 \quad \text{on } \left(0, \frac{\pi}{\sqrt{q}}\right). \tag{2.23}$$

Let

$$a = \sup\{x > 0 \colon v' > 0 \text{ on } (0, x)\}.$$

Plainly, $a < y$. We shall prove that $a \geqslant \pi/\sqrt{q}$. Suppose, to the contrary, that $a < \pi/\sqrt{q}$. Then

$$v''' + qv' = w' > 0 \quad \text{on } (0, a]$$

and hence

$$0 < \int_0^a \sin\left(\frac{\pi x}{a}\right)(v''' + qv')\,dx = \left(q - \frac{\pi^2}{a^2}\right)\int_0^a \sin\left(\frac{\pi x}{a}\right)v'(x)\,dx.$$

Since by assumption $q < \pi^2/a^2$, we have a contradiction.

Remembering from Lemma 2.5 that $\pi/\sqrt{q} > L$, we may now conclude that

$$u_2' > u_1' \quad \text{on } (0, L],$$

which shows that

$$\xi > L \quad \text{and} \quad u_2(\xi) > u_1(L).$$

Thus, since $v(L) > v(0)$ or $u_2(L) - u_1(L) > u_2(0) - u_1(0)$, it follows that

$$u_2(\xi) - u_1(L) > u_2(L) - u_1(L) > u_2(0) - u_1(0),$$

as we set out to prove.

In addition,

$$\begin{aligned}
u_2'''(\xi) - u_2'''(0) &= \int_0^\xi u_2^{\mathrm{iv}}(x)\,dx \\
&= -\int_0^\xi f\big(u_2(x)\big)\,dx \\
&= -\int_0^L f\big(u_2(x)\big)\,dx - \int_L^\xi f\big(u_2(x)\big)\,dx.
\end{aligned}$$

Since

$$\xi > L \quad \text{and} \quad u_2(x) > u_2(L) > u_1(L) > 0 \quad \text{for } L < x < \xi,$$

it follows that

$$-\int_L^\xi f\big(u_2(x)\big)\,dx > 0,$$

and because $u_2 > u_1$ on $(0, L)$, we have

$$-\int_0^L f\big(u_2(x)\big)\,\mathrm{d}x > -\int_0^L f\big(u_1(x)\big)\,\mathrm{d}x = u_1'''(L) - u_1'''(0) = 0.$$

Therefore

$$u_2'''(\xi) > u_2'''(0) \geqslant 0,$$

as required in (2.20).

Thus the inequalities in (2.20) have been proved. The inequalities in (2.21) can be proved in an identical manner; we shall omit the details.

This completes the proof of Theorem 2.2. $\qquad\square$

COROLLARY 2.1. *Let $u(x)$ be a periodic solution of equation* (1.1) *in which*

$$f(-s) = -f(s) \quad \text{for } s \in \mathbb{R}.$$

Suppose that $u(0) = 0$. Then

$$u(-x) = -u(x) \quad \text{for } x \in \mathbb{R}.$$

PROOF. Let $u(x)$ be a periodic solution of equation (1.1), shifted so that $u(0) = 0$, $u'(0) > 0$. Note that $u'(0) \neq 0$, because critical values of $u(x)$ are either positive (maxima) or negative (minima). Then the function $v(x) = -u(-x)$ satisfies the equation

$$v^{\mathrm{iv}} + qv'' - f(-v) = 0.$$

Remembering the asymmetry of f we find that

$$v^{\mathrm{iv}} + qv'' + f(v) = 0, \quad v(0) = 0, v'(0) > 0.$$

By uniqueness, $v = u$, so that

$$u(-x) = -u(x) \quad \text{for } x \in \mathbb{R}. \qquad\square$$

Theorems 2.1 and 2.2 together state that, for every $q \in \mathbb{R}$ and for every $E < 0$, there exists a unique nontrivial solution $u(x; q, E)$ of equation (1.1). It is symmetric with respect to its critical points. We define the *amplitude*

$$M(q, E) \stackrel{\text{def}}{=} \max\big\{|u(x; q, E)| : x \in \mathbb{R}\big\}. \tag{2.24}$$

In Figure 3 we show graphs of M when $f(s) = -s - s^3$. In the one versus q we have put $E = -1$ and in the one versus E we put $q = 0$.

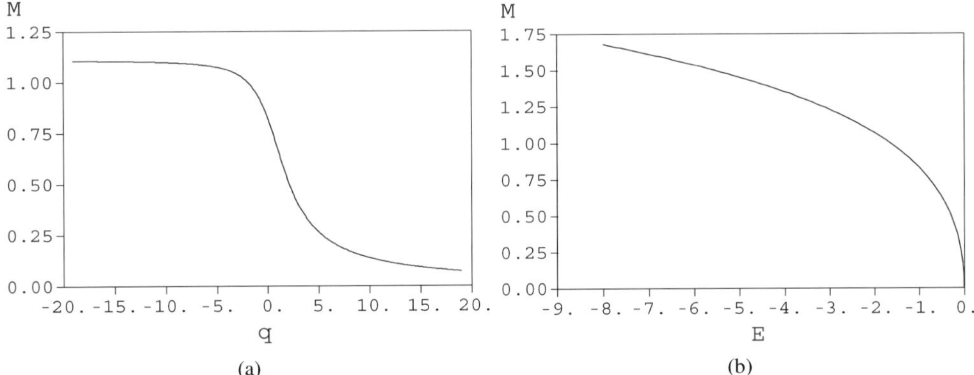

Fig. 3. Branches of periodic solutions of equation (1.1) with $f(s) = -s - s^3$: (a) M versus q ($E = -1$) and (b) M versus E ($q = 0$).

2.3. *Asymptotics*

In this subsection we study how the periodic solution evolves as the two parameters, q and E, become large, and when E becomes small. We first focus on the asymptotic properties of the periodic solution if $q \to -\infty$ and if $q \to +\infty$. As in comparable studies for the bi-stable nonlinearity [32], we find that as $q \to -\infty$, they converge to the periodic solution of the second-order equation with the given value of E, and as $q \to +\infty$ the periodic solutions shrink at a rate of $O(q^{-1})$. We shall do this analysis for the nonlinearity

$$f(s) = -s - s^3. \tag{2.25}$$

By symmetry, $c_+(E) = -c_-(E)$, and we shall write $c(E) = |c_\pm(E)|$.

2.3.1. *Large q.*   We first prove the following limit for $q \to -\infty$.

THEOREM 2.3. *Let $u(x; q)$ be the periodic solution for given $E < 0$ and $q \in \mathbb{R}$, shifted so that $u(0; q) = 0$ and $u'(0; q) > 0$. Then*

$$u(x; q) \sim V(y), \quad y = \frac{x}{\sqrt{|q|}} \text{ as } q \to -\infty, \tag{2.26}$$

where V is the solution of the equation

$$V'' + V + V^3 = 0,$$

with $V(0) = 0$, $V'(0) > 0$ and energy E, i.e.,

$$\max\{|V(y)| : y \in \mathbb{R}\} = \sqrt{\sqrt{1 + 4|E|} - 1}.$$

PROOF. We put $y = x/\sqrt{|q|}$ and write $u(x; q) = v(y; q)$. Then, when we transform (1.1) to the new variables y and v, we obtain, if $q < 0$,

$$\varepsilon^2 v^{\mathrm{iv}} - v'' - v - v^3 = 0, \quad \varepsilon = \frac{1}{|q|}, \tag{2.27}$$

and the energy identity becomes

$$\varepsilon^2 \left(v'v''' - \frac{1}{2}(v'')^2 \right) + \frac{1}{2}(v')^2 - \frac{1}{2}v^2 - \frac{1}{4}v^4 = E. \tag{2.28}$$

Since $|v(y; q)| < c(E)$ on $\mathbb{R}$ by Lemma 2.2, it follows from a simple Maximum Principle argument for $w = v''$, that

$$\left| v''(y; q) \right| < c(E) + c^3(E) \quad \text{for } y \in \mathbb{R}.$$

Let $\varphi \in C_0^\infty(\mathbb{R})$. Then, when we multiply (2.27) by φ, integrate over $\mathbb{R}$ and perform a few integrations by part, we find that

$$\varepsilon^2 \int_{\mathbb{R}} \varphi'' v'' + \int_{\mathbb{R}} \{\varphi' v' - \varphi(v + v^3)\} = 0.$$

Since the sets $\{v(\cdot; q)\}$ and $\{v'(\cdot; q)\}$ are both equicontinuous, we can let $q \to -\infty$ along a sequence to find that $v(y; q) \to V(y)$ and $v'(y; q) \to V'(y)$, where

$$V'' + V + V^3 = 0, \quad V(0) = 0, V'(0) > 0.$$

Doing the same in the energy identity, we find that

$$\frac{1}{2}(V')^2 - \frac{1}{2}V^2 - \frac{1}{4}V^4 = E.$$

These two relations together determine V uniquely, so that we may conclude that

$$\lim_{q \to \infty} v(y; q) = V(y).$$

This is the limit we set out to prove. $\square$

Next, let us consider the limit in the other direction, i.e., for $q \to \infty$.

THEOREM 2.4. *Let $u(x; q)$ be the periodic solution for given $E < 0$ and $q \in \mathbb{R}$ shifted so that $u(0; q) = 0$ and $u'(0; q) > 0$. Then*

$$u(x; q) \sim \frac{\sqrt{2|E|}}{q} \sin(x\sqrt{q}) \quad \text{as } q \to +\infty. \tag{2.29}$$

PROOF. We follow the proof of a comparable result in [32] (cf. Theorem 4.2.3) and begin with a preliminary estimate.

LEMMA 2.6. *Suppose that $q > 0$. Then*

$$|u(x; q)| < \sqrt{\frac{2|E|}{1 + q^2}} \quad \text{for } x \in \mathbb{R}.$$

PROOF. Write $w = u'' + qu$. Then

$$w'' = u^{\text{iv}} + qu'' = u + u^3 > 0 \quad \text{on } (0, \xi),$$

where ξ is the first positive zero of u'. Since $w'(\xi) = u'''(\xi) + qu'(\xi) = 0$, it follows that $w' < 0$ on $(0, \xi)$ and hence, because $w(0) = 0$, that

$$w(\xi) = u''(\xi) + qu(\xi) < 0,$$

so that

$$|u''(\xi)| > qu(\xi). \tag{2.30}$$

At ξ, the energy identity yields

$$\frac{1}{2}(u'')^2 = |E| - \frac{1}{2}u^2 - \frac{1}{4}u^4. \tag{2.31}$$

Using (2.30) to estimate the left-hand side of (2.31), we arrive at the required upper bound. $\qquad\square$

We continue with the proof of Theorem 2.4 and scale the variables as suggested by Lemma 2.5,

$$x = \frac{t}{\sqrt{q}}, \qquad u(x; q) = \frac{1}{q}v(t; q) \quad \text{and} \quad \xi = \frac{\tau}{\sqrt{q}}.$$

For v, we then obtain the equation

$$v^{\text{iv}} + v'' - \varepsilon^2 v - \varepsilon^4 v^3 = 0, \quad \varepsilon = \frac{1}{|q|},$$

and the energy identity

$$v'v''' - \frac{1}{2}(v'')^2 + \frac{1}{2}(v')^2 - \frac{\varepsilon^2}{2}v^2 - \frac{\varepsilon^4}{4}v^4 = E.$$

When we formally take the limit as $q \to \infty$ or $\varepsilon \to 0$, we find that $v(t; q) \to V(t)$ where V satisfies

$$V^{iv} + V'' = 0, \qquad V(0) = 0 \quad \text{and} \quad V'V''' - \frac{1}{2}(V'')^2 + \frac{1}{2}(V')^2 = E.$$

This implies that

$$V(t) = \sqrt{2|E|}\sin t,$$

as asserted. For the proof of this limit we refer to Section 4.2 of [32]. $\qquad\qquad\qquad\square$

2.3.2. *Small and large negative energy.*   It follows from Lemma 2.2 that

$$\left|u(x)\right| < c(E) = \sqrt{\sqrt{1 + 4|E|} - 1} < \sqrt{2|E|}. \tag{2.32}$$

This suggests that we scale $u(x)$ with $|E|^{1/2}$ for E small, and with $|E|^{1/4}$ for E large and negative.

Starting with small energy, we write $u(x) = |E|^{1/2}v(x)$ and substitute into (1.1) with f given by (2.25). This yields the equation

$$v^{iv} + qv'' - v - |E|v^3 = 0. \tag{2.33}$$

When we let $E \to 0$, and assume that $v(x) \to V(x)$, we find that V is a solution of the equation

$$V^{iv} + qV'' - V = 0. \tag{2.34}$$

The characteristic equation of (2.34) is

$$\lambda^4 + q\lambda^2 - 1 = 0$$

with roots $\lambda = \pm a$ and $\lambda = \pm ib$, where

$$a = \frac{1}{2}\left(\sqrt{q^2 + 4} - q\right) \quad \text{and} \quad b = \frac{1}{2}\left(\sqrt{q^2 + 4} + q\right). \tag{2.35}$$

Using arguments employed in establishing the limit for $q \to \pm\infty$ we obtain the following theorems.

THEOREM 2.5. *Let $q \in \mathbb{R}$ be fixed and let $u(x; E)$ be the periodic solution with energy $E < 0$, shifted so that $u(0; E) = 0$ and $u'(0; E) > 0$. Then*

$$u(x; E) \sim A(q)\sqrt{|E|}\sin(bx) \quad \text{as } E \to 0 \tag{2.36}$$

with

$$A(q) = \frac{2}{\sqrt{q^2 + q\sqrt{q^2 + 4} + 4}}.$$

Proceeding in an entirely similar manner we find the following limit for $E \to -\infty$.

THEOREM 2.6. *Let $q \in \mathbb{R}$ be fixed and let $u(x; E)$ be the periodic solution with energy $E < 0$, shifted so that $u(0; E) = 0$ and $u'(0; E) > 0$. Then*

$$u(x; E) \sim |E|^{1/4} V(y), \qquad y = |E|^{1/8} x \quad as \ E \to 0, \tag{2.37}$$

where V is the solution of the equation

$$V^{\text{iv}} - V^3 = 0, \quad V(0) = 0$$

with first integral

$$V'V''' - \frac{1}{2}(V'')^2 - \frac{1}{4}V^{\text{iv}} = 1.$$

The results of Theorems 2.3–2.6 are clearly borne out in the bifurcation graphs shown in Figure 3.

3. A superlinear bifurcation problem

Whereas the solution structure we found when $f(u)$ is decreasing turned out to be very simple – for every negative value of the energy E, we found a unique branch of periodic solutions extending from $q = -\infty$ to $q = +\infty$ – this no longer is the case for functions $f(u)$ which are increasing or nonmonotone. In this section we consider the relatively simple case of an increasing nonlinearity:

$$f(s) = s + s^p, \quad p > 1, \tag{3.1}$$

where we shall mean $s^p = |s|^{p-1}s$. Thus, we consider the equation

$$u^{\text{iv}} + qu'' + u^p + u = 0. \tag{3.2}$$

We shall find that results developed for (3.2) can be used to analyze more complicated nonlinearities. One such nonlinearity will be discussed in Section 4. We begin with some preliminaries, including a nonexistence theorem restricting the range of q in which we need look, then discuss an infinite family of branches of periodic solutions which branch off the trivial solution, and finally study their behavior as $q \to \infty$. We shall make frequent use of methods and results developed in [32].

3.1. *Preliminaries*

We begin with a nonexistence theorem.

THEOREM 3.1. *Equation (3.2) has no periodic solutions for $q \leqslant 2$.*

PROOF. We use an energy argument. Suppose that u is a nontrivial periodic solution with period $(0, L)$. We multiply equation (3.2) by u and integrate over $(0, L)$. Then, after some integrations by parts, we obtain

$$\int_0^L (u'')^2 \, dx - q \int_0^L (u')^2 \, dx + \int_0^L \left(u^2 + |u|^{p+1}\right) dx = 0. \tag{3.3}$$

But

$$2 \int_0^L (u')^2 \, dx = 2 \int_0^L u u'' \, dx \leqslant \int_0^L u^2 \, dx + \int_0^L (u'')^2 \, dx.$$

Using this interpolation inequality in (3.3) we obtain

$$(q - 2) \int_0^L (u')^2 \, dx \geqslant \int_0^L |u|^{p+1} \, dx. \tag{3.4}$$

Since u is nontrivial, the right-hand side in (3.4) is positive and it follows that $q > 2$. This implies that there can be no periodic solutions, with whatever energy, if $q \leqslant 2$. $\qquad\square$

We shall particularly focus on periodic solutions which bifurcate from the trivial solution $u = 0$. Thus, we need to inspect the linear equation, obtained from (3.2) by omitting the nonlinear term,

$$v^{\mathrm{iv}} + q v'' + v = 0. \tag{3.5}$$

The roots of its characteristic equation are

$$\lambda_{\pm} = \pm ia \quad \text{and} \quad \lambda_{\pm} = \pm ib, \tag{3.6}$$

where a and b are the positive roots of

$$a^2 = \frac{1}{2}\left(q + \sqrt{q^2 - 4}\right) \quad \text{and} \quad b^2 = \frac{1}{2}\left(q - \sqrt{q^2 - 4}\right). \tag{3.7}$$

Plainly, a and b are well defined if $q \geqslant 2$.

We shall find that branches of solutions bifurcate from $u = 0$ at values of q where *resonance* occurs, i.e., where the fraction a/b becomes rational, so that for some integers $n \geqslant 1$ and $m \geqslant 1$,

$$\frac{a}{b} = \frac{n}{m} \quad \Longleftrightarrow \quad q = q_{n,m} = \frac{n}{m} + \frac{m}{n}, \quad m \leqslant n. \tag{3.8}$$

The integers n and m will translate into specific geometric properties of the solutions. It will be convenient to write $q_n = q_{n,1}$. The sequence $\{q_n\}_{n=1}^{\infty}$ is seen to be increasing, starting from $q_1 = 2$.

We shall be studying single and multibump periodic solutions of equation (3.2) which are either even or odd. These solutions will often have more than two critical points per period; to keep track of them we introduce the following notation. Let $u(x)$ be a solution. We denote its local maxima on $\mathbb{R}^+$ by $\xi_1, \xi_2, \xi_3, \ldots$ and its local minima by $\eta_1, \eta_2, \eta_3, \ldots,$ where

$$\xi_k \leqslant \eta_k \leqslant \xi_{k+1} \leqslant \eta_{k+1}.$$

If $u'(x) > 0$ for small positive x, then the first maximum is ξ_1 and the first minimum is therefore η_1. If $u'(x) < 0$ for small $x > 0$, then the first minimum is η_1 and hence the first maximum is ξ_2. With the nonlinearity $f(s)$ defined by (3.1) it was shown in [32] that these sequences of critical points are infinite. Sometimes, it will be convenient to refer to these critical points collectively. We then denote them by ζ_j, where $\zeta_j \leqslant \zeta_{j+1}$, and ζ_1 is the first positive critical point.

In the analysis of multibump solutions of equation (3.2), the location and position of critical points of the zero energy solutions of the *linear* equation (3.5) play a pivotal role. In particular, we are interested in the solution $v(x)$ of the problem

$$\begin{cases} v^{\mathrm{iv}} + qv'' + v = 0, & x > 0, \\ v = 0,\, v' = 1,\, v'' = 0,\, v''' = -q/2 & \text{at } x = 0, \end{cases} \tag{3.9}$$

where the initial data have been chosen so that $\mathcal{E}(v) = 0$. We shall repeatedly use the following important property of solutions of problem (3.9).

LEMMA 3.1. *Let $n \geqslant 1$ and $q > q_{2n-1}$. Then*

$$v(\zeta_j) > 0 \quad \text{and} \quad v'''(\zeta_j) > 0 \quad \text{for } j = 1, 2, \ldots, n.$$

PROOF. The solution of problem (3.9) is given by

$$v(x) = \frac{1}{2}\left(\frac{\sin(ax)}{a} + \frac{\sin(bx)}{b}\right), \tag{3.10}$$

where a and b are given by (3.7). An elementary computation shows that a critical point ζ can be found as a root of the equation

$$\cos(a\zeta) + \cos(b\zeta) = 0, \tag{3.11}$$

and that

$$v'''(\zeta) = \frac{1}{2}\sqrt{q^2 - 4}\cos(b\zeta). \tag{3.12}$$

Using (3.10) and (3.12) to determine the value of v, respectively v''', at the roots of equation (3.11), we establish the desired properties. $\qquad\square$

In order to characterize the shape of solutions, we recall the notions of a *point of symmetry* and a *lap* of a function. We say a point $x_0 \in \mathbb{R}$ is a *point of symmetry* of the function $\phi \in C(\mathbb{R})$ if

$$\phi(x_0 + y) = \phi(x_0 - y) \quad \text{for } y \in \mathbb{R}.$$

By a *lap* of the function $\phi \in C^1(\mathbb{R})$ we mean an interval (x_1, x_2), where x_1 and x_2 are finite, such that

$$\phi'(x) \neq 0 \quad \text{for } x \in (x_1, x_2) \quad \text{and} \quad \phi'(x_1) = 0, \qquad \phi'(x_2) = 0.$$

Thus, the interval $(0, \pi)$ is a lap of the function $\cos(x)$.

3.2. *Multibump periodic solutions*

We begin with the construction of a family of relatively simple periodic solutions u of equation (3.2). Subsequently we turn to more complex periodic solutions. We first construct a family of odd periodic solutions.

THEOREM 3.2. *Let $n \geqslant 1$. For each $q > q_{2n-1}$, there exist an odd periodic solution $u_n(x)$ of equation (3.2) which has $2n - 1$ laps in each half-period, such that $u'_n(0) > 0$. Its first positive point of symmetry is ζ_n and its period is $4\zeta_n$. It has the following properties:*

$$
\begin{aligned}
u_n(x) &> 0 && \text{for } 0 < x < 2\zeta_n, \\
u'_{j+1}(0) &< u'_j(0) && \text{for } j = 1, 2, \ldots, n - 1 \ (n \geqslant 2), \\
u'''_n(\zeta_{j+1}) &< u'''_n(\zeta_j) && \text{for } j = 1, 2, \ldots, n - 1 \ (n \geqslant 2).
\end{aligned}
\tag{3.13}
$$

Since $u'''_n(\zeta_n) = 0$, the third inequality implies that $u'''_n(\zeta_j) > 0$ for all $j = 1, 2, \ldots, n - 1$.

In Figure 4 we show the first three solutions u_1, u_2 and u_3 of this theorem.

The proof of Theorem 3.2, much like those to follow, is based on a shooting argument. Let $u(x)$ be an odd solution. Then it can be viewed as a solution of the initial value problem

$$
\begin{cases}
u^{\mathrm{iv}} + q u'' + u^p + u = 0, & x > 0, \\
u = 0, u' = \alpha, u'' = 0, u''' = \beta & \text{at } x = 0,
\end{cases}
\tag{3.14}
$$

where α and β are suitable constants. Conversely, if $u(x)$ is a solution of problem (3.14) on $\mathbb{R}^+$ for some α and β, then by reflection we can extend this solution to one on the whole real line. Thus it suffices to find periodic solutions of problem (3.14) with the desired geometric properties.

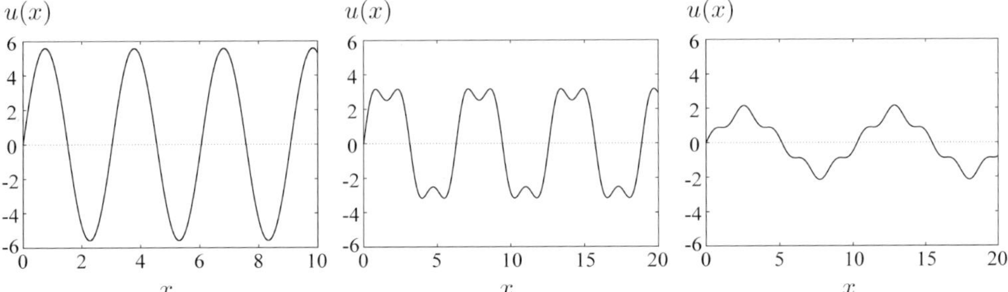

Fig. 4. The one-lap, three-lap and 5-lap solution of equation (3.2) for $p = 3$ (cf. Theorem 3.2).

Without loss of generality we may choose $\alpha \geqslant 0$ in (3.14). Evaluating the energy identity at the origin we find that α, β and E are related through

$$\alpha\beta + \frac{q}{2}\alpha^2 = 0, \tag{3.15}$$

where we have put $E = 0$. We can distinguish two types of solutions:

$$\text{Type I:} \quad \alpha > 0, \beta = -\frac{q}{2}\alpha \quad \text{and} \quad \text{Type II:} \quad \alpha = 0, \beta > 0.$$

Thus, in Type I, the parameter $\alpha > 0$ is arbitrary and in Type II, β is arbitrary. We denote the solution of problem (3.14) by respectively $u(x; \alpha)$ and $u(x; \beta)$. It is easily established that these solutions exist in a neighborhood of the origin and that they depend continuously on α, respectively β, on bounded intervals.

PROOF OF THEOREM 3.2. Since, by assumption, $\alpha > 0$, we seek a solution of Type I and we have the freedom to choose α. We select α so that $u(x; \alpha)$ becomes a solution with the desired geometric properties. This is done by carefully following the way the local maxima $u(\xi_1)$, $u(\xi_2)$, $u(\xi_3)$, ... and the local minima $u(\eta_1)$, $u(\eta_2)$, $u(\eta_3)$, ... vary with α, and tuning α so that they end up in the right positions. This is possible because $\xi_j \in C(\mathbb{R}^+)$ and $\eta_j \in C(\mathbb{R}^+)$ for all $j \geqslant 1$ [32].

We first prove the theorem for $n = 1$. For values $n \geqslant 2$ the result then follows upon iteration. In tuning α we specifically follow the value of $u'''(\xi_1(\alpha), \alpha)$. For convenience, we write

$$\varphi_1(\alpha) \stackrel{\text{def}}{=} u'''(\xi_1(\alpha), \alpha). \tag{3.16}$$

Suppose that $\varphi_1(\alpha) = 0$ for some $\alpha > 0$. Then $u' = 0$ and $u''' = 0$ at $\xi_1(\alpha)$ and so, $u(x)$ is symmetric with respect to ξ_1. Thus, by defining the function

$$w(x) = \begin{cases} u(x) & \text{for } 0 \leqslant x \leqslant \xi_1, \\ u(2\xi_1 - x) & \text{for } \xi_1 \leqslant x \leqslant 2\xi_1, \end{cases}$$

we obtain a solution of equation (3.2) on $(0, 2\xi_1)$: the first half of the periodic solution. The second half is obtained by translating $w(x)$ over a distance $2\xi_1$ and changing the sign. The function so constructed is *one* period of a periodic solution of equation (3.2) with period $4\xi_1$.

The essential ingredient in the construction above was the existence of a zero of φ_1. The existence of such a zero will be established by finding values of α for which φ_1 is positive and values for which it is negative. Continuity then ensures the existence of a zero.

We first determine the sign of $\varphi_1(\alpha)$ when α is large.

LEMMA 3.2. *Let $q \in \mathbb{R}$. Then there exists a constant $\alpha_+ > 0$ such that $\varphi_1(\alpha) < 0$ for $\alpha > \alpha_+$.*

PROOF. We use a scaling argument and introduce the new variables

$$t = \alpha^{(p-1)/(p+3)} x \quad \text{and} \quad v(t; \alpha) = \alpha^{-4/(p+3)} u(x; \alpha).$$

In terms of these variables problem (3.14) becomes

$$\begin{cases} v^{\text{iv}} + \alpha^{-2(p-1)/(p+3)} q v'' + \alpha^{-4(p-1)/(p+3)} v + v^p = 0, \\ v = 0, \, v' = 1, \, v'' = 0, \, v''' = -\frac{q}{2} \alpha^{-2(p-1)/(p+3)} \qquad \text{at } t = 0. \end{cases} \tag{3.17}$$

Let V be the solution of the limit problem, formally obtained by letting $\alpha \to \infty$,

$$\begin{cases} V^{\text{iv}} + V^p = 0, \\ V = 0, \, V' = 1, \, V'' = 0, \, V''' = 0 \quad \text{at } t = 0. \end{cases} \tag{3.18}$$

Because problem (3.17) is a regular perturbation of the limit problem (3.18), it follows that $v^{(j)}(t; \alpha) \to V^{(j)}(t)$ as $\alpha \to \infty$ uniformly on bounded intervals for $j = 0, 1, 2, 3, 4$. We see that $V''' < 0$ and $V'' < 0$ as long as $V > 0$. Hence

$$T = \sup\{t > 0 : V' > 0 \text{ on } (0, t)\}$$

is finite and $V''(T) < 0$ and $V'''(T) < 0$. Plainly,

$$\begin{cases} \xi_1(\alpha) \sim \alpha^{-(p-1)/(p+3)} T, \\ u'''(\xi_1(\alpha), \alpha) \sim \alpha^{(3p+1)/(p+3)} V'''(T) < 0, \end{cases} \quad \text{as } \alpha \to \infty,$$

so that $\varphi_1(\alpha) < 0$ for α large enough. $\qquad \square$

Next, we determine the sign of $\varphi_1(\alpha)$ when α is small.

LEMMA 3.3. *If $q > q_1 = 2$, there exists a constant $\alpha_- > 0$ such that $\varphi_1(\alpha) > 0$ for $0 < \alpha < \alpha_-$.*

PROOF. In [32] it was shown that for α sufficiently small, the solution $u(x)$ inherits the qualitative properties of the solution $v(x)$ of problem (3.9) on bounded intervals. In particular, because $n = 1$, we deduce from Lemma 3.1 that $u'''(\xi_1) > 0$, as required. $\qquad\square$

We conclude from Lemmas 3.2 and 3.3, and the continuity of $\varphi_1(\alpha)$ that there exists a point $\alpha_1^* \in (\alpha_-, \alpha_+)$ such that $\varphi_1(\alpha_1^*) = 0$. This implies that the function $u_1(x) = u(x; \alpha_1^*)$ is a periodic solution of equation (3.2). Plainly, all its critical points are points of symmetry and hence it is a *one-lap* solution, and $u(x; \alpha_1^*) > 0$ for $0 < x < 2\xi_1(\alpha_1^*)$. This concludes the proof of the case $n = 1$.

Next, let $n = 2$. We now look for a periodic solution which is symmetric with respect to its second critical point $\zeta_2 = \eta_1$. Lemma 3.1 states that if $q > q_3$, then

$$v(\eta_1) > 0 \quad \text{and} \quad v'''(\eta_1) > 0,$$

and hence, for $\alpha > 0$ small,

$$u(\eta_1) > 0 \quad \text{and} \quad u'''(\eta_1) > 0.$$

Let

$$\tilde{\alpha}_2 = \sup\{\alpha' > 0 \colon u(\eta_1) > 0 \text{ for } 0 < \alpha < \alpha'\}.$$

Since $u_1(\eta_1) = -u_1(\xi_1) < 0$ when $\alpha = \alpha_1^*$, it follows that $\tilde{\alpha}_2 \in (0, \alpha_1^*)$. Note that

$$u(\eta_1) = 0, \qquad u''(\eta_1) = 0 \quad \text{and} \quad u'''(\eta_1) < 0 \quad \text{at } \tilde{\alpha}_2. \tag{3.19}$$

The first equality is obvious and the second one follows from the energy identity. As to u''', by uniqueness, $u''' \neq 0$. This means that u''' must be negative. Hence, the function $\varphi_2(\alpha) = u'''(\eta_1(\alpha), \alpha)$ changes sign on the interval $(0, \tilde{\alpha}_2)$ so that there exists a point $\alpha_2^* \in (0, \tilde{\alpha}_2)$ such that $\varphi_2(\alpha_2^*) = 0$. Thus $u_2(x) = u(x; \alpha_2^*)$ is a *two-lap* periodic solution with period $4\eta_1(\alpha^*)$ such that

$$u_2(x) > 0 \quad \text{for} \quad 0 < x < 2\eta_1.$$

Note that $\alpha_2^* < \alpha_1^*$, which proves the second property.

The third property of u_2 follows from the following lemma.

LEMMA 3.4. *Let $u(x)$ be a solution of equation (1.1) and a and b are two critical points of $u(x)$ such that*

$$a < b \quad \text{and} \quad f(u) > 0 \quad \text{on } (a, b).$$

Then

$$u'''(a) > u'''(b).$$

PROOF. Integration of (1.1) over (a, b) yields

$$u'''(b) - u'''(a) + \int_a^b f(u(x)) \, dx = 0.$$

Since the integral is positive, the assertion follows. $\qquad\qquad\square$

In particular we see that $u_2'''(\zeta_1) > u_2'''(\zeta_2)$. This completes the proof of Theorem 3.2 for $n = 2$.

REMARK 3.1. Since by (3.19)

$$u(\eta_1) = 0 \quad \text{and} \quad u''(\eta_1) = 0 \quad \text{at } \tilde{\alpha}_2,$$

it follows that

$$u(\eta_1 + y) = -u(\eta_1 - y) \quad \text{for } y \in \mathbb{R}.$$

In particular,

$$u(2\eta_1) = 0 \quad \text{and} \quad u''(2\eta_1) = 0.$$

Therefore, the solution $\tilde{u}_2(x) = u(x; \tilde{\alpha}_2)$ is also periodic, with period $2\eta_1$, and

$$\tilde{u}_2(x) > 0 \quad \text{on } (0, \eta_1) \quad \text{and} \quad \tilde{u}_2(\eta_1) = 0.$$

Plainly, nowhere on $[0, 2\eta_1]$ does this solution have a point of symmetry, i.e., a point where both $\tilde{u}_2' = 0$ and $\tilde{u}_2''' = 0$.

Suppose that $n = 3$. Then we look for a periodic solution which is symmetric with respect to its third critical point $\zeta_3 = \xi_2$. Lemma 3.1 states that if $q > q_5$, then

$$u(\xi_2) > 0 \quad \text{and} \quad u'''(\xi_2) > 0 \quad \text{for } \alpha > 0 \text{ small.}$$

Since $u'''(\eta_1) = 0$ at α_2^*, it follows from Lemma 3.4 that

$$u'''(\xi_2) < 0 \quad \text{at } \alpha_2^*.$$

Thus, we have shown that

$$u''(\xi_2) > 0 \quad \text{for } \alpha \text{ small} \quad \text{and} \quad u'''(\xi_2) < 0 \quad \text{for } \alpha = \alpha_2^*.$$

It follows that there exist an $\alpha_3^* \in (0, \alpha_2^*)$ such that $u'''(\xi_2) = 0$ at α_3^*, and $u_3(x) = u(x; \alpha_3^*)$ is a periodic solution which is symmetric with respect to ξ_2. This establishes the existence of a *three-lap* periodic solution u_3. By construction, it has the properties

$$u_3(x) > 0 \quad \text{for } 0 < x < 2\xi_2, \quad u_3'(0) < u_2'(0) \quad \text{and} \quad u_3'''(\xi_1) > u_3'''(\eta_1) > 0,$$

where the last two inequalities follow from an application of Lemma 3.4.

REMARK 3.2. Note that when $\alpha = \alpha_2^*$, then $u(\eta_2) < 0$, and when $\alpha = \alpha_3^*$, then $u(\eta_2) > 0$. Hence

$$\tilde{\alpha}_3 = \sup\{\alpha' > \alpha_3^*\colon u(\eta_2) > 0 \text{ for } \alpha_3^* < \alpha < \alpha'\} < \alpha_2^*$$

and

$$u(\eta_2) = 0, \qquad u''(\eta_2) = 0 \quad \text{and} \quad u'''(\eta_2) < 0 \quad \text{at } \tilde{\alpha}_3.$$

Hence, $\tilde{u}_3(x) = u(x, \tilde{\alpha}_3^*)$ is a periodic solution with period $2\eta_2$, and

$$\tilde{u}_3(x) > 0 \quad \text{on } (0, \eta_2).$$

Continuing in this manner we can construct for every $n \geqslant 1$ a periodic solution u_n if $q > q_{2n-1}$, which is symmetric with respect to $x = \zeta_n$ and has n laps on $(0, \zeta_n)$, and which is positive on $(0, 2\zeta_n)$, and has the required properties.

This completes the proof of Theorem 3.2. $\qquad\qquad\qquad\qquad\qquad\qquad\square$

An easy corollary of the proof of Theorem 3.2, based on the reasoning given in the Remarks 3.1 and 3.2, is a *second* family of periodic solutions. Its properties are formulated in the next theorem.

THEOREM 3.3. *Let $n \geqslant 1$. For each $q > q_{2n+1}$, there exist an odd periodic solution $\tilde{u}_n$ of equation (3.2) which has $4n - 1$ laps in each period, such that $\tilde{u}_n'(0) > 0$. It has the properties*

$$\tilde{u}_n(x) = -\tilde{u}_n(2\eta_n - x) \quad \text{for } 0 < x < 2\eta_n,$$

$$\tilde{u}_n(x) > 0 \qquad\qquad\quad \text{for } 0 < x < \eta_n.$$

In Figure 5 we show graphs of $\tilde{u}_1$, $\tilde{u}_2$ and $\tilde{u}_3$ established in Theorem 3.3.

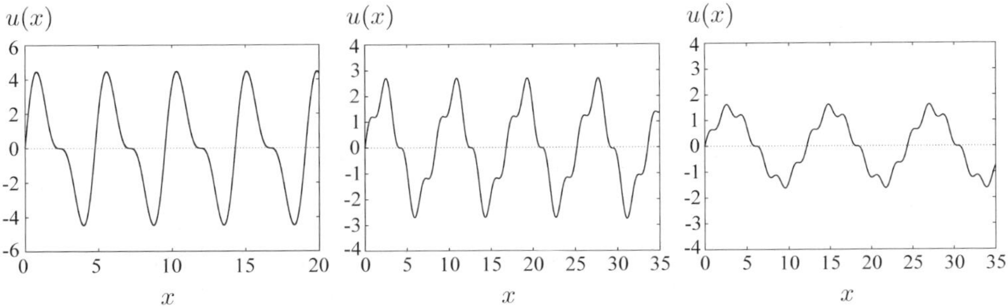

Fig. 5. The three-lap, 7-lap and 11-lap solution of equation (3.2) for $p = 3$ (cf. Theorem 3.3).

                                   L.A. Peletier

In an entirely analogous manner we can establish the existence of a Type II family of periodic solutions of equation (3.2). They are characterized by the fact that $\alpha = u'(0) = 0$. Specifically, we obtain the following theorem.

THEOREM 3.4. *Let $n \geqslant 1$. For each $q > q_{2n+1}$ there exist an odd periodic solution $\bar{u}_n$ of equation (3.2) which has $2n$ laps in each half-period, such that $\bar{u}'_n(0) = 0$. It has the properties*

$$\bar{u}_n(x) = \tilde{u}_n(2\eta_n - x) \quad \text{for } 0 < x < 2\eta_n,$$

$$\bar{u}_n(x) > 0 \qquad \qquad \text{for } 0 < x < \eta_n.$$

In Figure 6 we show graphs of $\bar{u}_1$, $\bar{u}_2$ and $\bar{u}_3$.

In Theorems 3.2–3.4 we have established the existence of odd periodic solutions for q in intervals of the form (q_{2n+1}, ∞), for $n = 1, 2, 3, \ldots$. They lie on branches which bifurcate from the trivial solutions at the *odd* eigenvalues q_{2n+1}. Similarly, there exist branches of even periodic solutions, which bifurcate from the *even* eigenvalues q_{2n} and extend al the way to $q = \infty$. The first three of these are shown in Figure 7.

Their existence is ensured by the following theorem.

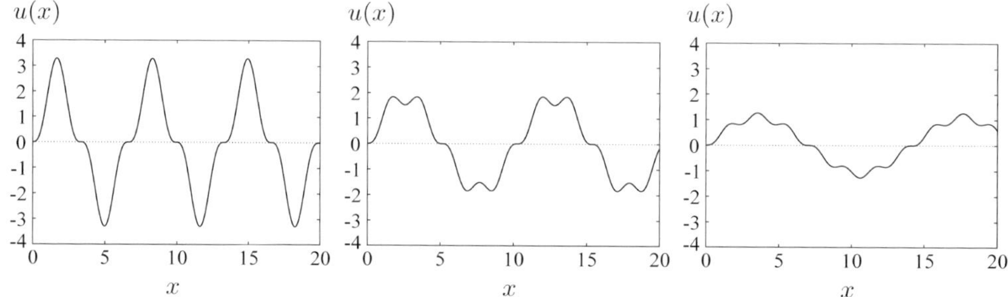

Fig. 6. The two-lap, four-lap and 6-lap solution of equation (3.2) for $p = 3$ (cf. Theorem 3.4).

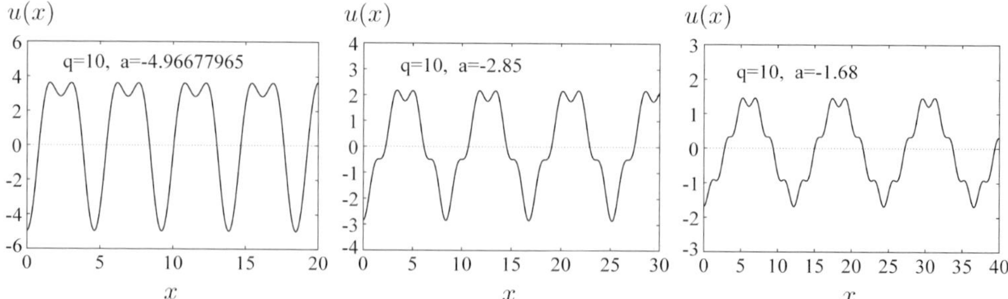

Fig. 7. The two-lap, four-lap and 6-lap solution of equation (3.2) for $p = 3$ (cf. Theorem 3.5).

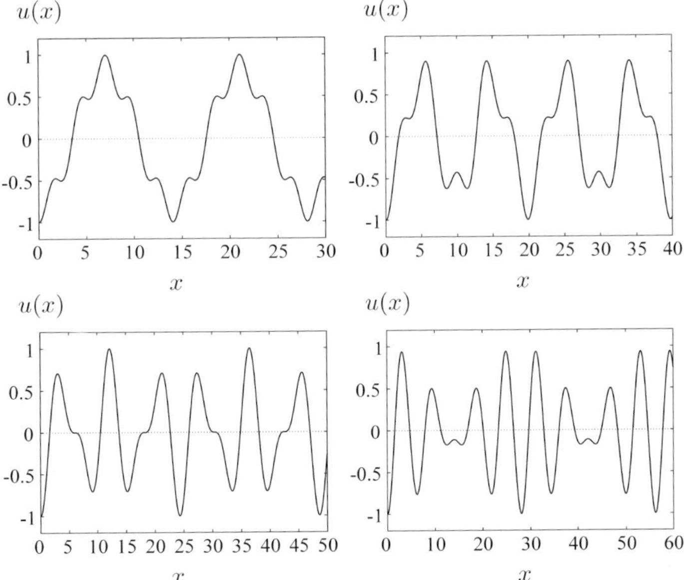

Fig. 8. Periodic solutions of the linear equation in (3.9) for (a) $q = q_{5,1}$, (b) $q = q_{5,2}$, (c) $q = q_{5,3}$ and (d) $q = q_{5,4}$.

THEOREM 3.5. *Let $n \geqslant 1$. For each $q > q_{2n}$, there exist an even periodic solution u_{2n} of equation (3.2) which has $2n$ laps in each half-period.*

So far we have discussed branches of solutions which bifurcate from the eigenvalues q_n, $n = 1, 2, 3, \ldots$. They all possess the property that they have precisely two zeros in each period. In [32] it has been shown that there are also branches of periodic solutions which bifurcate from the eigenvalues $q_{n,m}$ ($1 \leqslant m < n$). In Figure 8 we show graphs of solutions on branches which bifurcate from $q_{5,1}$, $q_{5,2}$, $q_{5,3}$ and $q_{5,4}$. Notice that they have respectively 2, 4, 6 and 8 zeros in each period. In fact, on each branch bifurcating from $q_{n,m}$ the solution can be shown to possess $2m$ zeros per period.

3.3. *Branches of periodic solutions*

In Section 3.2 we have constructed three countable families of odd periodic solutions, each type of solution existing on a half-line of values of q, which extends to infinity. In Figure 9 we show the branches of one-lap, three-lap, 5-lap and 7-lap solutions of the type obtained in Theorem 3.2 ($u'(0) > 0$) together with branches of two-lap, four-lap and 6-lap solutions of the type obtained in Theorem 3.4 ($u'(0) = 0$).

In the next theorem, which is due to van den Berg [41], we present an upper bound for bounded solutions of equation (3.2) on $\mathbb{R}$.

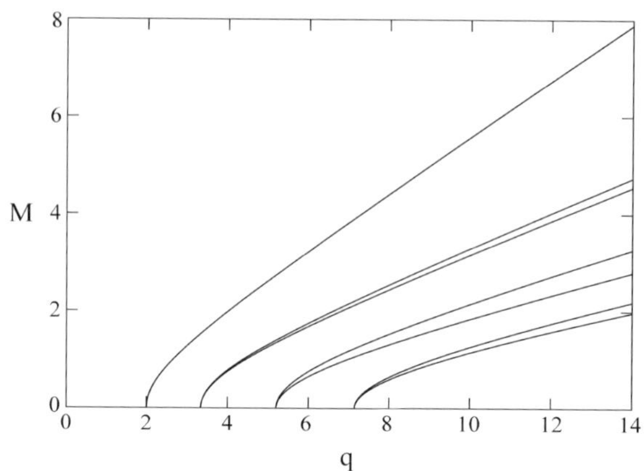

Fig. 9. Branches in the (q, M)-plane of one-lap, three-lap, 5-lap and 7-lap (per half-period) solutions of equation (3.2) obtained in Theorem 3.2 and branches of two-lap, four-lap and 6-lap (per half-period) solutions obtained in Theorem 3.4, when $p = 3$.

THEOREM 3.6. *There exists a positive constant K, which does not depend on q, such that any bounded solution $u(x, q)$ of equation (3.2) satisfies*

$$\|u(\cdot, q)\|_\infty \leqslant K(1 + q)^{2/(p-1)} \quad \text{for } q > 0.$$

PROOF. We proceed in two steps: we first fix $q \in \mathbb{R}^+$ and show that there exists a constant $C(q)$ such that $\|u(\cdot, q)\|_\infty \leqslant C(q)$, and then we show that $C(q)$ must be $O(q^{-2/(p-1)})$ as $q \to \infty$.

Thus, let $q > 0$ and suppose first that there exists a sequence of bounded solutions $\{u_n\}$ of equation (3.2) such that $\|u_n\|_\infty \to \infty$ as $n \to \infty$. Write

$$\mu_n = \|u_n\|_\infty^{-1}.$$

Then $\mu_n \to 0$ as $n \to \infty$. We now scale the solutions u_n and define the new variables

$$y = \mu^{-(p-1)/4}(x - x_n) \quad \text{and} \quad v_n(y) = \mu_n u_n(x), \tag{3.20}$$

where we have chosen the translations x_n in such a way, that

$$v_n(0) > \frac{1}{2} \quad \text{for every } n \geqslant 1. \tag{3.21}$$

When we transform equation (3.2) to these new variables we obtain

$$v_n^{\text{iv}} + q\mu_n^{(p-1)/2} v_n'' + v_n^p + \mu_n^{p-1} v_n = 0, \tag{3.22}$$

By construction, the sequence $\{v_n\}$ is bounded in $L^\infty(\mathbb{R})$; in fact

$$\|v_n\|_\infty = 1 \quad \text{for all } n \geqslant 1.$$

Hence, when we write (3.22) as

$$v_n^{\mathrm{iv}} + q \mu_n^{(p-1)/2} v_n'' = -\mu_n^{p-1} v_n - v_n^p, \tag{3.23}$$

we see that the right-hand side is uniformly bounded, so that all the derivatives of v_n are uniformly bounded on compact sets in $\mathbb{R}$. Therefore, there exists a subsequence, which we denote again by v_n, which converges in $C^4([-L, L])$ to a function V for any $L > 0$. Taking the limit in (3.22) we find that V satisfies the reduced equation

$$V^{\mathrm{iv}} + V^p = 0. \tag{3.24}$$

From (3.21) we conclude that $V(0) \geqslant 1/2$, so that V must be a bounded nontrivial solution of equation (3.24). But by [32], Exercise 3.2.1, equation (3.24) has no bounded nontrivial solutions. Therefore we have obtained a contradiction.

Next, we show that $C(q) < K(1+q)^{2/(p-1)}$ for some $K > 0$ when $q > 0$. Suppose to the contrary, that there exists a sequence $\{q_n\}$ tending to infinity as $n \to \infty$ with corresponding bounded solutions u_n such that

$$q_n^{-2/(p-1)} \|u_n\|_\infty \to \infty \quad \text{as } n \to \infty. \tag{3.25}$$

We now repeat the argument given in the first part of the proof. Because by (3.25),

$$\mu_n^{(p-1)/2} q_n \to 0 \quad \text{as } n \to \infty,$$

we obtain the same limit equation (3.24), and thus the same contradiction. This completes the proof of Theorem 3.6. $\qquad\square$

4. A sublinear bifurcation problem

In this section we investigate families of periodic solutions of the equation

$$u^{\mathrm{iv}} + q u'' + u - |u|^{p-1} u = 0, \quad p > 1, \tag{4.1}$$

i.e., we put $f(s) = s - |s|^{p-1}s$ in equation (1.1). In contrast to the superlinear equation (3.2), this equation has three constant solutions, $u = 0$ and $u = \pm 1$. The spectrum at $u = 0$ is the same as that of the trivial solution of the equation discussed in Section 3. At $u = \pm 1$ the spectrum consists of two real eigenvalues and two imaginary eigenvalues for any value of $q \in \mathbb{R}$. Specifically,

$$\lambda = \pm\alpha \quad \text{and} \quad \lambda = \pm i\beta, \tag{4.2}$$

where

$$\alpha = \frac{1}{\sqrt{2}}\sqrt{\sqrt{q^2 + 8} - q} \quad \text{and} \quad \beta = \frac{1}{\sqrt{2}}\sqrt{\sqrt{q^2 + 8} + q}. \tag{4.3}$$

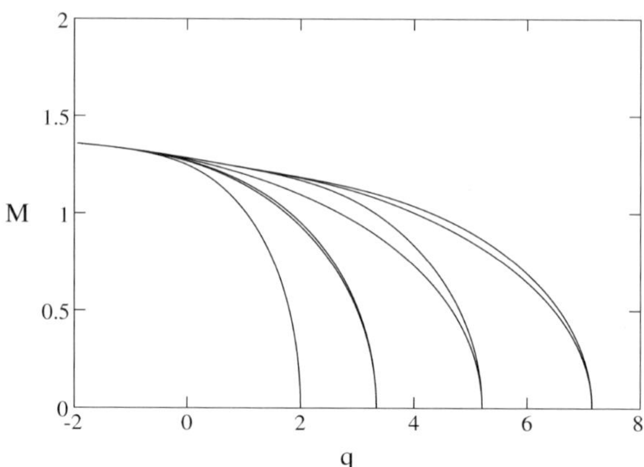

Fig. 10. Branches of one-lap, three-lap, 5-lap and 7-lap periodic solutions of equation (4.1) for $p = 3$.

As for (3.2), zero energy periodic solutions bifurcate from $u = 0$ at the critical values $q_{n,m}$. However, we shall see that here they bifurcate *subcritically*, whereas in Section 3 they bifurcated *supercritically*. In Figure 10 we show branches of solutions bifurcating from the values q_1, q_3, q_5 and q_7. As in Figure 7, one branch bifurcates from q_1, whilst from the values q_3, q_5 and q_7 two branches with different solutions bifurcate, one with $u'(0) > 0$ and one with $u'(0) = 0$ (see Figure 5).

As in Section 2, we can show that any periodic solution is bounded above by the unique positive root c_+ of the equation $F(s) = \frac{1}{2}s^2 + G(s) = 0$.

THEOREM 4.1. *Let u be a periodic solution of equation (4.1) with zero energy. Then*

$$\left|u(x)\right| < c_+ \overset{def}{=} \left(\frac{p+1}{2}\right)^{1/(p-1)} \quad \textit{for } x \in \mathbb{R}.$$

The proof is an easy consequence of the energy identity (1.9).

In the next two theorems we establish the existence of odd periodic solutions with a prescribed number of laps between nearest points of symmetry for different values of $q > -2$. We begin with values of $q \in (-2, 2)$.

THEOREM 4.2. *Let $q \in (-2, 2)$. Then for any odd number n there exists an odd periodic solution u_n of equation (4.1) with n laps between points of symmetry.*

PROOF. Clearly, linearization of equation (4.1) about $u = 0$ yields the same equation as we obtained for equation (3.2). Thus, we are led to the same linear problem as was discussed in Section 3.1, with solution $v(x)$. We now find that

$$-2 < q < 2 \quad \Longrightarrow \quad v'''(\xi_1) < 0, \tag{4.4}$$

so that

$$u'''(\xi_1(\alpha), \alpha) < 0 \quad \text{for } \alpha \text{ small.} \tag{4.5}$$

Let

$$\bar{\alpha}_1 = \sup\{\alpha > 0 \colon u(\xi_1) < c_+ \text{ on } (0, \alpha)\},$$

where c_+ has been defined in Theorem 4.1. As in Section 2, one can show that $\bar{\alpha}_1 < \infty$ and that
  (a) $\xi_1(\alpha) \in C([0, \bar{\alpha}_1])$,
  (b) $u(\xi_1(\bar{\alpha}_1), \bar{\alpha}_1) = c_+$, and
  (c) $u'''(\xi_1(\bar{\alpha}_1), \bar{\alpha}_1) > 0$.
It follows from (4.5) and properties (b) and (c) that the function

$$\varphi_1(\alpha) \stackrel{\text{def}}{=} u'''(\xi_1(\alpha), \alpha)$$

(i) is continuous and (ii) changes sign, on $(0, \bar{\alpha}_1)$. Let α_1^* be a zero of φ_1 on this interval. Then $u_1(x) = u(x; \alpha_1^*)$ can be continued to a one-lap periodic solution.

In order to construct a three-lap periodic solution, we notice that $u_1(\eta_1) = -u_1(\xi_1) < 0$. Let

$$\underline{\alpha}_1 = \sup\{\alpha > \alpha_1^* \colon u(\eta_1) < 0 \text{ on } (\alpha_1^*, \alpha)\}.$$

Then by continuity,

$$u(\eta_1(\underline{\alpha}_1), \underline{\alpha}_1) = 0 \quad \text{and} \quad u'''(\eta_1(\underline{\alpha}_1), \underline{\alpha}_1) < 0.$$

Since $\eta_1(\bar{\alpha}_1) = \xi_1(\bar{\alpha}_1)$ and $u'''(\xi_1(\underline{\alpha}_1), \underline{\alpha}_1) > 0$, it follows that the function

$$\varphi_3(\alpha) \stackrel{\text{def}}{=} u'''(\eta_1(\alpha), \alpha)$$

changes sign at some point $\alpha_3^* \in (\underline{\alpha}_1, \bar{\alpha}_1)$, and the function $u_3(x) = u(x; \alpha_3^*)$ is a three-lap periodic solution.
  Next, let

$$\bar{\alpha}_3 = \sup\{\alpha > \alpha_3^* \colon u(\xi_2) < 0 \text{ on } (\alpha_3^*, \alpha)\}.$$

Since $u_3(\xi_1) = u_3(\xi_2)$, this supremum is well defined. Plainly, $\bar{\alpha}_3 < \bar{\alpha}_1$ and as before,
  (a) $\xi_2(\alpha) \in C([\underline{\alpha}_1, \bar{\alpha}_3])$,
  (b) $u(\xi_2(\bar{\alpha}_3), \bar{\alpha}_3) = c_+$, and
  (c) $u'''(\xi_1(\bar{\alpha}_3), \bar{\alpha}_3) < 0$.
Recall that $\eta_1(\underline{\alpha}_1) = \xi_1(\underline{\alpha}_1)$, and hence $u'''(\xi_2) < 0$ at $\underline{\alpha}_1$. Hence, the function

$$\varphi_5(\alpha) \stackrel{\text{def}}{=} u'''(\xi_2(\alpha), \alpha)$$

changes sign on the interval $(\underline{\alpha}_1, \bar{\alpha}_3)$, say at α_5^*, and the function $u_5(x) = u(x; \alpha_5^*)$ is a 5-lap periodic solution.

Continuing in this manner we can successively construct periodic solutions with any odd number of laps between adjacent points of symmetry. $\qquad\square$

REMARK 4.1. As in Section 3, in the iteration process we also pick up a sequence of periodic solutions for which both u, u' and u'' vanish in the middle of the period (see Figure 4).

Next, we turn to values of $q \in (2, \infty)$. As we have seen in Lemma 3.1, the inequality (4.4) no longer holds for $q > q_1 = 2$. However, we do have

$$q_1 < q < q_3 \quad \Longrightarrow \quad v(\eta_1) < 0, \tag{4.6}$$

so that

$$u\big(\eta_1(\alpha), \alpha\big) < 0 \quad \text{for } \alpha \text{ small.} \tag{4.7}$$

This allows us to define $\underline{\alpha}_1$ again as in the proof of Theorem 4.2 and continue to construct a sequence of periodic n-lap solutions where n is odd and $n \geqslant 3$.

For $q > q_3$ we have

$$q_3 < q < q_5 \quad \Longrightarrow \quad v(\xi_2) > 0 \quad \text{and} \quad v'''(\xi_2) < 0, \tag{4.8}$$

and we can pick up the construction of periodic n-lap solutions at $n = 5$.

Summarizing we can prove the following existence theorem.

THEOREM 4.3. *Let n be any positive odd integer. Then for $-2 < q < q_n$, there exists an odd periodic solution of equation* (4.1) *of Type* I *with n laps in each half-period.*

PROOF. The above argument applied successively proves Theorem 4.3 for $q \neq q_k$, where $k < n$. However, since all the local extrema are nondegenerate, it follows from the continuous dependence of u, u' and u'' on q, that the statement remains true for the isolated values $\{q_k \colon 1 \leqslant k < n\}$. $\qquad\square$

REMARK 4.2. Let $n \geqslant 3$ and odd. Then for $-2 < q < q_n$ there also exist a periodic solution of equation (4.1) of Type II with $n - 1$ laps in each half-period.

For $q \leqslant 2$ the situation is quite different. In the next theorem we show that there are no periodic solutions which have points of symmetry and one zero between each of them, with respect to which they are odd.

THEOREM 4.4. *Let $q \leqslant 2$ and $E \geqslant 0$. Then there are no odd periodic solutions $u(x)$ of equation* (4.1) *which are positive on* $(0, L)$, *where $x = L$ is its first point of symmetry.*

PROOF. We use a maximum principle argument. This is made possible because the characteristic equation of the linear operator

$$\mathcal{L}(u) \overset{\text{def}}{=} u^{\text{iv}} + qu'' + u,$$

given by

$$\lambda^4 + q\lambda^2 + 1 = 0,$$

has real eigenvalues when $q \leqslant -2$. It can be factorized as follows

$$\left(\lambda^2 - \mu_+\right)\left(\lambda^2 - \mu_-\right) = 0, \quad \text{where } \mu_{\pm} = \frac{1}{2}\left(-q \pm \sqrt{q^2 - 4}\right).$$

Clearly, μ_+ and μ_- are real and positive whenever $q \leqslant -2$. For more details about this line of approach, we refer to [33,39,40] and [32].

Thus, suppose to the contrary that $u(x)$ is an odd periodic solution and that its first positive point of symmetry is located at $x = L$. Plainly, $u(x)$ is then a solution of the following boundary value problem:

$$\begin{cases} u^{\text{iv}} + qu'' + u - u^p = 0, & u > 0, 0 < x < L, \\ u(0) = 0, u''(0) = 0, u'(L) = 0, u'''(L) = 0. \end{cases} \tag{4.9}$$

In view of the factorization of the characteristic equation, we can also factorize $\mathcal{L}$ and write the fourth-oder differential equation in problem (4.9) as a system of two second-order differential equations

$$\begin{cases} u'' - \mu_+ u = v, \\ v'' - \mu_- v = u^p. \end{cases} \tag{4.10}$$

We note that

$$v(0) = 0 \quad \text{and} \quad v(L) = b(a) - \mu_+ a,$$

where we have written $u(L) = a$ and $u''(L) = b \overset{\text{def}}{=} \pm\sqrt{2\{F(a) - E\}}$. Note that since $E \geqslant 0$,

$$v(L) \leqslant \sqrt{2\{F(a) - E\}} - \mu_+ a \leqslant \sqrt{2F(a)} - \mu_+ a$$

$$= a\left\{\sqrt{1 - \frac{2a^{p-1}}{p+1}} - \mu_+\right\} < 0,$$

because $\mu_+ \geqslant 1$ when $q \leqslant -2$. Hence, by the Maximum Principle [34] applied to the second equation of (4.10), we conclude that $v(x) < 0$ for $0 < x \leqslant L$. Because

$$u(0) = 0 \quad \text{and} \quad u(L) \geqslant 0,$$

applying the Maximum Principle to the first equation of (4.10) yields that $u(x) > 0$ on $(0, L)$. Since $u(x)$ is a nontrivial solution, it follows that $u(L) > 0$ and by the Boundary Point Lemma [34], that $u'(L) > 0$, a contradiction. This completes the proof of Theorem 4.4. $\square$

It is interesting to note that when $q \leqslant -2$ and $E = 0$, then there exists a homoclinic orbit to the origin.

THEOREM 4.5. *Let $q \leqslant -2$ and $E = 0$. Then there exists an even solution $u(x)$ of equation (4.1), such that*

$$u(x) \to 0 \quad as \ x \to \pm\infty,$$

which is strictly increasing on $(-\infty, 0)$ and strictly decreasing on $(0, \infty)$. At the origin we have

$$m(q) < u(0) < c_+ = \left(\frac{p+1}{2}\right)^{1/(p-1)},$$

where

$$m(q) = \sup\left\{1 < \hat{s} < c_+ : \frac{f^2(s)}{F(s)} < \frac{q^2}{2} \ for \ 1 < s < \hat{s}\right\}. \tag{4.11}$$

PROOF. The existence of an even solution can be proved using a shooting argument much like that which has been used in Section 3. For further details we refer to [33].

The upper bound of $u(0)$ follows as in the proof of Theorem 4.1. In order to prove the lower bound we introduce the auxiliary function

$$H(x) \stackrel{\text{def}}{=} \frac{u'''(x)}{u'(x)} + \frac{q}{2}. \tag{4.12}$$

Let

$$x_1(\alpha) = \sup\{x > 0 : u(\cdot; \alpha) > 0 \text{ on } (0, x)\}$$

and let

$$\mathcal{A} = \{\tilde{\alpha} > 0 : u'''(\cdot; \alpha) > 0 \text{ on } (0, x_1(\alpha)) \text{ for } 0 < \alpha < \tilde{\alpha}\}.$$

We divide the proof into a series of steps.

Step 1. We have

$$(0, 1] \subset \mathcal{A}. \tag{4.13}$$

We note that since $q < 0$,

$$u^{\mathrm{iv}} = \frac{|q|}{2} u'' - f(u).$$

By assumption, we have $u''(0) < 0$. Moreover, because $0 < u(x) < \alpha \leqslant 1$ on $(0, x_1)$ and $f(u) > 0$ for $0 < u < 1$, it follows that $f(u(x)) > 0$ for $0 < x < x_1$. Therefore $u^{\mathrm{iv}}(x) < 0$ near $x = 0$, and $u'''(x) < 0$ and $u''(x) < 0$ for $0 < x \leqslant x_1$.

Step 2. Let $u(x)$ be a zero energy solution. Then

$$H\big(x_1(\alpha)\big) > 0 \quad \text{for } 0 < \alpha \leqslant \sqrt{2}. \tag{4.14}$$

Since $u(x)$ is a zero energy solution, we have

$$u'u''' - \frac{1}{2}\big(u''\big)^2 + \frac{q}{2}\big(u'\big)^2 + F(u) = 0.$$

We can write this as

$$\big(u'\big)^2 H - \frac{1}{2}\big(u''\big)^2 + F(u) = 0.$$

Since $F(u(x_1)) = 0$, it follows that

$$\big(u'\big)^2 H = \frac{1}{2}\big(u''\big)^2 \quad \text{at } x_1. \tag{4.15}$$

It is clear that $u''(x_1) \neq 0$, since otherwise we would have a periodic solution. Hence, $u'(x_1) < 0$ and it follows that $H(x_1)$ must be positive.

Step 3. Let $u(x)$ be a zero energy solution. Then

$$H(x) > 0 \quad \text{for } 0 < x < x_1 \text{ and for } 0 < \alpha \leqslant 1. \tag{4.16}$$

Fix $\alpha \in (0, 1]$. Then $0 < u(x) < 1$ for $0 < x < x_1$, and

$$\big(u'H\big)' = u^{\mathrm{iv}} + \frac{q}{2}u'' = \frac{|q|}{2}u'' - f(u) < 0.$$

Hence

$$u'(x)H(x) < u'(0)H(0) = 0 \quad \text{for } 0 < x < x_1.$$

Since $u'(x) < 0$, this implies that $H(x) > 0$ for $0 < x \leqslant x_1$.

Step 4. Let $u(x)$ be a zero energy solution. Then

$$H(0; \alpha) = \lim_{x \to 0} H(x) = -\frac{q}{2} + \frac{f(\alpha)}{\sqrt{2F(\alpha)}}. \qquad (4.17)$$

By l'Hôpital's rule we find that

$$\lim_{x \to 0} H(x) = \lim_{x \to 0} \frac{u^{\mathrm{iv}}(x)}{u''(x)} + \frac{q}{2}.$$

Since $u(x) \to \alpha$ and $u''(x) \to \beta(\alpha) = -\sqrt{2F(\alpha)}$ as $x \to 0$, the desired expression (4.17) for $H(0)$ follows.

REMARK 4.3. Note that if $q \leqslant -2$, then $H(0; \alpha) > 0$ for $\alpha \in (0, 1]$.

We are now ready to complete the proof of the lower bound for $\alpha = u(0)$. Let

$$\alpha^* = \sup \mathcal{A}.$$

Then it is clear that if $u(x; \alpha_0)$ is a positive solution such that $u(\pm\infty; \alpha_0) = 0$, then $\alpha_0 > \alpha^*$. We shall now prove that $\alpha^* \geqslant m(q)$.

Suppose to the contrary that $\alpha^* < m(q)$. Then there exists a point x^* such that

$$u'''(x^*; \alpha^*) = 0 \quad \text{and} \quad u'''(x^*; \alpha^*) \leqslant 0 \quad \text{for } 0 < x < x_1.$$

Since by assumption, $\alpha^* < m(q)$, it follows that $H(0; \alpha^*) > 0$, so that $x^* > 0$. Also since $H(x_1; a^*) > 0$, we must have $x^* < x_1$. Thus, x^* must be an interior point so that, by the definition of H,

$$H(x^*; \alpha^*) = \frac{q}{2} \leqslant -1.$$

By the result of Step 4, we can define the set

$$\widehat{\mathcal{A}} = \sup\{\tilde{\alpha} > 0 \colon H(\cdot; \alpha) > 0 \text{ on } [0, x_1], 0 < \alpha < \tilde{\alpha}\}.$$

At $\hat{\alpha} = \sup \widehat{\mathcal{A}}$ there exists a point $\hat{x} \in [0, x_1(\hat{\alpha})]$ such that $H(\hat{x}; \hat{\alpha}) = 0$. But since $\hat{\alpha} < \alpha^*$, and by assumption $\alpha^* \leqslant m(q)$, it follows that $H(0, \tilde{\alpha}) > 0$ and hence $\hat{x} > 0$. Similarly, since $H(x_1) > 0$, we deduce that $\hat{x} < x_1$. Therefore, $\hat{x}$ is an interior point and we must have

$$H(\hat{x}; \hat{\alpha}) = 0 \quad \text{as well as} \quad H'(\hat{x}; \hat{\alpha}) = 0.$$

From (4.15) we conclude that

$$\frac{1}{2}(u'')^2 = F(u) \quad \text{and} \quad u^{\mathrm{iv}} + qu'' = 0 \quad \text{at } x = \hat{x}, \alpha = \hat{\alpha}.$$

Using the differential equation we find that

$$\frac{2}{q^2} f^2(u) = F(u) \quad \text{at } x = \hat{x}, \alpha = \hat{\alpha}.$$

This means that $u(\hat{x}; \hat{\alpha}) \geqslant m(q)$. Since $u' < 0$ on $\mathbb{R}^+$, we conclude that $\hat{\alpha} = u(0; \hat{\alpha}) > u(\hat{x}; \hat{\alpha})$, and hence $\hat{\alpha} > m(q)$. Since $\hat{\alpha} < \alpha^*$ this is impossible, and we have a contradiction. This completes the proof of Theorem 4.5. $\qquad\square$

REMARK 4.4. For the function $f(s) = s - |s|^{p-1}s$ we find that, for $s > 0$,

$$\frac{f^2(s)}{F(s)} = 2\frac{(1 - s^{p-1})^2}{1 - (s/c_+)^{p-1}}.$$

Hence

$$m(q) = c_+\left\{1 - \frac{16}{(p-1)^3}q^{-2} + O\big(q^{-4}\big)\right\} \quad \text{as } q \to \infty. \tag{4.18}$$

We complete this section by stating a uniqueness theorem.

THEOREM 4.6. *Let $q \leqslant -2$. Then there exists at most one positive homoclinic orbit of equation (4.1) which tends to $u = 0$ as $x \to \pm\infty$.*

The proof of this theorem is given by means of maximum principle arguments. We refer to [33] or to [40].

5. A multiscale analysis

As we have seen in Sections 3 and 4, when q becomes large, periodic solutions emerge with an increasing number of bumps. In Figure 11 we show three zero energy periodic

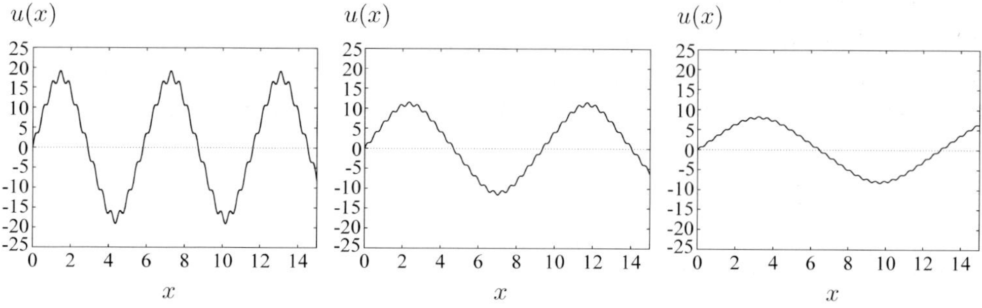

Fig. 11. n-lap periodic solutions for $n = 13$, $n = 21$, $n = 29$ when $q = 200$ and $E = 0$.

solutions of equation (5.1) when $f(s) = s + s^3$ and $q = 200$. They exhibit a typical multi-scale structure: a large scale periodic solution, the *Baseline solution*, with a high-frequency periodic solution superimposed on it. In this section we utilize this multiscale structure to study periodic solutions of the equation

$$u^{\mathrm{iv}} + qu'' + f(u) = 0 \tag{5.1}$$

when the number of bumps is large.

We follow [21] in which such methods were developed to study the shape of multibump periodic solutions of equation (5.1), such as shown in Figure 11 and to estimate the location of solution branches.

Throughout this section we make the following assumptions about the nonlinearity $f \in C^1(\mathbb{R})$,

$$f(s) = s + g(s) \quad \text{and} \quad g(-s) = -g(s) \quad \text{for } s \in \mathbb{R}. \tag{5.2}$$

In particular, we shall discuss the examples $g(s) = \pm s^3$.

In Section 5.1 we begin by deriving approximate expressions for multibump periodic solutions. Then, in Section 5.2, we use these approximations to obtain information about the corresponding solution branches.

5.1. *Multibump solutions*

To prepare for the multiscale analysis we rescale the independent variable and write

$$x^* = \frac{x}{\sqrt{q}} \quad \text{and} \quad u^*(x^*) = u(x). \tag{5.3}$$

Equation (5.1) then transforms into

$$\varepsilon^2 u^{\mathrm{iv}} + u'' + u + g(u) = 0, \tag{5.4}$$

where we have dropped the asterisks again.

In this subsection we focus on periodic solutions which are symmetric with respect to the origin, i.e., we require that

$$u'(0) = 0 \quad \text{and} \quad u'''(0) = 0. \tag{5.5}$$

In addition we require them to be odd with respect to some of their zeros. Let $L > 0$ be the smallest of such positive zeros. Then

$$u(L) = 0 \quad \text{and} \quad u''(L) = 0. \tag{5.6}$$

Now, let $u(x)$ be a solution of equation (5.4) which satisfies the initial conditions (5.5) as well as the symmetry condition (5.6) at $x = L$. Then we can continue u as an even function with respect to $x = 0$ and as an odd function with respect to $x = L$ and so construct a periodic solution of (5.4) on $\mathbb{R}$ with period $4L$.

Let us put

$$u(0) = \alpha \quad \text{and} \quad u''(0) = \beta. \tag{5.7}$$

Then the solution $u(x)$ of equation (5.4) is uniquely determined by the pair of constants (α, β). For (5.4), the energy identity is given by

$$\mathcal{E}(u) \stackrel{\text{def}}{=} \varepsilon^2 \left(u'u''' - \frac{1}{2}(u'')^2 \right) + \frac{1}{2}(u')^2 + \frac{1}{2}u^2 + G(u) = E,$$

$$G(u) = \int_0^u g(s)\,\mathrm{d}s. \tag{5.8}$$

Hence, using (5.5) and (5.7) to evaluate $\mathcal{E}(u)$ at the origin, we find that

$$E = -\frac{\varepsilon^2}{2}\beta^2 + \frac{1}{2}\alpha^2 + G(\alpha). \tag{5.9}$$

Thus, if we fix E, then apart from the sign, the value of β is determined through (5.9) by the value of α. If in addition we require (5.6) to be satisfied at some, as yet undetermined value of L, then the problem will be over determined. In fact, for a given value of $\alpha > 0$ we find distinct values of ε, or q, for which there exists a solution. These values correspond to the solution branches shown in Figure 1(a) for $g(s) = s^3$.

To identify the different scales, we put

$$\xi = \varepsilon^{-s}x \quad \text{and} \quad U(\xi) = u(x)$$

and we transform equation (5.4) to these new variables. This yields

$$\varepsilon^{2(1-2s)}U^{\mathrm{iv}} + \varepsilon^{-2s}U'' + U + g(U) = 0.$$

Plainly, by putting $s = 1$, we balance the first two terms and obtain

$$\varepsilon^{-2}\left(U^{\mathrm{iv}} + U''\right) + U + g(U) = 0.$$

Equating the coefficient of ε^{-2} to zero we find that

$$U^{\mathrm{iv}} + U'' = 0,$$

so that, if we assume that U is uniformly bounded on $\mathbb{R}$, it must be of the form

$$U(\xi) = A\cos\xi + C\sin\xi + B, \tag{5.10}$$

in which A, B and C are constants.

Thus, we distinguish two scales: a *slow* scale x of $O(1)$ in which the last three terms balance and a *fast* scale ξ of $O(\varepsilon^{-1})$ in which the first two terms balance and the last two terms are small. In the spirit of multiscale analysis [3,20], we follow [21], assume a solution of the form

$$u(x) = U(x, \xi) + B(x), \tag{5.11}$$

and treat ξ and x as independent variables. Writing $u' = u'(x)$ we then obtain $u' = U_x + \varepsilon^{-1} U_\xi + B_x$ and treat higher-order terms similarly. The functions $U(x, \xi)$ and $B(x)$ are expanded in powers of ε:

$$U(x, \xi) = U_0(x, \xi) + \varepsilon U_1(x, \xi) + \cdots, \tag{5.12}$$

$$B(x) = B_0(x) + \varepsilon B_1(x) + \cdots. \tag{5.13}$$

By substituting these expressions into the equation (5.4) and into the boundary conditions (5.5)–(5.7), we obtain expressions for the first few terms in these expansions. Specifically, we obtain the following proposition.

PROPOSITION 5.1. *Let α and E be $O(1)$ with respect to ε. Then the leading-order asymptotic approximation to the solution $u(x)$ of problem (5.4), (5.5) and (5.7), is given by*

$$u(x) = B_0(x) \pm \varepsilon\sqrt{\alpha^2 + 2G(\alpha) - 2E} \cos\frac{x}{\varepsilon} + O(\varepsilon^2) \quad as \ \varepsilon \to 0, \tag{5.14}$$

where the "+" sign applies when $u''(0) < 0$ and the "−" sign when $u''(0) > 0$. The baseline solution B_0 in (5.14) is the solution of the initial value problem

$$\begin{cases} B_0'' + B_0 + g(B_0) = 0, \\ B_0(0) = \alpha \mp \varepsilon\sqrt{\alpha^2 + 2G(\alpha) - 2E} \quad and \quad B_0'(0) = 0. \end{cases} \tag{5.15}$$

The first zero $L = L(\alpha, \varepsilon)$ of B_0 satisfies the condition

$$\frac{L(\alpha, \varepsilon)}{\varepsilon} = (2m + 1)\frac{\pi}{2}, \quad m = 0, 1, 2, \ldots. \tag{5.16}$$

REMARK 5.1. The integer m in (5.16) determines the number of laps n between points of symmetry through the relation

$$n = 2m + 1.$$

Before deriving this result, we note that the assumption that $E = O(1)$ implies that $U_0 = 0$ and hence that $U = O(\varepsilon)$. We give a brief heuristic explanation for this here; for a detailed derivation we refer to [21].

When we substitute (5.10) into the initial condition (5.7) we obtain for β,

$$u''(0) = \beta = B''(0) + \frac{1}{\varepsilon^2} U_{\xi\xi}(0, 0) + \frac{2}{\varepsilon} U_{x\xi}(0, 0) + U_{xx}(0, 0).$$

Assuming that B and U and their derivatives are O(1) or smaller at $x = 0$, and observing that $\beta = O(\varepsilon^{-1})$ by (5.9), we arrive at the conclusion that $U_{\xi\xi} = O(\varepsilon)$. In (5.10) we found that to leading order, $U(x,\xi) \sim A(x)\cos\xi$. This implies that $A = O(\varepsilon)$ and hence $U_0(x) \equiv 0$.

Thus, the leading-order term in $U(x,\xi)$ is $\varepsilon U_1(x,\xi)$. We substitute (5.11)–(5.13) into equation (5.4) and collect the coefficients of ε^j for $j = -1, 0, 1, \ldots$. The leading-order term yields the equation,

$$\mathrm{O}\big(\varepsilon^{-1}\big): \quad U_{1\xi\xi\xi\xi} + U_{1\xi\xi} = 0,$$

from which we conclude that U_1 must have the form

$$U_1(x,\xi) = A_1(x)\cos\xi + C_1(x)\sin\xi + D(x)\xi + E(x).$$

By symmetry $D(x) = 0$, and $E(x)$ can be absorbed into $B(x)$. This leading-order result suggests a more efficient form of the expansion for $U(x,\xi)$:

$$\begin{cases} U(x,\xi) = A(x)\cos(\lambda\xi) + C(x)\sin(\lambda\xi), \\ A(x) = \varepsilon A_1(x) + \varepsilon^2 A_2(x) + \cdots, \\ C(x) = \varepsilon C_1(x) + \varepsilon^2 C_2(x) + \cdots, \\ \lambda = 1 + \varepsilon\lambda_1 + \varepsilon^2\lambda_2 + \cdots. \end{cases} \tag{5.17}$$

Substituting (5.17) into (5.4), we obtain the O(1) equation,

$$\begin{aligned} \mathrm{O}(1): \quad B_0''(x) + B_0(x) + \gamma f\big(B_0(x)\big) \\ = -2\big\{\lambda_1 A_1 \cos(\lambda\xi) + \lambda_1 C_1 \sin(\lambda\xi) \\ + A_1'(x)\sin(\lambda\xi) - C_1'(x)\cos(\lambda\xi)\big\}. \end{aligned}$$

Following the multiscale assumption, we treat ξ and x as independent variables. This gives the following equations for B_0, A_1 and C_1:

$$B_0''(x) + B_0(x) + g\big(B_0(x)\big) = 0 \tag{5.18}$$

and

$$C_1'(x) = \lambda_1 A_1(x), \quad A_1'(x) = -\lambda_1 C_1(x). \tag{5.19}$$

From (5.19) we conclude that

$$A_1(x) = \rho_1 \cos(\lambda_1 x) \quad \text{and} \quad C_1(x) = \rho_1 \sin(\lambda_1 x), \tag{5.20}$$

where ρ_1 is a constant. Note that this yields

$$U_1(x,\xi) = \rho_1 \cos\left(\frac{x}{\varepsilon} + \lambda_1 x + \cdots - \lambda_1 x\right),$$

so that λ_1 cancels from the expression for $U_1(x, \xi)$. This is not surprising since variation of the solution on the scale of $\varepsilon\lambda_1\xi$ amounts to variation on the x scale, which is already captured in the coefficients A_j and C_j. Therefore, including λ_1 in the expansion yields no additional information about the solution. Thus, without loss of generality we may put $\lambda_1 = 0$ and hence set

$$A_1(x) = \rho_1 \quad \text{and} \quad C_1(x) = 0, \tag{5.21}$$

where ρ_1 is a constant that is yet to be determined. Combining (5.21) with the $O(\varepsilon^{-1})$ term in the boundary condition at $x = L$,

$$A_1(L)\cos\frac{L}{\varepsilon} + C_1(L)\sin\frac{L}{\varepsilon} = 0,$$

we conclude that

$$\cos\frac{L}{\varepsilon} = 0 \quad\Longrightarrow\quad \frac{L}{\varepsilon} = (2m+1)\frac{\pi}{2}, \quad m = 0, 1, 2, \ldots, \tag{5.22}$$

which establishes the condition (5.16) on α, ε and m.

To determine $B_0(x)$ we first consider the $O(1)$ terms in the boundary conditions. This yields for B_0

$$B_0'(0) = 0 \quad \text{and} \quad B_0(L) = 0. \tag{5.23}$$

Thus $B_0(x)$ solves the boundary value problem

$$\begin{cases} B_0'' + B_0 + g(B_0) = 0, & B_0 > 0, 0 < x < L, \\ B_0'(0) = 0, & B_0(L) = 0, \end{cases} \tag{5.24}$$

where L is given by (5.22).

Next, we relate B_0 – and hence its first zero L – to the initial conditions of $u(x)$ given in (5.7)

$$\alpha = B_0(0) + \varepsilon\rho_1, \qquad \beta = B_0''(0) - \frac{\rho_1}{\varepsilon}.$$

By the energy identity, we have

$$\beta = \pm\frac{1}{\varepsilon}\sqrt{\alpha^2 + 2G(\alpha) - 2E}.$$

Since $B_0''(0)$ is of lower order than ρ_1/ε, we obtain to first order $\rho_1 = \varepsilon\beta$, so that

$$B_0(0) = \alpha - \varepsilon\rho_1(\alpha, E) \quad \text{and} \quad \rho_1(\alpha, E) = \mp\sqrt{\alpha^2 + 2G(\alpha) - 2E}. \tag{5.25}$$

Thus, given E, the constants α and ε determine $B_0(x)$ and L uniquely, i.e., $L = L(\alpha, \varepsilon)$. Since L needs to satisfy (5.22), we obtain the relation (5.16) between α, ε and m.

We have seen that the high-frequency oscillations are also of $O(\varepsilon)$. Hence, we also need to include the first-order term $\varepsilon B_1(x)$. The equation for B_1 is obtained from the $O(\varepsilon)$ term in the equation

$$B_1''(x) + B_1(x) + g\big(B_0(x)\big)B_1(x)$$
$$= (1 + 2\lambda_2)A_1 \cos(\lambda\xi) - 3\gamma B_0^2 \rho_0 \cos(\lambda\xi), \tag{5.26}$$

and the corresponding boundary conditions at $O(\varepsilon)$,

$$B_1'(0) = 0 \quad \text{and} \quad B_1(L) = 0. \tag{5.27}$$

Treating ξ and x as independent variables in (5.26), we have obtained the following boundary value problem for the first-order correction to the baseline solution B_1:

$$\begin{cases} B_1''(x) + \big\{1 + g'\big(B_0(x)\big)\big\}B_1(x) = 0, \\ B_1'(0) = 0 \quad \text{and} \quad B_1(L) = 0. \end{cases} \tag{5.28}$$

The solution of this problem is given by

$$B_1(x) = B_0'(x)\left(c_1 + c_2 \int^x \big\{B_0'(s)\big\}^{-2} ds\right), \tag{5.29}$$

in which c_1 and c_2 are constants which we determine from the boundary conditions. Because $B_1'(0) = 0$ and $B_0'(0) = 0$ we find that $c_2 = 0$. This means that $B_1(L) = c_1 B_0'(L)$. Since $B_0'(L) \neq 0$, this implies that the boundary condition $B_1(L) = 0$ can only be satisfied if $c_1 = 0$. Therefore we conclude that $B_1(x) = 0$.

Thus, we have shown that if $E = O(1)$ as $\varepsilon \to 0$, then

$$B(x) = B_0(x) + O\big(\varepsilon^2\big) \quad \text{as } \varepsilon \to 0,$$

where $B_0(x)$ is the solution of problem (5.24). This completes the derivation of the expansion formulated in Proposition 5.1.

5.2. *Branches of multibump periodic solutions*

Multibump periodic solutions with a fixed number of high-frequency oscillations per half-period and a given energy E, lie on curves in the (ε, α)-plane. We denote these curves by $\mathcal{C}_m$, $m = 1, 2, \ldots$. When m is large a good approximation of the curves $\mathcal{C}_m$ is given by the relation between α, ε and m obtained in (5.16),

$$\frac{L(\alpha, \varepsilon)}{\varepsilon} = (2m + 1)\frac{\pi}{2}, \quad m = 1, 2, 3, \ldots. \tag{5.30}$$

We emphasize that in general, on branches $\mathcal{C}_m$ the number of bumps may change. Such a change takes place when a critical point becomes degenerate, i.e., at some $x_0 \in (0, L)$, where we have $u'(x_0) = 0$ and $u''(x_0) = 0$. It is clear from the energy identity (5.8) that if $g(s) = s^3$, then when $E = 0$, this can only happen when $u(x_0) = 0$ as well.

We can obtain an explicit expression for $L(\alpha, \varepsilon)$ by multiplying the differential equation in (5.24) by $2B_0'$ and integrating the resulting equation over the interval $(0, L)$. This yields after some rearrangement

$$\{B'(x)\}^2 = \gamma^2 - B^2(x) + 2G(\gamma) - 2G(B(x)),$$

where we have omitted the subscript "0". Taking the square root, we eventually find that

$$L(\alpha, \varepsilon) = \mathcal{J}(\gamma) \overset{\text{def}}{=} \int_0^\gamma \frac{ds}{\sqrt{\gamma^2 - s^2 + 2G(\gamma) - 2G(s)}}, \tag{5.31}$$

where

$$\gamma = \gamma(\alpha, \varepsilon) \overset{\text{def}}{=} \alpha - \varepsilon\rho(\alpha, E). \tag{5.32}$$

We discuss the two cases corresponding to the nonlinearities discussed in Sections 3 and 4,

$$\text{Case I}: \quad g(s) = +s^3 \quad \text{and} \quad \text{Case II}: \quad g(s) = -s^3.$$

5.2.1. *Case I: $g(s) = s^3$.* In this case $G(s) = \frac{1}{4}s^4$ and hence $\mathcal{J}(\gamma)$ becomes

$$\mathcal{J}(\gamma) = J^+(\gamma) \overset{\text{def}}{=} \frac{\sqrt{2}}{\gamma} \int_0^1 \frac{dt}{\sqrt{(1 - t^2)\{(2/\gamma^2) + 1 + t^2\}}}. \tag{5.33}$$

Plainly,

$$J^+(\gamma) = \frac{\sqrt{2}}{\gamma} \left\{ \int_0^1 \frac{dt}{\sqrt{1 - t^4}} + \mathcal{O}\left(\frac{1}{\gamma^2}\right) \right\} \quad \text{as } \gamma \to \infty. \tag{5.34}$$

An elementary computation, involving the transformation $t^4 = 1/(1 + r)$, allows us to compute the principal term explicitly,

$$\int_0^1 \frac{dt}{\sqrt{1 - t^4}} = \frac{1}{4} \int_0^\infty \frac{r^{-1/2}\, dr}{(1 + r)^{3/4}} = \frac{1}{4} B\left(\frac{1}{2}, \frac{1}{4}\right) = \frac{\sqrt{\pi}}{4} \frac{\Gamma(1/4)}{\Gamma(3/4)} = 1.3110\ldots. \tag{5.35}$$

Here $B(x, y)$ is the Beta function [1].

Using (5.35) in (5.34), (5.33), (5.31) and finally in (5.30), we obtain the following estimate for branches of multibump periodic solutions.

PROPOSITION 5.2. *Let $E = 0$. Then in the $(\varepsilon^{-1}, \alpha)$-plane, the two branches of even periodic solutions which are odd with respect to two zeros per period and have n-laps between points of symmetry, grow asymptotically as*

$$\alpha \sim \frac{1}{n\sqrt{2\pi}} \frac{\Gamma(3/4)}{\Gamma(1/4)} q \quad as \ q \to \infty, \quad q = \frac{1}{\varepsilon}. \tag{5.36}$$

In Figure 12 we compare the bifurcation curves $\mathcal{C}_m$ obtained numerically (solid) with the asymptotic approximation given above (dash-dotted), when $g(s) = s^3$. We compare these graphs on two different scales; as expected, the approximation improves with increasing m and the relative size of the error decreases with ε.

We can also use (5.30) in a different type of asymptotics, one in which α stays bounded and $\varepsilon \to 0$. Plainly, when α remains in a bounded interval, then by (5.32), $\gamma \to \alpha$, and

$$J^+(\gamma) \to J^+(\alpha) = \int_0^\alpha \frac{ds}{\sqrt{\alpha^2 - s^2 + 1/2(\alpha^4 - s^4)}} \quad as \ \varepsilon \to 0. \tag{5.37}$$

Note that $J^+(\gamma) < \pi/2$. In fact, an elementary computation shows that

$$J^+(\alpha) = \frac{\pi}{2}\left(1 - \frac{3}{8}\alpha^2 + \mathcal{O}(\alpha^4)\right) \quad as \ \alpha \to 0. \tag{5.38}$$

Thus, returning to (5.30), and using (5.31), (5.37) and (5.38), we obtain the following asymptotics:

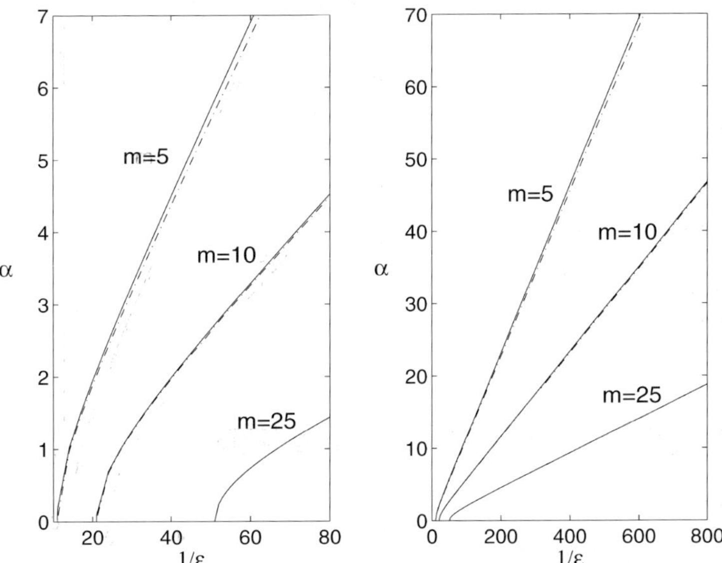

Fig. 12. Bifurcation curves $\mathcal{C}_m$ for $E = 0$ for the full equation (solid line) and asymptotic approximation (dash-dotted line); when $g(s) = s^3$.

PROPOSITION 5.3. *Let $E = 0$. Then, in the $(\varepsilon^{-1}, \alpha)$-plane, the branch of even periodic solutions which are odd with respect to two zeros per period and have n-laps between points of symmetry converges to*

$$q(\alpha) - n \sim n \left\{ \frac{\pi}{2J^+(\alpha)} - 1 \right\} \quad as \ n \to \infty, \quad q = \frac{1}{\varepsilon}, \tag{5.39}$$

on bounded intervals of α. From this expression we deduce that in the limit as $n \to \infty$, the local behavior near the branch points on the branch of trivial solutions is described by

$$\alpha(q) \sim \sqrt{\frac{8}{3n}} \sqrt{q - n}, \quad q > n. \tag{5.40}$$

5.2.2. *Case II: $g(s) = -s^3$.* In this case, $G(s) = -\frac{1}{4}s^4$ and we find from (5.31) that

$$\mathcal{J}(\gamma) = J^-(\gamma) \stackrel{\text{def}}{=} \frac{\sqrt{2}}{\gamma} \int_0^1 \frac{dt}{\sqrt{(1 - t^2)(\theta + 1 - t^2)}}, \quad \theta = \frac{2}{\gamma^2} - 2. \tag{5.41}$$

It is clear that in order for the integrand to be well defined, we need θ to be positive, i.e., we need to require that $\gamma < 1$. This means that α is bounded above so that only the type of asymptotics for $n \to \infty$ is possible.

In this case,

$$J^-(\alpha) = \int_0^\alpha \frac{ds}{\sqrt{\alpha^2 - s^2 - 1/2(\alpha^4 - s^4)}} > \frac{\pi}{2} \tag{5.42}$$

and in the limit as $\alpha \to 0$ we find

$$J^+(\alpha) = \frac{\pi}{2} \left(1 + \frac{3}{8}\alpha^2 + \mathcal{O}(\alpha^4) \right) \quad as \ \alpha \to 0, \tag{5.43}$$

and we obtain as before the following proposition.

PROPOSITION 5.4. *Let $E = 0$. Then in the $(\varepsilon^{-1}, \alpha)$-plane, the branch of even periodic solutions which are odd with respect to two zeros per period and have n-laps between points of symmetry converges to*

$$q(\alpha) - n \sim -n \left\{ 1 - \frac{\pi}{2J^-(\alpha)} \right\} \quad as \ n \to \infty, \quad q = \frac{1}{\varepsilon}, \tag{5.44}$$

on bounded intervals of α. From this expression we deduce that in the limit as $n \to \infty$, the local behavior near the branch points on the branch of trivial solutions is described by

$$\alpha(q) \sim \sqrt{\frac{8}{3n}} \sqrt{n - q}, \quad q < n. \tag{5.45}$$

Acknowledgements

It is a pleasure to thank B. Meulenbroek, whose MSc thesis [25] was the basis of Section 2, G.J.B. van den Berg, for making the bifurcation graphs in Sections 1–4 and R. Kuske for sharing with me the Matlab code for the numerical comparisons shown in Section 5.

References

[1] M. Abramowitz and I. Stegun, *Handbook of Mathematical Functions*, Dover, New York (1970).

[2] N.N. Akhmediev, A.V. Buryak and M. Karlsson, *Radiationless optical solitons with oscillating tails*, Opt. Commun. **110** (1994), 540–544.

[3] C.M. Bender and S.A. Orszag, *Advanced Mathematical Methods for Scientists and Engineers*, McGraw-Hill, New York (1978).

[4] D. Bonheure, *Multi-transition kinks and pulses for fourth order equations with a bistable nonlinearity*, Ann. Inst. H. Poincaré Anal. Non Linéaire **21** (2004), 319–340.

[5] D. Bonheure, *Multi-transition connections for fourth order equations with a bistable nonlinearity*, Differential Equations, F. Dumortier, H. Broer, J. Mahwin, A. Vanderbauwhede and S. Verduyn Lunel, eds, World Scientific, Singapore (2005).

[6] B. Buffoni, A.R. Champneys and J. Toland, *Bifurcation and coalescence of a plethora of homoclinic orbits for autonomous Hamiltonian systems*, J. Dyn. Differ. Equations **8** (1996), 221–279.

[7] A.R. Champneys and J. Toland, *Bifurcation of a plethora of multimodal homoclinic orbits for autonomous Hamiltonian systems*, Nonlinearity **6** (1993), 665–722.

[8] A.R. Champneys, M.D. Groves and P.D. Woods, *A global characterization of gap solitary-wave solutions to a coupled KdV system*, Phys. Lett. A **271** (2000), 178–190.

[9] J. Chaparova, *Existence and numerical approximations of periodic solutions of semilinear fourth-order differential equations*, J. Math. Anal. Appl. **273** (2002), 121–136.

[10] J.V. Chaparova, L.A. Peletier and S. Tersian, *Existence and nonexistence of nontrivial solutions of fourth and sixth order ordinary differential equations*, Adv. Differential Equations **8** (2003), 1237–1258.

[11] P. Collet and J.P. Eckmann, *Instabilities and Fronts in Extended Systems*, Princeton Series in Physics, Princeton Univ. Press (1990).

[12] G.T. Dee and W. van Saarloos, *Bistable systems with propagating fronts leading to pattern formation*, Phys. Rev. Lett. **60** (1988), 2641–2644.

[13] E. Doedel, A.R. Champneys, T.F. Fairgrieve, Y.A. Kutznetsov, B. Sandstede and X. Wang, *AUTO97: Continuation and bifurcation software for ordinary differential equations (with Hom. Cont.)* (1997); available at *http://indy.cs.concordia.ca/auto/*.

[14] B. Ermentrout, *Simulating, Analyzing, and Animating Dynamical Systems*, SIAM, Philadelphia, PA (2002); available at *www.math.pitt.edu/~bard/xpp/xpp.html*.

[15] P.C. Fife, *Pattern formation in gradient systems*, Handbook of Dynamical Systems II, Towards Applications, B. Fiedler, G. Iooss and N. Kopell, eds, Elsevier, Amsterdam–New York (2002), 677–722.

[16] P.C. Fife and M. Kowalsky, *A class of pattern forming models*, J. Nonlinear Sci. **9** (1999), 641–669.

[17] W.D. Kalies and R.C.A.M. van der Vorst, *Multi transition homoclinic and heteroclinic solutions to the Extended Fisher–Kolmogorov equation*, J. Differential Equations **131** (1996), 209–228.

[18] W.D. Kalies, J. Kwapisz, J.B. van den Berg and R.C.A.M. van der Vorst, *Homotopy classes for stable periodic and chaotic patterns in fourth-order Hamiltonian systems*, Comm. Math. Phys. **214** (2000), 573–592.

[19] W.D. Kalies, J. Kwapisz and R.C.A.M. van der Vorst, *Homotopy classes for stable connections between Hamiltonian saddle-focus equilibria*, Comm. Math. Phys. **193** (1998), 337–371.

[20] J. Kevorkian and J.D. Cole, *Multiple Scale and Singular Perturbation Methods*, Springer-Verlag, New York (1996).

[21] R. Kuske and L.A. Peletier, *A singular perturbation problem for a fourth order ordinary differential equation*, Nonlinearity **18** (2005), 1189–1222.

[22] P. Manneville, *Dissipative Structures and Weak Turbulence*, Academic Press, Boston, MA (1990).

[23] P.J. McKenna and W. Walter, *Nonlinear oscillations in a suspension bridge*, Arch. Ration. Mech. Anal. **87** (1987), 167–177.

[24] P.J. McKenna and W. Walter, *Traveling waves in a suspension bridge*, SIAM J. Appl. Math. **50** (1990), 703–715.

[25] B. Meulenbroek, *Periodic solutions of a fourth order equation*, Report MI 2001-05, Mathematical Institute, Leiden University (2001).

[26] V.J. Mizel, L.A. Peletier and W.C. Troy, *Periodic phases in second order materials*, Arch. Ration. Mech. Anal. **145** (1998), 343–382.

[27] M.A. Peletier, *Non-existence and uniqueness results for fourth-order Hamiltonian systems*, Nonlinearity, **12** (1999), 1555–1570.

[28] L.A. Peletier and W.C. Troy, *A topological shooting method and the existence of kinks of the Extended Fisher–Kolmogorov equation*, Topol. Methods Nonlinear Anal. **6** (1996), 331–355.

[29] L.A. Peletier and W.C. Troy, *Chaotic spatial patterns described by the Extended Fisher–Kolmogorov equation*, J. Differential Equations **129** (1996), 458–508.

[30] L.A. Peletier and W.C. Troy, *Spatial patterns described by the Extended Fisher–Kolmogorov (EFK) equation: Periodic solutions*, SIAM J. Math. Anal. **28** (1997), 1317–1353.

[31] L.A. Peletier and W.C. Troy, *Multibump periodic travelling waves in suspension bridges*, Proc. Roy. Soc. Edinburgh Sect. A **128** (1998), 631–659.

[32] L.A. Peletier and W.C. Troy, *Spatial Patterns: Higher Order Models in Physics and Mechanics*, Birkhäuser, Boston, MA (2001).

[33] L.A. Peletier, A.I. Rotariu-Bruma and W.C. Troy, *Pulse like patterns described by higher order model equations*, J. Differential Equations **150** (1998), 124–187.

[34] M.H. Protter and H.F. Weinberger, *Maximum Principles in Differential Equations*, Prentice-Hall International, Englewood Cliffs, NJ (1967).

[35] J.G. Ramsay, *A geologist's approach to rock deformation*, Inaugural Lectures, Imperial College of Science, Technology and Medicine, London (1967), 131–143.

[36] J.B. Swift and P.C. Hohenberg, *Hydrodynamic fluctuations at the convective instability*, Phys. Rev. A **15** (1977), 319–328.

[37] S. Tersian and J. Chaparova, *Periodic and homoclinic solutions of some semilinear sixth-order differential equations*, J. Math. Anal. Appl. **272** (2002), 223–239.

[38] J.F. Toland, *A necessary geometric condition for the existence of certain homoclinic orbits*, Math. Proc. Cambridge Philos. Soc. **100** (1986), 591–594.

[39] G.J.B. van den Berg, *Uniqueness of solutions of the extended Fisher–Kolmgorov equation*, C. R. Math. Acad. Sci. Paris **326** (1998), 447–452.

[40] G.J.B. van den Berg, *The phase plane picture for a class of fourth order differential equations*, J. Differential Equations **161** (2000), 110–153.

[41] G.J.B. van den Berg, *Dynamics and equilibria of fourth order differential equations*, Ph.D. Thesis, Leiden University (2000).

[42] G.J.B. van den Berg, Private communication.

[43] G.J.B. van den Berg, L.A. Peletier and W.C. Troy, *Global branches of multi bump periodic solutions of the Swift–Hohenberg equation*, Arch. Ration. Mech. Anal. **158** (2001), 91–153.

[44] A.C. Yew, *Localised solutions of a system of coupled nonlinear Schrödinger equations*, Ph.D. Thesis, Brown University (1998).

[45] A.C. Yew, A.R. Champneys and P.J. McKenna, *Multiple solitary waves due to second order harmonic generation in quadratic media*, J. Nonlinear Sci. **9** (1999), 33–52.

[46] A.C. Yew, B. Sandstede and C.K.R.T. Jones, *Instability of multiple pulses in coupled nonlinear Schrödinger equations*, Phys. Rev. E **61** (2000), 5886–5892.

Author Index

Roman numbers refer to pages on which the author (or his/her work) is mentioned. Italic numbers refer to reference pages. Numbers between brackets are the reference numbers. No distinction is made between the first author and co-author(s).

Abramowitz, M. 600, *603* [1]

Acerbi, E. 7, 8, 12, *97* [1]; *98* [2]; *98* [3]; *98* [4]; 127, 128, 130, 142, 151, 185, *208* [1]; *208* [2]; *208* [3]; *208* [4]

Adams, R.A. 15, *98* [5]; 325, *395* [1]

Adimurthi 284, 310, *312* [1]; *313* [2]; *313* [3]; *313* [4]; 376, *395* [2]

Agmon, S. 325, 339, 393, *395* [3]; 405, 407, 435, *461* [1]

Akhmediev, N.N. 556, *603* [2]

Alama, S. 375, *395* [4]

Alberti, G. 103, 117, 159, 166–169, 176, *208* [5]; *209* [6]; *209* [7]; *209* [8]; *209* [9]; *209* [10]; *209* [11]

Alessandrini, G. 512, 515, *551* [4]

Alicandro, R. 195, 201, 202, 206, 208, *209* [12]; *209* [13]; *209* [14]; *209* [15]

Alkhutov, Y. 12, *98* [6]

Alonso, A. 475, *551* [5]

Alt, H.W. 468, 469, 473, 475, *551* [1]; *551* [2]; *551* [3]

Amann, H. 321, 330, 333, 340, 341, 370, *395* [5]

Amar, M. 103, 178, 179, *209* [16]; *209* [17]

Ambrosetti, A. 320, 341, 369, 371, 374, *395* [6]

Ambrosio, L. 103, 109, 125, 157, 173, 188–190, *209* [18]; *209* [19]; *209* [20]; *209* [21]

Ansini, N. 147, 164, 165, 170, *209* [22]; *209* [23]; *209* [24]

Antontsev, S.N. 6, 8, 11, 12, 40, 49, 53, 67, 76, 95, *98* [7]; *98* [8]; *98* [9]; *98* [10]; *98* [11]; *98* [12]; *98* [13]; *98* [14]; *98* [15]; *98* [16]; *98* [17]

Anzellotti, G. 117, *209* [25]; *209* [26]

Aranda, C. 320, 335, 336, 353, 374, *395* [7]; *395* [8]; *395* [9]; *396* [10]

Attouch, H. 110, *209* [27]

Aubin, T. 219, 226, *313* [5]; 403, 441, *461* [2]

Badiale, M. 375, *396* [11]

Bahri, A. 218, 220, 231, *313* [6]; *313* [7]; *313* [8]

Baiocchi, C. 467–470, *551* [6]; *551* [7]; *551* [8]; *551* [9]; *551* [10]; *551* [11]

Baldo, S. 117, 171, 176, *209* [6]; *209* [25]; *209* [26]; *209* [28]

Ball, J.M. 127, *209* [29]

Balogh, Z. 451, *461* [3]

Bandle, C. 324, 364, 384, *396* [12]; 407, 451, 455, *461* [4]; *461* [5]; *461* [6]; *461* [7]

Baraket, S. 224, 285, 291, *313* [9]

Barbu, V. 475, 482, *551* [12]

Barenblatt, G. 10, *98* [18]

Barles, G. 12, *98* [19]

Barron, E.N. 133, *209* [30]

Bartolucci, D. 223, *313* [10]

Bear, J. 471, *551* [13]

Bellettini, G. 103, 158, 166, *209* [7]; *209* [8]; *209* [33]; *210* [34]

Ben Ayed, M. 222, 255, *313* [11]

Benci, V. 336, *396* [13]; 468, *551* [14]

Bender, C.M. 596, *603* [3]

Berestycki, H. 330, *396* [14]

Bers, L. 334, 393, *396* [15]; *461* [8]

Bertsch, M. 320, 324, 325, 329, 333, 355, 386, *396* [16]

Bethuel, F. 174, 175, *210* [35]

Bhattacharya, K. 103, 185, *209* [31]; *209* [32]

Blake, A. 206, *210* [38]

Blanc, X. 199, 203, *210* [36]; *210* [37]

Blanchard, P. 18, 78, *98* [20]

Bocea, M. 185, *210* [39]

Bodineau, T. 103, *210* [40]

Bonheure, D. 559, *603* [4]; *603* [5]

Bouchitté, G. 103, 119, 159, 167–169, *209* [9]; *209* [10]; *210* [41]; *210* [42]; *210* [43]

Bourdin, B. 191, *210* [44]

Braides, A. 103, 110, 112, 119, 120, 122,
    126–128, 130, 135, 138, 142, 144, 147, 151,
    153, 155, 157, 159, 162, 164, 165, 170,
    183–185, 188, 190, 191, 193–195, 197–199,
    202–208, *209* [12]; *209* [13]; *209* [14];
    209 [16]; *209* [18]; *209* [22]; *209* [23]; *209* [24];
    209 [31]; *210* [45]; *210* [46]; *210* [47]; *210* [48];
    210 [49]; *210* [50]; *210* [51]; *210* [52];
    210 [53]; *210* [54]; *210* [55]; *210* [56];
    210 [57]; *210* [58]; *210* [59]; *210* [60];
    210 [61]; *210* [62]; *211* [63]; *211* [64]; *211* [65]
Brandt, A. 406, *461* [9]; *461* [10]
Brezis, H. 174, 175, *210* [35]; 218, 219, 222, 223,
    269, 270, *313* [12]; *313* [13]; *313* [14];
    313 [15]; 320, 321, 324, 340, 341, 369, 371,
    374, 375, 385, 387, *395* [6]; *396* [17]; *396* [18];
    396 [19]; *396* [20]; 403, 404, 410, 426, 446,
    449, *461* [11]; *461* [12]; *461* [13]; 468, 473,
    475, *551* [15]; *552* [16]
Browder, F. 410, *461* [12]
Brown, K.J. 341, 356, *396* [21]
Brüning, E. 18, 78, *98* [20]
Buffoni, B. 556, *603* [6]
Buryak, A.V. 556, *603* [2]
Buttazzo, G. 103, 126, 142, 151, 185, *208* [1];
    208 [2]; *210* [42]; *210* [47]; *211* [66]; *211* [67]

Cabré, X. 374, *396* [22]; 406, 407, *461* [15];
    461 [17]
Caccioppoli, R. 410, *461* [14]
Caffarelli, L.A. 219, 223, *313* [16]; *313* [17]; 406,
    407, 451, *461* [15]; *461* [16]; *461* [17];
    461 [18]; 468, *552* [17]
Caglioti, E. 223, *313* [18]
Camar-Eddine, M. 134, 151, *211* [68]
Campanato, S. 406, *461* [19]; *461* [20]
Canino, A. 376, *396* [23]
Cao, D. 284, *313* [19]
Capelo, A. 470, *551* [10]
Capogna, L. 407, *462* [21]
Carriero, M. 188, *211* [92]
Carrillo, J. 468, 469, 473, 475, 507, 509, 521,
    551 [5]; *552* [18]; *552* [19]; *552* [20]
Cassandro, M. 103, 166, *209* [8]
Cazenave, T. 385, *396* [17]
Cerami, G. 320, 341, 369, 371, 374, *395* [6]
Chaljub-Simon, A. 407, 450, *462* [22]
Chambolle, A. 7, *98* [21]; 103, 191, 196, 206,
    210 [44]; *210* [48]; *211* [69]; *211* [70];
    211 [71]; *211* [72]
Champion, T. 133, *211* [73]
Champneys, A.R. 556, 559, 560, *603* [6]; *603* [7];
    603 [8]; *603* [13]; *604* [45]

Chandrasekhar, S. 217, 223, *313* [20]
Chang, K.C. 376, *396* [24]; *396* [25]
Chang, S.-Y.A. 223, *313* [21]
Chaparova, J.V. 556, *603* [9]; *603* [10]; *604* [37]
Chen, C.-C. 223, *313* [22]
Chen, F. 370, *396* [26]
Chen, Y. 7, *98* [22]; *100* [57]
Cheng, J. 349, 364, *400* [123]
Cherkaev, A.V. 141, *213* [119]
Chiadò Piat, V. 151, 155, 157, 164, 165, 188,
    208 [3]; *209* [23]; *210* [49]; *210* [50]
Chipot, M. 12, 26, 28, 40, 64, *98* [8]; *98* [23];
    98 [24]; 468–470, 473, 475, 507, 509, 521,
    552 [18]; *552* [19]; *552* [21]; *552* [22]; *552* [23]
Choi, Y.S. 366, 367, 370, *396* [27]; *396* [28];
    396 [29]
Choquet-Bruhat, Y. 407, 450, *462* [22]
Cicalese, M. 201, 202, 204, 206, 208, *209* [12];
    209 [13]; *209* [15]; *210* [51]
Cioranescu, D. 145, *211* [74]
Cirstea, F. 368, *396* [30]
Clapp, M. 266, *313* [23]
Clément, P. 324, 328, *396* [31]
Coclite, M.M. 320, 340, 368, 369, *396* [32];
    396 [33]; *396* [34]; *396* [35]
Cohen, D.S. 321, 324, 341, *396* [36]
Cole, J.D. 596, *603* [20]
Colesanti, A. 451, *462* [23]
Collet, P. 555, *603* [11]
Conti, S. 103, 172, 173, *211* [75]; *211* [76];
    211 [77]
Cordes, H.O. 407, *462* [24]; *462* [25]
Coron, J.M. 218, 221, 231, 255, *313* [7]; *313* [24]
Cortesani, G. 188, *211* [78]
Coscia, A. 197, *211* [79]
Courant, R. 325, *396* [37]; 403, *462* [26]
Crandall, M.G. 133, *211* [80]; 223, *313* [25]; 320,
    321, 324, 333–335, 341–344, 364, 385, *396* [38]
Cuoghi, P. 451, *462* [23]

Dacorogna, B. 127, 128, *211* [81]
Dai, Q. 364, *397* [39]
Dal Maso, G. 103, 110, 117, 120, 150, 151, 158,
    194, 207, 208, *208* [3]; *209* [33]; *210* [52];
    210 [53]; *211* [67]; *211* [72]; *211* [82];
    211 [83]; *211* [84]; *211* [85]; *211* [86]; *211* [87]
Damascelli, L. 489, *552* [24]
Dancer, E.N. 218, *313* [26]
Dautray, R. 330, *397* [40]
David, G. 188, *211* [88]
Davies, E.B. 357, *397* [41]
Dávila, J. 309, *313* [27]; *313* [28]; 322, 323, 366,
    367, 384–388, *397* [42]; *397* [43]; *397* [44];
    397 [45]; *397* [46]

Davini, A. 143, *211* [89]
de Figueiredo, D.G. 324, 328, *396* [31]
De Giorgi, E. 103, 106, 110, 120, 122, 128, 188, *211* [90]; *211* [91]; *211* [92]; *212* [93]; *212* [94]
de Giovanni, M. 376, *396* [23]
De Lellis, C. 173, *209* [19]
de Marsily, G. 10, *98* [25]
De Pascale, L. 133, *211* [73]
De Simone, A. 103, 173, *211* [75]; *212* [95]
de Thélin, F. 325, *397* [56]
Dee, G.T. 555, *603* [12]
Defranceschi, A. 103, 120, 122, 126–128, 135, 138, 142, 151, 153, 190, 208, *210* [54]; *210* [55]
del Pino, M. 221–224, 231, 262, 264–266, 270, 285, 293, 296–298, 301, 309, *313* [23]; *313* [27]; *313* [28]; *314* [29]; *314* [30]; *314* [31]; *314* [32]; *314* [33]; *314* [34]; *314* [35]; *314* [36]; 320, 321, 336, 342, 364, *397* [47]; *397* [48]
Díaz, G. 12, *98* [19]
Díaz, J.I. 5, 8, 11, 12, 49, 53, 95, *98* [9]; *98* [10]; *98* [19]; *99* [26]; *99* [27]; 319, 324, 341, 355, 366, 367, 383–385, *397* [49]; *397* [50]; *397* [51]
DiBenedetto, E. 12, *99* [28]; *99* [29]; 488, *552* [25]
Diening, L. 11, *99* [30]
Dieudonné, J. 329, 330, *397* [52]
Ding, W. 218, *314* [37]
Doedel, E. 560, *603* [13]
Dolbeault, J. 222, 223, 270, *314* [29]; *314* [30]
Dolzmann, G. 103, *211* [75]
D'Onofrio, L. 12, *99* [31]
Donsker, M. 330, *397* [53]
Douglis, A. 325, 339, 393, *395* [3]; 405–407, 435, *461* [1]; *462* [27]
Druet, O. 270, *314* [38]
Dugundji, J. 447, *462* [28]
Duzaar, F. 12, *99* [32]

Eckmann, J.P. 555, *603* [11]
Edmunds, D. 11, *99* [33]
Ehihara, Y. 364, *399* [104]
El Mehdi, K. 222, 255, 310, *313* [11]; *314* [39]
Entov, V. 10, *98* [18]
Ermentrout, B. 560, *603* [14]
Esposito, P. 310, 312, *314* [40]; *314* [41]; *314* [42]
Essén, M. 407, *461* [4]
Evans, L.C. 109, 129, 133, 144, *211* [80]; *212* [96]; *212* [97]; *212* [98]; 407, *461* [13]; *462* [29]

Fabricant, A. 333, *397* [54]
Fairgrieve, T.F. 560, *603* [13]
Fan, X.-L. 10–12, 35, 47, *99* [34]; *99* [35]; *99* [36]; *99* [37]; *99* [38]; *99* [39]; *99* [40]; *99* [41]; *99* [42]

Feireisl, E. 361, *397* [55]
Felli, V. 223, *314* [43]
Felmer, P. 221, 222, 231, 262, 264, 265, 285, *314* [31]; *314* [32]; *314* [33]; *314* [34]
Fife, P.C. 407, *462* [30]; 555, *603* [15]; *603* [16]
Fleckinger, J. 325, *397* [56]
Flucher, M. 178, 179, *212* [99]; *212* [100]; 451, *461* [5]
Focardi, M. 190, *212* [101]
Fonseca, I. 119, 125, 127, 130, 172, 173, 183–185, *210* [39]; *210* [43]; *210* [56]; *211* [76]; *212* [102]; *212* [103]; *212* [104]
Fowler, R.H. 273, *314* [44]
Fragalà, I. 12, *99* [43]; 142, *210* [47]
Francfort, G.A. 183, 184, 204, 205, *210* [56]; *210* [57]
Franzoni, T. 103, *212* [93]
Friedman, A. 468–470, *551* [11]; *552* [17]; *552* [26]; *552* [27]
Friesecke, G. 103, 185, 186, 204, *212* [105]; *212* [106]; *212* [107]; *212* [108]
Fu, Y. 10, *99* [44]
Fukushima, M. 133, *212* [109]
Fulks, W. 319, 321, 354, *397* [57]
Furusho, Y. 364, *399* [104]
Fusco, N. 103, 109, 125, 127, 128, 130, 188, 189, *208* [4]; *209* [20]

Gamba, I.M. 319, *397* [58]
Gariepy, R.F. 109, 133, *211* [80]; *212* [97]
Garroni, A. 132, 170, 178, 179, 207, *209* [17]; *210* [53]; *212* [100]; *212* [110]; *212* [111]; *212* [112]; *212* [113]
Gaucel, S. 370, *397* [59]
Gauss, C.F. 409, *462* [31]
Gazzola, F. 12, *99* [43]
Ge, Y. 223, *314* [45]
Gelfand, I.M. 223, *314* [46]
Gelli, M.S. 190, 195, 199, 202, 203, 206–208, *209* [14]; *210* [58]; *210* [59]; *210* [60]; *210* [61]; *212* [101]
Ghergu, M. 319, 335, 336, 349, 368, *396* [30]; *397* [60]; *397* [61]
Giacomoni, J. 376, *395* [2]; *397* [62]
Giaquinta, M. 406, *462* [32]
Gidas, B. 219, *313* [16]; 451, *461* [18]; *462* [33]
Gilardi, G. 468, *551* [3]
Gilbarg, D. 393, *397* [63]; *397* [64]; 403, 406, 407, 435, 451, *462* [34]; *462* [35]
Giraud, G. 410, *462* [36]; *462* [37]
Gobbino, M. 196, *212* [114]
Godoy, T. 335, 358, *395* [7]; *395* [8]; *397* [65]
Gomes, D. 144, *212* [98]

Gomes, S.N. 321, 336, 342, 348, 364, *397* [66]; *397* [67]
Gonçalves, J.V. 337, *398* [68]
Goulaouic, C. 407, 408, *462* [38]
Greenkorn, R.A. 468, 471, *552* [28]
Grossi, M. 222, 255, 310, 312, *312* [1]; *313* [11]; *314* [39]; *314* [40]
Groves, M.D. 559, *603* [8]
Gu, J. 364, *397* [39]
Gui, C. 284, *314* [47]; 321, 344, 349, *398* [69]
Gurney, W.S.C. 324, 384, *398* [70]
Gurtin, M.E. 324, 384, *398* [71]

Haitao, Y. 372, 374, 375, *398* [72]
Hale, J.K. 360, *398* [73]
Han, Q. 403, 406, 407, *462* [21]; *462* [39]; *462* [40]
Han, X. 11, *99* [34]
Han, Z.-C. 219, *314* [48]
Harjulehto, P. 10, 11, *99* [45]; *99* [46]; *99* [47]
Hästö, P. 10, 11, *99* [45]; *99* [46]; *99* [47]
Hélein, F. 174, 175, *210* [35]
Henry, D. 322, 353, 354, 356, 357, 360, *398* [74]
Hernández, G.E. 336, *397* [48]
Hernández, J. 320–323, 325, 330, 332, 339–341, 353, 355, 358, 362, 364, 366, 369, 370, 377, 383, *397* [50]; *397* [56]; *397* [65]; *398* [75]; *398* [76]; *398* [77]; *398* [78]; *398* [79]
Hess, P. 325, 341, 356, *396* [21]; *398* [80]
Hilbert, D. 325, *396* [37]; 403, *462* [26]
Hirano, N. 266, *314* [49]; 322, 376, *398* [81]
Hohenberg, P.C. 555, *604* [36]
Hölder, O. 406, 409, *462* [41]
Holopainen, I. 451, *461* [3]
Hopf, E. 406, 410, *462* [42]
Hörmander, L. 393, *397* [63]; 407, *462* [34]
Hua Lin, F. 321, 344, 349, *398* [69]
Huang, S.-Y. 469, *552* [27]
Hudzik, H. 12, *99* [48]

Ioffe, D. 103, *210* [40]
Ivanov, A.V. 12, *99* [49]
Iwaniec, T. 12, *99* [31]; *99* [50]

James, R.D. 103, 185, 186, *209* [32]; *212* [105]; *212* [106]; *212* [107]
Jerrard, R.L. 175, 176, *212* [115]; *212* [116]
Jing, R. 223, *314* [45]; *314* [50]
John, F. 334, 393, *396* [15]
Jones, C.K.R.T. 559, *604* [46]
Joseph, D.D. 223, *314* [51]
Jungel, A. 319, *397* [58]

Kalashnikov, A.S. 12, *99* [51]; *99* [52]; *99* [53]; *100* [54]
Kalies, W.D. 556, 559, *603* [17]; *603* [18]; *603* [19]
Kamin, S. 321, 324, 340, 341, *396* [18]
Kamynin, L.I. 393, *398* [82]
Karátson, J. 323, 377, 380, 383, *398* [76]; *398* [83]
Karlsson, M. 556, *603* [2]
Kato, T. 325, *398* [80]
Kaufmann, U. 358, *397* [65]
Kawohl, B. 12, *99* [43]
Kazdan, J. 218, *314* [52]
Kellogg, O.D. 410, *462* [43]
Kevorkian, J. 596, *603* [20]
Khimchenko, B.N. 393, *398* [82]
Khruslov, E.Ya. 145, *213* [120]
Kichenassamy, S. 403, 406–408, 435, 447, 450–452, 455, *462* [44]; *462* [45]; *462* [46]; *462* [47]; *463* [48]; *463* [49]; *463* [50]; *463* [51]
Kinderlehrer, D. 468, 473, 475, *552* [16]
Kohn, R.V. 173, *212* [95]
Koskenoja, M. 10, *99* [47]
Kováčik, O. 11, 12, *100* [55]
Kowalczyk, M. 224, 293, 296–298, 301, *314* [35]
Kowalsky, M. 555, *603* [16]
Krein, M.G. 404, 446, *463* [52]
Kufner, A. 356, *398* [84]
Kuske, R. 559, 594, 596, *603* [21]
Kutev, N. 333, *397* [54]
Kutznetsov, Y.A. 560, *603* [13]
Kwapisz, J. 556, 559, *603* [18]; *603* [19]

Ladyzhenskaya, O.A. 18, 24, 34, *100* [56]; 339, 387, 389, *398* [85]; 403, *463* [53]
Laetsch, T. 321, 324, 341, *396* [36]; *398* [86]
Lair, A.V. 336, 374, 375, *398* [87]
Lami-Dozo, E. 320, 335, 336, 353, 374, *395* [9]; *396* [10]
Langlais, M. 370, *397* [59]
Lazer, A.C. 321, 342, 343, 363, 364, 366, 367, 376, *396* [27]; *398* [88]
Le Bris, C. 199, 203, *210* [36]; *210* [37]; *212* [117]
Le Dret, H. 181, *212* [118]
Le Dung 488, 491, 516, *552* [29]
Leach, J.A. 319, 358, *398* [89]
Leaci, A. 188, *211* [92]
Leoni, G. 119, 125, 127, 172, 173, *210* [43]; *211* [76]; *212* [102]
Leray, J. 403, 404, 444, *463* [54]
Letta, G. 122, *212* [94]
Levine, S. 7, *98* [22]; *100* [57]
Lew, A.J. 207, 208, *210* [62]
Li, Y.Y. 220, 221, 223, 231, *313* [8]; *314* [53]; *314* [54]; *314* [55]

Lichtenstein, L. 410, *463* [55]
Lieberman, G.M. 512, 550, *552* [30]
Lin, C.-S. 223, 284, *313* [22]; *314* [47]; *315* [56]; *315* [57]; *315* [58]
Lin, F.H. 403, *462* [40]
Lions, J.-L. 12, *100* [58]; 330, *397* [40]; 477, *552* [31]
Lions, P.-L. 7, *98* [21]; 199, 203, *210* [36]; *210* [37]; *212* [117]; 223, *313* [18]; 321, 341, *398* [90]
Liouville, J. 223, *315* [59]
Loewner, C. 451, *463* [56]
López-Gómez, J. 325, 332, *398* [91]
Lundgren, T.S. 223, *314* [51]
Lurie, K. 141, *213* [119]
Lyaghfouri, A. 468, 469, 473, 475, 509, 521, *552* [20]; *552* [22]; *552* [23]; *552* [32]; *552* [33]; *552* [34]; *552* [35]

Ma, L. 223, *315* [60]
MacCamy, R.C. 324, 384, *398* [71]
Majer, P. 374, *396* [22]
Mancebo, F.J. 320–322, 325, 330, 332, 339–341, 353, 355, 362, 364, 366, 369, 370, *398* [77]; *398* [78]; *398* [79]
Mancini, G. 284, *313* [2]
Manes, A. 325, *398* [92]
Manneville, P. 555, *603* [22]
Mantegazza, C. 173, *209* [19]
Marcellini, P. 12, *100* [59]
March, R. 197, 198, *211* [63]
Marchenko, A.V. 145, *213* [120]
Marchioro, C. 223, *313* [18]
Marcus, M. 451, 455, *461* [6]; *461* [7]; *463* [57]
Martel, Y. 385, *396* [17]
Mascarenhas, L. 119, *210* [43]
Maybee, J.S. 319, 321, 354, *397* [57]
McKenna, P.J. 321, 336, 342, 343, 363, 364, 366, 367, 370, 376, *396* [27]; *396* [28]; *396* [29]; *398* [88]; *398* [93]; 557, 559, *604* [23]; *604* [24]; *604* [45]
Merle, F. 223, *313* [13]
Meulenbroek, B. 603, *604* [25]
Meyers, N.G. 406, *463* [58]
Michaille, M. 12, 26, 28, 64, *98* [23]; *98* [24]
Micheletti, A.M. 266, *314* [49]; *315* [61]; 325, 336, *396* [13]; *398* [92]
Mignot, F. 223, *315* [62]
Milton, G.W. 140, *213* [121]
Mingione, G. 7, 8, 12, *97* [1]; *98* [2]; *98* [3]; *98* [4]; *99* [32]
Miranda, C. *463* [59]
Mitidieri, E. 324, 328, *396* [31]

Mizel, V.J. 556, *604* [26]
Modica, L. 103, 108, 159, *211* [86]; *213* [122]; *213* [123]
Molle, R. 222, *315* [63]; *315* [64]
Monakhov, V. 10, *100* [60]
Montenegro, M. 322, 323, 366, 367, 384–388, *397* [43]; *397* [44]; *397* [45]; *397* [46]
Morel, J.M. 188, *213* [124]; 319, 355, 367, 385, *397* [51]
Morgan, F. 176, *213* [125]
Morrey, C.B. 127, *213* [126]; 403, 406, *463* [60]; *463* [61]
Mortola, S. 108, 159, *213* [123]
Mosco, U. 134, 150, 151, *211* [67]; *211* [87]; *213* [127]
Mugnai, L. 158, *210* [34]
Müller, S. 103, 117, 119, 130, 136, 138, 170, 173, 179, 185, 186, *209* [11]; *211* [75]; *212* [95]; *212* [100]; *212* [103]; *212* [104]; *212* [105]; *212* [106]; *212* [107]; *212* [111]; *212* [112]; *213* [128]; *213* [129]
Mumford, D. 158, 189, *213* [130]; *213* [131]
Murat, F. 145, *211* [74]; 223, *315* [62]
Musielak, J. 11, 12, *100* [61]
Musso, M. 221–224, 231, 253, 262–266, 270, 285, 293, 296–298, 301, 309, 310, 312, *313* [23]; *313* [27]; *313* [28]; *314* [29]; *314* [30]; *314* [32]; *314* [33]; *314* [34]; *314* [35]; *314* [36]; *314* [41]; *314* [42]; *315* [65]; *315* [66]

Nagasaki, K. 223, 224, *315* [67]
Namba, T. 324, 384, *398* [94]
Needham, D.J. 319, 358, *398* [89]
Nesi, V. 132, *212* [113]
Neumann, C. 409, *463* [62]
Ni, W.-M. 284, *315* [58]; *315* [68]; *315* [69]; *315* [70]
Nirenberg, L. 218, 222, *313* [14]; 325, 330, 339, 371, 375, 387, 393, *395* [3]; *396* [14]; *396* [19]; 403–407, 435, 451, *461* [1]; *462* [27]; *463* [56]; *463* [63]; *463* [64]
Nisbet, R.N. 324, 384, *398* [70]
Noussair, E.S. 284, *313* [19]

Oliveira, H. 8, *98* [9]
Orlandi, G. 176, *209* [6]
Orszag, S.A. 596, *603* [3]
Ortiz, M. 207, 208, *210* [62]
Oswald, L. 319, 324, 341, 355, 367, 385, *396* [20]; *397* [51]
Otto, F. 103, 173, *211* [75]; *212* [95]
Ouyang, T. 349, 353, 366, 383, *399* [95]

Pacard, F. 223, 224, 285, 291, *313* [9]; *314* [45];
    407, 450, *463* [65]
Pacella, F. 284, *313* [3]; *313* [4]
Paczka, S. 358, *397* [65]
Pagano, S. 203, *213* [132]
Pallara, D. 103, 109, 125, 188, 189, *209* [20]
Palmieri, G. 368, 369, *396* [35]
Pan, X.B. 284, *315* [68]
Pao, C.V. 321, 322, 340, 341, 370, *399* [96]
Paolini, M. 158, *209* [33]
Paroni, R. 203, *213* [132]
Passaseo, D. 218, 222, *315* [63]; *315* [64];
    315 [71]; *315* [72]; *315* [73]
Pedregal, P. 130, *212* [104]
Peletier, L.A. 219, 269, 270, *313* [15]; 556, 557,
    559, 561, 562, 569, 571–573, 575, 577, 579,
    583, 585, 589, 590, 593, 594, 596, *603* [10];
    603 [21]; *604* [26]; *604* [28]; *604* [29];
    604 [30]; *604* [31]; *604* [32]; *604* [33]; *604* [43]
Peletier, M.A. 556, *604* [27]
Percivale, D. 117, 151, 185, *208* [2]; *208* [3];
    209 [26]
Phillips, D. 386, *399* [97]
Piatnitski, A. 103, 155, 201, *210* [50]; *211* [64];
    213 [133]
Pistoia, A. 222, 253, 263, 266, 285, 310, 312,
    314 [36]; *314* [40]; *314* [41]; *314* [42];
    314 [49]; *315* [61]; *315* [65]; *315* [66]; *315* [74]
Pohozaev, S. 217, *315* [75]
Ponsiglione, M. 132, *212* [113]
Pozio, A. 324, 364, 384, *396* [12]
Presutti, E. 103, 166, *209* [8]
Prinari, F. 133, *211* [73]
Protter, M.H. 325, 394, *399* [98]; 589, 590,
    604 [34]
Pucci, C. 393, *399* [99]
Pucci, P. 6, 11, *100* [62]
Puel, J. 223, *315* [62]
Pulvirenti, M. 223, *313* [18]

Qiu, L. 370, *399* [100]

Rabinowitz, P.H. 223, *313* [25]; 320, 321, 324,
    333–335, 341–344, 364, 385, *396* [38];
    399 [101]; 403, 404, 444, 446, 447, *463* [66]
Radulescu, V. 319, 335, 336, 349, 368, *396* [30];
    397 [60]; *397* [61]
Raitums, U. 140, *213* [134]
Rajagopal, K. 7, *100* [63]
Rákosník, J. 11, 12, *99* [33]; *100* [55]
Ramiandrisoa, A. 385, *396* [17]
Ramsay, J.G. 555, *604* [35]
Rangelov, T. 333, *397* [54]

Rao, M. 7, *98* [22]
Raoult, A. 181, *212* [118]
Reichel, W. 336, *398* [93]
Remy, E. 201, *213* [133]
Ren, X. 310, *315* [76]; *315* [77]
Rey, O. 219, 220, 222, 231, 240, 255, 284, 285,
    313 [8]; *313* [11]; *315* [74]; *315* [78]; *315* [79];
    315 [80]; *315* [81]
Rieger, M.O. 190, *213* [135]
Rivière, T. 407, 450, *463* [65]
Rodrigues, J.F. 8, *98* [11]; 469, *552* [36]
Rogers, R.C. 167, *213* [136]
Rostamian, R. 320, 324, 325, 329, 333, 355, 386,
    396 [16]
Rotariu-Bruma, A.I. 556, 589, 590, 593, *604* [33]
Rutman, M.A. 404, 446, *463* [52]
Růžička, M. 7, *100* [63]; *100* [64]
Ryzhik, V. 10, *98* [18]

Saá, J.E. 12, *99* [27]
Saccon, C. 322, 376, *398* [81]
Samko, S.G. 11, *100* [65]; *100* [66]
Sandier, E. 174, 175, *213* [137]; *213* [138]
Sandstede, B. 559, 560, *603* [13]; *604* [46]
Santos, C.A.P. 337, *398* [68]
Šarapudinov, I.I. 12, *100* [67]
Sattinger, D.H. 321, 322, 340, 341, 354, 370,
    399 [102]; 403, 404, 446, 449, *463* [67];
    463 [68]
Sbordone, C. 12, *99* [50]
Schatzman, M. 324, 341, *399* [103]
Schauder, J. 403, 404, 406, 444, *463* [54];
    463 [69]; *463* [70]; *463* [71]
Schechter, M. 334, 393, *396* [15]
Schindler, I. 376, *397* [62]
Schweizer, B. 173, *211* [77]
Senba, T. 364, *399* [104]
Seppecher, P. 134, 151, 167–169, *209* [9];
    209 [10]; *211* [68]
Seregin, G.A. 8, *98* [4]
Serfaty, S. 174, 175, *213* [137]; *213* [138]
Serrin, J. 6, 11, *100* [62]
Shafrir, I. 223, *314* [55]
Shah, J. 189, *213* [131]
Shaker, A.W. 336, 374, 375, *398* [87]
Shaoping, W. 322, 369, 374, 375, *399* [117]
Shen, J. 11, *99* [35]
Shi, J. 340, 341, 349, 353, 366, 368, 383,
    399 [95]; *399* [105]; *399* [106]
Shimakura, N. 403, 407, 408, *462* [38]; *463* [72]
Shioji, N. 322, 376, *398* [81]
Shmarev, S. 6, 11, 12, 49, 53, 76, 95, *98* [10];
    98 [12]; *98* [13]; *98* [14]; *98* [15]; *98* [16]
Shteto, E. 336, *396* [13]

Shujie, L. 336, *399* [118]
Sigalotti, M. 202, 203, *210* [61]; 512, 515, *551* [4]
Simon, L. 406, *463* [73]
Simon, P.L. 323, 377, 380, 383, *398* [76]; *398* [83]
Simondon, F. 361, *397* [55]
Smoller, J. 321, 322, 340, 341, 354, 370, *399* [107]; 403, 404, 446, 447, *463* [74]
Solci, M. 167, 191, *210* [48]; *213* [139]
Solimini, S. 188, *213* [124]
Solonnikov, V.A. 339, 387, 389, *398* [85]
Soner, H.M. 176, *212* [116]
Sprekels, J. 336, 348, *397* [67]
Spruck, J. 219, *313* [16]; 324, 341, *399* [108]; 451, *461* [18]; *462* [33]
Stampacchia, G. 468, 473, 475, *552* [16]
Stanich, J. 7, *100* [57]
Stavre, R. 469, *552* [37]
Stegun, I. 600, *603* [1]
Stein, E.M. 406, 412, *464* [75]
Sternberg, P. 159, *213* [140]
Stroock, D. 405, *464* [76]
Struwe, M. 223, *315* [82]; 322, 370–372, *399* [109]
Stuart, C.A. 320, 335, 341, 369, *399* [110]
Sun, Y. 341, 368, *399* [111]
Suzuki, T. 223, 224, *315* [67]; *316* [83]
Swift, J.B. 555, *604* [36]

Takáč, P. 320, 322, 356–358, 360, 376, 394, *397* [62]; *399* [112]
Takagi, I. 284, *315* [58]; *315* [68]; *315* [69]; *315* [70]
Talenti, G. 219, 226, *316* [84]
Tarantello, G. 223, *313* [10]; *315* [82]; *316* [85]; 375, *396* [11]
Tartar, L. 140, 141, *213* [141]; 320, 321, 324, 333–335, 341–344, 364, 385, *396* [38]
Taylor, M. 405, *464* [77]
Terracini, S. 223, *314* [43]
Tersian, S. 556, *603* [10]; *604* [37]
Tesei, A. 324, 364, 384, *396* [12]
Theil, F. 204, *212* [108]
Tilli, P. 190, *213* [135]
Toader, R. 188, *211* [78]
Toland, J.F. 556, *603* [6]; *603* [7]; *604* [38]
Tortorelli, V.M. 190, *209* [21]
Triebel, H. 356, *399* [113]
Troianello, G. 407, *464* [78]
Troy, W.C. 556, 557, 561, 562, 569, 571–573, 575, 577, 579, 583, 585, 589, 590, 593, *604* [26]; *604* [28]; *604* [29]; *604* [30]; *604* [31]; *604* [32]; *604* [33]; *604* [43]
Trudinger, N.S. 393, *397* [64]; 403, 406, 407, 435, 451, *462* [35]; *464* [79]

Truskinovsky, L. 167, 207, *213* [136]; *213* [142]
Tsutsumi, M. 12, *100* [68]
Tyson, J.T. 451, *461* [3]

Ural'tseva, N.N. 18, 24, 34, *100* [56]; 339, 387, 389, *398* [85]; 403, *463* [53]
Urbano, J. 12, *99* [29]

Valente, V. 170, *209* [24]
van den Berg, J.B. 556, 565, 583, 589, 593, *603* [18]; *604* [39]; *604* [40]; *604* [41]; *604* [42]; *604* [43]
van der Vorst, R.C.A.M. 556, 559, *603* [17]; *603* [18]; *603* [19]
van Saarloos, W. 555, *603* [12]
Varadhan, S.R.S. 330, *396* [14]; *397* [53]; 405, *464* [76]
Varonen, S. 10, *99* [47]
Vázquez, J.L. 6, *100* [69]
Vega, J.M. 320–322, 325, 330, 332, 339–341, 353, 355, 362, 364, 366, 369, 370, *398* [77]; *398* [78]; *398* [79]
Velenik, Y. 103, *210* [40]
Vernescu, B. 469, *552* [37]
Véron, L. 450, 451, *463* [51]; *463* [57]
Verruijt, A. 471, *551* [13]
Vespri, V. 12, *99* [29]
Vitali, E. 103, 167, 190, *210* [55]; *213* [139]

Walter, W. 557, *604* [23]; *604* [24]
Wang, F.-Z. 10, *99* [36]
Wang, X.J. 284, *316* [86]; 560, *603* [13]
Warner, F. 218, *314* [52]
Wei, J. 220, 223, 285, 309, 310, *313* [27]; *315* [60]; *315* [76]; *315* [77]; *315* [80]; *315* [81]; *316* [87]
Weinberger, H.F. 325, 394, *399* [98]; 589, 590, *604* [34]
Weston, V.H. 223, *316* [88]
Wiegner, M. 335, 349, 354, *399* [114]; *399* [115]
Woods, P.D. 559, *603* [8]
Wu, H.-Q. 10, *99* [36]
Wu, S. 341, 368, *399* [111]; *399* [116]

Xie, Y. 40, *98* [8]

Yadava, S.L. 284, *313* [3]; *313* [4]
Yang, P. 223, *313* [21]
Yang, Y.-S. 223, *313* [17]
Yao, M. 340, 341, 349, 353, 366, 368, 370, 383, *396* [26]; *399* [95]; *399* [100]; *399* [105]; *399* [106]

Ye, D. 309, *316* [89]
Yew, A.C. 559, *604* [44]; *604* [45]; *604* [46]
Yijing, S. 322, 336, 369, 374, 375, *399* [117]; *399* [118]
Yiming, L. 322, 369, 374, 375, *399* [117]
Yu, J. 336, 366, *400* [124]

Zeppieri, C. 185, *211* [65]
Zhang, Q.H. 10, 11, *99* [37]; *99* [42]

Zhang, Z. 322, 323, 335, 336, 340, 349, 353, 364, 366, 376, *399* [119]; *399* [120]; *399* [121]; *399* [122]; *400* [123]; *400* [124]
Zhao, D. 11, 12, 35, 47, *99* [35]; *99* [38]; *99* [39]; *99* [40]; *99* [41]
Zhao, Y.Z. 11, *99* [42]
Zhikov, V.V. 8, 11, 12, 67, *98* [17]; *100* [70]; *100* [71]; *100* [72]; *100* [73]
Zhou, F. 309, *316* [89]
Ziemer, W. 109, *213* [143]
Zisserman, A. 206, *210* [38]

Subject Index

Γ-convergence, 110
– development by, 117
Γ^+-convergence, 177
Γ-limit, 110
– lower, 113
– upper, 113

A
$\mathcal{A}$-harmonic, 488, 512, 515
Ambrosio–Tortorelli energies, 191
anisotropic, 9
– diffusion, 56
– spaces, 14
anisotropy, essential, 73
anti-ferromagnetic interaction, 206
approximate gradient, 188
asymptotic analysis, 217

B
Baiocchi transformation, 467
barotropic gases, 9
baseline solution, 559, 594
beam equation, 555
Bernstein's inequality, 412
beta function, 600
bi-stable systems, 556
bifurcation, 446
– curves, 559
– problem, 573, 585
Blake–Zisserman approximation, 206
blow-up
– of positive solutions, 354
– technique, 119
borderline cases, 82
boundary
– behavior of solutions, 341
– blow-up, 451
– condition Dirichlet, 471, 491
– – leaky, 469, 471, 516
– – unified, 473
– essential, 156
branches, 573, 582, 583, 599, 600

Brouwer fixed-point theorem, 441
bubbling, 218–222, 224
BV-ellipticity, 157

C
Caccioppoli
– partition, 156
– set, 156
capacitary potential, 179
capacity, 145
Carathéodory functions, 16, 24
Cauchy–Born rule, 203
cell-problem homogenization formula, 137
central interaction, 199
closure of Riemannian metrics, 142
compact
– operator, 444
– support principle, 11
compressible fluids, 9
compression of a cone, 447
concentration–compactness, 176
convection
– terms, 32, 55
– functionals, 194
critical exponent, 218, 220, 223
criticality, 217
currents, 176

D
Darcy law, 9, 471
– linear, 510, 545
– nonlinear, 468, 471, 511
De Giorgi–Letta measure criterion, 122
De Giorgi rectifiability theorem, 156
diffusion–absorption
– balance, 43
– processes, 5
directional localization, 56
Dirichlet
– form, 133
– integral, 409
– problem, 409
distance function, 416

double-well energy, 158
doubly nonlinear, 10
dyadic decomposition, 412

E
elastica functional, 158
electrorheological fluids, 7
embedding theorems, 15
energy, 556, 560, 562, 572
– functions, 44, 79
– identity, 560, 570, 571, 577, 595, 600
– relation, 47
equation
– of anisotropic diffusion, 86
– of mixed type, 90
– with convective terms, 88
equicoercive sequence, 115
Euler–Lagrange equation, 556
Euler–Poisson–Darboux, 407
existence of a solution, 475
extension lemma, 151

F
fast diffusion–weak absorption, 5
ferromagnetic interaction, 205
finite-difference energies, 196
Fisher–Kolmogorov equation, extended, 555
fixed-point theorems, 443
flat convergence, 175
free boundary, 471, 490, 516
– continuity of, 495, 526
– solutions, 383
free-discontinuity problem, 186
Fuchsian, 452
– operator
– – of type (I), 435
– – of type (II), 435
– reduction, 452
function with bounded variation, 156
fundamental estimate, 122

G
G-convergence, 132
gamma function, 601
Gauss–Green formula, 156
generalized
– diffusion equation, 3, 24
– special functions of bounded variation, 189
geometric rigidity, 186
Gibbs phenomenon, 171
Ginzburg–Landau energy, 173

H
Hamiltonian structure, 556
harmonic center, 179
Hausdorff measure, 156
Hölder
– condition of order α, 409
– spaces, 411
– inequality, 13
– – inverse, 38
holomorphic semigroup, 358
homoclinic orbit, 593
homogenization
– of Hamilton–Jacobi equations, 143
– of networks, 204
– theorem, 135
– theory, 134
homogenized functional, 135

I
image recovery, 6
implicit function theorem, 563
initial value problem, 562, 576
internal normal, 156
interpolation inequalities, 415
Ising system, 166, 205
isolated singularities, 449, 450
isometric embedding, 186

J
Jacobian, distributional, 175
jump set, 156

K
Krein–Rutman theorem, 446

L
laminate, 139
Laplace
– equation, 408
– transform, 408
laps, 557, 561, 576, 584
lattice system, 199
Lax formula, 143
Lennard–Jones potential, 207
liminf inequality, 111
limsup inequality, 111
line-tension effect, 168
linearized stability, 353
Liouville
– equation, 410
– property, 450
Littlewood–Paley (LP), 412
local weak solutions, 43

localization, 5
– principle, 140
– property, 11
Loewner–Nirenberg equation, 451
lower bound, 104
log-continuous, 17
lower-semicontinuous envelope, 114

M
maximal solution, 337, 451, 502, 539
maximum principle, 393, 423, 449, 570, 589, 590, 593
mean curvature, 416, 451, 455
method of continuity, 440
minimal solution, 337, 502, 510, 539, 545
Minkowski content, 188
Modica–Mortola theorem, 159
monotone nonlinearities, 333
monotonicity property, 486
Moreau–Yosida transform, 112
multibump solution, 576, 594, 599
multiplicity of principal eigenvalues, 332
multiscale analysis, 558, 593, 594
Mumford–Shah functional, 189
Musielak–Orlicz space, 11

N
nearest neighbor, 200
Newtonian potential, 408, 419
next-to-nearest neighbor, 203
non-Newtonian fluids, 7, 8
nonhomogeneous, 9
nonoscillation result, 494, 520
nonstandard, 4
– growth condition, 4, 8
– – of (p_+, p_-) type, 8

O
obstacle problem, 151
optimal profile, 162
Orlicz–Sobolev space, 11

P
p-Laplacian, 3, 407, 450
pair potential, 199
perforated domain, 144
perimeter, 155
Perron, 410
Poincaré
– balayage method, 410
– inequality, 426
Poincaré–Perron, 435
point of density, 156
points of symmetry, 561, 576, 586

Poisson equation, 409
polyconvex function, 127
pool, 509
principal
– curvature, 416
– eigenvalue, 329, 330
$p(x)$-Laplace
– equation, 3, 7, 10
– – eigenvalue problem for, 34
– – generalized, 9, 16
$p(x)$-Laplacian, 3
– anisotropic, 89

Q
quasiconvex envelope, 128
quasiconvexity, 127

R
radial solutions
– multiplicity of, 377
– to singular problems, 377
– uniqueness of, 377
recovery sequence, 111
relaxation, 114
relaxed Dirichlet problem, 149
renormalization group, 207
renormalized unknown, 452
reservoirs-connected solution, 507–510, 544, 545
– uniqueness of, 510, 511, 545, 549
resonance, 574
Robin function, 179

S
S_3-connected solution, 507
SBV compactness theorem, 189
Schauder fixed-point theorem, 441
Schrödinger equation, 556, 558
Schwarz reflection principle, 434
semilinear
– elliptic boundary value problems, 217
– equation with nonlinear absorption term, 83
separation of scales, 207
set of finite perimeter, 156
shooting argument, 562
singular problems
– differenciability of, 350
– elliptic, 323
– – estimates for, 327, 334
– – sublinear, 340
– estimates for eigenvalues, 329
– variational methods, 370
slicing method, 122
slow diffusion–strong absorption, 5

Sobolev
– embedding, 176
– inequality, 14
– space, generalized, 10
solid–solid phase transitions, 172
space
– $L^{p(x)}(\Omega)$, 12
– $W_0^{1,p(x)}(\Omega)$, 12
spectrum of singular eigenvalue problem, 325
stabilization, 358
strong
– comparison principle, 489
– maximum principle, 11, 495, 520
sub- and supersolutions, 410, 448
subadditivity, 157
supersolution, 460
Swift–Hohenberg equation, 555, 557

T
tangential operator, 432
thermistor, 8
thermo-convective flows, 8

thin film, 181
trace-interpolation inequality, 50

U
unbounded domains, 72
uniqueness of principal eigenvalue, 332
upper bound, 104

V
vanishing absorption, 92
variational
– inequalities, 467
– problem, 477
– structure, 556
viscous fluid, 8

W
weak formulation, 473
– unified, 475
weighted norms, 413
wetting condition, 172

Y
Young inequality, 16